Physics of
OPTOELECTRONICS

OPTICAL ENGINEERING

Founding Editor
Brian J. Thompson
University of Rochester
Rochester, New York

Physics of
OPTOELECTRONICS

Michael A. Parker

Taylor & Francis
Taylor & Francis Group

Boca Raton   London   New York   Singapore

A CRC title, part of the Taylor & Francis imprint, a member of the
Taylor & Francis Group, the academic division of T&F Informa plc.

Published in 2005 by
CRC Press
Taylor & Francis Group
6000 Broken Sound Parkway NW, Suite 300
Boca Raton, FL 33487-2742

International Standard Book Number-10: 0-8247-5385-2 (Hardcover)
International Standard Book Number-13: 978-0-8247-5385-6 (Hardcover)
Library of Congress Card Number 2004059307

Library of Congress Cataloging-in-Publication Data

Parker, Michael A.
    Physics of optoelectronics / Michael A. Parker.
        p. cm. -- (Optical engineering ; 104)
    Includes bibliographical references and index.
    ISBN 0-8247-5385-2
    1. Optoelectronics--Materials. I. Title. II. Optical engineering (Marcel Dekker, Inc.) ; v. 104.

TA1750.P37 2004
621.381'045--dc22   2004059307

Taylor & Francis Group
is the Academic Division of T&F Informa plc.

Visit the Taylor & Francis Web site at
http://www.taylorandfrancis.com

and the CRC Press Web site at
http://www.crcpress.com

Preface

The study of optoelectronics examines matter, light and their interactions. The solid state and quantum theory provide fundamental descriptions of matter. The solid state shows the effect of crystal structure and departure from crystal structure on electronic transport. Classical and quantum electrodynamics describe the foundations of light and the interaction with matter.

The text introduces laser engineering physics in sufficient depth to make accessible recent publications in theory, experiment and construction. A number of well-known texts review present trends in optoelectronics while many others develop the theory. The *Physics of Optoelectronics* progresses from introductory material to that found in more advanced texts. Such a broad palette, however, requires the support of many sources as suggested by the reference sections after every chapter. The journal literature itself is dauntingly vast and best left to the individual texts for summary in any particular topical area. For this reason, the present text often overlaps many excellent references as a service to the reader to provide a self-contained account of the subject.

The *Physics of Optoelectronics* addresses the needs of students and professionals with a "standard" undergraduate background in engineering and physics. First- and second-year graduate students in science and engineering will most benefit, especially those planning further research and development. The textbook includes sufficient material for introducing undergraduates to semiconductor emitters and has been used for courses taught at Rutgers and Syracuse Universities over a period of six years. The students come from a variety of departments, but primarily from electrical and computer engineering. A subsequent course in optical systems and optoelectronic devices would be the most natural follow-up to the material presented herein.

The *Physics of Optoelectronics* focuses on the properties of optical fields and their interaction with matter. The laser, light emitting diode (LED) and photodetector perhaps represent the best examples of the interaction. For this reason, the book begins with an introduction to lasers and LEDs, and progresses to the rate equations as the fundamental description of the emission and detection processes. The rate equations exhibit the matter–light interaction through the gain terms. The remainder of the text develops the quantum mechanical expressions for gain and the optical fields. The text includes many of the derivation steps, and supplies figures to illustrate concepts in order to provide the reader with sufficient material for self-study.

The text summarizes and reviews the mathematical foundations of the quantum theory embodied in the Hilbert space. The mathematical foundations focus on the abstract form of the linear algebra for vectors and operators. These foundations supply the "pictures" often lacking in elementary studies of the quantum theory, that would otherwise make the subject more intuitive. A figure does not always accurately represent the mathematics but does help convey the meaning or "way of thinking" about a concept.

The quantum theory of particles and fields can be linked to the Lagrangian and Hamiltonian formulations of classical mechanics. A derivation of the field–matter interaction from first principles requires the electromagnetic field Lagrangian and Hamiltonian. A chapter on dynamics includes a brief summary and review of the formalism for discrete sets of particles and continuous media. The remainder of the discourse on dynamics covers topical areas in the quantum theory necessary for the study of optical fields, transitions and semiconductor gain. The chapter includes the density operator, time-dependent perturbation theory, and the harmonic oscillator from the operator point-of-view.

The description of lasers and LEDS would not be complete without a discussion of the fundamental nature of the light that these devices produce. In the best of circumstances, the emission approximates the classical view of a coherent state with well-defined phase and amplitude. However, this often-found description of the optical fields originating in Maxwell's equations does not provide sufficient detail to describe the quantum light field, nor to understand recent progress in the areas of quantum optics and low-noise communications. The text develops the "quantized" electromagnetic fields and discusses the inherent quantum.

The later portions of the book develop the matter–light interaction, beginning with the time-dependent perturbation theory and Fermi's golden rule. After reviewing density-of-states and Bloch wave functions from the solid state, the text derives the gain from Fermi's golden rule. The gain describes the matter–light interaction in optical sources and detectors. However, Fermi's golden rule does not fully account for the effects of the environment. The theory typically implements the density operator and develops the Liouville equation (master equation) to describe collision broadening and saturation effects. The book briefly examines the origin of the fluctuation–dissipation theorem and applies it to the master equation. The book naturally leads to further study areas, including quantum optics, nanoscale emitters and detectors, nonlinear optics, and standard studies of q-switched and mode-locked lasers, parametric amplifiers, gas and solid state lasers.

The typical first-year graduate course (28 classes with approximately 1.5 hours each) covers the introduction (1.1–1.7), laser rate equations (2.1–2.5), the wave equations and transfer matrices (3.1–3.2, 3.5–3.7), a brief summary of waveguiding (3.8–3.9), linear algebra (4.1–4.6, 4.8–4.10), basic quantum theory (5.6–5.8), especially time dependent perturbation theory and density operators (5.10–5.11), quantum dipole and Fermi's golden rule (7.1–7.3), the Liouville equation and gain (7.9–7.13), and Fermi's golden rule approach to gain (Chapter 8). Usually some material must be sacrificed between the Liouville equation and the Fermi's golden rule approach for a one-semester course. A follow-up short course (8–12 classes) can cover the introduction to quantum electro-dynamics (quantum optics — Chapter 6) with requisite material on quantum representations (5.9) and suggested material on noise from the beginning chapters (1.8, 2.6). Online lectures (with slides and audio) for a one-semester course are available free at www.crcpress.com. The topics for the one-semester course have been made independent of the other topics.

The author acknowledges the Rutgers, Cornell and Syracuse University programs in engineering and physics. The faculty, administration and staff at Rutgers have provided significant support for teaching and laboratory facilities. A number of individuals from the author's past have contributed to the author's view on semiconductor sources and detectors, especially C. L. Tang, P. D. Swanson, R. Liboff, E. A Schiff, P. Kornreich, R. J. Michalak, S. Thai, K. Kasunic, and J. S. Kimmet. The author thanks his wife Carol, for her assistance and her patience during the weekends and evenings over the past several years while the author prepared the courses, compiled the material, and wrote the textbooks. Thanks also go to the staffs at Marcel Dekker, CRC Press, and Taylor & Francis for their advice and efforts to bring the text to publication. Most of all, the author thanks his students for attending the courses and for their challenging questions and suggestions.

Contents

1

Introduction to Semiconductor Lasers

Semiconductor lasers have important applications in communications, signal processing and medicine, including optical interconnects, RF links, CD ROM, gyroscopes, surgery, printers, and photocopying (to name only a few). Compared with other optical sources, lasers have higher bandwidth and higher spectral purity; they function as bright coherent sources. These properties allow laser emission to be tightly focused, with minimum divergence.

The study of lasers encompasses perhaps the broadest array of subfields, including quantum theory, electromagnetics and optics, solid state engineering and physics, chemistry, and mathematics, as should be evident from the acronym *laser*, meaning "light amplification by stimulated emission of radiation." Quantum theory describes the "light amplification" process, whereby an optical field (with proper frequency) interacts with an ensemble of light emitters (i.e., atoms, molecules, etc.). The amplification occurs through the matter–field interaction (Figure 1.0.1), when the incident field "stimulates" the ensemble to emit more light with the same characteristics as the incident light, including the same frequency, direction of propagation, and phase (the emitted light is coherent).

Semiconductor lasers emit light when electrons and holes recombine. The microwave counterpart (maser) operates similarly. Traditionally, optics (including quantum optics) describes light and its properties using electromagnetic (EM) theory; this includes the manipulation of light by lenses and prisms, as well as dispersion and waveguiding. Solid state physics, engineering, and chemistry describe the composition of matter and its properties (including electrical properties). Mathematics comprises the most natural language for any field in science or engineering.

Semiconductor laser provides a natural laboratory for the study of matter, fields, and their interactions. This book is primarily concerned with the construction of laser and the characteristics of emitted optical energy. The first volume of the series, *Fundamentals of the Quantum Theory and the Solid State* (forthcoming), develops the prerequisite material on quantum theory and its natural language of linear algebra. That volume discusses the solid state, especially crystalline structure and implications for phonons, the origin bands, Bloch wavefunctions, and density of states.

The present volume extends the discussion to include applied electromagnetic fields perturbing the energy states of excited atoms in a gain medium to induce an electronic transition, so that the atom emits light. It also deals with Fermi's golden rule and the application of density operator theory. The chapter on light develops the quantum theory of the electromagnetic wave. It discusses the limits to measuring the fields, inherent noise, and the relation to the Poisson statistics. The last few chapters put it all together,

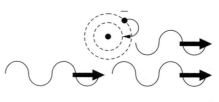

FIGURE 1.0.1

Pictorial representation of the stimulated emission process. An incident wave (left) induces the electron to make a transition to a lower energy level thereby producing a second wave (right).

and develop the quantum theory of gain, the rate equations, and apply the results to semiconductor lasers made from quantum wells and quantum dot.

This first chapter discusses the basic components for constructing and operating a laser. Feedback is a key ingredient for laser oscillation (lasing), which refers to sustained light emission from the ensemble of light-emitters without an *external* optical field for the induced emission process. This chapter introduces the concepts of semiconductors and band diagrams, which are necessary to understand the electrical and light-emission properties of the laser. It is good to have a strong grasp of the physical nature of the laser before proceeding to the abstract topics. In fact, you will notice that the text in this book progresses from physical to abstract and back to physical at the very end.

The main purpose of this book is to develop equations to predict the characteristics of the output signal in terms of the construction of the laser, the type of gain medium and the characteristics of the pump source. The rate equations are the most fundamental set of equations; they can be obtained phenomenologically or through detailed quantum mechanical considerations. These rate equations require the gain to be written in terms of basic material properties. It is the primary purpose of this book to develop the general theory and concepts, rather than list all the different types of lasers. GaAs serves as our prototype material system.

1.1   Basic Components and the Role of Feedback

First consider the basic components of the laser as depicted in Figure 1.1.1. There are four basic components necessary to obtain laser oscillation—a gain medium for amplification, a pump to add energy to the medium, positive feedback for oscillation, and a coupling mechanism to extract a signal. The gain medium contains active centers that emit light (or microwaves for masers). Light within the resonator (cavity), defined by a waveguide and two mirrors, interacts with these centers to produce additional light. An optical or electrical source (a pumping mechanism) supplies energy to the emission centers. As with any oscillator, the device must include a mechanism for returning the signal to the gain section (feedback). For the laser, mirrors provide the feedback that increases the interaction between the ensemble of gain centers and the signal. The output coupler consists of a partially reflective mirror (for a maser, the coupler might be a loop of wire or a small hole in a waveguide). Some of the light in the cavity escapes through the mirror. Lasers made of semiconductor material generally have two partially reflective mirrors, unless special reflective or antireflective coatings are used.

Lasers (and masers) produce coherent electromagnetic waves. The output wave can be pictured (to a good order of approximation) as a single sinusoidal wave, with a well-defined phase and amplitude traveling in a single direction. This is a consequence of

FIGURE 1.1.1
The gain medium, pump, feedback mechanism, and output coupler comprise the four basic components of a laser (or maser).

the ensemble of emission centers producing coherent stimulated emission. Classically, the same group of atoms can spontaneously emit without an impressed electromagnetic field. The spontaneous emission tends to propagate in all directions with random phases. We will later discuss how spontaneous emission can actually be attributed to random fluctuations of the electromagnetic field in the vacuum. We will also discuss how the output wave is never totally known, since a small uncertainty always exists in the phase and amplitude of the wave (refer to the chapter on quantum optics).

Now, we will discuss the importance of feedback by comparing a "ring" laser to a conventional op-amp electronic circuit having similar topology to the laser. All oscillators need to have a signal fed back into the input to regenerate the signal, and thereby compensate for power loss in the circuit. Figure 1.1.2 shows a triangular shaped cavity that closes on itself. Two of the mirrors are total internal reflection (TIR) mirrors that reflect 100% of any incident light on the inside of the waveguide. The output beam is labelled P_o (for power). A bias current "I" electrically pumps the gain section, while the rest of the waveguide remains unpumped. The quantity P_i represents an external input signal to be amplified. For the oscillator, the internal noise starts the oscillation (so that $P_i = 0$), but sometimes the noise can be represented by P_i if desired (spontaneous emission for lasers). We assume an asymmetry in the cavity so that the optical wave in the cavity propagates in a clockwise direction.

The electrical circuit equivalent to the ring laser has an amplifier with gain "g" in place of the gain section for the laser. The block $1/\alpha$ can represent "loss element" such as a resistor or another gain element. For the electrical circuit, we can write a set of equations that can be solved to determine a condition on the gain to achieve oscillation.

$$\left\{\begin{array}{l} P_o = gP_s \\ P_s = P_{in} + P_f \\ P_f = \dfrac{1}{\alpha}P_o \end{array}\right\} \quad \rightarrow \quad P_o = \frac{gP_{in}}{1 - (g/\alpha)} \qquad (1.1.1)$$

For a laser operating below the lasing threshold (the output light mostly consists of spontaneous emission rather than stimulated emission), increasing the pump power to the gain section causes the gain "g" to increase. The denominator of P_o becomes smaller

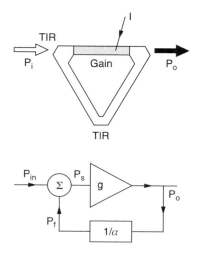

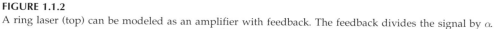

FIGURE 1.1.2
A ring laser (top) can be modeled as an amplifier with feedback. The feedback divides the signal by α.

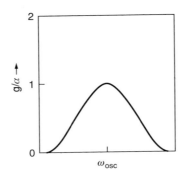

FIGURE 1.1.3
The oscillator operates at the frequency with highest net gain g/α.

as "g" increases. When $g = \alpha$ (gain equals loss), the loop gain becomes infinite and a sustained output (oscillation) is achieved, even though the input P_i is essentially zero. For a laser, α describes the optical energy lost from the cavity through mirrors, or scattering from imperfections in the waveguide or free carrier absorption.

Generally, for both the electrical circuit and the ring laser (as well as other oscillators), the gain depends on the angular frequency ω of the signal $g = g(\omega)$ as shown in Figure 1.1.3. For example, an electronic amplifier circuit might have some RC or RLC filters that provide a narrow resonance. For now, we assume α is independent of frequency. An oscillator normally operates at the peak of the gain curve $\omega = \omega_{osc}$ since that is where the condition "gain equals loss" $g = \alpha'$ occurs.

In practice, the gain can be slightly smaller than the loss $g/\alpha < 1$. This occurs because the difference $1 - (g/\alpha)$ is made up of a noise signal. The oscillator (and especially the laser) requires a certain amount of noise to start the oscillation; this noise must have a frequency component at the oscillation frequency. For a laser, the required noise consists of spontaneous emission from the emission centers in the gain medium. Without the coherent radiation produced by the laser, the atoms can only spontaneously emit (i.e., on their own—fluorescence), which is symbolized by P_i at one of the facets, even though the signal originates in the gain section. This spontaneously emitted light can be amplified just like the stimulated emission. As a result, the spontaneous emission reflects from the mirrors and induces transitions in the excited atoms comprising the gain medium (the light stimulates emission). Now both the spontaneous and stimulated emission can further induce transitions to produce the steady-state laser signal. The amount of spontaneous emission to initiate lasing sets the minimum required current (threshold current) and also sets the minimum achievable optical linewidth in the output spectrum. As a note, the optical signal from an LED (light emitting diode) consists entirely of spontaneous emission. In this case, spontaneous emission is not noise—it is the signal.

1.2   Basic Properties of Lasers

Lasers have many properties, which make them unique and highly applicable. The physical construction and material properties determine the operating wavelength and achievable output power. The "brightness," defined as the amount of output power per unit frequency, determines the spectral purity of the source.

1.2.1 Wavelength and Energy

Semiconductor lasers can be designed and fabricated with optical emission ranging from ultraviolet (UV) to infrared (IR). The UV lasers are particularly important for higher resolution work, since the smaller wavelengths can "see" smaller objects. Blue lasers can store roughly four times the amount of information on a standard size CD as red lasers. The IR lasers producing light with wavelengths of 1.3 and 1.5 microns (μm) have applications in fiber-based communication systems, since fibers have minimum dispersion and loss at these wavelengths. Low dispersion is important for maintaining pulse shape over large distances. Figure 1.2.1 shows how the emission wavelength varies with semiconductor composition. For example, pure GaAs emits nearly 0.85 μm while AlAs emits nearly 0.6 μm. The figure shows that the emission wavelength for a composition $Al_xGa_{1-x}As$ varies approximately linearly with "x" between the two extremes. For IR emission, a laser composed of a combination of InP and InAs should operate at either the minimum loss or dispersion wavelengths.

The emission wavelengths of some common laser systems can be found in Figure 1.2.2. The semiconductor lasers are labeled by "diode"; the double-headed arrow shows the possible emission range. Masers emit in the far infrared, with wavelengths larger than 30 μm.

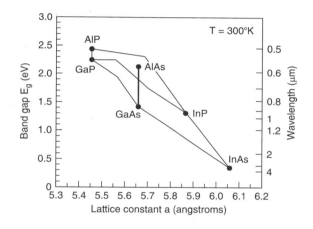

FIGURE 1.2.1
Relation between emission wavelength, semiconductor bandgap and the lattice spacing constant.

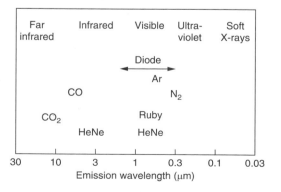

FIGURE 1.2.2
Common lasers and the emission wavelengths.

Some typical relations are

$$E = \hbar\omega = h\nu \quad \lambda = 2\pi/k \quad c = \nu\lambda = \omega/k$$

where the variables are defined in the order that they appear as photon energy E, $h/(2\pi)$, and "h" is Planck's constant, the angular frequency $\omega = 2\pi\nu$ and frequency ν in Hertz (Hz), wavelength λ, wave vector k, and speed of light c. These equations can be combined to provide an easy-to-remember relationship between the wavelength in nanometers (nm) and the energy in electron volts (eV) $E = 1240/\lambda$. For example, if an AlGaAs diode laser has a bandgap of approximately 1.45 eV, then the emission wavelength is nearly 850 nm.

1.2.2   Directionality

Laser beams can be highly directional. The longitudinal axis (along the length of the resonator) and the mirrors define a preferred direction for emission. Stimulated emission from atoms duplicates the characteristics of the light, causing the atom to emit. That is, the emitted light has the same wavelength and propagation direction as the perturbing (incident) light. Spontaneous emission does not behave in this way. However, diffraction effects are especially severe for semiconductor lasers. Waveguiding in in-plane lasers can confine the beam to within a few hundred nanometers. The light through the front facet can diverge at angles of 45 degrees or more. Vertical cavity lasers, on the other hand, have aperture sizes on the order of 10 µm and a laser beam has convergence angles smaller than 15 degrees.

1.2.3   Monochromaticity and Brightness

The spectra for semiconductor lasers consists of stimulated and spontaneous emission components. Spontaneously emitted light has a range of wavelengths covering many nanometers. Stimulated emission, on the other hand, has a typical width less than one angstrom. Figure 1.2.3 compares the two types. The wavelength spectral width can be related to the frequency spectral width by taking the differential of $\nu = c/\lambda$ to get $|\Delta\nu| = |\Delta\lambda|\, c/\lambda^2$.

One of the most important properties of the laser is that it can pack a lot of energy in a very narrow bandwidth ($\Delta\nu$ or $\Delta\lambda$). We can define the brightness[6] as the power per unit bandwidth that flows from a surface of area A at the source into a cone of steradian $\Delta\Omega$. Obviously, the more monochromatic the beam, the greater the brightness. Many lasers systems have frequency bandwidths smaller than 1 MHz. LEDs can have a spectral

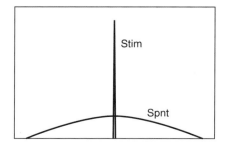

FIGURE 1.2.3
Comparison of spectrum for typical spontaneous and stimulated emission.

width as wide as 100 nm. However, the incandescent lamp (tungsten filament, for example) can span on the order of 1000 nm. For human vision requiring blue, green, and red spectral components, the LED (and especially the laser) can provide highly efficient sources.

1.2.4 Coherence Time and Coherence Length

The coherence time of the laser beam describes the length of time for two frequency components ν_1, ν_2 to become out of phase by a full cycle

$$\Delta \tau = \frac{1}{\Delta \nu}$$

For lasers, the spectral width $\Delta \nu$ can be related to the FWHM (full-width half-max) points of the spectrum. Later chapters relate the coherence time to the dipole dephasing time. The random phase model for the laser, as will be discussed in connection with the density operator for optical states, views the laser beam as monochromatic waves with the phase jumping randomly every so often.

Example 1.2.1

As an example, a laser with the bandwidth of 1 MHz produces a coherence time of 1 μs. Sunlight has a coherence time of approximately 2 fs. Spontaneous emission for the GaAs laser with the bandwidth of 4 nm produces a coherence time of 0.2 ps. This number is close to the dipole dephasing time, which is attributed to an average time between collisions within the laser gain medium.

1.3   Introduction to Emitter Construction

There are many types of light emitters. The gain medium can be a semiconductor such as GaAs or InP, a gas like HeNe, argon, CO_2, or a solid state material such as ruby or doped glass. The emitter can be made from a range of materials including optical fiber, semiconductor wafers, or bulk optical components. Generally, semiconductor devices have micron sizes and require a clean room and precision fabrication techniques (References 3 and 4). The basic emitter can be augmented with a number of optical components for special purpose applications.

1.3.1   In-Plane and Edge-Emitting Lasers

For semiconductor lasers, the cavity can be parallel to the semiconductor wafer (i.e., in the plane of the wafer) or perpendicular to the plane of the wafer as shown in Figures 1.3.1 and 1.3.2. The "edge emitting laser" or "in-plane laser" (IPL) refers to the parallel type, while "vertical cavity surface emitting laser" (VCSEL) refers to the perpendicular type. Hybrids can have the cavity within the plane while using mirrors to emit perpendicular to the plane.

The simplest IPL consists of a ridge with a metal electrode on the top along its length (see Figure 1.3.1). A second electrode runs across the bottom of the wafer. The ridge also forms part of an optical waveguide to confine the light wave as it moves

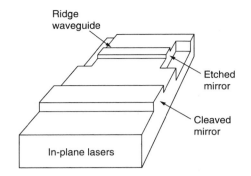

FIGURE 1.3.1

Representation of two ridge-guided lasers. The front one has cleaved mirrors while the rear one has mirrors made from a chemical etching process.

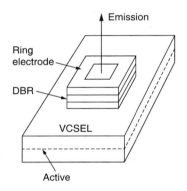

FIGURE 1.3.2

A vertical cavity laser. The top side often consist of p-type material. The bottom side (wafer side) is n-type. An electrode runs across the bottom side of the wafer.

between the mirrors. The lower portion of the top surface extends into the mode in the waveguide cladding region. The air–material interface produces an effective refractive index smaller in magnitude than that of the active region, in order to provide lateral waveguiding. For the IPLs shown, the mirrors on the left-hand and right-hand sides consist of nothing more than the air–semiconductor interfaces. The reflection coefficient for GaAs is approximately 34%. Such mirrors can be either cleaved similar to cutting glass, or etched by a chemical process (usually a gas etch).

1.3.2 VCSEL

Vertical Cavity Surface Emitting Lasers (VCSELs) were developed in the late 1980s as a result of advances in material growth and processing techniques. The VCSEL uses a series of layers of dissimilar refractive indices to produce a distributed Bragg reflector (DBR), which is also known as a mirror stack. Each layer in the DBR is a half-wavelength thick. The VCSEL wafers including the DBR mirrors and the active region with the quantum wells can be grown by Molecular Beam Epitaxy (MBE). There is a mirror stack above and below the active region, so as to define a vertical cavity. The simplest VCSELs have ring electrodes surrounding a window on the top side, where the laser beam emerges. Electrical current passing through the conducting mirrors pumps the gain medium (often called the active region).

1.3.3 Buried Waveguide Laser

Rather than making ridge waveguides as shown in Figure 1.3.3, the buried waveguide laser surrounds the higher index active region with lower index material. In this way, index differences provide waveguiding in the lateral and transverse directions (the two directions perpendicular to the length of the laser).

1.3.4 Lateral Injection Laser

The lasers in Figure 1.3.1 inject current through the top and bottom contacts in a direction perpendicular to the plane of the wafer. The lateral injection laser in Figure 1.3.3 injects current parallel to the internal layers of the heterostructure. This configuration helps provide a planar surface for fabrication.

1.3.5 The Light Emitting Diode

The LED consists of a *"pn"* or *"pin"* junction of light emitting material typically surrounded by a plastic lens, as shown in Figure 1.3.4. The lens helps to shape the output beam. Unlike the laser, the LED requires neither mirrors nor feedback to operate. Spontaneous recombination of electrons and holes produces spontaneous emission (fluorescence) for the output beam. The vacuum fields sufficiently perturb the energy levels to initiate the spontaneous emission.

1.3.6 Semiconductor Laser Amplifier

Semiconductor laser amplifiers resemble inplane lasers, except they do not have mirrors at the ends. In fact, manufacturers add antireflection coatings to prevent even the smallest amount of feedback. The electron and hole population provides the gain.

1.3.7 Gas Laser

The gas laser contains the gas in a gas tube, as shown in Figure 1.3.5. Two electrodes apply very high voltage to the gas in a manner similar to the "neon signs" used for outdoor advertising. The light oscillation builds up between a 100% reflective mirror and a partially reflective mirror. The partially reflective mirror provides feedback and an output beam. The Brewster windows allow only one polarization mode of light to

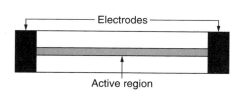

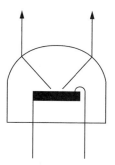

FIGURE 1.3.3
Electrodes allow current flow parallel to the internal layers for the lateral injection laser.

FIGURE 1.3.4
The light emitting diode.

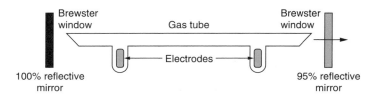

FIGURE 1.3.5
Block diagram of the gas laser.

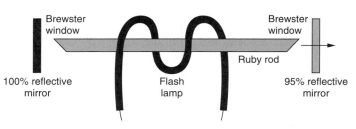

FIGURE 1.3.6
The ruby solid state laser.

pass through without any attenuation (100% transmittance), the other polarization mode reflects at an oblique angle and does not contribute to the laser output.

1.3.8   The Solid State Laser

An example of solid state laser is shown in Figure 1.3.6. The solid ruby rod provides the gain medium. A high intensity flash lamp pumps the ruby. Mirrors at either end provide the feedback. Unlike the continuously emitting gas laser, the ruby rod laser provides pulsed light.

1.4   Introduction to Matter and Bonds

Semiconductor light emitters require matter in one form or another for the physical form of the device and for producing light. The device construction depends on the type of matter used. Gas lasers look different from semiconductor lasers. The matter produces light using the matter–light interaction. The exact details depend on the type of matter.

The study of matter comprises the subject of solid state physics and chemistry (often termed condensed matter). The invention and engineering of new devices requires a thorough understanding of the solid state.

The present section reviews broad classifications of matter. Later chapters and sections use these concepts to develop the mathematical descriptions.

1.4.1   Classification of Matter

Gases, liquids, and solids represent three basic types of matter. Modern technology finds its grounding in solids in the form of crystals, polycrystals, and amorphous materials. The next section discusses the relation between the atomic configuration and the band diagrams.

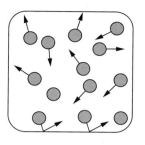

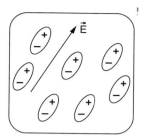

FIGURE 1.4.1
Gas molecules do not bind to one another and haven't any order.

FIGURE 1.4.2
An electric field can rotate molecules with a permanent dipole to create order.

Gases

Gases have atoms or molecules that do not bond to one another in a range of pressure, temperature, and volume (Figure 1.4.1). Argon consists of single atoms whereas hydrogen usually appears as H_2. These molecules haven't any particular order and move freely within a container.

Liquids and Liquid Crystals

Similar to gases, liquids haven't any atomic/molecular order and they assume the shape of the containers. Applying low levels of thermal energy can easily break the existing weak bonds.

Liquid crystals have mobile molecules, but a type of long range order can exist. Figure 1.4.2 shows molecules having a permanent dipole. Applying an electric field rotates the dipole and establishes order within the collection of molecules.

Solids

Solids consist of atoms or molecules executing thermal motion about an equilibrium position fixed at a point in space. Solids can take the form of crystalline, polycrystalline, or amorphous materials. Solids (at a given temperature, pressure, and volume) have stronger bonds between molecules and atoms than liquids. Solids require more energy to break the bonds.

Crystals have a long-range order as shown in Figure 1.4.3. Each lattice point in space has an identical cluster of atoms (atomic basis). Later chapters show how this order affects conduction and other properties. Silicon provides an example of a crystal with a two-atom basis set on a face centered cubic crystal.

Polycrystalline materials consist of domains. The molecular/atomic order can vary from one domain to the next. Polycrystalline silicon can be made from plasma enhanced chemical vapor deposition under the proper conditions; it has great technological uses in the area of MEMs. The material has medium range order that can extend over several microns. Figure 1.4.4 shows two domains with different atomic order. The interstitial material between the two domains has very little order. Many of the bonds remain unsatisfied, and hence there can be large voids. The growth process for polycrystalline materials can be imagined as follows. Consider a blank substrate placed inside a growth chamber. Crystals begin to grow at random locations with random orientation. Eventually, the clusters meet somewhere on the substrate. Because the clusters have different crystal orientations, the region where they meet cannot completely bond together. This results in the interstitial region.

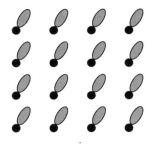

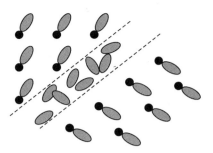

FIGURE 1.4.3
Crystals have identical clusters of atoms attached to lattice points in space.

FIGURE 1.4.4
A polycrystalline material showing two separate crystal phases separated by interstitial material.

Amorphous materials do not have any long-range order, but they have varying degrees of short-range order. Examples of amorphous materials include amorphous silicon, glasses, and plastics. Amorphous silicon provides the prototypical amorphous material for semiconductors. It has wide ranging and unique properties for use in solar cells and thin film transistors. The material can be grown by a number of methods including sputtering and plasma enhanced chemical vapor deposition (PECVD). The order of the atoms determines the quality of the material for conduction, and the order depends on the growth conditions.

In the amorphous state, the long-range order does not exist. The bonds for amorphous silicon all have essentially the same length and angle, but the dihedral angle can differ (a change in the dihedral angle occurs when two bonded atoms rotate with respect to each other about the bond axis, as shown in Figure 1.4.5. In some sense, a cluster of fully coordinated silicon atoms produces local order, but the distribution of dihedral angles yields variation in the spatial orientation of the clusters. Furthermore, some of the atoms have less than four-fold coordination and therefore have unsatisfied bonds. Under the proper preparation conditions, these dangling bonds terminate in hydrogen atoms to produce hydrogenated amorphous silicon (a-Si:H).

1.4.2   Bonding and the Periodic Table

Semiconductor materials generally fall in columns III through VI in the periodic table. Figure 1.4.6 shows a periodic table of elements. The letters *S, P, D, F* denote the bonding levels. The first two columns of the periodic table correspond to the *S* orbital, which

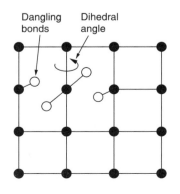

FIGURE 1.4.5
A rotation about the dihedral angle produces dangling bonds.

Periods	I A	II A	III B	IV B	V B	VI B	VII B	VIII			I B	II B	III A	IV A	V A	VI A	VII A	0
1	1.0079 H[1]																1.0079 H[1]	4.00260 He[2]
2	6.941 Li[3]	9.01218 Be[4]											10.81 B[5]	12.011 C[6]	14.0067 N[7]	15.9994 O[8]	18.9984 F[9]	20.179 Ne[10]
3	22.9898 Na[11]	24.305 Mg[12]											26.9815 Al[13]	23.086 Si[14]	30.9738 P[15]	32.06 S[16]	35.453 Cl[17]	39.948 Ar[18]
4	39.098 K[19]	40.08 Ca[20]	44.9559 Sc[21]	47.90 Ti[22]	50.9414 V[23]	51.996 Cr[24]	54.9380 Mn[25]	55.847 Fe[26]	58.9332 Co[27]	58.71 Ni[28]	63.546 Cu[29]	65.38 Zn[30]	69.72 Ga[31]	72.59 Ge[32]	74.9216 As[33]	78.96 Se[34]	79.904 Br[35]	83.80 Kr[36]
5	85.4678 Rb[37]	87.62 Sr[38]	88.9059 Y[39]	91.22 Zr[40]	92.9064 Nb[41]	95.94 Mo[42]	98.9062 Tc[43]	101.07 Ru[44]	102.9055 Rh[45]	106.4 Pd[46]	107.868 Ag[47]	112.40 Cd[48]	114.82 In[49]	118.69 Sn[50]	121.75 Sb[51]	127.60 Te[52]	126.9045 I[53]	131.30 Xe[54]
6	132.9054 Cs[55]	137.34 Ba[56]	[57-71]	178.49 Hf[72]	180.9479 Ta[73]	183.85 W[74]	186.2 Re[75]	190.2 Os[76]	192.22 Ir[77]	195.09 Pt[78]	196.9665 Au[79]	200.59 Hg[80]	204.37 Tl[81]	207.2 Pb[82]	208.9804 Bi[83]	(210) Po[84]	(210) At[85]	(222) Rn[86]
7	(223) Fr[87]	226.0254 Ra[88]	[89-103]	[104]	[105]	[106]	[107]	[109]										

FIGURE 1.4.6
The Periodic Table.

requires two electrons to be stable. For example, hydrogen has only one valence electron that occupies the spherically symmetric *S* orbital. Helium has two valence electrons in the *S* orbital. As an exception, helium appears in the last column of the periodic table to designate it as a stable noble gas. Columns III-A through VI-A (labeled at the top of the column) plus column *O* represent the *P* orbitals, which require six electrons for stability.

Example 1.4.1

Hydrogen needs a second electron for the *S* orbital to be filled. We therefore expect to see hydrogen molecules as H_2.

Example 1.4.2

Silicon in column 4 requires four extra electrons to fill the *P* level. However, silicon already has four electrons. We therefore expect one silicon atom to covalently bond to four other silicon atoms. Covalent bonds share valence electrons rather than completely transfering the electrons to neighboring atoms (as for ionic bonding).

Silicon represents a prototypical material for electronic devices. Similarly, amorphous silicon represents a prototypical material for amorphous semiconductors. Gallium arsenide (GaAs) represents a prototypical direct bandgap material for optoelectronic components. Aluminum and gallium occur in the same column of the table. We therefore expect to find compounds in which an atom of aluminum can replace an atom of gallium. Such compounds can be designated by $Al_xGa_{1-x}As$.

The most stable atomic bonds release the greatest amount of energy during the bonding process. Figure 1.4.7 shows the potential energy between two atoms as a function of the distance between them. The separation distance labeled as a_o yields a minimum in the energy. Moving the atoms closer than this distance increases the energy, as does moving them further apart. The binding energy $\mathcal{E}_b$ represents the minimum energy required to separate the two atoms, once bonding occurs.

Adding impurity atoms can affect the electronic and optical properties of a material. For example, doping can be used to control the conductivity of a host crystal. *n*-type dopants have one extra valence electron than the material itself. For example, we might expect phosphorus to be an *n*-type dopant for silicon (see Figure 1.4.8). Not all phosphorous valence electrons participate in bonding, and they can freely move about

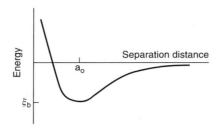

FIGURE 1.4.7
Total energy of two atoms as a function of their separation distance.

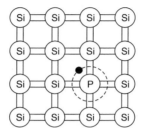

FIGURE 1.4.8
An *n*-type dopant atom embedded in a silicon host crystal. The electron is loosely bound to the dopant atom and free to roam about the crystal at room temperature.

the crystal. *p*-type dopants have one less electron in the valence shell than atoms in the host material. For example, boron is a *p*-type dopant for silicon.

The effects of doping on conduction can be easily seen for the *n*-type dopant in silicon. The "extra" fifth electron orbits the phosphorus nucleus similar to a hydrogen atom. However, the radius of the orbit must be much larger than the radius of a similar hydrogen orbit. Unlike the orbit shown in the figure, the electron orbit actually encloses many silicon atoms. The silicon atoms within the orbit can become polarized and screen the electrostatic force between the orbiting electron and the phosphorus ion. As a result, the electrons remain only weakly bonded to the phosphorus nucleus at low temperatures. These electrons break their bonds at room temperature and freely move about the crystal, thereby increasing the conductivity of the crystal. For GaAs, zinc and silicon provide a *p*-type and *n*-type dopant, respectively.

1.5   Introduction to Bands and Transitions

Semiconductor devices most often use the crystalline form of matter. The conduction and optical characteristics for emitters and detectors primarily depend on the band structure. This section introduces the bands and possible transitions between the bands. The matter–light interaction produces these transitions for lasers, light emitting diodes, and detectors.

1.5.1   Intuitive Origin of Bands

As previously discussed, a silicon atom has four valence electrons so that it can covalently bond to four other silicon atoms. Figure 1.5.1 shows a cartoon representation (at 0 Kelvin)

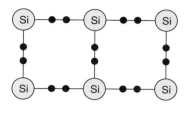

FIGURE 1.5.1
Cartoon representation of silicon crystal at 0 K.

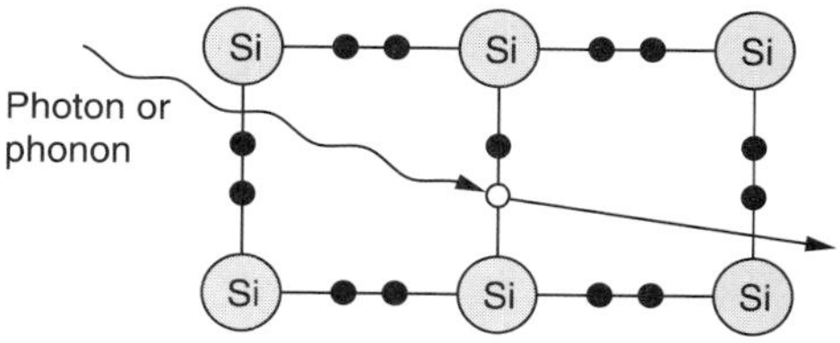

FIGURE 1.5.2
Cartoon representation of transition from VB to CB.

of the crystal and indicates adjacent atoms sharing two electrons. Adding energy to the crystal (Figure 1.5.2) frees the electrons from the bonds so that they can roam around the crystal lattice. This means that free electrons have larger energy than those electrons in the bonds. The band gap energy is the minimum energy required to liberate an electron. An electron that absorbs this minimum amount of energy must have a potential energy equal to the gap energy. If the electron absorbs more than the minimum, then it has not only the potential energy but also kinetic energy. The conduction band (CB) represents the energy of the free electrons (also known as conduction electrons). The vacancies left behind are "holes" in the bonding. The holes appear to move when electrons in neighboring bonds transfer to fill the vacancy. The transferred electron leaves behind another hole. The hole therefore appears to move from one location to the next.

The total energy of a conduction electron can be written as

$$E = \mathrm{PE} + \mathrm{KE} = E_g + \tfrac{1}{2}m_e v^2 \tag{1.5.1}$$

where the potential energy equals the gap energy. E_g using the momentum $p = m_e v$ we can rewrite the relation as

$$E = E_g + \frac{p^2}{2m_e} \tag{1.5.2}$$

where m_e denotes an effective mass for the electron. Therefore, as shown in Figure 1.5.3, the plot of the energy E vs momentum p must have a parabolic shape. If the electron receives just enough energy to surmount the band gap, then it does not have enough energy to be moving and the momentum must be $p = 0$. We will refer to these energy diagrams as band diagrams or dispersion curves.

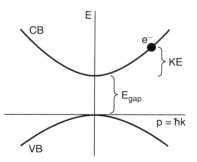

FIGURE 1.5.3
Band diagram showing a direct band gap for materials such as GaAs.

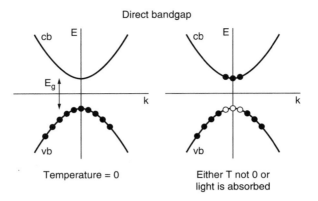

FIGURE 1.5.4
An elementary band diagram for gallium arsenide.

The promoted electron (conduction electron in the CB) leaves behind a hole at the Si–Si bond. Neighboring bonded electrons can tunnel into the empty state. The holes therefore move from one site to the next. This means that the holes can have kinetic energy. A plot of the kinetic energy vs momentum p or wave vector k also has a parabolic shape for the holes

$$E = \frac{p^2}{2m_h} \tag{1.5.3}$$

where m_h denotes the effective mass of the hole. The free holes live in the valence band (VB) and can participate in electrical conduction. The valence band has a parabolic shape similar to the conduction band.

Some of the features of the bands require a quantum mechanical analysis. Let's comment on the reason for referring to bands as dispersion curves. When atoms come close together to form a crystal, the energy levels for bonding split into many different energy levels. All of these split-levels from all of the atoms in the crystal produce the bands. "Bands" actually refer to a collection of *closely spaced* energy levels (see the circles in Figure 1.5.4). For example, the CB energies are very closely spaced to form the parabola. Sometimes people refer to these closely spaced states as "extended states" because the wave vector k indicates that electrons in these states are described by traveling plane waves.

The conduction and valence bands comprise the E vs k dispersion curve, where k denotes the electron (or hole) wave vector. We imagine that the electrons (and holes) behave as waves with wavelength $\lambda = 2\pi/k$. Using $p = \hbar k$, the band diagrams can be relabeled as in Figure 1.5.4. The band diagram gives the energy of the electrons (and holes) as a function of the wave vector (or momentum). The stationary particles have $k = 0$ and those moving have nonzero wave vector. The E vs k diagrams are similar to the frequency ω vs k diagrams used for optics (where ω is the angular frequency related to the frequency ν by $\omega = 2\pi\nu$).

For recombination, the electrons must give up excess energy. Electrons and holes recombine when they collide with each other and shed extra energy. They can emit photons and phonons. Regardless of the process, the total energy given up must equal or exceed the bandgap energy. The recombination of electrons and holes in direct bandgap materials produces photons (i.e., the electron loses energy and drops to the valence band, vb) in a direct bandgap material. These electron–hole pairs (sometime called excitons) are "emission centers" that can form the gain medium for a laser.

1.5.2 Indirect Bands, Light and Heavy Hole Bands

The material represented by Figure 1.5.4 has a direct band gap. A semiconductor has a direct bandgap when the *cb* minimum lines up with the *vb* maximum (GaAs is an example). A material having an indirect bandgap occurs (Figure 1.5.5) when the minimum and maximum do not have the same value of the wave-vector *k* (silicon is an example). For both direct and indirect bandgaps, the difference in energy between the minimum of the *cb* and the maximum of the *vb* gives the bandgap energy.

GaAs has light-hole and heavy-hole valence bands (see Figure 1.5.6). The effective mass of an electron or hole in one of the bands is proportional to the reciprocal of the band curvature according to

$$\frac{1}{m_{\text{eff}}} = \frac{1}{\hbar^2} \frac{\partial^2 E}{\partial k^2} \tag{1.5.4}$$

The heavy-hole band HH has holes with larger mass than the light-hole band LH. The light holes are a couple of orders of *smaller* magnitude than the free mass of an electron for GaAs. The effective mass m_e of a particle gives rise to the momentum according to $p = \hbar k = m_e v$. Both valence bands can contribute to the absorption and emission of light. For GaAs, the maximum of the two *vb*'s have approximately the same energy.

Adding indium to the GaAs causes strain in the lattice of gallium and arsenic atoms which forces them away from their normal equilibrium position in the lattice. Strain eliminates the degeneracy between the two valence bands at $k=0$ (separates them in energy). Strain also tends to increase the curvature of the HH band, reduces the mass of the holes in that band and therefore increases the speed of GaAs devices. It increases the gain for lasers. It also changes the bandgap slightly and therefore also the emission wavelength of the laser.

1.5.3 Introduction to Transitions

Consider two methods of adding energy to move electrons from the valence band to the conduction band. First, valence band electrons can absorb phonons. The phonon is the quantum of vibration of a collection of atoms about their equilibrium position. Second, the electron in the valence band can absorb a photon of light.

Figure 1.5.5 shows a full valence band at a temperature of $T=0$ K. If the semiconductor absorbs light or the temperature increases, some electrons receive sufficient energy

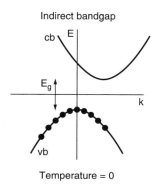

FIGURE 1.5.5
A semiconductor at zero degrees Kelvin with an indirect bandgap.

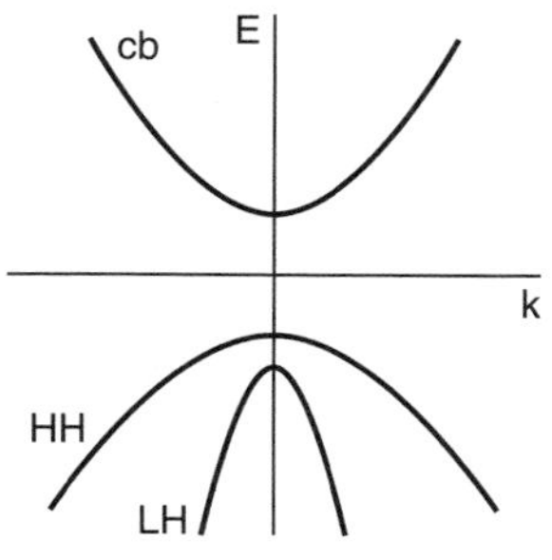

FIGURE 1.5.6
GaAs has a light LH and heavy HH hole valence band.

to make a transition from the valence to the conduction band. Those electrons in the conduction band *cb* and holes in the valence band *vb* become free to move and can participate in electrical conduction. Each value of "*k*," labels an available electron state in either the conduction or valence band. Notice that for nonzero temperatures, the electrons reside near the bottom of the conduction band and the holes occupy the top of the valence band. Carriers tend to occupy the lowest energy states because if they had higher energy, they would lose it through collisions.

Optical transitions between the valence and conduction bands require photons with energy larger than the bandgap energy. A photon has energy $E_\gamma = \hbar\omega_\gamma$ and momentum $p_\gamma = \hbar k_\gamma$, where the wavelength is $\lambda_\gamma = 2\pi/k_\gamma$ and the speed of the photon is $v = \omega_\gamma/k_\gamma$. We expect momentum and energy to be conserved when a semiconductor absorbs (or emits) a photon. The change in the *electron* energy and momentum must be $\Delta E = \hbar\omega_\gamma$ and $\Delta p = \hbar k_\gamma$ respectively. However, the momentum of the *photon* $p_\gamma = \hbar k_\gamma$ is small (but not the energy) and so $\Delta p \cong 0$. This means that $0 = \Delta p = \hbar\,\Delta k$ and, as a result $\Delta k = 0$, and so the transitions occur "vertically" in the band diagram.

Figure 1.5.7 shows an atom absorbing energy by promoting an electron to the *cb*. The absorbed photon has energy larger than the bandgap and the electron has nonzero wavevector *k*. Initially, the electron in the valence band had nonzero wavevector *k* (it was moving to the right). Now, the electron in the conduction band has nonzero wavevector (it also moves to the right with the same momentum it had in the valence band). However, now the electron has more energy than the minimum of the conduction band. The electron collides with the lattice (etc.) to produce *phonons* and drops to the minimum of the conduction band. The produced particles must be *phonons* because the settling process (a.k.a., thermalization) requires a large change in wavevector and therefore a large change in momentum. *Phonons* have small energy but large momentum whereas *photons* have large energy but small momentum. Any process that involves the phonon leads to a change in the electron wave vector; this explains why phonons are involved in transitions across the bandgap of *indirect* bandgap materials. As a side issue, notice the satellite valley on the conduction band in Figure 1.5.7 (i.e., the small dip on the right-hand side). Fast moving electrons (large *k*) can scatter into these valleys (inter-valley scattering) and constitutes an undesirable process in most cases.

1.5.4   Introduction to Band Edge Diagrams

We often describe the working of devices using band-edge diagrams. These diagrams plot energy vs position for the carriers inside a semiconductor. The next section uses this concept to explain the working of the pervasive *pn* junction.

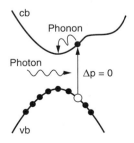

FIGURE 1.5.7

Optical transitions are "vertical" in the band diagram because the photon momentum is small. The electron can lose energy by phonon emission.

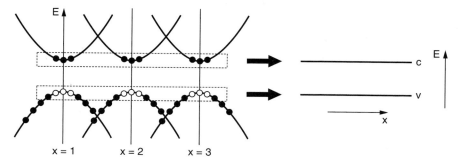

FIGURE 1.5.8

The states within an energy kT of the bottom of the conduction band or the top of the valence band form the levels in the band-edge diagram.

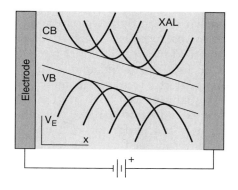

FIGURE 1.5.9

Bands bending between parallel plates connected to a battery.

The band-edge diagrams (spatial diagrams) can be found from the normal E–k band diagrams (dispersion curves). Recall that a *dispersion curve* has axes of E vs k and doesn't give any indication or information on how the energy depends on the position variable x. In fact, there must exist one dispersion curve for each value of "x" (we assume just one spatial dimension) in the material. *We group the states near the bottom of the E–k conduction band together to form the conduction band "c" for the band-edge diagram* (see Figure 1.5.8). Similarly, we group the topmost hole states in the E–k valence band to provide the valence band for the band-edge diagram. The width of the levels c and v are approximately 25 meV which is much smaller than the band gap. This is why the thin lines labeled "c" and "v" can represent the conduction and valence states in Figure 1.5.8.

Now consider the band bending effect. Imagine a semiconductor material embedded between two electrodes attached to a battery as shown in Figure 1.5.9. The electric field points from right to left inside the material. An electron placed inside the material would move towards the right under the action of the electric field. We must add energy to move an electron closer to the left-hand electrode (since it is negatively charged and naturally repels electrons). This means that *all* electrons have higher energy near the left-hand electrode and lower energy near the right-hand electrode.

For the situation depicted in Figure 1.5.9, all electrons have higher energy near the left hand electrode. The term "all electrons" refers to conduction and valence band electrons. This means that near the left electrode, the E–P diagrams must be shifted upwards to the higher energy levels. Once again grouping the states at the bottom of the conduction bands across the regions, we find a band-edge. Similarly, we group the tops of the

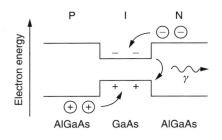

FIGURE 1.5.10
Band-edge diagram for heterostructure with a single quantum well.

valence bands. When we say that the conduction band *cb* (for example) bends, we are actually saying that the dispersion curves are displaced in energy for each adjacent point in *x*. Now we see that the electric field between the plates causes the electron energy to be larger on the left and smaller on the right. An electron placed in the crystal moves to the right to achieve the lowest possible energy. Stated equivalently, the electron moves opposite of the electric field towards the right-hand plate.

Band-edge diagrams can be used to understand a large number of opto-electronic components such as PIN photodetectors and semiconductor lasers. In fact, Figure 1.5.10 shows an example of a GaAs quantum well laser or LED having a PIN heterostructure. Actually the doping does not extend up to the well, but remains at least 500 nm away. The bands appear approximately flat under forward bias of approximately 1.7 V. The bandgap in $Al_xGa_{1-x}As$ is slightly larger than that for GaAs as can be seen from the approximate relation $E_g = 1.424 + 1.247x$ (eV) for $x < 0.5$. The semiconductor $Al_xGa_{1-x}As$ has a direct bandgap for $x < 0.5$ and becomes indirect for $x > 0.5$. Usually the clad layers (the layers right next to the well) with $x = 0.6$ are used which gives an approximate bandgap of 1.9 eV compared with 1.4 for GaAs. Applying a bias voltage to the structure causes carriers to be injected into the undoped GaAs region (well region) from the "*p*" and "*n*" regions. Electrons drop into the conduction band *cb* well, and holes drop into the valence band *vb* well. Sufficiently thin well regions form quantum wells that confine the carriers (holes and electrons) and enhance the radiative recombination process, producing photons γ.

1.5.5 Bandgap States and Defects

For perfect crystals, electrons can only occupy states in the valence and conduction bands (a similar statement holds for holes). The situation changes for doping and defects.

Consider the case for doping first. For simplicity, we specialize to *n*-type dopants such as phosphorus in silicon (refer to the previous section in connection with Figure 1.4.8). The electrons in Si–Si bonds require on the order of 1.1 eV of energy to break them free and promote them to the conduction band. Therefore, we know that the bonding electrons live in a band diagram with a band gap on the order of 1.1 eV (see the band-edge diagram in Figure 1.5.11). However, recall that a phosphorus dopant atom has 5 valence electrons but only needs 4 of them for bonding in the silicon crystal. The 5th electron remains only weakly bonded to the phosphorus nucleus at low temperatures. Small amounts of energy can ionize the dopant and promote the electron to the conduction band. Therefore, the dopant states must be very close to the conduction band as shown in the figure. At very low temperatures (below 70 K), we might expect all of the Si–Si bonding electrons to be in the valence band and most of the dopant electrons to be in the shallow dopant states. As the temperature increases, more of the dopant states empty their

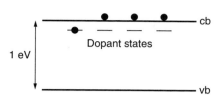

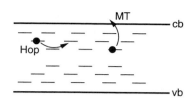

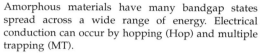

FIGURE 1.5.11

The *n*-type dopant states are very close to the conduction band.

FIGURE 1.5.12

Amorphous materials have many bandgap states spread across a wide range of energy. Electrical conduction can occur by hopping (Hop) and multiple trapping (MT).

electrons into the conduction band and the electrical conductivity must increase. By the way, the dopant states are localized states because electrons in the dopant states cannot freely move about the crystal—they orbit a nucleus in a fixed region of space.

The amorphous materials provide good examples for bandgap states arising from defects. Amorphous materials do not have perfect crystal structure. The material has many dangling bonds with 0, 1, or 2 electrons. The dangling bonds with 1 or 2 electrons require different amounts of energy to liberate an electron. For simplicity, consider dangling bonds with a single electron. These dangling bonds exist in a variety of conditions so that these electrons require a range of energy to be promoted to the conduction band (actually, for amorphous materials, the conduction band-edge becomes the "mobility edge"). The dangling bonds have very high density (i.e., the number of bonds per unit volume) and occupy a wide range of energy as shown in the band-edge diagram (Figure 1.5.12).

Electrical conduction can proceed by two mechanisms in the amorphous materials. Hopping conduction can take place between spatially and energetically close bandgap states. The electron quantum mechanically tunnels from one state to the next to produce current. Multiple trapping conduction takes place when conduction electrons repeatedly become trapped in the bandgap localized states and repeatedly absorb enough energy to become free again. Those electrons trapped closest to the center of the band gap require the greatest amount of energy to be freed. At room temperature, most phonons have an energy of approximately 25 meV. Fewer phonons have larger energy. Therefore, those electrons in the deeper traps must wait a longer amount of time to be released to the conduction band (i.e., above the mobility edge). We therefore see that the traps decrease the average mobility of the carriers by "freezing" them out for a period of time. With a little thought, you can see that the electrons tend to accumulate in the lower states. Also, these lower states near midgap tend to act as recombination centers. The electrons stay in the traps so long, that nearby holes almost certainly collide with them and recombine.

We therefore see another facet of the bandgap states. Some act purely as temporary traps and others as recombination centers. The function of the gap states, depends on the depth in the gap.

1.5.6 Recombination Mechanisms

The monomolecular, bimolecular, and Auger recombination mechanisms are especially important for light emitters. The bimolecular recombination produces spontaneous emission (radiative recombination) while the monomolecular and Auger recombination primarily involves phonons (nonradiative recombination). The recombination rates R (number of particles recombining per second per unit volume) can depend on the density of electrons n and holes p (number per volume), and on the density of bandgap defects.

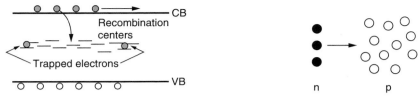

FIGURE 1.5.13
Recombination centers in the band gap.

FIGURE 1.5.14
Bimolecular recombination.

Monomolecular recombination involves a single type of carrier (at least initially). Monomolecular recombination occurs, for example, when electrons enter recombination states within the band gap (see Figure 1.5.13). They remain trapped (for the most part) until holes come along to recombine with the electrons. Even if the recombination is radiative, the emission would be at the wrong wavelength to contribute to the laser mode. The gap states remain relatively unoccupied. The carriers therefore become trapped in the gap states at a fairly constant rate. The lifetime τ_n describes the length of time that carriers (in the conduction band or valence band) can live before trapping-out in the gap states (in the absence of other processes).

The rate of decrease of carriers due to monomolecular recombination, given by n/τ_n, produces simple exponentials for the carrier distributions. For example, if we start at time $t = 0$ with a density of n_0 electrons in the conduction band without any pumping and any other form of decay, we could write

$$\frac{dn}{dt} = -\frac{n}{\tau_n} \qquad \text{to find that} \qquad n(t) = n_o \exp(-t/\tau_n) \qquad (1.5.5)$$

where the monomolecular recombination rate is $R_{\mathrm{mon}} = -dn/dt$. So τ_n represents a time constant that describes the length of time required for the carrier density n to drop to $1/e$ of the original population. The rate equations for the matter–light interaction incorporate this differential equation.

Bimolecular recombination involves both types of carriers and the rate depends on the number of each type. Figure 1.5.14 shows collisions between electrons and holes, which results in recombination. For radiative recombination, every recombination event produces a photon. The number of "collisions" must be proportional to the number of holes. The greater the number of holes and electrons in a finite volume, the greater the chance a hole and electron will collide. So we expect the number of recombination events to be proportional to the product "np." For intrinsic material as found in the active region of some light emitters, we assume equal numbers of holes and electrons ($n = p$) so that the recombination rate is proportional to n^2. We take B as the constant of proportionality. The radiative recombination rate becomes

$$R_{\mathrm{spont}} = Bn^2 \qquad (1.5.6)$$

Auger recombination is another nonradiative recombination mechanism and is important for lasers with emission wavelengths larger than 1 μm (small bandgap). For comparison, GaAs lasers generally emit between 800 to 860 nm. Semiconductor lasers made from InGaAsP can exhibit Auger recombination at relatively high power levels. There are several types of Auger recombination but they are all basically the same. Auger recombination involves collision between the same type of carrier (hole "collides" with a hole or an electron collides with an electron). This recombination channel requires phonons. Figure 1.5.15 shows an example where electron 1 collides and transfers its

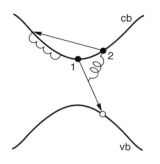

FIGURE 1.5.15
Example of Auger recombination.

energy to electron 2. Electron 1 recombines to lose its energy. Electron 2, having received the extra energy, moves higher up in the conduction band *cb*. Electron 2 cascades downward by transferring its energy to phonons, which heats up the crystal lattice. This is a nonradiative process that removes carriers that would otherwise contribute to the laser gain. It is therefore an unwanted process. Auger recombination usually occurs at larger optical power or at higher temperatures. The Auger recombination rate is proportional to n^3 because (in our example) the process involves two electrons and one hole (and $n = p$). The Auger recombination rate is

$$R_{\text{aug}} = Cn^3 \tag{1.5.7}$$

Nonradiative recombination occurs primarily through *phonon* processes. Materials with indirect bandgaps rely on the phonon process for carrier recombination. For direct bandgaps as in GaAs, the phonon processes are much less important since radiative recombination dominates the recombination channels. For nonradiative recombination, an energetic electron produces a number of phonons (rather than photons) and then recombines with a hole. The *phonon* processes reduce the efficiency of the laser because a portion of the pump must be diverted to feed these alternate (nonradiative) recombination channels. However, all lasers produce some phonons as the semiconductor heats up.

Combining all of the different types of recombination we can write the total rate of recombination R_T as

$$R_T = R_{\text{radiative}} + R_{\text{nonradiative}} = R_{\text{mon}} + R_{\text{spont}} + R_{\text{aug}} = An + Bn^2 + Cn^3 \tag{1.5.8}$$

where $A = 1/\tau_n$. The rates R have units of "number of recombination events per unit volume per second." The effective carrier lifetime τ_e, which depends on the carrier density "n," can be defined as

$$\frac{1}{\tau_e} = A + Bn + Cn^2 \tag{1.5.9}$$

Therefore, the total recombination rate must be

$$R_T = \frac{n}{\tau_e} \tag{1.5.10}$$

As we will see later, this turns out to be a nice way of writing the recombination rate since the carrier density will be approximately constant when the laser operates above threshold. For lasers made with "good" material, the term B dominates the recombination process. That means radiative recombination dominates the other recombination processes. If we restrict our attention to GaAs then C can be neglected. We will usually write

$$R_r = Bn^2 \qquad (1.5.11)$$

1.6   Introduction to the pn Junction for the Laser Diode

The semiconductor laser, light emitting diode (LED), and detector have *electronic* structures very similar to a semiconductor diode. The emitter and detector use adjacent layers of p and n type material or p, n and i (instrinsic or undoped) material. For the case of emitters, applying a forward bias voltage, controls the high concentration of holes and electrons near the junction and produces efficient carrier recombination for photon production. For the case of detectors, reverse bias voltages increase the electric field at the junction, which efficiently sweeps out (removes) any hole–electron pairs created by absorbing incident photons. The emitting and detecting devices operate only by virtue of the matter properties and the imposed electronic junction structure.

1.6.1   Junction Technology

The semiconductor *pn* junction (diode) has a special place in technology since it forms an integral part of most devices. The diode has "*p*" and "*n*" type regions as shown in Figure 1.6.1. Gallium arsenide (GaAs) serves as a prototypical material for light emitting devices. The *p*-type GaAs can be made using beryllium (Be) and zinc (Zn) as dopants whereas the *n*-type GaAs uses silicon (Si). The diode structure allows current to flow in only one direction and it exhibits a "turn-on" voltage. Some typical turn-on voltages are 1.5 for GaAs, 0.7 for Si, 0.5 for Ge. Typically, the light emitters have the *p*-type materials on the topside of the wafer where all of the fabrication takes place.

Forward or reverse bias voltages can be applied to the diode structure. The forward bias applies a field parallel to the direction of the triangle (Figure 1.6.1). In the case of GaAs, electrons and holes move into the active region where they recombine and emit light. Reverse bias voltages can be applied to the semiconductor diode, laser, and LED to use them as photodetectors. In reverse bias, photocurrent dominates the small amount of leakage current. Not all semiconductor junctions produce light under forward bias. Only the direct bandgap materials such as GaAs or InP efficiently emit light (a

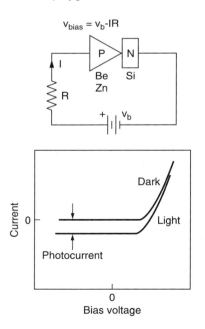

FIGURE 1.6.1
Forward biasing a GaAs laser diode (top). The *I–V* characteristics (bottom) show the photocurrent when the diode is reversed biased.

photon dominated process). The indirect bandgap materials like silicon support carrier recombination through processes involving *phonons* (lattice vibrations). Although indirect bandgap materials can emit some *photons*, the number of photons will be of orders of smaller magnitude than for the direct bandgap materials.

Semiconductor devices can be classified as homojunction or heterojunction depending on whether the laser diode consists of a single material or two (or more) distinct materials. For the emitter, the heterojunction provides better carrier and optical confinement at the active region of the device than the homojunction. Better confinement implies higher net gain and greater efficiency. The next topic discusses the formation and operation of the *pn* homojunction. Equilibrium statistics describe the carrier distributions in a diode without an applied voltage whereas nonequilibrium statistics describe the carrier distributions for forward bias.

1.6.2   Band-Edge Diagrams and the pn Junction

The doping and statistical characteristics of the material determine the properties of the *pn* junction. The *pn* diode consists of *n* and *p* type semiconductor layers. For the *n*-type material, the dopant atoms must have a weakly bound electron and the material must not have electrically active defects. Similar comments apply to the *p*-type material. Naturally the doped crystalline materials most easily satisfy these requirements. However, it is possible to form *pn* junctions in amorphous materials under appropriate conditions. The doping process "grows" mobile holes and electrons into the material. *Applying* an electric field causes the electrons in the *cb* to move from negative to positve (opposite to the direction of the applied field); holes move parallel to the applied field.

A cartoon representation of the conduction and valence bands vs *distance into a material* appears in Figure 1.6.2. The position of the Fermi level in the bandgap indicates the predominant type of carrier. For *p*-type, the Fermi level E_F has a position closer to the valence band and the material has a larger number of free holes than free electrons. Similarly, a Fermi level E_F closer to the conduction band implies a larger number of conduction electrons. When the *n*-type and *p*-type materials are isolated from each other,

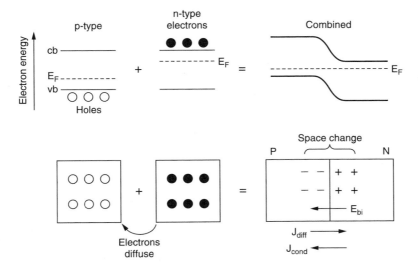

FIGURE 1.6.2
Combining two initially isolated doped semiconductors produces a PN junction with a built-in voltage (top). The built-in voltage is associated with a space charge region produced by drift and diffusion currents.

"excess" electrons in the *n*-type and holes in the *p*-type cannot come into equilibrium with each other and hence the Fermi levels (that represent statistical equilibrium) do not necessarily line up with each other.

Figure 1.6.2 shows an initial configuration for spatially separated and electrically isolated *p*-type and the *n*-type materials. Bringing the "*p*" and "*n*" type materials into contact forms a diode junction and forces the two Fermi energy levels to line up while approximately maintaining their position relative to each band except in the junction region. The final band diagram requires the conduction and valence bands to "bend" in the region of the junction. The "band" represents the energy of electrons or holes. So, to bend the band, energy must be added or subtracted in regions of space. We know from electrostatics that electric fields can change the energy.

What causes the electric field? When the two chunks of material are combined, the electrons can easily *diffuse* from the *n*-type material to the *p*-type material; similarly, holes diffuse from "*p*" to "*n*." This flow of charge maximizes the entropy and establishes equilibrium for the combined system. For example, the diffusion process might be pictured similar to the process occurring when a single blue drop and a single red drop of dye are spatially separated in a glass of water; each drop spreads out and eventually intermixes by diffusion. Unlike the dye drops, the holes and electrons carry charge and set up an electric field at the junction as they move across the interface. The diffusing electrons attach themselves to the *p*-dopants on the *p*-side but they leave behind positively charged cores. The separated charge forms a dipole layer. The direction of the built-in electric field prevents the diffusion process from indefinitely continuing. We define the diffusion current J_d to be the flow of positive charge due to diffusion alone (the figure shows positive charge diffuses to the right across the junction). We define the conduction current J_c to be the flow of charge in response to an electric field alone. Figure 1.6.2 shows that positive charge would flow from left to right under the action of the built-in field. Equilibrium occurs when $J_c = J_d$. The particles stop diffusing because of the established built-in field; an electrostatic barrier forms at the junction. Electrons on the *n*-side of the junction would be required to surmount the barrier to reach the *p*-side by diffusion; for this to occur, energy would need to be added to the electron. Diffusion causes the two Fermi levels to line-up and become flat. The Fermi energy E_F is really related to the probability that an electron will occupy a given energy level.

1.6.3 Nonequilibrium Statistics

The previous topic discussed how *n* and *p* type semiconductors when brought into contact establish a junction at statistical equilibrium. Applying forward bias to the diode produces a current and interrupts the equilibrium carrier population. Basically, any time the carrier population departs from that predicted by the Fermi-Dirac distribution, the device must be described by nonequilibrium statistics.

How should nonequilibrium situations be described? To induce current flow, we need to *apply* an electric field to reduce the electrostatic barrier at the junction so that diffusion can again occur as shown in Figure 1.6.3. The built-in electric field E_{bi} (for the equilibrium case) points from "*n*" to "*p*" and so we must apply an electric field E_{appl} that points from "*p*" to "*n*" to reduce the barrier. This requires us to connect the *p*-side of the diode to the positive terminal of a battery and the *n*-side to the negative terminal. Figure 1.6.3 shows that the applied voltage *V* reduces the built-in barrier and allows diffusion current to surmount the barrier. Notice also that the Fermi level is no longer flat in the junction region. The *applied* field is proportional to the gradient of the Fermi

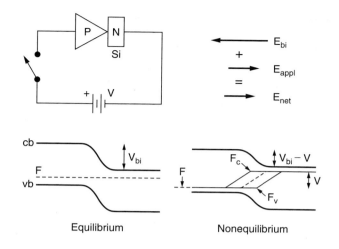

FIGURE 1.6.3
Band-edge diagrams for a PN diode in thermal equilibrium (no bias voltage) and one not in equilibrium (switch closed). The Fermi-level is flat for the case of equilibrium. However for the nonequilibrium case, the single Fermi level splits into two quasi-Fermi levels. The dotted line on the right-hand side shows the position dependent Fermi level.

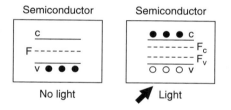

FIGURE 1.6.4
Light shining on a semiconductor (even without bias voltage) produces two quasi-Fermi levels. The quasi- Fermi levels show that we expect more electrons in the conduction band and more holes in the valence band than predicted by thermal equilibrium statistics (i.e., the Fermi-Dirac distribution).

energy E_F. The hole and electron density in the "n" and "p" regions are described by the quasi-Fermi energy levels F_v and F_c respectively. The quasi-Fermi levels describe nonequilibrium situations. We will see the importance of quasi-Fermi levels for obtaining a population inversion in a semiconductor to produce lasing. The separation between the two quasi-Fermi levels can be related to the applied voltage.

The absorption of light by a semiconductor (without any bias voltage) shows the reason for using quasi-Fermi levels. Consider Figure 1.6.4. The semiconductor absorbs photons with energy larger than the bandgap $E_g = E_c - E_v$ by promoting an electron from the valence band to the conduction band. Therefore, shining light on the material produces more electrons in the conduction band and more holes in the valence band. For the intrinsic semiconductor, the number of holes and electrons remain equal. However, if we insist on describing the situation with a single Fermi level F, then moving it closer to one of the bands increases the number of carriers in that band but reduces the number in the other. Therefore the single Fermi level must split into two in order to increase the number of carriers in both bands. The energy difference between the electron quasi-Fermi energy levels and the conduction band provides the density of electrons in the conduction band (a similar statement holds for holes and the valence band).

1.7 Introduction to Light and Optics

Semiconductor emitters produce light, and detectors absorb it. In order to describe these processes, it is first necessary to discuss the nature of light and to develop a mathematical framework. Light has both particle and wave properties. The quantum theory describes the particle nature of light. Maxwell's field equations describe the wave nature of light, the classical interaction of matter and light, and unifies all of electromagnetic (EM) phenomena (RF and optical). The classical matter–light interaction explains the refractive index, absorption/gain and nonlinear EM phenomena. These define the study of optics. In this book, we group the traditional study of optics with the study of light (including the quantum theory of light) and reserve the study of the "matter light interaction" for the quantum description of transitions.

1.7.1 Particle–Wave Nature of Light

Light and matter have both particle and wave properties. The early Greeks first proposed an "atomic" model of matter. An "atomic" model for light does not depart much from this earlier notion. In the 1600s, Newton favored the particle nature of light described by a corpuscular theory. At the same time, Huygens explained a number of light phenomena with the wave theory. In the early 1800s, Young demonstrated the interference of light beams and laid to rest the corpuscular theory. Maxwell collected all electromagnetic phenomena into the field equations, which unified the optical and RF phenomena and predicted the speed of light in vacuum. In the early 1900s, Planck proposed a new particle theory for energy transfer in order to explain the ultraviolet catastrophe of light. The quantum of energy for a wave having wavelength λ must be $E = \hbar\omega = hc/\lambda$, where $h = 2\pi\hbar$ is Planck's constant, $\omega = 2\pi f$ is the *angular* frequency corresponding to the frequency f (Hz), $f = c/\lambda$, and c is the speed of light in vacuum. Afterwards, Einstein explained the photoelectric effect with the new particle theory and later received the Nobel prize. Also about this time, Einstein developed the special theory of relativity making use of the constant speed of light in vacuum and thereby uniting space–time (and momentum–energy). Since that time, the wave–particle duality for both light and matter has become an everyday fact.

With such a long history, how do we picture light as both a particle and a wave? Most of the time, people say the act of observation (i.e., certain experiments) forces light to behave as either a particle or as a wave. Sometimes people say the particle aspect refers to quantities, such as energy and momentum, usually reserved to describe physical (nonlight) particles having mass. Today, since the advent of quantum electrodynamics (QCD—"the best theory we have"), the particle and wave properties appear in a single equation. For example, the equation for an EM plane wave $\mathscr{E} \sim \hat{b}\,e^{ikz - i\omega t}$ has both the wave aspect due to the classical plane wave $e^{ikz - i\omega t}$ (spatial-temporal mode) and the particle aspect because of the amplitude operator $\hat{b}$. A similar plane wave can be used to represent matter such as electrons (second quantization). This shows that the particle and wave aspects actually refer to separate aspects of the light; the wave aspect comes from the spatial-temporal mode (classical sinusoidal wave) and the particle aspect must be related to the amplitude. Although some aspect of light appears as a wave, only whole multiples of the quantum $E = \hbar\omega$ can be transferred.

Initially in the early 1900s, the quantum of light referred to the notion of particles of energy ($\mathscr{E}^*\mathscr{E}$). The theory later evolved to mean the quantum theory of electromagnetic fields $\mathscr{E}$. Therefore, we might surmise the quantum of energy should be recovered from the quantity $\mathscr{E}^*\mathscr{E} \sim \hat{b}^+\hat{b}$ where "+" represents the complex conjugate for operators. In fact, we will see that the quantity $\hat{b}^+\hat{b}$ represents the number operator that gives the number of photons in an EM mode.

We represent the amplitude of the light by an operator, which must operate on "something" to provide a value. This something is a vector space (i.e., a function space). The vectors in the space determine the exact nature of the plane wave. We consider Fock, coherent, and squeezed type vectors. A laser operated at sufficiently high power produces a sinusoidal wave most closely related to the coherent state. Low noise lasers and parametric amplifiers can produce the squeezed states. It turns out that repeated simultaneous measurements of the magnitude and phase of the amplitude do not yield a single number for the amplitude and a single number for the phase; the measurements interfere with each other. As a result, the light has an intrinsic statistical distribution for the photon number and the phase. In the coherent state, repeated measurement of the photon number produces a range of values. The photon number has a Poisson distribution.

1.7.2 Classical Method of Controlling Light

Maxwell's equations unify the electromagnetic phenomena using the framework of waves. It describes both the free fields and those interacting with matter. In the theory, matter can produce or absorb light using dipoles (in addition to other mechanisms). Electrical dipoles consist of a bound pair of charges of opposite sign as shown in Figure 1.7.1. As will be discussed in more detail later, the dipole is represented by

$$\vec{p} = q\vec{r} \tag{1.7.1}$$

where q is the magnitude of one of the charges and $\vec{r}$ is the separation between them. Usually, we are most interested in the induced dipoles which means they are formed by applying an electric field $\vec{\mathscr{E}}$. Although bound, the charges in the induced dipole can move (for example, imagine the two charges connect by a linear spring). The figure shows two charges capable of changing positions. The polarization $\vec{\mathscr{P}}$ describes the number of dipoles per unit volume. The induced polarization must be related to the field according to

$$\vec{\mathscr{P}} = \chi\vec{\mathscr{E}} \tag{1.7.2}$$

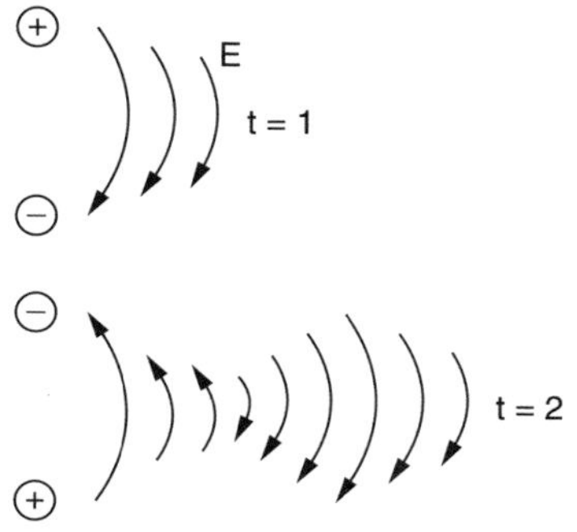

FIGURE 1.7.1
The oscillating charge produce an electric field that moves into space at speed c.

where χ represents the susceptibility and describes how easily the electric field can induce dipoles.

The dipoles produce light (i.e., produce gain), absorb light and provide an index of refraction. Actually, the gain and absorption can be related to the complex parts of a refractive index and a wave vector. First, let's see how the dipoles produce an EM wave. Suppose an induced dipole moment oscillates at frequency ω as shown in Figure 1.7.1. The top portion of the figure shows the electric field due to the two point charges. After a period of time, as shown in the bottom portion of the figure, the two charges have changed position and the electric field points in the opposite direction. As the field changes, the lines of force radiate into space with the speed of light. The moving charge produces current at the position of the dipole and therefore produces a magnetic field that also moves into space.

What happens if the radiated field from an oscillating dipole travels through a dielectric (i.e., a material capable of being polarized)? The EM wave can excite (induce) the oscillation of a collection of dipoles (the incident electric field forces the charges to separate). If the electric field interacts with a dielectric then its speed becomes c/n (where n is the refractive index). Basically, dipoles absorb the EM field and then re-radiate the field. The absorption-radiation sequence takes some time and slows down the progression of the EM wave. The colors (frequencies) that most closely match the resonant frequency of the dipoles therefore interact the strongest and should therefore propagate the slowest. The index of refraction must therefore be linked with the frequency response of the dipoles (i.e., the frequency response of the susceptibility). If the oscillators are damped (friction), then the absorbed light can be converted into heat and not re-radiated.

Elementary courses on optics show how the refractive index can be used to manipulate and control light. The refractive index of glass makes it possible to focus light using lenses. The dipole absorption can be used for color filters. The dipole emission properties produce gain in semiconductor lasers and optical amplifiers. As indicated in the next topic, the index makes it possible to control the position of the wave as it propagates through a semiconductor wafer (waveguiding). The laser would not be of much use without the waveguide.

Nonlinear optics uses the departure of dipoles from the simple linear relation $\vec{\mathscr{P}} = \chi \vec{\mathscr{E}}$ with χ a constant over the field range of interest. In some cases, larger electric fields stretch the "dipole spring" to where it no longer behaves linearly with field. In this case, the susceptibility χ changes with the field. We can now imagine applying a "steady-state" field to stretch the dipoles to set the value of χ. Then a small incident EM wave will experience a refractive index set by the polarization through χ. This nonlinear behavior can be used to make electrically controlled lenses and waveguide switches.

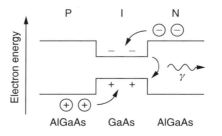

FIGURE 1.7.2
Band-edge diagram for an AlGaAs-GaAs heterostructure.

1.7.3 The Ridge Waveguide

The heterojunction GaAs laser has a PIN structure consisting of an undoped (i.e., intrinsic) layer sandwiched between the "*p*" and "*n*" layers as shown in Figure 1.7.2. The bands appear approximately flat under forward bias of approximately 1.7 V. The bandgap of $Al_xGa_{1-x}As$ is slightly larger than that for GaAs according to the approximate relation $E_g = 1.42 + 0.78x$ in electron volts for $x < 0.5$ (see Figure 1.2.1). The semiconductor $Al_xGa_{1-x}As$ has a direct bandgap for $x < 0.5$ and becomes indirect for $x > 0.5$. Usually the heterostructure uses $x = 0$ (pure GaAs) and $x = 0.5$ (50% aluminum) which gives bandgaps of 1.5 and 1.8 eV, respectively. Besides controlling the bandgap, the aluminum concentration also determines the refractive index of the material. A form of the Sellmeier equation gives the refractive index of undoped $Al_xGa_{1-x}As$ to within a few percent

$$n = \left[A + \frac{B}{\lambda^2 - C} - D\lambda^2 \right]^{1/2} \tag{1.7.3}$$

where $A = 13.5 - 15.4x + 11.0x^2$, $B = 0.690 + 3.60x - 4.24x^2$, $C = 0.154 - 0.476x + 0.469x^2$, $D = 1.84 - 8.18x + 7.00x^2$, and the vacuum wavelength λ has the range of 0.564 to 1.033 µm. Corrections to the index of refraction due to dielectric absorption and the conductivity of doped material are neglected. Equation (1.7.3) shows that increasing aluminum concentrations produce decreasing refractive indices.

The heterostructure plays two very important roles for the laser. First it provides an optical waveguide and second it can be used to make quantum wells. The optical waveguide confines the light to regions of high gain. Figure 1.7.3 shows a ridge guided laser (also see Figure 1.7.4) using two different waveguiding mechanisms for the transverse and longitudinal directions. Consider the transverse direction. The waveguide has a core region with a refractive index larger than that for the surrounding cladding. Equation (1.7.3) shows that the larger bandgap material has the smaller refractive index. The lower refractive index of the cladding ($Al_{0.5}Ga_{0.5}As$) confines the transverse optical mode to a width about 300 nm. Often the composition of the PIN structure is graded rather than the flat structure shown in the figure.

The efficiency of the semiconductor laser can be improved by keeping the optical mode away from the doping. The evanescent tail of the optical mode can extend as far as 500 nm or more depending on the difference between the refractive indices of the active and surrounding regions. Free carriers in the doped regions tend to absorb the fields and reduce the efficiency of the laser. The core resides between the cladding layers (transverse direction along x) which have smaller refractive indices.

The ridge waveguide provides an example of a waveguiding mechanism that confines the optical mode along the lateral direction. The ridge along the length of the laser (longitudinal direction) defines a waveguide with a length on the order of 100 to 1000 µm. The ridge and therefore the output beam are typically 5 µm wide. The ridge provides lateral confinement so long as the surface "*s*" next to the ridge is within approximately 150 nm of the active region (see Figure 1.7.4). The evanescent tail of the optical mode must press against the air—$Al_{0.5}Ga_{0.5}As$ interface. The effective index of the $Al_{0.5}Ga_{0.5}As$ decreases because the lower index of the air must be made part of the average. The *lateral* confinement (especially for gain or ridge guided lasers) can be quite weak and the evanescent tail can extend up to a micron along the lateral direction. Better lateral confinement can be obtained by using a buried heterojunction which places the optical mode in the "center" of the wafer with low index materials on all four sides along the length.

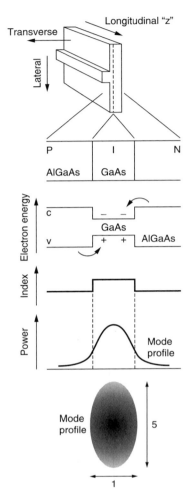

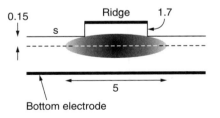

FIGURE 1.7.3
Construction of the laser heterostructure and the resulting mode profile.

FIGURE 1.7.4
Edge view of the mode.

The heterostructure can be used to form electron and hole quantum wells in the active region of the laser. The wells appear similar to those in Figure 1.7.3 but with a width on the order of 50 to 200 angstroms (Å). The quantum wells spatially confine the electrons and holes and thereby increase the recombination efficiency. In addition, the clad layers improve the overlap between the mode and the well region to further improve the efficiency.

1.7.4 The Confinement Factor

The confinement factor represents the fraction of power (or fraction of photons) confined to a volume; it often determines the performance of a device. Let $P(\vec{r}) \sim \tilde{\mathscr{E}}^* \mathscr{E}$ represent the optical power density at point $\vec{r}$. The fraction of the total power contained within a volume V can be written as

$$\Gamma = \frac{\text{Power in } V}{\text{Total Power}} = \frac{\int_V P(\vec{r}) \, dV}{\int_{\text{All Space}} P(\vec{r}) \, dV} \tag{1.7.4}$$

Two types of confinement factor are often encountered in optoelectronics. The first measures the percentage of optical power confined to the core of a waveguide (volume V_c). The second measures the percentage confined to an "active" region (volume V_a) such as the quantum wells in semiconductor laser.

The active region (volume V) contains the holes and electrons that recombine. For a laser, this region must interact with the optical mode to produce more light. The active region consists of intrinsic material for a number of semiconductor lasers. Electrical pumping (as indicated by the bias current I in Figure 1.7.5) or optical pumping can be used to initiate and maintain an electron and hole population. In the active region (the I region for the example in Figure 1.7.5), the holes and electrons recombine to produce light. Spontaneous emission initiates laser oscillation for sufficiently large pump levels. Once lasing begins, the laser light propagates back and forth between the two partially reflective mirrors. The escaping light provides the output laser signal. During laser operation, the active region continues to produce a small amount of spontaneous emission, which escapes through the mirrors and the sides of the laser.

The index differences between the core and adjacent cladding regions determine the confinement of the optical mode to the active region. The optical mode extends into the cladding so that the modal volume V_γ is larger than volume V of the active region as shown in Figure 1.7.5. The simplest model assumes that the optical power density P is uniformly distributed in the volume V_γ and zero outside the volume. The optical confinement factor Γ in Equation (1.7.4) gives the fraction of the optical mode that overlaps the gain region $\Gamma = V/V_\gamma$. In reality, the optical energy is not uniformly distributed along "x" (the transverse direction in this case). For example, the mode intensity drops off exponentially (this is the evanescent tail) in the cladding region. Therefore, the actual confinement factor can best be found by integrating the optical power density along the x-direction.

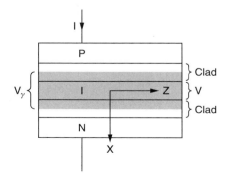

FIGURE 1.7.5
Structure of the semiconductor laser.

1.8 Introduction to Noise in Optoelectronic Components

Many types of noise can be found in optoelectronic devices and systems. There are many different contexts for the term noise. In one context, it might refer to random fluctuations in a signal and in another, it might refer to unwanted steady-state levels. Discussions of noise in optoelectronic systems often focus on shot, Johnson, low frequency, and spontaneous emission noise. The relative intensity noise (RIN) describes the fluctuations in the power emitted from a device. Other sources of noise exist such as the production of harmonic components by nonlinear devices or mode hopping for lasers.

Shot noise can be generally viewed as due to the fluctuating arrival times of randomly generated particles. Shot noise in optoelectronic components refers to the fluctuations in photocurrent due to the random generation of carriers or to fluctuations in optical power due to the random arrival of photons. This type of noise is often viewed as a basic limitation and the devices are termed shot noise limited. The random arrival of photons can best be understood in terms of the photon statistics of an electromagnetic wave in a coherent state as discussed in Chapter 5. Sub-shot noise components can be designed that use squeezed states of light.

Relative intensity noise (RIN) refers to a ratio of the noise (as measured by a standard deviation) to the signal power. If a variance characterizes the noise then the power squared characterizes the signal. The definition is quite general and describes a variety of noise sources including thermal, quantum mechanical, and spontaneous emission. Sometimes people apply the term RIN solely to spontaneous emission (since it dominates the others under many circumstances). For a laser, the coherent output beam comprises the signal and spontaneous emission comprises the noise. However, we should not consider spontaneous emission to be noise for all systems and devices. Light emitting diodes produce spontaneous emission as the signal and not as noise.

Thermal background noise occurs in semiconductors when phonons interact with the charges to produce thermal equilibrium. The Boltzmann distribution (a limiting case for the Fermi-Dirac statistics) gives the probable number of electrons in the conduction band (for thermal equilibrium). Electrons in the conduction band can also absorb thermal energy, which increases the kinetic energy of those electrons until other processes dissipate the energy. The noise appears in conduction processes. Background thermal noise can be controlled to some extent by using wide-bandgap materials and thermal coolers.

The mechanisms and analysis of noise, aside from being very interesting in their own right, often lead to new or improved devices and systems. The classical studies of noise lead to the study of noise in the quantum theory. The present section introduces noise in optoelectronic components including Johnson, low frequency, shot, and spontaneous emission noise.

1.8.1 Brief Essay on Noise for Systems

Communication systems normally require large signal-to-noise ratios (SNR) for clear and accurate transfer of information. Sometimes people define the SNR as the reciprocal of the RIN. At other times, it can be defined as the ratio of the average signal level to the standard deviation of the noise signal $SNR = \overline{P}/\sigma$. The SNR can be improved by either increasing the signal level or decreasing the noise. Increasing the signal level generally requires larger power and physically larger devices. However, larger power and larger-sized devices run contrary to the requirements and trends in modern VLSI design and space applications. Therefore reducing the noise level constitutes a desirable alternative approach.

Communication systems and links typically require large dynamic range. The dynamic range refers to the range of values that an output parameter can assume. Large signals sometimes induce nonlinear behavior in components and subsystems. Many systems limit the useful signal range to exclude this adverse behavior. For example, the signal range for an analog transistor amplifier should be limited to prevent the onset of saturation as the voltage swings near the supply rails. However, the swing from rail-to-rail can be desirable for digital systems. The "noise floor" represents the background noise always present in a given system. Obviously, the noise floor limits the dynamic range. By lowering the noise-floor, the dynamic range can be increased.

Noise can potentially be more detrimental to analog signals than to digital ones. An analog signal usually carries information on a continuously varying parameter (such as temperature or amplitude) and therefore, the noise determines the ultimate precision of the measurement or the quality of the impressed information. Noise as small as 0.1% can be significant for audio applications (for example). For a digital system, the impact of noise manifests itself somewhat differently. The digital system is designed with hysteresis and a threshold to provide a clear distinction between a logic "0" and "1." The effects of noise can be characterized by the "bit error rate." Of course, anyone who compares music from a vinyl record with that from a compact disk clearly understands the distinction between the digital and analog noise.

Photonic and RF systems use signals in the form of the electromagnetic field, which carries shot noise. Although present at all power levels, these nonclassical effects become most evident for small numbers of photons. The shot noise can be related to variations in the number of photons in a light beam. The power level for which the granularity of the field becomes significant depends on the frequency. High-frequency sources (GHz and larger), such as the laser or maser, require relatively few photons to give a specific power level as compared with low frequency sources, such as AM radio. Generally, low power levels translate directly to small photon numbers.

Communications and data transfer systems would most benefit from low power, small sized devices with high S/N ratios. Space platforms especially must have light launch-pad weight. The space platforms have limited power resources and power dissipation capabilities. The small devices have small numbers of atoms that can only produce small numbers of photons. Small systems tend to be more noise-prone (even for EM noise) than larger ones and therefore, the S/N ratios must generally be smaller. For small systems, noise must be a problem because small (and low power) components do not deal with many particles (electrons, holes, and photons) at one time. For low particle numbers, the uncertainty (or standard deviation) in the signal is roughly the same size as the magnitude of the quantity itself. Equivalently stated, the standard deviation of the number of particles (that represent the signal) is relatively large compared with the average number.

1.8.2 Johnson Noise

Johnson noise refers to random variations in voltage across a resistor even when left unbiased. The literature often terms this type of noise as resistor or Nyquist noise. The random fluctuations in the motion of charge carriers within the resistor produces random fluctuations in voltage or current. The noise originates in collisions between carriers and scattering centers (the mechanisms producing resistance). A thermal distribution sets the mean velocity. This type of noise occurs even when the number of electrons or holes remains constant.

The power of the noise can be calculated by several methods. We follow a method first developed by Nyquist. Consider two resistors in thermal equilibrium with each other

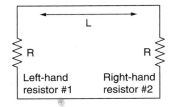

FIGURE 1.8.1

Two resistors in thermal equilibrium transfer noise power between each other.

and interconnected by a transmission line as shown in Figure 1.8.1. The two resistors produce identical amounts of noise power when they have the same temperature. We find the noise power from the left-hand resistor. For simplicity, assume the power from the left-hand resistor flows in a clockwise direction (and the power from the right-hand flows in the counter-clockwise direction). Therefore, the waves in the upper part of the loop have the form $e^{ikz-i\omega t}$. None of the power incident on the right-hand resistor will be reflected so long as the value of R matches the characteristic impedance of the transmission line. We could allow the power to flow toward the right-hand resistor along both the top and bottom branches so long as we realize that only half the power flows along each branch and, later, double the number of available modes.

Assume the closed loop supports propagation modes with wavelengths having sub-multiples of the length L. That is, assume

$$\lambda = \frac{L}{m} \quad \rightarrow \quad k = \frac{2\pi}{\lambda} = \frac{2m\pi}{L} \quad \text{where} \quad m = 1, 2, 3, \dots \quad (1.8.1)$$

Each k-value gives an allowed mode of the system (traveling wave). Similar to the procedure for density of states found in the solid state, the spacing between adjacent k-values must be $\Delta k = 2\pi/L$. The number of modes per unit k-length must be

$$g_k = \frac{1}{\Delta k} = \frac{L}{2\pi} \quad (1.8.2)$$

The number of modes in the interval $(0, k)$ must be

$$N(k) = g_k k = \frac{Lk}{2\pi} \quad (1.8.3)$$

This must be the same number of modes in the frequency interval $(0, \nu]$ where $\nu = c/\lambda = ck/(2\pi)$ in Hz, and c represents the speed of the wave on the transmission line (smaller than the speed of light in vacuum). The number of modes in this frequency interval can be written as

$$N(\nu) = \frac{Lk}{2\pi} = \frac{L\nu}{c}$$

The density of frequency states can be written

$$g_\nu = \frac{d}{d\nu} N(\nu) = \frac{2L}{c} \quad (1.8.4)$$

Notice the same result for the density of frequency states can most easily be found by the usual formula $g_\nu \, d\nu = g_k \, dk$.

The power flow for energy in the modes in the frequency interval $(\nu, \nu + \Delta\nu)$ can be written as

$$dP = \frac{\text{Energy}}{\text{Length}} \frac{\text{Length}}{\text{Second}} = \frac{1}{\text{Length}} \frac{\text{Energy}}{\text{Mode}} \frac{\text{Modes}}{\text{Hertz}} \frac{\text{Length}}{\text{Second}} d\nu = \frac{1}{L} \frac{\hbar\omega}{e^{\hbar\omega/kT} - 1} \frac{L}{c} c \, d\nu \quad (1.8.5)$$

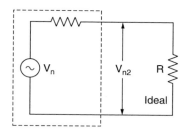

FIGURE 1.8.2
The left-hand resistor represented by a noise source and an ideal resistor R.

where T denotes the equilibrium temperature in degrees Kelvin, $\hbar\omega = h\nu$. The Nyquist paper clearly indicates the factor $1/L$ appears since energy transferred to the right-hand resistor only needs to move through the distance L. The exponential term can be approximated by

$$\hbar\omega\left[e^{\hbar\omega/kT} - 1\right]^{-1} \sim kT \quad \text{when} \quad \frac{\hbar\omega}{kT} \ll 1$$

We therefore find the power flowing to the right-hand resistor must be

$$dP = kT\,d\nu \tag{1.8.6}$$

Often times, the quantity $kT\,d\nu$ is termed the "available noise power." The same results can be derived by assuming a thermal distribution and calculating

Now suppose the right-hand resistor is ideal in the sense that it does not generate Johnson noise. For example, the right-hand resistor might be held at or near 0 K. Figure 1.8.2 shows a model for the left-hand resistor as a voltage source in series with an ideal resistor. The model defines an RMS noise voltage V_n. The RMS noise voltage across the right-hand resistor due to power flow from the left-hand side has the form $V_{n2} = V_n/2$. The RMS current through the right-hand resistor can be defined as $I_{n2} = V_{n2}/R = V_n/2R$. Therefore, the current through the ideal right-hand resistor must be

$$V_{n2}I_{n2} = dP = kT\,d\nu \quad \rightarrow \quad V_{n2}^2 = kTR\,d\nu \tag{1.8.7}$$

Substituting the relation between V_{n2} and V_n provides an expression for the RMS source V_n

$$V_n^2 = 4kTR\,d\nu \quad \rightarrow \quad V_n = \sqrt{4kTR\,d\nu} \tag{1.8.8}$$

The voltage per root Hertz is $\sqrt{4kTR}$. It might seem strange that the voltage increases with bandwidth $d\nu$. The bandwidth in the formula comes directly from the constant density of states.

We might intuitively understand the formulas in Equations (1.8.6) and (1.8.7) as follows. The resistance appears because it directly depends on the number of collisions that produce sudden changes of velocity and hence random changes in current. The rate of collision must depend on temperature since the thermal distribution determines the speed of the electron between collisions (also the density of phonons increases with temperature).

EXAMPLE 1.8.1
Find the RMS voltage at 300 K across a 1 kΩ resistor in a bandwidth of 1 Hz.
Solution: Use $4kT = 1.67 \times 10^{-20}$ to find $V_n = \sqrt{4kTR\,d\nu} = 4.1\,nV$ Volts$\sqrt{\text{Hz}}$

EXAMPLE 1.8.2
Find the noise power in dBm for the previous example.
Solution: The definition of noise power in dBm is

$$\text{dBm} = 10 \log_{10}\left(\frac{P}{P_o}\right) \tag{1.8.9}$$

where $P_o = 1\,\text{mW}$. This gives $\text{dBm} = -168$.

1.8.3 Low Frequency Noise

Low frequency noise (or excess noise) has a $1/f$ (where f denotes the frequency) power spectrum. This type of noise occurs only when current flows through a device in contradistinction to Johnson noise that does not require an applied voltage. In the case of $1/f$ noise, the charged particles move under the applied field and randomly encounter a scattering center or trap. The change in motion of the particle results in noise. The low frequency noise has a number of alternative names including excess noise, current noise, pink noise and semiconductor noise. This noise is in addition to the Johnson noise already present.

Composition carbon resistors exhibit significant amounts of excess noise compared with the metal film resistors. The carbon composition resistors consist of granules of carbon pressed together. Moving charges experience a nonuniform medium that results in random variations of the current flow. Some books [Reference 5] report excess resistor noise as high as $3\,\mu\text{V/decade}$.

The noise power has the form (power spectrum) $P(f) = C/f$ where C is a constant measured in watts. The total power has the form

$$P_{1/f} = \int_{f_L}^{f_H} df\,\frac{C}{f} = C\,\text{Ln}\frac{f_H}{f_L} \tag{1.8.10}$$

Therefore, each decade of frequency has the same power. As much noise power exist in the range 0.1 to 1 Hz as in 100 to 1000 Hz.

1.8.4 The Origin of Shot Noise

Shot noise originates in the random generation or arrival times of particles. A Poisson probability distribution describes the number of particles produced or detected in each interval of time. Examples abound for optical, electronic, and mechanical systems. For example, thermal generation of carriers within the depletion region of the *pn* or PIN junction and their subsequent sweep-out produces shot noise. The carriers suddenly appear according to a Poisson distribution; this random variation of charge density produces random changes in the current through the device.

For electronic components, Johnson noise differs from shot noise since Johnson noise does not require the number of carriers to change nor does it require any current. Although $1/f$ noise requires current, it does not require the generation of particles at random times. Therefore, the noise through an electronic component might simultaneously exhibit $1/f$ and shot noise. For optical components, Chapter 6 covering the quantum theory of electromagnetic fields will show that coherent states also exhibit Poisson statistics.

We first demonstrate the Poisson distribution characteristic of shot noise. For illustration, consider the random generation of electrons at the left-hand plate of

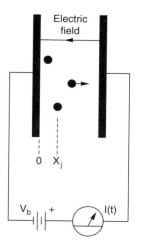

FIGURE 1.8.3
Electrons randomly generated move in the applied field.

a capacitor and their subsequent arrival at the right-hand side as shown in Figure 1.8.4. Similar arrangements apply to *pn* junctions and vacuum tubes. Assume t_j denotes the random electron-generation time and the x_j represents the position for the *j*th electron at time t. We assume a uniform distribution of generation times. We want to know the probability of finding the number n of electrons arriving at the right-hand electrode during a time interval t.

The calculation proceeds by using recursion relations. We first require a few preliminary relations that require the probability $P(n, \Delta t)$ of finding n electrons emitted in the small time interval $\Delta t \sim 0$. The probability of finding $n = 0$ electrons in the interval must be $P(0, 0) = \mathrm{Lim}_{\Delta t \to 0} P(0, \Delta t) = 1$. The probability of finding a single electron in the interval Δt must be proportional to the rate of generation (or arrival) r and the size of the time interval Δt according to

$$P(1, \Delta t) \approx r\,\Delta t \tag{1.8.11}$$

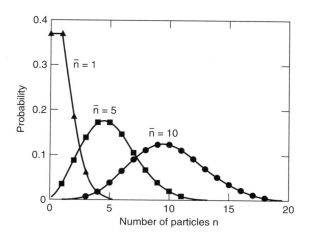

FIGURE 1.8.4
Example plots of the Poisson distribution.

We will later substantiate the claim of r being a rate (number of particles emitted per second). The probability of at least 1 electron arriving in Δt must be

$$P(0, \Delta t) + P(1, \Delta t) + P(2, \Delta t) + \cdots = 1$$

We assume a small enough time interval Δt that $P(n, \Delta t)$ remains negligible for $n \geq 2$. Therefore, we find

$$P(0, \Delta t) + P(1, \Delta t) = 1 \tag{1.8.12}$$

Now we find the functional form for the probability $P(n, t)$ of finding n electrons during the interval of time t. Consider the interval $(0, t + \Delta t)$. For sufficiently small times Δt, either 0 or 1 electrons might be emitted. This means the probability of finding n electrons in the time interval $(0, t + \Delta t)$ must be given by

$$P(n, t + \Delta t) = P(n, t; 0, \Delta t) + P(n-1, t; 0, \Delta t) = P(n, t)P(0, \Delta t) + P(n-1, t)P(1, \Delta t) \tag{1.8.13}$$

where the symbol $P(n, t + \Delta t)$ is the conditional probability of finding n electrons by time t given that 0 electrons are found during the time interval Δt. This last equation holds because the number of electrons m_Δ arriving in interval Δt must be independent of the number m arriving in the interval t so that $P(m, t; m_\Delta, \Delta t) = P(m, t)P(m_\Delta, \Delta t)$.

Use Equation (1.8.12) to eliminate $P(0, \Delta t)$ and use Equation (1.8.11) in place of $P(1, \Delta t)$ to find

$$\frac{P(n, t + \Delta t) - P(n, t)}{\Delta t} + r P(n, t) = r P(n-1, t) \tag{1.8.14a}$$

which becomes a differential equation in the limit $\Delta t \to 0$

$$\frac{dP(n, t)}{dt} + r P(n, t) = r P(n-1, t) \tag{1.8.14b}$$

This formula can be rewritten using an integrating factor as described in Appendix 1. We find the recursion relation with $P(n, 0) = 0$ to be

$$P(n, t) = P(n, 0) + r e^{-rt} \int_0^t d\tau\, e^{r\tau} P(n-1, \tau) = r e^{-rt} \int_0^t d\tau\, e^{r\tau} P(n-1, \tau) \tag{1.8.14c}$$

We now use the recursion relation to find $P(n, t)$. We need a starting function $P(0, t)$. This can be found from Equation (1.8.14b) since $P(n-1, t)$ would not be present in this case. We find

$$\frac{dP(0, t)}{dt} = -r P(0, t)$$

The solution can be found using the starting condition $P(0, 0) = 1$

$$P(0, t) = e^{-rt} \tag{1.8.15a}$$

Other cases can be found. The $n=1$ case comes from Equations (1.8.14c) and (1.8.15a)

$$P(1,t) = r\,\mathrm{e}^{-rt}\int_0^t d\tau\,\mathrm{e}^{r\tau}P(0,\tau) = r\,\mathrm{e}^{-rt}\int_0^t d\tau\,1 = rt\,\mathrm{e}^{-rt} \qquad (1.8.15b)$$

Similarly the recursion relation provides the desired results (proof by induction)

$$P(n,t) = \frac{(rt)^n\,\mathrm{e}^{-rt}}{n!} \qquad (1.8.16)$$

The last equation represents the Poisson distribution. The key assumptions include the independent nature of the emission events (or emission times or arrival times).

Some of the important parameters can be evaluated. The expected number of electrons emitted in the time t can be evaluated as follows (Problem 1.10)

$$\langle n \rangle \equiv \bar{n} = \sum_{n=0}^{\infty} n\,P(n,t) = \mathrm{e}^{-rt}\sum_{n=0}^{\infty} n\,\frac{(rt)^n}{n!} = rt \qquad (1.8.17)$$

This shows that r can be interpreted as the average rate of emission. The Poisson distribution can now be written as

$$P(n,t) = \frac{\bar{n}^n\,\mathrm{e}^{-\bar{n}}}{n!} \qquad (1.8.18)$$

The value $\bar{n}$ represents the average number of particles generated during the time t. In the case of emission, we therefore expect the number of particles per unit length to be

$$\rho_L = \frac{\text{number}}{\text{length}} = \frac{\text{number}}{\text{second}}\,\frac{\text{second}}{\text{length}} = r/v.$$

The standard deviation can be found using Equation (1.8.18)

$$\sigma^2 = \langle n^2 - \bar{n}^2 \rangle = \sum_{n=0}^{\infty}(n^2 - \bar{n}^2)\,P(n,t) = \bar{n} \qquad \rightarrow \qquad \sigma = \sqrt{\bar{n}} \qquad (1.8.19)$$

where σ^2 and σ represent the variance and the standard deviation, respectively (see Problem 1.11). Therefore, the average and standard deviation are not independent parameters for the Poisson distribution.

1.8.5   The Magnitude of the Shot Noise

For a diode (or LED), two sources of shot noise exist. Figure 1.8.5 shows an example band structure for a PIN structure near thermal equilibrium. Two sources of current can be identified. Diffusion causes electrons to randomly surmount the barrier into the P region. Generation randomly produces electron–hole pairs that surmount the bandgap, enter the bands, and separate under the action of the fields.

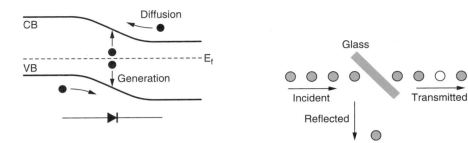

FIGURE 1.8.5
A PIN junction with diffusion and thermally generated current.

FIGURE 1.8.6
Imperfect reflecting surfaces induce partition noise.

Correlation studies show the shot noise current due to DC current I must be

$$I_{\text{shot}} = \sqrt{2qI\,\Delta f} \qquad (1.8.20)$$

For a reverse biased junction such as for a photodetector, thermally generated carriers and the resulting reverse saturation current I_s make up the predominate current. Under forward bias, the conduction current dominates

$$I = I_s\big[e^{qV/kT} - 1\big] \qquad (1.8.21)$$

An unbiased diode balances I_s and I so that the mean-square value of shot noise becomes

$$I_{\text{shot}} = \sqrt{4qI_s\,\Delta f} \qquad (1.8.22)$$

1.8.6 Introduction to Noise in Optics

Later chapters write the electric field in terms of quadratures. The single mode field has the form

$$\vec{E}(\vec{r},t) \sim \hat{Q}\,\sin\!\big(\vec{k}\cdot\vec{r} - \omega t\big) + \hat{P}\,\cos\!\big(\vec{k}\cdot\vec{r} - \omega t\big)$$

where the quadrature amplitude operators do not commute $[Q,P] = i$. This naturally requires a Heisenberg uncertainty relation and we cannot simultaneously and precisely know the components of the field. The commutation relations lead to quantum noise. The quadrature amplitudes must operate on a vector space. The particular vectors determine the specifics of an optical beam. Coherent states produce shot noise that follow a Poisson distribution for the photon number. Squeezed states produce sub-shot noise.

Other mechanisms produce noise. For example, partially reflective surfaces introduce noise as illustrated in Figure 1.8.6. A piece of glass, for example, reflects a portion of the incident photons. For every reflected photon, the output stream must be missing one (indicated by the open circle). The input stream has a standard deviation of zero. The reflected and transmitted beams have nonzero standard deviation. This example shows the noise added by the reflecting surface—partition noise. Interfaces not perfectly reflecting or transmitting add partition noise. Absorbers and multiple modes in a laser beam introduce similar sources of noise.

1.9   Review Exercises

1.1 Consider semiconductor lasers operating at 0.8 μm and 1.55 μm.

1. Find the bandgaps required for the lasers to operate at these wavelengths. Assume the electrons and holes occupy minimum energy states.
2. Use the graphs in the chapter to find the possible materials that will provide these wavelengths.

1.2 Explain how optical confinement and electron/hole confinement improve the efficiency of the laser.

1.3 Consider a PN junction. Sketch an approximate plot of the optical power emitted from the junction as a function of voltage. Explain any assumptions.

1.4 Suppose a homojunction LED is made from semiconductor with a bandgap of $E_g = 1.5\,\text{eV}$. Assume the LED produces approximately 2 mW of optical power at a bias current of 20 mA.

1. Explain why the bias voltage across the LED must be approximately 1.5 V to produce significant emission. Hint: Use a band-edge diagram for flat bands.
2. Draw a circuit diagram using a 10 V battery, a resistor and the LED that will make the LED glow. What value of bias resistor should be used to achieve approximately 2 mW of optical power?

1.5 Applying reverse bias to an LED allows it to operate as a photodetector.

1. What composition of $Al_xGa_{1-x}As$ will allow a homojunction device to absorb near 800 nm?
2. Show the circuit diagram for connecting a 10 V battery, resistor, and LED to make a detector circuit. Assume the output signal is taken across R.
3. If the LED in reverse bias has a response of 0.2 mA/mW, determine R to produce a signal gain of 200 V/W.

1.6 You have a parts box with a yellow-emitting LED, a red-emitting LED, a silicon NPN transistor with a current gain of $\beta \sim 200$, a resistor, and a 10 V battery.

1. Draw a circuit for a wavelength converter using the common emitter configuration. Use the yellow LED as a detector and the red LED as an emitter.
2. If the yellow LED produces 0.1 mA/mW as a detector and the red LED produces 0.2 mW/mA, then find the overall gain of the circuit from input to output as the ratio of output power to input power.

1.7 A student wants to build a semiconductor laser transmitter. She plans to connect the output of a radio (100 mV rms output signal) to the transmitter. Design a transistor laser driver and receiver using any assortment of other parts including resistors, capacitors, and LM-741 Op Amps. Assume the laser has a turn-on voltage of 1.5 V and a threshold current of 25 mA. Assume the laser power output is linear in the bias current and the transfer function has the magnitude of 0.2 mW/mA.

1.8 A student wants to observe the thermal noise from a resistor. He buys a low noise amplifier with a voltage gain up to 10,000. Assume the noise level for the amplifier is $1\,\text{nV}/\sqrt{\text{Hz}}$ referred to the input (i.e., the noise level must be multiplied by the gain to find the output noise). Assume the resistor connected to the input of the amplifier is held at 300 K. For a bandwidth of 100 kHz, what value of resistance R must be used so that the output noise of the resistor matches the output noise of the amplifier? Discuss any assumptions.

1.9 Consider a *pn* diode with $I_s = 10^{-12}$ at room temperature of 300 K.

1. Find the shot noise due to I_s.
2. Assume the junction has a bias voltage of 0.6 V. Assume $q/kT = 1/0.025$. Find the shot noise (in $A/\sqrt{\text{Hz}}$) due to the corresponding current.

1.10 Show $\langle n \rangle = \sum_{n=0}^{\infty} n\, P(n, t) = e^{-rt} \sum_{n=0}^{\infty} n((rt)^n/n!) = rt$ referred to in Equation (1.8.17). Hint: redefine dummy indices in the summation and recall the Taylor expansion of e^{rt}.

1.11 In a manner similar to the previous problem, demonstrate the relation for the variance of the Poisson distribution referred to in Equation (1.8.19)

$$\sigma^2 = \langle n^2 - \bar{n}^2 \rangle = \sum_{n=0}^{\infty} \left(n^2 - \bar{n}^2 \right) P(n, t) = \bar{n}$$

1.12 Consider a long narrow glass tube with a source of high-speed particles as shown in Figure P1.12. Assume the number of particles in time Δt obey a Poisson distribution with an average of r particles emitted per second.

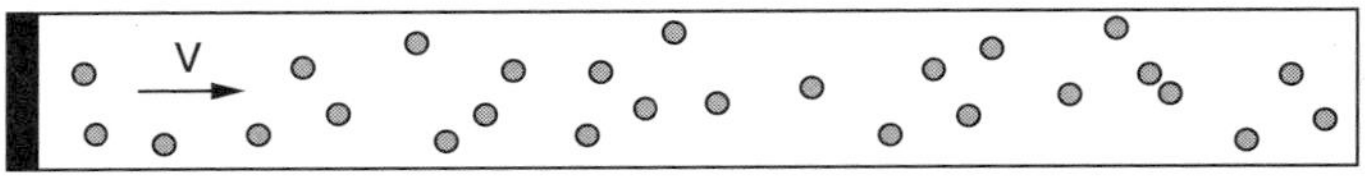

FIGURE P1.12
Particles produced randomly at the left side move to the right with speed ν.

1. Explain why the Poisson distribution can be written as

$$P(n, \Delta x) = \frac{1}{n!} \left(\frac{r}{v} \Delta x \right)^n e^{\Delta x r/v}$$

which gives the probability of n particles in the length Δx.
2. Assume $r = 1000$, $v = 100$, and $\Delta x = 1$. Using the figures in the chapter, what is the probability of finding 5 particles per unit length.

1.13 Read the "Amateur Scientist" column in the following editions of the *Scientific American* popular magazine (available in the archives of most libraries). Check the following editions: September 1964, December 1965, Febuary 1969, September 1971. List the lasers and basic components for construction including power supplies, any flash lamps and gases, and special optics.

1.14 The popular magazine *Nuts & Volts* starting in June 2003 shows how to construct a ruby rod laser. Draw several diagrams to show the circuits, construction, and optical train. Detail required voltages and currents. Discuss the expected optical power and whether the output is pulsed or steady state. Check the archives in your local library.

1.15 The popular magazine *Poptronics* from April 2002 has an article on laser pointers and driver circuits starting on page 46. Draw the circuit and explain how it works. Most libraries have the magazine in their archives.

1.16 Read and briefly summarize the following journal publication on high-power laser diode arrays. A university library has the journals and either a citation index or computer system for searches.
M. Sakamoto et al., "Ultrahigh power 38 W continuous-wave monolithic laser diode arrays," *Appl. Phys. Lett.* **52**, 2220 (1988).

1.17 Summarize the operating mechanisms for PIN and avalanche photodiodes. For the avalanche photodiodes, refer to the following publication.
Spinelli, et al., "Physics and numerical simulation of single photon avalanche diodes," *IEEE Transactions on Electron Devices* **44**, 1931 (1997).

1.10 Further Reading

The following list has references to interesting and informative reading material. The "easy reading" section has construction plans for various lasers and optical systems.

Easy Reading

1. Moore J.H., Davis C.C., Coplan M.A., *Building Scientific Apparatus, A Practical Guide to Design and Construction*, Addison-Wesley Publishing, London, 1983.
2. McComb G., *Lasers, Ray Guns, & Light Cannons*, Projects from the Wizard's Workbench, McGraw-Hill, New York, 1997.

Fabrication

3. Ralph Williams, *Modern GaAs Processing Methods*, Artech House, Boston, 1990.
4. Nishi Y. and Doering R., *Handbook of Semiconductor Manufacturing Technology*, Marcel Dekker, Inc., New York, 2000.

Noise

5. Motchenbacher C.D., Connelly J.A., *Low-Noise Electronic System Design*, John Wiley & Sons, New York, 1993.
6. Davenport W.B., Root W.L., *An Introduction to the Theory of Random Signals and Noise*, McGraw-Hill, New York, 1958.

Optoelectronics: Circuits

7. Marston R.M., *Optoelectronics Circuits Manual*, 2nd ed., Newnes, 1999.
8. Petruzzellis T., *Optoelectronics, Fiber Optics and Laser Cookbook*, More than 150 Projects and Experiments, McGraw-Hill, New York, 1997.

Principles and Systems

9. Kasap S.O., *Optoelectronics and Photonics, Principles and Practices*, Prentice Hall, Saddle River, 2001.

10. Kuhn K.J., *Laser Engineering*, Prentice Hall, Saddle River, 1998.
11. Jenkins, F.A., *Fundamentals of Optics*, 4th ed., H. E. White, McGraw-Hill, New York, 1976.

Reference Books

12. Miller J.L. and Friedman E., *Photonics Rules of Thumb: Optics, Electro-Optics, Fiber Optics and Lasers*, McGraw-Hill Professional, 1996.
13. Wang C.T., *Introduction to Semiconductor Technology, GaAs and Related Compounds*, Ed., John Wiley & Sons, New York, 1990.

Semiconductors

14. Streetman B.G., Banerjee S., *Solid State Electronic Devices*, 5th ed., Prentice Hall, Saddle River, 1999.
15. Kittel C., *Introduction to Solid State Physics*, 5th ed., John Wiley & Sons, New York, 1976.
16. Pankove J.I., *Optical Processes in Semiconductors*, Dover Publications, New York, 1971.
17. Sze S.M., *Physics of Semiconductor Devices*, 2nd ed., John Wiley & Sons, New York, 1981.

2

Introduction to Laser Dynamics

The construction of an emitter or detector ensures the interaction between matter and light. Maxwell's equations and the quantum theory furnish the details of the interactions, while the so-called rate equations provide the best summary. These equations represent a key result for optoelectronic devices by describing relatively complicated physical phenomena (covered in detail in the last part of this book). We primarily focus on the laser but show how the equations apply to the light emitting diode (LED) and the laser amplifier, both of which come from the laser geometry but with the appropriate output facets.

The rate equations describe how the gain, pump, feedback, and output coupler mechanisms affect the carrier and photon concentration in a device. The rate equations manifest the matter–light interaction through the gain term. The gain represents the mechanisms for stimulated emission and stimulated absorption which both require an incident photon field to operate. Later chapters will develop the quantum mechanics of this type of emission and absorption. The photon rate equation describes the effects of the output coupler and feedback mechanism through a relaxation term incorporating the cavity lifetime.

The rate equations provide a wealth of information and have great predictive power. These equations can determine the bandwidth, the threshold current, the emitted optical power versus bias current, and the noise content of the beam. This chapter introduces the simplest rate equations and relates its parameters to the physical construction. A great amount of engineering physics must be included from later chapters to make accurate models of the construction.

2.1 Introduction to the Rate Equations

Matter and light interact to produce a number of phenomena including optical emission and absorption. The interaction appears as a gain term embedded in rate equations that provide the most fundamental description of the laser. We can use some elementary reasoning to deduce phenomenological rate equations that track the number of electron–hole pairs and the number of photons, and relate these numbers to the pump rate and the parameters associated with the laser construction and the material properties. They express energy conservation in terms of the number of excited atoms or the number of carriers in an energy level. The equations describe the magnitude (and phase) of the optical signal. Although elementary reasoning and physical experience lead to these phenomenological equations, they can be (and will be) derived based on more fundamental physical principles.

We can state at least three "different" sets of rate equations with three corresponding sets of variables. The first set of equations uses the variables describing the density of photons γ (the number of photons per cm^3), the density of electrons n (#/cm^3) and the pump-current number density $\mathscr{J}$ (#carriers/s/cm^3). A second set of equations uses variables describing the optical power P (W) and current I (A). The first set using γ and $\mathscr{J}$ must be equivalent to the second set with P and I since the number of photons γ must be equivalent to the optical power P. The phase does not enter into these equations since the optical power does not depend on phase ($P \sim \tilde{\mathscr{E}}^*\tilde{\mathscr{E}}$). The third set involves the electric field $\tilde{\mathscr{E}}$ (amplitude and phase ϕ), and carrier density n. Notice how this set has additional information on the phase of the electric field. This last description applies, for example, to mode locking and injection locking.

Light emission and absorption has similar descriptions for a variety of gain media. However, the microscopic details vary. For example, an emitter using a gas plasma uses a collection of gas molecules with the constituent electrons making transitions between atomic levels. A semiconductor emitter uses closely arrayed atoms with electrons making transitions between bands. Often times the systems are treated as "two level atoms." For the generic system, saying the "electron makes a transition from one atomic level to another" or "it makes a transition from one band to another" conveys identical meaning. For semiconductors, the matter–light interaction often involves electrons and holes and we therefore refer to electron–hole pairs. Only later will we focus on the exact physical mechanism involved.

As just mentioned, the rate equations provide a primary description of light emission and absorption from a collection of atoms. We use them to describe the output optical power vs. input current (resulting in P–I curves), the modulation response to a sinusoidal bias current, and the operating characteristics for laser amplifiers.

2.1.1 The Simplest Rate Equations

Let us consider the rate of change of a number of carrier pairs in the active region of a semiconductor-based device. We assume an intrinsic semiconductor so that the number of electrons per unit volume matches the number of holes per unit volume ($n = p$). We assume the semiconductor can be viewed as two levels, one level for the conduction band and one for the valence band. We want to know what physical phenomena can change the number of electrons in the conduction and valence bands. These changes must be related to the number of photons produced (for a direct bandgap semiconductor).

The rate of change of the total number of electrons (or holes) $N = nV$ comes from electron–hole generation and recombination. We assume that the electrons and holes remained confined to the active region having volume V. The rate equation has the basic form

$$\frac{dN}{dt} = \text{Generation–Recombination} \tag{2.1.1a}$$

Generation processes such as pumping and absorption increase the total number of electron–hole pairs (i.e., increases the number of electrons in the conduction band). Recombination processes such as stimulated and spontaneous emission reduces the total number of electrons in the conduction band. These facts can be incorporated into the basic rate equation to write

$$\frac{dN}{dt} = -\left(\begin{array}{c}\text{Stimulated}\\\text{Emission}\end{array}\right) + \left(\begin{array}{c}\text{Stimulated}\\\text{Absorption}\end{array}\right) + \text{Pump} - \left(\begin{array}{c}\text{Non-Radiative}\\\text{Recombination}\end{array}\right) - \left(\begin{array}{c}\text{Spontaneous}\\\text{Recombination}\end{array}\right)$$
$$\tag{2.1.1b}$$

This last equation calculates the change in the number of carriers "nV" in the active region. Absorption and pumping increase the number while emission and recombination

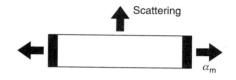

FIGURE 2.1.1
Two mechanisms included in the "optical loss" term.

decrease it. We will carefully examine each term as the section proceeds. The pump consists of either bias current or optical flux. The pump term describes the number of electron–hole pairs produced in the active volume V in each second. We therefore use the form

$$\text{Pump} = \mathcal{J}V \qquad (2.1.1c)$$

where the pump-current density J has units of # carriers/vol/sec.

Many of the processes that *decrease* the total number of carriers N must also *increase* the total number of photons $\Upsilon = \gamma V_\gamma$ in the modal volume V_γ. We can therefore write a photon rate equation as

$$\frac{d\Upsilon}{dt} = +\begin{pmatrix}\text{Stimulated} \\ \text{Emission}\end{pmatrix} - \begin{pmatrix}\text{Stimulated} \\ \text{Absorption}\end{pmatrix} - \begin{pmatrix}\text{Optical} \\ \text{Loss}\end{pmatrix} + \begin{pmatrix}\text{Fraction of} \\ \text{Spont. Emiss.}\end{pmatrix} \qquad (2.1.1d)$$

The "optical loss" term accounts for the optical energy lost from the cavity (see Figure 2.1.1). Some of the light scatters out of the cavity sidewalls and some passes through the mirrors. The light passing through the mirrors, although considered to be an "optical loss," comprises a useful signal. Notice that the pump-current number density J does not appear in the photon equation since it does not directly change the cavity photon number.

The rate equations provide relations between the photon density, carrier density and the pump current density. For now, we characterize the semiconductor material comprising the laser as having two energy levels. These two levels correspond to the conduction and valence band edges obtained from the effective density of states approximation.

The remainder of this section examines the pump, recombination, and optical loss terms in the basic rate equations. The next section continues with the gain and its relation to stimulated emission and absorption. The end of the next section will combine all of the terms into the rate equations.

2.1.2 Optical Confinement Factor

A block diagram of the physical construction of the typical laser diode appears in Figure 2.1.2. Semiconductor–air interfaces form two mirrors on the left-hand and right-hand side of the laser diode. The active region (i.e., gain region) has volume V, which is smaller than the modal volume V_γ containing the optical energy (refer to Section 1.7). The simplest model assumes that the optical power is uniformly distributed in V_γ and is zero outside the volume. The optical confinement factor Γ specifies the fraction of the optical mode that overlaps the gain region $\Gamma = V/V_\gamma$. In other words, the confinement factor gives the percentage of the total optical energy found in the active region V.

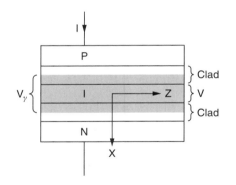

FIGURE 2.1.2
Structure of the semiconductor laser.

2.1.3 Total Carrier and Photon Rate

The simplest rate equations (2.1.1) describe the number of electron–hole pairs (i.e., carrier pairs) and the number of photons in the cavity. If we refer to intrinsic materials or if we confine our attention to the carriers created by optical absorption or electrical pumping, then the number of pairs must be identical to the number of electrons. Therefore, for the simplest model, we discuss only the number of electrons. If the holes and electrons do not distribute themselves evenly throughout the active region, it becomes necessary to describe each type of carrier by its own equation. Quantum well lasers for example do not generally have identical electron distributions in each well. For now, we assume charge neutrality ($n = p$) in all portions of the active region. Let's denote the electron density (number per volume) by "n" so that nV represents the total number of electrons in the active region. Similarly, let γ be the photon density (number of photons per volume). The total number of photons in the modal volume must be γV_γ and the total number of *photons* in the active region must be γV.

2.1.4 The Pump Term and the Internal Quantum Efficiency

The number of electron–hole pairs that contribute to the photon emission process in each unit of volume (cm^3) of the active region in each second can be related to the bias current I by

$$\mathscr{I} = \frac{\eta_i I}{qV} \tag{2.1.2}$$

where $\mathscr{I}$ represent the "pump-current number density," η_i is the internal quantum efficiency, the elementary charge "q" changes the units from Coulombs to the "number of electrons," and V represents the active volume. The upper-middle diagram in Figure 2.1.3 shows the pump increases the number of electrons and holes in the conduction band (*cb*) and valence band (*vb*), respectively.

The internal quantum efficiency η_i represents the fraction of terminal current I that generates carriers in the active region. Therefore, the quantity $\eta_i I$ provides the actual current absorbed in the active region. Well-designed lasers have internal quantum efficiency close to one. The internal efficiency can be smaller than one if some of the current I shunts around the junction as it travels between the "p" to the "n" materials. For example, current might flow along a surface exterior to the junction as shown in the upper-right diagram in Figure 2.1.3. The current might also flow through the active

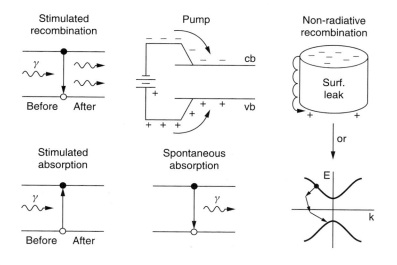

FIGURE 2.1.3
A number of mechanisms for changing the number of carriers and the number of photons.

region without producing photons (phonon process perhaps). Pumping (whether optical or electrical) initiates laser action when the carrier population reaches the "threshold" density (to be discussed later). Without carrier recombination, the pump would continuously increase the carrier population according to $dn/dt = \mathcal{J}$ (recall $n = p$).

2.1.5 Recombination Terms

The radiative and nonradiative processes comprise two broad categories of recombination mechanisms. The radiative process produces photons usually as spontaneous emission (a.k.a, fluorescence) from a semiconductor laser or LED. The stimulated emission process also involves carrier recombination; however, because it requires an incident photon field, it is usually studied as part of the laser gain.

Spontaneous recombination refers to the recombination of holes and electrons without an applied optical field (i.e., no incident photons—see lower-middle diagram in Figure 2.1.3). Spontaneous recombination produces photons by reducing the number of electrons in the conduction band and holes in the valence band. This spontaneous emission initiates laser action but decreases the efficiency of the laser. The three processes of stimulated recombination, spontaneous recombination, and nonradiative recombination all reduce the number of electrons in the conduction band and holes in the valence band. Absorption and pumping increases the number of electrons.

From the classical point of view, spontaneous emission occurs "on its own" without an applied optical field. Classically speaking, cause and effect do not appear to hold. Quantum theory shows that "vacuum fields" actually initiate the spontaneous emission. The actual magnitude of the emission can be calculated from knowledge of the quantum-mechanical vacuum fields and the self-reaction of the oscillating dipoles in the material. The vacuum fields can be pictured as electromagnetic waves that exist in all space; they represent a type of zero-point motion of the electric field. The reader might imagine a Universe without any light sources and without any photons. This very strange Universe still has sporadic electromagnetic fields in the available optical modes. These fields are the "vacuum fields." The "modes" refer to a type of physical "storage" mechanisms for photons (if the Universe has photons). For analogy, the modes for a string on a violin store phonons when someone plucks the string. The vacuum fields initiate spontaneous emission by perturbing the energy levels of the electrons and holes thereby causing recombination. It turns out that the spontaneous emission from a laser can be reduced by removing some of the possible modes or by "squeezing" the vacuum fields. Loosely speaking, these zero point fields "stimulate" the spontaneous recombination in a manner similar to the stimulated emission for laser action.

Nonradiative recombination occurs primarily through *phonon* processes. Materials with indirect bandgaps rely on the phonon process for carrier recombination. For direct bandgaps as in GaAs, the phonon processes are much less important since radiative recombination dominates the recombination channels. For nonradiative recombination, an energetic electron produces a number of phonons (rather than photons) and then recombines with a hole. The right-hand portion of Figure 2.1.3 shows an example of nonradiative recombination whereby an electron moves along an outside surface of a laser, and interacts with phonons until it recombines with a hole. The *phonon* processes reduce the efficiency of the laser because a portion of the pump must be diverted to feed these alternate (nonradiative) recombination channels. However, all lasers produce some phonons as the semiconductor heats up.

Monomolecular (nonradiative), bimolecular (radiative) and Auger recombination (nonradiative) comprise three important recombination mechanisms for laser operation (refer to Section 1.5.6). Recall that the nonradiative monomolecular recombination occurs when carriers "trap out" in midgap states and recombine. The rate of change of electron density (or holes since $n = p$) due to monomolecular recombination can be written as

$$\frac{dn}{dt} = -\frac{n}{\tau_n} = -An \tag{2.1.3a}$$

where $\tau_n = 1/A$ represents a lifetime. The total number of monomolecular recombination events in the active volume V can be written as

$$R_{\mathrm{mono}} V = AnV \tag{2.1.3b}$$

R_{mono} has units of "number of recombination events per volume per second." Monomolecular recombination reduces the efficiency of the laser by recombining holes and electrons without emitting photons into the lasing mode.

Bimolecular recombination produces spontaneous emission. Electrons and holes recombine without the need for bandgap states so that the spontaneous recombination rate R_{sp} is proportional to np. The total number of spontaneous emission recombination events in active volume V must be

$$R_{sp} V = Bn^2 V \tag{2.1.4}$$

Bimolecular recombination events can inject photons into the lasing mode to initiate laser oscillation but most of the energy escapes through the sides of the laser and reduces the laser efficiency and increases the threshold current (see the lower middle diagram in Figure 2.1.3).

The nonradiative Auger recombination occurs when carriers transfer their energy to other carriers, which interact with phonons to return to an equilibrium condition. Auger recombination is important for lasers (such as InGaAsP) with emission wavelengths larger than $1\,\mu m$ (small bandgap). For comparison, GaAs lasers generally emit between 800 to 860 nm. As discussed in Section 1.5.6, Auger recombination involves three charged particles and an energy transfer mechanism. The charged particles might be two electrons and a hole. The rate of Auger recombination in the active volume V can be written as

$$R_{\text{aug}}V = Cn^3 V \qquad (2.1.5)$$

where the power of 3 serves as a reminder of the three particles. This form of recombination reduces the efficiency of the laser since it recombines the carriers without producing photons for the laser mode.

Combining all the different types of recombination we can write the total rate of recombination R_r as

$$R_r = R_{\text{radiative}} + R_{\text{nonradiative}} = An + Bn^2 + Cn^3 \qquad (2.1.6)$$

where $A = 1/\tau_n$. The rates R have units of "number of recombination events per unit volume per second." Some people define an effective carrier lifetime τ_e which depends on the carrier density "n" as

$$\frac{1}{\tau_e} = A + Bn + Cn^2 \qquad (2.1.7)$$

so that the total recombination rate can be written as $R_r = n/\tau_e$. As we will see later, this turns out to be a nice way of writing the recombination rate since the carrier density will be approximately constant when the laser operates above threshold. For lasers made with "good" material, the B term (radiative recombination) dominates the recombination process. If we restrict our attention to GaAs then C can be neglected. We will usually write

$$R_r = Bn^2 \qquad (2.1.8)$$

2.1.6 Spontaneous Emission Term

The spectrum of the laser beam consists of nearly a single wavelength. Lasers can achieve linewidths (i.e., the width of the spectral line) on the order of 1 kHz. As discussed in Chapter 1, an oscillator operates at a single frequency because the gain equals the loss at that frequency. We expect the same to be true for a semiconductor laser (the mirrors provide the feedback path). It turns out that "homogenously broadened" lasers have one lasing frequency but "nonhomogeneously broadened" ones can lase at multiple frequencies (i.e., multiple modes within the cavity can be excited).

The number of photons in the lasing mode increases not only from stimulated emission but also from the spontaneous emission. Let's see how this happens. Excited atoms in the gain medium spontaneously emit photons in all directions. The wavelength range of spontaneously emitted photons cannot be confined to a narrow spectrum. Figure 2.1.4 compares the typical spectra for spontaneous and stimulated emssion (for GaAs). Some of the spontaneously emitted photons propagate in exactly the correct direction to enter the waveguide of the laser cavity. Of those photons that enter the waveguide, a fraction of them have exactly the right frequency to match that of the lasing mode. This small fraction of spontaneously emitted photons adds to the photon density γ of the cavity.

The rate of spontaneous emission into the cavity mode can be written as

$$V_\gamma R_{sp} = V_\gamma \beta Bn^2 \qquad (2.1.9)$$

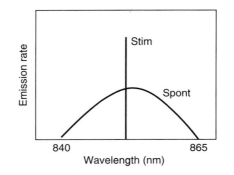

FIGURE 2.1.4
Comparing spectra for spontaneous and stimulated emission.

where B and Bn^2 are the same terms as previously found for the spontaneous recombination rate. The geometry factor β gives the fraction of the total spontaneously emitted photons that actually couple into the laser mode. The value of β typically ranges from 10^{-2} to 10^{-5}. R_{sp} has units of number per volume per second.

The small fraction of spontaneous photons coupling into the cavity with the right frequency start the lasing process. Above threshold, it wastes a significant fraction of the pump energy—raises the laser threshold current. The threshold current is the minimum pump current (into the semiconductor laser) required to initiate lasing. Similarly, optically pumped lasers have a threshold optical power.

2.1.7 The Optical Loss Term

The optical loss term describes changes in the photon density that can be linked with the optical components of the laser cavity. The reader can picture the cavity as the space bounded by two mirrors and the sidewalls as shown in Figure 2.1.5. The cavity retains the optical characteristics of the material such as a waveguide without a gain medium. As the photons bounce back and forth between the mirrors, some are lost through the mirrors and some are lost through the sides. Other loss mechanisms also influence the photon density. For example, free carriers can absorb light when the light waves drive the motion of the electrons and the surrounding medium damps this motion by converting the kinetic energy into heat. All of the optical losses contribute to an overall relaxation time τ_γ (*called the cavity lifetime*). The total number of photons in the modal volume $V_\gamma \gamma$ *decreases* because of these optical losses. Greater numbers of photons must be lost from the cavity for greater numbers of photons inside the cavity. Therefore, a simple differential equation expresses the dynamics in the absence of other sources or losses of photons

$$V_\gamma \frac{d\gamma}{dt} = -\frac{V_\gamma \gamma}{\tau_\gamma} \qquad (2.1.10a)$$

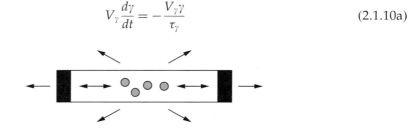

FIGURE 2.1.5
The Fabry-Perot cavity.

This last simple equation can be solved to give

$$\gamma(t) = \gamma_0 \exp\left(-\frac{t}{\tau_\gamma}\right) \tag{2.1.10b}$$

which shows that the initial photon density decays exponentially as the carriers are lost. If the cavities include a gain medium then two things can happen. An absorptive medium ($g < 0$) causes the photon density to relax faster than τ_γ. On the other hand, a medium with positive material gain ($g > 0$) causes the photon density to grow rather than decay. The physical ideas accompanying the solution of the simple differential equation above must be included in the full set of laser rate equations since they describe the basic optical properties of the cavity.

Let's examine the loss α in a little more detail. Light can be lost from a cavity due to either distributed or point-loss mechanisms. Distributed loss refers to energy loss along the length of a device. For example, distributed loss includes optical energy lost through the sidewalls and free carrier absorption. Sometimes this loss mechanism is also termed "internal loss," denoted by α_{int}, because it refers to light leaving the body of the laser in a manner other than through the end mirrors. The laser mirrors represent the second type of loss—the point loss. The light escapes the cavity at specific points. The loss is not distributed along the length of the laser. However, we still find it convenient to describe the mirror loss as if it were a "distributed loss" and give it the symbol α_m.

The cavity lifetime in Equations (2.1.10) describes a lumped device. In order to include a spatial dimension, we *define* an optical loss per unit length. As will be explained in the next section for the gain, we will find it useful to define the "optical loss" α with units of cm^{-1} as

$$\alpha = \tau_\gamma^{-1}/v_g \qquad \text{or} \qquad \frac{1}{\tau_\gamma} = \alpha v_g \tag{2.1.11}$$

where v_g represents the group velocity of the wave.

The optical loss (per unit length) α gives the number of photons lost in each unit length of cavity. As such, it is useful for describing physically extended systems (those having non-zero sizes). We picture the optical loss α as taking place along the length of the laser body. Appendix 2 discusses the use of an optical equation of continuity that accounts for both spatial and temporal variations in the photon number.

How should we picture the mirror loss α_m as distributed along the length of the cavity? To really answer this question, a partial differential equation with boundary conditions at the mirrors must be solved. However, for the purpose of the lumped model (rate equations depending only on time), the amount of energy lost at *both* mirrors will be averaged over the length of the cavity (as will be demonstrated later). The result will be

$$\frac{1}{\tau_m} = v_g \alpha_m = \frac{v_g}{L} Ln(1/R) \tag{2.1.12a}$$

which assumes that both mirrors have the same power reflectivity R (0.34 for GaAs). The loss per mirror must be $\alpha_m/2$. The reciprocal of the cavity lifetime becomes

$$\frac{1}{\tau_\gamma} = \frac{1}{\tau_{\text{int}}} + \frac{1}{\tau_m} = v_g \alpha = v_g(\alpha_{\text{int}} + \alpha_m) \tag{2.1.12b}$$

The internal loss and single mirror loss are typically on the order of 30/cm.

2.2   Stimulated Emission—Absorption and Gain

The matter–light interaction produces the optical emission and absorption in a material. The rate equations include this interaction in the gain that appears in the stimulated emission and absorption terms in Equations (2.1.1). However, the rate equations handle spontaneous emission (a.k.a., fluorescence) separately from the gain even though quantum mechanics shows it also originates in matter–light interaction. We proceed to define several types of gain and show how all the pieces fit together to make realistic rate equations.

2.2.1   Temporal Gain

Consider an ensemble of two level atoms or bands in a semiconductor material. Stimulated emission and absorption affect the number of electrons in the two energy levels. These processes help to determine the rate of change of the number of carriers in the active region and the number of photons in the modal volume. The process of stimulated emission appears in the upper-left portion of Figure 2.1.3 in the previous section. A photon perturbs the energy levels of atoms (i.e., electron–hole pairs or "excitons" for the semiconductor) and induces *radiative* recombination. In the case of Figure 2.1.3, the number of photons increases by one while the number of conduction electrons decrease by the same number. The CB electrons and VB holes produce "gain" in the sense that incident photons with the proper wavelength can stimulate carrier recombination and thereby produce more photons with the same characteristics as the incident ones. The figure indicates a gain of two by defining a generic form of gain as the ratio of the "output number of photons" to the "input number of photons."

The same ensemble of atoms can also absorb photons from the beam (as shown in the lower-left portion of Figure 2.1.3) by promoting a valence electron to the conduction band. The stimulated emission increases the number of photons in the laser while the stimulated absorption decreases the number. Therefore, the gain really should describe the difference between the emission and absorption rates. The stimulated emission and absorption terms in Equations (2.1.1) can be grouped together into a single term incorporating the gain.

The word "stimulated" means that a photon must be incident on the material before either stimulated emission or absorption can proceed. Therefore, the change in the total number of photons γV_γ in the modal volume V_γ must be proportional to the number of photons present

$$R_{\text{stim}} V_\gamma = V_\gamma \frac{d\gamma}{dt}\bigg|_{\text{stim}} \sim \gamma$$

where R_{stim} represents the net number of photons produced ($R_{\text{stim}} > 0$) or absorbed ($R_{\text{stim}} < 0$) in each unit volume in each second. However, only those photons in the *active* region (volume V) can stimulate additional photons since the electron–hole pairs are confined to that region. Therefore

$$R_{\text{stim}} V_\gamma = V_\gamma \frac{d\gamma}{dt}\bigg|_{\text{stim}} \sim V\gamma$$

Define the "temporal gain" g_t to be the constant of proportionality so that

$$R_{\text{stim}}V_\gamma = V_\gamma \frac{d\gamma}{dt}\bigg|_{\text{stim}} = V g_t \gamma \tag{2.2.1a}$$

or equivalently

$$R_{\text{stim}} = \frac{d\gamma}{dt}\bigg|_{\text{stim}} = \frac{V}{V_\gamma} g_t \gamma = \Gamma g_t \gamma \tag{2.2.1b}$$

where $\Gamma = V/V_\gamma$ is the confinement factor defined in Section 2.1.2. The temporal gain g_t must have units of "per second" since $R_{\text{stim}}V_\gamma$ has units of #events per second. As previously discussed, electron–hole pairs produce stimulated emission and therefore the temporal gain g_t must depend on the number of excited carriers n (in a semiconductor) or the number of excited atoms (in a gas) so that $g_t = g_t(n)$. The temporal gain describes the stimulated emission and absorption terms in Equations (2.1.1) from the previous section.

2.2.2 Single Pass Gain

The rate of stimulated emission in Equations (2.2.1) does not depend on the spatial coordinates. Therefore, these rate equations treat physical devices as lumped elements as if they have the size of a single point rather than occupying a finite volume of space. However, we would like to apply the rate equations to the case of optical energy propagating in an extended gain medium such as the laser amplifier shown in Figure 2.2.1. As a first step, consider the "single pass gain" produced by the collection of "two-level" atoms depicted in Figure 2.2.2. These atoms have only two possible energy levels for the electrons. The figure shows three photons incident on the left side of the gain medium. These photons enter the material and interact with the atoms. Five of the atoms emit photons (stimulated emission) while two of them absorb photons (absorption or sometimes called stimulated absorption) and two do nothing. The number of output photons is six which gives a single pass gain of $G = 6/3 = 2$. The gain describes only the stimulated emission and absorption processes and does not include photon losses through the side of the laser or through the mirrors. The single pass gain is defined as the ratio between the numbers of output and input photons. In the case of the laser amplifier shown in Figure 2.2.1, the "input" represents the number of the photons in the cavity at point z_1 (time t_1) and the "output" represents the number of photons at point z_2

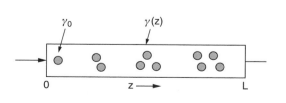

FIGURE 2.2.1
Block diagram of the laser amplifier. The number of photons increases as they travel along the gain medium. The gain can depend on position.

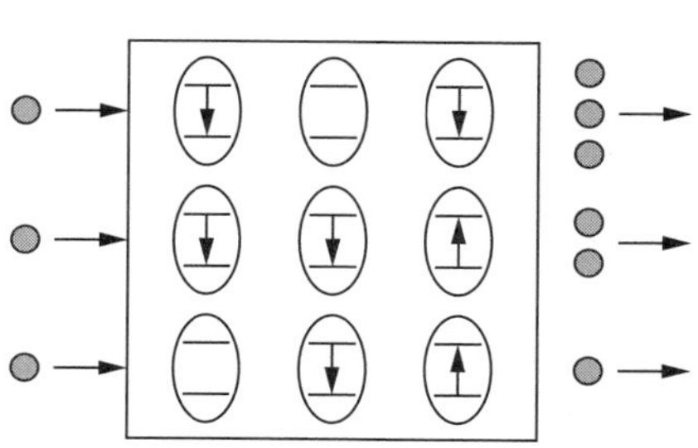

FIGURE 2.2.2
An example where two levels produce more photons than they absorb.

(time t_2). The laser amplifier shown in Figure 2.2.1 produces a single pass gain of four in a length $z_2 - z_1 = L$ of the material.

2.2.3 Material Gain

We want a gain that can represent the ability of a material to produce photons in each unit volume along the length of a device. The single pass gain represents the whole piece of material. For example, increasing the length L of the material must also increase the single pass gain. Therefore, we would like to eliminate the length dependence from the definition of *gain*. We define the material gain "g" in terms of the number of photons produced in the medium in each unit of length for each photon entering that unit length.

We can find the material gain from the temporal gain by changing the units of g_t from "per second" to those of the material gain g, namely "per unit length." The two gains "g and g_t" can be seen to be equivalent on an intuitive level. Consider the "extended" laser amplifier described by g as shown in Figure 2.2.1. Photons travel from one end to the other and spend a time $t = L/v_g$ in the amplifier. The quantity $v_g = dz/dt$ is the group velocity of the wave and, for a *single frequency laser*, it is given by c/n where n is the refractive index of the medium (see Appendix 3). The output from the laser amplifier is increased by the gain g. Now consider the case of the point-sized lumped device. The signal enters the lumped device and remains for a period of time t. The temporal gain g_t amplifies the number of photons during that time. The amplified signal then leaves the lumped device. For both the extended and lumped devices, the signals remain in contact with the gain medium for the same length of time t. We expect the output signal size to be the same and we find a relation of the form

$$g \sim \frac{1}{\text{Length}} \sim \frac{1}{\text{Sec}} \frac{\text{Sec}}{\text{Length}} \sim g_t \frac{t}{L} \sim \frac{g_t}{v_g}$$

Converting between length and time also converts between the two gains.

The typical demonstration of the equivalence between the two gains "g and g_t" starts with Equations (2.2.1).

$$\frac{d\gamma}{dt} = \Gamma g_t \gamma \tag{2.2.2a}$$

The chain rule for differentiation gives

$$\frac{d\gamma}{dt} = \frac{d\gamma}{dz}\frac{dz}{dt} = v_g \frac{d\gamma}{dz} \tag{2.2.2b}$$

Combining Equations (2.2.2a) and (2.2.2b) produces

$$\frac{d\gamma}{dz} = \Gamma g \gamma \tag{2.2.2c}$$

where $g = g(n) = g_t/v_g$ depends on the number of carriers n (per unit volume).

Essentially we are changing variables from "t" to "z" in the photon density γ. In the case of "t," we imagine that photons enter a "node" (i.e., a small box without spatial extent) and after a time t emerge from the node but with more of them (i.e., γ increases). In the case of "z," we imagine a steady state process where photons enter a spatially extended device (laser amplifier) and each subsequent unit length of material produces more photons than each previous unit length. It is possible of course, to imagine a situation where γ can depend on both z and t as in $\gamma = \gamma(z, t)$. For example, the bias current to the laser amplifier might be modulated.

We can lastly demonstrate the single pass gain. The stimulated emission from the laser amplifier in Figures 2.2.1 and 2.2.2 can be found from the net rate of stimulated emission given by Equation (2.2.2c) (assuming $g =$ constant) as

$$\frac{d\gamma}{dz} = \Gamma g \gamma \quad \rightarrow \quad \gamma(z) = \gamma(0)\, e^{\Gamma g z} \tag{2.2.3}$$

The material gain appears in the argument of the exponential. The single pass gain G can be written as

$$G = e^{\Gamma g z} \tag{2.2.4}$$

Since the material gain g in Equation (2.2.3) depends on the number of excited carriers n or the number of excited atoms in a gas, so g can produce either gain or absorption. The material gain $g(n)$ can increase (or decrease) the number of photons in each unit of length. Without any exited carriers $n = 0$, we expect incident photons to be absorbed which means $G < 1$ and therefore $g < 0$. For sufficiently large n, the material gain g becomes positive and produces stimulated emission. By the way, the assumption that $g =$ constant is equivalent to assuming that "n" is independent of length, which in turn is equivalent to assuming that the laser amplifier is far from "saturation." Near saturation, the carrier density decreases from its quiescent value in regions where the optical power density is large.

The gain often appears as a logarithm of the form shown in Figure 2.2.3

$$g(n) = g_o\, \mathrm{Ln}\left(\frac{n - n_\infty}{n_o - n_\infty}\right) \tag{2.2.5}$$

The n_∞ is negative and never attainable. The parameter n_o represents the transparency density.

Example 2.2.1

Suppose all of the atoms shown in Figure 2.2.2 are in the ground state (electrons in the lowest level). The three incident photons in the figure would most likely be absorbed and the single pass gain G would be 0. However, the equation $G = \exp(gL)$ indicates

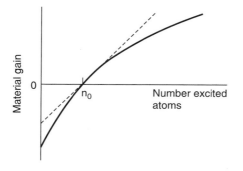

FIGURE 2.2.3
The material gain as a function of the number n of excited atoms.

that the material gain "g" must be negative. In fact, to produce $G=0$, we must have $g \sim -\infty$ and similarly $g_t \sim -\infty$. This situation is shown in Figure 2.2.3.

Example 2.2.2

If half of the atoms in Figure 2.2.2 are excited and half are still in the ground state, then for every photon that we put into the material, we might expect to get exactly one in the output. In such a case, the single pass gain is $G=1$ but the material gain is $g=0$; this condition is termed "material transparency."

2.2.4 Material Transparency

The semiconductor material becomes "transparent" ("material transparency") when the rate of absorption just equals the rate of stimulated emission. One incident photon produces exactly one photon in the output. This means that the single pass gain must be unity, i.e., $G=1$. The material gain in this case is $g=0$ as can be seen by setting the single pass gain equal to one in Equation (2.2.4). The transparency density n_o (number per unit volume) represents the number of excited atoms (or electrons per volume) required to achieve transparency.

The gain curves can be approximated by a straight line at n_o (refer to Figure 2.2.3) by making a Taylor expansion about the transparency density n_o to find $g = g(n) \cong g_o(n - n_o)$. The symbol $g_o = dg/dn$ is typically called the differential gain. It plays a predominant role in designing high efficiency and large bandwidth laser heterostructure.

The material gain required to achieve lasing will be much larger than zero since the gain must offset other losses besides stimulated absorption (typically $g = 150\,\text{cm}^{-1}$). The number of carriers required to achieve lasing will be larger than the transparency density (sometimes approximately double). We will spend considerable time learning about gain especially for the quantum mechanical treatment.

2.2.5 Introduction to the Energy Dependence of Gain

We now provide a simple argument as to why the gain depends on frequency ω (i.e., energy $E = \hbar\omega$) somewhat similar to the op-amp circuit discussed in the first chapter. We will also see how the pump level affects the peak gain and the bandwidth of the gain curve. Strictly speaking, the remainder of the book will examine this development in greater detail.

Suppose we connect a pn GaAs homojunction to a battery so that the forward bias places a nonequilibrium number of electrons in the conduction band cb and holes in the valence band vb as shown in the top portion of Figure 2.2.4. The quasi-Fermi levels F_c and F_v mark the approximate top of filled cb states and the bottom of empty vb states, respectively.

Assume a photon enters the semiconductor with energy E. We want to know the effect it will have on the population distribution. In particular, we want to know if it will induce a transition and if so, what type. The energy $E = E_1$ is smaller than the bandgap energy E_g and cannot connect a state in the vb with one in the cb. Furthermore, no states exist within the bandgap. Therefore the photon does not induce emission or absorption. The lower portion shows the gain must be zero at energy E_1. Stimulated photons can only be produced when the energy of the incoming photon matches the energy difference between filled cb states and the corresponding empty vb states. Therefore, we expect incident photons with energy $E_g \leq E \leq E_2$ to produce stimulated emission and hence, positive gain. The energy $E_2 = F_c - F_v$ corresponds to the difference

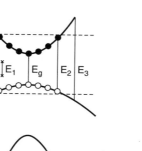

FIGURE 2.2.4
Top portion shows filled conduction and empty valence states. Bottom portion shows gain as a function of energy.

in quasi-Fermi levels which was essentially set by the bias voltage. When the incident photon has energy $E > F_c - F_v$ such as for E_3, then the photon connects a filled valence state with an empty conduction state. In such a case, the photon can only be absorbed and hence the gain must be negative.

As a result, the figure shows that the bias voltage sets the width of positive gain in addition to determining the peak value at E_p. Notice that the gain reaches a peak at an energy other than the bandgap energy E_g. This means that the average photon energy will be slightly larger than the gap energy. The phenomena of shifting the photon to larger energy (i.e., shorter wavelength) occurs because of the band filling effect through the quasi-Fermi levels. For the region of positive gain, the semiconductor emits photons and can be used as a laser or led. For the region of negative gain, the semiconductor absorbs the photon. For the bias levels determining the gain curve in Figure 2.2.4, the semiconductor can absorb short wavelength light to optically pump the semiconductor and then emit the absorbed energy at a longer wavelength.

If we reverse bias the pn junction, then the gain curve remains negative for all energy. In such a case, the semiconductor only absorbs light and can be used as a photodetector.

2.2.6 The Phenomenological Rate Equations

Now we combine all of the individual terms into the laser rate equations. The equation for the number of excited atoms (number of electrons in the conduction band)

$$V\frac{dn}{dt} = -\left(\begin{array}{c}\text{Stimulated}\\\text{Emission}\end{array}\right) + \left(\begin{array}{c}\text{Stimulated}\\\text{Absorption}\end{array}\right) + JV - \left(\begin{array}{c}\text{Non-Radiative}\\\text{Recombination}\end{array}\right) - \left(\begin{array}{c}\text{Spontaneous}\\\text{Recombination}\end{array}\right)$$

becomes

$$V\frac{dn}{dt} = -V v_g g\,\gamma + JV - \frac{n}{\tau_e}V \qquad (2.2.6a)$$

Referring to earlier equations we have

$$\frac{1}{\tau_e} = A + Bn + Cn^2 \qquad \text{and} \qquad A = 1/\tau_n \qquad (2.2.6b)$$

For good material, $A = C = 0$ and the recombination term equals the spontaneous recombination rate

$$\frac{dn}{dt} = -v_g g \gamma + J - Bn^2 \qquad (2.2.7)$$

Next consider the photon rate equation given by

$$V_\gamma \frac{d\gamma}{dt} = +\left(\begin{array}{c}\text{Stimulated} \\ \text{Emission}\end{array}\right) - \left(\begin{array}{c}\text{Stimulated} \\ \text{Absorptions}\end{array}\right) - \left(\begin{array}{c}\text{Optical} \\ \text{Loss}\end{array}\right) + \left(\begin{array}{c}\text{Fraction of} \\ \text{Spont. Emiss.}\end{array}\right)$$

We can now rewrite the photon rate equation as

$$V_\gamma \frac{d\gamma}{dt} = +V v_g g(n) \gamma - V_\gamma \frac{\gamma}{\tau_\gamma} + \beta Bn^2 V_\gamma \qquad (2.2.8)$$

or, using the optical confinement factor of $\Gamma = V/V_\gamma$, we have the second rate equation

$$\frac{d\gamma}{dt} = +\Gamma v_g g(n) \gamma - \frac{\gamma}{\tau_\gamma} + \beta Bn^2 \qquad (2.2.9)$$

For convenience, let's write the rate equations together

$$\frac{dn}{dt} = -v_g g(n) \gamma + J - Bn^2 \qquad (2.2.10a)$$

$$\frac{d\gamma}{dt} = +\Gamma v_g g(n) \gamma - \frac{\gamma}{\tau_\gamma} + \beta Bn^2 \qquad (2.2.10b)$$

These equations describe the laser system (except for the phase of the EM wave). The rate equations describe all that a person wants to know about a number of optoelectronic devices. The rate equations mainly exist to find the output power (also cavity power) as a function of the bias current. We can also use them for a small signal analysis of time response of the beam to small changes in the bias current. The reader should realize that the laser rate equations are quite nonlinear especially since g depends on "n." Also keep in mind that the rate equations should really be generalized to a partial differential equation that includes a spatial coordinate (refer to Appendix 2).

EXAMPLE 2.2.3

Find the number of electron–hole pairs when only the pump operates. Do not include the stimulated emission and recombination terms.

  Solution: The n rate equation reduces to

$$\frac{dn}{dt} = \mathscr{J}$$

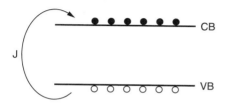

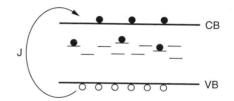

FIGURE 2.2.5
The pump continuously increases the number of electron–hole pairs.

FIGURE 2.2.6
A semiconductor with pump and monomolecular recombination.

The number of electron–hole pairs must be

$$n = \mathscr{J}t$$

Figure 2.2.5 shows how the pump increases the number.

EXAMPLE 2.2.4
Find the number of electrons in the conduction band when only monomolecular recombination and the pump operate.
  Solution: Figure 2.2.6 shows how the pump increases the number of electron–hole pairs while the gap states trap out the electrons. Eventually, but with a different time constant, the holes will be reduced as they recombine with the electrons in the traps. Let n refer solely to the electrons in the conduction band. The rate equation provides

$$\frac{dn}{dt} = \mathscr{J} - \frac{n}{\tau_e} \tag{2.2.11}$$

The equation can be easily solved using an integrating factor (Appendix 1) or by Laplace transforms.
  Let $\tilde{n}(s)$ be the Laplace transform of $n(t)$. The Laplace transform of Equation (2.2.11) produces

$$s\tilde{n} + \frac{\tilde{n}}{\tau_e} = \frac{\mathscr{J}}{s} + n_o \quad \text{or} \quad \tilde{n} = \frac{J + sn_o}{s(s + 1/\tau_e)}$$

where n_o represents the initial number of electrons.
  Using partial fractions and basic results for Laplace transforms, we find

$$n(t) = \mathscr{J}\tau_e\left(1 - e^{-t/\tau_e}\right) + n_o\,e^{-t/\tau_e}$$

The second term shows that the initial number of electrons decays as they trap out. The first term shows the pump increases the number of *cb* electrons while the trapping tends to decrease the number. As a result, the number approaches an asymptote $\mathscr{J}\tau_e$ set by the interplay between the pump and the recombination term.

2.3 The Power–Current Curves

The relation between optical output power and the pump strength provides the most fundamental information on the operation of light-emitting devices. The rate equations provide power versus current curves for semiconductor lasers and light emitting diodes. The power–current curves are alternately termed *P–I* or *L–I* curves. A computer provides the most accurate solutions to these highly nonlinear equations. However, the most important results can be found using some very insightful and highly accurate approximations. Separate approximations must be applied to the lasing and nonlasing regimes of operation.

2.3.1 Photon Density versus Pump-Current Number Density

We solve the rate equations for the steady-state photon density γ inside the laser cavity as a function of the steady state pump-current number density J. The rate equations are

$$\frac{dn}{dt} = -v_g\, g(n)\,\gamma + \mathscr{J} - Bn^2 \tag{2.3.1}$$

$$\frac{d\gamma}{dt} = +\Gamma v_g\, g(n)\,\gamma - \frac{\gamma}{\tau_\gamma} + \beta Bn^2 \tag{2.3.2}$$

A system attains steady state when all of the time derivatives become zero. We assume that the laser has been operating for a long time compared with the time constants τ_γ and τ_e. We define the effective carrier lifetime τ_e by $\tau_e = 1/(Bn)$ as discussed in Section 2.1.5. For these sufficiently long times, the rate equations become the steady-state equations

$$0 = -v_g g(n)\gamma + \mathscr{J} - Bn^2 \tag{2.3.3}$$

$$0 = +\Gamma v_g g(n)\gamma - \frac{\gamma}{\tau_\gamma} + \beta Bn^2 \tag{2.3.4}$$

The steady-state condition imposed on Equations (2.3.1) and (2.3.2) in order to arrive at Equations (2.3.3) and (2.3.4) has nothing to do with the time-dependent sinusoidal variation of the electromagnetic waves inside the cavity. The above equations describe the photon density, which refers to the optical power density. Equation (2.3.4) requires the amplitude of the EM waves to be independent of time. The power contained in the EM waves neither grows nor decays with time.

Case 1   *Below Lasing Threshold*

The phrase "below lasing threshold" implies that the laser has insufficient gain to support oscillation. Small value of current density $\mathscr{J}$ implies small values for the carrier density "n" and the photon density γ. As discussed in Case 2 below there exists a "threshold" pump-current number density $\mathscr{J}_{\text{thr}}$ for which $\mathscr{J} > \mathscr{J}_{\text{thr}}$ produces lasing and $\mathscr{J} < \mathscr{J}_{\text{thr}}$ produces only spontaneous emission. For case 1 considered here, we assume that $\mathscr{J} < \mathscr{J}_{\text{thr}}$. For this case, the photon density γ in the cavity remains relatively small compared with that achieved for lasing. Therefore, we drop the stimulated emission/absorption terms $v_g g(n)\gamma$ in Equations (2.3.3) and (2.3.4). The second steady-state equation (2.3.4) provides

$$\gamma = \beta\tau_\gamma Bn^2 \tag{2.3.5a}$$

for the photon density for spontaneous emission. The first steady-state equation (2.3.3) provides the expression

$$Bn^2 = \mathscr{J} \tag{2.3.5b}$$

Equations (2.3.5) can be combined to yield the $\gamma - \mathscr{J}$ relation

$$\gamma = \beta\tau_\gamma \mathscr{J} \tag{2.3.6}$$

This is the photon density *in the cavity* (i.e., between the two mirrors) due to spontaneous emission. Notice that the photon density is linear in the pump-current number density J. The factor β accounts for the geometry factors describing the coupling of spontaneous emission to the cavity mode.

Some books include a coupling coefficient η_{ex} and show the linear relation between the photon density (proportional to the optical power) and the pump current (see Coldren's book). Our Equation (2.3.6) above describes the photon density inside the laser; the wave vectors point along the longitudinal axis (i.e., the long axis). These "spontaneous" photons can be emitted through the mirrors. We will find an expression for the emitted optical power later.

Case 2 Above Lasing Threshold

The phrase "above lasing threshold" refers to the situation of sufficiently large pump current (or pump power) to produce stimulated emission in steady state ($J > J_{thr}$). We assume that stimulated emission provides the primary source of cavity photons whereas the number of spontaneously emitted photons remains relatively small. The ratio of spontaneous to stimulated photons in the lasing mode is further reduced by the coupling coefficient β. We therefore neglect the term βBn^2 in Equation (2.3.4); this can later be justified by a self-consistency argument. The steady-state laser equations become

$$0 = -v_g g(n)\gamma + \mathscr{J} - R_{rec} \tag{2.3.7}$$

$$0 = +\Gamma v_g g(n)\gamma - \frac{\gamma}{\tau_\gamma} \tag{2.3.8}$$

where the spontaneous recombination term has the form

$$R_{rec} = \frac{n}{\tau_e} = \frac{n}{\tau_n} + Bn^2 + Cn^3 \cong Bn^2$$

Equation (2.3.8) can be solved for $v_g g(n)$ to obtain a most remarkable equation!

$$v_g g(n) = \frac{1}{\Gamma\tau_\gamma} \tag{2.3.9}$$

The left-hand side depends on the carrier density "n" (or number of excited atoms) but the right-hand side is independent of "n"! *This requires "n" to be a constant. The "threshold* *carrier density*" n_{thr} *represent the approximate value of the carrier density n to produce laser oscillation* $n \cong n_{thr}$. According to this very good approximation, the carrier density remains fixed regardless of the magnitude of the current above lasing threshold. In fact, since the term $\beta B n^2$ in Equation (2.3.4) must always be positive, *the carrier density n must be slightly smaller than the threshold density.* Below lasing threshold, the approximation $n \cong n_{thr}$ does not hold since Case 1 shows that the device produces mostly spontaneous emission; consequently, the spontaneous emission term in the photon rate equation cannot be ignored. The value of the gain at lasing threshold can obviously be written as $g_{thr} = g(n_{thr})$. If we write the cavity lifetime in terms of the loss coefficients from Equation (2.1.11)

$$\frac{1}{\tau_\gamma} = v_g \alpha \tag{2.3.10}$$

then Equation (2.3.9) becomes

$$\Gamma g_{thr} = \alpha \tag{2.3.11}$$

Similar to the op-amp oscillator in Chapter 1, this last equation clearly shows that the gain equals the loss when the laser oscillates (i.e., lases). Keep in mind that the material gain, just like the carrier density, remains approximately fixed for currents larger than the threshold current!

We can substitute Equation (2.3.9) into Equation (2.3.7) to obtain the equation for the $\gamma - \mathscr{J}$ curve

$$\gamma = \Gamma \tau_\gamma (\mathscr{J} - R_{rec}) \cong \Gamma \tau_\gamma (\mathscr{J} - B n_{thr}^2) = m(\mathscr{J} - \mathscr{J}_{thr}) \tag{2.3.12a}$$

where

$$J_{thr} = A n_{thr} + B n_{thr}^2 + C n_{thr}^3 \cong B n_{thr}^2 \tag{2.3.12b}$$

Notice that the carrier density "n" has been replaced with its threshold value n_{thr}. Equation (2.3.12a) describes a straight line that passes through the threshold density $\mathscr{J}_{thr}$ and has slope $m = \Gamma \tau_\gamma$ as shown in Figure 2.3.1. Actually, the threshold current density is defined to be the point where an extrapolated straight line with slope

$$m = \Gamma \tau_\gamma = \frac{\Gamma}{v_g(\alpha_m + \alpha_{int})} \tag{2.3.13}$$

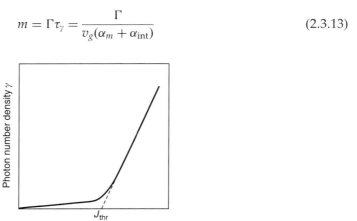

FIGURE 2.3.1
Example of $\gamma - \mathscr{J}$ characteristics for a laser operating below threshold $\mathscr{J} < \mathscr{J}_{thr}$ and above threshold $\mathscr{J} > \mathscr{J}_{thr}$.

intersects the $\mathscr{J}$-axis (a similar definition holds for the bias current I). As shown by Equation (2.3.12b), the threshold pump current number density $\mathscr{J}_{thr} = Bn_{thr}^2$ becomes larger for materials (and laser designs) with greater tendency to spontaneously emit (since B is larger). So even though spontaneous emission initiates laser oscillation, it also shifts the threshold currents to larger values. Larger rates of spontaneous emission therefore generally waste power and increase heating.

2.3.2 Comment on the Threshold Density

The previous topic mentioned the most remarkable phenomenon whereby the number of electrons remains independent of the pump current above lasing threshold. This means that we can increase the pump current and hence the optical output as much as we like without altering the gain! How can we understand this? It turns out that the stimulated emission and feedback mechanisms conspire to produce this result.

Consider a simple analogy that helps illustrate the relation. Suppose we represent the electrons in the conduction band by molecules in a water bucket as shown in Figure 2.3.2. The first bucket corresponds to the laser below threshold. The pumped water flows through the small leaks; this corresponds to spontaneous emission. As the water level rises as for the right-hand bucket, the water eventually flows out of the large slot. Increasing the pump only slightly increases the water level since the slightly higher pressure ensures more water flow through the large slot. Therefore the output water flow increases, as the pump strength increases without raising the water level in the bucket. The slot provides an additional channel for water to escape the bucket, but it only becomes active once the level reaches the height of the slot.

Now consider the case of the laser. As the pump strength increases, the number of electron–hole pairs increases in the active region and produce spontaneous emission. At some point, the laser oscillates and the number of electron–hole pairs reaches the threshold value. Increasing the pump further slightly increases the number of pairs but greatly increases the number of photons. The feedback mechanism causes these photons to produce even more through stimulated emission, which therefore increases the pair recombination rate. This negative feedback lowers the number of pairs in opposition to the effect of the pump, which very accurately mains the total number of pairs. The stimulated emission represents an alternate channel for photon production,

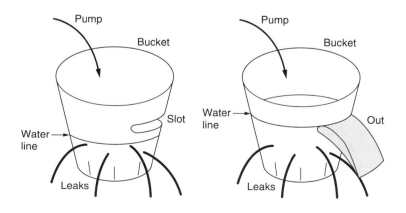

FIGURE 2.3.2
The pump increases the water level until the water line encounters the large slot in the bucket.

but it can only become active once the number of electron–hole pairs reaches a certain level similar to the water-bucket analogy.

2.3.3 Power versus Current

It's nice to use photon and current number densities when deriving the basic relations but they are not very useful in the laboratory. We need quantities like Watts and Amps. Simple scaling factors can be applied to the γ–$\mathscr{J}$ equations in Section 2.3.2 to provide convenient units for power and current. We first write down an expression for power in terms of the photon density. Consider the dimensional analysis argument for the power passing through both laser mirrors (with equal reflectivity)

$$\frac{\text{Power out}}{\text{two mirrors}} = P_o = \frac{\text{Energy}}{\text{Sec}} = \left(\frac{\text{Energy}}{\text{Photon}} * \frac{\text{Photons}}{\text{Volume}} * \text{Mode volume}\right) * \frac{1}{\tau_m} \quad (2.3.14)$$

The energy of each photon is hc/λ_o where λ_o is the wavelength in vacuum. The photons in the laser mode propagate in two directions, so the number of photons striking a single mirror must be proportional to $\gamma/2$ and for both mirrors must be proportional to γ. The modal volume is V_γ and the mirror time constant is $1/\tau_m = v_g \alpha_m$. Equation (2.3.14) treats the laser cavity as a reservoir of photons that can "leak" out with a time constant τ_m. As a comment, Equation (2.3.14) uses a constant photon density γ but the density is not necessarily uniform along the cavity. We should use an average since the photon density near the mirrors might differ from that near the interior. Substituting all of the expressions, the power from *both* mirrors of the laser must be

$$P_o = \gamma \frac{hc}{\lambda_o} V_\gamma v_g \alpha_m \quad (2.3.15)$$

Actually, the optical power through a mirror can take on a number of different forms. We next consider two separate cases for currents above and below threshold.

Case 1   P–I Below Threshold

Now we can find the output power from *both* mirrors as a function of the bias current I for a laser operating below threshold $I < I_{\text{thr}}$ (LED regime). The P–I curves can be found from the number density relation (Equation (2.3.6)) below threshold

$$\gamma = \beta \tau_\gamma \mathscr{J}$$

by substituting into Equation (2.3.15), which is

$$P_o = \gamma \frac{hc}{\lambda_o} V_\gamma v_g \alpha_m$$

to obtain

$$P_o = \left(\beta \tau_\gamma \mathscr{J}\right) \frac{hc}{\lambda_o} V_\gamma v_g \alpha_m \quad (2.3.16)$$

Next, writing the cavity lifetime τ_γ in terms of the losses α_m and α_{int}

$$\frac{1}{\tau_\gamma} = \frac{1}{\tau_m} + \frac{1}{\tau_{\text{int}}} = v_g(\alpha_m + \alpha_{\text{int}})$$

and writing the pump-current number density J in terms of I, we obtain

$$P_o = \beta \tau_\gamma \frac{\eta_i I}{qV} \frac{hc}{\lambda_o} V_\gamma v_g \alpha_m = \beta \frac{\eta_i}{q\Gamma} \frac{hc}{\lambda_o} \frac{\alpha_m}{\alpha_m + \alpha_{int}} I \tag{2.3.17}$$

The output power below threshold is linear in the bias current I. The modal coupling coefficient β causes the output power to be of smaller magnitude than the power for the same laser above threshold.

Case 2 *P–I Above Threshold*

Now we find the output power from *both* mirrors as a function of the bias current I for a laser operating above threshold $I > I_{thr}$. Substitute Equation (2.3.15), and the expression for the pump-current number density (2.3.2), namely $J = \eta_i I/(qV)$, into Equation (2.3.12a), namely $\gamma = \Gamma \tau_\gamma (\mathscr{J} - \mathscr{J}_{thr})$ to get

$$\frac{P_o}{(hc/\lambda_o)\gamma V_\gamma v_g \alpha_m} = \Gamma \tau_\gamma \left(\frac{\eta_i I}{qV} - \frac{\eta_i I_{thr}}{qV} \right) \tag{2.3.18}$$

Using the relation for the optical confinement factor $\Gamma = V/V_\gamma$ and the relation for the cavity lifetime

$$\frac{1}{\tau_\gamma} = \frac{1}{\tau_m} + \frac{1}{\tau_{int}} = v_g(\alpha_m + \alpha_{int})$$

we obtain the *P–I* curve (for currents larger than the threshold current I_{thr} and for light emitted from both mirrors)

$$P_o = \eta_i \frac{hc}{q\lambda_o} \frac{\alpha_m}{\alpha_{int} + \alpha_m} (I - I_{thr}) \tag{2.3.19}$$

The equation for the *P–I* relation above threshold represents a straight line with an intercept of I_{thr} as shown in Figure 2.3.3. The mirror loss and internal loss determine the slope of the line. Smaller mirror reflectivity gives larger loss α_m and also, therefore, larger output power. With some effort the threshold current $I_{thr} = (qV/\eta_i)\mathscr{J}_{thr}$ can be obtained by using $J_{thr} = Bn_{thr}^2$.

2.3.4 Power versus Voltage

The power output versus bias current is linear for this simple model. The semiconductor laser has the basic form of a PIN diode. The current–voltage relation for the diode has a form similar to

$$I = I_o\left(e^{qV/kT} - 1\right) \sim I_o\, e^{qV/kT} \tag{2.3.20}$$

Therefore, the plot of power versus bias voltage does not follow a simple linear relation. For best linearity, the laser should be driven with a current source.

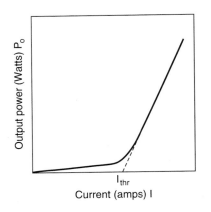

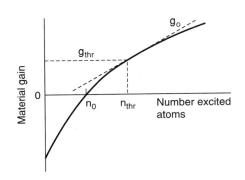

FIGURE 2.3.3
Output power versus bias current.

FIGURE 2.3.4
Example of the gain vs. the number of excited atoms.

2.3.5 Some Comments on Gain

The results of this section allow us to show the most important points for the laser gain curve. Figure 2.3.4 shows the transparency density n_o must be smaller than the threshold density n_{thr}. Above threshold, the gain doesn't vary much from $g_{thr} = g(n_{thr})$. One step to linearize the rate equations consists of Taylor expanding the gain g around n_{thr}. The *differential gain* is the *slope* of the gain $g(n)$

$$g_o(n) = \frac{dg(n)}{dn}$$

where for lasing, the differential gain is evaluated at the threshold density n_{thr}. The lowest order Taylor series approximation centered on the transparency density n_o is $g(n) = g_o(n - n_o)$. The differential gain is g_o. The gain at threshold must be $g_{thr} = g(n_{thr}) = g_o(n_{thr} - n_o)$.

2.4 Relations for Cavity Lifetime, Reflectances and Internal Loss

In order to apply the rate equations to "real world" situations, we need to relate the parameters appearing in the rate equation to the physical construction of the device. We can most easily find a relation between the mirror loss and the mirror reflectance. The free carrier absorption and optical scattering for the internal (distributed) losses require the wave equation developed in the next chapter. Here we demonstrate the relation for the reflection of optical power without interference effects. We then state the output power for mirrors with significantly unequal reflectance.

2.4.1 Internal Relations

We need to relate the cavity lifetime τ_y to the reflectance of the two mirrors R_1 and R_2 on the ends of the laser, and to the internal losses such as sidewall scattering and free carrier absorption. The term "reflectance" refers to the amount of optical *power* reflected from a mirror whereas reflectivity usually refers to the electric field (power is essentially the square of the field). The method presented in this topic uses a laser operating at steady state above threshold and requires the power in the beam to have the same magnitude after a round trip as it did at the start. The procedure again shows that the gain must equal the optical losses above threshold. We should also require the phase of the waves to agree after the round trip, but we will include this effect after discussing the optical scattering and transfer matrices.

Let us demonstrate the following basic relation for the cavity lifetime

$$\frac{1}{\tau_\gamma} = v_g \alpha_{int} + v_g \alpha_m = v_g \alpha_{int} + \frac{v_g}{2L} \text{Ln}\left(\frac{1}{R_1 R_2}\right) \tag{2.4.1}$$

where R_1 and R_2 denote the reflectivity of the two mirrors and L represents the length of the cavity. Equation (2.4.1) describes the case of the cavity without the gain or absorption normally encountered for semiconductor materials. The equation relates the time required for optical energy to escape from the cavity to the time constants for the internal and mirror loss. The internal loss includes the scattering and free carrier absorption. If the cavity has semiconductor material, we can define an *effective* cavity lifetime τ_{eff} that can be much larger than the cavity lifetime τ_γ since the material can have gain that produces photons thereby compensating for those photons lost. The second term on the right-hand side of Equation (2.4.1) describes the mirror loss. The factor $1/L$ provides an average loss over the length L of the laser. The logarithm term describes the "fractional loss" at the mirrors. The time required for light to travel from one mirror to the other L/v_g must be the same as the time interval during which the light attempts to escape from one mirror. The factor of $1/2$ occurs because the light makes a round trip.

The relation (2.4.1) can be derived by requiring the optical power within the cavity to maintain steady state. This means that a beam starting at $z=0$ (with power P_o), reflecting from the mirror at $z=L$, and finally reflecting from the mirror at $z=0$, must have the same power with which it started. The number of photons starting at $z=0$ must be the same as the number returning to $z=0$ after a round trip.

We must calculate the increase and decrease of the optical energy as it propagates from $z=0$ to $z=L$ and back to $z=0$. We first consider the exponential growth of the wave as it propagates across the gain medium. Consider photons starting at the left mirror and traveling to the right mirror across the length L of the gain medium. These photons encounter distributed internal loss α_{int} and gain, but not mirror loss. The photon rate equation (2.3.2) without spontaneous recombination can be written as

$$\frac{d\gamma}{dt} = +\Gamma v_g g(n)\gamma - \frac{\gamma}{\tau_\gamma} \tag{2.4.2}$$

where the distributive losses produce the cavity lifetime of

$$\frac{1}{\tau_\gamma} = \frac{1}{\tau_{int}} = v_g \alpha_{int} \tag{2.4.3}$$

The mirror-loss term does not appear in Equation (2.4.3) because we first consider an EM wave that only propagates between mirrors. Setting the confinement factor equal to one $\Gamma = 1$ for convenience, and changing variables from time "t" to distance "z," Equations (2.4.2) and (2.4.3) provide

$$\frac{d\gamma}{dt} = v_g \frac{d\gamma}{dz} = v_g g\gamma - \gamma v_g \alpha_{int} \quad \text{or} \quad \frac{d\gamma}{dz} = g\gamma - \gamma \alpha_{int} = g_{net}\gamma \tag{2.4.4}$$

where the net gain

$$g_{net} = g - \alpha_{int} \tag{2.4.5}$$

accounts for the material gain and distributed losses. The photon density is proportional to the power $P(z)$ traveling through the medium so that Equation (2.4.4) can also be written as

$$\frac{dP}{dz} = g_{net}P \tag{2.4.6}$$

Assuming the carrier density is constant along z, the solution to this simple differential equation is

$$P(z) = P_o \exp(g_n z) \tag{2.4.7}$$

So the power grows exponentially as it propagates from $z=0$ to $z=L$. Just before the right-hand mirror in Figure 2.4.1, the power must be

$$P(L) = P_o \exp(g_n L) \tag{2.4.8}$$

Now calculate the power after a round trip by repeatedly using Equation (2.4.8). The power at $z=L$ for a beam starting at $z=0$ is $P_o \exp(g_n L)$.
The reflectance R of mirror R_2 decreases the power to

$$R_2 P_o \exp(g_n L)$$

Reflectance refers to the ratio of the reflected to incident *power*; it is the square of the *reflectivity* $R = r^2$ (reflectivity refers to the fields). The power in the beam increases exponentially as it travels from $z=L$ back to $z=0$

$$R_2 P_o \exp(2g_n L)$$

Finally, mirror R_1 reduces the beam power to produce the power

$$R_1 R_2 P_o \exp(2g_n L)$$

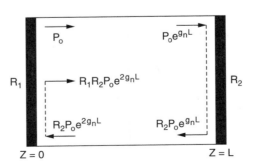

FIGURE 2.4.1
The effect of the gain medium on the optical power of a beam making a round-trip in the laser Fabry-Perot cavity.

just to the right of the mirror at $z = 0$. At this time, the beam has made a complete round trip. For steady state, the initial power P_o must be the same as the final power

$$P_o = R_1 R_2 P_o \exp(2g_n L)$$

which yields the relation

$$g_{net} = \frac{1}{2L} \text{Ln}\left(\frac{1}{R_1 R_2}\right) \tag{2.4.9a}$$

where

$$g_{net} = g - \alpha_{int} \tag{2.4.9b}$$

The material gain g in Equation (2.4.9b) can be rewritten by appealing to Equation (2.3.11) which says that the gain must equal the loss at steady state $\Gamma g = \alpha$ or, for unity confinement $\Gamma = 1$, we have $g = \alpha$. We can also appeal to the equivalent form in terms of the cavity lifetime and the temporal gain g_t

$$v_g g = g_t = \frac{1}{\tau_\gamma} \tag{2.4.9c}$$

Combining Equations (2.4.9) provides the cavity lifetime (without the gain or absorption due to semiconductor material)

$$\frac{1}{\tau_\gamma} = v_g \alpha_{int} + \frac{v_g}{2L} \text{Ln}\left(\frac{1}{R_1 R_2}\right) \tag{2.4.10}$$

This last equation provides the lifetime for the case when energy escapes from the cavity through "internal loss" and through *both* mirrors. If the two mirrors have the same reflectivity (as is typical for a semiconductor laser with cleaved or etched facets), then $R = R_1 = R_2$ and

$$\frac{1}{\tau_\gamma} = v_g \alpha_{int} + \frac{v_g}{L} \text{Ln}\left(\frac{1}{R}\right) \tag{2.4.11}$$

The power reflectivity for facets cleaved in GaAs is nearly 0.34. Typically $\alpha_m \approx 100 \, \text{cm}^{-1}$ and $\alpha_{int} \approx 50 \, \text{cm}^{-1}$.

2.4.2 External Relations

In this topic, we illustrate the output power from mirrors with reflectance R_1 and R_2. Often times one mirror receives a high reflectance coating to increase the power out of the other one. The photon density (intracavity power) varies significantly from a constant value near the mirrors. A higher reflectance mirror produces a higher photon density just inside that cavity than does a lower reflectance one. By calculating the average photon density in the laser body (Reference 5, the Agrawal and Dutta book) and applying the appropriate boundary conditions, we find the power P_1 and P_2 through the mirrors with reflectance R_1 and R_2, respectively

$$P_1 = \frac{(1 - R_1)\sqrt{R_2}}{\left(\sqrt{R_1} + \sqrt{R_2}\right)\left(1 - \sqrt{R_1 R_2}\right)} P_o \tag{2.4.12a}$$

$$P_2 = \frac{(1 - R_2)\sqrt{R_1}}{(\sqrt{R_1} + \sqrt{R_2})(1 - \sqrt{R_1 R_2})} P_o \qquad (2.4.12b)$$

where P_o gives the total power through *both* mirrors given in Equation (2.3.19). The relation between the reflectance and the loss coefficients is given by Equation (2.4.11). We can easily show that $P_1 + P_2 = P_o$, which indicates the output power divides itself between the two facets. If the facets are identical, i.e., $R_1 = R_2 = R$, then each facet handles half the total since $P_1 = P_2 = P_o/2$.

2.5   Modulation Bandwidth

The first part of this chapter discusses the laser rate equations that provide a wealth of information on the semiconductor laser. For example, the previous sections demonstrate the all-important *P–I* curves. The *P–I* curves represent the laser system at steady state. However, the rate equations can also provide information on the transient and small signal behaviors of the laser. The present section discusses the modulation bandwidth.

We start by asking a question. What is the optical response of a laser to small sinusoidal changes of bias current? That is, what is the transfer function? This question has very important implications for modern communications and data transfer systems. Fiber communications systems couple the semiconductor laser to one end of the fiber and position the receiver many kilometers away. The laser must have large bandwidth in order to transmit the greatest amount of information. Most laser systems modulate the output laser beam by modulating the bias current to the laser. We must have some knowledge of its frequency response to successfully implement the semiconductor laser in the communication system. We would like to have the changes in amplitude of the output beam due to changes in the bias current to be independent of frequency. However, as with any real device, the response function usually depends on frequency. For example, electronic circuits have parasitic capacitance that cause the frequency to roll-off at high frequencies. We will see that the output response is relatively flat up to a resonant frequency that is determined by the rate equations.

2.5.1   Introduction to the Response Function and Bandwidth

Figure 2.5.1 shows a conceptual experiment to determine the bandwidth. The laser is biased with the battery, which provides a steady-state current $\bar{I}$ to the laser. The signal generator applies a *small* sinusoidal bias current $I = I(t)$ (where $I \ll \bar{I}$). The left-hand coils

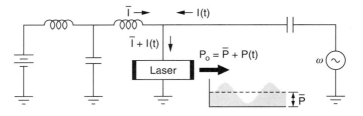

FIGURE 2.5.1
A semiconductor laser with DC bias and AC modulation.

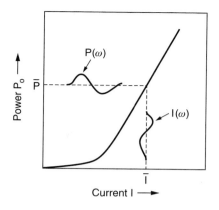

FIGURE 2.5.2
Output power vs bias current.

and capacitor prevent the small AC signal from shunting through the battery; however, they allow the DC current to flow from the battery to the laser. The right-hand capacitor prevents the DC current from conducting through the signal source but allows the AC signal to reach the laser. The input current I' to the laser consists of the sum of a large DC bias and a small AC signal with angular frequency ω

$$I' = \bar{I} + I(t) \tag{2.5.1}$$

The small signal current is given in phasor notation as $I(t) = I\,e^{i\omega t}$ where $I = I(\omega)$ represents the amplitude. Notice that we have temporarily changed notation from that in the earlier sections. The "primes" indicate totals as required by Equations (2.2.10). The "bars" indicate the steady-state quantities given in Equations (2.3). The output signal consists of the optical beams emitted through both mirrors. The total output power P_o must be the sum of the steady state power $\bar{P}$ and the small sinusoidal signal $P(t)$. Equation (2.5.1) has a prime to indicate the total current (it does not mean derivative). Figure 2.5.2 shows how the steady-state $\bar{P}$–$\bar{I}$ transfer function changes the small sinusoidally varying bias current into a small sinusoidally varying output power P. For larger modulation frequency ω, we expect the small changes in output power P to decrease in amplitude.

We will perform a small signal analysis of the γ'–$\mathcal{J}'$ rate equations. This means that we will find two sets of equations. The first set describes the steady-state photon density $\bar{\gamma}$ in response to the steady-state current-number pump density $\bar{\mathcal{J}}$. The second set determines the small sinusoidal changes in the photon density γ as a function of the small changes in the bias current $\mathcal{J}$. In terms of photons, we can imagine the average optical power to be represented by an average number of photons $\bar{\gamma}$ and the modulation to be represented by small changes in this number as indicated in Figure 2.5.3. Notice how the photon density corresponds to the amplitude of the carrier. The carrier is the electric field of the light wave from the laser. The carrier has a very high frequency on the order of 10^{15} Hz. The power is essentially the square of the electric field.

The results of the small signal analysis appear in Figure 2.5.4 as a plot of the response (output divided by input) versus the modulation frequency ω. The modulation signal decreases rapidly at high modulation frequencies. The response curve develops a peak which we label as the resonant frequency ω_{res}. We take the resonant frequency as a measure of the bandwidth. We will see that the bandwidth depends on the construction

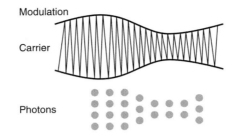

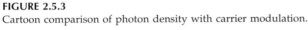

Modulation

Carrier

Photons

FIGURE 2.5.3
Cartoon comparison of photon density with carrier modulation.

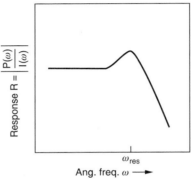

ω_{res}

Ang. freq. ω ⟶

FIGURE 2.5.4
Resonant frequency of the response curve is roughly equal to the bandwidth.

of the laser through the loss terms (i.e., $1/\tau_\gamma$) the differential gain g_o, and the pump level (through the average photon density $\bar{\gamma}$ within the cavity) according to

$$\omega_{\text{res}} = \sqrt{\frac{v_g g_o \bar{\gamma}}{\tau_\gamma}} \tag{2.5.2}$$

where v_g is the speed of the wave in the gain medium. Recall that g_o comes from the lowest order Taylor approximation $g(n) \approx g_o(n - n_o)$, and g_o represents the slope of the gain curve near $n = n_{\text{thr}}$.

2.5.2 Small Signal Analysis

To demonstrate the equation for resonance, we must perform a small signal analysis of the laser rate equations. A small signal analysis requires us to consider both steady state quantities, which will be denoted by a "bar" over the quantity (for example, $\bar{\gamma}$ should remind the reader of an "average"), and the sinusoidal varying terms. Let the quantities in the rate equation be denoted by primes (the prime indicates the total of the DC and AC quantities). For simplicity, we linearize the gain by making an approximation for the carrier density n near the transparency density n_o of $g' = g(n') \cong g_o(n' - n_o)$. The recombination term is linearized by using n'/τ_n (with τ_n a constant). Keep in mind that the variables in the rate equation denoted by primes represent the total of the DC and AC quantities

$$\frac{dn'}{dt} = -v_g g_o(n' - n_o)\gamma' + J' - \frac{n'}{\tau_n} \tag{2.5.3a}$$

$$\frac{d\gamma'}{dt} = +\Gamma v_g g_o(n' - n_o)\gamma' - \frac{\gamma'}{\tau_\gamma} \tag{2.5.3b}$$

that is, the primed quantities are defined by

$$n' = \bar{n} + n\,e^{i\omega t} \qquad \gamma' = \bar{\gamma} + \gamma\,e^{i\omega t} \qquad J' = \bar{J} + J\,e^{i\omega t} \tag{2.5.4}$$

The quantities n, γ and J are the small signal quantities (i.e., the small Fourier amplitudes).

The steady-state quantities $\{\bar{n}, \bar{\gamma}, \bar{J}\}$ satisfy the steady-state equations obtained from Equations (2.5.3) by setting the derivatives to 0

$$0 = -v_g g_o(\bar{n} - n_o)\bar{\gamma} + \bar{J} - \frac{\bar{n}}{\tau_n} \tag{2.5.5a}$$

$$0 = +\Gamma v_g\,g_o(\bar{n} - n_o)\bar{\gamma} - \frac{\bar{\gamma}}{\tau_\gamma} \tag{2.5.5b}$$

The actual derivation of the resonant frequency requires a great deal of algebra. We outline the procedure as follows. Defining $\bar{g} = g_o(\bar{n} - n_o)$ and substituting the primed quantities from Equation (2.5.4) into the rate equations (2.5.3) provides

$$i\omega n\,e^{i\omega t} = -v_g\big(\bar{g} + g_o n\,e^{i\omega t}\big)\big(\bar{\gamma} + \gamma\,e^{i\omega t}\big) + \bar{J} - \frac{\bar{n}}{\tau_n} + J\,e^{i\omega t} - \frac{n}{\tau_n}e^{i\omega t}$$

$$i\omega\gamma\,e^{i\omega t} = \Gamma v_g\big(\bar{g} + g_o n\,e^{i\omega t}\big)\big(\bar{\gamma} + \gamma\,e^{i\omega t}\big) - \frac{\gamma\,e^{i\omega t}}{\tau_\gamma} - \frac{\bar{\gamma}}{\tau_\gamma} \tag{2.5.6}$$

Use the steady-state equations (2.5.5) to cancel some of the terms with the steady-state quantities $\{\bar{n}, \bar{\gamma}, \bar{J}\}$ to get

$$i\omega n\,e^{i\omega t} = -v_g\big[g_o n\,e^{i\omega t}\,\bar{\gamma} + \bar{g}\gamma\,e^{i\omega t} + g_o\gamma n\,e^{i\omega t}\big] + J\,e^{i\omega t} - \frac{n}{\tau_n}e^{i\omega t}$$

$$i\omega\gamma\,e^{i\omega t} = \Gamma v_g\big[\bar{\gamma}g_o n\,e^{i\omega t} + \bar{g}\gamma\,e^{i\omega t} + g_o n\gamma\,e^{2i\omega t}\big] - \frac{\gamma\,e^{i\omega t}}{\tau_\gamma} \tag{2.5.7}$$

Drop the second-order nonlinear terms such as γn, n^2; this procedure also removes terms such as $e^{2i\omega t}$. Then cancel the exponentials $e^{i\omega t}$ from both sides

$$i\omega n = -v_g g_o n\bar{\gamma} - v_g\bar{g}\gamma + J - \frac{n}{\tau_n}$$

$$i\omega\gamma = \Gamma v_g g_o n\bar{\gamma} + \Gamma v_g\bar{g}\gamma - \frac{\gamma}{\tau_\gamma} \tag{2.5.8}$$

Rewrite Equations (2.5.8) to obtain

$$v_g\bar{g}\gamma = J - \left(v_g g_o\bar{\gamma} + \frac{1}{\tau_n} + i\omega\right)n$$

$$\Gamma v_g g_o\bar{\gamma}n = \left(\frac{1}{\tau_\gamma} - \Gamma v_g\bar{g} + i\omega\right)\gamma \tag{2.5.9}$$

Solve the second of Equations (2.5.9) for n and substitute into the first of Equations (2.5.9). Multiply out the terms to find

$$\gamma = \frac{\Gamma v_g g_o \bar{\gamma} J}{\Gamma v_g^2 g_o \bar{g} \bar{\gamma} + \left(v_g g_o \bar{\gamma} + \frac{1}{\tau_n} + i\omega\right)\left(\frac{1}{\tau_\gamma} - \Gamma v_g \bar{g} + i\omega\right)} \tag{2.5.10}$$

Multiply out the denominator and define the following terms

$$\frac{1}{\tau} = \left(\frac{1}{\tau_\gamma} - \Gamma v_g \bar{g} + v_g g_o \bar{\gamma} + \frac{1}{\tau_n}\right) \tag{2.5.11}$$

$$\omega_o^2 = \Gamma v_g^2 g_o \bar{g} \bar{\gamma} + \left(v_g g_o \bar{\gamma} + \frac{1}{\tau_n}\right)\left(\frac{1}{\tau_\gamma} - \Gamma v_g \bar{g}\right) \tag{2.5.12}$$

to obtain the *transfer* function

$$\frac{\gamma}{J} = \frac{\Gamma v_g g_o \bar{\gamma}}{(\omega_o^2 - \omega^2) + (i\omega/\tau)} \tag{2.5.13}$$

Defined the *response* function

$$R_J = \left|\frac{\gamma}{J}\right|^2 \tag{2.5.14}$$

to show that

$$R_J = \frac{(\Gamma v_g g_o \bar{\gamma})^2}{(\omega_o^2 - \omega^2)^2 + (\omega^2/\tau^2)} \tag{2.5.15}$$

We find the peak of the response function by setting its derivative with respect to the angular frequency equal to 0

$$\frac{dR_J}{d\omega} = 0 \tag{2.5.16}$$

This gives a condition on the resonant frequency

$$\omega_r^2 = \omega_o^2 - \frac{1}{2\tau^2} \tag{2.5.17}$$

If $1/\tau^2$ is very small, then $\omega_r \cong \omega_o$. Earlier portions of this chapter show that, above threshold, the gain essentially locks to the loss

$$\frac{1}{\tau_\gamma} = \Gamma v_g \bar{g} \quad \rightarrow \quad \Gamma v_g \bar{g} \bar{\gamma} = \frac{\bar{\gamma}}{\tau_\gamma} \tag{2.5.18}$$

Combining Equations (2.5.17) and (2.5.18) with Equation (2.5.12) provides the desired results

$$\omega_r = \sqrt{\frac{v_g g_o \bar{\gamma}}{\tau_\gamma}} \tag{2.5.19}$$

We can also write the resonant frequency in terms of the mirror and internal loss. The cavity lifetime in Equation (2.5.19) can be replaced by the internal loss and the mirror loss

$$\frac{1}{\tau_\gamma} = v_g(\alpha_{int} + \alpha_m)$$

to obtain

$$\omega_{res} = \sqrt{g_o v_g^2 \bar{\gamma} \left[\alpha_{int} + \frac{1}{L} \text{Ln}\left(\frac{1}{R}\right) \right]} \qquad (2.4.20)$$

Only a single reflectance R appears in the equation since we assume the optical power emits through two identical mirrors. As the length of the cavity L or the mirror reflectivity R increases, the resonant frequency decreases. Bandwidth essentially describes how fast the laser can respond to changes in current. A large cavity lifetime means it will take a long time for the light in the cavity to leak out and thereby change the optical energy in the cavity. A large mirror reflectance increases the cavity lifetime and therefore decreases the resonant frequency. On the other hand, the length of the cavity determines the amount of time that it takes for the light to travel between the two mirrors. If there were not any internal loss, only the mirror loss term would be present. In such a case, the cavity loses light only when the light strikes the mirrors. The amount of time required to travel across the cavity is $t = v_g/L$, where L is the length of the cavity. This means that the laser can respond no faster than the time t. Also notice that the resonant frequency depends on the number of photons in the cavity. As bias current to the laser increases, the number of photons in the laser cavity also increases and hence the resonant frequency of a laser must also increase. Therefore higher powers also provide larger bandwidths. By using the results of Equation (2.5.15) and using the expression (2.5.20) for the resonant frequency in place of ω, we can calculate the peak of a resonant curve. As mentioned, longer lasers have lower bandwidth and also lasers with higher reflectance mirrors also have lower bandwidth. A VCSEL might have a cavity length of approximately 5 μm and the mirror reflectivity might be as high as 95%.

2.6 Introduction to RIN and the Weiner–Khintchine Theorem

The study of noise in optoelectronic systems has an increasingly important position with the advent of small sized devices and small signals. The present section discusses the relative intensity noise (RIN) for the photon density and optical power. The RIN measures the expected fluctuation of the signal. The laser rate equations, when supplemented with Langevin noise terms, provide the main equations to predict the noise from the semiconductor laser. The Langevin noise terms represent the external influence of the environment on the lasing process such as for the pump and mirrors. The RIN can most generally be stated as an autocorrelation which requires the calculation of the correlation between Langevin sources. The section discusses the Kronecker and impulse correlation for discrete and continuous processes, respectively. The Weiner–Khintchine formula relates the correlation function to the spectral density.

The subsequent section shows how the response to the noise sources can be deduced from the laser rate equations. Appendix 4 discusses many of the preliminary concepts in probability theory and statistics.

2.6.1 Basic Definition of Relative Noise Intensity

Relative Intensity Noise (RIN) measures the amount of noise relative to the size of the signal. Consider a beam characterized by either the power P or the photon density γ. At first thought, we might define the RIN as the ratio of the deviation σ of the signal to the average signal $\bar{P}$. However, for convenience, people define RIN as the ratio of the noise variance σ^2 to the average power squared $\bar{P}^2$. Figure 2.6.1 shows an optical signal (with average power $\bar{P}$) with superimposed noise. The figure assumes that the noise, represented by the excursions of the signal from the average, obeys a Gaussian distribution. Let the symbol σ denote the standard deviation for the distribution. The RIN for a DC signal can be equally well expressed in terms of either power or photon density

$$\text{RIN} = \frac{\sigma_P^2}{\bar{P}^2} \quad \text{RIN} = \frac{\sigma_\gamma^2}{\bar{\gamma}^2} \tag{2.6.1}$$

where Appendix 4 shows the average and variance in Equation (2.6.1) of the form

$$\langle z \rangle = \int dz\, z\, f(z) \qquad \sigma^2 = \big\langle (z - \bar{z})^*(z - \bar{z}) \big\rangle_{\text{Real}} = \big\langle (z - \bar{z})^2 \big\rangle \tag{2.6.2}$$

The second version in Equation (2.6.1) most conveniently uses the photon density found in the laser rate equations. In either case, one must be careful to distinguish between the photon density inside or outside of the cavity since partition noise (due to the mirrors for example) modifies the RIN.

2.6.2 Basic Assumptions

Equation (2.6.1) and Figure 2.6.1 make use of some implicit assumptions (Reference 6, Mandel and Wolf, Chapter 2). First, we assume that measuring the random variable $P(t)$

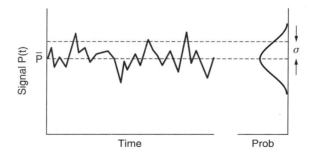

FIGURE 2.6.1
A signal as a function of time. A large amount of noise is superimposed on the average signal.

produces discrete points on the graph. We think of the time t as an index. Each sequence of points $P(t)$ (i.e., each possible graph like Figure 2.6.1) represents a "realization" of a random process. Sometimes we refer to the collection of points $P(t)$ as a random process. Let t_1 be a specific time. The random variable $P(t_1)$ representing the power at time t_1 can take on any number of values. For example, we might find $P(t_1) = 2.1$ or $P(t_1) = 1.8$ or $P(t_1) = 2.4$. Therefore at time t_1, the random variable $P(t_1)$ takes on a range of values and has an average and standard deviation. For every time $t_1, t_2, \ldots$, there exists a random variable $P_1 = P(t_1)$, $P_2 = P(t_2), \ldots$. The collection of all random variables form the random process $P(t)$. As discussed in Appendix 4, the joint probability distribution $f(P_1, t_1; P_2, t_2)$ represents the probability per unit time (squared) that $P = P_1$ at time $t = t_1$ *and* $P = P_2$ at time $t = t_2$; this probability density is also called the "two time probability." Notice that the two-time probability refers to two-separate times for the same process. The multi-time probabilities contain information on correlation.

As a second assumption, we require the *noise* process to be "stationary" in the sense that its characteristics do not change with time. For example, the standard deviation (for the noise) cannot depend on time. We usually characterize the stationary process by the time-dependence of the probability distribution. The probability distribution for a single random variable such as the power $P(t)$ at a fixed time t cannot depend on time for a stationary process. However, a joint probability distribution $f(P_1, t_1; P_2, t_2; P_3, t_3 \ldots)$ describing a number of random variables P_i (which denotes the power in the beam at time "t_i"), depends only on a *difference* in time

$$f(P_1, t_1; P_2, t_2; P_3, t_3 \ldots) = f(P_1, 0; P_2, t_2 - t_1; P_3, t_3 - t_1 \ldots)$$

If the joint distribution could not be written in this form, the results of our measurements would depend on when we start the clock. In this case, the fundamental nature of the system must be changing.

Our third assumption concerns the "ergodic" property of the distribution. For some processes, an average $\bar{P} = \langle P(t) \rangle$ can be found by using either an ensemble average or by a time average. An ergodic process assumes that the average of a function $f(t)$ can be computed by either a time or ensemble average

$$\langle y(t) \rangle = \operatorname*{Lim}_{\tau \to 0} \frac{1}{\tau} \int_0^\tau dt\, y(t) \quad \text{or} \quad \langle y \rangle = \int dy\, y f(y) \tag{2.6.3}$$

where $f(y)$ denotes the probability density. For the average over time, the interval τ should be long compared with any correlation time, but short compared with the time scale of interest. Strictly speaking, a process is ergodic if every realization contains exactly the same statistical information as the ensemble. In this case, the realizations don't all need to start at the same time.

Example 2.6.1

A small pressure sensor is glued to the bottom of a tin can filled with water. Suppose the sensor produces noise in proportion to the pressure; i.e., the noise is a fixed percentage of the total signal (say 1%). The pressure at the bottom of the can must be proportional to the amount of water in the can. Suppose we place a hole at the bottom of the tin can and the water slowly drains away. The pressure changes with time. Therefore, the noise content depends on when the first measurement occurs. The distribution "f" cannot be written as in the previous equation

$$f(P_1, t_1; P_2, t_2) \neq f(P_1, 0; P_2, t_2 - t_1)$$

2.6.3 The Fluctuation-Dissipation Theorem

We should introduce concepts leading up to the discussion of the Langevin noise sources before continuing with the RIN for a semiconductor laser. There are three main questions to be answered. (1) How are random noise sources included in the rate equations? (2) How should we think of the noise sources? (3) What physical mechanism leads to the noise sources?

First, let's write the final set of rate equations (neglecting the phase). Let δn and $\delta \gamma$ be the differential response of the carrier and photon number density to the Langevin noise sources $F_n(t)$ and $F_p(t)$, respectively. The symbols δn and $\delta \gamma$ refer to the small differences between the quantity and its average $\delta n = n(t) - \bar{n}$ and $\delta \gamma = \gamma - \bar{\gamma}$ similar to Section 2.5 where the "bar" indicates the average quantity. The next section shows that the rate equations with noise terms have the form

$$\frac{d}{dt} \delta n = -v_g \bar{g}_n \bar{\gamma} \, \delta n - v_g \bar{g} \, \delta \gamma - \bar{R}_n \, \delta n + F_n(t) \tag{2.6.4}$$

$$\frac{d}{dt} \delta \gamma = \Gamma v_g \bar{g}_n \bar{\gamma} \, \delta n + \Gamma v_g \bar{g} \, \delta \gamma - \frac{\delta \gamma}{\tau_\gamma} - \beta \bar{R}_n \delta n + F_\gamma(t) \tag{2.6.5}$$

The subscripts "n" indicate a derivative with respect to "n." Classically, the noise sources are a phenomenological addition to the equations. These noise sources are intimately related to the damping terms in Equations 2.6.4 and 2.6.5 (for example, γ/τ_γ) through the so-called "fluctuation-dissipation" theorem. Figure 2.6.2 shows how the functions might be pictured. Adjacent points have absolutely no relation between them (delta function correlated).

The Langevin noise sources arise from the interaction of the laser with so-called *reservoirs*. The pumping mechanisms can be considered to be reservoirs and they can supply an infinite number of charges (i.e., current * time). The phonons in the crystal are part of another reservoir—a thermal reservoir—since their distribution in energy corresponds to a Boltzmann distribution with a specific (controlled) temperature. The mirrors provide "windows" into a third type of reservoir—a "light" reservoir. Every external influence on the laser system can be related to a reservoir. Subsequent sections show that a reservoir induces rapid fluctuations (Langevin noise) as well as

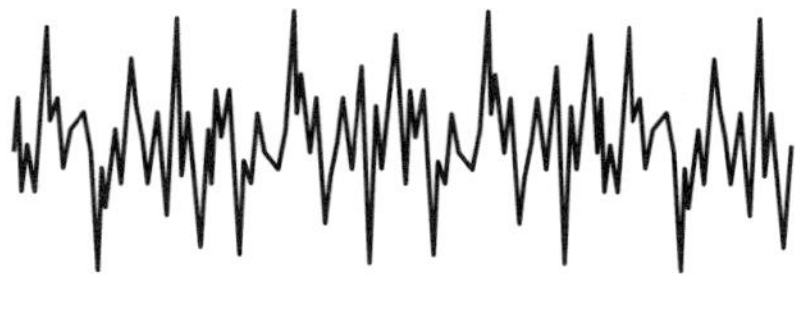

FIGURE 2.6.2
An example of the Langevin function $F_n(t)$ or $F_\gamma(t)$. The functions have random values and cannot be specified by a formula.

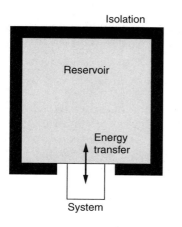

Isolation

Reservoir

Energy
transfer

System

FIGURE 2.6.3
The reservoir can exchange energy with the system under study.

damping. This section introduces the notion of a reservoir and discusses the associated fluctuation-dissipation theorem.

To see how fluctuations arise in a number density, we discuss the thermal reservoir. The laser and its environment (i.e., the external influences) are divided into a small system under study and a collection of reservoirs. The reservoirs are large systems that provide equilibrium for the smaller system. A reservoir is a system with an extremely large number of degrees of freedom. For example, a reservoir of two-level atoms or harmonic oscillators necessarily contains a large number of atoms or oscillators. A reservoir of light consists of a set of modes where the number of such modes is extremely large. Typically, we assume a specific energy distribution exists in the reservoir. For example, if the reservoir consists of point particles (such as gas molecules) then one might assume a Boltzmann distribution for the energy.

Let's bring the reservoir into contact with the small system so that energy can flow between the system and the reservoir as indicated in Figure 2.6.3. The reservoir has such a large number of degrees-of-freedom that any energy transferred to/from the small system has negligible affect on the energy distribution in the reservoir.

For a concrete example, suppose the small system consists of a single gas molecule and the reservoir has a large number of molecules all at thermal equilibrium (i.e., a Boltzmann energy distribution). The temperature of the small system will eventually match the temperature of the larger system. However, temperature measures the average kinetic energy. Therefore, to say that the reservoir and system have equal temperatures means that the average kinetic energy of the single molecule matches the average kinetic energy of all the molecules in the reservoir.

Suppose the molecule in the small system has a much larger than average kinetic energy (maybe a factor of 10). The extra energy eventually transfers to the reservoir. This extra energy distributes to all of the molecules in the reservoir, which makes negligible changes in the total reservoir distribution. Essentially the small system loses the initial packet of energy to the large system; the initial packet distributes over many degrees of freedom. In effect, the reservoir has "absorbed" the "extra" system energy and the motion of the single molecule "damps."

The reservoir energy distribution defines average quantities for the reservoir. The contact between the two systems brings the small system into equilibrium with the reservoir, which therefore defines the average quantities for the small system. Suppose the single atom in the small system is initially in equilibrium with the reservoir. Occasionally, a large chunk of energy will be transferred from the reservoir to the small system—a thermal fluctuation. As a result, the single atom will have more energy than its equilibrium value. Eventually, the energy of the atoms damps when the energy transfers back to the reservoir. We assume any correlation between fluctuations occurring on short times scales to be negligible. The process of transferring energy between the small system and the reservoir provides an example of the fluctuation-dissipation theorem. The theorem basically states that a reservoir both damps the small system and induces fluctuations in the small system. *The two processes go together and cannot be separated.* Often, on a phenomenological level, the fluctuations are included in rate equations through the Langevin functions.

2.6.4 Definition of Relative Intensity Noise as a Correlation

The relative intensity noise (RIN) can be measured in terms of either the power P or the photon density γ. The RIN can be defined through an autocorrelation function Γ

$$\text{RIN} \equiv \frac{\Gamma_\gamma(t, t + \tau)}{\bar{\gamma}^2} \equiv \frac{\langle \delta\gamma^*(t)\, \delta\gamma(t + \tau)\rangle}{\bar{\gamma}^2} \quad \text{or} \quad \text{RIN} = \frac{\Gamma_P(t, t + \tau)}{\bar{P}^2} = \frac{\langle \delta P^*(t)\, \delta P(t + \tau)\rangle}{\bar{P}^2} \quad (2.6.6)$$

where the RIN appears to depend on $(t, t + \tau)$ and the "variance" is replaced by an autocorrelation function Γ (not to be confused with the confinement factor). We will see later that RIN depends only on the time difference τ for a stationary probability distribution. The definition explicitly shows the complex conjugate in spite of the fact that the photon density and power are real. The quantity $\delta\gamma$ specifies the difference between the photon density $\gamma(t)$ at time t and the average photon density $\bar{\gamma}$. The RIN in Equation (2.6.6) and RIN in Equation (2.6.1) must be the same for $\tau = 0$. The RIN in Equation (2.6.6) contains additional information on the correlation of the fluctuations and on the bandwidth of the fluctuation spectrum.

To understand the relation between the correlation and the standard deviation, consider the simplest case of the discrete process x_i with zero average. The discrete sequence of numbers x_i can be Kronecker-delta function correlated (refer to Appendix 4). Consider two sub-sequences of N numbers from the same sequence x_i (assume large N and zero average). We might take a sequence starting at number "i," specifically $x_i, x_{i+1}, \ldots, x_{i+N}$ and another sequence displaced by n terms, specifically $x_{i+n}, x_{i+n+1}, \ldots, x_{i+n+N}$. We assume that x_i corresponds to the noise and has an expectation value equal to zero $\langle x_i \rangle = 0$ (where we either average over "i" or use the ensemble average). The correlation between the two sub-sequences must be given by

$$\Gamma(i, n) = \langle x_i x_{i+n} \rangle = \frac{1}{N} \sum_{i=1}^{N} x_i x_{i+n} \quad (2.6.7)$$

If the adjacent (and succeeding) values of "x" are truly random so that x_i does not have any relation what-so-ever to x_{i+1}, then we expect the quantity $\langle x_i x_{i+n} \rangle$ to be zero, specifically $\langle x_i x_{i+n} \rangle = 0$ for any $n > 0$. On the other hand, for $n = 0$, we see that the right-hand side of Equation (2.6.7) reduces to the variance of "x" which does not necessarily produce zero $\langle x_i x_i \rangle = \sigma_x^2 \neq 0$. Therefore, it is possible to have Kronecker-delta correlated discrete sequences

$$\Gamma(n) \equiv \Gamma(i, i + n) = \sigma_x^2\, \delta_{n,0}$$

This depends on the fact that adjacent values of "x" are not related. Notice the correlation Γ depends on "n" and not i. That is, Γ depends only on the difference $(i + n) - i = n$, which has the mark of a stationary process.

Continuous processes $x(t)$, such as for the Langevin noise terms, can be Dirac-delta function correlated (assume x has zero average). For two times t, t' (possibly arbitrarily close), we consider quantities such as $\langle x(t) x(t') \rangle = \langle x(t) x(t + \tau) \rangle$ where τ represents a displacement that takes the place of "n" above. We assume that the two values $x(t)$ and

$x(t+\tau)$ do not have any relation to each other regardless of the size of τ (except for $\tau = 0$). This says that arbitrarily large values of the frequency ω must be contained in the Fourier transform of a realization $x(t)$, since "x" at t and t' can make wild swings regardless of the proximity of t and t'. On the other hand for $\tau = 0$, $x(t)$ and $x(t')$ must be the same sequence and $\langle x(t)\,x(t')\rangle = \langle x(t)\,x(t+\tau)\rangle = \langle x(t)\,x(t)\rangle$ must be related to the standard deviation. The Langevin noise terms must be Dirac delta function correlated

$$\langle x(t)\,x(t+\tau)\rangle = S_x\,\delta[(t+\tau)-t] = S_x\,\delta(\tau - 0) \tag{2.6.8}$$

where S_x represents the correlation strength.

In analogy with the discrete case, a continuous process can be delta-function correlated according to Equation (2.6.8). For the Ergodic process, Equation (2.6.8) can be written as

$$S_x\,\delta(\tau) = \langle x(t)\,x(t+\tau)\rangle = \lim_{T\to\infty}\frac{1}{T}\int_0^T dt\,x(t)\,x(t+\tau) \cong \lim_{N\to\infty}\frac{1}{N}\sum_i x(t_i)\,x(t_i+\tau) \tag{2.6.9}$$

where the time interval T is divided into the number N of small intervals Δt. We can define the time τ also in terms of a discrete index. In the continuous case, we can write $\tau = \eta\Delta t$ where η represents the continuous counterpart of the discrete index n. The right-hand side of Equation (2.6.9) becomes $\lim_{N\to\infty}(1/N)\sum_i x_i\,x_{i+n}$. However, the left side has the Dirac delta function that can be rewritten as $\delta(\tau) = \delta(\eta\Delta t) = \Delta t\,\delta(\eta)$. The Dirac delta can be expressed as Kronecker-delta function by $\delta(\eta - 0) \cong \delta_{n,0}/\Delta t$ where η is approximately the integer n and Δt is the cell width. Equation (2.6.9) then becomes

$$S_x\,\delta_{n,0} = \lim_{N\to\infty}\frac{1}{N}\sum_i x_i\,x_{i+n} \tag{2.6.10}$$

One problem arises with regard to taking the Fourier transform of a stationary process $z(t)$ as normally done to find the spectral density $S_z(\omega)$. The function $z(t)$ must be square integrable

$$\int_{-\infty}^{\infty} dt\,|z(t)|^2 < \infty \tag{2.6.11}$$

so that the function z has finite "length." However, a stationary process z does not have finite length since the fluctuations away from the average (zero in this case) do not change in magnitude with time since the standard deviation does not depend on time. The expected value of $|z(t)|^2$ must be nonzero for any time and therefore the summation over all times must become infinite. The Weiner–Khintchine formula in the next topic circumvents the issue of whether or not the Fourier integrals exist. The derivation uses the correlation in time, which easily shows any impulse-function character of the transform for z^2 (i.e., γ^2 or δP^2 for the RIN).

The next topic discusses the correlation in more detail. In preparation, we make a brief note on the noise for a Fourier transform. Suppose $P(t)$ represents the optical power in a light beam. Assume the signal consists of a steady state part and superimposed noise. The Fourier integral representation of $P(t)$ becomes

$$P(t) = \int_{-\infty}^{\infty} d\omega\,\tilde{P}(\omega)\,\frac{e^{i\omega t}}{\sqrt{2\pi}}$$

Any noise in $P(t)$ must appear in the Fourier transform $\tilde{P}(\omega)$. The noise cannot be in $e^{i\omega t}$ since it is a formula and ω is not a random variable. The RF spectrum analyzer provides a measurement of the noise by displaying $S_{\delta P}(\omega)$ and not $\tilde{P}(\omega)$. One must remember that the RF spectrum analyzer is not an oscilloscope and any small variation in the plot does not indicate the noise in the signal (optical power in this case). The displacement of the trace above the bottom of the plot indicates the noise.

2.6.5   The Weiner–Khintchine Theorem

The Weiner–Khintchine theorem states that a stationary process $z(t)$ has a power spectrum $S(\omega) = \sqrt{2\pi}\,\Gamma(\omega)$ that is essentially the Fourier transform $\Gamma(\omega)$

$$\Gamma(\omega) = \int_{-\infty}^{\infty} d\tau\, \Gamma(\tau)\, \frac{e^{-i\omega\tau}}{\sqrt{2\pi}} \tag{2.6.12}$$

of the correlation function $\Gamma(\tau) = \langle z(t)^* z(t+\tau)\rangle$. In addition

$$\langle z^*(\omega')\, z(\omega)\rangle = S(\omega)\delta(\omega - \omega') \quad \text{or} \quad S(\omega) = \int_{-\infty}^{\infty} d\omega'\, \langle z^*(\omega')\, z(\omega)\rangle \tag{2.6.13}$$

We can demonstrate the Weiner–Khintchine Theorem by first noting that the stationary processes produce autocorrelation functions that only depend on the difference in time

$$\Gamma(t_1, t_2) = \langle z^*(t_1)\, z(t_2)\rangle = \iint dz_1 dz_2\, z_1^* z_2\, f(z_1, t_1; z_2, t_2) \tag{2.6.14}$$

where "f" represents the joint probability density. Recall the definition of a stationary process as one for which the density function is independent of the origin of time. Let's shift the origin of time by t_1 to obtain

$$f(z_1, t_1; z_2, t_2) = f(z_1, t_1 - t_1; z_2, t_2 - t_1) = f(z_1, 0; z_2, \tau) \tag{2.6.15}$$

where $\tau = t_2 - t_1$. Therefore, the average in Equation (2.6.14) depends only on the difference $\tau = t_2 - t_1$

$$\Gamma(t_1, t_2) = \iint dz_1 dz_2\, z_1^* z_2\, f(z_1, 0; z_2, t_2 - t_1) = \langle z^*(0)\, z(t_2 - t_1)\rangle \equiv \Gamma(t_2 - t_1) = \Gamma(\tau)$$

The same reasoning shows $\Gamma(\tau) = \Gamma(t_1, t_2) = \Gamma(-\tau)$.

The expectation value $\langle z^*(\omega_1)\, z(\omega_2)\rangle$ can be written in terms of the Fourier transform

$$z(t) = \int_{-\infty}^{\infty} d\omega\, \tilde{z}(\omega)\, \frac{e^{i\omega t}}{\sqrt{2\pi}} \quad \text{or} \quad \tilde{z}(\omega) = \int_{-\infty}^{\infty} dt\, z(t)\, \frac{e^{-i\omega t}}{\sqrt{2\pi}} \tag{2.6.16}$$

as

$$\langle z^*(\omega_1)\, z(\omega_2)\rangle = \int_{-\infty}^{\infty} dt_1 \int_{-\infty}^{\infty} dt_2\, \langle z^*(t_1)\, z(t_2)\rangle\, \frac{e^{+i\omega_1 t_1}}{\sqrt{2\pi}}\, \frac{e^{-i\omega_2 t_2}}{\sqrt{2\pi}} \tag{2.6.17}$$

Substitute the correlation coefficient, set $t = t_2 - t_1$, and separate the integrals to find

$$\langle z^*(\omega_1)\, z(\omega_2) \rangle = \int_{-\infty}^{\infty} dt_1 \int_{-\infty}^{\infty} dt_2\, \Gamma(t_2 - t_1)\, \frac{e^{+i\omega_1 t_1}}{\sqrt{2\pi}} \frac{e^{-i\omega_2 t_2}}{\sqrt{2\pi}} = \int_{-\infty}^{\infty} dt\, \Gamma(t)\, e^{-i\omega_2 t} \int_{-\infty}^{\infty} dt_1\, \frac{e^{+i(\omega_1 - \omega_2)t_1}}{2\pi}$$

Substitute the Dirac delta function to find

$$\langle z^*(\omega_1)\, z(\omega_2) \rangle = \int_{-\infty}^{\infty} dt\, \Gamma(t)\, e^{-i\omega_2 t}\, \delta(\omega_1 - \omega_2) = \sqrt{2\pi}\, \Gamma(\omega_2)\, \delta(\omega_1 - \omega_2) \qquad (2.6.18)$$

where $\Gamma(\omega)$ is the Fourier transform of the autocorrelation function $\Gamma(\tau)$. This last relation shows the delta function correlation in frequency. Solving this last equation for $\Gamma(\omega_2)$ provides

$$\Gamma(\omega) = \int_{-\infty}^{\infty} dt\, \Gamma(t)\, \frac{e^{-i\omega t}}{\sqrt{2\pi}} \qquad (2.6.19)$$

Now, defining the power spectral density as $S(\omega) = \sqrt{2\pi}\Gamma(\omega)$, we find the second result from Equation (2.6.18)

$$\langle z^*(\omega_1)\, z(\omega_2) \rangle = S(\omega_2)\, \delta(\omega_1 - \omega_2) \qquad (2.6.20)$$

2.6.6 Alternate Derivations of the Weiner–Khintchine Formula

The development of the Weiner–Khintchine formula in the previous topic circumvents the issue of whether or not the Fourier integrals exist. The derivation does not explicitly require the convergence properties. However, to demonstrate the physical interpretation of the correlation function, we need to use an alternative approach that explicitly makes use of the convergence properties of the integrals. In particular, the function $z(t)$ must be square integrable

$$\int_{-\infty}^{\infty} dt\, |z(t)|^2 < \infty \qquad (2.6.21)$$

so that the function z has finite "length." However, a stationary process z does not have finite length since the fluctuations away from the average (zero in this case) do not change in magnitude (the standard deviation does not depend on time). This means that the expected excursion of z at $t = 0$ must be the same as for any other time. The expected value of $|z(t)|^2$ must be nonzero for any time and therefore the summation over all times must become infinite. The correlation function in the previous topic circumvents this issue by considering nonzero time delays τ.

In order to bring out the physical nature of the correlation function (especially for a delay of $\tau = 0$), we consider a sample of $z_T(t)$ over a finite interval $(-T/2, T/2)$ and define $z_T(t) = 0$ for $|t| > T/2$. At the end of the discussion, we will be interested in letting T grow without bound. The Fourier transform of the process $z_T(t)$ becomes

$$z_T(\omega) = \int_{-T/2}^{T/2} dt\, z(t)\, \frac{e^{-i\omega t}}{\sqrt{2\pi}} \qquad (2.6.22)$$

For now, it should be emphasized that the finite interval where $z(t) \neq 0$ produces the finite limits on the integral so as to circumvent issues of convergence. The procedure is equivalent to using a convergence factor $e^{-|\alpha| t}$, which would produce an integrand of the form $e^{-|\alpha| t - i \omega t}$. At the end of the procedure, we would take $\alpha \to 0$ to produce the result for the Fourier transform. The inverse transform becomes

$$z(t) = \mathop{\mathrm{Lim}}_{T \to \infty} \int_{-T/2}^{T/2} d\omega \, z_T(\omega) \frac{e^{i\omega t}}{\sqrt{2\pi}} \tag{2.6.23}$$

We now reproduce the Weiner–Khintchine theorem. Consider the correlation function for a stationary process defined as

$$\Gamma(\tau) = \langle z^*(t) \, z(t + \tau) \rangle = \mathop{\mathrm{Lim}}_{T \to \infty} \frac{1}{T} \int_{-T/2}^{T/2} dt \, z^*(t) \, z(t + \tau) \tag{2.6.24}$$

We assume that τ remains small compared with T so that the region where $z = 0$ does not significantly affect the value of the integral. Substituting Equation (2.6.23) into Equation (2.6.24) produces

$$\Gamma(\tau) = \langle z^*(t) \, z(t + \tau) \rangle = \mathop{\mathrm{Lim}}_{T \to \infty} \frac{1}{T} \int_{-T/2}^{T/2} dt \int_{-\infty}^{\infty} \int_{-\infty}^{\infty} d\omega' \, d\omega \, z_T^*(\omega) \, z_T(\omega') \frac{e^{-i\omega t}}{\sqrt{2\pi}} \frac{e^{i\omega'(t+\tau)}}{\sqrt{2\pi}} \tag{2.6.25a}$$

Interchanging integrals and separating the exponentials provides

$$\Gamma(\tau) = \langle z^*(t) \, z(t + \tau) \rangle = \mathop{\mathrm{Lim}}_{T \to \infty} \frac{1}{T} \int_{-\infty}^{\infty} \int_{-\infty}^{\infty} d\omega' \, d\omega \, z_T^*(\omega) \, z_T(\omega') \, e^{i\omega'\tau} \int_{-T/2}^{T/2} dt \frac{e^{i(\omega' - \omega)t}}{2\pi} \tag{2.6.25b}$$

Use the Dirac delta function

$$\mathop{\mathrm{Lim}}_{T \to \infty} \int_{-T/2}^{T/2} dt \frac{e^{i(\omega' - \omega)t}}{2\pi} = \delta(\omega' - \omega)$$

to find the correlation function

$$\Gamma(\tau) = \langle z^*(t) \, z(t + \tau) \rangle = \mathop{\mathrm{Lim}}_{T \to \infty} \frac{1}{T} \int_{-\infty}^{\infty} d\omega \, |z_T(\omega)|^2 e^{i\omega\tau} \tag{2.6.26a}$$

Therefore the correlation function and the power spectrum is

$$\Gamma(\tau) = \langle z^*(t) \, z(t + \tau) \rangle \underset{\substack{\text{Fourier} \\ \text{Transform}}}{\longleftrightarrow} S(\omega) = \mathop{\mathrm{Lim}}_{T \to \infty} \frac{\sqrt{2\pi}}{T} |z_T(\omega)|^2 \tag{2.6.26b}$$

As an important note, other references use different expressions for the power spectrum. A common one is

$$S(\omega) = \mathop{\mathrm{Lim}}_{T \to \infty} \frac{4\pi}{T} |z_T(\omega)|^2$$

2.6.7 Langevin Noise Terms

The Langevin noise terms for the rate equations can be pictured as impulse correlated (delta function correlated). The noise source function $F(t)$ can be pictured as a sequence of random numbers albeit an infinite continuous sequence (see Figure 2.6.2). The correlation is imagined to be extremely short (the Markovian approximation)—shorter than any time scale of interest. The two values $F(t_1)$ and $F(t_2)$ do not have any relation to each other regardless of the proximity of t_2 to t_1. Alternatively, the amplitude at frequency ω_1 does not have any relation to the amplitude at frequency ω_2 regardless of the proximity of ω_1 and ω_2. We can show

$$\left\langle F(t_1)\, F^*(t_2) \right\rangle = S_F\, \delta(t_2 - t_1) \qquad \left\langle \tilde{F}(\omega_1)\, \tilde{F}^*(\omega_2) \right\rangle = \tilde{S}_F\, \delta(\omega_1 - \omega_2) \qquad (2.6.27a)$$

Often different noise sources i and j are correlated according to

$$\left\langle F_i(t_1)\, F_j^*(t_2) \right\rangle = S_{ij}\, \delta(t_2 - t_1) \qquad \left\langle \tilde{F}_i(\omega_1)\, \tilde{F}_j^*(\omega_2) \right\rangle = \tilde{S}_{ij}\, \delta(\omega_1 - \omega_2) \qquad (2.6.27b)$$

The Weiner–Khintchine theorem shows that a stationary process has impulse correlated frequency components $\left\langle F^*(\omega_1)\, F(\omega_2) \right\rangle = \sqrt{2\pi}\, \Gamma(\omega_2)\, \delta(\omega_1 - \omega_2)$. Using this frequency correlation and the Fourier transform provides Equation (2.6.27a)

$$\left\langle F^*(t_1)\, F(t_2) \right\rangle = \int_{-\infty}^{\infty} d\omega_1 \int_{-\infty}^{\infty} d\omega_2 \left[\sqrt{2\pi}\, \Gamma(\omega_2)\, \delta(\omega_1 - \omega_2) \right] \frac{e^{-i\omega_1 t_1}}{\sqrt{2\pi}} \frac{e^{+i\omega_2 t_2}}{\sqrt{2\pi}}$$

Eliminating the Dirac delta function produces

$$\left\langle F^*(t_1)\, F(t_2) \right\rangle = \int_{-\infty}^{\infty} d\omega_1 \sqrt{2\pi}\, \Gamma(\omega_1)\, \frac{e^{i\omega_1(t_2 - t_1)}}{2\pi}$$

Assume that $\Gamma(\omega_1)$ is fairly independent of ω_1, we find the desired results

$$\left\langle F^*(t_1)\, F(t_2) \right\rangle = \sqrt{2\pi}\, \Gamma(\omega)\, \delta(t_2 - t_1)$$

2.6.8 Alternate Definitions for RIN

The following list shows a variety of definitions for RIN.

1. $\text{RIN}(\tau) = \dfrac{\left\langle \delta P^*(t)\, \delta P(t + \tau) \right\rangle}{\bar{P}^2}$ 2. $\text{RIN}(\omega) = \dfrac{S_{\delta p}(\omega)}{\bar{P}^2}$

3. $\text{RIN}(\omega) = \dfrac{\left\langle \delta P(\omega) \delta P^*(\omega) \right\rangle}{\bar{P}^2}$ 4. $\dfrac{\text{RIN}}{\Delta f} = \dfrac{2 S_{\delta p}(\omega)}{\bar{P}^2}$

The first two definitions are interrelated by the Weiner–Khintchine formula where S is the Fourier transform of the autocorrelation function $\Gamma(\tau) = \left\langle \delta P^*(t)\, \delta P(t + \tau) \right\rangle$.

The third relation as a shorthand notation can be related to the first and second ones as follows. Start with

$$\text{RIN}(\tau) \equiv \frac{\Gamma_P(t, t + \tau)}{\bar{P}^2} \equiv \frac{\left\langle \delta P^*(t)\, \delta P(t + \tau) \right\rangle}{\bar{P}^2} \qquad (2.6.28)$$

Substituting the indicated Fourier transforms provides

$$\mathrm{RIN}(\tau) = \frac{1}{\bar{P}^2} \int_{-\infty}^{\infty} \int_{-\infty}^{\infty} d\omega \, d\omega' \, \left\langle \delta\tilde{P}^*(\omega) \, \delta\tilde{P}(\omega') \right\rangle \frac{e^{-i\omega t}}{\sqrt{2\pi}} \frac{e^{i\omega'(t+\tau)}}{\sqrt{2\pi}}$$

The Weiner–Khintchine theorem shows the term $\left\langle \delta\tilde{P}^*(\omega) \, \delta\tilde{P}(\omega') \right\rangle$ has imbedded Dirac delta functions $\delta(\omega - \omega')$, which can be used to eliminate one integral, and requires $\omega = \omega'$. The last equation becomes

$$\mathrm{RIN}(\tau) = \frac{1}{\bar{P}^2} \int_{-\infty}^{\infty} d\omega \, \frac{1}{\sqrt{2\pi}} \left\langle \delta\tilde{P}^*(\omega) \, \delta\tilde{P}(\omega) \right\rangle \frac{e^{i\omega\tau}}{\sqrt{2\pi}} \qquad (2.6.29)$$

where now the symbol $\langle \delta\tilde{P}^*(\omega) \, \delta\tilde{P}(\omega) \rangle$ does not include the Dirac delta function. Substituting $\mathrm{RIN}(\omega)$ into Equation (2.6.29)

$$\mathrm{RIN}(\tau) = \frac{1}{\bar{P}^2} \int_{-\infty}^{\infty} d\omega \, \frac{1}{\sqrt{2\pi}} \mathrm{RIN}(\omega) \frac{e^{i\omega\tau}}{\sqrt{2\pi}} \quad \text{or} \quad \mathrm{RIN}(\omega) = \frac{S_{\delta P}(\omega)}{\bar{P}^2} \qquad (2.6.30)$$

Often to indicate the change of notation (missing delta function) the density is stated as

$$S_{\delta P}(\omega) = \int_{-\infty}^{\infty} d\omega' \, \left\langle \delta P^*(\omega') \, \delta P(\omega) \right\rangle$$

Case 3 reduces to case 2.

The fourth case accommodates RF spectrum analyzers. Assume the filter in the RF spectrum analyzer has the band pass transfer function $F = 1$ over the frequency range Δf centered at ω_o. The analyzer then measures

$$\mathrm{RIN} = \frac{\sigma^2}{\bar{P}^2} = \frac{1}{\bar{P}^2} \langle \delta P^*(t) \, \delta P(t) \rangle = \frac{1}{\bar{P}^2} \int_{-\infty}^{\infty} \int_{-\infty}^{\infty} d\omega_1 d\omega_2 \langle \delta P^*(\omega_1) \, \delta P(\omega_2) \rangle \frac{e^{i(\omega_2 - \omega_1)t}}{2\pi} F^*(\omega_1) F(\omega_2)$$

Use the relation from the Weiner–Khintchine theorem

$$\left\langle \delta P^*(\omega_1) \, \delta P(\omega_2) \right\rangle = S_{\delta P}(\omega_1) \delta(\omega_1 - \omega_2)$$

to find

$$\frac{1}{\bar{P}^2} \langle \delta P^*(t) \, \delta P(t) \rangle = \frac{1}{\bar{P}^2} \int_{-\infty}^{\infty} d\omega \, S_{\delta P}(\omega) \frac{|F|^2}{2\pi} = \frac{S_{\delta P}(\omega)}{\bar{P}^2} 2\Delta f$$

where the factor of 2 accounts for both positive and negative frequency accepted by the filter. The spectral density $S_{\delta P}(\omega)$ replaces the variance.

2.7   Relative Intensity Noise for the Semiconductor Laser

This section determines the relative intensity noise (RIN) for the semiconductor laser by using rate equations that include the Langevin noise sources. The noise terms model the noise effects induced by the pump and optical reservoirs. Fluctuations induced by the quantum mechanical vacuum produce RIN usually identified with spontaneous emission. Small fluctuations associated with the pump produce small fluctuations in the carrier density that lead to very large effects in the photon density. Both types of noise produce noise in the emitted beam. The fluctuations in the carrier density produce a greater amount of noise in the output beam. Not surprisingly then, the noise in the output beam becomes maximum near the laser resonance frequency discussed in Section 2.4. The reason is that the response curve near resonance relates small changes in the carrier population to the power in the output beam.

2.7.1 Rate Equations with Langevin Noise Sources and the Spectral Density

We are interested in finding the response of the laser (or LED) to noise. In particular, we define the relative intensity noise RIN in Equation (2.6.12) as

$$\frac{\text{RIN}}{\Delta f} = \frac{2S_{\delta P}(\omega)}{\bar{P}^2} \tag{2.7.1a}$$

where the correlation strength has the form

$$S_{\delta P}(\omega) = \int_{-\infty}^{\infty} d\omega' \, \langle \delta P^*(\omega') \, \delta P(\omega) \rangle \tag{2.7.1b}$$

The laser rate equations can predict the response provided they incorporate the Langevin noise sources. These sources model the effect of random external influences on the lasing process. We include one Langevin noise source in the electron–hole rate equation; it has an effect similar to a randomly time-varying pump current. We place another Langevin noise source in the photon rate equation; it provides a randomly varying photon generation rate. Figure 2.7.1 shows a conceptual view of the response of the output power to Langevin noise sources. The noise sources provide sudden spikes that randomly change the photon and carrier density. The spectral densities of the sources contain components at all frequencies. We expect the noise in the output power to be enhanced near the resonant frequency of the laser as predicted by the laser rate equations.

We consider small changes in both n and γ as $\delta n = n(t) - \bar{n}$ and $\delta \gamma = \gamma(t) - \bar{\gamma}$. We insert these into the laser rate equations given by Equations (2.1.19) and (2.1.20)

$$\frac{dn}{dt} = -v_g g \gamma + J - R_{sp} \qquad \frac{d\gamma}{dt} = \Gamma v_g g \gamma - \frac{\gamma}{\tau_\gamma} + \beta R_{sp} \tag{2.7.2}$$

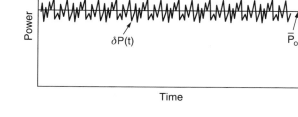

FIGURE 2.7.1
Representation of the noise in the optical power from a light emitter.

where n, γ represent the electron and photon density, respectively, Γ, β denote the optical confinement and coupling factors, and $g = g(n)$, $\mathscr{J}$, R represent the material gain, the pump-number current density, and the spontaneous recombination. Assume that g denotes the unsaturated gain (does not depend on the photon density). The variation in the carrier number and photon number can be found by performing variations similar to $\delta_n f(n, \gamma) = (\partial f / \partial n)|_{\bar{n}, \bar{\gamma}} \delta n \equiv f_n \, \delta n$. The procedure duplicates that for finding the bandwidth

$$\frac{d}{dt} \delta n(t) = -v_g \bar{g}_n \bar{\gamma} \, \delta n - v_g \bar{g} \, \delta \gamma - \bar{R}_n \delta n + F_n(t)$$

$$\frac{d}{dt} \delta \gamma(t) = \Gamma v_g \bar{g}_n \bar{\gamma} \, \delta n + \Gamma v_g \bar{g} \, \delta \gamma - \frac{\delta \gamma}{\tau_\gamma} + \beta \bar{R}_n \delta n + F_\gamma(t)$$

(2.7.3)

The subscript "n" indicates a derivative with respect to "n" except on the Langevin terms F_n and F_γ. The term $\bar{R}_n$ indicates the derivative of the nonradiative recombination with respect to "n" and evaluated at the steady-state value of $\bar{n}$ (i.e., the threshold value). The "bars" on top indicate a steady-state (i.e., average) value.

Since we are most interested in the RIN power spectrum in Equation (2.7.1), we substitute the Fourier transforms $\tilde{n}(\omega) = \delta n(\omega)$ and $\tilde{\gamma}(\omega) = \delta \gamma(\omega)$ into both sides of Equations (2.7.3) to produce

$$i\omega \tilde{n} = -v_g \bar{g}_n \bar{\gamma} \, \tilde{n} - v_g \bar{g} \, \tilde{\gamma} - \bar{R}_n \tilde{n} + \tilde{F}_n$$

$$i\omega \tilde{\gamma} = \Gamma v_g \bar{g}_n \bar{\gamma} \, \tilde{n} + \Gamma v_g \bar{g} \, \tilde{\gamma} - \frac{\tilde{\gamma}}{\tau_\gamma} + \beta \bar{R}_n \tilde{n} + \tilde{F}_\gamma$$

(2.7.4)

where we assume the pump-current number density $\mathscr{J}$ doesn't vary in time. We can alternatively write $g_o = \bar{g}_n$, $\tau_n^{-1} = \bar{R}_n$.

Collecting terms in Equation (2.7.4) produces the matrix equation

$$\begin{bmatrix} i\omega + v_g \bar{g}_n \bar{\gamma} + \bar{R}_n & v_g \bar{g} \\ -\Gamma v_g \bar{g}_n \bar{\gamma} - \beta \bar{R}_n & i\omega - \Gamma v_g \bar{g} + (1/\tau_\gamma) \end{bmatrix} \begin{bmatrix} \tilde{n} \\ \tilde{\gamma} \end{bmatrix} = \begin{bmatrix} \tilde{F}_n \\ \tilde{F}_\gamma \end{bmatrix}$$

(2.7.5)

Denoting the determinant by $\Delta_M = \text{Det}\, \underline{M}$ where $\underline{M}$ represents the 2×2 matrix in the last equation, the solution to Equation (2.7.5) is

$$\begin{bmatrix} \tilde{n} \\ \tilde{\gamma} \end{bmatrix} = \frac{1}{\text{Det}\, \underline{M}} \begin{bmatrix} i\omega - \Gamma v_g \bar{g} + \frac{1}{\tau_\gamma} & -v_g \bar{g} \\ \Gamma v_g \bar{g}_n \bar{\gamma} + \beta \bar{R}_n & i\omega + v_g \bar{g}_n \bar{\gamma} + \bar{R}_n \end{bmatrix} \begin{bmatrix} \tilde{F}_n \\ \tilde{F}_\gamma \end{bmatrix}$$

(2.7.6)

The zeros (or approximate zeros) of the determinant give the resonances. The determinant is

$$\text{Det}\, \underline{M} = \left(i\omega + v_g \bar{g}_n \bar{\gamma} + \bar{R}_n \right) \left(i\omega - \Gamma v_g \bar{g} + \frac{1}{\tau_\gamma} \right) + v_g \bar{g} \left(\Gamma v_g \bar{g}_n \bar{\gamma} + \beta \bar{R}_n \right)$$

(2.7.7)

Similar to Equations (2.5.11) and (2.5.12), define two terms

$$\frac{1}{\tau} = \frac{1}{\tau_\gamma} - \Gamma v_g \bar{g} + v_g \bar{g}_n \bar{\gamma} + \bar{R}_n$$

(2.7.8)

$$\omega_o^2 = v_g \bar{g}\left(\Gamma v_g \bar{g}_n \bar{\gamma} + \beta \bar{R}_n\right) + \left(v_g \bar{g}_n \bar{\gamma} + \bar{R}_n\right)\left(\frac{1}{\tau_\gamma} - \Gamma v_g \bar{g}\right) \tag{2.7.9}$$

so that the determinant becomes

$$\Delta = \mathrm{Det}\,\underline{M} = \left(\omega_o^2 - \omega^2\right) + \frac{i\omega}{\tau} \tag{2.7.10}$$

where $\omega_r^2 = \omega_o^2 - (1/2\tau^2) \sim \omega_o^2$ gives the resonant frequency and ω/τ represents a damping term. Recall the resonant frequency has the form $\omega_r = \sqrt{v_g \bar{g}_o \bar{\gamma}/\tau_\gamma}$. The solution for the photon density $\tilde{\gamma}$ (inside the cavity) comes from Equation (2.7.6)

$$\tilde{\gamma}(\omega) = \frac{\left(\Gamma v_g \bar{g}_n \bar{\gamma} + \beta \bar{R}_n\right)\tilde{F}_n(\omega) + \left(i\omega + v_g \bar{g}_n \bar{\gamma} + \bar{R}_n\right)\tilde{F}_\gamma(\omega)}{\left(\omega_o^2 - \omega^2\right) + i\omega/\tau} = \frac{C_{\gamma n}\tilde{F}_n(\omega) + C_{\gamma\gamma}\tilde{F}_\gamma(\omega)}{\Delta} \tag{2.7.11a}$$

where $C_{\gamma n}$ is real and $\omega_o^2 \sim \omega_r^2$ and

$$C_{\gamma n} = \Gamma v_g \bar{g}_n \bar{\gamma} + \beta \bar{R}_n \quad C_{\gamma\gamma} = i\omega + v_g \bar{g}_n \bar{\gamma} + \bar{R}_n \quad \Delta = \left(\omega_o^2 - \omega^2\right) + i\omega/\tau \tag{2.7.11b}$$

As discussed in the previous section, we can calculate spectral density $S_{\delta P}(\omega)$ in Equation (2.7.1b) to find the RIN. We start by calculating the frequency correlation

$$\left\langle \tilde{\gamma}(\omega)\,\tilde{\gamma}^*(\omega')\right\rangle = \frac{|C_{\gamma n}|^2}{|\Delta|^2}\left\langle \tilde{F}_n(\omega)\tilde{F}_n^*(\omega')\right\rangle + \frac{C_{\gamma n}C_{\gamma\gamma}^*}{|\Delta|^2}\left\langle \tilde{F}_n(\omega)\tilde{F}_\gamma^*(\omega')\right\rangle$$

$$+ \frac{C_{\gamma\gamma}C_{\gamma n}^*}{|\Delta|^2}\left\langle \tilde{F}_\gamma(\omega)\tilde{F}_n^*(\omega')\right\rangle + \frac{|C_{\gamma\gamma}|^2}{|\Delta|^2}\left\langle \tilde{F}_\gamma(\omega)\tilde{F}_\gamma^*(\omega')\right\rangle$$

and then form the spectral density

$$S_\gamma(\omega) = \int_{-\infty}^{\infty} d\omega' \left\langle \tilde{\gamma}(\omega)\,\tilde{\gamma}^*(\omega')\right\rangle = \frac{|C_{\gamma n}|^2}{|\Delta|^2}\left\langle \tilde{F}_n \tilde{F}_n\right\rangle + \frac{2\,\mathrm{Re}\left\{C_{\gamma n}C_{\gamma\gamma}^*\right\}}{|\Delta|^2}\left\langle \tilde{F}_n \tilde{F}_\gamma\right\rangle + \frac{|C_{\gamma\gamma}|^2}{|\Delta|^2}\left\langle \tilde{F}_\gamma \tilde{F}_\gamma\right\rangle \tag{2.7.12}$$

where we *define* symbols of the form $\left\langle \tilde{F}_i \tilde{F}_j\right\rangle$ to mean

$$\left\langle \tilde{F}_i \tilde{F}_j\right\rangle = \int_{-\infty}^{\infty} d\omega' \left\langle \tilde{F}_i(\omega)\,\tilde{F}_j^*(\omega')\right\rangle \tag{2.7.13}$$

and so on. The cross correlation terms were combined since we will find $\langle \tilde{F}_n \tilde{F}_\gamma$ is real and that $\langle \tilde{F}_n \tilde{F}_\gamma\rangle = \langle \tilde{F}_\gamma \tilde{F}_n\rangle$. We can substitute for $C_{\gamma n}$ and $C_{\gamma\gamma}$ in order to write $\mathrm{Re}\{C_{\gamma n}C_{\gamma\gamma}^*\} = \left(\Gamma v_g \bar{g}_n \bar{\gamma} + \beta \bar{R}_n\right)\left(v_g \bar{g}_n \bar{\gamma} + \bar{R}_n\right)$.

To find the RIN, we need to find the correlation strengths $\langle \tilde{F}_n \tilde{F}_n\rangle$, $\langle \tilde{F}_n \tilde{F}_\gamma\rangle$, and $\langle \tilde{F}_\gamma \tilde{F}_\gamma\rangle$. Also, we have primary interest in the output power P_o and therefore want to know the noise in the output power. As discussed in Section 1.8, the output facets induce partition noise beyond the noise already present inside the cavity. Figure 2.7.2 illustrates the point. A piece of glass, for example, reflects a portion of the incident photons. For every reflected photon, the output stream must be missing one (indicated by the open circle). In this example, the input stream has a standard deviation of zero. The reflected and

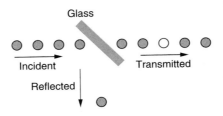

FIGURE 2.7.2
Imperfect reflecting surfaces induce partition noise.

transmitted beams have nonzero standard deviation. Interfaces not perfectly reflecting or transmitting add partition noise.

2.7.2   Langevin–Noise Correlation

This topic presents a simple method for calculating the correlation strength between Langevin noise terms. The discussion follows Coldren's book and his references to Lax and McCumber. The method replaces the quantum treatment of noise and assumes the noise originates in the shot noise associated with the transport of particles into and out of particle reservoirs.

For shot noise, the correlation strengths have the form

$$\langle F_i F_i \rangle = \sum R_i^+ + \sum R_i^- \quad \langle F_i F_j \rangle = -\sum R_{ij} - \sum R_{ji} \qquad (2.7.14)$$

The symbols R_i^+, R_i^- represent the rate of particle flow (# particles/time) into and out of a particle reservoir, respectively. The symbol R_{ij} refers to the particle flow between reservoirs i and j. In order to find the correct results, the noise functions must be converted to numbers per unit time. For example, we must convert F_n in Equations (2.7.3) from units of "number per volume per second" to "number per second." Examples for the method appear in the ensuing calculations.

We can apply Equations (2.7.14) to the sources in the laser rate equations (2.7.1)

$$\frac{dn}{dt} = -\left(R_{\substack{\text{Stim} \\ \text{Emiss}}} - R_{\substack{\text{Stim} \\ \text{Abs}}} \right) + \mathcal{J} - R_{sp} \qquad \frac{d\gamma}{dt} = \Gamma\left(R_{\substack{\text{Stim} \\ \text{Emiss}}} - R_{\substack{\text{Stim} \\ \text{Abs}}} \right) - \frac{\gamma}{\tau_\gamma} + \beta R_{sp} \qquad (2.7.15)$$

where Γ represents the optical confinement factor V_a/V_γ. Let the symbols R_{SE} and R_{SA} represent the stimulated emission and absorption rates.

Sometimes the rate of spontaneous emission in the photon equation is redefined as

$$\beta R_{sp} = \Gamma R'_{sp}$$

The noise sources F_n and F_γ have units of "number per volume per time" in keeping with the units in Equations (2.7.1). The rate equations show how the recombination mechanisms affect the reservoir populations. We need units of "number per unit time." Multiply Equation (2.7.15) by the volume of the active region V_a and the second of the

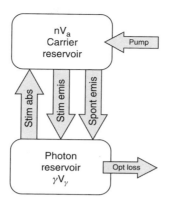

FIGURE 2.7.3
The photon and carrier reservoirs.

equations by the modal volume V_γ to obtain

$$\frac{d(V_a n)}{dt} = -(R_{SE} - R_{SA})V_a + \mathscr{J} V_a - V_a R_{sp}$$

$$\frac{d(V_\gamma \gamma)}{dt} = +(R_{SE} - R_{SA})V_a - \frac{V_\gamma \gamma}{\tau_\gamma} + V_a R'_{sp}$$

(2.7.16)

One more point, the stimulated term must be divided into stimulated emission and stimulated absorption terms since they affect the population of the photon and carrier reservoirs as suggested by Figure 2.7.3. We redefine the noise sources to be

$$F'_n = V_a F_n \quad F'_\gamma = V_\gamma F_\gamma$$

(2.7.17)

First, consider the electron noise source. We think of $V_a n$ as the number of pairs in the carrier reservoir. We identify the following rates assuming steady state except for the noise

$$\sum R_i^+ + \sum R_i^- = \left[\bar{R}_{SA} V_a + J V_a \right] + \left[\bar{R}_{SE} V_a + \bar{R}_{sp} V_a \right]$$

The electron correlation strength becomes

$$\langle F'_n F'_n \rangle = V_a^2 \langle F_n F_n \rangle = \sum R_i^+ + \sum R_i^- = \left[\bar{R}_{SA} V_a + J V_a \right] + \left[\bar{R}_{SE} V_a + \bar{R}_{sp} V_a \right]$$

Therefore

$$\langle F_n F_n \rangle = \frac{\bar{\mathscr{J}}}{V_a} + \frac{\bar{R}_{SA}}{V_a} + \frac{\bar{R}_{SE}}{V_a} + \frac{\bar{R}_{sp}}{V_a}$$

(2.7.18a)

Next, consider the photon noise source. Thinking of $V_\gamma \gamma$ as the number of particles in a reservoir, we find

$$\langle F'_\gamma F'_\gamma \rangle = V_\gamma^2 \langle F_\gamma F_\gamma \rangle = \sum R_i^+ + \sum R_i^- = \left[\bar{R}_{SE} V_a + \bar{R}'_{sp} V_a \right] + \left[\bar{R}_{SA} V_a + \frac{V_\gamma \bar{\gamma}}{\tau_\gamma} \right]$$

Therefore the photon correlation strength must be

$$\langle F_\gamma F_\gamma \rangle = \frac{V_a}{V_\gamma^2}\left[\bar{R}_{SE} + \bar{R}_{SA} + \bar{R}'_{sp}\right] + \frac{\bar{\gamma}}{V_\gamma \tau_\gamma} \tag{2.7.18b}$$

Consider the cross correlation strength between the electrons and photons

$$\left\langle F'_n F'_\gamma \right\rangle = V_a V_\gamma \langle F_n F_\gamma \rangle = -\sum R_{ij} - \sum R_{ji} = -\left[\bar{R}_{SA} + \bar{R}_{SE} + \bar{R}'_{sp}\right]V_a$$

This last equation provides

$$\langle F_n F_\gamma \rangle = -\frac{\bar{R}_{SA} + \bar{R}_{SE} + \bar{R}'_{sp}}{V_\gamma} \tag{2.7.18c}$$

Notice how the stimulated terms "SE" and "SA" both have "plus" signs which prevents us from substituting the usual expression $v_g \bar{g} \bar{\gamma} = \bar{R}_{SE} - \bar{R}_{SA}$. Instead, we use this relation for $\bar{R}_{SA}$ to write $\bar{R}_{SA} = \bar{R}_{SE} - v_g \bar{g} \bar{\gamma}$. As we will see in a subsequent chapter, the term $\bar{R}_{SE}$ can be related to the spontaneous emission through $\bar{R}_{SE} = \bar{R}'_{sp} \bar{\gamma} V_\gamma$. The results for the correlation strengths in Equations (2.7.18) become (refer to Coldren's book for alternate expressions that include nonradiative recombination)

$$\langle F_n F_n \rangle = \frac{2\bar{R}'_{sp}\gamma}{\Gamma} - \frac{v_g \bar{g} \bar{\gamma}}{V_a} + \frac{\bar{\mathscr{I}} + \bar{R}_{sp}}{V_a} = \frac{2\bar{R}'_{sp}\bar{\gamma}}{\Gamma}\left[1 + \frac{1}{2\bar{\gamma}V_\gamma}\right] - \frac{v_g \bar{g} \bar{\gamma}}{V_a} + \frac{\bar{\mathscr{I}} + \bar{\mathscr{I}}'_{thr}}{V_a} \tag{2.7.19a}$$

$$\langle F_\gamma F_\gamma \rangle = 2\Gamma \bar{R}'_{sp} \bar{\gamma}\left[1 + \frac{1}{\bar{\gamma}V_\gamma}\right] \tag{2.7.19b}$$

$$\langle F_\gamma F_n \rangle = -2\bar{R}'_{sp}\bar{\gamma}\left[1 + \frac{1}{2\bar{\gamma}V_\gamma}\right] + \frac{v_g \bar{g} \bar{\gamma}}{V_\gamma} \tag{2.7.19c}$$

where $\bar{\mathscr{I}}'_{thr} = \bar{R}_{sp} - \bar{R}'_{sp} \cong \bar{R}_{sp}$ and $1/\bar{\gamma}V_\gamma \ll 1$ above threshold.

As mentioned in the previous topic, the mirrors introduce noise into the beams transmitted through the mirror and reflected from it. However, the term $\bar{\gamma}/\tau_\gamma$ already accounts for the reflected term. We need to find the noise added to the output signal described by the output power P_o. Starting with the relation between the photon density and the output power similar to Equation (2.3.19)

$$P_o = \gamma \frac{hc}{\lambda_o} V_\gamma v_g \alpha_m = \frac{hc}{\lambda_o} V_\gamma \frac{\gamma}{\tau_m} \tag{2.7.20a}$$

where $v_g \alpha_m = 1/\tau_m$ refers to the mirror loss. Using the cavity lifetime instead of the mirror loss would include the combined optical loss through mirrors, sidewalls, and free carrier absorption. The relative size of these losses depends on the reflectance of the mirrors. Including the Langevin noise term in Equation (2.7.20a) and focusing on the deviation from the average produces

$$\delta P_o = \frac{hcV_\gamma}{\lambda_o \tau_m}\delta\gamma + F_o(t) \tag{2.7.20b}$$

Converting to number per unit time

$$\left(\frac{hc}{\lambda_o}\right)^{-1} \delta P_o = \frac{V_\gamma}{\tau_m} \delta \gamma + \left(\frac{hc}{\lambda_o}\right)^{-1} F_o$$

Setting

$$F'_o = \left(\frac{hc}{\lambda_o}\right)^{-1} F_o$$

The correlation strength for the new function then must be related to the number of photons per second $V_\gamma \gamma / \tau_m$ leaving the photon reservoir through the mirror

$$\left\langle \tilde{F}'_o \tilde{F}'_o \right\rangle = \left(\frac{hc}{\lambda_o}\right)^{-2} \left\langle \tilde{F}_o \tilde{F}_o \right\rangle = \frac{V_\gamma \gamma}{\tau_m} \quad \rightarrow \quad \left\langle \tilde{F}_o \tilde{F}_o \right\rangle = \left(\frac{hc}{\lambda_o}\right)^2 \frac{V_\gamma \gamma}{\tau_m} = \frac{hc}{\lambda_o} P_o \qquad (2.7.21)$$

The cross correlation $\langle \tilde{F}_o \tilde{F}_n \rangle$ can be taken as zero by assuming that fluctuations in the *output* light does not have anything to do with fluctuations in the carrier density

$$\left\langle \tilde{F}_o \tilde{F}_n \right\rangle \qquad (2.7.22)$$

The cross correlation term $\langle \tilde{F}_o \tilde{F}_\gamma \rangle$ can be related to the particle flow from the internal to the external reservoir as required by Equation (2.7.14). Equation (2.7.20a) shows that F_o has units of Watts, while Equation (2.7.3) indicates that F_γ has units of "number per volume per second." Define the new sources as

$$F'_o = \left(\frac{hc}{\lambda_o}\right)^{-1} F_o \quad \text{and} \quad F'_\gamma = V_\gamma F_\gamma$$

The rate of flow from the internal to external reservoir must be $\gamma V_\gamma / \tau_m$

$$\left\langle F'_o F'_\gamma \right\rangle = \left(\frac{hc}{\lambda_o}\right)^{-1} V_\gamma \langle F_o F_\gamma \rangle = -\frac{V_\gamma \gamma}{\tau_m} \quad \rightarrow \quad \langle F_o F_\gamma \rangle = -\frac{hc}{\lambda_o} \frac{\gamma}{\tau_m} = -P_o / V_\gamma \qquad (2.7.23)$$

2.7.3 The Relative Intensity Noise

We want to know the correlation strength defining the relative noise intensity (RIN) defined in Equation (2.7.1)

$$\frac{\text{RIN}}{\Delta f} = \frac{2 S_{\delta P}(\omega)}{\bar{P}^2} \quad S_{\delta P}(\omega) = \int_{-\infty}^{\infty} d\omega' \left\langle \delta \tilde{P}^*(\omega') \, \delta \tilde{P}(\omega) \right\rangle \equiv \langle \delta P \, \delta P \rangle \qquad (2.7.24)$$

where $\bar{P}$ is the average output power $P_o(t)$, noise causes the time dependence in $P_o(t)$, and the difference $\delta P = P_o - \bar{P}$ can be attributed to the noise. This last equation needs the Fourier transform of Equation (2.7.20b)

$$\delta P_o(\omega) = \frac{hc V_\gamma}{\lambda_o \tau_m} \delta \gamma(\omega) + F_o(\omega) \qquad (2.7.25)$$

Substituting into Equations (2.7.24) provides

$$\bar{P}^2 \text{RIN}/\Delta f = \langle \delta \tilde{P} \, \delta \tilde{P} \rangle = \left(\frac{hcV_\gamma}{\lambda_o \tau_m}\right)^2 \langle \delta \tilde{\gamma} \, \delta \tilde{\gamma} \rangle + \frac{hcV_\gamma}{\lambda_o \tau_m} \langle \tilde{F}_o \, \delta \tilde{\gamma}^* + \tilde{F}_o^* \, \delta \tilde{\gamma} \rangle + \langle \tilde{F}_o \tilde{F}_o \rangle$$

$$= \left(\frac{hcV_\gamma}{\lambda_o \tau_m}\right)^2 \langle \delta \tilde{\gamma} \, \delta \tilde{\gamma} \rangle + \frac{2hcV_\gamma}{\lambda_o \tau_m} \text{Re} \langle \tilde{F}_o \, \delta \tilde{\gamma} \rangle + \langle \tilde{F}_o \tilde{F}_o \rangle$$

(2.7.26)

The complex conjugate appears in the middle term in the top line in order to show the term is real contrary to the standard compact notation setup in Equation (2.7.13). Now we substitute the correlation relations found in the first topic.

Equation (2.7.26) requires a number of correlation strengths. The embedded correlations in the first term $\langle \delta \tilde{\gamma} \, \delta \tilde{\gamma} \rangle$ can be found in Equations (2.7.19). The third term $\langle \tilde{F}_o \tilde{F}_o \rangle$ appears in Equation (2.7.21).

The second term $\langle \tilde{F}_o^* \, \delta \tilde{\gamma} \rangle$ uses Equations (2.7.11)

$$\langle \tilde{F}_o^* \, \delta \tilde{\gamma} \rangle = \left\langle \tilde{F}_o^* \, \frac{C_{\gamma n} \tilde{F}_n(\omega) + C_{\gamma \gamma} \tilde{F}_\gamma(\omega)}{\Delta} \right\rangle = \frac{C_{\gamma n}}{\Delta} \langle \tilde{F}_o^* \, \tilde{F}_n \rangle + \frac{C_{\gamma \gamma}}{\Delta} \langle \tilde{F}_o^* \, \tilde{F}_\gamma \rangle$$

(2.7.27)

where the second term abuses the compact notation defined in Equation (2.7.13) in order to show the subdivision of the integral into the two last terms. The complex conjugate should be removed to correspond to the notation in Equation (2.7.13). The correlation in Equation (2.7.27) appear in Equations (2.7.22) and (2.7.23).

We can now write the RIN as

$$\frac{P_o^2 \text{RIN}}{\Delta f} = \left(\frac{hcV_\gamma}{\lambda_o \tau_m}\right)^2 \langle \delta \tilde{\gamma} \, \delta \tilde{\gamma} \rangle + \frac{2hcV_\gamma}{\lambda_o \tau_m} \text{Re} \frac{C_{\gamma \gamma}}{\Delta} \langle \tilde{F}_o \tilde{F}_\gamma \rangle + \langle \tilde{F}_o \tilde{F}_o \rangle$$

(2.7.28)

Substitute Equations (2.7.12) for the term $\langle \delta \tilde{\gamma} \, \delta \tilde{\gamma}^* \rangle$ and Equations (2.7.19), (2.7.21), and (2.7.23) for the resultant correlation strengths to find

$$\frac{\text{RIN}}{\Delta f} = \frac{E_\gamma}{\bar{P}} \left[1 + \frac{a_1 + a_2 \omega^2}{|\Delta|^2} \right]$$

$$a_1 = \frac{8\pi(\Delta\nu)\bar{P}}{E_\gamma \tau_{\Delta n}^2} \left[1 + \frac{1}{\bar{\gamma} V_\gamma} \right] + \omega_R^4 \left[\frac{\eta_i(I + I_{\text{thr}})}{I_{st}} - 1 \right] - 2\omega_\Delta^2 \left[\omega_R^2 + \frac{1}{\tau_{\Delta n} \tau_\gamma} \right]$$

$$a_2 = \frac{8\pi(\Delta\nu)\bar{P}}{E_\gamma \tau_{\Delta n}^2} \left[1 + \frac{1}{\bar{\gamma} V_\gamma} \right] - \frac{8\pi(\Delta\nu)}{\tau_\gamma}$$

where we assume negligible change in the gain with photon density and the symbols E_γ, $(\Delta\nu)$, I_{st}, ω_R^2, ω_Δ^2, $\tau_{\Delta n}$ mean as

$$E_\gamma = hc/\lambda_o \qquad (\Delta\nu) = \Gamma \bar{R}'_{sp}/(4\pi\bar{\gamma}) \qquad \Gamma \bar{R}'_{sp} = \beta R_{sp} \qquad 1/\tau_{\Delta n} = \bar{R}_n$$

$$I_{st} = q\bar{P}/E_\gamma \qquad \omega_R^2 = v_g \bar{g}_n \bar{\gamma}/\tau_\gamma \qquad \omega_\Delta^2 = \gamma_{NP}\gamma_{PN} + \gamma_{NN}\gamma_{PP} - \omega_R^2$$

$$\gamma_{NN} = v_g \bar{g}_n \bar{\gamma} + \bar{R}_n \qquad \gamma_{NP} = v_g \bar{g} = \frac{1}{\Gamma \tau_\gamma} - \frac{\beta \bar{R}_{sp}}{\Gamma \bar{\gamma}}$$

$$\gamma_{PN} = \Gamma v_g \bar{g}_n \bar{\gamma} + \beta \bar{R}_n \qquad \gamma_{PP} = \frac{1}{\tau_\gamma} - \Gamma v_g \bar{g}$$

2.8 Review Exercises

2.1 A semiconductor has recombination centers in the middle of the bandgap as shown in Figure P2.1. The symbols n, p, N_r, n_r, p_r represent the density of electrons in the conduction band, the density of holes in the valence band, the density of recombination centers, density of electrons trapped in the recombination centers, and the density of recombination centers without an electron, respectively. The cross-section refers to the area of a disk; a carrier falling within the area is most likely captured by the trap. Larger cross-sections mean that the traps more easily capture the carrier. We have $N_r = n_r + p_r$. The electron and hole lifetimes can be written as

$$\tau_n = \frac{1}{s_n v p_r} \qquad \text{and} \qquad \tau_p = \frac{1}{s_p v n_r}$$

where v, s_n, s_p represent the thermal velocity, trap capture cross-section for electrons, and cross-section for holes, respectively. Neglect spontaneous and stimulated recombination. Consider a pump current and only recombination represented by the two lifetimes.

1. Explain why the lifetimes depend on p_r, n_r.
2. Explain why the following rate equations hold

$$\frac{dn}{dt} = \mathcal{J} - \frac{n}{\tau_n} \qquad \frac{dp}{dt} = \mathcal{J} - \frac{p}{\tau_p} \qquad \frac{dn_r}{dt} = \frac{n}{\tau_n} - \frac{p}{\tau_p}$$

3. Assume $n, n_r \ll N_r$. Find the electron density n at steady state in terms of $\mathcal{J}$ and N_r.

2.2 Reconsider Problem 2.1 for the case of high injection levels defined by $n, p \gg N_r$.

1. Explain why $n \cong p$
2. Use the steady state solutions to show

$$\frac{n_r}{p_r} = \frac{s_n}{s_p}$$

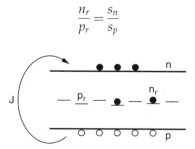

FIGURE P2.1
Semiconductor with recombination centers.

3. Using the total density of recombination centers N_r, show

$$n_r = N_r \frac{s_n}{s_n + s_p}$$

4. Finally show and explain the meaning of the results

$$\tau_n = \tau_p = \begin{cases} \dfrac{1}{N_r v s_n} & s_n \ll s_p \\[2ex] \dfrac{1}{N_r v s_p} & s_p \ll s_n \end{cases}$$

2.3 A laser researcher measures the power versus current curve from an inplane semiconductor laser shown in Figure P2.3. Assume the following constants.

$$L = 200\,\mu m, \quad R = 0.34, \quad c = 3 \times 10^{10}\,cm/s$$

$$\lambda = 850\,nm, \, q = 1.6 \times 10^{-19}\,Coul$$

$$h = 6.64 \times 10^{-34}\,J\,s, \, \eta_i = 1$$

$$\frac{hc}{\lambda} = 2.35 \times 10^{-19}\,J$$

Assume area of the top of the semiconductor is 200 μm × 5 μm. Consider the curve above threshold.

1. Using $R_1 = R_2 = 0.34 = R$, calculate the mirror loss α_m in units of cm^{-1}.
2. Find the internal optical loss α_i in units of cm^{-1}. Hint: Use a ruler to find the slope.
3. Find the cavity lifetime τ_γ. Assume $n = 3.5$ in $v_g = c/n$.
4. Calculate the photon density in the cavity when the total output power is $P = 4\,mW$. Assume that the mode occupies the volume $V_\gamma = 200\,\mu m \times 5\,\mu m \times 0.4\,\mu m$.
5. Calculate the optical confinement factor Γ if the thickness of the active region is 0.1 μm.
6. If the differential gain is $g_o = 5.1 \times 10^{-15}\,cm^2$, calculate the resonant frequency ω_r and $f_r = \omega_r/2\pi$.
7. Using Figure P2.3, calculate the geometry factor β for spontaneous emission. Hint: Measure the slope in Figure P2.3 and use the results from Section 2.3.

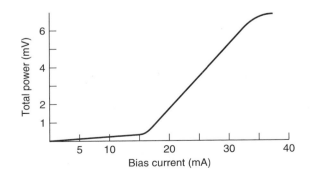

FIGURE P2.3
Power from both mirrors vs. bias current.

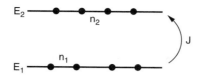

FIGURE P2.4
Pump removes electrons from level 1 and places them in level 2.

2.4 Suppose a system has two energy levels E_2 and E_1 (Figure P2.4) with N electrons distributed between the two levels. At any given time, level 1 has n_1 electrons and level 2 has n_2 electrons. Consider the following definitions.
$J=$ pump current number density,
number of electrons per volume per second
removed from level 1 and placed in level 2.
$n_i =$ density of electrons in level E_i
$R_{sp} =$ spontaneous recombination rate
$R_{nr} =$ nonradiative recombination rate
$g'n_2\gamma =$ rate of stimulated recombination $E_2 \rightarrow E_1$
$g'n_1\gamma =$ rate of stimulated absorption $E_1 \rightarrow E_2$
$g' =$ constant

1. Explain why the electron rate equations have the form

$$\frac{dn_2}{dt} = -g'n_2\gamma + g'n_1\gamma + J - R_{sp} - R_{nr}$$

$$\frac{dn_1}{dt} = g'n_2\gamma - g'n_1\gamma - J + R_{sp} + R_{nr}$$

2. Note $N = n_1 + n_2$ must be constant and define $N_\delta = n_2 - n_1$. Use the two equations in "part 1" to show

$$\frac{dN_\delta}{dt} = -2g'N_\delta\gamma + 2J - 2R_{sp} - 2R_{nr}$$

3. Write N_δ in terms of n_2 and N, and show

$$\frac{dn_2}{dt} = -2g'\left(n_2 - \frac{N}{2}\right)\gamma + J - R_{sp} - R_{nr}$$

4. Explain why $d\gamma/dt = g'n_2\gamma - g'n_1\gamma - \gamma/\tau_\gamma + \beta R_{sp}$
5. Show $d\gamma/dt = 2g'(n_2 - N/2)\gamma - \gamma/\tau_\gamma + \beta R_{sp}$
6. How do the results of parts 3 and 5 compare with the rate equations discussed in Chapter 2 with $g(n) = g_o(n - n_o)$?

2.5 A laser amplifier can be made from an inplane laser by evaporating anti-reflective coatings on the two mirrors. These coatings prevent positive feedback. The rate equations become

(1) $$\frac{dn}{dt} = -v_g g\gamma + J - \frac{n}{\tau_n}$$

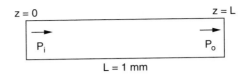

FIGURE P2.5
The laser amplifier.

(2) $\quad \dfrac{d\gamma}{dt} = \Gamma v_g g \gamma - \dfrac{\gamma}{\tau_{\text{int}}}$

Assume that the power in the amplifier can be written as

(3) $\quad P = \dfrac{hc}{\lambda_o}\gamma A_p v_g$

where $V_p = A_p L$ and A_p denotes the cross-sectional area. Assume $\eta_i = 1$, $\tau_{\text{int}} = (v_g \alpha_{\text{int}})^{-1}$ and λ_o denotes the vacuum wavelength.

1. Show that the two rate equations can be written as

$$\dfrac{dn}{dt} = -v_g g' P + \dfrac{I}{qV_a} - \dfrac{n}{\tau_n} \qquad \dfrac{dP}{dt} = \Gamma v_g g P - v_g \alpha_{\text{int}} P$$

   where $g' = (\lambda_o/hc)(g/A_p v_g)$, V_a denotes the volume of the active region, and Γ denotes the optical confinement factor.

2. Find the output power P_o at $Z = L$ using the second equation in Part 1 and the constants $\Gamma = 0.3$, $g = 400\ \text{cm}^{-1}$, $\alpha_{\text{int}} = 50\ \text{cm}^{-1}$, $L = 1\ \text{mm}$, $P_i = 1\ \text{mW}$.

3. For steady-state $dn/dt = 0$, use Equation (1) to show that

$$g = g_o(\tau_n J - n_o) \Big/ \left(1 + \dfrac{\gamma}{\gamma_s}\right) = g_o\left(\dfrac{\tau_n I}{qV_a} - n_o\right) \Big/ \left(1 + \dfrac{P}{P_s}\right)$$

   where $\gamma_s = (v_g g_o \tau_n)^{-1}$ and $P_s = (1/g_o\tau_n)(hc/\lambda_o)A_p$. Hint: Solve Equation (1) for "n" and substitute into $g \cong g_o(n - n_o)$.

4. Part 3 shows that the gain actually decreases with increasing optical power in the waveguide. Calculate the gain g at $P = 1\ \text{mW}$ and at $100\ \text{mW}$.
   Assume $v_g = c/n_g$, $c = 3 \times 10^{10}\ \text{cm/s}$, $n_g = 3.5$, $q = 1.6 \times 10^{-19}$, $g_o = 5 \times 10^{-16}\ \text{cm}^2$, $\tau_n = 10^{-9}\ \text{s}$, $hc/\lambda_o = 2.35 \times 10^{-19}\ \text{J}$, $L = 1\ \text{mm}$, $A_p = 0.3 \times 5\ \mu\text{m}^2$, $V_a = 0.1 \times 5 \times 1000\ \mu\text{m}^3$, $n_o = 10^{18}\ \text{cm}^{-3}$ and $J = 1.8 \times 10^{27}/\text{cm}^3\ \text{s}$.

2.6 Fill in the missing steps in the derivation of bandwidth in Section 2.5.

2.7 A beam of photons travels in air along the $+z$ axis toward the flat surface of a large semiconductor material. It meets the surface at a $90°$ angle. Some of the photons reflect and some enter the material. The material has refractive index n and the surface has reflectance R for the photon number. The vacuum wavelength is λ_o.

1. Assume γ_o photons per volume strike the surface. How many photons per unit area per second strike the surface? Call this number the photon current density $\mathscr{J}_\gamma$ (similar to a current density).

2. What photon current reflects from the surface and what photon current density passes into the material?

3. Develop a formula for the incident power in terms of the photon current. Assume the photons are all confined to the cross-sectional area A.

4. What is the power inside the semiconductor in terms of the power P_o incident on the surface?

5. Using the previous results, show the photon density inside the semiconductor must be

$$\gamma_{\text{inside}} = n(1 - R)\gamma_{\text{outside}}$$

Explain why this formula makes sense.

2.8 Reconsider the previous problem. Assume the material is large so that the photons never encounter the sides and we can neglect any optical losses and we can set the confinement factor to one. Assume negligible spontaneous and nonradiative recombination. Omit any pump. Find the absorbed power as a function of distance into the material when the incident power is P_o. Use the following procedure.

1. Using the γ–$\mathcal{J}$ rate equations, find the photon density inside the material as a function of distance. Use $\gamma_{0,\text{in}}$ as the photon density just inside the surface (see Figure P2.8). Assume the gain g is independent of position.

2. Explain why the gain g must be negative.

3. Combine the results from this problem and the last problem to show

$$P(z) = P_{\text{outside}}(1 - R)\,e^{-\alpha_{\text{abs}}z}$$

where $\alpha_{\text{abs}} = -|g|$.

2.9 Determine the photon density inside the cavity of a 850 nm laser with 34% mirror reflectance for both mirrors and emitting 1 mW of power through one of the mirrors.

2.10 Determine the photon density inside the cavity of a 1300 nm semiconductor laser with one mirror having 1% reflectance and the other having 99% reflectance. Assume the power from the low reflectivity mirror is 1 mW.

2.11 Repeat Example 2.2.4 and show the math.

2.12 Suppose a GaAs–AlGaAs heterostructure with five quantum wells absorbs low intensity light within a distance of 100 μm, which corresponds to e^{-1}. Find the value of the material gain assuming it's constant in distance. Neglect scattering loss, mirror loss, and spontaneous emission.

2.13 Show the following relation holds for the correlation function $\Gamma(\tau)$ of a real process y (i.e., y is real)

$$\Gamma(\tau) = \Gamma(t_1, t_2) = \Gamma(-\tau)$$

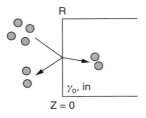

FIGURE P2.8 (See Problem 2.8, Part 1)
Photons reflecting from the surface.

2.14 Explain why the correlation function satisfies $\Gamma(0) \geq 0$.

2.15 Show that the correlation function for a real process y (i.e., y is real) satisfies $\Gamma(0) \geq |\Gamma(\tau)|$. Hint: Use $\langle [y(t+\tau) \pm y(t)]^2 \rangle \geq 0$ and for a stationary process

$$\langle y^2(t+\tau) \rangle = \langle y^2(t) \rangle = \Gamma(0) \quad \text{and} \quad \langle y(t)y(t+\tau) \rangle = \Gamma(\tau)$$

2.16 Show correlation functions for complex stationary processes satisfy $\Gamma^*(-\tau) = \Gamma(\tau)$.

2.17 Let $\Gamma(t)$ be a correlation function for a stationary complex process Z and let $\Gamma(\omega)$ be its Fourier transform. Show $(1/\sqrt{2\pi}) \int_{-\infty}^{\infty} d\omega \, S_z(\omega) = \Gamma_z(t=0)$ where S_z denotes the spectral density for z.

2.18 Show the equations found in Section 2.7

$$\frac{d}{dt} \delta n(t) = -v_g \bar{g}_n \bar{\gamma} \, \delta n - v_g \bar{g} \, \delta \gamma - \bar{R}_n \delta n + F_n(t)$$

$$\frac{d}{dt} \delta \gamma(t) = \Gamma v_g \bar{g}_n \bar{\gamma} \, \delta n + \Gamma v_g \bar{g} \, \delta \gamma - \frac{\delta \gamma}{\tau_\gamma} + \beta \bar{R}_n \delta n + F_\gamma(t)$$

Use the procedure found in Section 2.5.

2.19 Repeat the derivation of the bandwidth using the matrix methodology in Section 2.7.

2.20 Show $\tilde{F}_n(-\omega) = \tilde{F}_n(\omega)$ requires $\tilde{F}_n(\omega)$ to be real when $F_n(t)$ is real.

2.21 Suppose DC current I leaves a region of space (such as a capacitor plate). Show that the shot noise must be given by $\langle \delta i^2 \rangle = 2qI\Delta f$, where $I(t) = I + F(t)$ and F represents a Langevin noise source and $i(t) = I(t) - I$.

2.22 Suppose a steady state beam of light with γ photons per volume leaves a region of space. Imagine that the photons are uniformly spaced across the cross-sectional area A and that they travel at speed c. Starting with the equation with the Langevin noise term $P(t) = P + F$, show that the shot noise must be given by $\langle \delta P^2 \rangle = (hc/\lambda_o)P \, 2\Delta f$, where P represents the steady state power in the beam.

2.23 The transient response (i.e., large signal response) of lasers and diodes can be more important than the small signal response. Read the following journal papers and summarize your findings. Check for more recent publications on the same topic but any author by using the citation indices or computer resources at the local university library.
D. Marcuse *et al.*, *IEEE J. Quant. Electr.* **QE-19**, 1397 (1983).

2.24 Semiconductor lasers with two gain regions can exhibit pulsations in an otherwise steady-state output beam. Read the following journal papers and summarize your findings. Check for more recent publications on the same topic of any author by using the citation indices or computer resources at the local university library.
M. Ueno, R. Lang, "Conditions for self-sustained pulsation and bistability in semiconductor lasers," *J. Appl. Phys.* **58**, 1689 (1985).
R. W. Dixon, W. B. Joyce, "A possible model for sustained oscillations (pulsations) in (Al,Ga)As Double-Heterostructure Lasers," *IEEE J. Quant. Electr.* **QE-15**, 470 (1979).
C. Harder *et al.*, "Bistability and pulsations in semiconductor lasers with inhomogeneous current injection," *IEEE J. Quant. Electr.* **QE-18**, 1351 (1982).

2.25 Similar to the situation described in Problem 2.24, aging lasers also exhibit self-pulsation. Read the following journal papers and summarize your findings. Check for more recent publications on the same topic of any author by using the citation indices or computer resources at the local university library.
R.L. Hartman *et al.*, "Pulsations and absorbing defects in (Al,Ga)As injection lasers,"

J. Appl. Phys. **50**, 4616 (1979).

C.H. Henry, "Theory of defect-induced pulsations in semiconductor injection lasers," *J. Appl. Phys.* **51**, 3051 (1980).

2.26 Some solutions have been investigated to the aging problems described in previous problem. Read the following articles and summarize them. Check for more recent publications on the same topic.

F.U. Herrmann *et al.*, "Reduction of mirror temperature in GaAs/AlGaAs quantum well laser diodes with segmented contacts," *Appl. Phys. Lett.* **58**, 1007 (1991).

W.C. Tang, "Comparison of the facet heating behavior between AlGaAs single quantum-well lasers and double heterostructure lasers," *Appl. Phys. Lett.* **60**, 1043 (1992).

2.27 Read how to measure the transparency current density in the following paper and summarize your findings.

T. R. Chen *et al.*, "Experimental determination of transparency current density and estimation of the threshold current of semiconductor quantum well lasers," *Appl. Phys. Lett.* **56**, 1002 (1990).

2.28 A variety of methods have been proposed to make mirrors ranging from coatings, gratings to total internal reflection. Read the following papers and summarize your findings. Find some others using the citation indices or computer system at the university library.

M. Hagberg *et al.*, "Single-ended output GaAs/AlGaAs single quantum well laser with a dry-etched corner reflector," *Appl. Phys. Lett.* **56**, 1934 (1990).

F. Shimokawa *et al.*, "Continuous-wave operation and mirror loss of a U-shaped GaAs/AlGaAs lasser diode with two totally reflecting mirrors," *Appl. Phys. Lett.* **56**, 1617 (1990).

T. Takamori *et al.*, "Lasing characteristics of a continuous-wave operated folded-cavity surface emitting laser," *Appl. Phys. Lett.* **56**, 2267 (1990).

S. Ou *et al.*, "High-power cw operation of InGaAs/GaAs surface-emitting lasers with 45 degree intracavity micro-mirrors," *Appl. Phys. Lett.* **59**, 2085 (1991).

2.29 Read how to measure the mirror reflectance for semiconductor lasers. Read the following journal papers and summarize your findings. What differences do you see?

J. Johnson *et al.*, "Precise determination of turning mirror loss using GaAs/AlGaAs lasers with up to ten 90° intracavity turning mirrors," *IEEE Phot. Tech. Lett.* **4**, 24 (1992).

H. Appelman *et al.*, "Self-aligned chemically assisted ion-beam-etched GaAs/(Al,Ga). As turning mirrors for photonic applications," *J. Lightwave Techn.* **8**, 39 (1990).

2.30 A variety of methods exist for modulating semiconductor lasers for optical links and interconnects. Read the following paper that compares and constrast several methods and report on your findings.

*Cox *et al.*, "Techniques and performance of intensity-modulation, direct-detection analog optical links", *IEEE Trans. Microwave Thy. and Techniq.* **45**, 1375 (1997).

2.9   Further Reading

The following list contains references pertinent to the material discussed in the chapter.

Introduction

1. Kuhn K.J., *Laser Engineering*, Prentice Hall, Saddle River, 1998.

General References

2. Coldren L.A., *Diode Lasers and Photonic Integrated Circuits*, John Wiley & Sons, New York, 1995.
3. Davis C.C., *Lasers and Electro-Optics, Fundamentals and Engineering*, Cambridge University Press, Cambridge, 1996.
4. Verdeyen J.T., *Laser Electronics*, 2nd ed., Prentice Hall, Englewood Cliffs, 1989.
5. Agrawal G.P., Dutta N.K., *Semiconductor Lasers*, 2nd ed., Van Nostrand Reinhold, New York, 1993.

Stochastic Processes and Statistical Theory

6. Mandel L., Wolf E., *Optical Coherence and Quantum Optics*, Cambridge University Press, Cambridge, 1995.
7. Mood A.M., Graybill F.A., Boes D.C., *Introduction to the Theory of Statistics*, 3rd ed., McGraw-Hill, New York, 1963.

3

Classical Electromagnetics and Lasers

The first two chapters illustrate the basic construction of semiconductor lasers. The construction incorporates the four fundamental components of the gain, pump, output coupler and feedback mechanisms. The phenomenological rate equations describe the operation of the laser in terms of fundamental mathematical quantities that represent the basic components (for example, the mirror loss α_m or bimolecular recombination coefficient B).

The present chapter delves deeper into the construction of the laser by discussing the dynamics of optical waveguiding and the flow of optical power though complicated optical systems. Maxwell's equations play a central role for those topics and for a classical description of the material gain. Not too surprising, the material gain can be described in terms of the polarization and susceptibility. Later chapters use the quantum theory to describe the material gain. The polarization and susceptibility provide the link between the classical and quantum mechanical treatments of lasers.

The first section in the present chapter reviews basic electromagnetic theory for Maxwell's equations. It then develops the wave equation and applies it to a classical gain medium in order to develop classical expressions for the gain, absorption, and index in terms of the susceptibility. The chapter shows how the internal energy of matter changes when it absorbs energy from electromagnetic waves. The absorbed energy can be (1) dissipated as heat, (2) stored as internal electric and magnetic fields, (3) stored in polarized atoms and molecules, and (4) stored in the magnetization of the material (however, we assume negligible magnetization). The chapter next discusses the boundary conditions necessary to solve the wave equation and applies the results to reflecting surfaces. The chapter continues the review of electromagnetic theory by discussing the Poynting vector in some detail and then applies it to the flow of optical power through complicated optical systems using the scattering and transfer matrices. The transfer matrices lead to the laser gain conditions, longitudinal modes, and threshold conditions. The chapter finishes with the electromagnetic theory of waveguiding in rectangularly shaped waveguides. The transverse modes are discussed.

The following two chapters include the groundwork for advanced studies of the electromagnetic field and for the matter–light interaction. The next chapter reviews 4-vector notation in Minkowski space and the psuedo inner product developed to describe the "warping of space–time" encountered in the special theory of relativity. The subsequent chapter develops the connection between Maxwell's equation and the vector potential. It also develops the Lagrangian and Hamiltonian for the electromagnetic field, shows how they reproduce Maxwell's equations, and how they yield the total energy of a system including the free field energy, particle energy and the matter–field interaction energy.

3.1   A Brief Review of Maxwell's Equations and the Constituent Relations

The present section reviews relevant concepts in electromagnetic theory. First we discuss Maxwell's equations and the constituent relations. The electric dipole receives special attention because of its importance for polarization and hence, optical gain. The section discusses boundary conditions especially suited for applications of Maxwell's equations. We use Maxwell's equations in subsequent sections to (1) find a complex wave vector k_n to describe gain/absorption and refractive index, (2) find the Poynting vector for electromagnetic power flow, (3) develop scattering and linear systems theory for optical devices, and (4) develop the theory of optical waveguides.

3.1.1   Discussion of Maxwell's Equations and Related Quantities

Maxwell's equations in differential form can be written as

$$\nabla \times \vec{\mathscr{E}} = -\frac{\partial \vec{\mathscr{B}}}{\partial t} \qquad \nabla \cdot \vec{\mathscr{D}} = \rho_{\text{free}}$$

$$\nabla \times \vec{\mathscr{H}} = \vec{\mathscr{J}} + \frac{\partial \vec{\mathscr{D}}}{\partial t} \qquad \nabla \cdot \vec{\mathscr{B}} = 0 \tag{3.1.1}$$

In addition to Maxwell's equations, there are three constitutive relations among (1) the displacement field $\vec{\mathscr{D}}$ and the electric field $\vec{\mathscr{E}}$, (2) the magnetic field $\vec{\mathscr{H}}$ and the magnetic induction $\vec{\mathscr{B}}$, and (3) the current density $\vec{\mathscr{J}}$ and the electric field $\vec{\mathscr{E}}$. We must discuss two types of charge density. The free charge can move around in the material. The bound charge does not appear in Maxwell's equations but instead, appears in the displacement field $\vec{\mathscr{D}}$ in terms of the polarization. The current density has the usual units of amps per unit cross-sectional area. As a note, the symbols in Maxwell's equations written in "script" signify that they are functions of both position and time. "Block style" characters will be used to represent the amplitude of the quantities that can be functions of position but not functions of time such as $\mathscr{E}(\vec{r}, t) = E(\vec{r})e^{i\omega t}$ where the position vector has the form $\vec{r} = x\tilde{x} + y\tilde{y} + z\tilde{z}$. The quantities of the form $\tilde{x}$ (etc.) denote unit vectors; the "twiddle" distinguishes the quantity from the quantum mechanical operators such as the x-position operator $\hat{x}$.

For a review, first consider the relationship between the electric field and the displacement field

$$\vec{\mathscr{D}} = \varepsilon_0 \vec{\mathscr{E}} + \vec{\mathscr{P}} \tag{3.1.2}$$

where $\vec{\mathscr{P}}$ represents the polarization of the medium. The displacement field is important because some of the optical energy can be stored in the polarization of the material rather than in the fields. For example consider the capacitor and battery shown in Figure 3.1.1. The capacitor has a dielectric between the two plates. The external field induces the formation of electric dipoles (the figure shows three dipoles) and thereby polarizes the dielectric. The dipoles consist of two oppositely charged particles separated by a distance d (see Figure 3.1.2). We sometimes imagine that the applied field stretches the molecules or atoms to form the dipoles. Separating the two charges in space requires energy; in this case, the work done on the dipole appears as potential energy (because of the electrostatic attraction between the opposite charges). The greater the number of dipoles per unit

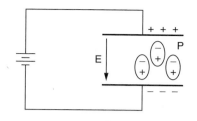

FIGURE 3.1.1
The electric field between the capacitor plates induces dipoles.

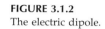

FIGURE 3.1.2
The electric dipole.

volume, the greater the stored energy per unit volume. The dipole moment $\vec{p} = q\vec{d}$ points from the negative charge to the positive charge where $\vec{d}$ extends from the negative to the positive charge and its magnitude gives the distance between the two charges. As a note, we discuss two types of electric dipoles in studies of emitters and detectors. Permanent dipoles consist of two opposite charges permanently separated by a distance d. The induced dipoles consist of two overlapping charge distributions ($d=0$) that separate under the action of an applied electric field. The induced type of dipole leads to gain and absorption.

Now consider the relation between the applied fields, the dipoles and the stored energy. Consider again the capacitor example as depicted by Figure 3.1.3. The battery places charge on the top and bottom capacitor plates and induces dipoles within the bulk of the dielectric. The figure divides the dielectric into three regions denoted by A, B, C. Region A appears closest to positively charged top plate. The figure shows that the two negative charges in region A effectively cancel two of the positive charges on the top plate; the same considerations hold for region C near the bottom capacitor plate. For region B in the interior of the dielectric, the positive and negative tails of the dipole effectively cancel and do not affect the electric field. Therefore, only the regions near the top and bottom plates alter the interior electric field. The total energy stored within the capacitor now has two sources (1) the actual electric field (only four of the six charges on the plate in the figure contribute to the field) and (2) the electric dipoles (the figure shows four dipoles).

The polarization $\vec{\mathcal{P}}(\vec{r}, t)$ denotes the total dipole moment per unit volume at the position $\vec{r}$ at time t. The polarization and the dipole moment are related by

$$\vec{P} = \frac{\#\,\text{dipoles}}{\text{vol}}\,\vec{p} \tag{3.1.3}$$

if an electromagnetic wave travels through free space and encounters a chunk of polarizable material, then inside the material, the electric field decreases and the material

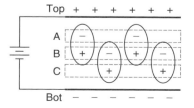

FIGURE 3.1.3
Dipoles store energy and lowers the electric field between the capacitor plates.

becomes polarized. The polarized atoms or molecules gain the energy lost by the electric field.

We can relate the induced polarization to the electric field at the location of the dipole, which is not necessarily the same as the applied field. The most general relation requires tensors but we assume the medium to be isotopic. We further assume a linear relation between the induced polarization and the electric field. The constitutive relation between the polarization and the electric field can be written as

$$\vec{\mathscr{P}} = \varepsilon_o \chi(\omega)\vec{\mathscr{E}} \qquad (3.1.4)$$

where $\chi(\omega)$ denotes the (complex) susceptibility and the constant ε_o represents the permittivity of free space. In principle, the susceptibility can also depend on electric field so that the material has a nonlinear response to an applied electric field. Basically we can think of the susceptibility as the polarization—the susceptibility measures the polarization per unit electric field and describes the ease with which a material can be polarized.

A free charge density ρ_{free} can produce both an electric field and polarization according to Gauss' law

$$\nabla \cdot \vec{\mathscr{D}} = \rho_{\text{free}} \rightarrow \varepsilon_o \nabla \cdot \vec{\mathscr{E}} + \nabla \cdot \vec{\mathscr{P}} = \rho_{\text{free}}$$

the "+" indicates that energy stored as polarization decreases the energy stored in the field within a dielectric (since the sum of the two terms equals the constant free charge). The constitutive relation between the magnetic induction $\vec{\mathscr{B}}$ and the magnetic field $\vec{\mathscr{H}}$ can be written as

$$\vec{\mathscr{B}} = \mu_o \vec{\mathscr{H}} + \mu_o \vec{\mathscr{M}} \qquad (3.1.5)$$

where μ_o represents the permeability of free space.

We neglect any material magnetization $\vec{M}$ in the typical semiconductor used for semiconductor emitters and detectors; we assume the material cannot be magnetized. The magnetization measures the number of magnetic dipoles per volume; these magnetic dipoles can be pictured as microscopic bar magnets. The magnetic induction and the magnetic field differ for reasons very similar to the reasons that the electric and displacement fields differ. The magnetic material can form magnetic dipoles that super-impose their magnetic fields with $\vec{\mathscr{H}}$. That is, the magnetic induction $\vec{\mathscr{B}}$ includes both $\vec{\mathscr{H}}$ and $\vec{\mathscr{M}}$. To discuss this in more detail, consider steady state conditions for a magnetic material. One of Maxwell's equations provides

$$\nabla \times \vec{\mathscr{H}} = \vec{\mathscr{J}} + \frac{\partial \vec{\mathscr{D}}}{\partial t} \rightarrow \nabla \times \vec{\mathscr{H}} = \vec{\mathscr{J}} \rightarrow \nabla \times \vec{\mathscr{B}} - \mu_o \nabla \times \vec{\mathscr{M}} = \mu_o \vec{\mathscr{J}}$$

The last equation basically says that as the current increases, both the magnetization and the magnetic induction also increase. This shows that the number of magnetic field lines $\vec{\mathscr{B}}$ consist of the field lines $\vec{\mathscr{H}}$, due solely to $\vec{\mathscr{J}}$, and to the field lines due to the magnetization $\vec{\mathscr{M}}$. The magnetization $\vec{\mathscr{M}}$ originates in the magnetic dipoles lining up due to the field $\vec{\mathscr{H}}$ already present. We can then say that the current $\vec{\mathscr{J}}$ produces a field $\vec{\mathscr{H}}$ which in turn lines up the magnetic dipoles to produce the magnetization $\vec{\mathscr{M}}$. The magnetization produces additional field lines. The magnetic induction $\vec{\mathscr{B}}$ describes the total field consisting of $\vec{\mathscr{H}}$ along with those due to $\vec{\mathscr{M}}$. Basically, Maxwell's equation

$\nabla \times \vec{\mathscr{H}} = \vec{\mathscr{J}}$ says the current $\vec{\mathscr{J}}$ produces only magnetic field $\vec{\mathscr{H}}$. The field $\vec{\mathscr{B}}$ measures the field lines due to $\mathscr{H}$ and $\mathscr{M}$. Inside a magnetic material, we therefore have $\vec{\mathscr{B}} = \mu_o \vec{\mathscr{H}} + \mu_o \vec{\mathscr{M}}$. Outside the magnetic material, we must have $\vec{\mathscr{B}} = \mu_o \vec{\mathscr{H}}$.

As a final relation, we can relate the current density $\vec{\mathscr{J}}$ to the electric field $\vec{\mathscr{E}}$ by

$$\vec{\mathscr{J}} = \sigma \vec{\mathscr{E}} \tag{3.1.6}$$

where σ denotes the conductivity of the material. The reader will recognize this last relation as Ohm's law.

3.1.2 Relation between Electric and Magnetic Fields in Vacuum

As every reader knows, the electromagnetic wave consists of an electric and magnetic field. The vector $\vec{\mathscr{E}} \times \vec{\mathscr{H}}$ points in the same direction as the wave vector $\vec{k}$, which in turn points in the propagation direction of the wave. Figure 3.1.4 shows the energy propagating towards the right. Subsequent sections relate the magnitude and direction of energy flow to the Poynting vector.

We can find the relation between the plane-wave electric and magnetic fields $\vec{\mathscr{E}}, \vec{\mathscr{H}}$ in vacuum by using the plane wave versions

$$\vec{\mathscr{E}}(z,t) = E_o\, e^{ik_o z - i\omega t}\, \hat{x} \tag{3.1.7}$$

$$\vec{\mathscr{H}}(z,t) = H_o\, e^{ik_o z - i\omega t}\, \hat{y} \tag{3.1.8}$$

The fact that H has only the y-component will be verified below. We want the relation between E_o and H_o for a wave in free space. We need one of Maxwell's equations, namely

$$\nabla \times \vec{\mathscr{H}} = \vec{\mathscr{J}} + \frac{\partial \vec{\mathscr{D}}}{\partial t} \tag{3.1.9}$$

where the current density $\vec{\mathscr{J}}$ has units of amperes per area, and in free space $\vec{\mathscr{J}} = 0$. We also need the constituent relation for the displacement field $\vec{\mathscr{D}}$ in terms of the electric field $\vec{\mathscr{E}}$ and the polarization $\vec{\mathscr{P}}$

$$\vec{\mathscr{D}} = \varepsilon_o \vec{\mathscr{E}} + \vec{\mathscr{P}}$$

However, the polarization must be zero for free space $\vec{\mathscr{P}} = 0$. Maxwell's equation (3.1.9) reduces to

$$\nabla \times \vec{\mathscr{H}} = \varepsilon_o \frac{\partial \vec{\mathscr{E}}}{\partial t} \tag{3.1.10}$$

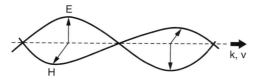

FIGURE 3.1.4
The electromagnetic wave.

Now calculate the various terms in Equation (3.1.10). The cross product can be written as

$$\nabla \times \vec{\mathscr{H}} = \begin{vmatrix} \tilde{x} & \tilde{y} & \tilde{z} \\ \dfrac{\partial}{\partial x} & \dfrac{\partial}{\partial y} & \dfrac{\partial}{\partial z} \\ \mathscr{H}_x & \mathscr{H}_y & \mathscr{H}_z \end{vmatrix} = \tilde{x}\left(\dfrac{\partial \mathscr{H}_z}{\partial y} - \dfrac{\partial \mathscr{H}_y}{\partial z}\right) - \tilde{y}\left(\dfrac{\partial \mathscr{H}_z}{\partial x} - \dfrac{\partial \mathscr{H}_x}{\partial z}\right) + \tilde{z}\left(\dfrac{\partial \mathscr{H}_y}{\partial x} - \dfrac{\partial \mathscr{H}_x}{\partial y}\right)$$

Therefore, since the magnetic field only has the y-component, the cross product reduces to

$$\nabla \times \vec{\mathscr{H}} = -\tilde{x}\frac{\partial \mathscr{H}_y}{\partial z} = -i\tilde{x}k_o H_o e^{+ik_o z - i\omega t}$$

where the direction of the wave vector (which has magnitude k_o) parallels the z-axis $\hat{z}$. The time derivative in Equation 3.1.10 provides

$$\frac{\partial}{\partial t}\varepsilon_o \vec{\mathscr{E}}(z,t) = -i\omega\varepsilon_o \tilde{x}E_o\, e^{ik_o z - i\omega t}$$

Substituting these last two results back into Maxwell's equation (3.1.10) yields

$$k_o H_o = \omega\varepsilon_o E_o \quad \rightarrow \quad H_o = \frac{\omega\varepsilon_o}{k_o}E_o \qquad (3.1.11)$$

in vacuum. The definition for magnetic induction gives $B_o = \mu_o H_o$ where μ_o symbolizes the permeability of free space. Next, recall the relation among the speed of light c in vacuum, the permitivity, and the permeability, namely $c^2 = (\varepsilon_o \mu_o)^{-1}$. Equation (3.1.11) becomes

$$B_o = \frac{E_o}{c} \qquad (3.1.12)$$

3.1.3 Relation between Electric and Magnetic Fields in Dielectrics

The relation between the electric and magnetic fields changes in a polarizable medium. Let's ignore any absorption (so that k must be real) and define the following electric and magnetic fields inside the material

$$\vec{\mathscr{E}} = E_1\, e^{ikz - i\omega t}\, \tilde{x} \qquad \vec{\mathscr{H}} = H_1\, e^{ikz - i\omega t}\, \tilde{y}$$

The magnitude of the wave vector k depends on the real index of refraction "n" according to

$$k = \frac{2\pi}{\lambda_n} = \frac{2\pi}{\lambda_o/n} = k_o n \qquad (3.1.13)$$

where λ_n, λ_o represent the wavelengths in the medium and in vacuum, respectively. Assuming there aren't any free charges and free currents, Maxwell's equation can now be written as

$$\nabla \times \vec{\mathscr{H}} = \frac{\partial}{\partial t}\vec{\mathscr{D}} = \frac{\partial}{\partial t}\left(\varepsilon_o\vec{\mathscr{E}} + \vec{\mathscr{P}}\right) = \frac{\partial}{\partial t}\left(\varepsilon_o\vec{\mathscr{E}} + \varepsilon_o\chi\vec{\mathscr{E}}\right) \qquad (3.1.14)$$

The last term can be simplified by defining the permittivity ε of the material in terms of the free space permittivity ε_o and the susceptibility χ of the medium (note that $\chi = 0$ for the vacuum)

$$\varepsilon = \varepsilon_o(1 + \chi) \tag{3.1.15}$$

Maxwell's Equation (3.1.14) for an electromagnetic wave in the medium becomes

$$\nabla \times \vec{\mathcal{H}} = \frac{\partial}{\partial t}\left(\varepsilon \vec{\mathcal{E}}\right) \tag{3.1.16}$$

The cross product and derivative can be performed in the same manner as above to obtain

$$-i\tilde{x}kH_1\, e^{ikz-i\omega t} = -i\tilde{x}\omega \varepsilon E_1\, e^{ikz-i\omega t}$$

which provides

$$H_1 = \frac{\varepsilon \omega}{k_o n} E_1 \tag{3.1.17}$$

where $k = k_o n$. A complex index shifts phase but does not affect the phase velocity. Again using $B = \mu_o H$, which neglects the magnetization (i.e., $M = 0$), we obtain

$$B_1 = \frac{\omega \mu_o \varepsilon}{k_o n} E_1 \tag{3.1.18}$$

This last expression can be rewritten using the speed of light in the medium

$$v = \frac{1}{\sqrt{\mu_o \varepsilon}} = \frac{c}{n} \quad \rightarrow \quad \mu_o \varepsilon = \frac{n^2}{c^2}$$

and using the expression for the speed of light in terms of the angular frequency and the wave vector

$$c = \frac{\omega}{k_o}$$

to provide the new expression

$$B_1 = \frac{c\mu_o \varepsilon}{n} E_1 = \frac{c(n/c)^2}{n} E_1 = \frac{n}{c} E_1 = \frac{E_1}{v} \tag{3.1.19}$$

where $v = c/n$ gives the speed of light in the dielectric material. Equation (3.1.19) relates the magnetic and electric fields inside the dielectric.

3.1.4 General Form of the Complex Traveling Wave

The transverse traveling wave can depend on position according to

$$\vec{\mathcal{E}} = \tilde{z}\, u(x,y)\, e^{ik_n z - i\omega t}$$

A slowly varying amplitude in time can be written as $u(x, y, t)$.

3.2 The Wave Equation

The gain term appears in the rate equations as one of the primary quantities of interest for light emitters and detectors. The gain most fundamentally represents the quantum mechanical, matter–light interaction. However, the classical theory links the gain to the familiar (and easy to picture) polarization and susceptibility. Both the classical and quantum approaches rely on results obtained from Maxwell's wave equation. Therefore, we derive the wave equation for an electromagnetic (EM) wave traveling through a conductor or dielectric material. The results show how the gain and absorption come from the motion of the electric dipole moments, which gives rise to the complex permittivity, refractive index, and wave vector. The analysis includes conductive media in order to show the origin of free-carrier absorption. Subsequent sections discuss how the boundary conditions arise from the Maxwell differential equations to produce reflection and Snell's law.

3.2.1 Derivation of the Wave Equation

Maxwell's equations for the electric and magnetic field

$$\nabla \times \vec{\mathcal{E}} = -\frac{\partial \vec{\mathcal{B}}}{\partial t} \tag{3.2.1}$$

$$\nabla \times \vec{\mathcal{H}} = \vec{\mathcal{J}} + \frac{\partial \vec{\mathcal{D}}}{\partial t} \tag{3.2.2}$$

can be combined. Taking the curl $\nabla \times$ of the top equation (3.2.1), we obtain

$$\nabla \times \nabla \times \vec{\mathcal{E}} = -\frac{\partial}{\partial t} \nabla \times \vec{\mathcal{B}} = -\frac{\partial}{\partial t} \nabla \times (\mu_o \vec{\mathcal{H}})$$

Substituting Equation (3.2.2) and using $\vec{\mathcal{J}} = \sigma \vec{\mathcal{E}}$, we obtain

$$\nabla \times \nabla \times \vec{\mathcal{E}} = -\mu_o \frac{\partial}{\partial t} \left(\vec{\mathcal{J}} + \frac{\partial \vec{\mathcal{D}}}{\partial t} \right) = -\mu_o \frac{\partial \vec{\mathcal{J}}}{\partial t} - \mu_o \frac{\partial^2 \vec{\mathcal{D}}}{\partial t^2} = -\mu_o \sigma \frac{\partial \vec{\mathcal{E}}}{\partial t} - \mu_o \frac{\partial^2}{\partial t^2} \left(\varepsilon_o \vec{\mathcal{E}} + \varepsilon_o \chi \vec{\mathcal{E}} \right)$$

Using the relation between the susceptibility and the permittivity $\varepsilon = \varepsilon_o(1 + \chi)$, we find

$$\nabla \times \nabla \times \vec{\mathcal{E}} = -\mu_o \sigma \frac{\partial \vec{\mathcal{E}}}{\partial t} - \mu_o \frac{\partial^2}{\partial t^2} \left(\varepsilon \vec{\mathcal{E}} \right)$$

For now, we ignore any spatial dependence of the permittivity and also ignore any possibility of modulating it with externally applied voltages. We can use the "bac-cab" rule

$$\vec{A} \times \vec{B} \times \vec{C} = \vec{B}(\vec{A} \cdot \vec{C}) - \vec{C}(\vec{A} \cdot \vec{B})$$

in the form appropriate for a differential operator

$$\nabla \times \nabla \times \vec{\mathcal{E}} = \nabla(\nabla \cdot \vec{\mathcal{E}}) - \nabla^2 \vec{\mathcal{E}}$$

Requiring the divergence of the electric field to be zero $\nabla \cdot \vec{E} = 0$ equivalently says that the net "free charge" must be negligible $\nabla \cdot \vec{D} = \rho_{\text{free}} = 0$. Now the wave equation takes on the form

$$\nabla^2 \vec{\mathcal{E}} = \mu_o \sigma \frac{\partial \vec{\mathcal{E}}}{\partial t} + \mu_o \varepsilon_o (1 + \chi) \frac{\partial^2 \vec{\mathcal{E}}}{\partial t^2} \tag{3.2.3}$$

For a wave equation, the coefficient of the second derivative (with respect to time) can be related to the speed of the wave in the medium. Therefore, examining Equation (3.2.3) shows the susceptibility must be related to the index of refraction. The first derivative of the electric field can be related to damping. In mechanics, this term would be related to frictional forces.

3.2.2 The Complex Wave Vector

The present topic shows how the complex wave vector produces the absorption/gain and real refractive index. We start by substituting a plane wave into the wave equation. This procedure yields an expression for the complex wave vector in terms of susceptibility and conductivity. Any electromagnetic wave can be written as a sum of plane waves using the Fourier transform. Let's assume that the electric field consists of a single traveling plane wave

$$\vec{\mathcal{E}} = \hat{e} E_o \exp(i k_n z - i \omega t) \tag{3.2.4}$$

where $k_n = 2\pi/\lambda_n = 2\pi n/\lambda_o$ denotes a complex wave vector, λ_o denotes the wavelength in free space, n represents the complex refractive index and $\tilde{e}$ symbolizes a unit vector along the direction of polarization. We discover the meaning of the complex wave vector $k_c = k_r + i k_i = \text{Re}(k_c) + i \, \text{Im}(k_c)$ and find speed of the wave by substituting the plane wave into the wave equation for the electric field. The substitution provides

$$-k_c^2 + i \mu_o \sigma \omega + \mu_o \varepsilon_o \omega^2 (1 + \chi) = 0 \tag{3.2.5}$$

To continue, we need to combine this last expression with two different expressions of the speed of light in vacuum

$$c = \frac{1}{\sqrt{\varepsilon_o \mu_o}} \qquad c = \frac{\omega}{k_o} \rightarrow k_o = \frac{\omega}{c}$$

where ε_o, μ_o, ω, k_o represent the permittivity, permeability, angular frequency, and the magnitude of the wave vector in free space, respectively. Substituting these terms into Equation (3.2.5) provides another expression for the wavevector

$$k_c^2 = \frac{\omega^2}{c^2}(1 + \chi) + i \mu_o \sigma \omega = k_o^2 (1 + \chi) + i \mu_o \sigma \omega$$

Rearranging terms

$$k_c^2 = k_o^2(1 + \chi) + i \frac{k_o^2}{k_o^2} \mu_o \sigma \omega = k_o^2(1 + \chi) + i k_o^2 \frac{\mu_o \sigma \omega}{(\omega/c)^2} = k_o^2(1 + \chi) + i k_o^2 \frac{\sigma}{\varepsilon_o \omega}$$

and, by factoring out the common term of k_o, we find an expression for the complex wave vector

$$k_c^2 = k_o^2 \left[1 + \chi + i \frac{\sigma}{\varepsilon_o \omega} \right] \tag{3.2.6}$$

where $i = \sqrt{-1}$ and χ can be complex. We see that the wave vector k_c consists of the sum of real and imaginary parts. For emphasis, the real and imaginary parts can be explicitly written

$$k_c = k_o \left[1 + \text{Re}(\chi) + i \left(\text{Im}(\chi) + \frac{\sigma}{\varepsilon_o \omega} \right) \right]^{1/2} \tag{3.2.7}$$

We explicitly find the square root on the right-hand side by writing the argument under the square root in phasor form $r e^{i\theta}$ and then setting the square root to $\sqrt{r} e^{i\theta/2}$. The results show the complex wave vector has both real and imaginary components. We will return to this equation after a few definitions and a discussion of the meaning of the complex wave vector.

3.2.3 Definitions for Complex Index, Permittivity and Wave Vector

Before continuing with the complex wave vector k_c, we make some definitions for the complex refractive index and the complex permittivity. Define the complex refractive index as

$$n_c = n_r + i n_i \tag{3.2.8a}$$

which can be related to the complex wave vector

$$k_c = k_o n_c \tag{3.2.8b}$$

Comparing Equation (3.2.8b) with Equation (3.2.7) shows the complex index has the form

$$n_c^2 = 1 + \text{Re}(\chi) + i \left(\text{Im}(\chi) + \frac{\sigma}{\varepsilon_o \omega} \right) \tag{3.2.9}$$

The meaning will become clear shortly. We also define a complex permittivity as

$$\varepsilon_c = \varepsilon_r + i \varepsilon_i \tag{3.2.10}$$

By convention, we often write the real part of the wave vector, index, and permittivity without subscripts as $k = k_r$, $n = n_r$, and $\varepsilon = \varepsilon_r$. The reader should recall that the real permittivity produces the real refractive index (as usually stated in optics) according to

$$n^2 = \varepsilon/\varepsilon_o \quad \text{or} \quad n = \sqrt{\varepsilon/\varepsilon_o} \tag{3.2.11}$$

This last relation makes it clear that the index of refraction must be related to the dynamics of the dipoles because the permittivity ε can be related to the polarization. We assume that Equation (3.2.11) also holds for the complex index of refraction and the complex permittivity

$$n_c = \sqrt{\frac{\varepsilon_c}{\varepsilon_o}} = \sqrt{\frac{\varepsilon_r}{\varepsilon_o} + i\frac{\varepsilon_i}{\varepsilon_o}} \qquad (3.2.12)$$

Comparing Equation 3.2.12 with Equation 3.2.9 gives a relation for the complex permittivity as

$$\frac{\varepsilon_r}{\varepsilon_o} + i\frac{\varepsilon_i}{\varepsilon_o} = 1 + \text{Re}(\chi) + i\left(\text{Im}(\chi) + \frac{\sigma}{\varepsilon_o\omega}\right) \qquad (3.2.13a)$$

This last equation provides the relations for the permittivity

$$\frac{\varepsilon_r}{\varepsilon_o} = 1 + \text{Re}(\chi) \qquad \frac{\varepsilon_i}{\varepsilon_o} = \text{Im}(\chi) + \frac{\sigma}{\varepsilon_o\omega} \qquad (3.2.13b)$$

So we see that the real permittivity and (hence) the real part of the refractive index are related to the real part of the susceptibility. Likewise the imaginary part of the permittivity is related to both the imaginary part of the susceptibility and the conductivity. The conduction mechanism (expressed through the conductivity) absorbs part of the electromagnetic wave. Finally, as another definition (to be explained later), the complex wave vector can be written in the following way

$$k_c = k_o n_c = k_o n + i\frac{\alpha}{2} = k_o n - i\frac{g_n}{2} \qquad (3.2.14)$$

where α and g_n represent the absorption and the gain, respectively. Notice that the gain and absorption terms differ by a minus sign. The absorption coefficient and gain g_n do not agree with the material gain discussed in Chapter 2 because it includes the free carrier loss term (through the conductivity).

3.2.4 The Meaning of k_n

The complex wave vector plays a central role in determining the gain or absorption of a material. We devote this topic to exploring the immediate consequences of Equations (3.2.7) and (3.2.14).

Assume that an electromagnetic wave strikes a chunk of material as shown in Figure 3.2.1. We can see that the wave vector $\text{Re}(k_c)$ must be larger inside than the wave vector k_o on the outside due to the real part of the index of refraction,

$$\text{Re}(k_c) = k_o n_r > k_o \qquad \rightarrow \qquad \lambda_{\text{medium}} < \lambda_{\text{vacuum}}$$

Therefore the wavelength inside of the material must be smaller than outside $n_r = \text{Re}(n_c)$. The real part of the wave vector provides information on the wavelength.

Now we will see how the imaginary part of the wave vector leads to an exponential increase or decrease of the electric field (or power) depending on whether the material exhibits gain or absorption, respectively. Assuming an unpumped material, the electric

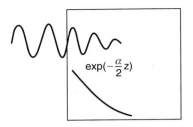

FIGURE 3.2.1
An incident electric field decays as it travels through an absorptive medium.

field inside the material exponentially decays due to absorption. Substituting Equation (3.2.14) into the equation for the plane wave (3.2.4) shows this behavior

$$E = E_o \exp(ik_c z) = E_o \exp[i(k_o n + i\alpha/2)z)] = E_o \exp(-z\alpha/2)\exp[ik_o nz] \qquad (3.2.15)$$

where the time dependence has been omitted. This last equation makes it clear that the absorption α causes an exponential decrease of the electric field. We now see the reason for the factor of 2 in the definition of the complex wave vector in Equation (3.2.14). The power has the form of $P \sim E^* E \sim \exp(-\alpha z) = \exp(+gz)$ and the factor of 2 does not appear (refer to Section 3.5 on the Poynting vector). The absorption (or gain) coefficient for the field is $\alpha/2$ (or $g/2$) while that for the power is α (or g).

The imaginary part of the complex wave vector in Equation (3.2.7) has the term $i\sigma/\varepsilon_o c^2$ which represents the free carrier absorption. The mobile carriers in a metal oscillate in response to the incident wave and absorb some of the energy. The oscillating charges transfer the energy to the metal as heat. As a result, the free charge attenuates the electromagnetic wave. The same thing happens for a doped semiconductor. The doping increases the number of mobile electrons or mobile holes. These carriers can then absorb any electromagnetic field that happens to be incident on the doped material. The free carrier absorption is part of the internal optical loss α_{int} for a laser.

3.2.5 Approximate Expression for the Wave Vector

Now to find approximate expressions for the complex wave vector we return to Equation 3.2.7, namely

$$k_c = k_o n_c = k_o \left[1 + \mathrm{Re}(\chi) + i\left(\mathrm{Im}(\chi) + \frac{\sigma}{\varepsilon_o \omega}\right)\right]^{1/2} \qquad (3.2.16a)$$

As discussed in the previous topic, the imaginary part of the wave vector gives the exponential decay or growth (absorption or gain, respectively) of the traveling wave. For simplicity, substitute the real index $n_r^2 = 1 + \mathrm{Re}(\chi)$ into Equation (3.2.16a) and factor it from the square root to get

$$k_c = k_o n_c = k_o n_r \left[1 + \frac{i}{n_r^2}\left(\mathrm{Im}(\chi) + \frac{\sigma}{\varepsilon_o \omega}\right)\right]^{1/2} \qquad (3.2.16b)$$

We assume the imaginary term remains small. We apply a Taylor series expansion of the form

$$\sqrt{1+y} \approx 1 - \frac{y}{2}$$

to find

$$k_c = k_o n_c = k_o n_r + i \frac{k_o}{2n_r} \left(\text{Im}(\chi) + \frac{\sigma}{\varepsilon_o \omega} \right) \tag{3.2.16c}$$

Now we can find a very important result for the gain/absorption by comparing Equation 3.2.15, namely $k_c = k_o n + i\alpha/2$ with Equation (3.2.16c). The absorption α of the material can be written as

$$\alpha = \frac{k_o}{n_r} \left(\text{Im}(\chi) + \frac{\sigma}{\varepsilon_o \omega} \right) = -g_n \tag{3.2.17}$$

where $n = n_r$ and

$$\alpha = -g_n$$

This shows that the imaginary part of the susceptibility can produce loss or gain. As we will see, the strength of the pump determines the value of the susceptibility. For the quantum wall laser, the susceptibility increases when the number of excitons (electron-hole pairs increases). The reason is simple—more excitons mean more dipoles. Let's next examine the form of Equation (3.2.17).

The absorption term in Equation (3.2.17) consists of two parts. The term $\alpha_{\text{stim}} = k_o \, \text{Im}(\chi)/n$ represents stimulated absorption. If $\text{Im}\chi < 0$ then the term provides the material gain g from Chapter 2. The second term $\alpha_{fc} = k_o \sigma/(n_r \varepsilon_o \omega)$ represents free-carrier absorption (see Review Exercise 3.1). Therefore the full absorption coefficient has the form of $\alpha = \alpha_{\text{stim}} + \alpha_{fc}$. If we set $\alpha = -g_n$ and $\alpha_{\text{stim}} = -g$ then the absorption equation takes the form of $g_n = g - \alpha_{fc}$ so that g_n represents the net gain. Notice the net gain does not agree with the net gain found for a laser since the wave equation does not include scattering losses and mirror losses.

3.2.6 Approximate Expressions for the Refractive Index and Permittivity

Now we state relations for the refractive index and the permittivity. The refractive index involves a square root as shown in Equation (3.2.12)

$$n_c = \sqrt{\frac{\varepsilon_c}{\varepsilon_o}} = \sqrt{\frac{\varepsilon_r}{\varepsilon_o} + i \frac{\varepsilon_i}{\varepsilon_o}}$$

We can find a simple approximate expression for the complex refractive index by using the binomial expansion on this last equation

$$\tilde{n} = \sqrt{\frac{\varepsilon_r}{\varepsilon_o}} \left[1 + i \frac{\varepsilon_i/\varepsilon_o}{\varepsilon_r/\varepsilon_o} \right]^{1/2} \cong n_r \left[1 + i \frac{\varepsilon_i/\varepsilon_o}{2\varepsilon_r/\varepsilon_o} \right] = n_r \left[1 + i \frac{\varepsilon_i/\varepsilon_o}{2n_r^2} \right] = n_r + i \frac{\varepsilon_i/\varepsilon_o}{2n_r} \tag{3.2.18}$$

where we assume the imaginary part of ε is small. The complex index $\tilde{n} = n_r + i n_i$ has real and imaginary parts given by

$$n_r = \sqrt{\frac{\varepsilon_r}{\varepsilon_o}} \qquad n_i = \frac{\varepsilon_i/\varepsilon_o}{2n_r} \tag{3.2.19}$$

Equation (3.2.19) in conjunction with Equation (3.2.16c) provides the complex index as

$$n_r = \sqrt{1 + \mathrm{Re}(\chi)} \qquad n_i = \frac{1}{2n_r}\left(\mathrm{Im}(\chi) + \frac{\sigma}{\varepsilon_o\omega}\right) \qquad (3.2.20)$$

and therefore

$$\varepsilon_i = \varepsilon_o\,\mathrm{Im}(\chi) + \frac{\sigma}{\omega} \qquad (3.2.21)$$

3.2.7 The Susceptibility and the Pump

The polarization induced by an electromagnetic field traveling through a medium has real and imaginary parts just like the susceptibility since $\vec{\mathscr{P}} = \varepsilon_o\chi\vec{\mathscr{E}}$. The oscillating electric field forces the dipoles to also oscillate which, according to classical electromagnetic theory, produces more electromagnetic waves (the dipole oscillation consists of the periodic exchange of the positive and negative charges). The real part of the susceptibility leads to the index of refraction while the imaginary part leads to absorption or gain as can be seen from the main two results

$$n = n_r = [1 + \mathrm{Re}(\chi)]^{1/2} \qquad (3.2.22)$$

$$\alpha = \frac{k_o}{n_r}\left(\mathrm{Im}(\chi) + \frac{\sigma}{\varepsilon_o\omega}\right) = -g_n \qquad (3.2.23)$$

The portion of the polarization corresponding to the real part of the susceptibility will be in-phase with the driving electric field. Similarly, the portion of the polarization corresponding to the imaginary part of the susceptibility will be out of phase with the driving field. More on this topic appears in subsequent sections and chapters.

Question: The pump mechanism adds energy to the laser. Which quantities depend on the pumping? That is, which quantities depend on the extra number of carriers added to the semiconductor due to the pumping? It is the susceptibility that changes with pumping. We should think of susceptibility as being very similar to polarization since the susceptibility is essentially the polarization per unit electric field ($P = \varepsilon_o\chi E$). Adding carriers through the pump mechanism increases the number of possible dipoles. Some books divide the susceptibility into a background term and a pump term as $\chi = \chi_b + \chi_p$. The background term describes the number of possible dipoles already present in the material. The pump adds carriers to the semiconductor gain medium (which contributes to χ_p). Figure 3.2.2 shows a cartoon representation of how electrons and holes in an electric field give rise to dipoles. Both the background and the pump susceptibility respond to an incident electromagnetic field.

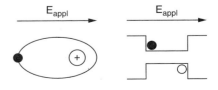

FIGURE 3.2.2
Adding atoms or molecules (left) to a material increases the number of dipoles. Increasing the number of electrons–holes to a quantum well (right) increases the number of dipoles.

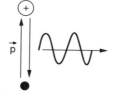

FIGURE 3.2.3
Oscillating dipole produces an EM field.

To show the effect of the pump on the absorption and index of refraction, we substitute the background and pump susceptibility into their respective equations

$$\alpha = \frac{k_o}{n_r}\left(\text{Im}(\chi_b) + \text{Im}(\chi_p) + \frac{\sigma}{\varepsilon_o\omega}\right) = \frac{k_o}{n_r}\frac{\sigma}{\varepsilon_o\omega} + \frac{k_o}{n_r}\text{Im}(\chi_b + \chi_p) = \alpha'_{int} - g$$

where α'_{int} is related to the term containing the conductivity σ and "g" is related to the term containing the susceptibility (Figure 3.2.3). We have seen similar equations to this before. When the material gain g, which includes stimulated emission and stimulated absorption, is larger than the free carrier absorption term α'_{int}, the net absorption α will be negative and the material will therefore exhibit gain. The reader should realize that the loss α as defined in the previous equation does not include other optical losses such as that through the mirrors and sidewalls of the laser. Next, consider the (real) refractive index

$$n = n_r = \left[1 + \text{Re}(\chi_b) + \text{Re}(\chi_p)\right]^{1/2}$$

Define the background refractive index as $n_b = \sqrt{1 + \text{Re}(\chi_b)}$. Use a Taylor expansion to rewrite the (real) refractive index as

$$n = n_b\left[1 + \frac{\text{Re}(\chi_p)}{n_b^2}\right]^{1/2} \cong n_b\left[1 + \frac{\text{Re}(\chi_p)}{2n_b^2}\right] = n_b + \frac{\text{Re}(\chi_p)}{2n_b}$$

We see that the refractive index is smaller than the background refractive index when the real part of the pump susceptibility is negative. The pump susceptibility changes the index of refraction.

Example 3.2.1 Laser Frequency and Refractive Index
Changes in the refractive index can lead to changes in the operating wavelength of the laser. Consider the Fabry-Perot cavity shown in Figure 3.2.4 with the half-integral number of wavelengths. Recall that the wavelength of light in a material with refractive index n is given by $\lambda_n = \lambda/n$. Let m be the number of half wavelengths that exactly fits in the cavity $L = m(\lambda_n/2)$. The wavelength in air must be $\lambda_o = \frac{nL}{m}$.

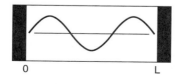

0   L

FIGURE 3.2.4
A half-integral number of wavelengths fit into the laser cavity.

In order to keep λ_o constant, any changes in n must be accompanied by an equal change in m. However for fixed λ_o, m can only change by an integer. Therefore, some changes in n must cause the operating wavelength to change.

3.3 Boundary Conditions for the Electric and Magnetic Fields

Boundary conditions play a key role for solving partial differential equations. They determine the form of the basis set used in the expansion of the general solution. The sets of eigenvalues and basis functions can be either continuous or discrete (or a combination) depending on the nature of the boundary conditions. We will need boundary conditions when we solve Maxwell's equations for waveguides, reflection coefficients and Snell's law. The boundary conditions considered in the present section consist of those that describe how the electric and magnetic fields behave as they move across interfaces between different materials.

3.3.1 Electric Field Perpendicular to an Interface

Two cases apply to finding the electric field perpendicular to an interface. In the first case, we assume that an interface between materials doesn't have any free charge. The second case includes the free charge. As a point worth remembering, the index of refraction of a material (used to find the speed of light $v = c/n$) can be written in terms of the permittivity of the material as $n = \sqrt{\varepsilon/\varepsilon_o}$ where ε_o denotes the permittivity in free space. Therefore, either the refractive index n or the permittivity characterizes the different materials.

Case 1   No Free Surface Charge

Suppose an interface separates two materials with dissimilar refractive indices $n_1 = \sqrt{\varepsilon_1/\varepsilon_o}$ and $n_2 = \sqrt{\varepsilon_2/\varepsilon_o}$ as shown in Figure 3.3.1. We assume the interface doesn't have any free charge. How do we relate the two displacement fields $\vec{\mathcal{D}}_1$ and $\vec{\mathcal{D}}_2$?

Without free charges, Maxwell's equation for the displacement field can be written as

$$\nabla \cdot \vec{\mathcal{D}} = 0 \qquad\qquad (3.3.1)$$

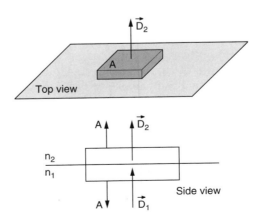

FIGURE 3.3.1
Interface separates two media with different refractive indices.

Integrating over the volume of a small box (Figure 3.3.1) we find the integral

$$\int_V \nabla \cdot \vec{\mathscr{D}} \, dV = 0 \quad \rightarrow \quad \int_{A_T} \vec{\mathscr{D}} \cdot d\vec{a} = 0 \tag{3.3.2}$$

from the divergence theorem. The symbol A_T represents the total surface area of the entire box. For Case 1, assume the displacement fields to be perpendicular to the interface, which means they must be perpendicular to the top and bottom of the box and parallel to the vertical sides. The dot product therefore produces two nonzero integrals, one over the top and another over the bottom of the box

$$0 = \int_{A_T} \vec{\mathscr{D}} \cdot d\vec{a} = \int_{A \atop \text{top}} \mathscr{D}_2 \, da - \int_{A \atop \text{bottom}} \mathscr{D}_1 \, da \tag{3.3.3}$$

where the minus sign occurs because the displacement field points opposite the area vector on the bottom side (see the bottom portion of Figure 3.3.1). For small enough boxes, the displacement fields must be approximately constant over the top and bottom surfaces and can be removed from the integrals. As a result, we find

$$\vec{\mathscr{D}}_2 = \vec{\mathscr{D}}_1 \tag{3.3.4}$$

Substituting the definition of the displacement field in terms of the permittivity and electric field

$$\vec{\mathscr{D}} = \varepsilon \vec{\mathscr{E}}$$

for each displacement field provides

$$\mathscr{E}_2 = \frac{\varepsilon_1}{\varepsilon_2} \mathscr{E}_1 = \left(\frac{n_1}{n_2}\right)^2 \mathscr{E}_1 \tag{3.3.5}$$

Equation (3.3.4) indicates that the *displacement* fields must be continuous across the *dielectric* interface whereas Equation (3.3.5) shows that the *electric* fields cannot be continuous. Let's examine the reason as to why the electric fields have this discontinuity. Assume that the electric field due to a traveling wave points upward similar to E_{wave} in Figure 3.3.2. The fields from electric dipoles *inside* the medium tend to cancel (refer to the discussion in connection with the capacitor in Figure 3.1.3). The fields due to induced dipoles near the surface tend not to cancel. For example, the figure shows two negative charges compared with four positive charges at the interface; the interface must have a net charge of +2. The resulting sheet charge at the interface (i.e., say the $+q$ part of the dipole charges) tends to produce fields that point upward and downward.

FIGURE 3.3.2
Dipole fields produce a discontinuity in the electric fields on either side of the interface.

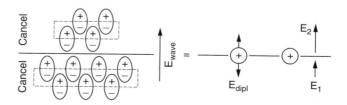

The total electric field must be the sum of the traveling wave field and that due to the dipoles

$$\left(\mathscr{E}_{\text{wave}} + \mathscr{E}_{\text{dipole}}\right)_{\text{bottom}} = \mathscr{E}_1 \quad \text{and} \quad \left(\mathscr{E}_{\text{wave}} + \mathscr{E}_{\text{dipole}}\right)_{\text{top}} = \mathscr{E}_2$$

In the region where the dipole field points downward, the electric field decreases and where the dipole field points upward, the electric field increases. Therefore across the interface, there must be a discontinuity in the electric field.

Case 2   With Free Surface Charges

For this case, assume the interface supports a surface charge σ_{free} (charge per unit area). The volume integral of Gauss's law now provides (replacing Equation (3.3.2))

$$\nabla \cdot \vec{\mathscr{D}} = \rho = \sigma_{\text{free}} \delta(z) \quad \rightarrow \quad \int_V \nabla \cdot \vec{\mathscr{D}} \, dV = \int_{A_T} \sigma_{\text{free}} \, da \quad \rightarrow \quad \int_{A_T} \vec{\mathscr{D}} \cdot d\vec{a} = \int_{A_T} \sigma_{\text{free}} \, da$$

where $\delta(z)$ represents the Dirac delta function and $z = 0$ gives the position of the surface charge (the z-axis is perpendicular to the surface in Figure 3.3.1).

Following the remainder of the development in case 1, we find the result

$$\vec{\mathscr{D}}_2 = \vec{\mathscr{D}}_1 + \sigma_{\text{free}} \quad \rightarrow \quad \varepsilon_2 \vec{\mathscr{E}}_2 = \varepsilon_1 \vec{\mathscr{E}}_1 + \sigma_{\text{free}}$$

3.3.2   Electric Fields Parallel to the Surface

This topic asks whether or not *electric* fields parallel to a dielectric interface (without free charge or free currents) must be continuous or discontinuous (Figure 3.3.3). For this case, we use another of Maxwell's equations

$$\nabla \times \vec{\mathscr{E}} = -\frac{\partial \vec{\mathscr{B}}}{\partial t} \tag{3.3.6}$$

We assume there aren't any free currents or changing magnetization so that we can rewrite the Maxwell equation as

$$\nabla \times \vec{\mathscr{E}} = 0 \tag{3.3.7}$$

Integrating over the area of the loop and then converting the integral to a path integral produces

$$\int_A \nabla \times \vec{\mathscr{E}} \cdot d\vec{a} = 0 \quad \rightarrow \quad \oint \vec{\mathscr{E}} \cdot d\vec{s} = 0$$

where we have used the curl theorem (Stokes theorem). Note that the dot product must be zero for the left and right sides of the loop because the path makes a 90° angle with respect to the fields. The fields are parallel and antiparallel to the path directions on the top and bottom respectively. Assume the upper and lower paths have length L. The path integral can be expanded to write

$$0 = \oint \vec{\mathscr{E}} \cdot \vec{ds} = \underbrace{\mathscr{E}_2 L}_{\text{Top}} + \underbrace{-\mathscr{E}_1 L}_{\text{Bottom}} \quad \rightarrow \mathscr{E}_2 = \mathscr{E}_1 \tag{3.3.8}$$

Therefore we see that electric fields parallel to the interface must be continuous across the interface. If the EM wave *propagates* perpendicular to the interface (i.e., the wave vector $\vec{k}$ perpendicular to the interface with the electric fields tangent to the interface), the condition $\mathscr{E}_2 = \mathscr{E}_1$ must refer to not only the incident and transmitted fields but also to the reflected fields. That is, one of the fields must contain the incident and reflected fields (for example, $\vec{\mathscr{E}}_1 = \vec{\mathscr{E}}_{\text{inc}} + \vec{\mathscr{E}}_{\text{refl}}$) while the other contains the transmitted field (for example, $\vec{\mathscr{E}}_2 = \vec{\mathscr{E}}_{\text{trans}}$). We will use these equations to find the reflection coefficients between media.

Another point, in Section 3.5.3 we discuss the flow of power across a boundary. We find that the tangential electric field increases when it exits the dielectric. This seems to contradict the results in Equations (3.3.8). There are some points worth considering. (1) For Section 3.5.3, energy must be conserved so that the fields must grow when the wave exits. The dipoles in the dielectric store some of the energy while only the EM field stores the energy in the vacuum. (2) Any calculation of power must include both reflected and incident power except in the case of an antireflective (AR) coating. (3) The calculation for Equation (3.3.8) neglects changing magnetic fields. This is a good approximation because we would calculate the magnetic field through the total area bounded by the loop to be $\mathscr{B}A$. We can make the short sides of the loop (those that pass through the interface) infinitesimally small so that the area is $A \sim 0$.

3.3.3   The Boundary Conditions for the General Electric Field

For an electric field neither parallel nor perpendicular to a boundary, the field decomposes into the tangential and perpendicular pieces. The perpendicular piece can be discontinuous at the interface while the tangential piece must be continuous across the interface. In the final topic for the present section, we will write the general condition in a condensed form.

3.3.4   The Tangential Magnetic Field

Some authors state boundary conditions for waveguides using the magnetic fields instead of electric fields. Therefore for completeness, we will show how to find the boundary conditions for the magnetic field $\vec{\mathscr{H}}$.

We wish to find a relation between the magnetic field $\vec{\mathscr{H}}_1$ just below the interface and the magnetic field $\vec{\mathscr{H}}_2$ just above the interface. Let's assume interfacial current flows along the interface through the loop as shown in Figure 3.3.4. Notice the current $\vec{\mathscr{J}}$ will produce a curling magnetic field. We therefore expect the fields $\vec{\mathscr{H}}_1$ and $\vec{\mathscr{H}}_2$ below and above the interface, respectively, to differ. To see this, start with Maxwell's equation

$$\nabla \times \vec{\mathscr{H}} = \vec{\mathscr{J}} + \frac{\partial \vec{\mathscr{D}}}{\partial t} \tag{3.3.9}$$

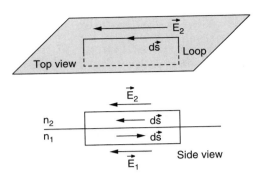

FIGURE 3.3.3
The geometry for fields parallel to the interface.

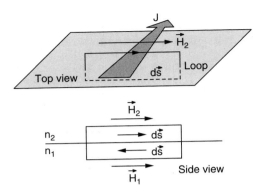

FIGURE 3.3.4
Geometry for the tangential magnetic fields.

Consider a small loop of area A as shown in Figure 3.3.4. When we integrate Equation (3.3.9) over the area enclosed by the loop, we will have a term of the form $\mathscr{D}A$. We can make the small sides of the loop (those passing through the interface) infinitesimally small so that $A=0$ and the displacement field doesn't make any contribution to the magnetic field. Notice that we also have a term $\mathscr{J}A$. We cannot neglect this term because we assume the current runs along the interface (without any depth into the material) and through the small loop of area A. Regardless of how small we make the small sides, we still enclose the current. We only need to consider Equation (3.3.9) in the form

$$\nabla \times \vec{\mathscr{H}} = \vec{\mathscr{J}} \tag{3.3.10}$$

Integrating the last equation over the area bounded by the loop and then passing to the line integral, we find

$$\int_A \nabla \times \vec{\mathscr{H}} \cdot d\vec{a} = \int_A \vec{\mathscr{J}} \cdot d\vec{a} = \int_0^L K\,ds \quad \rightarrow \quad \oint \vec{\mathscr{H}} \cdot d\vec{s} = \int_0^L K\,ds \tag{3.3.11}$$

where we made a new definition for the surface current $K=$ amps/length. The magnetic field is perpendicular to the loop along the short sides and therefore the integrals over these sides don't make any contribution. The path integral becomes

$$\oint \vec{\mathscr{H}} \cdot d\vec{s} = \int_0^L K\,ds \quad \rightarrow \quad \underbrace{\mathscr{H}_2 L}_{\text{top}} - \underbrace{\mathscr{H}_1 L}_{\text{bottom}} = KL \quad \rightarrow \quad \mathscr{H}_2 = \mathscr{H}_1 + K \tag{3.3.12}$$

The minus sign occurs for the bottom path because of the opposite direction of the magnetic field and the path.

Equation (3.3.12) shows that the tangential magnetic fields can be discontinuous provided there exists a surface current K. In the absence of a surface current, we see that the tangential fields must be continuous across the boundary.

3.3.5 Magnetic Field Perpendicular to the Interface (Without Magnetization)

The final boundary condition deals with a magnetic field perpendicular to a boundary. We assume for simplicity that the material is purely a dielectric so that the magnetization does not affect the fields

$$\nabla \cdot \vec{\mathscr{B}} = 0 \quad \rightarrow \quad \nabla \cdot \vec{\mathscr{H}} = 0$$

where we have assumed that the magnetic induction is linear in the magnetic field. We can construct a Gaussian volume as we did in Section 3.3.1 to see that the magnetic field perpendicular to the boundary must be continuous.

3.3.6 Arbitrary Magnetic Field

With surface current present, the tangential component of the magnetic field will be discontinuous while the perpendicular part will be continuous. Without surface currents, both fields will be continuous.

3.3.7 General Relations and Summary

We can summarize and generalize the expressions for the boundary conditions. The notation takes care of both the TE and TM cases. Maxwell's equations in integral form can be written as

$$\int \vec{\mathscr{D}} \cdot \tilde{n} \, da = \int \rho_{\text{free}} \, dV = \int \sigma_{\text{free}} \, dA$$

$$\int \vec{\mathscr{B}} \cdot \tilde{n} \, da = 0$$

$$\oint_C \vec{\mathscr{H}} \cdot d\vec{s} = \int \vec{\mathscr{J}} \cdot \tilde{n} \, da + \int \frac{\partial \vec{\mathscr{D}}}{\partial t} \cdot \tilde{n} \, da \qquad (3.3.13)$$

$$\oint_C \vec{\mathscr{E}} \cdot d\vec{s} = -\frac{\partial}{\partial t} \int \vec{\mathscr{B}} \cdot \tilde{n} \, da$$

where σ_{free} is the free surface charge (not related to the conductivity).

Figure 3.3.5 shows arbitrarily oriented surfaces, Gaussian boxes, and loops. The first integral can be evaluated over the small Gaussian box to give

$$\vec{\mathscr{D}}_2 \cdot \tilde{n}_2 + \vec{\mathscr{D}}_1 \cdot \tilde{n}_1 = \sigma_f$$

where the subscript "2" indicates the top of the box and "1" indicates the bottom.

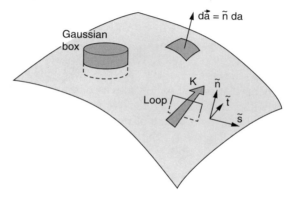

FIGURE 3.3.5
The small Gaussian box and loop necessary to evaluate volume and surface integrals. Side 1 = Bottom, Side 2 = Top.

Taking $\hat{n} = \hat{n}_2 = -\hat{n}_1$ we find

$$\left(\vec{\mathscr{D}}_2 - \vec{\mathscr{D}}_1\right) \cdot \tilde{n} = \sigma_f \quad \rightarrow \quad \left(\varepsilon_2 \vec{\mathscr{E}}_2 - \varepsilon_1 \vec{\mathscr{E}}_1\right) \cdot \tilde{n} = \sigma_f \tag{3.3.14}$$

The electric field perpendicular to a dielectric interface is discontinuous (even when there isn't any free surface charge).

Consider the fourth integral in Equations (3.3.13). Look at the loop in Figure 3.3.5. The area integral over the magnetic field can be made arbitrarily small just by shrinking the vertical sides of the loop (i.e., the sides that penetrate into the surface). We are left with

$$\oint \vec{E} \cdot d\vec{s} = 0$$

Evaluating the integral along the top and bottom of the loop provides

$$\left(\vec{\mathscr{E}}_2 - \vec{\mathscr{E}}_1\right) \cdot \tilde{s} = 0 \tag{3.3.15a}$$

for all directions $\tilde{s}$ along the surface. This equation can alternatively be written as

$$\left(\vec{\mathscr{E}}_2 - \vec{\mathscr{E}}_1\right) \cdot \tilde{t} = 0 \tag{3.3.15b}$$

where $\tilde{t}$ stands for a unit vector tangent to the surface.

Equations 3.3.15 say that the tangential component of the electric field must always be continuous across an interface. We have finished with the electric field. Let's move on to the magnetic field.

The second integral in Equations (3.3.13) can be evaluated over the small Gaussian box to give

$$\int \vec{\mathscr{B}} \cdot d\vec{a} = 0 \quad \rightarrow \quad \vec{\mathscr{B}}_2 \cdot \tilde{n} - \vec{\mathscr{B}}_1 \cdot \tilde{n} = 0 \quad \rightarrow \quad \left(\vec{\mathscr{B}}_2 - \vec{\mathscr{B}}_1\right) \cdot \tilde{n} = 0$$

For *nonmagnetic material*, this last result can be written as

$$\left(\vec{\mathscr{H}}_2 - \vec{\mathscr{H}}_1\right) \cdot \tilde{n} = 0 \tag{3.3.16}$$

This result says that the normal component of the magnetic field is continuous across an interface (for nonmagnetic materials). Finally, the third of the integral relations in Equations (3.3.13)

$$\oint_C \vec{\mathscr{H}} \cdot d\vec{s} = \int \vec{\mathscr{J}} \cdot \tilde{n}\, da + \int \frac{\partial \vec{\mathscr{D}}}{\partial t} \cdot \tilde{n}\, da$$

can be evaluated using a loop such as that in Figure 3.3.5. First note that we should consider surface currents rather than volume currents J. The reason is that we can shrink the vertical size of the loop and eliminate all current except that moving along the interface. Therefore, the integral can be written as

$$\oint_C \vec{\mathscr{H}} \cdot d\vec{s} = \int \vec{K} \cdot \tilde{t}\, ds + \int \frac{\partial \vec{\mathscr{D}}}{\partial t} \cdot \tilde{n}\, da$$

where ds is just the length of one of the long sides of the loop. The area integral over the displacement field can be taken as zero because we can shrink the vertical sides of the loop and make the area approach zero. Making the top and bottom sides small, we find

$$\left(\vec{\mathcal{H}}_2 - \vec{\mathcal{H}}_1\right) \cdot \tilde{s} = \vec{K} \cdot \tilde{t}$$

If there isn't any surface currents $K = 0$ we find for all directions $\hat{s}$

$$\left(\vec{\mathcal{H}}_2 - \vec{\mathcal{H}}_1\right) \cdot \tilde{s} = 0 \qquad (3.3.17a)$$

Given the lack of surface currents, we can also write this as

$$\left(\vec{\mathcal{H}}_2 - \vec{\mathcal{H}}_1\right) \cdot \tilde{t} = 0 \qquad (3.3.17b)$$

The tangential component of the magnetic field must be continuous across an interface without surface currents.

The following list summarizes the general boundary conditions without free surface charge or surface currents in the absence of magnetic media

$$\left(\varepsilon_2 \vec{\mathcal{E}}_2 - \varepsilon_1 \vec{\mathcal{E}}_1\right) \cdot \tilde{n} = 0$$

$$\left(\vec{\mathcal{E}}_2 - \vec{\mathcal{E}}_1\right) \cdot \tilde{t} = 0$$

$$\left(\vec{\mathcal{H}}_2 - \vec{\mathcal{H}}_1\right) \cdot \tilde{n} = 0$$

$$\left(\vec{\mathcal{H}}_2 - \vec{\mathcal{H}}_1\right) \cdot \tilde{t} = 0$$

3.4 Law of Reflection, Snell's Law and the Reflectivity

The present section uses the boundary conditions for Maxwell's equations to derive the law of reflection ($\theta_i = \theta_r$), Snell's law ($n_1 \sin \theta_1 = n_2 \sin \theta_2$), and expressions for the Fresnel reflection and transmission coefficients.

3.4.1 The Boundary Conditions

The previous section shows that the boundary conditions for Maxwell's equation can be written as

$$\left(\varepsilon_2 \vec{\mathcal{E}}_2 - \varepsilon_1 \vec{\mathcal{E}}_1\right) \cdot \tilde{n} = 0 \qquad (3.4.1a)$$

$$\left(\vec{\mathcal{E}}_2 - \vec{\mathcal{E}}_1\right) \cdot \tilde{t} = 0 \qquad (3.4.1b)$$

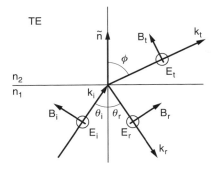

FIGURE 3.4.1
Transverse electric fields point into the page. i = incident, r = reflected, and t = transmitted.

$$\left(\vec{\mathscr{H}}_2 - \vec{\mathscr{H}}_1 \right) \cdot \tilde{n} = 0 \tag{3.4.1c}$$

$$\left(\vec{\mathscr{H}}_2 - \vec{\mathscr{H}}_1 \right) \cdot \tilde{t} = 0 \tag{3.4.1d}$$

where $\tilde{n}$ and $\tilde{t}$ are unit vectors perpendicular and parallel to the interface, respectively. We must keep in mind that the fields in these equations represent the *total* field on either side of the boundary. Assume region 2 refers to the topside of the interface while region 1 refers to the bottom side. Figure 3.4.1 shows an example of transverse electric fields. Notice, right next to the interface on the bottom side, there exists two electric fields. We make the following definitions

$$\vec{\mathscr{E}}_1 = \vec{\mathscr{E}}_i + \vec{\mathscr{E}}_r \qquad \vec{\mathscr{E}}_2 = \vec{\mathscr{E}}_t \qquad \vec{\mathscr{H}}_1 = \vec{\mathscr{H}}_i + \vec{\mathscr{H}}_r \qquad \vec{\mathscr{H}}_2 = \vec{\mathscr{H}}_r$$

The subscripts i, r, and t refer to incident, reflected, and transmitted, respectively. The figure shows the electric fields parallel to the interface and pointing into the plane of the page.

With the definitions for the electric and magnetic fields, we can now rewrite the first two boundary conditions in a form suitable for optical activity at an interface.

$$\left[\varepsilon_2 \vec{\mathscr{E}}_t - \varepsilon_1 \left(\vec{\mathscr{E}}_i + \vec{\mathscr{E}}_r \right) \right] \cdot \tilde{n} = 0 \tag{3.4.2a}$$

$$\left[\vec{\mathscr{E}}_t - \left(\vec{\mathscr{E}}_i + \vec{\mathscr{E}}_r \right) \right] \times \tilde{n} = 0 \tag{3.4.2b}$$

Notice that we converted the "$\cdot\tilde{t}$" into "$\times\tilde{n}$" (refer to Figure 3.3.5 in Section 3.3), and noting "$\cdot\tilde{t}$" gives the component parallel to the interface as does "$\times\tilde{n}$" because $\tilde{n}$ is perpendicular to the interface (use the right-hand rule).

We would like to restate the last two boundary conditions in Equations (3.4) in terms of the electric field for convenience. We will need the relation between the magnitude of the magnetic and electric fields in a dielectric

$$\mathscr{H} = \frac{\mathscr{E}}{\mu_0 v_g}$$

We only need to include the direction of the fields. The cross-product vector $\vec{\mathscr{E}} \times \vec{\mathscr{H}}$ points in the same direction as the wave vector $\vec{k}$ as will be further discussed in connection with the Poynting vector in Section 3.5. Therefore we can write

$$\vec{\mathscr{H}} = \tilde{k} \times \frac{\vec{\mathscr{E}}}{\mu_o v_g} = \frac{\vec{k}_n}{k_n} \times \frac{\vec{\mathscr{E}}}{\mu_o v_g} = \frac{\vec{k}_n}{k_o n} \times \frac{\vec{\mathscr{E}}}{\mu_o v_g} = \frac{\vec{k}_n \times \vec{\mathscr{E}}}{k_o \mu_o c}$$

where k_n is the wave vector in a dielectric with refractive index n, k_o is the wave vector in vacuum, v_g is the speed of light in the dielectric, and c is the speed of light in vacuum.

Let's, drop the subscript "n" for simplicity.

$$\vec{\mathscr{H}} = \frac{\vec{k} \times \vec{\mathscr{E}}}{k_o \mu_o c}$$

This last relation can be used to rewrite the remaining two boundary conditions in Equations (3.4.1). We find

$$\left(\vec{k}_t \times \vec{\mathscr{E}}_t - \vec{k}_i \times \vec{\mathscr{E}}_i - \vec{k}_r \times \vec{\mathscr{E}}_r \right) \cdot \tilde{n} = 0 \tag{3.4.2c}$$

$$\left(\vec{k}_t \times \vec{\mathscr{E}}_t - \vec{k}_i \times \vec{\mathscr{E}}_i - \vec{k}_r \times \vec{\mathscr{E}}_r \right) \times \tilde{n} = 0 \tag{3.4.2d}$$

Let's group the boundary conditions together to write

$$\left[\varepsilon_2 \vec{\mathscr{E}}_t - \varepsilon_1 \left(\vec{\mathscr{E}}_i + \vec{\mathscr{E}}_r \right) \right] \cdot \tilde{n} = 0 \tag{3.4.2a}$$

$$\left[\vec{\mathscr{E}}_t - \left(\vec{\mathscr{E}}_i + \vec{\mathscr{E}}_r \right) \right] \times \tilde{n} = 0 \tag{3.4.2b}$$

$$\left(\vec{k}_t \times \vec{\mathscr{E}}_t - \vec{k}_i \times \vec{\mathscr{E}}_i - \vec{k}_r \times \vec{\mathscr{E}}_r \right) \cdot \tilde{n} = 0 \tag{3.4.2c}$$

$$\left(\vec{k}_t \times \vec{\mathscr{E}}_t - \vec{k}_i \times \vec{\mathscr{E}}_i - \vec{k}_r \times \vec{\mathscr{E}}_r \right) \times \tilde{n} = 0 \tag{3.4.2d}$$

3.4.2 The Law of Reflection

Now we show that the angle of incidence equals the angle of reflection. Using the plane wave form of the electric field

$$\vec{\mathscr{E}} = \vec{\mathscr{E}}_o \, e^{i\vec{k}\cdot\vec{r} - i\omega t} \tag{3.4.3}$$

the boundary conditions give equations of the form (use Equation (3.4.2a) for example)

$$\varepsilon_2 \vec{\mathscr{E}}_{ot} \cdot \tilde{n} \, e^{i\vec{k}_t\cdot\vec{r} - i\omega t} = \varepsilon_1 \vec{\mathscr{E}}_{oi} \cdot \tilde{n} \, e^{i\vec{k}_i\cdot\vec{r} - i\omega t} + \varepsilon_1 \vec{\mathscr{E}}_{or} \cdot \tilde{n} \, e^{i\vec{k}_r\cdot\vec{r} - i\omega t} \tag{3.4.4a}$$

The driving frequency is the same on both sides of the interface.

$$\varepsilon_2 \vec{\mathscr{E}}_{ot} \cdot \tilde{n} \; e^{i\vec{k}_t \cdot \vec{r}} = \varepsilon_1 \vec{\mathscr{E}}_{oi} \cdot \tilde{n} \; e^{i\vec{k}_i \cdot \vec{r}} + \varepsilon_1 \vec{\mathscr{E}}_{or} \cdot \tilde{n} \; e^{i\vec{k}_r \cdot \vec{r}} \qquad (3.4.4b)$$

We assume that the interface is at $z = 0$. As $\vec{r}$ varies along the interface, only the exponentials change. Therefore to keep the equality in this last equation, we must require

$$e^{i\vec{k}_t \cdot \vec{r}} = e^{i\vec{k}_i \cdot \vec{r}} = e^{i\vec{k}_r \cdot \vec{r}} \qquad (3.4.5)$$

We can also see this by recognizing the exponentials form a basis set (refer to Chapter 2 in Volume 1); consequently Equation (3.4.4b) must have the same basis vector in each term or else the coefficient of each term would need to be zero. This last equation can only hold so long as

$$i\vec{k}_t \cdot \vec{r} = i\vec{k}_i \cdot \vec{r} = i\vec{k}_r \cdot \vec{r} \quad \text{for } z = 0 \quad (\text{i.e., } \vec{r} \text{ is confined to } x{-}y \text{ plane}) \qquad (3.4.6)$$

The dot product gives the projection of $\vec{k}$ onto the $x{-}y$ plane. Figure 3.4.1 tells us to use the sine of the indicated angles to $\vec{k}$ onto the $x{-}y$ plane.

$$\vec{k}_t \cdot \vec{r} = \vec{k}_i \cdot \vec{r} = \vec{k}_r \cdot \vec{r} \;\; \rightarrow \;\; k_t r \sin \phi = k_i r \sin \theta_i = k_r r \sin \theta_r$$

$$\rightarrow \;\; k_t \sin \phi = k_i \sin \theta_i = k_r \sin \theta_r .$$

The term

$$k_i \sin \theta_i = k_r \sin \theta_r$$

can be rewritten by noting that $k_i = k_o n_1 = k_r$ so that we must have

$$\theta_i = \theta_r \quad (\text{Law of Reflection})$$

The term

$$k_t \sin \phi = k_i \sin \theta_i$$

can be rewritten by noting that $k_i = k_o n_1$ and that $k_t = k_o n_2$ to get

$$n_2 \sin \phi = n_1 \sin \theta_i \quad (\text{Snell's law})$$

3.4.3   Fresnel Reflectivity and Transmissivity for TE Fields

Now we derive the Fresnel reflectivity and transmissivity from the boundary conditions. The Fresnel reflectivity and transmissivity apply to electric fields rather than power. We want to write the reflected field in terms of the incident field and then the transmitted field in terms of the incident field. We have three variables but only need to solve for two in terms of the third (the incident field). We therefore require two equations in the three variables.

As shown in Figure 3.4.1, the electric field points into the page in a direction perpendicular to the unit vector $\tilde{n}$. Therefore, the first two boundary conditions in Equations

(3.4.2a) and (3.4.2c) don't provide any useful information. However Equations (3.4.2b) and (3.4.2d) provide the reflectivity and transmissivity.

$$\left[\vec{\mathcal{E}}_t - \left(\vec{\mathcal{E}}_i + \vec{\mathcal{E}}_r \right) \right] \times \tilde{n} = 0 \tag{3.4.2b}$$

$$\left(\vec{k}_t \times \vec{\mathcal{E}}_t - \vec{k}_i \times \vec{\mathcal{E}}_i - \vec{k}_r \times \vec{\mathcal{E}}_r \right) \times \tilde{n} = 0 \tag{3.4.2d}$$

For Equation (3.4.2b), notice that the fields are all perpendicular to the unit vector so that we must require

$$\mathcal{E}_t - (\mathcal{E}_i + \mathcal{E}_r) = 0 \tag{3.4.7}$$

For Equation (3.4.2d), we must manipulate the terms a little bit. Look at the first term and use the BAC–CAB rule to evaluate the triple cross product.

$$\vec{k}_t \times \vec{\mathcal{E}}_t \times \tilde{n} = \vec{\mathcal{E}}_t \left(\vec{k}_t \cdot \tilde{n} \right) - \tilde{n} \left(\vec{k}_t \cdot \vec{\mathcal{E}}_t \right) = \vec{\mathcal{E}}_t \left(\vec{k}_t \cdot \tilde{n} \right) \tag{3.4.8a}$$

The electric field is everywhere perpendicular to the wave vector so the second term gives zero.

$$\vec{k}_t \times \vec{\mathcal{E}}_t \times \tilde{n} = \vec{\mathcal{E}}_t \left(\vec{k}_t \cdot \tilde{n} \right) \tag{3.4.8b}$$

Projecting the wave vector onto the unit vector requires the cosine. We find

$$\vec{k}_t \times \vec{\mathcal{E}}_t \times \tilde{n} = \vec{\mathcal{E}}_t \left(\vec{k}_t \cdot \tilde{n} \right) = \vec{\mathcal{E}}_t k_t \cos \phi = \vec{\mathcal{E}}_t k_o n_2 \cos \phi$$

The other two terms in Equation (3.4.2d) work the same way. Equation (3.4.2d) can now be written as

$$\mathcal{E}_t k_o n_2 \cos \phi - \mathcal{E}_i k_o n_1 \cos \theta + \mathcal{E}_r k_o n_1 \cos \theta = 0 \tag{3.4.9}$$

where we have taken the magnitude of the fields. Notice the sign of the last term has been changed because the reflected k-vector makes an angle of $\pi - \theta$ with respect to the vertically pointing unit vector.

We now have two equations to solve,

$$\mathcal{E}_t - (\mathcal{E}_i + \mathcal{E}_r) = 0 \tag{3.4.10a}$$

$$\mathcal{E}_t n_2 \cos \phi - (\mathcal{E}_i - \mathcal{E}_r) n_1 \cos \theta = 0 \tag{3.4.10b}$$

Solving the first equation for the transmitted field and substituting into the second one provides

$$r_{\text{TE}} = \frac{\mathcal{E}_r}{\mathcal{E}_i} = \frac{n_1 \cos \theta - n_2 \cos \phi}{n_1 \cos \theta + n_2 \cos \phi} \tag{3.4.11a}$$

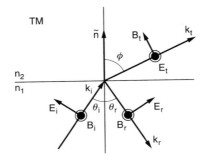

FIGURE 3.4.2
Definitions for the TM fields.

The ϕ angles can be eliminated by using Snell's law. Solving the second equation for the reflected field and substituting gives

$$t_{\text{TE}} = \frac{\mathscr{E}_t}{\mathscr{E}_i} = \frac{2n_1 \cos\theta}{n_1 \cos\theta + n_2 \cos\phi} \tag{3.4.11b}$$

3.4.4 The TM Fields

We can perform the same analysis using the TM fields depicted in Figure 3.4.2. This time we will need Equations (3.4.2a,c). You can show

$$r_{\text{TM}} = \frac{E_r}{E_i} = \frac{n_1 \cos\phi - n_2 \cos\theta}{n_2 \cos\theta + n_1 \cos\phi} \tag{3.4.12a}$$

$$t_{\text{TM}} = \frac{E_t}{E_i} = \frac{n_1}{n_2}(1 + r_{\text{TM}}) \tag{3.4.12b}$$

3.4.5 Graph of the Reflectivity Versus Angle

The relations in Equations (3.4.11a) and (3.4.12a) appear in Figure 3.4.3 for glass with refractive index of 1.5. Notice how the TM reflectivity becomes zero near 60° (the Brewster angle).

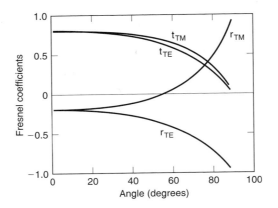

FIGURE 3.4.3
The reflectivity and transmissivity versus the angle of incidence for $n_1 = 1$ and $n_2 = 1.5$.

3.5   The Poynting Vector

As systems, light emitters, and detectors produce and receive electromagnetic power. The present section provides a basic understanding of electromagnetic power flow and the mechanisms for storing electromagnetic energy in a material. The Poynting vector describes the energy flow from the material. The relation between power flow and the change in stored energy can be described by an equation of continuity. We first discuss the calculation of energy flow for real electromagnetic fields and then indicate how the mathematical description changes for complex notation. Two examples show how power flows across an interface with an antireflective coating.

3.5.1   Introduction to Power Transport for Real Fields

The Poynting vector $\vec{S} = \vec{E} \times \vec{H}$ describes the power (per unit surface area) carried by an electromagnetic wave. The energy flow includes the fields and polarization. Figure 3.5.1 shows the cross-sectional area of a waveguide. The magnitude of the Poynting vector gives the power (as a function of time) flowing through each unit area. In general, the Poynting vector has two sources of time dependence: one at the optical frequency and another due to an impressed modulation. For a steady-state source, we only need to consider the optical carrier varying at the optical frequency. The next example shows that the Poynting vector has this sinusoidal time dependence.

We can easily calculate the Poynting vector for a plane wave. The magnitude provides units of Watts/area and its direction parallels the propagation vector. Suppose the fields are given by

$$\vec{\mathscr{E}}(z, t) = E_o \sin(k_o z - \omega t)\tilde{x} \qquad \vec{\mathscr{H}}(z, t) = H_o \sin(k_o z - \omega t)\tilde{y} \tag{3.5.1}$$

Then the Poynting vector must be

$$\vec{\mathscr{S}} = \vec{\mathscr{E}} \times \vec{\mathscr{H}} = \tilde{z} E_o H_o \sin^2(k_o z - \omega t) \tag{3.5.2}$$

where S is a script S.

Notice that the Poynting vector $\vec{S}$ points in the $\tilde{z}$-direction.

The result of the basic calculation for the Poynting vector in Equation (3.5.2) shows that the power fluctuates with angular frequency ω on the order of 10^{16}. Figure 3.5.2 shows the rapid fluctuation versus time for a specific point z. Equation (3.5.2) is a

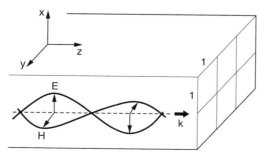

FIGURE 3.5.1
The Poynting vector gives the instantaneous power flowing through a surface.

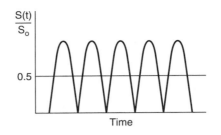

FIGURE 3.5.2
The Poynting vector vs. time and the average over a half cycle.

perfectly fine expression and represents the power versus time. Often people take an average over an optical cycle: measurement equipment does not detect such fast variations (neglecting interference effects between fields). Any modulation sometimes included in the coefficient $E_o H_o$ (such as modulating a laser for optical communications) is orders of magnitude slower than the optical variation. As a result, averaging over a single cycle does not affect most calculations. If time coherence affects are important, the fields must be first added and then the power calculated from the Poynting vector. And then a single cycle average can be made. We will see that the complex version of the Poynting vector automatically includes the averaging procedure.

The averaging procedure can be most easily seen using Equation (3.5.2) and Figure 3.5.1. Assume that the surface is located at $z = z_o$ and that the wave has uniform magnitude across the surface of area A (i.e., a plane wave). The instantaneous power $P(z, t)$ (Watts through the surface) depends on both the position z and time t

$$P(z, t) = \vec{S} \cdot \vec{A} = \vec{S} \cdot A\tilde{z}$$

where $\vec{A} = A\tilde{z}$ represents a vector that points out of the volume having the side with area $|\vec{A}| = A$. For us the area vector $\vec{A} = A\tilde{z}$ points out of the volume toward the right. Substituting the fields provides the power

$$P(z_o, t) = \vec{S} \cdot A\tilde{z} = AE_oH_o \, \mathrm{Sin}^2(k_oz_o - \omega t)$$

The average power (averaged over a cycle) becomes

$$\left\langle \vec{P}(z, t) \right\rangle_t = \frac{1}{T} \int_0^T dt \, \vec{P}(z, t)$$

where T represents the period of the wave given by $T = 2\pi/\omega$.

The average produces the results

$$\left\langle P(z, t) \right\rangle_t = \tfrac{1}{2} AE_oH_o$$

This last calculation gives the average power transmitted through a surface by a plane wave. The Poynting vector can still depend on time, the amplitudes E_o and H_o depend on time with frequency less than $100\,\mathrm{GHz}$.

We formally calculate the optical power that flows into a volume. Let P_{in} and P_{out} be the power flowing "into" and "out of" the volume (Figure 3.5.3) with $P_{\mathrm{in}} = -P_{\mathrm{out}}$. The power flowing into the volume must be related to the power leaving the volume

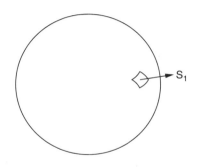

FIGURE 3.5.3
Power flows through a patch of area *da*.

across a patch of differential area $d\vec{a}$ in accordance with $P_{in} = -P_{out}$. The total power leaving the surface must be the sum of the power through each patch

$$P = \sum_{\text{surface}} \vec{S}(\vec{r}, t) \cdot d\vec{a}_{\vec{r}} = \int_{\text{surface}} \vec{S}(\vec{r}, t) \cdot d\vec{a} \qquad (3.5.3)$$

Example 3.5.1

If the average Poynting vector is $\vec{S} = \tilde{r}/r^2$, what is the average power flowing through a sphere of radius R_o centered at $z = 0$? Refer to Figure 3.5.4.

Solution: For a sphere, the patch of area is $d\vec{a} = \hat{r}\, da$, where the magnitude da can be written as

$$da = r^2 \sin\theta \, d\theta \, d\varphi$$

The power must be given by

$$P = \int \vec{S} \cdot d\vec{a} = \int\int \frac{\tilde{r}}{R_o^2} \cdot \hat{r} R_o^2 \sin\theta \, d\theta \, d\varphi = \int_{\varphi=0}^{2\pi} \int_{\theta=0}^{\pi} \sin\theta \, d\theta \, d\varphi = 4\pi$$

3.5.2 Power Transport and Energy Storage Mechanisms

Now we calculate how the power leaving a volume affects the amount of energy stored within the volume. Electromagnetic energy can be stored by a number of mechanisms.

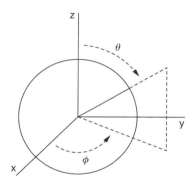

FIGURE 3.5.4
Spherical coordinates.

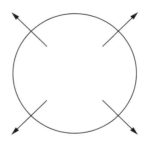

FIGURE 3.5.5
Energy diverges away from the volume.

First the electric and magnetic fields store energy. Second, electric and magnetic dipoles store energy as previously discussed in the first section of this chapter. Third, the interaction between currents and fields can generate energy. Often we associate this third type as Joule heating but it can also be related to the production of electromagnetic energy. We want to relate the power leaving the volume to the change in energy stored by these various mechanisms. Therefore we will need to calculate the Poynting vector by using Maxwell's equations and the constituent relations.

The power leaving a volume as described by Equation (3.5.3) can be written using the divergence theorem (Figure 3.5.5)

$$\int_{\text{volume}} \nabla \cdot \vec{S}\, dV = \int_{\text{surface}} \vec{S} \cdot d\vec{a} = P_{\text{out}} \tag{3.5.4}$$

As previously mentioned, the energy within the volume must be stored as a combination of electromagnetic fields, currents $\vec{\mathscr{J}}$ (current density), polarization $\vec{\mathscr{P}}$, and magnetization $\vec{M}$. In order to calculate the power flowing out through the surface in Equation (3.5.4), we first calculate the divergence using a vector identity

$$\nabla \cdot \vec{S} = \nabla \cdot \left(\vec{\mathscr{E}} \times \vec{\mathscr{H}} \right) = \vec{\mathscr{H}} \cdot \nabla \times \vec{\mathscr{E}} - \vec{\mathscr{E}} \cdot \nabla \times \vec{\mathscr{H}} \tag{3.5.5}$$

We can substitute for the two curls appearing in Equation (3.5.5) by using Maxwell's equations

$$\left\{ \begin{array}{l} \nabla \times \vec{\mathscr{H}} = \vec{\mathscr{J}} + \frac{\partial \vec{\mathscr{D}}}{\partial t} \\[2mm] \nabla \times \vec{\mathscr{E}} = -\frac{\partial \vec{\mathscr{B}}}{\partial t} \end{array} \right\} \quad \text{with} \quad \left\{ \begin{array}{l} \vec{\mathscr{D}} = \varepsilon_o \vec{\mathscr{E}} + \vec{\mathscr{P}} \\[2mm] \vec{\mathscr{B}} = \mu_o \vec{\mathscr{H}} + \mu_o \vec{M} \end{array} \right\} \tag{3.5.6}$$

The fields $\vec{\mathscr{E}}$ and $\vec{\mathscr{H}}$ exist inside the medium. Substituting Equations (3.5.6) into Equations (3.5.5) provides

$$\nabla \cdot \vec{S} = \vec{\mathscr{H}} \cdot \left(\nabla \times \vec{\mathscr{E}} \right) - \vec{\mathscr{E}} \cdot \nabla \times \vec{\mathscr{H}} = \vec{\mathscr{H}} \cdot \left(-\frac{\partial \vec{\mathscr{B}}}{\partial t} \right) - \vec{\mathscr{E}} \cdot \left(\vec{J} + \frac{\partial \vec{\mathscr{D}}}{\partial t} \right) \tag{3.5.7}$$

Using the constituent relations in Equation 3.5.6, we find

$$\nabla \cdot \vec{S} = -\vec{\mathscr{H}} \cdot \frac{\partial}{\partial t} \left(\mu_o \vec{\mathscr{H}} + \mu_o \vec{M} \right) - \vec{\mathscr{E}} \cdot \vec{\mathscr{J}} - \vec{\mathscr{E}} \cdot \frac{\partial}{\partial t} \left(\varepsilon_o \vec{\mathscr{E}} + \vec{\mathscr{P}} \right) \tag{3.5.8}$$

Multiply by –1 and observe that rewriting the derivative as

$$\vec{\mathcal{H}} \cdot \frac{\partial \vec{\mathcal{H}}}{\partial t} = \frac{\partial}{\partial t} \frac{\vec{\mathcal{H}} \cdot \vec{\mathcal{H}}}{2} \quad \text{and} \quad \vec{\mathcal{E}} \cdot \frac{\partial \vec{\mathcal{E}}}{\partial t} = \frac{\partial}{\partial t} \frac{\vec{\mathcal{E}} \cdot \vec{\mathcal{E}}}{2}$$

changes Equation (3.5.8) into

$$-\nabla \cdot \vec{S} = \frac{\partial}{\partial t} \left(\frac{\mu_0}{2} \mathcal{H}^2 + \frac{\varepsilon_0}{2} \mathcal{E}^2 \right) + \vec{\mathcal{E}} \cdot \vec{\mathcal{J}} + \mu_0 \vec{\mathcal{H}} \cdot \frac{\partial \vec{\mathcal{M}}}{\partial t} + \vec{\mathcal{E}} \cdot \frac{\partial \vec{P}}{\partial t} \tag{3.5.9}$$

From Equation (3.5.4), the power leaving the volume must be

$$P_{\text{out}} = \int \nabla \cdot \vec{S} \, dV$$

and therefore the power flowing into the volume must be

$$P_{\text{in}} = -\int \nabla \cdot \vec{S} \, dV \tag{3.5.10}$$

Substituting for the divergence of the Poynting vector gives

$$P_{\text{in}} = \underbrace{\int \vec{\mathcal{E}} \cdot \vec{\mathcal{J}} \, dV}_{\text{damping}} + \underbrace{\frac{\partial}{\partial t} \int dV \left(\frac{\mu_0}{2} \mathcal{H}^2 + \frac{\varepsilon_0}{2} \mathcal{E}^2 \right)}_{\text{field energy}} + \underbrace{\int dV \vec{\mathcal{E}} \cdot \frac{\partial \vec{\mathcal{P}}}{\partial t}}_{\text{polarization energy}} + \underbrace{\int dV \, \mu_0 \vec{\mathcal{H}} \cdot \frac{\partial \vec{\mathcal{M}}}{\partial t}}_{\text{magnetization energy}} \tag{3.5.11}$$

Now examine the four individual terms. The damping term describes how a conductive material absorbs the electromagnetic field to produce currents. Recall Ohm's law $\vec{\mathcal{J}} = \sigma \vec{\mathcal{E}}$. Alternatively, the first term in Equation (3.5.11) can be viewed as describing the electromagnetic power produced by the currents. The "field term" describes how the energy can be stored as electromagnetic fields inside the volume. For power flowing into the volume, the electromagnetic *energy density*

$$\rho_{\text{em}} = \frac{\mu_0}{2} \mathcal{H}^2 + \frac{\varepsilon_0}{2} \mathcal{E}^2 \tag{3.5.12}$$

must increase. The polarization power shows that energy flowing into the volume can appear as an increase in the polarization of the medium; i.e., the number or strength of the dipoles can increase. Similarly power flowing into the volume can increase the strength or number of the magnetic dipoles. Notice that Equation (3.5.11) does not include $\mathcal{B}$ because the equation identifies each individual influence on the power flow whereas $\mathcal{B}$ includes the fields due to $\mathcal{J}$ and due to $\mathcal{M}$. Using both $\mathcal{B}$ and $\mathcal{M}$ would double count the energy stored as magnetization $\mathcal{M}$.

As a note, Equation 3.5.11 can be used to show the equation of continuity for the electromagnetic field namely $\partial_t \rho_{\text{em}} + \nabla \cdot \vec{S} = -\vec{\mathcal{J}} \cdot \vec{\mathcal{E}}$. Reconsider Equation 3.5.11 without the integrals in order to work with energy density (energy per volume). We want all of the terms to have the form $\partial \rho_{\text{em}} / \partial t$. Only two terms need to be changed. We assume a linear, homogeneous, isotropic medium. First consider the polarization term

$$\vec{\mathcal{E}} \cdot \frac{\partial \vec{\mathcal{P}}}{\partial t} = \vec{\mathcal{E}} \cdot \frac{\partial}{\partial t} \left(\varepsilon_0 \chi \vec{\mathcal{E}} \right) = \frac{\varepsilon_0 \chi}{2} \frac{\partial}{\partial t} \mathcal{E}^2$$

Similarly, the magnetization term becomes

$$\mu_o \vec{\mathcal{H}} \cdot \frac{\partial \vec{\mathcal{M}}}{\partial t} = \mu_o \beta \vec{\mathcal{H}} \cdot \frac{\partial \vec{\mathcal{H}}}{\partial t} = \frac{\mu_o \beta}{2} \frac{\partial \mathcal{H}^2}{\partial t}$$

We therefore see that the density ρ_{em} has the form

$$\rho_{\text{em}} = \left(\frac{\mu_o}{2} \mathcal{H}^2 + \frac{\varepsilon_o}{2} \mathcal{E}^2 \right) + \left(\frac{\varepsilon_o \chi}{2} \mathcal{E}^2 \right) + \left(\frac{\mu_o \beta}{2} \mathcal{H}^2 \right)$$

as required.

3.5.3 Poynting Vector for Complex Fields

The Poynting vector for complex fields can be defined as

$$\vec{S} = \tfrac{1}{2} \vec{\mathcal{E}} \times \vec{\mathcal{H}}^* \tag{3.5.13}$$

which already has the average over time (for the optical frequencies). Notice the complex conjugate on one of the fields ($\vec{\mathcal{H}}$ in this case) and the $\frac{1}{2}$ out front. The Poynting vector $\vec{S}$ refers to the real quantity "power." How did we arrive at Equation (3.5.13)? To make the Poynting vector $\vec{S}$ real, we need a quantity such as $\mathcal{E}\mathcal{E}^* = |\mathcal{E}|^2$. For the plane wave $\mathcal{E} \sim e^{ikz - i\omega t}$ we see that the time dependence of $\mathcal{E}\mathcal{E}^* = |\mathcal{E}|^2$ disappears. The most reasonable thing is to multiply by the factor of $\frac{1}{2}$ and realize that the Poynting vector for complex fields gives the average power flow. The amplitudes might still depend on time by impressing a slower modulation.

Example 3.5.2

For the plane waves given in Section 3.5.1, specifically

$$\vec{\mathcal{E}}(z,t) = E_o \sin(k_o z - \omega t)\tilde{x} \qquad \vec{\mathcal{H}}(z,t) = H_o \sin(k_o z - \omega t)\tilde{y}$$

calculate $\vec{S}$ using complex fields.

Solution: The fields can be written in complex notation as

$$\vec{\mathcal{E}} = E_o e^{ikz - i\omega t} e^{i\varphi} \tilde{x} \quad \text{and} \quad \vec{\mathcal{H}} = H_o e^{ikz - i\omega t} e^{i\varphi} \tilde{y}$$

Notice the ease of including an extra phase factor $e^{i\varphi}$. Therefore the Poynting vector can be written

$$\vec{S} = \tfrac{1}{2} \vec{\mathcal{E}} \times \vec{\mathcal{H}}^* = \tfrac{1}{2} \tilde{z} \, E_o \, e^{ikz - i\omega t} \, e^{i\varphi} H_o^* \, e^{-ikz + i\omega t} \, e^{-i\varphi} = \tilde{z} \, \tfrac{1}{2} E_o H_o^*$$

This last expression identically agrees with the average Poynting vector for real fields as can be seen as follows.

$$\left\langle \vec{S} \right\rangle = \frac{1}{T} \int_0^T dt \, \vec{\mathcal{E}} \times \vec{\mathcal{H}} = \frac{1}{T} \int_0^T dt \, \tilde{z} \, E_o \sin(kz - \omega t) \, H_o \sin(kz - \omega t) = \tfrac{1}{2} \tilde{z} \, E_o H_o$$

3.5.4   Power Flow Across a Boundary

The present topic clearly demonstrates the role of polarization for storing energy. We consider two examples of an electromagnetic wave initially traveling in a dielectric and passing through a surface into vacuum. The previous section demonstrates that reflections occur at an interface separating two dissimilar optical materials. Consider Figure 3.5.6. The total field $\vec{E}_1$ inside the dielectric consists of an incident field E_i moving toward the right and a reflected field E_r moving toward the left so that $\vec{E}_1 = \vec{E}_i + \vec{E}_r$. The field E_2 consists of a transmitted field E_o moving toward the right. The examples considered in the present section neglect the reflected field. The situation corresponds to an interface with an antireflection coating as would be appropriate for laser amplifiers or for good waveguide–air coupling.

As a first example, we wish to find the electric field inside and outside a dielectric (neglecting reflections) by using the Poynting vectors. Subscript the fields inside the dielectric with a "i" while those outside of the dielectric subscript with "o." See Figure 3.5.6.

The Poynting vectors give the power flowing inside and outside the dielectric

$$\vec{S}_i = \tfrac{1}{2}\vec{\mathscr{E}}_i \times \vec{\mathscr{H}}_i^* \quad \text{and} \quad \vec{S}_o = \tfrac{1}{2}\vec{\mathscr{E}}_o \times \vec{\mathscr{H}}_o^* \tag{3.5.14}$$

as given by Equation (3.5.13). We assume steady-state conditions so that the energy lost from within the dielectric must appear on the outside, that is $\vec{S}_i = \vec{S}_o$ (neglecting any variation with area). Next using $\mathscr{B} = \mu_o \mathscr{H}$ (no magnetization) and Equation (3.1.19), namely $\mathscr{B} = \mathscr{E}/v_g$ where v_g represents the speed of light in the material, we find that Equation (3.5.14) becomes (for $\vec{S}_i = \vec{S}_o$)

$$\vec{\mathscr{E}}_i \times \vec{\mathscr{H}}_i^* = \vec{\mathscr{E}}_o \times \vec{\mathscr{H}}_o^* \quad \rightarrow \quad \mathscr{E}_i\!\left(\frac{\mathscr{E}_i^*}{\mu_o v_g}\right) = \mathscr{E}_o\!\left(\frac{\mathscr{E}_o^*}{\mu_o v_g}\right) \quad \rightarrow \quad |\mathscr{E}_o|^2 = \frac{c}{v_g}|\mathscr{E}_i|^2$$

Using the definition of the real index of refraction n, namely $v_g = c/n$, and taking the square root we find

$$|\mathscr{E}_o| = \sqrt{n}\,|\mathscr{E}_i|$$

This last expression says that the electric field grows as it leaves the dielectric. Although the expression correctly states the relations between the fields, if we assume the power to be proportional to the square of the field, then we find that the power stored in the electromagnetic fields grows according to $P_o = nP_i$. Apparently energy has been created! This cannot be. The power must include the energy stored in the polarization of the medium. The electric field increases once it leaves the dielectric since all of the energy must be stored in the fields in vacuum whereas only a fraction of the energy is stored in the fields in the dielectric.

The second example calculates an identical result explicitly using the energy conservation equations considered at the beginning of this section. Figure 3.5.7 shows optical

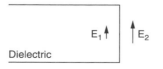

FIGURE 3.5.6
Electromagnetic wave travels to the right. Field outside dielectric is larger than inside.

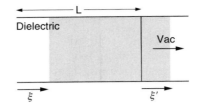

FIGURE 3.5.7
Optical energy travels to the right across a dielectric–vacuum interface.

energy in a volume $V = AL$ with length L and cross-sectional area A. The volume travels from a dielectric into vacuum. These plane waves have penetrated a distance ξ' past the interface while moving a distance ξ within the dielectric. The two distances ξ, ξ' differ because the speed of light in the two media differs. Equation (3.5.11) gives the rate of energy leaving a volume (for negligible magnetization and currents)

$$P = \frac{\partial}{\partial t} \int dV \left(\frac{\mu_0}{2} \mathcal{H}^2 + \frac{\varepsilon_0}{2} \mathcal{E}^2 \right) + \int dV \, \vec{E} \cdot \frac{\partial \vec{\mathscr{P}}}{\partial t} \tag{3.5.15}$$

Let "i" refer to quantities inside the dielectric and "o" refer to quantities in vacuum.

First calculate the terms in Equation (3.5.15) for the optical wave in the dielectric medium. Using $\mathscr{B} = \mu_0 \mathcal{H}$ (no magnetization) and Equation (3.1.19), namely $\mathscr{B} = \mathcal{E}/v_g$ where v_g is the speed of light in the material, we can write

$$\mathcal{H}_i = \frac{\mathcal{E}_i}{\mu_0 v_g} = \frac{n}{\mu_0 c} \mathcal{E}_i \tag{3.5.16}$$

Assuming a linear medium, the polarization in Equation (3.5.15) can be written in terms of the electric field as $\vec{\mathscr{P}} = \varepsilon_0 \chi \mathcal{E}$. Therefore the integrand of the second integral in Equation 3.5.15 can be rewritten as

$$\vec{\mathcal{E}}_i \cdot \frac{\partial}{\partial t} \vec{P}_i = \varepsilon_0 \chi \vec{\mathcal{E}}_i \cdot \frac{\partial \vec{\mathcal{E}}_i}{\partial t} = \tfrac{1}{2} \varepsilon_0 \chi \frac{\partial \mathcal{E}_i^2}{\partial t} \tag{3.5.17}$$

Combining Equations (3.5.15), (3.5.16), and (3.5.17) provides

$$P = \frac{\partial}{\partial t} \int dV \left[\frac{\mu_0}{2} \left(\frac{n}{\mu_0 c} \right)^2 \mathcal{E}_i^2 + \frac{\varepsilon_0}{2} \mathcal{E}_i^2 \right] + \int dV \, \tfrac{1}{2} \varepsilon_0 \chi \frac{\partial \mathcal{E}_i^2}{\partial t} = \frac{\partial}{\partial t} \int dV \left[\frac{n^2}{2\mu_0 c^2} \mathcal{E}_i^2 + \frac{\varepsilon_0}{2} \mathcal{E}_i^2 + \tfrac{1}{2} \varepsilon_0 \chi \mathcal{E}_i^2 \right]$$

Next using the fact that the permittivity in the dielectric can be written as $\varepsilon = \varepsilon_0 (1 + \chi)$ and that

$$\frac{\mu_0}{2} \left(\frac{n}{\mu_0 c} \right)^2 = \frac{\varepsilon}{2}$$

since $c^{-2} = \varepsilon_0 \mu_0$, the power becomes

$$P_i = \frac{\partial}{\partial t} \int dV \, \varepsilon \mathcal{E}_i^2 \tag{3.5.18}$$

Second, the power flowing outside the dielectric (i.e., in vacuum) is simply found by substituting $\varepsilon = \varepsilon_o$ into Equation (3.5.18) to get

$$P_o = \frac{\partial}{\partial t} \int dV \, \varepsilon_o \mathscr{E}_o^2 \qquad (3.5.19)$$

To evaluate the integrals in Equations (3.5.18) and (3.5.19), consider the following generic procedure. Assume the fields have the form $\mathscr{E} = E \sin(kz - \omega t)$. Elementary integral calculus provides a definition for the average $\langle \varepsilon_o \mathscr{E}^2 \rangle = \frac{1}{V} \int_{V'} dV \, \varepsilon_o \mathscr{E}^2$. However, we know the average must be $\langle \varepsilon_o \mathscr{E}^2 \rangle = \varepsilon_o E^2/2$. Therefore, the integral over the volume $\int_{V'} dV \, \varepsilon_o \mathscr{E}^2$ can be evaluated in terms of the average $\int_{V'} dV \, \varepsilon_o \mathscr{E}^2 = V' \varepsilon_o E^2/2$.

We now evaluate the integrals in Equations (3.5.18) and (3.5.19) by using these last results for the average. We assume the electric fields have constant amplitudes over the regions occupied by the beam. The two integrals become

$$P_i = \frac{\partial}{\partial t} \int_{V_i} dV \, \varepsilon \mathscr{E}_i^2 = \frac{\partial}{\partial t} \, V_i \, \frac{\varepsilon E_i^2}{2} \qquad P_o = \frac{\partial}{\partial t} \int_{V_o} dV \, \varepsilon_o \mathscr{E}_o^2 = \frac{\partial}{\partial t} \, V_o \, \frac{\varepsilon E_o^2}{2} \qquad (3.5.20)$$

The volumes depend on time although the amplitudes do not.

$$P_i = \frac{\varepsilon \mathscr{E}_i^2}{2} \frac{\partial V_i}{\partial t} \quad \text{and} \quad P_o = \frac{\varepsilon_o \mathscr{E}_o^2}{2} \frac{\partial V_o}{\partial t} \qquad (3.5.21)$$

Figure 3.5.7 provides

$$\frac{dV_i}{dt} = A \frac{d}{dt}(L - \xi) = -A \frac{d\xi}{dt} \quad \text{and} \quad \frac{dV_o}{dt} = A \frac{d\xi'}{dt}$$

Note that the derivatives in the last set of equations are related to the speed of the beam in the medium. They can be rewritten as

$$\frac{dV_i}{dt} = -A v_g \quad \text{and} \quad \frac{dV_o}{dt} = +Ac \qquad (3.5.22)$$

Equations (3.5.21) interpret P_i and P_o as the rate of change of energy in the media. The negative sign indicates a decrease of energy. The plus sign indicates an increase of vacuum energy at the expense of the dielectric energy. Therefore, we have

$$-P_i = P_o \qquad (3.5.23)$$

Substituting Equations (3.5.23) and (3.5.22) into Equations (3.5.21), we find

$$\mathscr{E}_o^2 = n \mathscr{E}_i^2$$

as found in the first example for this topic.

Given the results of these two examples, the Poynting vector accounts for not only the energy stored in the fields but also the energy stored in the polarization.

3.6 Electromagnetic Scattering and Transfer Matrix Theory

Many emitters and detectors use multiple optical elements as part of the device structure. For example, the vertical cavity lasers (VCSELs) use multiple layers of dissimilar optical materials to form a "tuned" mirror (a distributed Bragg reflector). Simple in-plane lasers (IPL) have two parallel mirrors. The behavior of the light wave within the laser can be easily modeled by representing each interface and layer by a transfer matrix. Multiple layers can then be represented as a product of the corresponding matrices.

In this section, we examine linear systems theory for optical elements. We primarily focus on the theory of reflection and transmission through multiple optical elements. The reflected and transmitted optical power and phase can be easily calculated using the scattering and transfer matrices. The theory has equal applications to RF and quantum mechanical devices.

We first discuss the scattering theory in general terms and then derive the power-amplitudes from the Poynting vector. These amplitudes serve as the input and output for the optical system. The matrices transform the amplitudes in a manner that mimics the transformation of the optical beams. We then discuss the reflection and transmission coefficients using scattering matrices. Although the scattering matrix equation relates the amplitude of an output beam to the amplitude of an input beam, it does not provide the most convenient representation of a system with multiple optical elements. Each optical element can be represented by a transfer matrix obtained from the corresponding scattering matrix. The product of transfer matrices has the same order as the sequence of optical elements. Subsequent sections will apply the basic theory to the Fabry-Perot laser.

3.6.1 Introduction to Scattering Theory

For lasers, we have great interest in finding the reflected and transmitted waves from various optical elements. These elements provide optical feedback and introduce optical loss. Figure 3.6.1 shows a simple glass plate with a single incident plane-wave but with multiple reflected and transmitted plane waves. The amplitude and phase of the input beam along with the index and thickness of the optical element determines the amplitude and phase relations for the reflected and transmitted light. The figure shows a quarter-wave plate with thickness $\lambda_n/4$ where λ_n represents the wavelength in the glass (with refractive index n_g). For dielectrics, the wavelength in the material can be related to the vacuum wavelength λ_o by $\lambda_n = \lambda_o/n_g$. The incident beam strikes the plate at point "a." Because the index of air is less than that of the glass ($n_a < n_g$), there must be a 180° phase shift for the reflected light in beam 2. The portion of light entering the glass propagates to point "b" where it partially reflects and partially transmits through the right-hand interface. However at point "b," the reflected signal does not undergo a phase shift since $n_g > n_a$. The signal reflected from point "b" travels to point "c" where another reflection occurs. The quarter wave thickness of the plate (as measured in the glass) ensures the light phase shifts by 180° for the total trip from point "a" to point "b" to point "c." In passing from the glass to air, the beam does not have a phase shift. As a result, beams 2 and 4 emerge in-phase and they constructively add together to produce a wave with larger amplitude. In this case, the quarter-wave plate functions as a fairly good mirror. As a result, we see the vital importance of both the *phase* and *amplitude* of the incident, reflected and transmitted electric fields for the function of the optical system.

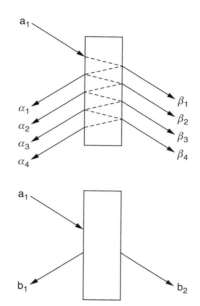

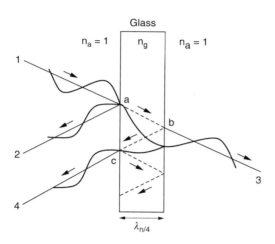

FIGURE 3.6.1
Wave picture of reflected and transmitted beams.

FIGURE 3.6.2
The reflect α_i and transmitted β_i amplitudes add together to produce b_1 and b_2, respectively.

The multiple reflected and transmitted beams in Figure 3.6.1 can be schematically represented as in Figure 3.6.2. The bottom portion of the figure shows a block diagram where a_1 represents the complex amplitude of the input beam (i.e., the magnitude and phase of the electric field). The top portion of the figure shows that the total reflected complex amplitude b_1 and transmitted complex amplitude b_2, respectively, must consist of the summation of the individual complex amplitudes α_i of the reflected beams and the complex amplitudes β_i of the transmitted beams.

$$b_1 = \sum_i \alpha_i \qquad b_2 = \sum_i \beta_i$$

In addition to magnitude and phase, the amplitudes must also contain information on the polarization of the field. We assume a single polarization. The next topic carefully defines the complex amplitudes using the Poynting vector. The phase can be affected by the thickness of the plate, the type of material, and the reflection and transmission coefficients at an interface.

It should be clear that we can linearly relate the two output amplitudes b_1, b_2 to the input amplitude a_1 for a linear optical system. We can write

$$b_1 = S_{11}a_1$$

$$b_2 = S_{21}a_1$$

where S_{ij} symbolizes the scattering matrix and NOT the Poynting vector. The scattering matrix describes the particular optical element. Most importantly, the output is linearly related to the input.

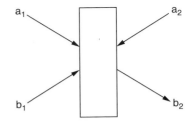

FIGURE 3.6.3
A beam strikes the glass plate from either side. The figure can be further generalized by having any number of beams from left or right.

Obviously, the situation can be generalized by including two input beams as illustrated in Figure 3.6.3. Once again the two outputs must be linearly related to the two inputs according to

$$b_1 = S_{11}a_1 + S_{12}a_2$$

$$b_2 = S_{21}a_1 + S_{22}a_2$$

or, in matrix notation

$$\begin{pmatrix} b_1 \\ b_2 \end{pmatrix} = \underline{S} \begin{pmatrix} a_1 \\ a_2 \end{pmatrix}$$

Having developed the general notion of the scattering matrix, we now discuss how the Poynting vector leads to the power amplitudes. These amplitudes allow one to retain the phase information necessary for intereference effects while describing the magnitude in terms of power. We do not need to convert from fields to power at the end of the calculation.

3.6.2 The Power-Amplitudes

The general complex electromagnetic fields can be written as

$$\vec{\mathscr{E}} = \tilde{x}\, E_o\, u(x,y) \exp(ik_n z - i\omega t)$$

$$\vec{\mathscr{H}} = \tilde{y}\, H_o\, u(x,y) \exp(ik_n z - i\omega t)$$

(3.6.1)

where the symbol k_n denotes the wave vector (that points in the z-direction) in a medium with refractive index "n."

The function $u(x, y)$ can be normalized such that

$$\int dx\, dy\, |u(x,y)|^2 = 1$$

(3.6.2)

The constant amplitude E_o adjusts the overall magnitude of the wave amplitude. The function $u(x, y)$ allows the magnitude of the electric field to depend on position (such as a plane wave with greatest intensity near the center of the beam). We will soon see the reason for requiring $u(x, y)$ to have "unit magnitude." Often we simply take "$u = $ constant" for convenience. The phasor representations for the electric and magnetic fields become

$$E(z) = E_o\, u\, \exp(ik_n z)$$

$$H(z) = H_o\, u\, \exp(ik_n z)$$

(3.6.3)

The argument $k_n z$ provides the z-dependent phase.

We would like to define "power amplitudes" $a(z)$ such that the total power in the beam can be written as

$$P = aa^* \tag{3.6.4}$$

(without additional constants). We can easily calculate the power in the beam (the usual quantity of interest) once we have calculated these generalized complex amplitudes. The power amplitudes in Equation (3.6.4) differ from the electric field by some constants. We need to retain the phase information $\exp(ik_n z)$ in the amplitudes so that the equations properly take into account the optical thickness of the optical elements and summations of phasors properly account for coherency between the waves. Absorption and gain change the constant amplitude E_o.

To find the amplitude "a," we need the Poynting vector for complex fields

$$\vec{S} = \tfrac{1}{2}\left(\vec{\mathscr{E}} \times \vec{\mathscr{H}}^*\right) \tag{3.6.5}$$

where, for a polarizable medium, we recall from Section 3.1 that the magnitude of the magnetic field can be written as

$$H = \frac{E}{\mu_o v} \quad \text{and} \quad v = c/n$$

from Section 3.1. Using our definitions for the electric and magnetic field, the Poynting vector for complex fields becomes

$$\vec{S} = \tfrac{1}{2}\hat{z}\, EH^*|u|^2 = \tfrac{1}{2}\hat{z}\, E\frac{E^*}{\mu_o v}|u|^2 = \frac{|E|^2|u|^2}{2\mu_o v}\hat{z} = \frac{\varepsilon v}{2}|E|^2|u|^2\hat{z}$$

where $\varepsilon = n^2 \varepsilon_o$ provides the permittivity of the medium.

We want the total power through a surface in the x–y plane to be given by

$$P = a(z)a^*(z) = \int dx\,dy\, \hat{z}\cdot\vec{S} = \frac{\varepsilon v}{2}|E|^2\int dx\,dy\,|u|^2 = \frac{\varepsilon v}{2}|E|^2$$

Using a relation for the speed of light

$$v_g = c\frac{1}{n} = \frac{1}{\sqrt{\varepsilon_o\mu_o}}\sqrt{\frac{\varepsilon_o}{\varepsilon}} = \frac{1}{\sqrt{\varepsilon\mu_o}}$$

we find that the total power depends on the index according to

$$P = a(z)a^*(z) = \frac{\varepsilon v}{2}|E|^2 = \frac{\sqrt{\varepsilon}}{2\sqrt{\mu_o}}|E|^2 = \tfrac{1}{2}\sqrt{\frac{\varepsilon_o}{\mu_o}}\sqrt{\frac{\varepsilon}{\varepsilon_o}}|E|^2 = \frac{\varepsilon_o c}{2}\, nE(z)E^*(z)$$

Therefore, the power amplitude can be taken as

$$a(z) = \sqrt{\frac{\varepsilon_o c}{2}}\, n\, E_o e^{ik_n z} \tag{3.6.6}$$

The complex amplitude in Equation (3.6.6) includes $e^{ik_n z}$ to account for changes in the phase of the wave due to propagation. Notice that $e^{ik_n z}$ does not contribute to the power when calculating $P = aa^*$. The reader will recognize that these power amplitudes are the same as those used in the introduction (denoted by a_i and b_i).

The next two topics develop the scattering and transfer matrix theory. Scattering matrices facilitate the identification of the reflectance and transmittance of simple optical elements. The transfer matrices are especially suited for stacked optical elements. They can be easily found from the scattering matrices.

3.6.3 Reflection and Transmission Coefficients

We first provide a brief summary for the reflectivity at a dielectric interface. Section 3.4 provides more detail and shows how their values can be found from boundary conditions. Alternatively, Grant Fowles' book titled *Introduction to Modern Optics*, published by Dover Books, contains a review of the reflection and transmission coefficients. The reader will also find alternate formulations for the scattering and transfer matrix theory. The reflectivity and transmissivity can be complex quantities for metal films, but remain real for interfaces between dielectrics. For our discussion, *we assume* real reflection and transmission coefficients.

Right from the start we need to be careful when handling the power amplitudes used for the scattering and transfer matrices. The Fresnel reflectivity and transmissivity refer to electric fields and not the power amplitudes. The refractive index makes the most important difference between the power amplitudes and the fields

$$a(z) = \sqrt{\frac{\varepsilon_o c}{2}}\, n\, E_o e^{ik_n z}$$

because the interface separates materials of differing index.

The Fresnel reflectivity *(for electric fields)* depends on the direction of the electric field with respect to the plane of the boundary between the two media (Figures 3.6.4). A transverse magnetic (TM) wave has the magnetic field parallel to the interface (transverse to the plane of incidence) while a transverse electric (TE) wave has the electric field parallel to the interface (transverse to the plane of incidence). The reflection coefficients can be written as follows

$$r = \frac{n_1 \cos\theta - n_2 \cos\phi}{n_1 \cos\theta + n_2 \cos\phi} \quad \text{TE}$$

$$r = \frac{-n_2 \cos\theta + n_1 \cos\phi}{n_2 \cos\theta + n_1 \cos\phi} \quad \text{TM}$$

(3.6.7)

where θ represents the angle of incidence and ϕ represents the angle of refraction.

Sometimes the TE and TM modes are called "s" and "p" respectively. For normally incident beams (i.e., $\theta = 0$ and $\phi = 0$), the two reflection coefficients are equal and given by

$$r = \frac{n_1 - n_2}{n_1 + n_2}$$

(3.6.8)

Notice that Equation (3.6.8) gives the correct sign for $n_1 < n_2$. Snell's law relates the angles of incidence and refraction by $n_1 \sin\theta = n_2 \sin\phi$ so that Equation (3.6.7) can

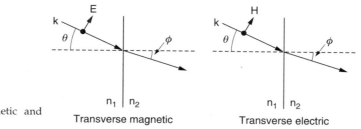

FIGURE 3.6.4
Definitions for Transverse Magnetic and Electric waves.

be written in terms of θ, n_1 and n_2 if desired. Equation (3.6.7) can also be written in terms of the angles alone (using Snell's law to the equations of the index of refraction). The transmissivity for the TE and TM modes can also be found in Section 3.4.

$$t = \frac{2\cos\theta\sin\phi}{\sin(\theta + \phi)} \qquad \text{TE}$$

$$t = \frac{2\cos\theta\sin\phi}{\sin(\theta + \phi)\cos(\theta - \phi)} \qquad \text{TM}$$

For normal incidence ($\theta = 0 = \phi$), both expressions for the transmissivity must agree. Working with the TE version, setting

$$\sin(\theta + \phi) = \sin\theta\,\cos\phi - \sin\phi\,\cos\theta$$

Setting $\theta = 0$ and then substituting Snell's law $n_1 \sin\theta = n_2 \sin\phi$ provides

$$t_{1\to 2} = \frac{2n_1}{n_1 + n_2} \qquad \text{(Normal Incidence)}$$

Notice we have added the $1 \to 2$ to indicate that the wave propagates from n_1 to n_2. Question: How do the Fresnel coefficients relate to the power reflection and transmission coefficients? These can be found from the previous relations for the electric field. Consider first the power-amplitude reflectivity r_{pa} using Figure 3.6.5 and Equation (3.6.6)

$$r_{pa} = \frac{b_1}{a_1} = \frac{\sqrt{n_1}E_{b1}e^{-ik_n z}}{\sqrt{n_1}E_{a1}e^{ik_n z}}\bigg|_{z=0} = \frac{E_{b1}}{E_{a1}} = r \qquad (3.6.9a)$$

where constants have cancelled and the minus sign appears in the exponential since the reflected wave travels along the minus z direction. We assume the interface occurs at $z = 0$ for simplicity. Notice that the reflectivity for the power amplitudes has the same value as for the fields. We drop the "pa" subscript from now on. The two reflectivity agree since the incident and reflected waves travel in the same refractive media and the refractive indices therefore cancel.

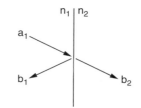

FIGURE 3.6.5
Amplitudes for input and output beams from an interface.

We can likewise find the transmissivity for the power-amplitudes. The transmissivity for the power-amplitudes for a wave traveling from media 1 into media 2, denoted by $\tau_{1\to2}$ can be written as (see Figure 3.6.5)

$$\tau_{1\to2} = \frac{b_2}{a_1} = \left.\frac{\sqrt{n_2}E_{b2}e^{ik_{n2}z}}{\sqrt{n_1}E_{a1}e^{ik_{n1}z}}\right|_{z=0} = \sqrt{\frac{n_2}{n_1}}\frac{E_{b2}}{E_{a1}} = \sqrt{\frac{n_2}{n_1}}t_{1\to2} \qquad (3.6.9b)$$

This time the power-amplitude and Fresnel transmissivity do not agree simply because the incident and transmitted waves travel in different media. The power-amplitude transmissivity can be written in terms of the refractive indices by substituting for $t_{1\to2}$. We think of the power-amplitude as the electric field amplitude except it includes all necessary constants to find the power.

So far we have discussed the field and power *amplitudes*. Question: What should we use to find reflected and transmitted power? The reflectance R and transmittance T refer to the power reflection and transmission coefficients. First, let's find the reflectance R

$$R = \frac{P_{\text{refl}}}{P_{\text{inc}}} = \frac{b_1 b_1^*}{a_1 a_1^*} = \left(\frac{b_1}{a_1}\right)\left(\frac{b_1}{a_1}\right)^* = rr^* = |r|^2 \qquad (3.6.10a)$$

where we have used Equation (3.6.9a). Second, we can find the transmittance T in a similar manner

$$T = \frac{P_{\text{trans}}}{P_{\text{inc}}} = \frac{b_2 b_2^*}{a_1 a_1^*} = \left(\frac{b_2}{a_1}\right)\left(\frac{b_2}{a_1}\right)^* = \tau_{1\to2}\tau_{1\to2}^* = |\tau_{1\to2}|^2 \qquad (3.6.10b)$$

Finally, we can write energy conservation using the reflectance R and transmittance T

$$P_{\text{inc}} = P_{\text{refl}} + P_{\text{trans}}$$

substituting the definitions for the reflected and transmitted power

$$P_{\text{inc}} = RP_{\text{inc}} + TP_{\text{inc}}$$

and therefore

$$R + T = 1 \qquad (3.6.10c)$$

In summary, the following relations can be shown.

$$R = |r|^2 \qquad T = \frac{n_2}{n_1}|t_{1\to2}|^2 = |t|^2 \qquad (3.6.11a)$$

where $t^2 = t_{1\to2}t_{2\to1}$.

For normal incidence, the TE and TM power-amplitude reflectivity and transmissivity are

$$r = -\frac{n_2 - n_1}{n_2 + n_1} \qquad \tau_{1\to2} = \sqrt{\frac{n_2}{n_1}}t_{1\to2} = \frac{2\sqrt{n_1 n_2}}{n_1 + n_2} \qquad (3.6.11b)$$

So that

$$1 = R + T = |r|^2 + |\tau|^2$$

where $|\tau|^2 = \tau_{1\to2}\tau_{1\to2}^*$ and also for real index

$$\tau^2 = t_{1\to2}t_{2\to1} \tag{3.6.11c}$$

3.6.4 Scattering Matrices

As mentioned in the introductory discussion in Section 3.6.1, the scattering matrix relates the output amplitudes b_i to the input amplitudes a_i. Here the a_i represent the beams actually incident on the optical element (see Figure 3.6.6). The b_i represent the beams (after superposition of reflected and transmitted components) actually leaving the element. The scattering matrix represents the effect of the optical element upon the incident beams. We term the amplitudes a_i and b_i the "physical" input and output for the optical element because they represent the actual beams that travel "into" and "out of" this element. These physical inputs and outputs are not the most convenient quantities when working with stacks of multiple optical elements. Later topics in this section show how the transfer matrix uses the most convenient input quantities—mathematical inputs and outputs.

To begin, consider a general optical element as a black box (Figure 3.6.6) with multiple input beams denoted by "a_j" and multiple output beams denoted by "b_j." These symbols represent the power amplitudes of the input and output beams, respectively. The introduction to scattering theory (Section 3.6.1) shows how output beams can be the result of many signal transformations (such as reflections) within the optical element. The optical element can be a waveguide, a lens, or an interface between two media. In a sense, the optical element "operates" on the input to produce the output.

"Operators" represent the operations that the optical element performs on optical beams. For linear systems, matrices provide these operators.

$$\begin{pmatrix} b_1 \\ b_2 \end{pmatrix} = \underline{S} \begin{pmatrix} a_1 \\ a_2 \end{pmatrix} = \begin{pmatrix} S_{11} & S_{12} \\ S_{21} & S_{22} \end{pmatrix} \begin{pmatrix} a_1 \\ a_2 \end{pmatrix} = \begin{pmatrix} S_{11}a_1 + S_{12}a_2 \\ S_{21}a_1 + S_{22}a_2 \end{pmatrix} \tag{3.6.12a}$$

$$\begin{pmatrix} b_1 \\ b_2 \end{pmatrix} = \begin{pmatrix} S_{11}a_1 + S_{12}a_2 \\ S_{21}a_1 + S_{22}a_2 \end{pmatrix} \tag{3.6.12b}$$

The matrix $\underline{S}$ symbolizes the scattering matrix and should *not* be confused with the Poynting vector. The word "linear" in "linear systems" means that the output must be linearly related to the input (doubling the input, doubles the output). However, the reader should realize that the matrices can depend on other parameters (such as frequency) and these matrices may be nonlinear in these other parameters. Also notice how the physical input appears on the right hand-side of Equation (3.6.11a) while

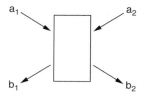

FIGURE 3.6.6
Representing an optical element as a black box.

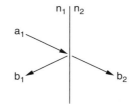

FIGURE 3.6.7
The simplest optical element is the interface.

the physical output appears on the left-hand side. This separation of input and output essentially defines the scattering matrix.

Let us consider the simplest optical element consisting of the boundary between two media with refractive indices n_1 and n_2 (with $n_1 < n_2$) as shown in Figure 3.6.7. As a simple example, let us assume only one incident beam from the left and none from the right. The scattering matrix equation becomes

$$\begin{pmatrix} b_1 \\ b_2 \end{pmatrix} = \underline{S} \begin{pmatrix} a_1 \\ 0 \end{pmatrix} = \begin{pmatrix} S_{11}a_1 \\ S_{21}a_1 \end{pmatrix}$$

Apparently, the matrix element $S_{11} = b_1/a_1$ must represent the Fresnel reflectivity for the front surface by definition of the reflectivity (see the comment on the minus sign below)

$$S_{11} = -r \qquad 0 \le |r| \le 1$$

Similarly, the matrix element $S_{21} = b_2/a_1$ must represent the Fresnel transmissivity

$$S_{21} = \tau_{1 \to 2}$$

We should make a few comments on the reflectivity r and the transmissivity τ important for the scattering matrix $\underline{S}$. The reflectivity and transmissivity are the power-amplitude reflectivity and transmissivity directly related to the field reflectivity "r" and transmissivity "t" as given by Equations (3.6.11). The reflectance R and transmittance T describe the reflected and transmitted *optical power*. We know the relations $R = |r|^2$ and $T = |\tau|^2$ since the power is proportional to the square of the electric field. The power-amplitude reflectivity has the same value as the electric field reflectivity. For example, the Fresnel reflectivity for a GaAs-air interface is 0.58 while the reflectance is 0.34.

We finally comment on the negative sign for the reflectivity "r" in S_{11}. For example, if $n_2 > n_1$ then the reflectivity $S_{11} = b_1/a_1$ must be negative while the reflectivity $S_{22} = b_2/a_2$ must be positive as found from the relations in the previous topics. We set $S_{11} = b_1/a_1 = -r$ to explicitly account for the phase change of approximately 180° for a wave reflecting from a medium with larger refractive index. We don't really need to include this "minus" sign since the formulas for reflectivity and transmissivity in previous topics take care of this for us. As shown in Figure 3.6.9, the Fresnel reflectivity for the electromagnetic wave incident on the n_1 side of the boundary must be the negative of the reflectivity for the wave traveling in the n_2 material and incident on the boundary. For example, suppose $a_1 = 0$ but that $a_2 \ne 0$. The scattering matrix becomes

$$\begin{pmatrix} b_1 \\ b_2 \end{pmatrix} = \begin{pmatrix} S_{12}a_2 \\ S_{22}a_2 \end{pmatrix}$$

with $S_{21} = \tau$ and $S_{22} = +r$.

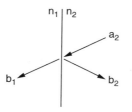

FIGURE 3.6.8
Beam enters on the right-hand side.

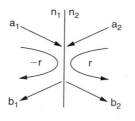

FIGURE 3.6.9
Beams propagating right and left are incident on an interface.

To illustrate the principal of superposition, consider waves incident on the interface from the right and from the left as shown in Figure 3.6.8. The scattering matrix equation becomes

$$\begin{pmatrix} b_1 \\ b_2 \end{pmatrix} = \underline{S} \begin{pmatrix} a_1 \\ a_2 \end{pmatrix} = \begin{pmatrix} S_{11} & S_{12} \\ S_{21} & S_{22} \end{pmatrix} \begin{pmatrix} a_1 \\ a_2 \end{pmatrix} = \begin{pmatrix} -r & \tau_{2\to1} \\ \tau_{1\to2} & r \end{pmatrix} \begin{pmatrix} a_1 \\ a_2 \end{pmatrix} \qquad (3.6.13)$$

Note the convention of "$-r$" and "$+r$" in the figure makes $r>0$ for $n_1<n_2$. If in reality $n_1>n_2$, then the sign of "r" will be reversed; either way, the equations work out ok.

An optical element of finite thickness with two interfaces provides a slightly more complicated example. Assume that an object imbedded in air has refractive index n as indicated by the block diagram in Figure 3.6.10. We should consider three different scattering matrices—one for each boundary and one for the material between the boundaries. The matrices for the two interfaces involve the Fresnel reflection and transmission coefficients. However, viewing the object as a black box, we can define *effective* reflection and transmission coefficients. For example, the effective reflectivity would be $-r_{\text{eff}}=b_1/a_1$. *The effective reflectivity and transmissivity account for multiple reflections within the object (i.e., multiple reflected beams contribute to the two output beams).* The scattering matrix equation can be written as

$$\begin{pmatrix} b_1 \\ b_2 \end{pmatrix} = \underline{S} \begin{pmatrix} a_1 \\ a_2 \end{pmatrix} = \begin{pmatrix} S_{11} & S_{12} \\ S_{21} & S_{22} \end{pmatrix} \begin{pmatrix} a_1 \\ a_2 \end{pmatrix} = \begin{pmatrix} -r_{\text{eff}} & t_{\text{eff}} \\ t_{\text{eff}} & -r_{\text{eff}} \end{pmatrix} \begin{pmatrix} a_1 \\ a_2 \end{pmatrix}$$

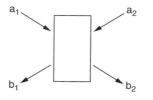

FIGURE 3.6.10
Input and output beams.

The reader will recognize this as an application of the principle of superposition. We have not really solved the problem but instead we have hidden the problem in the effective reflectivity and transmissivity. The effective reflectivity and transmissivity can be complex numbers because they contain information on the relative phases. We will see that transfer matrices are better suited for complicated problems such as this one. In fact, we can actually calculate the effective reflectivity and transmissivity.

3.6.5   The Transfer Matrix

The scattering matrix describes the reflected and transmitted amplitudes at an interface. However, the scattering matrix equation

$$\begin{pmatrix} b_{01} \\ b_{02} \end{pmatrix} = \begin{pmatrix} r & \tau_{2\to1} \\ \tau_{1\to2} & -r \end{pmatrix} \begin{pmatrix} a_{01} \\ a_{02} \end{pmatrix} \qquad \text{or} \qquad \underline{b} = \underline{S}\,\underline{a}$$

puts both physical inputs (a_1, a_2) on the right-hand side and both physical outputs (b_1, b_2) on the left-hand side. The scattering matrix does not provide a convenient formulation for solving more complicated problems. For example, consider Figure 3.6.11, which shows an optical device consisting of multiple optical elements and interfaces. We could find the output amplitudes $\underline{b}$ if we knew the input amplitudes $\underline{a}$. However, we must know the effect of the right-most and left-most elements before we can find the $\underline{a}$ amplitudes. We can write a matrix equation to include the two outer optical elements, but there's a simpler method. It would be nice to look at a figure such as Figure 3.6.11 and just write a matrix for each optical element in the order that it occurs. That is, we would really like to just look at a figure with a series of optical elements and write one matrix for each optical element; these matrices would be written in the same order as each optical element appears in the figure. This is where the transfer matrix comes into play.

Figure 3.6.12 shows an expanded view of the stacked optical elements. The top portion of the figure shows three optical elements. The middle portion of the figure separates the elements and labels the amplitudes of the input and output beams. The bottom portion shows how the transfer-matrix equation corresponds to each element. *We want to use beams A_2 and B_2 as the **input** to optical element No. 1. We want to interpret the amplitudes A_1 and B_1 as the **output** from the optical element.* In this way, the amplitudes A_2 and B_2 can be interpreted as the input to the first optical element and as the output from the second optical element. We picture the inputs to an optical element as residing on the right-hand side while the outputs from an element are on the left-hand side.

In matrix notation, the first two elements of the optical train produce an equation of the form

$$\begin{pmatrix} A_1 \\ B_1 \end{pmatrix} = \underline{T}_1 \begin{pmatrix} A_2 \\ B_2 \end{pmatrix} = \underline{T}_1\underline{T}_2 \begin{pmatrix} A_3 \\ B_3 \end{pmatrix} \qquad (3.6.14)$$

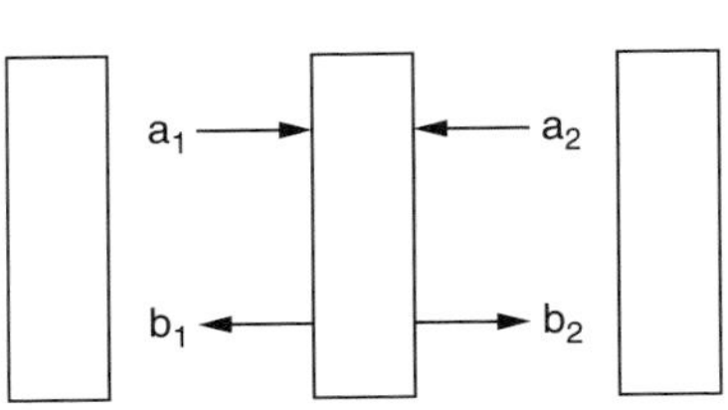

FIGURE 3.6.11
A multi-element electronic device.

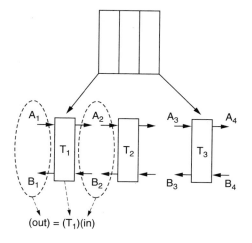

FIGURE 3.6.12
Stacked optical element. The transfer matrix equation is shown for the first element.

Clearly, an optical train such as in the figure can easily be represented by a series of multiplied transfer matrices. As with the scattering matrix S, the transfer matrix T represents the effects of the optical element.

The right-hand side of a transfer-matrix equation (such as Equation (3.6.14)) has the "input" column vectors while the left-hand side has the "output" column vectors. Here lies the important distinction between the scattering and transfer matrices. Consider the first optical element. Physically speaking, the A_1 amplitude represents an incident beam (i.e., an input beam) but it appears on the output side of the matrix equation. A similar comment applies to the *output* amplitude A_2 appearing on the input side of the matrix equation. We see that the mathematical inputs to the transfer matrix equation are different than the physical inputs to the scattering matrix equation. For the transfer matrix, amplitudes referring to a single side of an optical element appear in the same column vector (regardless of whether they represent physical inputs or not).

For the transfer-matrix equation, how can "output" variables be used on the "input" side of the equation? The answer comes from the fact that we are using linear systems. The variables in the scattering matrix equation (Eq. 3.6.13 for example) can be rearranged to give the variables for the transfer matrix equation. Consider the first optical element in Figure 3.6.12. For the scattering matrix, we denote the "input" amplitudes by a_i and the "output" amplitudes by b_i. Figure 3.6.13 shows that the scattering and transfer variables must be related by $A_1 = a_1$, $A_2 = b_2$, $B_1 = b_1$, and $B_2 = a_2$. A relation can be found between the scattering and transfer matrices. Start with the scattering matrix equation

$$b_1 = S_{11}a_1 + S_{12}a_2$$
$$b_2 = S_{21}a_1 + S_{22}a_2$$

(3.6.15)

A_1 = a_1 → □ → A_2 = b_2

B_1 = b_1 ← □ ← B_2 = a_2

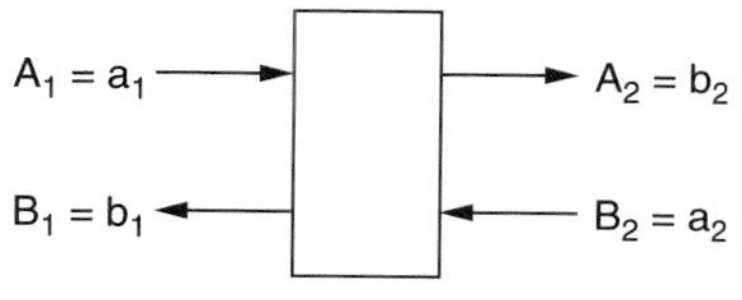

FIGURE 3.6.13
Relation between the scattering and transfer variables.

Next eliminate the scattering variables in favor of the transfer variables.

$$B_1 = S_{11}A_1 + S_{12}B_2$$
$$A_2 = S_{21}A_1 + S_{22}B_2$$

(3.6.16)

Equation 3.6.16 must be compared with the defining relation for the transfer-matrix equation (3.6.14). Equation 3.6.16 needs to be rearranged. We move A_2 and B_2 to the right-hand side and A_1 and B_1 to the left-hand side. The coefficients of A_2 and B_2 will be the elements of the transfer matrix. We find

$$A_1 = \frac{1}{S_{21}}A_2 - \frac{S_{22}}{S_{21}}B_2$$
$$B_1 = S_{11}A_1 + S_{12}B_2 = \frac{S_{11}}{S_{21}}A_2 - \frac{S_{11}S_{22} - S_{12}S_{21}}{S_{21}}B_2$$

The right-hand side of the previous two lines provides the elements of the transfer matrix

$$T = \frac{1}{S_{21}}\begin{pmatrix} 1 & -S_{22} \\ S_{11} & -\text{Det}(S) \end{pmatrix}$$

(3.6.17)

where Det stands for the determinant. We could just as easily demonstrate the scattering matrix in terms of the transfer matrix

$$S = \frac{1}{T_{11}}\begin{pmatrix} T_{21} & \text{Det}\,T \\ 1 & -T_{12} \end{pmatrix}$$

(3.6.18)

3.6.6 Examples Using Scattering and Transfer Matrices

Perhaps the best way to understand the scattering and transfer matrices is to consider several examples. The last example introduces the Fabry-Perot cavity. The next section uses the results to discuss the longitudinal modes present in a laser cavity.

Example 3.6.1 The Simple Interface

Reconsider an interface at $z = 0$ separating two media with refractive indices n_1 and n_2. The plane waves "A" travel toward the right ($+z$ direction) and have the form

$$A_i(z) = A_{oi}\,e^{ik_i z}$$

while the "B" waves travel toward the left and have the form

$$B_i(z) = B_{oi}\,e^{-ik_i z}$$

The wave vectors carry a subscript to indicate the index of refraction of the material $k_i = k_o n_i (n_i$ is real) through which the wave travels. The scattering matrix is

$$\underline{S} = \begin{pmatrix} -r & \tau_{2\to 1} \\ \tau_{1\to 2} & r \end{pmatrix}$$

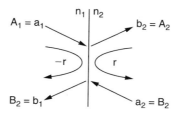

FIGURE 3.6.14
The simple interface.

Therefore the transfer matrix in

$$\begin{pmatrix} A_1(z) \\ B_1(z) \end{pmatrix}_{z=0} = T \begin{pmatrix} A_2(z) \\ B_2(z) \end{pmatrix}_{z=0} \rightarrow \begin{pmatrix} A_{01} \\ B_{01} \end{pmatrix} = T \begin{pmatrix} A_{02} \\ B_{02} \end{pmatrix}$$

must be given by

$$T = \frac{1}{\tau_{1\to2}} \begin{pmatrix} 1 & -r \\ -r & rr + \tau^2 \end{pmatrix} \tag{3.6.19}$$

by Equation 3.6.17 where $\tau^2 = \tau_{1\to2}\tau_{2\to1}$. Energy conservation gives $-\text{Det}(\underline{S}) = r^2 + \tau^2 = 1$ (refer to the previous topics). Two important notes: (1) if we exchange $-"r"$ and $"r"$ in the figure, then they must also be exchanged in the scattering and transfer matrices; (2) if we leave the minus signs as shown in Figure 3.6.14 but consider the right-hand side to be the output (and the left-hand side to be the input) for the transfer matrix, then r and $-r$ must be interchanged in the transfer matrix.

A single interface is not very interesting. Real optical devices have at least two interfaces and an interior. The simplest example consists of a beam of light propagating from one interface to the other. We take into account only the phase factor e^{ikz}.

Example 3.6.2 The Simple Waveguide

Suppose a wave travels toward the right and another travels toward the left inside a dielectric. We want to know how the amplitudes change when the one wave moves from a point z_o to a point $z_o + L$ or the other wave moves in the opposite direction. These interfaces at z_o and $z_o + L$ are not real and do not produce reflections; they both exist inside the semiconductor and do not indicate discontinuity in the refractive index. We can make a real waveguide by including actual interfaces at z_o and $z_o + L$; for this real waveguide, we must use three transfer matrices. For now, we focus on the simplest case of a wave propagating from one point to another.

Assume the two waves represented by power amplitudes a_1 and a_2 pass through the left-hand and right-hand interfaces inside the material with refractive index $"n"$ as indicated by Figure 3.6.15. The waves do not reflect from these imaginary interfaces. Assume that the beams propagate straight through the material. The forward propagating wave (from left to right) has the form $a_1 \sim E_o \exp(ik_n z)$ while the backward propagating wave has the form $a_2 \sim E_o \exp(-ik_n z)$. The amplitude b_2 at $z_o + L$ must be related to the amplitude a_1 at z_o by a phase factor

$$b_2 = E_o \exp[ik_n(z_o + L)] = a_1 \exp(ik_n L)$$

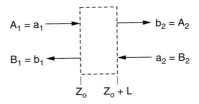

FIGURE 3.6.15
Block diagram for the simple waveguide.

The backward propagating wave with amplitude b_1 at z_o must be related to the wave with amplitude a_2 at z_o+L

$$a_2 = E_o \exp[-ik_n(z_o + L)] = b_1 \exp(-ik_nL)$$

or, in other words,

$$b_1 = a_2 \exp(ik_nL)$$

We can write the scattering matrix as

$$\begin{pmatrix} b_1 \\ b_2 \end{pmatrix} = \begin{pmatrix} S_{11} & S_{12} \\ S_{21} & S_{22} \end{pmatrix} \begin{pmatrix} a_1 \\ a_2 \end{pmatrix} = \begin{pmatrix} 0 & \exp(ik_nL) \\ \exp(ik_nL) & 0 \end{pmatrix} \begin{pmatrix} a_1 \\ a_2 \end{pmatrix}$$

Therefore, the transfer matrix in

$$\begin{pmatrix} A_1 \\ B_1 \end{pmatrix} = T \begin{pmatrix} A_2 \\ B_2 \end{pmatrix}$$

must be given by Equation (3.6.17)

$$T = \frac{1}{S_{21}} \begin{pmatrix} 1 & -S_{22} \\ S_{11} & -\text{Det}(S) \end{pmatrix} = \frac{1}{\exp(ik_nL)} \begin{pmatrix} 1 & 0 \\ 0 & \exp(2ik_nL) \end{pmatrix} = \begin{pmatrix} \exp(-ik_nL) & 0 \\ 0 & \exp(ik_nL) \end{pmatrix}$$

$$(3.6.20)$$

Now we can discuss a Fabry-Perot cavity. We must include two physical interfaces and the interior between them.

Example 3.6.3 The Fabry-Perot Cavity

The semiconductor laser has a Fabry-Perot cavity as depicted in Figure 3.6.16. The gain section corresponds to the region with index n_2. To model the laser, consider a slab of material with refractive index "n_2" embedded within another material of refractive index "n_1." The region corresponding to n_1 usually consists of air for a cleaved-facet (in-plane) semiconductor laser. Assume that reflections occur at each of the two parallel boundaries. We assume the only input beam comes from the left and so $B_1=0$. Notice the reflectivity is assumed positive for waves reflecting off the inner surfaces. Examples 3.6.1

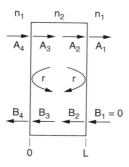

FIGURE 3.6.16
The Fabry-Perot cavity.

and 3.6.2 provide the transfer matrices. Starting with the right-hand interface we find,

$$\begin{pmatrix} A_2 \\ B_2 \end{pmatrix} = \frac{1}{\tau_{21}} \begin{pmatrix} 1 & r \\ r & 1 \end{pmatrix} \begin{pmatrix} A_1 \\ 0 \end{pmatrix}$$

We gave the reflectivity a "positive" sign in the transfer matrix contrary to that in Example 3.6.1. We did this since the output side (left side) of the $z = L$ boundary has $+r$ reflectivity rather than $-r$ reflectivity in Example 3.6.1. The subscript "21" on τ_{21} indicates the right-hand interface with a wave moving from medum "2" to "1." These subscripts provide a bookkeeping tool. In principle, there may be three separate refractive indices but not in this case. The waveguide (excluding the interfaces) has a transfer matrix given by Example 3.6.2

$$\begin{pmatrix} A_3 \\ B_3 \end{pmatrix} = \begin{pmatrix} e^{-i\phi} & 0 \\ 0 & e^{i\phi} \end{pmatrix} \begin{pmatrix} A_2 \\ B_2 \end{pmatrix}$$

where $\phi = k_n L$. The transfer matrix for the left-hand side is similar to that for Example 3.6.1 and is given by

$$\begin{pmatrix} A_4 \\ B_4 \end{pmatrix} = \frac{1}{\tau_{12}} \begin{pmatrix} 1 & -r \\ -r & 1 \end{pmatrix} \begin{pmatrix} A_3 \\ B_3 \end{pmatrix}$$

Multiplying the three individual matrices provides the total transfer matrix

$$\begin{pmatrix} A_4 \\ B_4 \end{pmatrix} = \frac{1}{\tau_{12}} \begin{pmatrix} 1 & -r \\ -r & 1 \end{pmatrix} \begin{pmatrix} e^{-i\phi} & 0 \\ 0 & e^{+i\phi} \end{pmatrix} \frac{1}{\tau_{21}} \begin{pmatrix} 1 & r \\ r & 1 \end{pmatrix} \begin{pmatrix} A_1 \\ 0 \end{pmatrix}.$$

Calculating the product, we find the total transfer matrix to be

$$\begin{pmatrix} A_4 \\ B_4 \end{pmatrix} = \frac{1}{\tau^2} \begin{pmatrix} e^{-i\phi} - r^2 e^{+i\phi} & re^{-i\phi} - re^{+i\phi} \\ -re^{-i\phi} + re^{+i\phi} & -r^2 e^{-i\phi} + e^{+i\phi} \end{pmatrix} \begin{pmatrix} A_1 \\ B_1 = 0 \end{pmatrix} \tag{3.6.21}$$

The phase $\phi = k_n L$ can incorporate the complex wave vector. Recall that the complex part of k_n describes absorption and gain. The laser uses a basic Fabry-Perot resonator filled with semiconductor to produce gain when pumped. We won't need to worry about the complex portion of the refractive index for the expression of the power amplitudes.

$$a(z) = \sqrt{\frac{\varepsilon_o c}{2}} \, n \, E_o \, e^{ik_n z}$$

Retracing the relation between E and H in Section 3.1 indicates the real part of the index determines the phase speed while the imaginary part affects the amplitude. The refractive index occurs in the coefficient of the power amplitude to provide the phase velocity in the medium. The next section discusses how the presence of gain (or loss) affects the transfer matrix and the resulting characteristics of the Fabry-Perot cavity.

3.7 The Fabry-Perot Laser

The in-plane laser has two parallel mirrors forming a Fabry-Perot resonator (i.e., Fabry-Perot etalon). The results for the transfer matrix describing the simple waveguide with two interfaces demonstrate (1) the gain condition (i.e., the gain must equal the loss for a laser operating above threshold), (2) the existence of longitudinal modes, (3) the approximate line shape of the emitted laser light, and (4) line narrowing. The "line-shape" function describes the shape of the curve representing the output power as a function of wavelength. The power spectrum attains a peak value and decreases on either side of the peak. The quality (finesse) of the cavity and the spontaneous emission rate determine the width of the power spectrum—often described by the "line width." For a semiconductor laser, the gain compensates for the optical loss of the cavity and thereby effectively increases the finesse of the resonator and decreases the line width. Indeed, increasing the gain from below-threshold to above-threshold levels decreases the line width as the spectrum transitions from spontaneous to stimulated emission. The decrease in line width along with a sudden jump in output power and the appearance of "speckle pattern" signifies the onset of lasing. Ultimately, spontaneous emission limits the linewidth for a laser.

3.7.1 Implications of the Transfer Matrix for the Fabry-Perot Laser

We now calculate the optical power leaving the Fabry-Perot resonator as a function of the incident optical power. The magnitude of the output power depends on the pumping level to the gain medium. When the gain compensates the losses, the resonator begins to lase and produces an output beam independent of the input beam similar to the ring laser discussed in Section 1.1. The interior of the Fabry-Perot etalon normally contains material with a refractive index n_2 larger than the refractive index n_1 ($n_2 > n_1$) of the surrounding medium. For simplicity, we assume the surrounding medium consist of air with $n_1 = 1$ and, for convenience, set $n = n_2$. Also assume "normal incidence" for all the beams (i.e., the beams propagate perpendicular to the interfaces); therefore, the reflectivity for TE and TM waves must be the same and denoted by r.

The previous section demonstrates the transfer matrix for a Fabry-Perot resonator with a single input beam incident on the left-hand side (Example 3.6.3). Equation (3.6.20) provides

$$\begin{pmatrix} A_4 \\ B_4 \end{pmatrix} = \frac{1}{\tau^2} \begin{pmatrix} e^{-i\phi} - r^2 e^{i\phi} & re^{-i\phi} - re^{i\phi} \\ -re^{-i\phi} + re^{i\phi} & -r^2 e^{-i\phi} + e^{i\phi} \end{pmatrix} \begin{pmatrix} A_1 \\ B_1 = 0 \end{pmatrix}$$

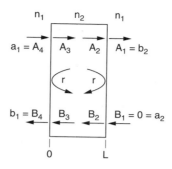

FIGURE 3.7.1

The amplitudes for the scattering matrix (lower case letters) and for the transfer matrix (upper case letters).

with the transfer matrix given by

$$T = \begin{pmatrix} T_{11} & T_{12} \\ T_{21} & T_{22} \end{pmatrix} = \frac{1}{\tau^2} \begin{pmatrix} e^{-i\phi} - r^2 e^{i\phi} & r e^{-i\phi} - r e^{i\phi} \\ -r e^{-i\phi} + r e^{i\phi} & -r^2 e^{-i\phi} + e^{i\phi} \end{pmatrix} \tag{3.7.1}$$

where the phase $\phi = k_n L$ uses the complex wavevector. Although the transfer matrix is a very useful mathematical abstraction, we eventually require the output amplitudes for physical lasers. We can better use the scattering matrix for this purpose. Recall the basic definition of the scattering matrix

$$\begin{pmatrix} A_4 \\ B_4 \end{pmatrix} = \frac{1}{\tau^2} \begin{pmatrix} e^{-i\phi} - r^2 e^{i\phi} & r e^{-i\phi} - r e^{i\phi} \\ -r e^{-i\phi} + r e^{i\phi} & -r^2 e^{-i\phi} + e^{i\phi} \end{pmatrix} \begin{pmatrix} A_1 \\ B_1 = 0 \end{pmatrix}$$

The various amplitudes for the scattering and transfer matrices appear in Figure 3.7.1. Equation (3.6.17) in the previous section gave the relation between the two types of matrices.

$$S = \frac{1}{T_{11}} \begin{pmatrix} T_{21} & \text{Det } T \\ 1 & -T_{12} \end{pmatrix} = \frac{\tau^2}{e^{-i\phi} - r^2 e^{i\phi}} \begin{pmatrix} T_{21} & \text{Det } T \\ 1 & -T_{12} \end{pmatrix} \tag{3.7.2}$$

For the laser oscillator, we are interested in the output signal as a function of the input signal. We can use either b_1 or b_2 as the output signal. Consider b_1 and write

$$b_1 = S_{11} a_1 \tag{3.7.3}$$

Equation (3.7.1) and (3.7.2) provide the relevant transfer function

$$\frac{\text{output}}{\text{input}} = \frac{b_1}{a_1} = S_{11} = \frac{-r e^{-i\phi} + r e^{i\phi}}{e^{-i\phi} - r^2 e^{i\phi}} = -r \frac{1 - e^{2i\phi}}{1 - r^2 e^{2i\phi}} \tag{3.7.4}$$

We assume both mirrors (i.e., the interfaces between the two media) have the same reflectivity. Equations (3.7.3) and (3.7.4) can be compared with the op-amp circuit discussed in Chapter 1. In particular, for the op-amp, the relation between the output power P_o and input power P_{in} has the form

$$P_o = \frac{g P_{\text{in}}}{1 - g/\alpha} \tag{3.7.5}$$

The parameters g and α replace the material gain and loss. Compare the op-amp equation (3.7.5) with Equation (3.7.4). Both denominators might become zero. The op-amp circuit begins to oscillate for small values of the denominator. Similarly, we expect Equation (3.7.4) to produce denominators close to zero when the Fabry-Perot laser begins to lase.

The power flowing into and out of the Fabry-Perot cavity must be proportional to the square of the power amplitudes. Equation (3.7.3) provides the relation between the reflected power P_{ref} and the incident power P_{in}

$$P_{\text{ref}} = |S_{11}|^2 P_{\text{in}} \qquad (3.7.6)$$

The "reflected" power actually originates from two sources. The first source consists of light produced within the resonator (such as from pumping) and passing from the interior to the exterior across the left-hand interface. The second source consists of a fraction of incident power P_{in} reflecting from the left-hand interface. The word "reflected" appears in quotes because the beam b_1 actually consists of the superposition of many beams from within the etalon. Calculating the square of the complex transfer function S_{11} we find

$$|S_{11}|^2 = S_{11} S_{11}^* = r^2 \frac{\left(1 - e^{2i\phi}\right)\left(1 - e^{2i\phi}\right)^*}{(1 - r^2 e^{2i\phi})(1 - r^2 e^{2i\phi})^*} \qquad (3.7.7)$$

Sometimes people call the matrix element S_{11} (or any of the S_{ij}) a transfer function because it relates an output variable to an input variable. The term "transfer function" is a standard term in engineering.

We need a few definitions at this point. The complex phase factor $\phi = k_n L$ contains the complex wave vector

$$k_n = k_o n - i \frac{g_{\text{net}}}{2}$$

so that the real and imaginary parts of the phase factor can be written

$$\phi = \phi_r + i\phi_i = k_n L = k_o n L - i \frac{g_{\text{net}}}{2} L$$

The gain g_{net} describes the material gain "g" and the distributed internal loss α_{int} as represented by the familiar relation

$$g_{\text{net}} = \Gamma g - \alpha_{\text{int}}$$

The power reflectivity R is related to the Fresnel reflectivity r (assumed real) by

$$R = r^2$$

We define an *effective* reflectivity $\mathscr{R}$ by

$$\mathscr{R} = \mathscr{R}\exp(-2\phi_i) \qquad (3.7.8)$$

Also, the complex nature of $e^{i\phi}$ must be properly handled during complex conjugation since imaginary parts occur in two places $(e^{i\phi})^* = e^{-i\phi^*}$.

Returning to the power transfer function, we find

$$|S_{11}|^2 = R\frac{1 + \exp(-4\phi_i) - 2\exp(-2\phi_i)\cos(2\phi_r)}{1 + R^2\exp(-4\phi_i) - 2R\exp(-2\phi_i)\cos(2\phi_r)} = R1 + \frac{\frac{\mathscr{R}^2}{R^2} - 2\frac{\mathscr{R}}{R}\cos(2\phi_r)}{1 + \mathscr{R}^2 - 2\mathscr{R}\cos(2\phi_r)}$$

Using the cosine expansion, $\cos(2\phi_r) = \cos^2(\phi_r) - \sin^2(\phi_r) = 1 - 2\sin^2(\phi_r)$, we find

$$|S_{11}|^2 = R\frac{[1 - \frac{\mathscr{R}}{R}]^2 + 4\frac{\mathscr{R}}{R}\sin^2\phi_r}{[1 - \mathscr{R}]^2 + 4\mathscr{R}\sin^2\phi_r} \tag{3.7.9}$$

Using Equation (3.7.6) and (3.7.9), the relation between the "reflected" power and the input power must be given by

$$P_{\text{ref}} = |S_{11}|^2 P_{\text{in}} = R\frac{[1 - \frac{R}{R}]^2 + 4\frac{R}{R}\sin^2\phi_r}{[1 - R]^2 + 4R\sin^2\phi_r}P_{\text{in}} \tag{3.7.10}$$

Example 3.7.1 A Fabry-Perot Etalon Without Material Gain or Internal Loss

For a Fabry-Perot etalon, let's plot the reflected and transmitted power as a function of the optical frequency $\omega = 2\pi\nu$ of the electromagnetic wave. Assume that an electromagnetic wave enters the etalon from the left, bounces around a bit, and emerges from (1) the left-hand facet as a "reflected" beam and from (2) the right-hand facet as a "transmitted" beam. For the present example, assume that the material comprising the Fabry-Perot etalon hasn't any material gain/absorption nor any internal distributed losses (such as free carrier absorption or optical scattering through the side walls). The phase factor and the complex wave vector must be real

$$\phi = \phi_r + i\phi_i = k_n L = k_o n L - i\frac{g_{\text{net}}}{2}L = k_o n L$$

The "reflected" power P_{ref} given by Equation (3.7.10) becomes (with $R = R$ for real ϕ; i.e., $\phi_i = 0$)

$$P_{\text{ref}} = |S_{11}|^2 P_{\text{in}} = R\frac{4\sin^2(k_o n L)}{[1 - R]^2 + 4R\sin^2(k_o n L)}P_{\text{in}}$$

where $k_o = 2\pi/\lambda_o$ and $\lambda_o = c/\nu = 2\pi c/\omega$ is the wavelength of the electromagnetic wave in air and ν is the frequency in Hz. Although not written out here, a similar relation can be found for the power transmitted through the etalon to the other side. However, power conservation requires that the transfer function be

$$P_{\text{trans}} = |S_{21}|^2 P_{\text{in}} = [1 - |S_{11}|^2]P_{\text{in}}$$

We can now plot the transfer functions for the reflected and transmitted powers. Figure 3.7.2 shows $P_{\text{ref}}/P_{\text{in}}$ for three different values of the power reflectivity all

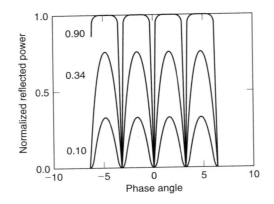

FIGURE 3.7.2
The "reflected" power through the Fabry-Perot resonator.

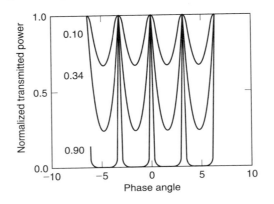

FIGURE 3.7.3
The transmitted power.

plotted against the phase angle $k_o nL$. The value of $R = 0.34$ corresponds to cleaved facets for GaAs lasers. Notice that larger facet reflectance R gives larger finesse (narrower line widths). This means greater optical loss must lower the finesse of the cavity. Different reflected powers can be obtained by changing the index n, the wavelength or the physical thickness of the etalon. The finesse can be exceedingly large (narrow line widths). The output power from the etalon depends on the wavelength through $k_o nL = 2\pi nL/\lambda_o$. Figure 3.7.3 shows the transmitted power. Only a very narrow band of wavelengths can propagate through the etalon thereby producing a very narrow bandpass filter.

3.7.2 Longitudinal Modes and the Threshold Condition

The previous topic shows the amount of power leaving a Fabry-Perot resonator as a function of the power in an incident beam. Pumping the resonator can initiate lasing. Above threshold, the laser produces considerably more power than it receives from the input beam. In this case, the effective transmittance and reflectance must become infinite as discussed in Section 1.1 concerning the ring laser and op-amp oscillator. These terms become infinite when the denominators of S_{11} or $|S_{11}|^2$ (etc) equals to zero. Although it might be more appropriate to work with "transmitted" power for a laser, either the "reflected" or "transmitted" power can be used since they have the same denominator. The results identify the longitudinal modes for the cavity and demonstrate the threshold condition for the gain (i.e., gain equals loss).

Although Equation (3.7.10) can be used, we return to the simpler forms given in Equations (3.7.3) and (3.7.4)

$$b_1 = S_{11}a_1 \quad \rightarrow \quad P_{\text{ref}} = |S_{11}|^2 P_{\text{in}} \tag{3.7.3}$$

$$S_{11} = \frac{-r\,e^{-i\phi} + r\,e^{i\phi}}{e^{-i\phi} - r^2\,e^{i\phi}} = -r\frac{1 - e^{2i\phi}}{1 - R\,e^{2i\phi}} \tag{3.7.4}$$

where $R = r^2$. As typical for linear systems, the pole of the transfer function determines the characteristics of the oscillation. For a laser, we do not usually inject optical power through the mirrors ($P_{\text{in}} = 0$) except for special cases of injection locking or optical pumping. Spontaneous emission within the laser gain medium, sometimes associated with P_{in}, initiates the laser oscillation. In what follows, let D represent the denominator of the transfer function S_{11} for the "reflected" power

$$D = 1 - R\,e^{i2\phi} \tag{3.7.11}$$

where we can now write the denominator as

$$D = 1 - R\,\exp(g_{\text{net}}L)\,\exp(i2k_onL) = 1 - R\,\exp(g_{\text{net}}L)\,[\cos(2k_onL) + i\,\sin(2k_onL)] \tag{3.7.12}$$

For the denominator to be zero, we require both the real and imaginary parts to be simultaneously zero.

As an important note, the index of refraction of the cavity material can change due to pumping. For this reason, the index of refraction "n" that appears in the formulas must include both the so-called background refractive index n_b and the pumping refractive index n_p.

Case 1 *Imaginary Part Produces the Wavelength of the Longitudinal Modes*

Setting the imaginary part of the denominator (given in Equation (3.7.12)) equal to zero provides

$$0 = \sin(2k_onL) = 2\sin(k_onL)\cos(k_onL)$$

We might try to make either the sine or the cosine term equal to zero. However, if we choose the cosine term to be zero, we would not necessarily find the *real* part of the denominator to be zero. Setting the sine term equal to zero gives a condition on the wave vector and the wavelength

$$k_onL = m\pi \quad (m = 1, 2, 3, \dots)$$

or equivalently,

$$\lambda_o = \frac{2nL}{m} \quad \text{or} \quad \lambda_n = \frac{2L}{m}$$

The corresponding optical frequencies must be

$$\nu_m = \frac{mc}{2nL} \quad (m = 1, 2, \dots)$$

FIGURE 3.7.4
The $m = 3$ longitudinal mode.

These expressions provide the allowed wavelengths and frequencies of the electromagnetic wave within the Fabry-Perot cavity. A sinusoidal wave with exactly one of these
wavelengths constitutes a so-called longitudinal mode. Figure 3.7.4 shows an $m = 3$
longitudinal mode. These modes derive their name "longitudinal" from the fact that
multiple waves fit along the length of the resononator (longitudinal direction). We will
investigate the transverse modes in connection with the slab waveguide later in this
chapter.

We can find the spacing between adjacent frequencies of the longitudinal modes using
$vn = mc/(2L)$. Let $\Delta m = 1$ and calculate

$$\Delta(vn) = \frac{\Delta mc}{2L}$$

$$n\Delta v + v\Delta n = \frac{c}{2L} \quad \rightarrow \quad \Delta v \left[n + v\frac{\Delta n}{\Delta v} \right] = \frac{c}{2L}$$

Define the group index $n_g = n + v\dfrac{\partial n}{\partial v}$ and substitute to find

$$\Delta v = \frac{c}{2Ln_g} \cong \frac{c}{2Ln_g}$$

The last expression gives the frequency difference between adjacent longitudinal
modes. For relatively frequency-independent refractive index, the group index reduces to
the ordinary refractive index ($n_g = n$). The difference in wavelength between adjacent
lines can also be found by using $v\lambda_o = c$ so that

$$\lambda_o \Delta v + v\Delta \lambda_o = 0 \qquad \text{or} \qquad |\Delta \lambda_o| = \frac{\lambda_o c}{2Ln}$$

Before continuing with the second case that demonstrates the gain relation, we should
make some comments on the physical significance of the longitudinal modes. These
modes are actually standing waves. If all of the EM energy in a resonator occupies
a longitudinal mode then none of the energy would escape through the mirrors to
form a useful signal. We must surmise the EM energy can occupy two types of
modes. One type constitutes traveling plane waves that can pass through the mirrors. The
other type makes up the longitudinal modes. Photons in the standing waves occupy
the longitudinal modes. These two types of modes produce two types of effects. The
longitudinal modes produces narrow line widths and well-defined laser frequencies.
The traveling wave produces a useful signal and broadens the laser line.

First consider the longitudinal modes. We will find in the following chapters that
the gain depends on the optical frequency (i.e., the wavelength) somewhat similar
to that shown in Figure 3.7.5. The figure also shows the allowed frequencies for the
Fabry-Perot cavity. Obviously, those modes under the center portion of the gain curve
will be the most highly excited and have the largest amplitudes. Those longitudinal
modes near the edges of the gain curve will not experience any amplification at all
and will not be present in the output spectra. The actual line shape pattern appears

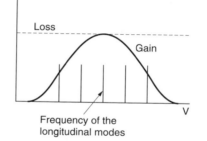

FIGURE 3.7.5

Delta-like functions mark the resonant frequencies (i.e., the frequencies of the longitudinal modes). A typical laser gain curve is shown.

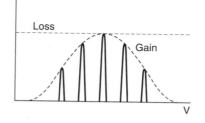

FIGURE 3.7.6

Homogeneously broadened lasers operate in the mode for which gain = loss.

in Figure 3.7.6. The gain essentially consists of the semiconductor gain multiplied by the resonator transfer function. For homogeneously broadened lasers, the gain equals the loss for only one longitudinal mode as shown. This one longitudinal mode will provide the output beam. The wavelength of the output light will be equal to the wavelength of this one particular longitudinal mode. Inhomogenously broadened lasers can simultaneously amplify all of the longitudinal modes and multiple wavelengths appear on the output beam.

Next consider the traveling modes. Previous discussion shows larger mirror reflectivity leads to smaller output power and narrower lines that span smaller ranges of wavelengths. Therefore, mirrors with large reflectivity produce very little traveling wave through the mirror and increase the proportion of the electromagnetic waves in the longitudinal modes. Ignoring material gain and distributed internal loss, a cavity with perfectly reflecting mirrors would have a power spectrum (a plot of power vs. wavelength) consisting of delta functions (i.e., functions with zero width as shown in Figure 3.7.5). The optical loss of the resonator widens the lines because the lower Q (lower finesse, wider lines) of the resonator allows frequencies adjacent to the peak to be amplified. This is just a restatement of the fact that not all of the electromagnetic wave in the cavity occupies one of the longitudinal modes. In effect, because the resonator is a lossy system, the longitudinal modes are not the exact eigenmodes for the total system (the total system, in this case, includes both the laser and the external environment). We will see shortly that the gain of the medium helps to offset the loss and tends to narrow linewidths.

An extremely important point concerns the role of homogeneously vs. inhomogeneously broadened lasers. A homogeneous broadened laser will have at most one lasing longitudinal mode. For an inhomogeneously broadened laser, any number of longitudinal modes can lase. The reason for these different behaviors has to do with the way the resonator produces the gain. An ensemble of identical atoms produces the so-called homogeneously broadened line shapes. Atoms that are affected differently from one another by their environment produce the inhomogeneously broadened line shapes. For example one or two atoms might experience a different strain than another. The Doppler effect can also produce inhomogeneous broadening. For Doppler broadening, each atom radiates at a frequency slightly different from an average frequency because of its random motion. We will see later in the book that saturation effects can cause multiple longitudinal modes to lase (even for homogeneous broadening).

Case 2   *The Real Part Produces the Threshold Condition*

We return to the expression for the denominator (Equation (3.7.12)) and consider the real part

$$\mathrm{Re}(D) = 1 - R \exp(g_{net}L) \cos(2k_o nL) \tag{3.7.13}$$

We have already chosen values for the argument of the cosine term namely

$$k_o nL = m\pi \qquad (m = 1, 2, 3 \ldots) \tag{3.7.14}$$

so that

$$\cos(2k_o nL) = 1$$

In order for the real part of the denominator to be zero, we must require

$$0 = 1 - R \exp(g_{net}L) \tag{3.7.15}$$

The gain g_{net} contains the material gain "g" and the distributed internal losses α_{int} as represented by the familiar relation

$$g_{net} = \Gamma g - \alpha_{int} \tag{3.7.16}$$

We insert the net gain into the analysis because the transfer matrix equations for Fabry-Perot cavity do not directly incorporate the laser rate equations (which require the net gain). We can solve for the net gain in Equation (3.7.15) to find

$$g_{net} = \frac{1}{L} Ln\left(\frac{1}{R}\right) \qquad \text{or} \qquad \Gamma g = \alpha_{int} + \frac{1}{L} Ln\left(\frac{1}{R}\right) \tag{3.7.17}$$

as found previously for a laser operating above threshold. We know that the laser must be operating above threshold because we explicitly require the denominator to be zero. Below threshold this denominator would not be small.

3.7.3   Line Narrowing

Consider a Fabry-Perot laser. For low levels of pumping, the "lines" in the power spectra have large linewidth as indicated in Figure 3.7.7. As the level of pumping increases, these "lines" become narrower. The present topic shows how the material gain compensates for the losses of the mirrors and the various distributed losses. When the gain equals the loss, it's almost as if the cavity has perfect mirrors and as if the medium hasn't any absorption.

Some books define the finesse $\mathscr{F}$ in order to discuss the quality of the Fabry-Perot resonator. For convenience, we define an effective reflectivity $\mathscr{R}_o$ as

$$\mathscr{R}_o = R \exp(g_{net}L) \tag{3.7.18}$$

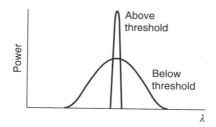

FIGURE 3.7.7
Compare line shapes above and below lasing threshold.

Grant Fowle's book shows that the finesse can be written as

$$\mathscr{F} = \frac{4\mathscr{R}_o^2}{(1 - \mathscr{R}_o)^2} \tag{3.7.19}$$

We can think of the finesse as the reciprocal of the linewidth. The "finesse" essentially measures the quality of the Fabry-Perot cavity as the number of round trips that a light pulse can make from one mirror to the other and back again before it dissipates away. Therefore large gain should produce large finesse and small line widths. The following discussion shows how this happens.

The spectrum of emitted power versus wavelength can be found by plotting the power transfer function vs. wavelength (k_o in the transfer function depends on wavelength). Recall that the denominator for the transfer function (Equation (3.7.12)) appears as

$$D = 1 - R \exp(g_{net}L) \exp(i2k_onL) = 1 - R \exp(g_{net}L) \cdot [\cos(2k_onL) + i \sin(2k_onL)] \tag{3.7.20}$$

Near resonance, we have $\lambda_n \sim 2L/m$ and so $2k_onL = 2k_nL = 4\pi L/\lambda_n \sim 2\pi m$. Therefore the sine term approximates 0 and the cosine term approximates 1 according to

$$\sin(2k_onL) \sim \sin(2\pi m) = 0 \quad \text{and} \quad \cos(2k_onL) \sim \cos(2\pi m) = 1$$

where m and λ_n represent the mode number and the wavelength in the medium, respectively. Substituting these into Equation (3.7.20) shows the relevant form of the denominator near resonance.

$$D \cong 1 - R \exp(g_{net}L) \tag{3.7.21}$$

Recall the effective reflectivity $\mathscr{R}_o = R \exp(g_{net}L)$ in Equation (3.7.18) to see the denominator in Equation (3.7.21) appears in the denominator of the finesse $\mathscr{F} = 4\mathscr{R}_o^2/(1 - \mathscr{R}_o)^2$ in Equation (3.7.19).

Now we show the finesse becomes large and the linewidth decreases as the pumping level increases. To show the denominator term becomes small, we need an accurate expression for the gain—more than just saying gain equals loss. We need to show that as the photon density in the cavity becomes large (due to lasing) then the gain approaches the loss. To this end, we use the steady-state photon laser rate equation

$$0 = \frac{d\gamma}{dt} = v_g\Gamma g\gamma - \gamma v_g(\alpha_{\text{int}} + \alpha_m) + \beta R_{sp}$$

Solving for the net gain (excluding the mirror term since its included in R),

$$g_{net} = \Gamma g - \alpha_{int} = \alpha_m + \frac{\beta v_g R_{sp}}{\gamma}$$

As the photon density γ increases, the term containing the spontaneous emission decreases. Substituting g_{net} into the denominator term $(1-R_o)$ yields

$$1 - \mathscr{R}_o = 1 - R\,\exp(g_{net}L) = 1 - R\exp(\alpha_m L)\exp\left(\frac{\beta v_g R_{sp}}{\gamma}\right)$$

Fortunately, we already know that the mirror loss term is given by

$$\alpha_m = \frac{1}{L}\,\mathrm{Ln}\left(\frac{1}{R}\right)$$

which can be rearranged to show that

$$R\,\exp(\alpha_m L) = 1$$

As a result

$$1 - \mathscr{R}_o = 1 - R\,\exp(g_{net}L) = 1 - \exp\left(\frac{\beta v_g R_{sp}}{\gamma}\right)$$

Therefore, for photon densities above threshold, the exponential approaches 1 and the denominators approach zero. The width of each line therefore approaches zero. Just for completeness, the finesse can be written as

$$\mathscr{F} = \frac{4\mathscr{R}^2}{(1 - \mathscr{R})^2} \cong \frac{4}{\left[1 - \exp\left(\frac{\beta v_g R_{sp}}{\gamma}\right)\right]^2} \rightarrow \infty \quad \text{as} \quad \text{gain} \rightarrow \text{loss}$$

However, as a very important note, the gain is always infinitesimally smaller than the loss due to the presence of spontaneous emission. It should be clear at this point that spontaneous emission prevents the linewidth from actually becoming zero.

3.8 Introduction to Waveguides

Many devices incorporate optical waveguides as a means of transporting an optical signal from one point to another. Waveguides can be monolithically integrated on semiconductor wafers to channel light from one optically active component to another. Electrically active waveguides make possible Mach Zender interferometers, optical switches and lasers. Semiconductor lasers use waveguides to confine optical energy as it travels back and forth between end mirrors. The communications industries make extensive use of the waveguide in the form of an optical fiber.

Waveguides consist of a core material with refractive index n_2 embedded within a cladding material having smaller index n_1. The difference in refractive index controls the amount of light confined to the core. The majority of the optical electromagnetic (EM) wave flows within the higher index core material. A portion of the wave extends beyond the core into the cladding as the evanescent tail. Decreasing the difference in index $n_2 - n_1$ increases the evanescent tail.

We explore two complimentary approaches to waveguiding. The geometric optics approach uses rays to represent the waves and provides the most convenient visual picture. The law of reflection and Snell's law play dominant roles. The physical optics approach in the following section solves Maxwell's equations and provides the most detailed description of waveguiding. Many excellent books can be found covering the waveguides. The following books treat waveguiding in a manner similar to the present book: (1) the 5th ed. of Yariv's book on *Quantum Electronics* published by John Wiley and Sons and (2) the third edition of R. G. Hunsperger's book *Integrated Optics: Theory and Technology* which is published by Springer-Verlag with the copyright date of 1991.

3.8.1 Basic Construction

The present section explores elementary optical waveguiding appropriate for semiconductor lasers that require optical confinement along both the transverse and lateral directions (see Figure 3.8.1). These structures most often achieve transverse confinement using index guiding. The higher index core material contains the active region while the lower index material forms the cladding. The lateral waveguiding can be achieved by three methods. First, the necessary change in refractive index can be achieved by etching away some of the material to form a ridge waveguide so that an air–semiconductor interface slightly lowers the effective index. A second method, gain guiding, uses the injected current to change the carrier density, which very slightly changes the refractive index. The third method uses regrowth to place lower index material next to the active region. Regardless of the exact mechanism, all waveguiding requires a difference in refractive index. The real challenge consists of finding materials that can be compatibly combined, at a low cost, and that do not decrease the performance of the device.

3.8.2 Introduction to EM Waves for Waveguiding

Consider a symmetric slab waveguide composed of three transverse layers as shown at the top of Figure 3.8.2. The middle layer has larger refractive index than both the lower and upper ones ($n_2 > n_1$). A symmetric waveguide has two outside clad layers with the same refractive index. The middle portion of the figure shows an enlarged view with

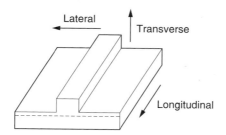

FIGURE 3.8.1
Directions relative to the waveguide and substrate.

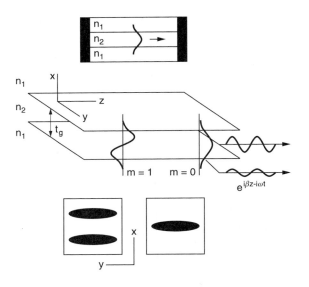

FIGURE 3.8.2
The $m=0$ and $m=1$ transverse modes.

waves propagating along the z-direction (longitudinal direction) and showing "standing waves" along the x-direction (transverse). As the wave propagates along the longitudinal direction, some of the optical energy penetrates into the low index regions. The curves labeled as $m=0$ and $m=1$ represent the amplitude of the electric field at a point (x, y) in the x–y plane; the exponential factor $e^{i\beta z - i\omega t}$ multiplies this amplitude. At the far right of the figure, the two arrows pointing toward the right represent the (effective) wave vector $\vec{\beta}$ (which is related to the complex wave vector that we discussed previously). Notice that, for the $m=0$ mode, the amplitude of the wave is largest near the center and smallest near the tails. The tails of both the $m=0$ and $m=1$ curves represent the evanescent fields that propagate to the right but decay exponentially with the distance into the n_1 material. Sines or cosines describe the center portion of the curves. The bottom portion of Figure 3.8.2 shows the power distribution versus position on the output facet (i.e., we are looking at the right-hand side of the middle portion of the figure). The lobes represent the "brightest" portion of the beam. The size of the small vertical widths of the lobes roughly correspond to the distance t_g between the two interfaces. The horizontal width of the lobe corresponds to the horizontal size of the two slabs. The fact that the beams have finite width along y indicates that there must be a mechanism to confine the beam in that direction even though we have not shown it in the figure. The $m=0$ transverse mode has a single maximum of the power distribution within the lobe. The $m=1$ transverse mode has two peaks in the optical power distribution. The peaks correspond to bright spots when viewing the laser output on a screen. Notice that the number of bright spots must equal the number of maxima for the electric field amplitude; this occurs because the power in the beam is proportional to the square of the electric field.

The following topics introduce waveguiding through the geometric optics approach.

3.8.3 The Triangle Relation

Consider a plane wave traveling the length of a symmetric waveguide as shown in Figure 3.8.3. Assume real refractive index "n_2" for convenience. In terms of ray optics, the wave bounces between the interfaces as it travels down the waveguide so long as the angle of incidence at each interface exceeds the critical angle (total internal reflections).

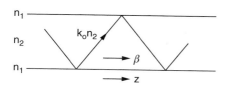

FIGURE 3.8.3

Net motion of the wave in the waveguide is to the right.

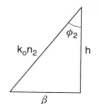

FIGURE 3.8.4

The triangle relation for the wave vectors.

The magnitude of the actual wave vector must be $k_o n_2$. The wave has net motion toward the right that can be represented by an effective wave vector $\vec{\beta}$. The magnitude of effective propagation constant β differs from the propagation constant $k_o n_2$ previously used. We can define an effective index n_{eff} so that the effective propagation constant becomes $\beta = k_o n_{\text{eff}}$ and the average speed of the wave must be c/n_{eff}. The waveguiding structure produces a speed along the length of the waveguide (i.e., the z direction) between that for the cladding and the core. This result appears counter intuitive since $\beta \leq k_o n_2$ from the triangle relation.

Now we discuss a triangle relation for the actual and effective wave vectors. Figure 3.8.3 shows that the wave vector $k_o n_2$ is not parallel to the effective wave vector β. The figure indicates the effective wave vector must be smaller than the actual wave vector $\beta \leq k_o n_2$. The triangle relation appears in Figure 3.8.4. As shown, the vector β represents propagation along the length of the waveguide while the vector "h" represents propagation perpendicular to the length of the waveguide. The quantity "h" must also be a wave vector. The triangle diagram provides the relation

$$\beta^2 + h^2 = (k_o n_2)^2 \tag{3.8.1}$$

The "h" represents the magnitude of a wave vector $\vec{h}$ for waves propagating perpendicular to the interface. In the physical optics approach, we will see that the perpendicular motion sets up standing waves as shown in Figure 3.8.5. The wave vector "h" in Equation (3.8.1) represents the "m" waves for the transverse direction in Figure 4.1.2. Only certain values of "h" produce standing waves. That is, there are certain allowed wavelengths for the transverse modes approximately given by $\lambda_{\text{trans}} = 2\pi/h = 2t_g/(m+1)$. The fact that "$h$" takes on discrete values means that only certain values of ϕ_2 are allowed and therefore only certain values of β.

3.8.4 The Cut-off Condition from Geometric Optics

A waveguide can only propagate electromagnetic waves with certain values for the effective wave vector. The previous topic indicates that the angle the wave vector makes

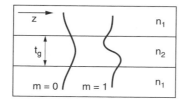

FIGURE 3.8.5

The wavelength is approximately equal to $2t_g$ and to t_g for $m = 0, 1$ respectively.

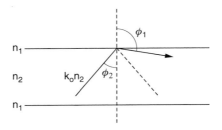

FIGURE 3.8.6
A slab waveguide with a beam undergoing total internal reflection.

with respect to the normal (see Figure 3.8.2) determines the effective wave vector β. Only a narrow range of angles produces a wave confined to the waveguide. The triangle relation requires $\beta \leq k_o n_2$. This topic shows how Snell's law and the existence of a critical angle for total internal reflection (TIR) leads to a minimum value for the propagation constant $k_o n_1 \leq \beta$. The smallest value of β is the cutoff value.

Consider Figure 3.8.6 depicting a symmetric slab waveguide. Suppose a wave (with wave vector $k_o n_2$) travels in a direction making the angle ϕ_2 with respect to the normal. Small angle ϕ_2 produces a substrate mode whereby the wave propagates across the interface into the interior of material n_1. The substrate mode is not a guided mode and not very desirable in most circumstances.

Next consider the condition for waveguiding. As the angle ϕ_2 increases, the angle ϕ_1 increases past 90 degrees and the wave undergoes total internal reflection. The wave bounces back and forth between the interfaces as it travels along the length of the waveguide. Snell's law ($n_2 \sin\phi_2 = n_1 \sin\phi_1$) provides a relationship between the various refractive indices and wave vectors. Setting $\phi_1 = 90°$, we require $\sin\phi_2 \geq n_1/n_2$ for total internal reflection. However, the triangle relation in Figure 3.8.4 shows that $\sin\phi_2 \geq \beta/k_o n_2$ and as a result, for waveguiding we must have

$$\beta \geq k_o n_1 \tag{3.8.2}$$

This last relation gives the cutoff condition for the waveguide. Effective wave vectors smaller than $k_o n_1$ will not propagate along the length of the guide. Some waveguides have cladding layers with differing refractive indices on either side of the core. In such a case, the situation becomes more complicated; very interesting switching mechanisms can be realized. However, this more complicated situation will not be covered here.

Combining the cutoff condition and the triangle relation, we find that the effective wave vector must have a magnitude within the range given by

$$\beta_{\min} = k_o n_1 \leq \beta \leq k_o n_2 = \beta_{\max} \tag{3.8.3}$$

The $m=0$ transverse (Figure 3.8.7) mode corresponds to the condition $\beta \cong k_o n_2$ since the smallest $h \sim \pi/t_g$ produces the largest $\beta = \sqrt{(k_o n_2)^2 - h^2} \cong k_o n_2$. Higher-order modes correspond to larger "m" which means smaller wavelength along the direction perpendicular to the interfaces which, in turn, means larger wave vectors $h \sim 2\pi/\lambda_{\text{perp}}$. Higher-order modes therefore have smaller effective wave vectors β. The waveguide allows only certain values of β since the transverse direction (perpendicular to the layers in Figure 3.8.7) accommodates only certain wavelengths λ_{perp}. If the wave must be guided then only those effective wave vectors in the range

$$k_o n_1 \leq \cdots \beta_2 \leq \beta_1 \leq \beta_0 \leq k_o n_2 \tag{3.8.4}$$

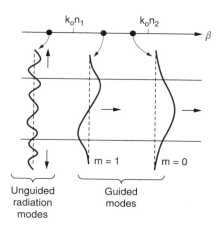

FIGURE 3.8.7
The characteristics of the various possible regions of β.

$k_o n_1 \leq \ldots \beta_2 \leq \beta_1 \leq \beta_0 \leq k_o n_2$ will work. Figure 3.8.7 shows the transverse modes for the various regions of the effective wave vector.

3.8.5   The Waveguide Refractive Index

We see that the confinement by the waveguide requires us to characterize the wave motion by an effective propagation constant β rather than the wave vector $k_n = k_o n_2$. We might expect the speed of the wave along the length of the guide to be smaller than usual since the ray does not travel straight down the waveguide. However, the structure of the waveguide forces the wave to "sample" two different materials with index n_1 and n_2 so that the effective average must be between the two $n_1 \leq n_{\text{eff}} \leq n_2$.

A plane wave traveling along a path making an angle ϕ_2 as shown in Figure 3.8.4 has the form

$$\mathscr{E} = E_o \, e^{i\vec{k}_n \cdot \vec{r} - i\omega t} \tag{3.8.5}$$

as represented schematically in Figure 3.8.5. Along the lines representing the wavefronts, the field has constant phase $\phi = \vec{k} \cdot \vec{r}$. Consider the effective propagation vector β and the velocity of the wave v_z along the length of the waveguide. We define an effective (guide) index n_{eff} by the relation $v_z = c / n_{\text{eff}}$ and

$$\beta = k_o n_{\text{eff}} \tag{3.8.6}$$

where $\vec{\beta} = \vec{k}_n \sin \phi_2$.

The index n_{eff} must be smaller than the usual propagation index n_2 since the triangle relation gives

$$\beta \leq k_o n_2 \quad \rightarrow \quad n_{\text{eff}} \leq n_2 \tag{3.8.7a}$$

This last expression therefore indicates the speed of the wave along the length of the waveguide (z direction) must be larger than a wave propagating in the medium n_2 without the wave guide structure.

$$v_z = c / n_{\text{eff}} \geq c / n_2 \tag{3.8.7b}$$

The effective index n_{eff} includes both the effects of n_2 and the zig-zag motion along the guide. It must be smaller than n_2 since the wave penetrates into the cladding region with the smaller refractive index n_1. The cutoff condition gives

$$k_o n_1 \leq \beta \;\rightarrow\; v_z = \frac{c}{n_{\text{eff}}} \leq \frac{c}{n_1} \qquad (3.8.7c)$$

The value of n_{eff} depends on the size of the wave vector h. Large "h" and therefore small β correspond to a wave propagating nearly perpendicular to the length of the waveguide.

Similar to phase velocity, we can define a group velocity for a wave propagating along the longitudinal z direction as

$$v_{gw} = \frac{d\omega}{d\beta} \qquad (3.8.8)$$

The guided group velocity v_{gw} is

$$v_{gw} = \frac{d\omega}{d\beta} = \frac{d\omega}{dk_n}\frac{dk_n}{d\beta} = v_g \frac{dk_n}{d\beta} \qquad (3.8.9)$$

where the subscripts "g, w" refer to "group" and the "waveguiding" case, v_g is the group velocity without waveguiding, and k_n (assumed real) appears in Equation (3.8.1). Using the triangle in Figure 3.8.4, we have $\beta = k_n \sin \phi_2$ and so

$$v_{gw} = v_g \frac{dk_n}{d\beta} = \frac{v_g}{\sin \phi_2} \geq v_g$$

3.9 Physical Optics Approach to Waveguiding

Although the geometric optical approach to waveguiding provides some insight into the nature of waveguiding, the physical optics approach forms a more predictive system covering a greater diversity of cases. The physical optics approach uses Maxwell's equations to model the waveguide structure. Combining these equations with the appropriate boundary conditions produces the transverse modes, evanescent fields and the allowed propagation constants. In a waveguide with mirrors at either end, the waveguide supports both longitudinal and transverse modes. The transverse mode produce regions of higher and lower intensity in the beam. The section first introduces the waves, next finds the solutions to Maxwell's equations, and then discusses some applications.

3.9.1 The Wave Equations

The basic symmetric slab waveguide appears in Figure 3.9.1. The wave propagates toward the right. The thickness of the core (n_2) is t_g (the subscript "g" make sense if you think of the core as made of glass). The figure shows an $m = 0$ transverse mode. The wave penetrates approximately a distance "$1/p$" into the cladding with refractive indices n_1 (where $n_1 < n_2$); this penetrating wave defines the evanescent field.

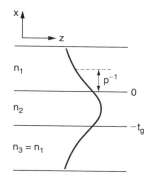

FIGURE 3.9.1
A slab waveguide with the penetration depth $1/p$.

A number of devices rely on evanescent fields for proper operation. For example, consider two closely spaced waveguides. Evanescent coupling occurs when the cores of the two waveguides come within the approximate distance $1/p$. The wave propagating in one guide can "leak" into the other one since the evanescent fields couple the energy between them. Electrically controlled optical switches make use of such arrangements. Applying a voltage to material separating the two waveguides can alter the refractive index and thereby switch "on" or "off" the cross coupling. As another example, consider the evanescent fields in a semiconductor laser. The doping should be as close as possible to the active region (i.e., core region with thickness t_g) in order to reduce the electrical resistance as much as possible. However, the optical mode extends a distance of approximately "$1/p$" from the core; in many cases, this distance is $p^{-1} \sim 200$ nm. Free carriers within this distance "$1/p$" can "absorb" the light (i.e., free carrier absorption). Therefore it is best to keep the doping more than distance p away from the core.

Let us continue with the slab waveguide in Figure 3.9.1. We divide the waveguide into three sections with differing indices of refraction. The z-direction is along the length of the waveguide, the x-direction is perpendicular to the layers (transverse direction), and the y-direction points upward out of the page (lateral direction). The upper interface for the n_2 material defines $x = 0$ while the lower interface defines $x = -t_g$.

Sections 3.1 and 3.2 show how Maxwell's equations lead to the EM wave equation and the meaning of the complex wave vector. In the present section, we assume real indices of refraction and we assume non-conductive media (negligible free carrier absorption $\sigma = 0$). Each of the three regions has an associated wave equation with very similar form; the only difference concerns the various possible values of the index of refraction n_j.

$$\nabla^2 \vec{\mathscr{E}} - \frac{n_j^2}{c^2} \frac{\partial^2 \vec{\mathscr{E}}}{\partial t^2} = 0 \tag{3.9.1}$$

We assume real refractive indices n_j ($j = 1, 2, 3$) and real corresponding wave vectors. The phase velocity of the light in any of the materials must be $v_j = c/n_j$, and the refractive indices do not include waveguiding effects. Each region produces a wave equation as in Equation (3.9.1). Regions 1 and 3 have the same refractive index and therefore have the same wave equation. We should consider both transverse electric TE and transverse magnetic TM polarization for the electromagnetic wave. The TE case places the electric field parallel to the interface while the TM case places the magnetic field parallel to the interface. Figure 3.9.2 considers the TE case with the electric field parallel to the y direction (the lateral direction). Often lasing starts in the TE mode for semiconductor lasers since the TM mode often require slightly higher threshold current.

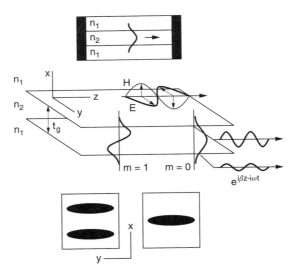

FIGURE 3.9.2
The TE mode shown in relation to the layers and the x, y, z directions.

3.9.2 The General Solutions

We already know that the solutions to the wave equation for the three regions have the form

$$\vec{\mathscr{E}} = \tilde{y}E_y(x)\exp(i\beta z - i\omega t) \tag{3.9.2}$$

where we consider β to be known. Each of the three regions produce a solution similar to Equation (3.9.2). Notice that we allow the amplitude to vary only along the x-direction (perpendicular to the layers) as described by $E_y(x)$; the electric field points in the y-direction as shown by the unit vector $\tilde{y}$ and the subscript y on E_y. Substituting into the wave equation we find

$$\frac{\partial^2 E_y}{\partial x^2} + (k_o^2 n_i^2 - \beta^2)E_y = 0$$

where

$$k_o n_j = \frac{n_j \omega}{c} \tag{3.9.3}$$

We can solve the second-order differential equation (with constant coefficients) in Equation (3.9.2) for E_y. The solutions have the form

$$E_y(x) \sim \exp\left(\pm i x \sqrt{k_o^2 n_j^2 - \beta^2}\right)$$

The solutions are easily seen to be sinusoidal or exponential according to

$$\text{Sinusoidal} \quad k_o^2 n_j^2 > \beta^2$$

$$\text{Exponential} \quad k_o^2 n_j^2 < \beta^2$$

By using the triangle relation ($\beta < k_o n_2$) and the relation for cut-off ($\beta > k_o n_1$), we see that Region 1 has exponential solutions while Region 2 has sinusoidal solutions.

We assume specific boundary conditions, but in particular for $x \to \pm\infty$, we require the solutions to remain finite. The solutions for the three regions then have the general form

$$\text{Region 1} \quad E_y = A \exp(-px)$$

$$\text{Region 2} \quad E_y = B \cos(hx) + C \sin(hx)$$

$$\text{Region 3} \quad E_y = D \exp[p(x + t_g)]$$

where

$$p = \sqrt{\beta^2 - k_o^2 n_1^2} \qquad h = \sqrt{k_o^2 n_2^2 - \beta^2} \qquad (3.9.4)$$

In order to determine the allowed β and the unspecified constants A, B, and C, we need to match the solution in Region 1 with the solution in Region 2 and so on. Therefore boundary conditions must be specified at the two interfaces. We expect at least one arbitrary constant in the final solution since the overall field strength (i.e., the beam power) has not been specified.

3.9.3 Review of the Boundary Conditions

The boundary conditions for electromagnetics can be found in Section 3.3. First consider the boundary condition on magnetic fields tangent to a boundary as shown in Figure 3.9.3. Physical currents can change tangential magnetic fields H but do not affect magnetic fields perpendicular to the interface. Therefore, magnetic fields perpendicular to an interface must be continuous across the interface. The electric field must also satisfy boundary conditions. The polarization or free charge at a surface causes a discontinuity in the value of an electric field polarized perpendicular to the interface (refer to Figure 3.9.4). A transverse electric field must be continuous across the interface. We assume that the partial derivatives of the transverse electric field are also continuous.

3.9.4 The Solutions

Let's repeat the general solutions for convenience

$$\text{Region 1} \quad E_y = A \exp(-px)$$

$$\text{Region 2} \quad E_y = B \cos(hx) + C \sin(hx) \qquad (3.9.5a)$$

$$\text{Region 3} \quad E_y = D \exp[p(x + t_g)]$$

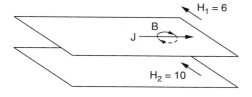

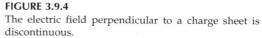

FIGURE 3.9.3
Tangential magnetic fields are discontinuous across sheets of current.

FIGURE 3.9.4
The electric field perpendicular to a charge sheet is discontinuous.

where

$$p = \sqrt{\beta^2 - k_o^2 n_1^2} \qquad h = \sqrt{k_o^2 n_2^2 - \beta^2} \qquad (3.9.5b)$$

The following table lists the boundary conditions and the results of applying those boundary conditions to the solutions for the three regions given above.

Condition	Result
1. E_y continuous at $x = 0$	$A = B$
2. $\dfrac{\partial E_y}{\partial x}$ continuous at $x = 0$	$C = -\dfrac{P}{h} A$
3. E_y continuous at $x = -t_g$	$D = A \cos(h\, t_g) + \dfrac{pA}{h} \sin(h t_g)$
4. $\dfrac{\partial E_y}{\partial x}$ continuous at $x = -t_g$	$\tan(h t_g) = \dfrac{2ph}{h^2 - p^2}$

The first three results can be used to write the electric field as a function of position within the waveguide. Substituting into the coefficients from the table into Equations 3.9.5a for the electric fields produces the results

Region 1: $\quad \vec{\mathscr{E}} = \tilde{y} E_y(x)\, e^{i\beta z - i\omega t} = A\, \tilde{y}\, e^{-px} e^{i\beta z - i\omega t}$

Region 2: $\quad \vec{\mathscr{E}} = \tilde{y} E_y(x)\, e^{i\beta z - i\omega t} = A\, \tilde{y} \left[\cos(hx) - \dfrac{p}{h} \sin(h t_g) \sin(hx) \right] e^{i\beta z - i\omega t}$

Region 3: $\quad \vec{\mathscr{E}} = \tilde{y} E_y(x)\, e^{i\beta z - i\omega t} = A\, \tilde{y} \exp[p(x + t_g)] \left[\cos(h t_g) + \dfrac{p}{h} \sin(h t_g) \right] e^{i\beta z - i\omega t}$

where

$$p = \sqrt{\beta^2 - k_o^2 n_1^2} \qquad h = \sqrt{k_o^2 n_2^2 - \beta^2}$$

Notice that the allowed parameters β, p, and h, have not yet been determined; we will address this issue shortly.

As an essential fact concerning the solutions in regions 1 and 3, the waves propagate along the z-direction because of the factor exp($i\beta z - i\omega t$) even though they decay along the x-direction (transverse direction). Therefore, lasers should have good mirrors not only for the core region (i.e., the n_2 region with thickness t_g), but also extending into the cladding for the distance "$1/p$." The value of "$1/p$" is typically 0.2 μn for GaAs lasers.

How do we find this penetration depth "$1/p$" for the evanescent field? And what are the values for β and h? These all follow from the relation (given in the table)

$$\tan(h t_g) = \frac{2ph}{h^2 - p^2} \qquad (3.9.6)$$

by writing "h" and "p" in terms of β using

$$p = \sqrt{\beta^2 - k_o^2 n_1^2} \qquad h = \sqrt{k_o^2 n_2^2 - \beta^2} \qquad (3.9.7)$$

and then solving for the effective propagation constant β. One generally obtains a large number of allowed values for β. An easy way to find the allowed values of the effective propagation constant consists of simultaneously plotting the left and right sides of

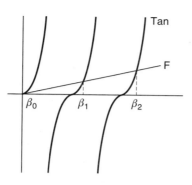

FIGURE 3.9.5
Intersection points give allowed values of the effective wave vector.

Equation (3.9.6)

$$\tan\left[t_g h\right] = \tan\left[t_g \sqrt{k_o^2 n_2^2 - \beta^2}\right] \quad \text{and} \quad F = 2ph/(h^2 - p^2)$$

versus β on the same set of axes. The intersection points of the two sets of curves provide the allowed values of the effective propagation constant β (see Figure 3.9.5). However, not all intersection points provide allowed values for β since the triangle relation and propagation cut-off place limits on the allowed range according to

$$\beta_{\text{cut-off}} = k_o n_1 \leq \beta \leq k_o n_2 = \beta_{\max}$$

Once having found the values of β, the values of "p" and "h" can also be found. The "p" parameter gives the penetration depth $(1/p)$ of the evanescent field into the cladding surrounding the core of the waveguide.

The "h" parameter occurs in the solution for region 2 given by the following equation.

$$\text{Region 2}: \vec{\mathscr{E}} = \tilde{y} E_y(x)\, e^{i\beta z - i\omega t} = A\,\tilde{y}\left[\cos(hx) - \frac{p}{h}\sin(ht_g)\,\sin(hx)\right] e^{i\beta z - i\omega t}$$

It is easy to see that "h" controls the period of the transverse sine or cosine wave within the core of the waveguide; it controls the mode structure (i.e., $m=0$, $m=1$ etc). We therefore see that the acceptable values of "h" must approximate the number of half-integral wavelengths that fit within the distance t_g. In other words to lowest order approximation, if $\lambda_n' = t_g/m$ represents the wavelength within a dielectric, then $h \cong 2\pi/\lambda_n' = 2\pi m/t_g$. However, this is only approximate because the wave actually extends part way into the cladding. Furthermore, we cannot allow all integer values of "m" because of the cut-off condition.

Example 3.9.1 Find the propagation constant and penetration depth for a GaAs-Al$_{0.6}$Ga$_{0.4}$As slab waveguide with $t_g = 200$ nm, $n_2 = 3.63$, $n_1 = 3.25$ using $\lambda_o = 850$ nm.

Solution: Figure 3.9.5 shows a plot near the intersection point $\beta = 25.5$. Other books show an easier graphical method to find the intersection points by plotting $R^2 = (pt_g)^2 + (ht_g)^2$ and $\tan[t_g h]$ on the same set of axis. Equation (3.9.7) gives the penetration depth of approximately 116 nm.

3.9.5 An Expression for Cut-off

In this topic, we find a relationship between (i) the wavelengths of light capable of propagating through a waveguide with core thickness t_g, (ii) the difference in indices $\Delta n = n_2 - n_1$, and (iii) the sum of indices $n_2 + n_1$ which can often be approximated by either $2n_2$ or $2n_1$ for semiconductor-type waveguides. At cut-off, Equation (3.9.5) provides $\beta \sim k_o n_1$ so that the equations

$$p = \sqrt{\beta^2 - k_o^2 n_1^2} \qquad h = \sqrt{k_o^2 n_2^2 - \beta^2}$$

become

$$p = 0 \qquad h = \sqrt{k_o^2 (n_2^2 - n_1^2)}$$

From $p = 0$, we see that the penetration depth $= 1/p$ extends a long way into the cladding region. Although it might seem paradoxical at first, an electromagnetic wave propagating along the waveguide experiences greater optical scattering (and hence greater distributed loss) when "p" becomes smaller.

Let's continue with the calculation. The tangent function becomes

$$\tan(h t_g) = \frac{2ph}{h^2 - p^2} = 0 \quad \Rightarrow \quad h t_g = m\pi$$

where $m = 1, 2, 3 \ldots$ denotes the same mode index as used previously. The solutions for the three regions become

Region 1: $\vec{\mathscr{E}} = \tilde{y} E_y(x) e^{i\beta z - i\omega t} = A\, \tilde{y}\, e^{i\beta z - i\omega t}$

Region 2: $\vec{\mathscr{E}} = \tilde{y} E_y(x) e^{i\beta z - i\omega t} = A\, \tilde{y}\, \cos(hx)\, e^{i\beta z - i\omega t} = A\, \tilde{y}\, \cos\left(\dfrac{m\pi x}{t_g}\right) \rightleftharpoons e^{i\beta z - i\omega t}$

Region 3: $\vec{\mathscr{E}} = \tilde{y} E_y(x) e^{i\beta z - i\omega t} = A\, \tilde{y}\, \cos(h t_g) e^{i\beta z - i\omega t} = A\, \tilde{y}\, (-1)^m e^{i\beta z - i\omega t}$

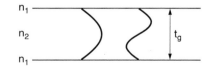

FIGURE 3.9.5
Approximate transverse modes in a slab waveguide.

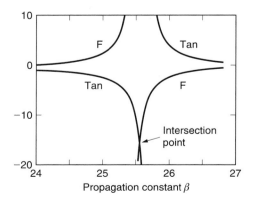

FIGURE 3.9.6
Example plot for 850 nm and $t_g = 200$ nm.

Notice that the wave has the same size in both regions 1 and 3.
The triangle relation provides

$$h^2 = k_o^2 n_2^2 - \beta^2 \leq k_o^2 (n_2^2 - n_1^2)$$

where the cut-off condition $\beta \sim \beta_{\text{cut-off}} = \beta_{\min} = k_o n_1$ was used. The "less than or equal sign" occurs because we require $\beta \geq \beta_{\text{cut-off}} = k_o n_1$. Substituting the relation $h t_g = m\pi$, $k_o = 2\pi / \lambda_o$, and $\Delta n = n_2 - n_1$ we find

$$\frac{(m\pi)^2}{t_g^2} \leq \left(\frac{2\pi}{\lambda_o}\right)^2 \Delta n (n_2 + n_1)$$

So that only modes with

$$\lambda_o^2 \leq \Delta n (n_2 + n_1) \frac{4 t_g^2}{m^2}$$

will propagate. Notice that the lowest order mode $m = 0$ will always propagate in a *symmetric* waveguide. An *asymmetric* waveguide has three different refractive indices for the three regions. Even though electromagnetic waves with arbitrary wavelength will propagate, the confinement will be poor.

3.10   Dispersion in Waveguides

The rate at which light propagates along a waveguide depends on the frequency of the wave and upon the construction of the waveguide. We discuss intermodal and intramodal dispersion and how they limit the bandwidth of communication systems. The term "mode" appears in a number of contexts. The waveguide mode refers to the particular zig-zag path along which the beam can propagate. This is equivalent to specifying the transverse wave pattern embodied by the "h" wave vectors in previous sections. Alternately, the mode can be specified by the pattern of bright spots observed on an output screen.

"Dispersion in waveguides" refers to the spreading of a pulse of light as it travels the length of the waveguide. We will use rectangular optical fiber as our prototype waveguide. We consider two basic mechanisms responsible for the spreading. The first concerns the construction of the waveguide. Light can follow a zig-zag path with some paths longer than others. Also, some light penetrates into the material with lower index and therefore travels faster than the light not penetrating as far. The second mechanism concerns the index of refraction. Material dispersion refers to the fact that light with different frequencies (i.e., different colors) travel at differing speeds. Although not considered here, we might expect the speed of light to depend on polarization as well.

We can distinguish between intermodal and intramodal dispersion. *Intermodal* dispersion refers to light, once injected into the fiber, traveling in multiple waveguide modes at the same time. As mentioned above, the different modes have different path lengths and lead to varying penetration into the low index cladding. Therefore, various parts of the wave travel at various speeds and the pulse must broaden. *Intramodal* dispersion refers to light that travels in exactly one waveguide mode (such as for a single mode fiber). In this case, we eliminate any delays due to light propagating in multiple modes (different path lengths for example). However, the waveguide group velocity can still depend on construction. For example, consider light made of multiple frequencies propagating in a single mode. In this case, light with longer wavelength penetrates further into the lower-index cladding and therefore travels faster. Also, the refractive index depends on frequency.

3.10.1 The Dispersion Diagram

In general, the dispersion diagram displays ω vs k or E vs k, where $E = \hbar\omega$ represents energy. The slope of the curves in the dispersion diagram gives the group velocity of EM waves. This topic applies the same ideas for a wave propagating along the length of the waveguide.

The dispersion diagram for a waveguide shows the relation between the angular frequency ω and the effective propagation constant β. The slope provides the group velocity of the wave along the length of the waveguide. Figure 3.10.1 shows an example (Reference 10, Kasap's book). Some points should be noted. First, the diagram shows that for a given ω, only certain values of β are allowed (as found in previous sections); these values are found by drawing a horizontal line through the chosen value of ω. Second, if ω (the color) varies continuously so does β for values past cut-off. Third, the two dotted lines give the maximum and minimum waveguide group velocities. Previous sections demonstrate the minimum and maximum values of β according to

$$\beta_{\min} = k_o n_1 \leq \beta \leq k_o n_2 = \beta_{\max}$$

where

$$k_o = \frac{2\pi}{\lambda_o} = \frac{\omega}{c}$$

and k_o is the wave vector in vacuum. Therefore, we can find the minimum and maximum waveguide group velocity according to (ignoring any dependence of n on ω)

$$v_{wg}^{(\min)} = \frac{\partial \omega}{\partial \beta_{\max}} = (\partial \beta_{\max}/\partial \omega)^{-1} = (\partial k_o n_2/\partial \omega)^{-1} = \left(\frac{\partial}{\partial \omega} \frac{\omega n_2}{c} \right)^{-1} = \frac{c}{n_2}$$

and similarly for the maximum

$$v_{wg}^{(\max)} = \frac{c}{n_1}$$

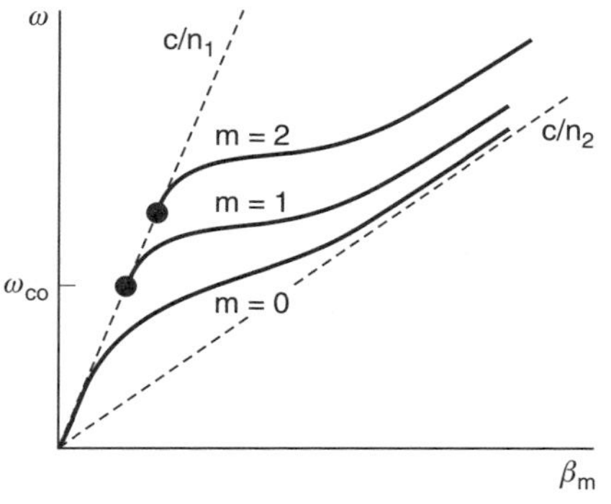

FIGURE 3.10.1
Dispersion curves for fiber in TE modes (after Kasap).

The maximum and minimum phase velocity serve as fiduciaries. Fourth, the different modes have different cutoff frequencies. The $m = 0$ mode propagates for all frequencies. Near cutoff, each of the modes has very large group velocity indicating that the cladding layer carries the greater portion of the mode. We therefore expect large penetration into the cladding layer. The smallest frequency and largest wavelength occur at cutoff. For large frequencies, the group velocity asymptotically approaches the lower limit of c/n_2. Apparently away from cutoff, the core of the waveguide carries the majority of the mode where the wave travels slowest. Also for fixed ω, the group velocity at the allowed β tends to be larger for higher mode numbers m because of greater penetration into the cladding.

3.10.2 A Formula for Dispersion

Dispersion causes waves with different frequencies or composed of different waveguide modes to travel at different speeds. This causes the waves to broaden as they travel the length of the waveguide. Figure 3.10.2 shows a pulse that starts fairly narrow but broadens as it travels along. We would find a range of wavelengths in the Fourier decomposition of the pulses. The dispersion measures the amount of "spreading" per unit length of waveguide (or material). The "spread" can either be measured as distance or as a time. For the distance measure, we can write

$$\sigma_{\text{final}} - \sigma_{\text{initial}} = v\,\Delta\tau$$

where v represents average wave speed, and $\Delta\tau$ denotes the time required to spread from an initial width σ_{initial} to the final width σ_{final}. Equivalently, we can say $\Delta\tau$ measures the spreading of the pulse in time. The time method is preferable because it does not require an average velocity.

We can write the dispersion as a formula (dispersions add to first-order perturbation theory).

$$\frac{\text{Spread}}{\text{length}} = \frac{\Delta\tau}{\text{length}} = (D_m + D_w)\,\Delta\lambda$$

where

$$D_m = -\frac{\lambda}{c}\left(\frac{d^2 n}{d\lambda^2}\right)$$

$$D_w = \frac{1.984 N_{g1}}{(\pi t_g)^2 2 c n_1^2}$$

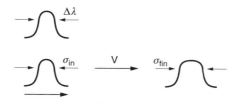

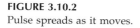

FIGURE 3.10.2
Pulse spreads as it moves.

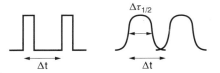

FIGURE 3.10.3
Spreading pulses limit the bandwidth.

and where the symbols m and w stand for material and waveguide dispersion, respectively, and N_{g1} represents the group index for material n_1. The group index can be found as follows (the first equality defines N_g).

$$\frac{c}{N_g} = v_g = \left(\frac{\partial k_n}{\partial \omega}\right)^{-1} = \left(\frac{\partial}{\partial \omega} \frac{\omega n}{c}\right)^{-1} = \left(\frac{n}{c} + \frac{\omega}{c} \frac{\partial n}{\partial \omega}\right)^{-1} = \frac{c}{n + \omega \frac{\partial n}{\partial \omega}}$$

so that

$$N_g = n + \omega \frac{\partial n}{\partial \omega} = n - \lambda \frac{\partial n}{\partial \lambda}$$

3.10.3 Bandwidth Limitations

Communications systems have transmitters that modulate a laser and inject the signal into an optical fiber. At the other end of the fiber, a detector circuit receives the signal. Digital transmitters send pulses of light, which represent 0, 1. Suppose that R is the repitition rate for the pulses. R has units of #pulses/sec so that the time between a point on one pulse to the identical point on an adjacent one must be $\Delta t = 1/R$. Assume for simplicity that the pulses are very narrow.

The pulse spreads as it moves as shown in Figure 3.10.3. At some point along the fiber, the pulses will start to overlap. We can estimate the maximum possible bit rate $B = R$ by insisting that the pulses remain separated by about $2\,\Delta\tau_{1/2}$. Therefore we can write

$$B = \frac{1}{\Delta t} = \frac{0.5}{\Delta\tau_{1/2}}$$

3.11 The Displacement Current and Photoconduction

Collisions between carriers and scattering centers determine the steady state current flow and therefore the resistance of a material. These models visualize current as electrons moving past a fixed point in the material or wire. However, the motion of charge not confined to a wire as part of a circuit, such as between the plates of a capacitor, also produces current in the circuit. The motion of this charge produces a changing electric field at the position of the plates, which produces conduction current in the circuit. This displacement current does not require a conductive medium nor does it require Ohmic contacts. Figure 3.11.1 shows an example whereby light absorbed at the surface of a semiconductor produces a layer of photocarriers that move under the action of an applied field. The current returns to zero once the charge reaches the lower electrode

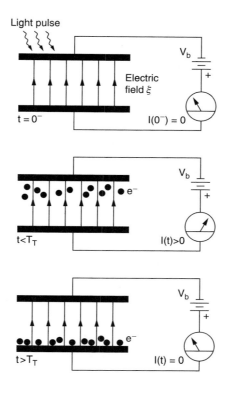

FIGURE 3.11.1
The motion of charge between the plates produces current in the external circuit.

at the "transit time" T_T. The displacement current finds common applications in AC conduction through capacitors, PIN photodetectors, electron time-of-flight experiments, and noise measurements.

3.11.1 Displacement Current

The displacement and physical currents comprise the current density between the electrodes shown in Figure 3.11.2. For simplicity, the figure shows a sheet of negative charge density σ (where $Q = \sigma A \ll CV$) moving with speed v under the action of the applied field $\mathscr{E}$. The calculation provides the same results when using a point charge rather

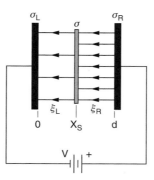

FIGURE 3.11.2
A sheet of electrons moves from left to right under the influence of an applied field.

than the sheet charge. Current flows through the battery due to the moving sheet even before it reaches the right-hand electrode because of the displacement current. The current density can be written as the sum of the displacement $J_d = \partial \mathscr{E}(x, t)/\partial t$ and physical currents $J_{pc} = v\sigma\, \delta(x - vt)$

$$J_{\text{Between}} = J_d + J_{pc} = \varepsilon \frac{\partial \mathscr{E}(x, t)}{\partial t} + v\sigma\, \delta(x - vt) \tag{3.11.1}$$

where σ and ε denote the conductivity and the permittivity of the medium, respectively, and $\delta(x - vt)$ represents the charge density at the position of the sheet. Regions outside of the sheet have only the displacement current such as for the right-hand plate, for example, before the carriers arrive.

We can show the conventional current flow in the right-hand plate J_c equals the displacement current J_d by elementary electrodynamics. Let J be the current due to flowing charge in one of Maxwell's equation

$$\nabla \times \vec{\mathscr{H}} = \vec{J} + \frac{\partial \vec{D}}{\partial t} \tag{3.11.2}$$

where the displacement field $\vec{\mathscr{D}}$ and the electric field $\vec{\mathscr{E}}$ can be related through the permittivity by $\vec{\mathscr{D}} = \varepsilon\vec{\mathscr{E}}$. Using the relation $\nabla \cdot \nabla \times \vec{\mathscr{H}} = 0$, the last equation becomes

$$\nabla \cdot \vec{J} = -\varepsilon \frac{\partial \vec{\mathscr{E}}}{\partial t} \tag{3.11.3}$$

For a Gaussian box with one side at d^+ (just inside the right-hand plate) and another to the left of the plate surface by a distance Δx, Equation (3.11.3) can be approximated by

$$\frac{J(d^+, t) - J(d - \Delta x, t)}{\Delta x} = -\varepsilon \frac{\partial}{\partial t} \frac{\mathscr{E}(d^+, t) - \mathscr{E}(d - \Delta x, t)}{\Delta x}$$

Using $J(d - \Delta x, t) = 0$, $J(d^+, t) = J_c$, and $\mathscr{E}(d^+, t) = 0$, we find the result $J_d = -J_c$.

The current produced in the circuit (i.e., through the battery) due to moving charge between the plates can be calculated by either of two methods. The first method has the advantage of clearly illustrating how the motion produces the current. The results provide a clear indication of the origin of noise as discussed in the next section. Figure 3.11.2 shows the plates with surface charge density $\sigma_R = Q/A$, σ_L, separated by distance d with voltage difference V given by

$$\mathscr{E}_R(d - x_s) + \mathscr{E}_L x_s = V \tag{3.11.4a}$$

The integral form of Gauss' law applied to the left plate, the right plate and the sheet charge provides

$$\mathscr{E}_L = \sigma_L/\varepsilon \qquad \mathscr{E}_R = -\sigma_R/\varepsilon \qquad \mathscr{E}_R - \mathscr{E}_L = -\sigma/\varepsilon \tag{3.11.4b}$$

Solving for $\mathscr{E}_L$ and $\mathscr{E}_R$ using Equation (3.11.4a) and the last of Equations (3.11.4b) yields

$$\mathscr{E}_L = \frac{V}{d} + \frac{\sigma}{\varepsilon}\left(1 - \frac{x_s}{d}\right) \quad \text{and} \quad \mathscr{E}_R = \frac{V}{d} - \frac{\sigma x_s}{\varepsilon d} \tag{3.11.5}$$

Then using $J_d = \varepsilon \partial \mathscr{E}(x, t)/\partial t$ we find the current in the circuit must be

$$J_c = -J_d = -\frac{\partial \mathscr{E}_R(x, t)}{\partial t} = \frac{\sigma v}{\varepsilon d} \tag{3.11.6a}$$

where the exact functional dependence on time depends on the speed of the sheet as it moves from one electrode to the next according to $v = dx_s/dt$. For a single electron $q = \sigma A = -e$, we have the current $I = J_c A$

$$I = -\frac{ev}{\varepsilon d} \qquad (3.11.6b)$$

The last relation produces two interesting effects. First, any variation in the speed of the electron in moving from one electrode to the other will induce a time dependence in the current I. Collisions between the electron and phonons in particular will induce noise in the current I. Second, calculating the photocurrent in a photodetector requires one to add up all of the charge moving entirely across the capacitor from one electrode to another. The case of moving holes and electrons does not multiply the result for electrons by two as can be seen as follows. A hole and electron must be produced together at the same point between the electrodes and if the electrons move a distance x, then the holes moves a distance $d-x$. The motion is equivalent to a single charge moving through the distance d.

3.11.2 The Power Relation

As a second method, the photocurrent induced in a circuit due to the motion of injected carriers can be deduced from energy considerations (refer to Section 3.5). The power expended by the battery in moving the carrier sheet is

$$P_{\text{Batt}} = V I(t) \qquad (3.11.7a)$$

As discussed in Section 3.5, the power absorbed per unit length (for this one-dimensional problem) by the medium is

$$\frac{d P_{\text{Medium}}}{dx} = J \cdot \mathscr{E} = e\, n(x,t)\, v\, \mathscr{E} \qquad (3.11.7b)$$

where n has units of # per unit length. Therefore, energy conservation requires

$$P_{\text{Batt}} = P_{\text{Medium}} \;\to\; I(t) = \frac{1}{V} \int_0^d e\, n(x,t)\, v\, \mathscr{E} \qquad (3.11.8)$$

where d symbolizes the separation of the electrodes as in Figure 3.11.2. In cases where only a single charge carrier moves, such as electrons, the velocity must be related to the drift mobility μ_e according to $v = \mu_e \mathscr{E}$. Further assuming a constant field, we can write $V = \mathscr{E}\, d$. Equation (3.11.8) can be written as

$$I(t) = \frac{e\mu_e \mathscr{E}}{d} \int_0^d n(x,t)\, dx \qquad (3.11.9)$$

3.11.3 Voltage Induced by Moving Charge

Once again assume a sheet of charge moves between two capacitor plates so that the position of the sheet depends on time $x_s = x_s(t)$. This time, the capacitor plates remain unconnected to any circuit and we calculate the voltage difference between the plates as a function of time. Starting with Equation (3.11.4a)

$$V = \mathscr{E}_R(d - x_s) + \mathscr{E}_L x_s \qquad (3.11.10)$$

The last of Equations (3.11.4b) for positive sheet charge

$$\mathscr{E}_R - \mathscr{E}_L = \sigma/\varepsilon \qquad (3.11.11)$$

allows us to rewrite Equation (3.11.10) as

$$V = -\frac{\sigma}{\varepsilon} x_s + \frac{\sigma d}{2\varepsilon} \qquad (3.11.12)$$

A point charge instead of the sheet charge requires us to use an image charge to calculate the fields and hence the voltage. Equation (3.11.12) shows the position of the sheet charge determines the voltage between two points in space. In particular, if the position of the sheet is random, say due to thermal fluctuations, then the voltage at the location of the electrodes will be random.

3.12 Review Exercises

3.1 Consider the wave equation $\nabla^2 f = \dfrac{1}{c^2} \dfrac{\partial^2 f}{\partial t^2}$

1. Show that any function $f(z - ct)$ satisfies the wave equation the 1-D equation so long as f can be differentiated.
2. Show the spherical traveling wave $e^{i(kr - \omega t)}/r$ satisfies the 3-D wave equation where $r = \sqrt{x^2 + y^2 + z^2}$. Use spherical coordinates.

3.2 Find the group velocity when the refractive index has the form $n = A + B/\lambda^2$ where λ denotes the wavelength in vacuum and A and B are constants. If $A = 1.5$ and $B = 4 \times 10^4 \, \text{nm}^2$ then find the group velocity at 850 nm.

3.3 A converging lens appears in Figure P3.3 with three primary rays. There are two focal points, one on either side of the lens.

* A ray traveling parallel to the optic axis deflects through the focal point.
* A ray initially traveling through the focal point deflects parallel to the optical axis.
* A ray passing through the center of the lens is not deflected.

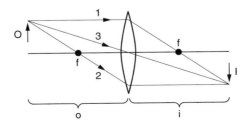

FIGURE P3.3
Three primary light rays for the converging lens.

A real image forms where the three rays intersect. O, I represent the object and image height. The focal length is positive for the converging lens.

1. Prove the lens formula $\dfrac{1}{f} = \dfrac{1}{i} + \dfrac{1}{o}$ based on Figure P3.3

2. If the magnification is given by $M = \dfrac{I}{O}$, then prove $M = \dfrac{f}{o-f}$

3. Show how to place two converging lenses (with focal lengths f_1 and f_2) to make a Galilean telescope. A beam of light (with diameter D_1) enters one lens and emerges from the second as a beam (with diameter D_2). The input and output beam both have parallel sides. Show the ratio of diameters must be given by $d_1/d_2 = f_1/f_2$. Hint: overlap two of the focal points.

3.4 Based on the previous problem, explain the following.

1. A point emitter placed at the focal point will produce uniform illumination on the other side of the converging lens.

2. A flat 2-D circular emitter placed a distance f from the lens will produce uniform illumination over some circular area on the other side of the lens.

3.5 Figure P3.5 shows a diverging lens with three primary rays.

* A ray traveling parallel to the optic axis deflects such that it appears to come from the focus.

* A ray traveling toward a focus deflects parallel to the optic axis.

* A ray traveling through the center of the lens passes through without deflection.

Show how to place a converging and diverging lens (with focal lengths f_1 and f_2) to make a Galilean telescope. A beam of light (with diameter D_1) enters one lens and emerges from the second as a beam (with diameter D_2). The input and output beam both have parallel sides. Show the ratio of diameters must be given by $d_1/d_2 = f_1/f_2$. Hint: overlap two of the focal points.

3.6 Using Snell's law, $n_1 \sin \theta_1 = n_2 \sin \theta_2$, find the critical angle for $n_1 = 3.5$ and $n_2 = 1$ when the incident beam initially travels in medium #1.

3.7 An engineering student wants to find the real and imaginary part of the susceptibility χ of a material for infrared light with vacuum wavelength of 850 nm. Assume the material is a dielectric with negligible conductivity $\sigma = 0$. The student performs two experiments. First, she allows the light to propagate in the material (surrounded by vacuum) and finds that the power drops to approximately 1/3 the original amount $P(z)/P(0) = e^{-1}$ in a distance of 100 μm. Next, she varies the incident angle of the light

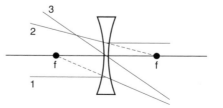

FIGURE P3.5
Three primary rays for the diverging lens.

and watches it leave the material from one of the facets. She finds a critical angle of 17°. Find the real and imaginary parts of χ.

3.8 Find the penetration depth $1/\alpha$ of an electromagnetic wave with wavelength λ in a material having $\mathrm{Im}(\chi) = 0$ and index $n = 4$. Use the following data. A resistor made from the material has the shape of a cube with 1 cm sides and has a resistance of 1 Ω.

3.9 Consider a cylinder of height L and radius R_o enclosed in free space except for a very thin layer generating plane waves given by

$$\vec{\mathcal{E}} = E\,e^{+ikz-i\omega t}\tilde{x} \quad z > 0$$

$$\vec{\mathcal{E}} = E\,e^{-ikz-i\omega t}\tilde{x} \quad z < 0$$

1. Calculate the time averaged total power leaving the generator in both directions using SA where S is the Poynting vector and A is the area of the generator.

2. Recalculate the time-averaged power leaving the generator by calculating the total power passing through the cylinder surface.

3.10 Consider an interface separating a dielectric with refractive index n from the vacuum with refractive index 1 as shown in Figure P3.10. An electromagnetic wave with field E_i strikes the interface. Some of the wave transmits across the interface with field E_t and some reflects back into the dielectric with field E_r.

Show $S_i = S_r + S_t$ where S is the magnitude of the Poynting vector and i, r, t refer to the incident, reflected, and transmitted waves respectively. Use the complex version of the Poynting vector and use the reflectivity r and the transmissivity t given in Section 3.6.3.

3.11 Repeat Problem 3.10 using the boundary conditions given in Section 3.3 for $\mathcal{E}$ and $\mathcal{H}$.

3.12 An engineering student plans to make a laser amplifier and needs to place antireflective (AR) coatings on the end facets. In fact to eliminate all reflections

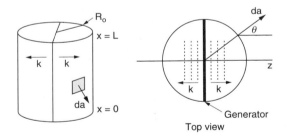

FIGURE P3.9
Thin EM wave generator in a hollow cylinder.

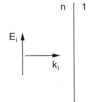

FIGURE P3.10
Wave in a dielectric strikes the interface.

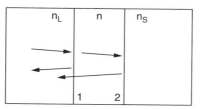

FIGURE P3.12
Layer with refractive index n functions as an AR coating.

in the optical system, he plans to place AR coatings on all of the lenses. As shown in Figure P3.12, the middle layer with refractive index n serves as the AR coating. Assume the following order for the refractive indices $n_L > n > n_S$. Assume the wavelength of interest is λ in vacuum.

1. What should be the smallest thickness of the coating so that the wave reflected from interface 1 will be 180° phase shifted from the wave reflected from interface 2 and passing through interface 1? Write your answer in terms of $\lambda_n = \lambda/n$.

2. Show to lowest-order approximation that $n = \sqrt{n_L n_S}$ as follows. Calculate the reflectivity r_1 for interface 1 and r_2 for interface 2. Require $r_1 = r_2$.

3.13 Repeat Problem 3.12 using the scattering—transfer matrix formalism. Use the following notation. The symbols r_1 and r_2 refer to the reflectivity of interface 1 and 2 respectively. The phase $\phi = k_o nL$ for the AR coating is real.

1. What should be the smallest thickness of the coating?
2. Show $n = \sqrt{n_L n_S}$

3.14 An optoelectronics student wants to make an antireflective coating similar to that discussed in Problems 3.8 and 3.9. However, he does not have a material giving the correct value of n. While working problems for a certain laser course, he suddenly thinks about adding some atoms to the AR coating that can provide gain or absorption. He then thinks that maybe the gain or absorption would change the value of n required to make the AR coating. For simplicity, assume the complex part of n_2 in r_1 and r_2 remains small. Use the complex expression for $\phi = \phi_r + i\phi_i = k_n L = k_o nL - ig_{net}L/2$.
Using the notation in Problems 3.12 and 3.13, find a relation for n_2 in terms of n_1 and n_3. Note and hint: unlike the chapter where we set $\sin\phi = 0$, you will need to set $\cos\phi = 0$; use the lowest value of m.

3.15 In Topic 3.6.3, show the formula $t_{1\rightarrow 2} = \dfrac{2n_1}{n_1 + n_2}$ and in Equation (3.6.10c), show the relation $\tau^2 = t_{1\rightarrow 2} t_{2\rightarrow 1}$

$$t_{1\rightarrow 2} = \frac{2n_1}{n_1 + n_2}$$

$$\tau^2 = t_{1\rightarrow 2} t_{2\rightarrow 1}$$

3.16 Starting with Equation (3.7.7), derive Equation (3.7.9).

3.17 Explain how the mirrors on the VCSEL work.

3.18 A student places a layer of glass (refractive index $n_2 = 1.5$) on a very thick piece of undoped AlGaAs (refractive index $n = 3.5$). A gas etchant (index $n_1 = 1$) removes

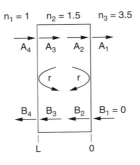

FIGURE P3.18

A glass layer.

the glass layer at a steady rate. As the layer etches, a laser beam strikes the wafer at normal incidence (perpendicular to the surface). Assume the laser has a wavelength of 700 nm, and the starting thickness of the glass layer is 4 μn. Determine the ratio B_4/A_4 as a function of time (see Figure P3.18) using transfer matrices.

3.19 Show $\nabla \cdot \nabla \times \vec{G} = 0$ for $\vec{G}$ differentiable. Show $\int \nabla \times \vec{F} \cdot d\vec{a} = 0$ using $\vec{G} = \nabla \times \vec{F}$.

3.20 Find the magnetic induction field $\vec{B}$ due to current I in a thin wire embedded in a magnetic material. Start with the appropriate Maxwell equation in differential form. Assume $\vec{M} = \beta \vec{H}$ and $\vec{D} = 0$.

3.21 Find the electric field and polarization at a distance R from a point charge $+Q$. Start with Maxwell's equations in differential form. Write the final answer in terms of the susceptibility, distance R and charge Q.

3.22 An optical beam enters a fiber as shown in Figure P3.22. The beam waveguides so long as θ remains larger than the critical angle θ_c. For $n_1 = 1.6$ and $n_2 = 1.7$, find the maximum acceptance angle α_{max} so that the beam will be waveguided. Assume the surrounding medium consists of air with refractive index $n_0 = 1$.

3.23 If $\alpha_{max} = 20$ in the previous problem, then what focal length lens gives α_{max} at the fiber (Figure P3.23)? Assume the input beam has parallel sides and a diameter of 2 mm.

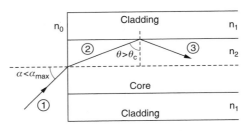

FIGURE P3.22

Beam enters a waveguide.

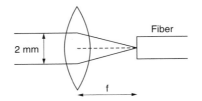

FIGURE P3.23

A lens focuses light into a fiber.

3.24 In Section 3.9.4, show that the boundary conditions and the general solutions for the three regions lead to the results for the constants A, B, C, D shown in Table 3.9.1.

3.25 Consider a laser made from GaAs and AlGaAs that emits light with a wavelength of 850 nm. Find the allowed values of the effective propagation constant. Assume the core has a thickness of $t = 0.5$ μm and a refractive index of $n_2 = 3.6$. The cladding has a refractive index of $n_1 = 3.4$.

1. Find the range of allowed β using the triangle relation.
2. Use a computer or graphical method to find some of the allowed values of β.
3. For the fundamental mode, find the penetration depth $1/p$.
4. Discuss the similarity and differences between the wavelength, the propagation constant k_n and the effective propagation constant β.

3.26 Write the transmissivity in Section 3.6.3 in terms of the two refractive indices n_1 and n_2.

3.27 Find the waveguide solutions for the TM polarization. Note that you will need the boundary conditions appropriate for the TM polarization.

3.28 Consider a GaAs-Al$_{0.6}$Ga$_{0.4}$As slab waveguide with $t_g = 200$ nm, $n_2 = 3.63$, $n_1 = 3.25$ using $\lambda_o = 850$ nm. Using the results of Example 3.9.1, find the angle θ for the triangle relation.

3.29 Consider a GaAs-Al$_{0.6}$Ga$_{0.4}$As slab waveguide with $t_g = 200$ nm, $n_2 = 3.63$, $n_1 = 3.25$ using $\lambda_o = 850$ nm.

1. Find the allowed modes (i.e., propagation constants β).
2. Find the penetration depths of the evenescent fields for each mode.
3. Find the transverse wave vector h for each mode.
4. Find the angle θ for the triangle relation for each mode.
5. Find the waveguide phase velocity for each mode.

3.30 Starting with $N_g = n + \omega \dfrac{\partial n}{\partial \omega}$, show $N_g = n - \lambda \dfrac{\partial n}{\partial \lambda}$.

3.31 Show the photocurrent in the electrodes attached to the photoconductor in Figure P3.31

$$I(t) = \frac{e\mu\mathscr{E}}{d} \int_0^d n(x, t)\, dx$$

must reduce to the photocurrent due to a moving point charge

$$I = \frac{ev}{\varepsilon d}$$

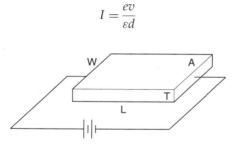

FIGURE P3.31
Photoconductor absorbs power through area A.

3.32 A photoconductor shown in Figure P3.31 absorbs P watts (photons) through the area A with negligible dark current. Find the photocurrent produced as a function of wavelength.

3.33 A beam of photons, with density γ, strikes a reverse biased photodiode which absorbs all of the photons. Find the photocurrent as a function of the incident optical power.

3.34 A beam of photons, with density γ, strikes a photocell consisting of an undoped semiconductor. Assume the photocell absorbs all of the light. Find the photocurrent as a function of the optical power and the bias voltage. Assume the carriers have a lifetime τ that is small compared with the transit time.

3.13   Further Reading

The following list contains references relevant to the chapter material.

Electromagnetics (easiest to more difficult)

1. Hayt W.H., Buck J.A., *Engineering Electromagnetics*, 6th ed., McGraw-Hill Higher Education, New York, 2001.
2. Percell E.M., *Electricity and Magnetism*, Berkeley Physics Course Volume 2, McGraw-Hill, New York, 1965.
3. Reitz J.R., Milford F.J., *Foundations of Electromagnetic Theory*, 2nd ed., Addison-Wesley Publishing, Reading, MA, 1967.
4. Jackson J.D., *Classical Electrodynamics*, 2nd ed., John Wiley & Sons, New York, 1975.

General

5. Agrawal G.P., Dutta N.K., *Semiconductor Lasers*, 2nd ed., Van Nostrand Reinhold, New York, 1993.
6. Coldren, L.A., *Diode Lasers and Photonic Integrated Circuits*, John Wiley & Sons, New York, 1995.

Optics

7. Hecht E., Zajac A., *Optics*, 4th ed., Addison-Wesley Publishing, Reading, MA, 1987.
8. Fowles G.R., *Introduction to Modern Optics*, Dover Publications, Mineola, NY, 1989.
9. Saleh B.E.A., Teich M.C., *Fundamentals of Photonics*, Wiley Interscience, New York, 1991.

Optical Fiber

10. Kasap S.O., *Optoelectronics and Photonics, Principles and Practices*, Prentice Hall, Saddle River, 2001.
11. Keiser G., *Optical Fiber Communications*, 3rd ed., McGraw-Hill Higher Education, 2000.

Waveguides and Optical Filters

12. Hunsperger R.G., *Integrated Optics: Theory and Technology*, 3rd ed., Springer-Verlag, New York, 1991.
13. Chuang S.H., *Physics of Optoelectronic Devices*, John Wiley & Sons, New York, 1995.
14. Yariv A., *Quantum Electronics*, 3rd ed., John Wiley & Sons, New York, 1989.
15. Yariv A., *Optical Electronics in Modern Communications*, 5th ed., Oxford University Press, New York, 1997.
16. Madsen C.K., Zhao J.H., *Optical Filter Design and Analysis: A Signal Processing Approach*, John Wiley & Sons, New York, 1999.
17. Pollock C.R., *Fundamentals of Optoelectronics*, Irwin, Chicago, 1995.

4

Mathematical Foundations

Linear algebra is the natural mathematical language of quantum mechanics. For this reason, the present chapter starts with a review of Hilbert spaces for vectors and operators. We introduce vector and Hilbert spaces along with inner products and metrics. The Dirac notation is developed for the Euclidean vector spaces as a starting point for the concepts of complete orthonormal sets of vectors, closure, dual vector spaces, and adjoint operators. The Dirac delta function in various forms and the principal part is introduced in Appendix 5 as an essential tool. The chapter turns to the main use of Dirac notation for function spaces; the concepts of norm, inner product, and closure are discussed. Fourier, Cosine, and Sine series are discussed as examples of expansions in complete orthonormal sets of functions.

Although Hilbert spaces are interesting mathematical objects with important physical applications, the study of linear algebra remains incomplete without a study of linear operators (i.e., linear transformations). In fact, the set of linear transformations itself forms a vector space and therefore has a basis set. The basis set for the operator is linked with the basis sets for the spaces that it operates between. The linear operator can be discussed as an abstract operator or through an isomorphism as a matrix or as a generalized expansion in operator space.

A Hermitian (a.k.a., *self*-adjoint) operator produces a basis set within a Hilbert space. The basis set comes from the eigenvector equation for the particular operator. The fact that a Hermitian operator produces a complete set (of orthornomal vectors) has special importance for quantum mechanics. Observables such as energy or momentum correspond to Hermitian operators. Complete sets make it possible to represent every possible result of a measurement of the observable by an object (vector) in the theory. The Hermitian operators have real eigenvalues which represent the results of the measurement.

4.1 Vector and Hilbert Spaces

Linear algebra starts with the definition of the vector space. An inner product space consists of a vector space with an inner product defined on it. The Hilbert space often refers to an inner product space of functions. However, this section uses the term Hilbert and inner product spaces interchangeably.

4.1.1 Definition of Vector Space

A vector space consists of a set F with a defined binary operation "+" and a scalar multiplication (SM) over the field of numbers $\mathcal{N}$ such that (assuming f, f_1, f_2 are in F and α, β are in $\mathcal{N}$) the following relations hold.

Closure:	$f_1 + f_2$ is in F and αf is in F
Associative:	$(f_1 + f_2) + f_3 = f_1 + (f_2 + f_3)$
Commutative:	$f_1 + f_2 = f_2 + f_1$
Zero:	There exists a zero vector $\mathcal{O}$ such that $\mathcal{O} + f = f$
Negatives:	For every f in F, there exists $(-f)$ in F such that $f + (-f) = \mathcal{O}$
SM Associative:	$(\alpha\beta)f = \alpha(\beta f)$
SM Distributive:	$\alpha(f_1 + f_2) = \alpha f_1 + \alpha f_2$
SM Distributive:	$(\alpha + \beta)f = \alpha f + \beta f$
SM Unit:	$1f = f$

If "F" is a set of functions then "F" is sometimes called a function space. For complex functions F, the number field $\mathcal{N}$ must be the set of complex numbers $\mathcal{C}$ while, for real functions F, the number field $\mathcal{N}$ consists of the real numbers $\mathcal{R}$. For example, if F represents the set of real functions but the number field consists of complex numbers, then objects such as $c_1 f(x)$ (where c_1 is complex) cannot be in the original vector space because the function $g(x) = c_1 f(x)$ has complex values. Therefore, for this example, closure cannot be satisfied contrary to the requirements of the definition for the vector space.

4.1.2 Inner Product, Norm, and Metric

An *inner product* $\langle \bullet | \bullet \rangle$ in a (real or complex) vector space F is a scalar valued function that maps $F \times F \to C$ (where C is the set of complex numbers) with the properties

1. $\langle f | g \rangle = \langle g | f \rangle^*$ with f, g elements in F and where "*" denotes complex conjugate.
2. $\langle \alpha f + \beta g | h \rangle = \alpha^* \langle f | h \rangle + \beta^* \langle g | h \rangle$ and $\langle h | \alpha f + \beta g \rangle = \alpha \langle h | f \rangle + \beta \langle h | g \rangle$ where f, g, h are elements of F and α, β are elements in the complex number field $\mathcal{C}$.
3. $\langle f | f \rangle \geq 0$ for all vectors f. The inner product can be zero $\langle f | f \rangle = 0$ if and only if $f = 0$ (except at possibly a few points for functions).

The *norm* or "length" of a vector f is defined to be $\|f\| = \langle f | f \rangle^{1/2}$.

A *metric* $d(f, g)$ is a relation between two elements f and g of a set F such that

1. $d(f, g) \geq 0$ and $d = 0$ only when $f = g$ (except at possibly a few points for piecewise continuous functions $C_p[a,b]$). Recall that two functions are equal only when $f(x) = g(x)$ for all "x" in the domain of definition.
2. $d(f, g) = d(g, f)$.
3. $d(f, g) \leq d(f, h) + d(h, g)$ where h is any third element of F.

The metric measures the distance between two elements of the space. The properties of the inner product are very similar to those of the metric. In fact, if $d(f, g)$ is a metric then it can be written as

$$d(f, g) = \langle f - g | f - g \rangle^{1/2}$$

Consider $\mathscr{R}^2$ which is the set of Euclidean vectors in the $x - y$ plane. Assume $\vec{r}_1$ and $\vec{r}_2$ are two vectors in R^2 with $\vec{r}_1 = x_1 \tilde{x} + y_1 \tilde{y}$ and $\vec{r}_2 = x_2 \tilde{x} + y_2 \tilde{y}$. Simple vector analysis provides the following relations.

Inner product $\qquad \langle \vec{r}_1 \, |. \, \vec{r}_2 \rangle = \vec{r}_1 \cdot \vec{r}_2 = x_1 x_2 + y_1 y_2$

Norm $\qquad \|\vec{r}_1\| = \langle \vec{r}_1 \mid \vec{r}_1 \rangle^{1/2} = (x_1^2 + y_1^2)^{1/2}$

Metric $\qquad d(\vec{r}_1, \vec{r}_2) = \|\vec{r}_1 - \vec{r}_2\| = [(\vec{r}_1 - \vec{r}_2) \cdot (\vec{r}_1 - \vec{r}_2)]^{1/2} = \sqrt{(x_1 - x_2)^2 + (y_1 - y_2)^2}$

The inner product can be defined for functions as follows.

Inner product
Norm $\qquad \|f(x)\| = \langle f \mid f \rangle^{1/2} = \left[\int_a^b dx \, f(x)^* f(x) \right]^{1/2} = \left[\int_a^b dx \, |f(x)|^2 \right]^{1/2}$

Example 4.1.1
Find the length of $f(x) = x$ for $x \in [-1, 1]$

$$\|f\| = \langle f \mid f \rangle^{1/2} = \left[\int_{-1}^1 dx \, x^* x \right]^{1/2} = \left[\int_{-1}^1 dx \, x \cdot x \right]^{1/2} = \left[\int_{-1}^1 dx \, x^2 \right]^{1/2} = \sqrt{\frac{2}{3}}$$

where we use the fact that $f(x) = x$ is real. If we were to divide the function by the norm and write $g(x) = f(x)/\|f\|$ then the length of $g(x)$ would be unity. In general, we normalize a function $f(x)$ to one by dividing the norm of $f(x)$.

4.1.3   Hilbert Space

We define a Hilbert space H to be a vector space with an inner product defined on the space. Some books reserve the term "Hilbert space" for vector spaces of *functions* with an inner product; they sometimes denote the inner product by (f_1, f_2). For function spaces, the functions must be square integrable in the sense that the following integral must exist for $f \in H$

$$\int_a^b dx \, |f(x)|^2$$

Sometimes the term "inner product space" refers to a vector space (regardless of whether it is a Euclidean or function space) having a defined inner product. This book doesn't make any distinction between the function or Euclidean vector spaces and assumes all of the inner products exist (such as the previous integral).

4.2   Dirac Notation and Euclidean Vector Spaces

The present section introduces a notation created by P. A. M. Dirac during the early 20th century. Professor Dirac, a mathematician and physicist, was intimately familiar with linear algebra and quantum theory. For our purposes, the Dirac notation helps to unify Euclidean and function spaces and those with discrete and continuous sets of basis functions. The notation appears for the vector space spanned by the basis set of unit vectors $\{\tilde{x}, \tilde{y}, \tilde{z}\}$. We then discuss the concepts of closure and completeness.

4.2.1   Kets, Bras, and Brackets for Euclidean Space

The basis vectors for 3D Euclidean space $\{\tilde{x}, \tilde{y}, \tilde{z}\}$ can also be written in "ket" $|\ \rangle$ notation. The vector $\vec{v}$ can be written as $|v\rangle$ and the basis vectors as

$$\tilde{x} \leftrightarrow |1\rangle \quad \tilde{y} \leftrightarrow |2\rangle \quad \tilde{z} \leftrightarrow |3\rangle$$

A general basis vector appears as $\tilde{e}_n \leftrightarrow |n\rangle$. For example, the vector $\vec{v} = 3\tilde{x} - 4\tilde{y} + 10\tilde{z}$ can be written as $|v\rangle = 3|1\rangle - 4|2\rangle + 10|3\rangle$. Sometimes the vector sum and scalar product are written as $|v_1\rangle + |v_2\rangle \equiv |v_1 + v_2\rangle$. and $|\alpha v\rangle = \alpha|v\rangle$, respectively.

We define a "bra" $\langle|$ to be a projection operator. The bras $\langle 1|$, $\langle 2|$, $\langle 3|$ represent operators that project a vector $\vec{v}$ onto the unit vectors $\tilde{x}$, $\tilde{y}$, $\tilde{z}$, respectively. For example, if $|v\rangle = 3|1\rangle - 4|2\rangle + 10|3\rangle$ then the projection operators provide the components $\langle 1|\vec{v} = 3$, $\langle 2|\vec{v} = -4$, and $\langle 3|\vec{v} = 10$. Here the bra $\langle 1|$, for example in Figure 4.2.1, operates on $\vec{v}$ to give the component of $\vec{v}$ along the $\tilde{x}$ axis. We would do better to write the combination of projection operators and vectors as $\langle 1|\vec{v} = \langle 1|v\rangle$. This combination of the "bra" + "ket" gives the "braket" (or bracket).

In general, $\langle w|$ represents the operator that projects an arbitrary vector onto the vector $\vec{w}$. The linear operator $\langle w|$ corresponds to "$\vec{w}\cdot$" where the dot refers to the usual dot product $\langle w|v\rangle = (\vec{w}\cdot)\vec{v} = \vec{w} \cdot \vec{v}$. We see that the bracket must be an inner product (the same inner product defined earlier). If "n" represents an integer corresponding to one of the basis vectors then $\langle n|v\rangle$ represents a component of the vector. The bras are linear operators and can be distributed across a sum.

$$\langle w| \quad [|v_1\rangle + |v_2\rangle] = \langle w \mid v_1\rangle + \langle w \mid v_2\rangle$$

As a note, some books call the bras "projectors" and they call objects like $|\bullet\rangle\langle\bullet|$ projection operators. We consider objects like $|\bullet\rangle\langle\bullet|$ to be more complicated compound objects.

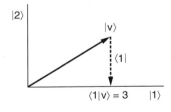

FIGURE 4.2.1
Projection of $\vec{v} = 3\tilde{x} + 5\tilde{y}$ onto $|1\rangle, |2\rangle$.

4.2.2 Basis, Completeness, and Closure for Euclidean Space

A basis set must be orthonormal and complete. Two vectors $|m\rangle, |n\rangle$ are orthonormal when

$$\langle m|n\rangle = \delta_{m,n} = \begin{cases} 1 & m = n \\ 0 & m \neq n \end{cases} \tag{4.2.1}$$

The Kronecker delta function $\delta_{m,n}$ expresses orthonormality for countable (or discrete) basis set (i.e., the elements of the basis set are in one-to-one correspondence with a subset of integers—could be an infinite subset). A set of vectors $B = \{\,|1\rangle, |2\rangle, \ldots, |N\rangle\,\}$ is *orthonormal* if for any two vectors $|m\rangle$, $|n\rangle$ in B, the inner product between them satisfies $\langle m|n\rangle = \delta_{m,n}$. For cases where a set of basis functions has a one-to-one relation with a continuous subset of the real numbers, the Dirac Delta Function (i.e., the impulse function) $\delta(x - x')$ replaces the Kronecker delta function $\delta_{m,n}$. If a vector space has the basis set $B = \{|1\rangle, |2\rangle, \ldots, |N\rangle\}$ then the notation it spans is V.

A linear combination of "N" orthonormal vectors $B = \{|1\rangle, |2\rangle, \ldots, |N\rangle\}$ has the form

$$|v\rangle = \sum_{i=1}^{N} C_i |i\rangle \tag{4.2.2}$$

where $\{C_i\}$ can be complex numbers. The collection of all such vectors $V = \{\,|v\rangle\,\}$ forms a vector space and the set B must be a basis set. The set B *spans* the vector space $V = Sp\,(B)$, which has dimension $\mathrm{Dim}(V) = N$. Since every vector in V can be found by a suitable choice of the C_i, the set B is said to be *complete*. On the other hand, given a vector space V then a set of orthonormal vectors is complete in V if every vector in the space V can be written as a linear combination of the form (4.2.2). Such a set of vectors forms a basis set.

Next we demonstrate the closure (i.e., completeness) relation. The components of the vector, namely C_i in Equation (4.2.1), can be written in terms of "brackets" by projecting the vector $|v\rangle$ onto each basis vector $|m\rangle$.

$$\langle m \mid v\rangle = \langle m| \sum_{i=1}^{n} C_i |i\rangle = \sum_{i=1}^{n} C_i \langle m \mid i\rangle = \sum_{i=1}^{n} C_i \delta_{i,m} = C_m \tag{4.2.3}$$

The results from Equation (4.2.3), written as $C_i = \langle i \mid v\rangle$, can be substituted into Equation (4.2.2) to obtain $|v\rangle = \sum_{i=1}^{n} C_i |i\rangle = \sum_{i=1}^{n} [\langle i \mid v\rangle]|i\rangle$. Then

$$|v\rangle = \sum_{i=1}^{n} |i\rangle\langle i \mid v\rangle \quad \text{or} \quad |v\rangle = \left(\sum_{i=1}^{n} |i\rangle\langle i| \right)|v\rangle \tag{4.2.4}$$

Consider the quantity in parenthesis to be an operator and realize that the equation must hold for all vectors $|v\rangle$ in the vector space V. Consequently, Equation (4.2.4) becomes

$$\sum_{i=1}^{n} |i\rangle\langle i| = 1 \tag{4.2.5}$$

for the vector space V spanned by the basis $B = \{|1\rangle, |2\rangle, \ldots, |n\rangle\}$. The "1" that appears in Equation (4.2.5) actually represents an operator and not the number "1." Although not demonstrated here, it is possible to show that the closure relation is equivalent to the completeness of a basis set.

Example 4.2.1

The completeness relation for $\mathscr{R}^3$ using $\langle w| = \vec{w}\cdot$ is

$$1 = |1\rangle\langle 1| + |2\rangle\langle 2| + |3\rangle\langle 3| \quad \text{so} \quad 1 = \tilde{x}\tilde{x}\cdot + \tilde{y}\tilde{y}\cdot + \tilde{z}\tilde{z}\cdot$$

Note that the unit vectors are written next to each other without an operator between them.

4.2.3 The Euclidean Dual Vector Space

The previous topic shows that a bra $\langle w|$ projects an arbitrary vector onto the vector $\vec{w}$. The linear operator $\langle w|$ maps a vector space V into the complex numbers $\mathscr{C}$ (i.e., $\langle w|: V \to \mathscr{C}$). These projection operators form a vector space—a vector space of linear operators. For Euclidean vector $\vec{v} = |v\rangle$, the corresponding bra is the operator $|v\rangle^+ = \langle v| = \vec{v}\cdot$.

The set V^+ consisting of all bra operators $\langle w|$ defines the "dual" of the vector space V. For each ket $|w\rangle$, there exists a bra $\langle w|$ and vice versa so that the original vector space V must be in 1-1 correspondence with the "dual vector space V^+." Mathematically, the two vector spaces $V = \{ |v\rangle \}$ and $V^+ = \{ \langle w| \}$ are related by an antilinear 1-1 (isomorphic) map denoted by the dagger superscript. The isomorphic map $+ : V \leftrightarrow V^+$ is called the Hermitian conjugate (or adjoint operator).

$$\langle \cdot| \quad \underset{+}{\leftrightarrow} \quad |\cdot\rangle \quad \text{or as} \quad |w\rangle^+ = \langle w|$$

If $\alpha, \beta \in \mathscr{C}$ (the complex numbers) then the antilinearity property can be written as

$$[\alpha|v\rangle + \beta|w\rangle]^+ = [\alpha|v\rangle]^+ + [\beta|w\rangle]^+ = \alpha^*\langle v| + \beta^*\langle w|$$

where "*" indicates complex conjugate. Part of the reason for taking the complex conjugate of the coefficients has to do with finding the magnitude of a "complex" vector.

The adjoint operator maps a basis set for V into a corresponding basis set for V^+. If $\{|i\rangle: i = 1, \ldots, n\}$ comprises a basis set for V then $\{ \langle i| : i = 1, \ldots, n \}$ must be a basis set for V^+. Therefore the dual basis set consists of operators that project an arbitrary vector onto the set of basis vectors of the vector space V. The dual basis allows us to write an arbitrary bra as $\langle v| = \sum_n A_n \langle n|$.

Example 4.2.2

Find the vector dual to $|2\rangle = \hat{y}$.
The dual vector is $\langle 2| = \hat{y}\cdot$ which is an operator that projects an arbitrary vector $\vec{v}$ onto $\hat{y}$. We can explicitly represent the result of the projection as the y-component of $\vec{v}$:

$$\langle 2 \mid v \rangle = \hat{y} \cdot \vec{v} = v_y$$

Example 4.2.3

Some relations can be demonstrated for $\vec{v} = |v\rangle = a|1\rangle + b|2\rangle$ where $\{ |1\rangle, |2\rangle \}$ spans R^2.

1. $\langle v| = |v\rangle^+ = [a|1\rangle + b|2\rangle]^+ = \left[|1\rangle^+ a^+ + |2\rangle^+ b^+\right] = a^*\langle 1| + b^*\langle 2|$

2. $\langle v \mid 1\rangle = [a^*\langle 1| + b^*\langle 2|]|1\rangle = a^*$ and $\langle 1 \mid v\rangle = \langle 1|[a|1\rangle + b|2\rangle] = a$

3. $\langle 1 \mid v\rangle = a = (a^*)^* = \langle v \mid 1\rangle^*$ Note that $\langle v \mid 1\rangle^+ = \langle 1 \mid v\rangle = \langle v \mid 1\rangle^*$.

The adjoint reverses the order of operators. Suppose the linear operators $\hat{L}, \hat{L}_1, \hat{L}_2$ act on the vector space V which has basis vectors $\{|1\rangle, |2\rangle, \ldots, |n\rangle\}$. For example, consider $\hat{L} = \hat{L}_1, \hat{L}_2$ and $\langle w|\hat{L}|v\rangle$ where $|v\rangle, |w\rangle \in V$ and $\langle w| \in V^+$. The adjoint operator reverses the direction of all the objects and adds the "+" to each operator.

$$\langle v|L_1 L_2|w\rangle^+ = \langle w|L_2^+ L_1^+|v\rangle.$$

4.2.4 Inner Product and Norm

Assume $\{|i\rangle : i = 1, 2, 3\}$ is a basis set for a 3D vector space. The norm (or length) of a vector is found by taking the square root of the inner product.

$$\|\vec{v}\|^2 = \langle v \mid v \rangle = \left(\sum_{i=1}^{3} v_i|i\rangle\right)^+ \left(\sum_{j=1}^{3} v_j|j\rangle\right) = \sum_{i=1}^{3} \langle i|v_i^* \sum_{j=1}^{3} v_j|j\rangle = \sum_{i,j=1}^{3} \langle i|v_i^* v_j|j\rangle = \sum_{i,j=1}^{3} v_i^* v_j \langle i \mid j\rangle$$

The last step follows since $v_i^* v_j$ is just a number and so it can be moved outside the brackets. Now use the orthonormality property for unit vectors to write

$$\|v\|^2 = \sum_{i,j=1}^{3} v_i^* v_j \delta_{i,j} = \sum_{i=1}^{3} v_i^* v_i = \sum_{i=1}^{3} |v_i|^2$$

where $|v_i|$ is the magnitude of the complex number. Notice how this is equivalent to the usual method of taking inner products $\langle v|v\rangle = [\langle v|]|v\rangle = [\vec{v}\cdot]\vec{v} = \vec{v} \cdot \vec{v}$ which has the usual dot product.

4.3 Hilbert Space

A Hilbert space consists of a vector space of functions with a defined inner product. We define the Hilbert space to include the Euclidean space defined in the previous section. The vector space of functions can have either a countable and uncountable number of vectors in the basis set. A function $f(x)$ in the space can be represented as an abstract vector $|f\rangle$ with components formed by projecting them onto basis functions $\langle \phi_n|f\rangle$ or onto Dirac delta functions (refer to Appendix 5). The Dirac delta functions produce the coordinate representation $\langle \delta(x - x_0)|f\rangle = \langle x_0|f\rangle = f(x_0)$.

The first topic develops the notation for those Hilbert spaces (of functions) that have a discrete basis set. The results quite straightforwardly generalize the notation and concepts for the Euclidean vectors. In fact, if readers were not warned ahead of time, they might think they were reading about Euclidean vectors all over again. Next we begin the study of function spaces with an uncountably infinite number of basis vectors that produces the coordinate representation. The study completes the interpretation for the Hilbert space with the discrete (but perhaps infinite) basis set and introduces the Hilbert space with an uncountably infinite basis set (i.e., the "continuous" basis set).

4.3.1 Hilbert Space of Functions with Discrete Basis Vectors

Functions in a set $F = \{\phi_0, \phi_1, \phi_2, \ldots, \phi_n\}$ are linearly independent if for complex constants $c_i(i = 0, \ldots, n)$, the sum

$$\sum_{i=0}^{n} c_i \phi_i(x) = 0$$

can only be true when all of the complex constants are zero $c_i = 0$. Functions in the set $F = \{\phi_0, \phi_1, \phi_2, \ldots, \phi_n\}$ are orthonormal if $\langle \phi_i | \phi_j \rangle = \delta_{ij}$ for every integer i, j in the set $\{0, 1, 2, \ldots, n\}$. An orthonormal set of functions must be linearly independent. A linearly independent set of functions $F = \{\phi_0, \phi_1, \phi_2, \ldots, \phi_n\}$ is complete if every function $f(x)$ in the space can be written as

$$f(x) = \sum_{i=0}^{n} c_i \phi_i(x) \quad \text{or} \quad |f\rangle = \sum_{i=0}^{n} c_i |\phi_i\rangle \tag{4.3.1}$$

(except at possibly a few points) for some choice of complex numbers c_i (Figure 4.3.1). If the set $\{\phi_i\}$ is "complete and orthonormal" then the functions ϕ_i can be chosen as basis functions (or basis vectors) to span the function space. A complete orthonormal set of functions $F = \{\phi_0, \phi_1, \phi_2, \ldots, \phi_n\}$ forms a basis for a Hilbert space $\mathcal{H}$. The basis functions can be written using Dirac notation as $\{|\phi_0\rangle, |\phi_1\rangle, \ldots\}$ or, more conveniently as $\{|0\rangle, |1\rangle, \ldots\}$. In some cases, there might be a *countably* infinite number of basis vectors in which case the infinite series

$$|f\rangle = \sum_{i=0}^{\infty} c_i |i\rangle \tag{4.3.2}$$

must properly converge. Assume that the series has the appropriate convergence properties so that it can be integrated or differentiated as necessary. Notice the similarity between these formulas and those for the Euclidean space.

The components of the vector $|f\rangle$ (i.e., the expansion coefficients c_i) can be found from Equation (4.3.2) by operating with the bra $\langle j|$ as follows

$$\langle j | f \rangle = \langle \phi_j | f \rangle = \langle j | \sum_{i=0}^{\infty} c_i |i\rangle = \sum_{i=0}^{\infty} c_i \langle j | i \rangle = \sum_{i=0}^{\infty} c_i \delta_{ij} = c_j \tag{4.3.3}$$

so, just like Euclidean vectors, the vector components must be $c_j = \langle j | f \rangle$. The projection of the function on the ith axis produces the inner product between the two complex functions "ϕ_i" and "f" over the range (a, b)

$$\langle \phi_i | f \rangle = \int_a^b dx\, \phi_i^*(x) f(x) \tag{4.3.4}$$

The components $c_i = \langle \phi_i | f \rangle = \langle i | f \rangle$ can be used to demonstrate the closure relation by substituting into Equation (4.3.2).

$$|f\rangle = \sum_{i=0}^{\infty} c_i |i\rangle = \sum_{i=0}^{\infty} \langle i | f \rangle |i\rangle = \sum_{i=0}^{\infty} |i\rangle \langle i | f \rangle = \left(\sum_{i=0}^{\infty} |i\rangle \langle i| \right) |f\rangle \tag{4.3.5}$$

The vector $|f\rangle$ is an arbitrary member of the Hilbert space. Recall that two operators $\hat{A}, \hat{B}$ are equal if and only if $\hat{A}|v\rangle = \hat{B}|v\rangle$ for all vectors $|v\rangle$ in the Hilbert space. Therefore, by definition of equality between operators, Equation (4.3.5) yields

$$\sum_{i=0}^{\infty} |i\rangle \langle i| = 1 \tag{4.3.6}$$

The closure relation ensures completeness of the basis set and vice versa.

The bra for functions can be written in terms of an operator as

$$\langle f| = \int dx\, f^*(x)\circ$$

where the circle serves as a reminder to insert a function in place of the circle.

4.3.2 The Continuous Basis Set of Functions

Now we discuss the continuous basis set of functions. Let $B = \{\phi_k\}$ be a set of basis vectors with one such vector for each real number "k" in some interval $[a, b]$, which could also be infinite. The orthonormality relation has the form

$$\langle \phi_K \mid \phi_k \rangle = \delta(k - K) \tag{4.3.7}$$

where the inner product between two general functions has the form

$$\langle f \mid g \rangle = \int dx\, f^*(x)g(x) \tag{4.3.8}$$

A general vector $|f\rangle$ has an integral expansion since there are more basis vectors than a conventional summation can handle.

$$|f\rangle = \int_a^b dk\, c_k |\phi_k\rangle \tag{4.3.9}$$

The subscript on the coefficient c resembles the index used in the summation over discrete sets. The expansion coefficients c_k can be written as a function $c_k = c(k)$ and can be viewed as the components of the vector or as the transform of the function f with respect to the particular continuous basis (such as the Fourier transform). Figure 4.3.2 shows the function $|f\rangle$ projected onto two of the many basis vectors. If desired the coordinate projection operator $\langle x|$ can be applied to both sides to obtain

$$f(x) = \int_a^b dk\, c_k\, \phi_k(x) \tag{4.3.10}$$

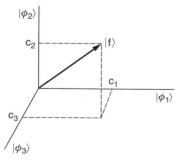

FIGURE 4.3.1
The function f projected onto the basis set of functions.

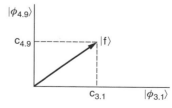

FIGURE 4.3.2
A function projected onto two of the many basis vectors.

The integral appearing in Equations (4.3.9) and (4.3.10) replaces the summation used for the discrete basis vectors. The quantities c_k and u_k can also be written in functional form as $c_k = c(k)$ and $\phi_k(x) = \phi(x, k)$.

Continuing to work with Equation (4.3.9), the component c_k can be found by operating on the left with $\langle \phi_K |$ and using the orthonormality relation (note the index of capital K) to get

$$\langle \phi_K \mid f \rangle = \int_a^b dk\, c_k \langle \phi_K \mid \phi_k \rangle = \int_a^b dk\, c_k \delta(k - K) = c_K \qquad (4.3.11)$$

which assumes that $K \in (a, b)$. Note that when computing inner products such as $\langle \phi_K \| \phi_k \rangle$, the integral is over a spatial coordinate "x" and has the form

$$\langle \phi_K \mid \phi_k \rangle = \int dx\, \phi_K^*(x)\, \phi_k(x) = \delta(k - K)$$

The closure relation can be found by using $c_k = \langle \phi_k | f \rangle$ as follows

$$|f\rangle = \int dk\, c_k |\phi_k\rangle = \int dk\, \langle \phi_k \mid f \rangle\, |\phi_k\rangle = \int dk\, |\phi_k\rangle\, \langle \phi_k \mid f \rangle$$

This last relation holds for arbitrary functions $|f\rangle$ in the Hilbert space so that

$$\int dk\, |\phi_k\rangle \langle \phi_k| = \underline{1} \qquad (4.3.12)$$

by definition of operator equality. Equation 4.3.12 provides the closure relation for a continuous set of basis vectors.

4.3.3 Projecting Functions into Coordinate Space

Recall that Euclidean vector $\vec{v}$ in a Hilbert space has components v_i. The components are really functions of the index "i" as in $v(i) = v_i = \langle i | v \rangle$. This is equivalent to projecting the vector $\vec{v}$ on to the ith coordinate. The index "i" is thought of similar to the x-axis, for example, except that "i" refers to the integer subset of the reals.

As shown in the previous topics, the symbol $|f\rangle$ denotes the function "f" in a vector space. We regard the function "f" (i.e., $|f\rangle$) as the most fundamental object and not the component $f(x)$. The reason is that the function "f" can equally well be represented, for example, as $f(x)$, or as the Fourier transform $f(k)$ or as a series expansion. We shall see how projecting the function "f" onto the xth coordinate produces the component $f(x)$. However, projecting "f" into k-space produces the Fourier transform $\langle k | f \rangle = f(k)$. The same "$f$" appears in $f(k)$ and $f(x)$ with the understanding that the explicit form of the two functions cannot be the same (i.e., $f(k)$ cannot be found by replacing "x" with "k").

Functions such as $f(x)$ can be thought of as vectors $|f\rangle$ projected onto the x-axis. To set the stage, recall how functions can be described as a collection of ordered pairs (x, f). We can consider "x" to be an index and write $f(x) = f_x$ where "x" takes on values in the domain. The only real difference between $f(x)$ and $v(i)$ is that $v(i)$ has a domain with a countable number of "x components" symbolized by i.

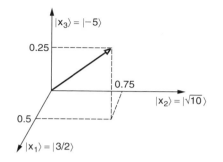

FIGURE 4.3.3
The function *f* projected onto several coordinates.

The function $f(x)$ can be pictured as a vector in *coordinate space* defined by $\{|x\rangle\}$. Imagine projecting a function "f" onto the coordinate "x" as $\langle x|f\rangle = f(x)$. Each real x is considered to be a basis vector. There can be an uncountably infinite number of "*vectors*" $|x\rangle$. Figure 4.3.3 shows a conceptual view of an example function f with values $f(3/2) = 0.5$, $f(\sqrt{10}) = 0.75$, $f(-5) = 0.25$. The components of the vector f must be $\langle 3/2|f\rangle = 0.5$, $\langle\sqrt{10}|f\rangle = 0.75$, $\langle -5|f\rangle = 0.25$. The figure shows three axes but there must be as many axes as coordinates x. Quantities of the form $\langle f|x\rangle$ can now be defined using the adjoint operator as

$$\langle f \mid x\rangle = \langle x \mid f\rangle^{+} = \langle x \mid f\rangle^{*} = \left[f(x)\right]^{*} = f^{*}(x) \tag{4.3.13}$$

What does it mathematically mean to project a function "f" into coordinate space to find an inner product $\langle x|f\rangle$? We already know that functions like $f \sim |f\rangle$ can be projected into function space (i.e., Hilbert space) to form inner products between functions such as $\langle f|g\rangle$. The coordinate basis $\{|x_0\rangle\}$ really consists of a set of Dirac delta functions

$$\left\{|x_0\rangle \equiv \left|\delta(x - x_0)\right\rangle \sim \delta(x - x_0)\right\}$$

as suggested by Figure 4.3.4. The bra $\langle x_0| \equiv \langle\delta(x - x_0)|$ is a projection operator that projects $|f\rangle$ onto the Dirac delta function $\delta(x - x_0)$. The projection of $f(x)$ onto the coordinate x_0 becomes

$$\langle x_0|f\rangle = \langle\delta(x - x_0|f(x)\rangle = \int_{-\infty}^{\infty} dx\, \delta(x - x_0)f(x) = f(x_0) \tag{4.3.14}$$

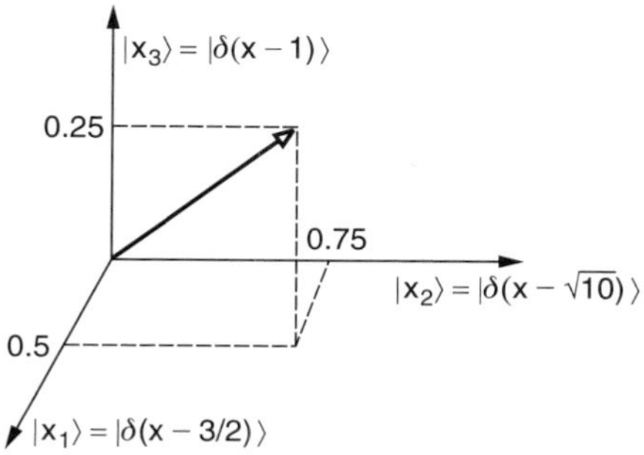

FIGURE 4.3.4
The coordinate space basis vectors are actually the Dirac delta functions.

We can demonstrate the orthonormality relation for the coordinate space. Let $|\xi|$ and $|\eta\rangle$ be two of the uncountably many coordinate kets. Using Equation (4.3.8) for the inner product, we can write

$$\langle \xi \mid \eta \rangle = \langle \delta(x - \xi) \mid \delta(x - \eta) \rangle = \int_{-\infty}^{\infty} dx \, \delta(x - \xi) \, \delta(x - \eta) = \delta(\xi - \eta) \qquad (4.3.15)$$

Therefore rather than have an orthonormality relation involving the Kronecker delta function as for Euclidean vectors, we see that the coordinate space uses the Dirac delta function.

Basis sets need to be complete in the sense that any function can be expanded in the set similar to Equation (4.3.9). Let $f(x)$ be an arbitrary element in the function space. Consider the expansion

$$|g\rangle = \int dx' \, |x'\rangle f(x')$$

If this is a legitimate expansion of $f(x)$ we should be able to show that $g(x)$ equals $f(x)$. To this end, operate on this last equation with $\langle x|$ to find

$$g(x) = \langle x \mid g \rangle = \int dx' \langle x \mid x' \rangle f(x') = \int dx' \, \delta(x' - x) \, f(x') = f(x)$$

So now we can think of the decomposition of a vector $\vec{f} = |f\rangle$ in function or "coordinate" basis sets (actually the same though).

Next, let's examine the closure relation for coordinate space. The table below shows how to replace the indices for the Euclidean vector and the summation by the coordinate x and integral, respectively.

$$\sum_{i=1}^{n} |i\rangle\langle i| = 1 \qquad \rightarrow \qquad \int |x\rangle dx \langle x| = 1$$
$$\langle m \mid n \rangle = \delta_{mn} \qquad \rightarrow \qquad \langle x' \mid x \rangle = \delta(x - x')$$
$$m, n \in \text{integers} \qquad\qquad x, x' \in \mathcal{R}$$

Note that the Dirac delta function replaces the Kronecker delta function for the continous basis set $\{|x\rangle\}$. Also notice that an integral replaces the discrete summation for the continuous basis.

Let's demonstrate the closure relation for the coordinate basis set. First consider the inner product between any two elements $|f\rangle$ and $|g\rangle$ of the Hilbert space. Using the fact that $\langle x|g\rangle$ is a complex number so that $g^*(x) = \langle x|g\rangle^* = \langle x|g\rangle^+$ and also $\langle x|g\rangle^+ = \langle g|x\rangle$, we have

$$\langle g \mid f \rangle = \int dx \, g^*(x) \, f(x) = \int dx \, \langle x \mid g \rangle^+ \langle x \mid f \rangle = \int dx \, \langle g \mid x \rangle \langle x \mid f \rangle = \langle g | \left(\int |x\rangle \, dx \, \langle x| \right) | f \rangle$$

$$(4.3.16)$$

However, the unit operator $\hat{1}$ does not change the vector $|f\rangle$, that is $\hat{1}|f\rangle = |f\rangle$, so that the inner product can be generally written as $\langle g|f\rangle = \langle g|\hat{1}|f\rangle$. Comparing this last expression with Equation (4.3.16) shows

$$\langle g|\hat{1}|f\rangle = \langle g| \left\{ \int |x\rangle \, dx \langle x| \right\} |f\rangle$$

This last relation must hold for all vectors $|f\rangle$ and $|g\rangle$ and therefore the operators on either side must be the same

$$\int |x\rangle\, dx \langle x| = \hat{1} \qquad (4.3.17)$$

Working Equation (4.3.16) in reverse, we can now see the reason for the definition of the inner product between two arbitrary functions $|f\rangle$ and $|g\rangle$ in the Hilbert space.

$$\langle g \mid f\rangle = \langle g|1|f\rangle = \langle g| \int |x\rangle dx \langle x||f\rangle = \int \langle g \mid x\rangle dx \langle x \mid f\rangle$$

Again using $g^*(x) = \langle g|x\rangle$, we find

$$\langle g \mid f\rangle = \int dx\, g^*(x) f(x)$$

as expected for the basic definition of inner product.

Further, we can see the connection between the inner product for the discrete basis sets and those for coordinate space. Recall for Euclidean vectors that

$$\langle g \mid h\rangle = \sum_i \langle g \mid i\rangle\langle i \mid h\rangle = \sum_i g_i^* h_i$$

where $\langle g|i\rangle = \langle i|g\rangle^+ = g_i^*$ since the inner product $\langle i|g\rangle$ is a complex number. Now suppose that g and f are functions so that the index "i" is replaced by the index "x." The inner product might then be written as

$$\langle g \mid f\rangle \sim \sum_x \langle g \mid x\rangle\langle x \mid f\rangle \sim \sum_x g^*(x) f(x) \sim \sum_x g_x^* f_x \sim \int dx\, g_x^* f_x \sim \int dx\, g^*(x)\, f(x)$$

Therefore, for functions, the inner product

$$\langle g \mid f\rangle = \int dx\, g^*(x) f(x) \qquad (4.3.18)$$

is viewed as a sum over components similar to the case for Euclidean vectors.

A later section shows that different sets of basis vectors F leads to different representations of the Dirac delta function. We can see this by considering any basis set $\{\phi_i(x)\}$ for an arbitrary function space so that

$$\delta(x - x') = \langle x \mid x'\rangle = \langle x|1|x'\rangle = \langle x| \left[\sum_{i=0}^{\infty} |\phi_i\rangle\langle\phi_i| \right] |x'\rangle$$

$$= \sum_{i=0}^{\infty} \langle x \mid \phi_i\rangle\langle\phi_i \mid x'\rangle = \sum_{i=0}^{\infty} \phi_i^*(x)\phi_i(x') \qquad (4.3.19)$$

Continuous basis sets can be similarly handled. If $\{|\phi_k\rangle\}$ has uncountably many basis vectors indexed by the continuous parameter k, then operating on the closure relation

$$\hat{1} = \int dk|\phi_k\rangle\langle\phi_k|$$

produces

$$\delta(x - x') = \langle x \mid x' \rangle = \langle x | \hat{1} | x' \rangle = \left\langle x \left| \left\{ \int dk | \phi_k \rangle \langle \phi_k | \right\} \right| x' \right\rangle = \int dk \, \phi_k^*(x') \, \phi_k(x) \qquad (4.3.20)$$

Equations (4.3.19) and (4.3.20) show that any complete orthonormal set of functions gives a representation of the Dirac delta function.

Example 4.3.1

Is the set $\{1, x\}$ orthonormal on the interval $[-1, 1]$?

Note that the "1" and "x" represent functions and not coordinates. Therefore, define functions $f = 1$ and $g = x$. These functions are orthogonal on the interval as can be seen

$$\langle f \mid g \rangle = \int_{-1}^{1} dx \, f^* g = \int_{-1}^{1} dx \, 1 \cdot x = 0$$

Neither function is normalized (unit length) since

$$\|f\|^2 = \langle f \mid f \rangle = \int_{-1}^{1} 1 \, dx = 2 \quad \text{and} \quad \|g\|^2 = \langle g \mid g \rangle = \int_{-1}^{1} dx \, x^2 = \frac{2}{3}$$

In general, any function $h(x)$ can be normalized by redefining it as $h \to h/\|h\|$. An orthonormal set can be formed by dividing each function by its length. The orthonormal set is

$$\left\{ \frac{1}{\sqrt{2}}, \sqrt{\frac{3}{2}} x \right\}$$

4.3.4 The Sine Basis Set

The sine functions provide another basis set for functions defined on the interval $x \in (0, L)$

$$B_s = \left\{ \sqrt{\frac{2}{L}} \sin\left(\frac{n\pi x}{L}\right) \quad n = 1, 2, 3 \ldots \right\} = \{\psi_n(x) : \ n = 1, 2, 3 \ldots\} \qquad (4.3.21)$$

The Hilbert space can be expanded to include functions that repeat every $2L$ along the x-axis. The normalization of $\sqrt{2/L}$ depends on the width of the interval L and on the fact that the sine function has "$n\pi x/L$" in the argument (where "n" is an integer).

A function in the vector space spanned by B_s can be written as a summation over the basis vectors

$$|f\rangle = \sum_{m=1}^{\infty} c_m |\psi_m\rangle \quad \text{or} \quad f(x) = \sum_{n=1}^{\infty} c_n \sqrt{\frac{2}{L}} \sin\left(\frac{n\pi x}{L}\right) \qquad (4.3.22)$$

The expansion coefficients are found by projecting the function onto the basis vectors

$$\langle \psi_n \mid f \rangle = \langle \psi_n | \left\{ \sum_m c_m |\psi_m\rangle \right\} = \sum_m c_m \langle \psi_n \mid \psi_m \rangle = c_n$$

These components can be evaluated

$$c_n = \langle \psi_n \mid f \rangle = \left\langle \sqrt{\frac{2}{L}} \sin\left(\frac{n\pi x}{L}\right) \,\middle|\, f(x) \right\rangle = \sqrt{\frac{2}{L}} \int_0^L dx\, f(x) \sin\left(\frac{n\pi x}{L}\right) \tag{4.3.23}$$

4.3.5 The Cosine Basis Set

The set of functions

$$B_c = \left\{ \frac{1}{\sqrt{L}}, \sqrt{\frac{2}{L}} \cos\left(\frac{n\pi x}{L}\right), \ \ldots \ \text{for } n = 1, 2, 3 \ldots \right\} = \{\phi_0, \phi_1, \ldots\} \tag{4.3.24}$$

is orthonormal on the interval $x \in (0, L)$. The functions in B_c form a basis set for piecewise continuous functions on $(0, L)$. The function space can be enlarged to include functions that repeat every L along the entire x-axis. An arbitrary function $f \in Sp(B_c)$ can be written as a summation

$$|f\rangle = \sum_{n=0}^{\infty} c_n |\phi_n\rangle \tag{4.3.25a}$$

Operating on both sides with $\langle x|$ provides

$$f(x) = \frac{c_0}{\sqrt{L}} + \sum c_n \sqrt{\frac{2}{L}} \cos\left(\frac{n\pi x}{L}\right) \tag{4.3.25b}$$

The normalization $\sqrt{2/L}$ depends on the interval endpoint L in $(0, L)$ and also upon the fact that the "$n\pi x/L$" occurs as the argument of the cosine function with "n" being an integer.

The expansion coefficients $c_0, c_1, \ldots$ (i.e., the components of the vector) in Equations (4.3.25) can be found from the inner product of "f" with each of the basis vectors $\cos(nx)$

$$c_0 = \langle \phi_0 \mid f \rangle = \left\langle \frac{1}{\sqrt{L}} \,\middle|\, f(x) \right\rangle = \frac{1}{\sqrt{L}} \int_0^L dx\, f(x) \tag{4.3.26}$$

and

$$c_n = \langle \phi_n \mid f \rangle = \left\langle \sqrt{\frac{2}{L}} \cos\left(\frac{n\pi x}{L}\right) \,\middle|\, f(x) \right\rangle = \sqrt{\frac{2}{L}} \int_0^L dx\, f(x) \cos\left(\frac{n\pi x}{L}\right) \tag{4.3.27}$$

where this expression for c_n holds for $n > 0$.

4.3.6 The Fourier Series Basis Set

For the Hilbert space of periodic, piecewise continuous functions on the interval $(-L, L)$, there exists a very important set of basis functions.

$$B = \left\{ \frac{1}{\sqrt{2L}} \exp\left(i\frac{n\pi x}{L}\right) \quad n = 0, \pm 1, \pm 2 \ldots \right\} \tag{4.3.28}$$

The orthonormality relation and the orthonormal expansion become

$$\left\langle \frac{1}{\sqrt{2L}}\exp\left(i\frac{n\pi x}{L}\right) \,\middle|\, \frac{1}{\sqrt{2L}}\exp\left(i\frac{m\pi x}{L}\right)\right\rangle = \delta_{nm}$$

and

$$f(x) = \sum_{n=-\infty}^{\infty} \frac{D_n}{\sqrt{2L}}\exp\left(i\frac{n\pi x}{L}\right) \tag{4.3.29}$$

Notice how this expansion in the complex exponential begins to look like a Fourier transform. The coefficients D_n can be complex. These equations can be reduced to the typical Fourier series.

For periodic boundary conditions encountered for traveling waves, the basis set is often restated in terms of the repetition length L. The wave is required to repeat itself every length L instead of $2L$ given above. In this case the basis becomes

$$B = \left\{\frac{1}{\sqrt{L}}\exp\left(i\frac{2n\pi x}{L}\right) \quad n = 0, \ \pm 1, \ \pm 2 \quad \cdots\right\} \tag{4.3.30}$$

For three dimensions, the periodic boundary conditions provide

$$B = \left\{\frac{1}{\sqrt{V}}\exp\left(i\vec{k}\cdot\vec{r}\right)\right\} \tag{4.3.31}$$

where $V = L_x L_y L_z$ and $k_x = (2\pi m/L_x)$, $k_y = (2\pi n/L_y)$, $k_z = (2\pi p/L_z)$ with $m, n, p = 0$, $\pm 1, \pm 2, \ldots$

The 3-D case has the Kronecker delta function orthonormality.

4.3.7 The Fourier Transform

The complete orthonormal basis set for a Hilbert space of bounded functions defined over the real x-axis is

$$\left\{\frac{e^{ikx}}{\sqrt{2\pi}}\right\} \tag{4.3.32}$$

For this section, the generalized expansion is defined as the integral over k.

$$f(x) = \int_{-\infty}^{\infty} dk\, \alpha(k)\frac{e^{ikx}}{\sqrt{2\pi}} \tag{4.3.33}$$

Define $\{|k\rangle\}$ to be the basis set

$$\left\{|k\rangle = |\phi_k\rangle = \left|\frac{1}{\sqrt{2\pi}}e^{ik0}\right\rangle \quad \rightarrow \quad \phi_k(x) = \langle x \mid k\rangle = \frac{1}{\sqrt{2\pi}}\exp\left(ikx\right)\right\} \tag{4.3.34}$$

where k is real and "0" provides a place for the variable x when the function is projected into coordinate space. We can demonstrate orthonormality for the basis set by substituting any two of the functions into the definition of the inner product.

$$\langle K \mid k \rangle = \int_{-\infty}^{\infty} dx \frac{e^{-iKx}}{\sqrt{2\pi}} \frac{e^{ikx}}{\sqrt{2\pi}} = \int_{-\infty}^{\infty} dx \frac{e^{i(k-K)x}}{2\pi} = \delta(k - K) \tag{4.3.35}$$

This expression agrees with the derivation for the Dirac delta function found in an appendix.

The closure relation

$$\hat{1} = \int_{-\infty}^{\infty} |k\rangle dk \langle k| \tag{4.3.36}$$

comes from the definition of completeness of the continuous basis set $\{|k\rangle = |\phi_k\rangle\}$. The projection of the closure relation into coordinate space and its dual produces a Dirac delta function. Operate on Equation (4.3.36) with $\langle x'|$ and $|x\rangle$ where x and x' represent spatial coordinates

$$\langle x' \mid x \rangle = \langle x' | \left[\int dk |k\rangle \langle k| \right] |x\rangle = \int_{-\infty}^{\infty} \langle x' \left| \frac{1}{\sqrt{2\pi}} e^{iko} \right\rangle \left\langle \frac{1}{\sqrt{2\pi}} e^{iko} \right| x \rangle dk$$

which can also be written as

$$\delta(x - x') = \int_{-\infty}^{\infty} dk \frac{e^{+ikx'}}{\sqrt{2\pi}} \frac{e^{-ikx}}{\sqrt{2\pi}} = \int_{-\infty}^{\infty} dk \frac{e^{-ik(x-x')}}{2\pi}$$

which agrees with the results in Appendix 5.

Projecting $|f\rangle$ into coordinate space produces $\langle x|f\rangle = f(x)$. Projecting $|f\rangle$ into k-space produces the Fourier transform $\langle k|f\rangle = \tilde{f}(k)$.

TABLE 4.3.1

Summary of Results

	Euclidean Vectors	Functions-Discrete Basis	Functions-Continuous Basis
Basis	$\{|n\rangle : n = 1, 2, 3 \ldots\} \sim \{\tilde{x}, \tilde{y}, \tilde{z} \ldots\}$ $n = $ Integer	$\{|n\rangle = |u_n\rangle \tilde{u}_n(x)\}$ $n = $ Integer	$\{|k\rangle = |\phi_k\rangle \tilde{\phi}_k(x)\}$ $k = $ Real
Projector	$\langle w| = \vec{w} \cdot$	$\langle f| = \int dx\, f^*(x) \circ$	$\langle f| = \int dx\, f^*(x) \circ$
Orthonormality	$\langle m \mid n \rangle = \delta_{m,n}$	$\langle u_m \mid u_n \rangle = \delta_{mn}$	$\langle \phi_K \mid \phi_k \rangle = \delta(k - K)$
Complete	$|v\rangle = \sum_n c_n |n\rangle$	$|f\rangle = \sum_n c_n |u_n\rangle$ $f(x) = \sum_n c_n u_n(x)$	$|f\rangle = \int dk\, c_k |\phi_k\rangle$ $f(x) = \int dx\, c_k\, \phi_k(x)$
Components	$c_n = \langle n \mid v \rangle$	$c_n = \langle u_n \mid f \rangle$	$c_k = \langle \phi_k \mid f \rangle$
Inner Product	$\langle v \mid w \rangle = \sum_n v_n^* w_n$	$\langle f \mid g \rangle = \int dx\, f^*(x)\, g(x)$	$\langle f \mid g \rangle = \int dx\, f^*(x)\, g(x)$
Closure	$\sum_n |n\rangle \langle n| = \hat{1}$	$\sum_n |u_n\rangle \langle u_n| = \hat{1}$ $\delta(x - x') = \sum_n u_n^*(x')\, u_n(x)$	$\int dk\, |\phi_k\rangle \langle \phi_k| = \hat{1}$ $\delta(x - x') = \int dk\, \phi_k^*(x')\, \phi_k(x)$

4.4 The Grahm–Schmidt Orthonormalization Procedure

The Grahm–Schmidt orthonormalization procedure transforms two or more independent functions (or vectors) into two or more orthogonal functions (or vectors). The Grahm–Schmidt procedure starts with a vector space and then develops a basis set.

Let two functions be represented as vectors $|f\rangle$ and $|g\rangle$ in a Hilbert space. Assume the function $|g\rangle$ is normalized to unity $\langle g|g\rangle = 1$ and choose $|g\rangle$ as one of the basis vectors as shown in Figure 4.4.1. We look for a function $h(x)$ in order to form a basis set $\{|g\rangle, |h\rangle\}$ for the space so that

$$|f\rangle = c_1|h\rangle + c_2|g\rangle \qquad (4.4.1)$$

Operating with $\langle g|$ on both sides of the equation for "f," we find an expression for the component c_2

$$\langle g|f\rangle = c_1\langle g|h\rangle + c_2\langle g|g\rangle = c_2$$

where we have used the orthogonality of "g" and "h," namely $\langle g|h\rangle = 0$, and the fact that "g" is normalized to 1. Now Equation (4.4.1) for "f" can be rewritten as

$$|h\rangle = |f\rangle - c_2|g\rangle = |f\rangle - |g\rangle\langle g\,|\,f\rangle \qquad (4.4.2)$$

where we have set $c_1 = 1$ but we will need to normalize $|h\rangle$ to 1.

The functional form $h(x)$ can be recovered by operating on Equation 4.4.2 with $\langle x|$ to find $h(x) = f(x) - g(x)\langle g|f\rangle$ or $h(x) = f(x) - g(x)\int_a^b dx\, g^*(x)f(x)$. We can easily prove that "h" and "g" are orthogonal by using Equation (4.4.2) and operating with $\langle g|$ as follows

$$\langle g|h\rangle = \langle g|\{|f\rangle - |g\rangle\langle g|f\rangle\} = \langle g|f\rangle - \langle g|g\rangle\langle g|f\rangle = 0$$

as required. In order for the set $\{|h\rangle, |g\rangle\}$ to be orthonormal, we need to normalize the function $|h\rangle$. Therefore define a normalized function $h' = h(x)/\|h(x)\|$. The basis set becomes $\{g, h'\}$.

We can easily include three or more independent vectors in the initial set. Assume that the Grahm–Schmidt procedure has been used to make two of the vectors ϕ_1, ϕ_2 orthonormal. Assume f to be independent of ϕ_1, ϕ_2. There must be a third basis function $h(x)$ for the set $\{\phi_1, \phi_2, f\}$ to be independent. Therefore, set $|f\rangle = |h\rangle + c_1|\phi_1\rangle + c_2|\phi_2\rangle$. The constants c_1 and c_2 are found similar to above. We can write $|f\rangle = |h\rangle + |\phi_1\rangle\langle\phi_1|f\rangle + |\phi_2\rangle\langle\phi_2|f\rangle$. Therefore the function $h(x)$ can be found by projecting $|h\rangle = |f\rangle - |\phi_1\rangle\langle\phi_1|f\rangle - |\phi_2\rangle\langle\phi_2|f\rangle$ onto coordinate space. It also needs to be normalized to serve as a basis function.

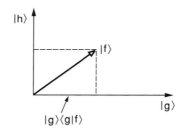

FIGURE 4.4.1
The relation between $|f\rangle, |g\rangle, |h\rangle$.

4.5 Linear Operators and Matrix Representations

Linear operators have a central role in many areas of mathematics, science, and engineering. This section discusses the linear operator and shows its relation to the matrix. Every linear operator $\hat{T}$ can be represented as a matrix $\underline{T}$. However, linear operators map vectors into other vectors whereas matrices map the set of components of the one vector into a set of components of the other vector.

4.5.1 Definition of a Linear Operator and Matrices

A linear operator $\hat{T}$ maps one Hilbert space V into another Hilbert space W according to $\hat{T} : V \rightarrow W$. For complex numbers c_1 and c_2 and vectors $|v_1\rangle, |v_2\rangle \in V$, the linear operator has the defining property of $\hat{T}\{c_1|v_1\rangle + c_2|v_2\rangle\} = c_1\hat{T}|v_1\rangle + c_2\hat{T}|v_2\rangle$. As will become evident from the matrices, if we know how a linear transformation $\hat{T}$ maps the basis vectors $|\phi_i\rangle$ then we know how it maps all vectors in the space.

We now define the matrix of a linear transformation $\hat{T} : V \rightarrow W$ that maps one Hilbert space $V = Sp\{|\phi_j\rangle : j = 1, 2, \ldots, M\}$ into another $W = Sp\{|\psi_i\rangle : i = 1, 2, \ldots, N\}$ as shown in Figure 4.5.1. The two spaces do not necessarily have the same dimension. The matrix for $\hat{T}$ with respect to the basis sets is

$$\underline{T} = \underline{T}_{ij} = \begin{bmatrix} T_{11} & T_{12} & \cdots & T_{1M} \\ T_{21} & T_{22} & \cdots & T_{2M} \\ \vdots & & & \vdots \\ T_{N1} & T_{N2} & \cdots & T_{NM} \end{bmatrix}$$

where the matrix elements are defined to be the coefficients in

$$\hat{T}|\phi_j\rangle = \sum_{i=1}^{N} T_{ij}|\psi_i\rangle \qquad (4.5.1)$$

for $j = 1, \ldots, M$. Figure 4.5.1 shows that the operator maps the basis vector $|\phi_1\rangle$ into a vector $|w\rangle$. This image vector must be a linear combination of the basis vectors for W. Equation (4.5.1) shows the transformation $\hat{T}$ can also be defined by how it affects each of the basis vectors in V.

Recall that each Hilbert space has a dual space. The basis set for $W^+ = \text{Dual}(W)$ consists of projection operators $\{\langle\psi_a|\}$. Now because $\hat{T}|\phi_i\rangle$ must be a vector in W, we can

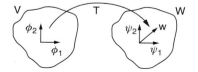

FIGURE 4.5.1

The linear operator T maps between vector spaces. The figure shows that the operator maps the basis vector ϕ_1 into the vector $|w\rangle$ which must be a linear combination of basis vectors in W.

operate on Equation (4.5.1) with say $\langle \psi_a |$ to find

$$\langle \psi_a | \hat{T} | \phi_j \rangle = \langle \psi_a | \sum_t T_{ij} | \psi_i \rangle = \sum_t T_{ij} \langle \psi_a \mid \psi_i \rangle = \sum_t T_{ib} \delta_{ai} = T_{aj}$$

The Dirac notation provides a compact expression for the matrix of an operator

$$T_{ab} = \langle \psi_a | \hat{T} | \phi_b \rangle \qquad (4.5.2)$$

This last expression makes it clear that matrix elements come from the inner products of operators between *basis* vectors. The values of the matrix elements depend on the basis vectors.

Dirac notation treats Euclidean and function spaces the same. As seen previously, there exists some slight distinction between discrete and continuous basis sets. Discrete basis sets require summations for generalized expansions and Kronecker delta functions for the orthonormality relation. Continuous basis sets require integrals for the generalized summations and Dirac delta functions for the orthonormality relations. It should be kept in mind that functions can have either discrete or continuous basis sets regardless of whether the function itself is continuous or not.

Example 4.5.1

Let $\hat{T} : V \rightarrow V$ and suppose that $\hat{T}$ is the unit operator; that is, $\hat{T} = 1$. The elements of the matrix with respect to the basis $B_v = \{|\phi_j\rangle = |j\rangle : j = 1, 2, \ldots, N\}$ must be

$$T_{ab} = \langle a | \hat{T} | b \rangle = \langle a | \left\{ \hat{1} | b \rangle \right\} = \langle a \mid bh \rangle = \delta_{ab}$$

The diagonal elements are 1 and all the others are zero.

4.5.2 A Matrix Equation

This topic shows how to write the matrix equation from the operator equation

$$|w\rangle = \hat{T} |v\rangle \qquad (4.5.3)$$

where $\hat{T} : V \rightarrow W$ and $W = Sp\{|\psi_j\rangle\}$ and $V = Sp\{|\phi_i\rangle\}$. Assume

$$|w\rangle = \sum_m y_m |\psi_m\rangle \quad \text{and} \quad |v\rangle = \sum_n x_n |\phi_n\rangle \qquad (4.5.4)$$

Start by inserting a unit operator between $\hat{T}$ and $|v\rangle$, and then replace it by the closure relation for the vector space V

$$|w\rangle = \hat{T} 1 |v\rangle = \hat{T} \left[\sum_b |\phi_b\rangle \langle \phi_b| \right] |v\rangle = \sum_b \hat{T} |\phi_b\rangle \langle \phi_b \mid v \rangle.$$

Operating on the last equation with $\langle \psi_a |$ produces

$$\langle \psi_a \mid w \rangle = \sum_b \langle \psi_a | \hat{T} | \phi_b \rangle \langle \phi_b \mid v \rangle = \sum_b T_{ab} \langle \phi_b \mid v \rangle$$

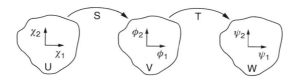

FIGURE 4.5.2
Three vector spaces for the composition of functions.

However, Equations 4.5.4 provide $\langle \psi_a | w \rangle = y_a$ and $\langle \phi_b | v \rangle = x_b$ so that

$$y_a = \sum_b T_{ab} x_b \quad \text{or} \quad \underline{T}\,\underline{x} = \underline{y}$$

which can be written in the usual form of

$$\begin{bmatrix} T_{11} & T_{12} & \cdots \\ T_{21} & \cdots & \\ \vdots & & \end{bmatrix} \begin{bmatrix} x_1 \\ x_2 \\ \vdots \end{bmatrix} = \begin{bmatrix} y_1 \\ y_2 \\ \vdots \end{bmatrix} \tag{4.5.5}$$

The expansion coefficients of the vectors appear in the column matrices.

4.5.3 Composition of Operators

Suppose $\hat{S} : U \to V$ and $\hat{T} : V \to W$ are two linear operators and U, V, and W are three distinct vector spaces (Figure 4.5.2) with the following basis sets

$$B_u = \{|\chi_i\rangle\} \; B_v = \{|\phi_j\rangle\} \; B_w = \{|\psi_k\rangle\}$$

The composition (i.e., product) $\hat{R} = \hat{T}\hat{S}$ first maps the space U to the space V and then maps V to W. The matrix of $\hat{R} = \hat{T}\hat{S}$ must involve the basis vectors B_u and B_w. The operator $\hat{R} = \hat{T}\hat{S}$ corresponds to the product of matrices.

$$R_{ab} = \langle \psi_a | \hat{R} | \chi_b \rangle = \langle \psi_a | \hat{T}\hat{S} | \chi_b \rangle$$

Inserting between $\hat{T}$ and $\hat{S}$ the closure relation for the space V gives

$$R_{ab} = \langle \Psi_a | \hat{T}\,\hat{1}\,\hat{S} | \chi_b \rangle = \langle \Psi_a | \hat{T} \left(\sum_c |\phi_c\rangle\langle\phi_c| \right) \hat{S} | \chi_b \rangle = \sum_c \langle \Psi_a | \hat{T} | \phi_c \rangle \langle \phi_c | \hat{S} | \chi_b \rangle = \sum_c T_{ac} S_{cb}$$

Notice that the closure relation corresponds to the range of $\hat{S}$ and the domain of $\hat{T}$. This last equation shows that the composition of operators corresponds to the multiplication of matrices $\underline{R} = \underline{T}\underline{S}$.

4.5.4 Introduction to the Inverse of an Operator

If $T : V \to W$ operates between spaces or even within one space, the function T must be "1-1" and "onto" to have an inverse. The property "1-1" requires every vector in V to have a unique image in W. The property "onto" requires every vector $|w\rangle \in W$ to have a preimage $|v\rangle \in V$ such that $\hat{T}|v\rangle = |w\rangle$.

The null space (also known as the kernel) provides a means for determining if a linear operator $\hat{T} : V \to W$ can be inverted. We define the null space to be the set of vectors

$N = \{|v\rangle\}$ such that $\hat{T}|v\rangle = 0$. Obviously, if the null space contains more than a single element (i.e., an element other than zero), the operator hasn't any inverse since an element of the range has multiple preimages. Furthermore, the chapter review exercises demonstrate the relation

$$\text{Dim}(V) = \text{Dim}(W) + \text{Dim}(N)$$

for $\hat{T} : V \rightarrow W$ where $W = \text{Range}(\hat{T})$. In this case, if the $\text{Dim}(N) > 0$ then the operator $\hat{T}$ hasn't any inverse since (1) the operator $\hat{T} : V \rightarrow W = \text{Range}(\hat{T})$ is already "onto," (2) the condition $\text{Dim}(N) = 0$ ensures the 1-1 property of the operator. Alternatively, we can require the determinant to be nonzero $\det(\hat{T}) \neq 0$ for the operator to be invertible.

4.5.5 Determinant

The *determinant of an operator* is defined to be the determinant of the corresponding matrix

$$\det(\hat{T}) = \det(\underline{T})$$

Generally, we assume for simplicity that the operator $\hat{T}$ operates within a single vector space. The determinant can be written in terms of the $\varepsilon_{ijk...}$ as

$$\det(\underline{T}) = \sum_{i,j,k...} \varepsilon_{ijk...} T_{1i} T_{2j} T_{3k} \cdots \quad \text{where} \quad \varepsilon_{ijk...} = \begin{cases} +1 & \text{Even permutations of } 1,2,3,\ldots \\ -1 & \text{Odd permutation of } 1,2,3,\ldots \\ 0 & \text{if any of } i = j = k \text{ holds} \end{cases}$$

$$(4.5.6)$$

For example $\varepsilon_{132} = -1$, $\varepsilon_{312} = +1$, and $\varepsilon_{133} = 0$.

Here's a couple of useful properties (see review exercises). The last proof can easily be handled using the unitary change-of-basis operator.

1. $\text{Det}(ABC) = \text{Det}(A)\,\text{Det}(B)\,\text{Det}(C)$
2. $\text{Det}(cA) = c^N\,\text{Det}(A)$ where $A : V \rightarrow V$, $N = \text{Dim}(V)$ and c is a complex number
3. $\text{Det}(A^T) = \text{Det}(A)$ where T signifies transpose
4. The $\det(T)$ is independent of the particular basis chosen for the vector space

4.5.6 Trace

The trace of an operator $\hat{T} : V \rightarrow V$ is the trace of the corresponding matrix (which is assumed square). The trace of a matrix is found by summing the diagonal elements of the matrix. If the basis for V is $B_v = \{|n\rangle\}$, then the trace of an operator can also be written as

$$Tr\left(\hat{T}\right) \equiv \sum_n \langle n|\hat{T}|n\rangle = \sum_n T_{nn} \tag{4.5.7}$$

The trace of an operator $\hat{T}$ is the sum of the diagonal elements of the matrix $\underline{T}$.

The trace can also be defined for coordinate space. Starting with the definition of trace, inserting a unit operator in two places, and then the closure relation in coordinate space gives

$$Tr\hat{A} = \sum_n \langle n|\hat{A}|n\rangle = \sum_n \langle n|1\hat{A}\,1|n\rangle = \iint dxdx' \sum_n \langle n\mid x\rangle\langle x|\hat{A}|x'\rangle\langle x'\mid n\rangle$$

The matrix elements are numbers that can be rearranged to give

$$Tr\hat{A} = \iint dxdx' \sum_n \langle x|\hat{A}\,|x'\rangle\langle x'|n\rangle\langle n|x\rangle = \iint dxdx'\langle x|\hat{A}|x'\rangle\langle x'|x\rangle = \iint dxdx'\langle x|\hat{A}|x'\rangle\delta(x - x')$$

where the closure relation is used for $|n\rangle$ and the Dirac delta function is substituted. Performing the final integration gives

$$Tr\hat{A} = \int dx\langle x|\hat{A}|x'\rangle \tag{4.5.8}$$

Here are some important properties for the Trace (see review exercises). Assume that the operators $\hat{A}, \hat{B}, \hat{C}$ have a domain and range within a single vector space V with basis vectors $B_v = \{|a\rangle\}$

1. $Tr(\hat{A}\hat{B}) = Tr(\hat{B}\hat{A})$
2. $Tr(\hat{A}\hat{B}\hat{C}) = Tr(\hat{B}\hat{C}\hat{A}) = Tr(\hat{C}\hat{A}\hat{B})$
3. The trace of the operator $\hat{T}$ is *independent* of the chosen basis set.

4.5.7 The Transpose and Hermitian Conjugate of a Matrix

The transpose operation means to interchange elements across the diagonal. For example

$$\begin{bmatrix} 1 & 2 & 3 \\ 4 & 5 & 6 \\ 7 & 8 & 9 \end{bmatrix}^T = \begin{bmatrix} 1 & 4 & 7 \\ 2 & 5 & 8 \\ 3 & 6 & 9 \end{bmatrix}$$

This is sometimes written as

$$\left(R^T\right)_{ab} = R_{ba}$$

Note the interchange of the indices "a" and "b." Sometimes this is also written as

$$R^T_{ab} = R_{ba}$$

The Hermitian conjugate (i.e., the adjoint) of the matrix requires the complex conjugate so that $R^+_{ab} = R^*_{ab} = R^*_{ba}$.

4.5.8 Basis Vector Expansion of a Linear Operator

The set of linear operators forms a vector space, which has a basis set. We will see the basis vectors have the form $|a\rangle\langle b|$. We begin by demonstrating how linear operators can be represented by sums over the basis vectors for the direct and dual spaces.

Consider an operator $\hat{T}: V \to W$ acting between two spaces $V = Sp\{|\phi_i\rangle\}$ and $W = Sp\{|\psi_j\rangle\}$. Starting with the definition of matrix elements

$$\hat{T}|\phi_b\rangle = \sum_a T_{ab}|\psi_a\rangle$$

multiplying by $\langle\phi_b|$ from the right, and summing over the index "b" provides

$$\hat{T}\sum_b |\phi_b\rangle\langle\phi_b| = \sum_{a,b} T_{ab}|\psi_a\rangle\langle\phi_b|$$

Substituting the closure relation provides the desired results.

$$\hat{T} = \sum_{a,b} T_{ab}|\psi_a\rangle\langle\phi_b| \tag{4.5.9a}$$

An operator $\hat{T}: V \to V$ produces the basis vector expansion

$$\hat{T} = \sum_{a,b} T_{ab}|\phi_a\rangle\langle\phi_b| \tag{4.5.9b}$$

These basis vector representations of an operator $\hat{T}: V \to V$ have a form very reminiscent of the closure relation. In fact, we can recover the closure relation if the operator $\hat{T}$ is taken as the unit operator $\hat{T} = 1$ so that the matrix elements are $T_{ab} = \delta_{ab}$.

Example 4.5.2
For the linear operator $\hat{T}: V \to V$ find an operator that maps the basis vectors as follows

$$|1\rangle \to |2\rangle \quad \text{and} \quad |2\rangle \to -|1\rangle \tag{4.5.10}$$

The solution can be found by noting that $|2\rangle\langle 1|$ can operate on the unit vector $|1\rangle$ and it gives $|2\rangle\langle 1|1\rangle = |2\rangle$. Similarly notice that $(-|1\rangle\langle 2|)|2\rangle = -|1\rangle\langle 2|2\rangle = -1$. The reader should show the desired operator is $\hat{T} = |2\rangle\langle 1| - |1\rangle\langle 2|$ by showing it reproduces the relations in Equation (4.5.10). The transformation T describes a rotation by $90°$.

4.5.9 The Hilbert Space of Linear Operators

The set of linear operators in the set $L = \{\hat{T}: V \to W\}$ forms a vector space with basis $B_L = \{Z_{ab} = |\psi_a\rangle\langle\phi_b| = |a\rangle\langle b|\}$ where for convenience, assume $V = Sp\{|\phi_a\rangle\}$ and $W = Sp\{|\psi_a\rangle\}$ have the same size $a = 1, 2, \ldots N = \text{Dim}(V)$. The vector space

$$L = \left\{\hat{T}: V \to V\right\} = Sp(B_L) = Sp\{Z_{ab} = |\psi_a\rangle\langle\phi_b|\} \tag{4.5.11}$$

can be made a Hilbert space by defining the inner product

$$\langle\hat{S} \mid \hat{T}\rangle = \text{Trace}\left(\hat{S}^+\hat{T}\right) \tag{4.5.12}$$

where $\hat{S}, \hat{T}$ are any two elements of L. A similar definition can be made for the set of linear operators $L = \{\hat{T} : V \to V\}$.

We can prove the properties (see Section 2.1.2) required of an inner product. Assume $\hat{A}, \hat{B}, \hat{C} \in L$ have basis expansions

$$\hat{A} = \sum_{aa'} A_{aa'} |\psi_a\rangle\langle\phi_{a'}| \qquad \hat{B} = \sum_{bb'} B_{bb'} |\psi_b\rangle\langle\phi_{b'}|$$

We prove the first required property of $\langle\hat{A}|\hat{B}\rangle = \langle\hat{B}|\hat{A}\rangle^*$. For simplicity, set indices a, b to refer to space W and indices a', b' to refer to space V.

$$\left\langle\hat{A} \mid \hat{B}\right\rangle^* = Tr\left\{\left[\sum_{aa'} A_{aa'} |a\rangle\langle a'|\right]^{+}\left[\sum_{bb'} B_{bb'} |b\rangle\langle b'|\right]\right\}^* = Tr\left\{\sum_{\substack{aa'\\bb'}} A_{aa'}^* B_{bb'} |a'\rangle\langle a \mid b\rangle\langle b'|\right\}^*$$

$$= \left[\sum_{\substack{aa'\\bb'}} A_{aa'}^* B_{bb'} \langle a \mid b\rangle\langle b' \mid a'\rangle\right]^* = \sum_{\substack{aa'\\bb'}} A_{aa'} B_{bb'}^* \langle a \mid b\rangle^*\langle b' \mid a'\rangle^* = \sum_{\substack{aa'\\bb'}} B_{bb'}^* A_{aa'} \langle b \mid a\rangle\langle a' \mid b'\rangle$$

$$= Tr\sum_{\substack{aa'\\bb'}} B_{bb'}^* |b'\rangle\langle b| A_{aa'} |a\rangle\langle a'| = Tr\left[\sum_{bb'} B_{bb'} |b\rangle\langle b'|\right]^{+}\left[\sum_{aa'} A_{aa'} |a\rangle\langle a'|\right] = \left\langle\hat{B} \mid \hat{A}\right\rangle$$

Notice that the adjoint operator switches $|a\rangle\langle a'| \to |a'\rangle\langle a|$ and places the complex conjugate on $A_{aa'}$ for the last term in the first line. The third result on the second line uses the fact that $\langle a|b\rangle^* = \langle b|a\rangle$ since $\langle a|b\rangle$ is an inner product.

The second property requires $\langle\hat{A}|\alpha\hat{B} + \beta\hat{C}\rangle = \alpha\langle\hat{A}|\hat{B}\rangle + \beta\langle\hat{A}|\hat{C}\rangle$ for the complex number field. This can easily be proved because the trace of the sum equals the sum of the traces.

The third property of $\langle f|f\rangle \geq 0 \forall f$ and $\langle f|f\rangle = 0$ if $f = 0$ follows using the following. Again set indices a, b to refer to space W and indices a', b' to refer to space V.

$$\left\langle\hat{A} \mid \hat{A}\right\rangle = Tr\left\{\sum_{aa'} A_{aa'}^* |a'\rangle\langle a| \sum_{bb'} A_{bb'} |b\rangle\langle b'|\right\} = \sum_{\substack{aa'\\bb'}} A_{aa'}^* A_{bb'} \delta_{ab}\delta_{a'b'} = \sum_{ab} |A_{ab}|^2 \geq 0$$

Now we can show the basis set B_L must be orthonormal. Let Z_{ab} and Z_{cd} be two basis vectors in Equation (4.5.11). Then Equation (4.5.12) provides

$$\langle Z_{ab} \mid Z_{cd}\rangle = Tr\left\{(|\psi_a\rangle\langle\phi_b|)^{+}(|\psi_c\rangle\langle\phi_d|)\right\} = Tr\{|\phi_b\rangle\langle\phi_d|\}\delta_{ac} = \sum_n \langle\phi_d \mid \phi_b\rangle\delta_{ac} = \delta_{ac}\delta_{bd}$$

4.5.10 A Note on Matrices

As a note, writing $\hat{T}$ as a sum over basis vectors is essentially the same as writing a matrix as a sum of "unit" matrices. For example, a 4×4 matrix can be written as

$$\begin{bmatrix} a & b \\ c & d \end{bmatrix} = a\begin{bmatrix} 1 & 0 \\ 0 & 0 \end{bmatrix} + b\begin{bmatrix} 0 & 1 \\ 0 & 0 \end{bmatrix} + c\begin{bmatrix} 0 & 0 \\ 1 & 0 \end{bmatrix} + d\begin{bmatrix} 0 & 0 \\ 0 & 1 \end{bmatrix}$$

So for real matrices

$$T = \begin{bmatrix} a & b \\ c & d \end{bmatrix}$$

the "basis set" consists of

$$\left\{ \begin{bmatrix} 1 & 0 \\ 0 & 0 \end{bmatrix}, \begin{bmatrix} 0 & 1 \\ 0 & 0 \end{bmatrix}, \begin{bmatrix} 0 & 0 \\ 1 & 0 \end{bmatrix}, \begin{bmatrix} 0 & 0 \\ 0 & 1 \end{bmatrix} \right\}$$

4.6 An Algebra of Operators and Commutators

The set of linear operators forms a vector space. The vector space properties do not include operator multiplication (i.e., composition). Operator multiplication satisfies the properties for an algebra, which does not include a property for the commutation of operators. This topic explores the effects of the noncommutivity of the operators. The linear isomorphism $M : \hat{T} \to T$ (i.e., it is "1-1" and "onto") between operators and matrices ensures identical properties for both the operators and the matrices.

Linear operators form an algebra which satisfy the properties

1. There exists a zero operator "0" such that $\hat{A}0 = 0\hat{A} = 0$
2. There exists a "unit" operator "I" such that $\hat{A}I = I\hat{A} = \hat{A}$
3. The distributive law holds $\hat{A}(\hat{B} + \hat{C}) = \hat{A}\hat{B} + \hat{A}\hat{C}$
4. The associative law holds $\hat{A}(\hat{B}\hat{C}) = (\hat{A}\hat{B})\hat{C}$
5. Scalar multiplication is defined $a\hat{A} = \hat{A}a$ where "a" is a complex number.

Properties 1–5 use the definition that two operators $\hat{A}$ and $\hat{B}$ are equal $\hat{A} = \hat{B}$ if $\hat{A}|v\rangle = \hat{B}|v\rangle$ for every vector $|v\rangle$ in the vector space V.

The algebraic properties for the multiplication of operators do not require them to commute. Two operators $\hat{A}$ and $\hat{B}$ commute when $\hat{A}\hat{B} = \hat{B}\hat{A}$ or equivalently $\hat{A}\hat{B} - \hat{B}\hat{A} = 0$. We represent the quantity $\hat{A}\hat{B} - \hat{B}\hat{A}$ by the *commutator* $[\hat{A}, \hat{B}] = \hat{A}\hat{B} - \hat{B}\hat{A}$. Therefore two operators $\hat{A}$ and $\hat{B}$ commute when $[\hat{A}, \hat{B}] = 0$. Our world vitally depends on the commutivity and noncommutivity of operators. It underlies all of quantum mechanics. It explains the differences between the classical and quantum views of the world.

Example 4.6.1

Show $[x, d/dx] \neq 0$. The commutator must be treated as an operator since it contains operators. Therefore, when calculating the commutator, it must operate on a function $f(x)$!

$$\left[x, \frac{d}{dx} \right] f = \left(x\frac{d}{dx} - \frac{d}{dx}x \right) f(x) = x\frac{df}{dx} - \frac{d}{dx}(xf) = x\frac{df}{dx} - \frac{dx}{dx}f - x\frac{df}{dx} = -f \neq 0$$

Notice that the derivative with respect to "x" operates on everything to the right.

The commutators satisfy the following properties where $\hat{A}, \hat{B}, \hat{C}$ represent operators and c denotes a complex number.

0. $[\hat{A}, \hat{B}] = \hat{A}\hat{B} - \hat{B}\hat{A}$	1. $[\hat{A}, \hat{A}] = 0$	2. $[c, \hat{A}] = 0$
3. $[\hat{A}, \hat{B}] = -[\hat{B}, \hat{A}]$	4. $[\hat{A}, \hat{B} + \hat{C}] = [\hat{A}, \hat{B}] + [\hat{A}, \hat{C}]$	5. $[\hat{A} + \hat{B}, \hat{C}] = [\hat{A}, \hat{C}] + [\hat{B}, \hat{C}]$
6. $[\hat{A}, \hat{B}\hat{C}] = [\hat{A}, \hat{B}]\hat{C} + \hat{B}[\hat{A}, \hat{C}]$	7. $[\hat{A}\hat{B}, \hat{C}] = [\hat{A}, \hat{C}]\hat{B} + \hat{A}[\hat{B}, \hat{C}]$	8. $f = f(\hat{A}) \rightarrow [f(\hat{A}), \hat{A}] = 0$

Properties 1 through 7 can be easily proven by expanding the brackets and using the definition of the commutator. For example, Property 5 is proved as follows

$$\left[\hat{A}, \hat{B}\right]\hat{C} + \hat{B}\left[\hat{A}, \hat{C}\right] = \left(\hat{A}\hat{B} - \hat{B}\hat{A}\right)\hat{C} + \hat{B}\left(\hat{A}\hat{C} - \hat{C}\hat{A}\right) = \hat{A}\hat{B}\hat{C} - \hat{B}\hat{C}\hat{A} = \left[\hat{A}, \hat{B}\hat{C}\right]$$

Functions of operators are defined through the Taylor expansion. Propertiy 8 can be proved by Taylor expansion of the function. The Taylor expansion of a function of an operator has the form

$$f\left(\hat{A}\right) = \sum_n c_n \hat{A}^n \quad \text{so that} \quad \left[f\left(\hat{A}\right), \hat{A}\right] = \left[\sum_n c_n \hat{A}^n, \hat{A}\right] = \sum_n c_n \left[\hat{A}^n, \hat{A}\right] = 0$$

where c_n can be a complex number and "n" is a nonnegative integer.

The following list of theorems can be proved by appealing to the properties of commutators, derivatives and functions of operators.

THEOREM 4.6.1. Operator Expansion Theorem

$$\hat{O} = e^{x\hat{A}}\hat{B}e^{-x\hat{A}} = \hat{B} + x\left[\hat{A}, \hat{B}\right] + \frac{x^2}{2!}\left[\hat{A}, \left[\hat{A}, \hat{B}\right]\right] + \cdots$$

THEOREM 4.6.2 $\quad e^A\hat{B}e^{-A} = \hat{B} + [\hat{A}, \hat{B}] + \frac{1}{2!}[\hat{A}, [\hat{A}, \hat{B}]] + \ldots$

THEOREM 4.6.3 $\quad$ If $[\hat{A}, \hat{B}] = c$ for $e^{x\hat{A}}\hat{B}e^{-x\hat{A}} = \hat{B} + cx$ where "c" is a complex number.

THEOREM 4.6.4 $\quad$ Product of Exponentials: Campbell–Baker–Hausdorff Theorem

$$e^{x(\hat{A}+\hat{B})} = e^{x\hat{A}}e^{x\hat{B}}e^{-x^2\left[\hat{A}, \hat{B}\right]/2} \quad \text{when} \quad \left[\hat{A}, \left[\hat{A}, \hat{B}\right]\right] = 0 = \left[\hat{B}, \left[\hat{A}, \hat{B}\right]\right]$$

If the operators commute, then the ordinary law of multiplication of exponentials holds.

THEOREM 4.6.5: $\quad [e^{x\hat{A}}\hat{B}e^{-x\hat{A}}]^n = e^{x\hat{A}}\hat{B}^n e^{-x\hat{A}}$

Theorem 4.6.1 can be proven by writing a Taylor expansion of $\hat{O}(x)$ as

$$\hat{O}(x) = \hat{O}(0) + \frac{\partial\hat{O}}{\partial x}\bigg|_{x=0} x + \frac{1}{2!}\frac{\partial^2\hat{O}}{\partial x^2}\bigg|_{x=0} x^2 + \ldots$$

where the two first terms have the form

$$\hat{O}(0) = e^{x\hat{A}}\hat{B}e^{-x\hat{A}}\Big|_{x=0} = \hat{B}$$

Higher-order derivatives can be similarly calculated

$$\frac{\partial\hat{O}}{\partial x}\Big|_{x=0} = \frac{\partial}{\partial x}(e^{x\hat{A}}\hat{B}e^{-x\hat{A}})_{x=0} = \hat{A}e^{x\hat{A}}\hat{B}e^{-x\hat{A}} - e^{x\hat{A}}\hat{B}e^{-x\hat{A}}.\hat{A}]_{x=0} = [\hat{A},\hat{B}]$$

Theorem 4.6.2 follows from the first by setting $x = 1$.

Theorem 4.6.5 uses the fact that $e^{x\hat{A}}e^{-x\hat{A}} = e^{x\hat{A}-x\hat{A}} = 1$ where the exponents can be combined because they commute. Then

$$\left[e^{x\hat{A}}\hat{B}e^{-x\hat{A}}\right]^{n} = \left(e^{x\hat{A}}\hat{B}e^{-x\hat{A}}\right)\left(e^{x\hat{A}}\hat{B}e^{-x\hat{A}}\right)\left(e^{x\hat{A}}\hat{B}e^{-x\hat{A}}\right)\dots = e^{x\hat{A}}\hat{B}^{n}e^{-x\hat{A}}$$

4.7   Operators and Matrices in Tensor Product Space

The tensor product space combines two or more vector spaces into one space. A variety of tensor product spaces can be formed. In this section, we simply place basis vectors next to each other and then build the algebra. This construction has applications to the quantum theory of multiple particles and spins as well as to group theory.

4.7.1   Tensor Product Spaces

Vector spaces V and W can be combined into a tensor product space (i.e., direct product space) with vectors $|v, w\rangle \equiv |v\rangle|w\rangle \in V \otimes W$. The vectors in the corresponding dual space have the form $|v, w\rangle^{+} = \langle v, w| \equiv \langle v|\langle w| \in [V \otimes W]^{*} = V^{*} \otimes W^{*}$. Suppose the vector spaces V and W have the the basis set $B_v = \{|\phi_i\rangle\}$ and $B_w = \{|\psi_j\rangle\}$, respectively. Then the product space and its dual have the basis sets

$$V \otimes W = \{|\phi_i\rangle|\psi_j\rangle = |\phi_i, \psi_j\rangle\} \quad \text{and} \quad V^{*} \otimes W^{*} = \{\langle\phi_i, \psi_j| = \langle\phi_i|\langle\psi_j|\}$$

and both have the dimension of $\text{Dim}(V)\,\text{Dim}(W)$.

Next, consider inner products on the product space. Inner products can only be formed between V^{*} and V, and also between W^{*} and W. So if $|v_1\rangle, |v_2\rangle \in V$ and $|w_1\rangle, |w_2\rangle \in W$ then the inner product can be written as

$$\langle v_1 w_1 \mid v_2 w_2 \rangle = \langle v_1 \mid v_2 \rangle \langle w_1 \mid w_2 \rangle \tag{4.7.1}$$

Now we can specify the standard properties for the Hilbert space. The basis vectors must satisfy an orthonormality relation of the form

$$\langle \phi_a \psi_b \mid \phi_c \psi_d \rangle = \langle \phi_a \mid \phi_c \rangle \langle \psi_b \mid \psi_d \rangle = \delta_{ac}\delta_{bd} \tag{4.7.2}$$

Every vector in the space $V \otimes W$ has a basis vector expansion with components

$$|\Psi\rangle = \sum_{ab} \beta_{ab} |\phi_a \psi_b\rangle \qquad \beta_{ab} = \langle \phi_a \psi_b \mid \Psi \rangle \qquad (4.7.3)$$

The closure relation has the form

$$\sum_{ab} |\phi_a \psi_b\rangle\langle \phi_a \psi_b| = \hat{1} \qquad (4.7.4)$$

Notice that the general vector in the tensor product space cannot generally be decomposed into the product of two vectors

$$|\Psi\rangle = \sum_{ab} \beta_{ab} |\phi_a \psi_b\rangle \neq \sum_{ab} \beta_a |\phi_a\rangle \beta_b |\psi_b\rangle = |\Phi\rangle|\Psi\rangle$$

since the components β_{ab} cannot be uniquely factored.

4.7.2 Operators

Operators $\hat{O}$ operate either between direct product spaces such as $\hat{O} : V \otimes W \to X \otimes Y$ or within a given direct product space such as $\hat{O} : V \otimes W \to V \otimes W$. For simplicity, we consider the second case in this section.

One type of direct product operator consists of the direct product of two operators $\hat{O}^{(V)} : V \to V$ and $\hat{O}^{(W)} : W \to W$, denoted by $\hat{O} = \hat{O}^{(V)} \otimes \hat{O}^{(W)}$. To find the image of the $\hat{O}|v\rangle|w\rangle$, we just need to remember that $\hat{O}^{(V)}$ operates only on vectors in V and $\hat{O}^{(W)}$ operates only on vectors in W. Therefore, we have

$$\hat{O}|v\rangle|w\rangle = \hat{O}^{(V)}\hat{O}^{(W)}|v\rangle|w\rangle = \hat{O}^{(V)}|v\rangle \hat{O}^{(W)}|w\rangle = |x\rangle|y\rangle$$

where $|x\rangle|y\rangle \in V \otimes W$. The inner product behaves in a similar manner

$$\langle q|\langle r|\hat{O}|v\rangle|w\rangle = \langle q|\langle r|\hat{O}^{(V)}\hat{O}^{(W)}|v\rangle|w\rangle = \langle q|\hat{O}^{(V)}|v\rangle \langle r|\hat{O}^{(W)}|w\rangle$$

where $|v\rangle \in V$, $|w\rangle \in W$, and $\langle q|\langle r|$ is a projector in the dual space $V^* \otimes W^*$. Not all operators can be subdivided in such a way that one part operates solely on V while another operates solely on W.

Another notation is quite common in the literature. It helps to distinguish between ordinary multiplication and the direct product type; this distinction becomes especially important for writing the matrix of a vector in the direct product space. If we have an operator $\hat{O}^{(V)} : V \to V$ then we can use the unit operator on W to write $\hat{O}^{(V)} \otimes \hat{1} : V \otimes W \to V \otimes W$ then $\{\hat{O}^{(V)} \otimes \hat{1}\}\{|v\rangle \otimes |w\rangle\} = \{\hat{O}^{(V)}|v\rangle\} \otimes \{\hat{1}|w\rangle\}$. More generally, we can write

$$\left\{\hat{O}^{(V)} \otimes \hat{O}^{(W)}\right\}\{|v\rangle \otimes |w\rangle\} = \left\{\hat{O}^{(V)}|v\rangle\right\} \otimes \left\{\hat{O}^{(W)}|w\rangle\right\}$$

What about the addition of two operators?

$$\left\{\hat{O}^{(V)} + \hat{O}^{(W)}\right\}\{|v\rangle \otimes |w\rangle\} \equiv \left\{\hat{O}^{(V)} \otimes \hat{1} + \hat{1} \otimes \hat{O}^{(W)}\right\}\{|v\rangle \otimes |w\rangle\}$$

Distributing terms gives

$$\left\{\hat{O}^{(V)} + \hat{O}^{(W)}\right\}\{|v\rangle \otimes |w\rangle\} = \left\{\hat{O}^{(V)} \otimes \hat{1}\right\}\{|v\rangle \otimes |w\rangle\} + \left\{\hat{1} \otimes \hat{O}^{(W)}\right\}\{|v\rangle \otimes |w\rangle\}$$

Simplifying gives

$$\left\{\hat{O}^{(V)} + \hat{O}^{(W)}\right\}\{|v\rangle \otimes |w\rangle\} = \left\{\hat{O}^{(V)}|v\rangle\right\} \otimes |w\rangle + |v\rangle \otimes \left\{\hat{O}^{(W)}|w\rangle\right\}$$

as expected. The notation helps signify that the addition between vectors must be on the direct product space.

4.7.3 Matrices of Direct Product Operators

The operators $\hat{O}$ acting on the direct product space $V \otimes W$ map one basis vector into another. Assume the basis vectors for the spaces can be written as

$$B_v = \{|\phi_1\rangle, |\phi_2\rangle\} \qquad B_w = \{|\psi_1\rangle, |\psi_2\rangle\} \qquad B_{V \otimes W} = \{|\phi_a\rangle|\psi_b\rangle\}$$

The matrix of $\hat{O}$ can be defined by the $O_{ab,cd}$ in $\hat{O}|\phi_c\psi_d\rangle \maltese \sum_{a,b} O_{a,b;c,d}|\phi_a\psi_b\rangle$, which produces the basis vector expansion of $\hat{O}$ as

$$\hat{O} \maltese \sum_{abcd} O_{ab;cd}|\phi_a, \psi_b\rangle\langle\phi_c, \psi_d| = \sum_{abcd} O_{ab,cd}|\phi_a\rangle|\psi_b\rangle\langle\phi_c|\langle\psi_d| \tag{4.7.5}$$

While this definition works, another grouping of the indices makes the direct product matrix easier to calculate when taking the direct product of two other matrices. To this end, rearrange the basis vectors and dummy indices in Equation 4.7.4 and write

$$\hat{O} = \sum_{abcd} O_{ac,bd}[|\phi_a\rangle\langle\phi_b|][|\psi_c\rangle\langle\psi_d|] \tag{4.7.6}$$

When necessary, we make the index convention that, for each "a" and "b," the summation is performed first over "d" and then over "c." The object $O_{ab,cd}$ is a single number (an element of a matrix). The collection of $O_{ac,bd}$ of complex numbers forms a matrix that cannot, most of the time, be divided into the product of two matrices.

Case 1: $\hat{O} = \hat{O}^{(V)} \otimes \hat{O}^{(W)} = \hat{O}^{(V)}\hat{O}^{(W)}$

This case supposes that the operator $\hat{O}$ operating on the direct product space $V \otimes W$ comes from two operators $\hat{O} = \hat{O}^{(V)}\hat{O}^{(W)}$ where $\hat{O}^{(V)} : V \to V$ and $\hat{O}^{(W)} : W \to W$. For simplicity, assume $\mathrm{Dim}(V) = \mathrm{Dim}(W) = 2$. The individual operators can be written as basis vector expansions

$$\hat{O}^{(V)} = \sum_{ab} O_{ab}^{(V)}|\phi_a\rangle\langle\phi_b| \qquad \text{and} \qquad \hat{O}^{(W)} = \sum_{cd} O_{cd}^{(W)}|\psi_c\rangle\langle\psi_d|$$

The operator $\hat{O} = \hat{O}^{(V)}\hat{O}^{(W)}$ can now be written as

$$\begin{aligned}
\hat{O} = \hat{O}^{(V)}\hat{O}^{(W)} &= \sum_{ab} O_{ab}^{(V)}|\phi_a\rangle\langle\phi_b| \sum_{cd} O_{cd}^{(W)}|\psi_c\rangle\langle\psi_d| \\
&= \sum_{abcd} O_{ab}^{(V)} O_{cd}^{(W)}[|\phi_a\rangle\langle\phi_b|][|\psi_c\rangle\langle\psi_d|]
\end{aligned} \tag{4.7.7}$$

For each a, b, there exists a set of matrix elements O_{cd}. A comparison of Equations 4.7.7 and 4.7.6 shows the matrix elements of $\hat{O} = \hat{O}^{(V)}\hat{O}^{(W)}$ must be related to those for $\hat{O}^{(V)}$ and for $\hat{O}^{(W)}$ by $O_{ac,bd} = O_{ab}^{(V)}O_{cd}^{(W)}$. In matrix notation, this becomes

$$\underline{O} = \underline{O}^{(V)} \otimes \underline{O}^{(W)} = \begin{bmatrix} O_{11}^{(v)} & O_{12}^{(v)} \\ O_{21}^{(v)} & O_{22}^{(v)} \end{bmatrix} \otimes \begin{bmatrix} O_{11}^{(w)} & O_{12}^{(w)} \\ O_{21}^{(w)} & O_{22}^{(w)} \end{bmatrix}$$

This is not the usual *matrix* multiplication! The matrix on the right-hand side is multiplied into each element of the matrix on the left-hand side.

$$\underline{O} = \underline{O}^{(V)} \otimes \underline{O}^{(W)} = \begin{bmatrix} O_{11}^{(v)}\underline{O}^{(W)} & O_{12}^{(v)}\underline{O}^{(W)} \\ O_{21}^{(v)}\underline{O}^{(W)} & O_{22}^{(v)}\underline{O}^{(W)} \end{bmatrix} = \begin{bmatrix} O_{11}^{(V)}O_{11}^{(W)} & O_{11}^{(V)}O_{12}^{(W)} & O_{12}^{(V)}O_{11}^{(W)} & O_{12}^{(V)}O_{12}^{(W)} \\ O_{11}^{(V)}O_{21}^{(W)} & O_{11}^{(V)}O_{22}^{(W)} & O_{12}^{(V)}O_{21}^{(W)} & O_{12}^{(V)}O_{22}^{(W)} \\ O_{21}^{(V)}O_{11}^{(W)} & O_{21}^{(V)}O_{12}^{(W)} & O_{22}^{(V)}O_{11}^{(W)} & O_{22}^{(V)}O_{12}^{(W)} \\ O_{21}^{(V)}O_{21}^{(W)} & O_{21}^{(V)}O_{22}^{(W)} & O_{22}^{(V)}O_{21}^{(W)} & O_{22}^{(V)}O_{22}^{(W)} \end{bmatrix}$$

Of course each entry $O_{ab}^{(V)}O_{cd}^{(W)}$ is just a single number found by ordinary multiplication between numbers. The above matrix illustrates the convention for the indices of $\underline{O}$.

Case 2: The operator $\hat{O}$ cannot be divided

The last matrix given in case 1 provides a clue as to how $\underline{O}$ should be written for the general case, namely

$$\underline{O} = \begin{bmatrix} O_{11,11} & O_{11,12} & O_{12,11} & O_{12,12} \\ O_{11,21} & O_{11,22} & O_{12,21} & O_{12,22} \\ O_{21,11} & O_{21,12} & O_{22,11} & O_{22,12} \\ O_{21,21} & O_{21,22} & O_{22,21} & O_{22,22} \end{bmatrix}$$

With the index convention, matrices in direct product space can be multiplied together as usual.

4.7.4 The Matrix Representation of Basis Vectors for Direct Product Space

Now let's show how the matrices multiply and define the unit vectors in the cross product space. Again for simplicity consider two 2-D Hilbert spaces V and W and use the product of two operators $\hat{O} = \hat{A}_v\hat{B}_w$ where the v and w indices refer to the original Hilbert space in $V \otimes W$. Let's convert the operator equation

$$\hat{A}_v\hat{B}_w|v_\otimes\rangle = |v'_\otimes\rangle$$

where the subscript $\otimes$ indicates the vector comes from $V \otimes W$. Operating with $\langle a_v|\langle b_w|$ and inserting the closure relation $\sum_{c,d} |c_vd_w\rangle\langle c_vd_w| = \hat{1}$ produces

$$\sum_{c,d} \langle a_v|\langle b_w|\hat{A}_v\hat{B}_w|c_vd_w\rangle\langle c_vd_w \mid v_\otimes\rangle = \langle a_vb_w \mid v'_\otimes\rangle$$

We can write this in matrix notation as

$$\sum_{c,d} A_{a_v c_v} \underbrace{B_{b_w d_w}}_{2} \underbrace{V_{c_v d_w}}_{2} = V'_{a_v b_w}$$

The numbers 1 and 2 under c, d indicate that we first sum over d and then over c. Writing this in matrix notation gives us

$$
\begin{bmatrix}
A_{11}^{(V)}B_{11}^{(W)} & A_{11}^{(V)}B_{12}^{(W)} & A_{12}^{(V)}B_{11}^{(W)} & A_{12}^{(V)}B_{12}^{(W)} \\
A_{11}^{(V)}B_{21}^{(W)} & A_{11}^{(V)}B_{22}^{(W)} & A_{12}^{(V)}B_{21}^{(W)} & A_{12}^{(V)}B_{22}^{(W)} \\
A_{21}^{(V)}B_{11}^{(W)} & A_{21}^{(V)}B_{12}^{(W)} & A_{22}^{(V)}B_{11}^{(W)} & A_{22}^{(V)}B_{12}^{(W)} \\
A_{21}^{(V)}B_{21}^{(W)} & A_{21}^{(V)}B_{22}^{(W)} & A_{22}^{(V)}B_{21}^{(W)} & A_{22}^{(V)}B_{22}^{(W)}
\end{bmatrix}
\begin{bmatrix} v_{11} \\ v_{12} \\ v_{21} \\ v_{22} \end{bmatrix}
=
\begin{bmatrix} v'_{11} \\ v'_{12} \\ v'_{21} \\ v'_{22} \end{bmatrix}
\tag{4.7.8}
$$

Notice the order of the factors and the order of the indices in Equation (4.7.8). The column vectors must come from the direct product of two individual matrices. If $|v_\otimes\rangle = |r_v\rangle|s_w\rangle$ then we see

$$
\begin{bmatrix} v_{11} \\ v_{12} \\ v_{21} \\ v_{22} \end{bmatrix}
=
\begin{bmatrix} r_1 s_1 \\ r_1 s_2 \\ r_2 s_1 \\ r_2 s_2 \end{bmatrix}
=
\begin{bmatrix} r_1 \begin{pmatrix} s_1 \\ s_2 \end{pmatrix} \\ r_2 \begin{pmatrix} s_1 \\ s_2 \end{pmatrix} \end{bmatrix}
=
\begin{pmatrix} r_1 \\ r_2 \end{pmatrix} \otimes \begin{pmatrix} s_1 \\ s_2 \end{pmatrix}
\tag{4.7.9}
$$

We therefore realize that the basis vectors can be represented by

$$
|1\rangle = |1\rangle_v|1\rangle_w \equiv \begin{pmatrix} 1 \\ 0 \end{pmatrix} \otimes \begin{pmatrix} 1 \\ 0 \end{pmatrix} = \begin{pmatrix} 1\begin{pmatrix} 1 \\ 0 \end{pmatrix} \\ 0\begin{pmatrix} 1 \\ 0 \end{pmatrix} \end{pmatrix} = \begin{pmatrix} 1 \\ 0 \\ 0 \\ 0 \end{pmatrix}
$$

$$
|2\rangle = |1\rangle_v|2\rangle_w \equiv \begin{pmatrix} 1 \\ 0 \end{pmatrix} \otimes \begin{pmatrix} 0 \\ 1 \end{pmatrix} = \begin{pmatrix} 1\begin{pmatrix} 0 \\ 1 \end{pmatrix} \\ 0\begin{pmatrix} 0 \\ 1 \end{pmatrix} \end{pmatrix} = \begin{pmatrix} 0 \\ 1 \\ 0 \\ 0 \end{pmatrix}
$$

$$
|3\rangle = |2\rangle_v|1\rangle_w \equiv \begin{pmatrix} 0 \\ 1 \end{pmatrix} \otimes \begin{pmatrix} 1 \\ 0 \end{pmatrix} = \begin{pmatrix} 0\begin{pmatrix} 1 \\ 0 \end{pmatrix} \\ 1\begin{pmatrix} 1 \\ 0 \end{pmatrix} \end{pmatrix} = \begin{pmatrix} 0 \\ 0 \\ 1 \\ 0 \end{pmatrix}
$$

$$
|4\rangle = |2\rangle_v|2\rangle_w \equiv \begin{pmatrix} 0 \\ 1 \end{pmatrix} \otimes \begin{pmatrix} 0 \\ 1 \end{pmatrix} = \begin{pmatrix} 0\begin{pmatrix} 0 \\ 1 \end{pmatrix} \\ 1\begin{pmatrix} 0 \\ 1 \end{pmatrix} \end{pmatrix} = \begin{pmatrix} 0 \\ 0 \\ 0 \\ 1 \end{pmatrix}
$$

4.8 Unitary Operators and Similarity Transformations

Unitary and orthogonal operators map one basis set into another. These operators do not change the length of a vector nor do they change the angle between vectors. Unitary operators act on abstract Hilbert spaces. Orthogonal operators, a subset of unitary operators, act on Euclidean vectors. This section also discusses the rotation of functions.

4.8.1 Orthogonal Rotation Matrices

Orthogonal operators rotate real Euclidean vectors. The word "orthogonal" implies that the length of a vector remains unaffected under rotations. The orthogonal operator can be most conveniently defined through its matrix.

$$\underline{R}^{-1} = \underline{R}^{T} \tag{4.8.1}$$

This relation is independent of the basis set chosen for the vector space as it should be since the effect of the *operator* does not depend on the chosen basis set. Recall the definition of the transpose

$$\left(R^{T}\right)_{ab} = R_{ba} \quad \text{or} \quad R^{T}_{ab} = R_{ba} \tag{4.8.2}$$

The defining relation in Equation (4.8.1) can be used to show $\text{Det}(\hat{R}) = 1$

$$1 = \text{Det}(\underline{1}) = \text{Det}\left(\hat{R}\hat{R}^{T}\right) = \text{Det}\,\hat{R}\,\text{Det}\,\hat{R}^{T} = \text{Det}\,\hat{R}\,\text{Det}\,\hat{R} = \text{Det}\left(\hat{R}\right)^{2}$$

and therefore $\text{Det}\,\hat{R} = 1$ by taking the positive root. The above string of equalities uses the unit operator (unit matrix) defined by $\underline{1} = [\delta_{ab}]$. The discussion shows later that the orthogonal matrix leaves angles and lengths invariant.

Recall that rotations can be viewed as either rotating *vectors* or the *coordinate system*. We take the point of view that operators rotate the vectors as suggested by Figure 4.8.1. Consider rotating all two-dimensional vectors by θ (positive when counter clockwise). We find the operator and then the matrix. The rotation operator provides $\hat{R}|1\rangle = |1'\rangle$ and $\hat{R}|2\rangle = |2'\rangle$. Reexpressing $|1'\rangle$ and $|2'\rangle$ in terms of the original basis vectors $|1\rangle$ and $|2\rangle$ then provides the matrix elements according to $\hat{R}|1\rangle = R_{11}|1\rangle + R_{21}|2\rangle$ and $\hat{R}|2\rangle = R_{21}|1\rangle + R_{22}|2\rangle$. Figure 4.8.1 provides

$$|1'\rangle = \hat{R}|1\rangle = \cos\theta|1\rangle + \sin\theta|2\rangle = R_{11}|1\rangle + R_{21}|2\rangle$$
$$|2'\rangle = \hat{R}|2\rangle = -\sin\theta|1\rangle + \cos\theta|2\rangle = R_{12}|1\rangle + R_{22}|2\rangle \tag{4.8.3}$$

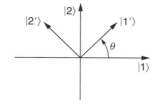

FIGURE 4.8.1
Rotating the basis vectors and re-expressing them in the original basis set.

where the coefficients are obtained from Figure 2. The results can be written as

$$\hat{R} = R_{11}|1\rangle\langle 1| + R_{12}|1\rangle\langle 2| + R_{21}|2\rangle\langle 1| + R_{22}|2\rangle\langle 2|$$

$$= \cos\theta|1\rangle\langle 1| - \sin\theta|1\rangle\langle 2| + \sin\theta|2\rangle\langle 1| + \cos\theta|2\rangle\langle 2| \tag{4.8.4}$$

The operator $\hat{R}$ is most correctly interpreted as associating a new vector $\vec{v}'$ (in the Hilbert space) with the original vector $\vec{v}$. A *rotation* implies an element of time ... time is not involved with these particular operators. The matrix $\underline{R}$ changes the components of a vector $|v\rangle = x|1\rangle + y|2\rangle$ into $|v'\rangle = x'|1\rangle + y'|2\rangle$ according to

$$\begin{bmatrix} x' \\ y' \end{bmatrix} = \begin{bmatrix} \cos\theta & -\sin\theta \\ \sin\theta & \cos\theta \end{bmatrix}\begin{bmatrix} x \\ y \end{bmatrix} = \begin{bmatrix} x\cos\theta - y\sin\theta \\ x\sin\theta + y\cos\theta \end{bmatrix} \quad \text{where} \quad \underline{R} = \begin{bmatrix} \cos\theta & -\sin\theta \\ \sin\theta & \cos\theta \end{bmatrix}$$

$$\tag{4.8.5}$$

This last relation easily shows $R^T R = 1$ so that $\underline{R}^{-1} = \underline{R}^T$ as required for an orthogonal operator $\hat{R}$ and matrix $\underline{R}$.

We can now see that the example rotation matrix transforms one basis into another. Equation (4.8.5) shows that the length of a vector does not change under a rotation by calculating the length

$$\|\vec{v}'\|^2 = (x')^2 + (y')^2 = (x\cos\theta - y\sin\theta)^2 + (x\sin\theta + y\cos\theta)^2 = x^2 + y^2 = \|\vec{v}\|^2$$

Therefore orthogonal matrices do not shrink or expand vectors. The same conclusion can be verified by using Dirac notation

$$\|v'\|^2 = \langle v' \mid v'\rangle = \langle v|\hat{R}^+\hat{R}|v\rangle = \langle v|\hat{R}^T\hat{R}|v\rangle = \langle v|1|v\rangle = \langle v \mid v\rangle = \|v\|^2$$

where the fourth term uses the fact that $\hat{R}$ is real. The "rotation" operator $\hat{R}$ does not change the angle between two vectors $|v'\rangle = \hat{R}|v\rangle$ and $|w'\rangle = \hat{R}|w\rangle$. The angle can be defined through the dot product relation $\langle v'|w'\rangle = \vec{v}' \cdot \vec{w}' = v'\,w'\cos\theta'$.

$$\cos\theta' = \frac{1}{v'w'}\langle v' \mid w'\rangle = \frac{1}{v\,w}\langle v|R^T R|w\rangle = \frac{1}{v\,w}\langle v \mid w\rangle = \cos\theta$$

The "rotation" operator $\hat{R}$ is called orthogonal because it does not affect the orthonormality of basis vectors $\{|1\rangle, |2\rangle, \ldots\}$ in a real vector space. The set $\{\hat{R}|1\rangle, \hat{R}|2\rangle, \ldots\}$ must also be a basis set.

4.8.2 Unitary Transformations

A unitary transformation is a "rotation" in the generalized Hilbert space. The set of orthogonal operators forms a subset of the unitary operators. A unitary operator "$\hat{u}$" is defined to have the property that

$$\hat{u}^+ = \hat{u}^{-1} \quad \text{or} \quad \hat{u}\hat{u}^+ = 1 = \hat{u}^+\hat{u} \tag{4.8.6}$$

The unitary operator therefore satisfies $|\text{Det}(u)|^2 = 1$ since

$$1 = \text{Det}(\hat{1}) = \text{Det}(\hat{u}\hat{u}^+) = \text{Det}(\hat{u})\,\text{Det}(\hat{u}^+) = \text{Det}(\hat{u})\,\text{Det}^*(\hat{u}) = |\text{Det}(\hat{u})|^2$$

which used the property of determinants $\text{Det}(u^T) = \text{Det}(u)$. We can write $\text{Det}(\hat{u}) = e^{i\phi}$. The relation $\hat{u}^+ = \hat{u}^{-1}$ therefore provides the determinant to within a phase factor. We customarily choose the phase to be zero ($\phi = 0$). Therefore, as an alternate definition of a unitary operator, require $\text{Det}(\hat{u}) = 1$.

The unitary transformations can be thought of as "change of basis operators" similar to the rotation operator $\hat{R}$ in the previous topic. That is, if $B_v = \{|a\rangle\}$ forms a basis set then so does $B'_v = \{\hat{u}|a\rangle = |a'\rangle\}$. The operator $\hat{u}$ maps the vector space V into itself $\hat{u} : V \to V$. Unitary operators preserve the orthonormality relations of the basis set.

$$\langle a' | b' \rangle = (\hat{u}|a\rangle)^+ (\hat{u}|b\rangle) = \langle a|\hat{u}^+\hat{u}|b\rangle = \langle a|1|b\rangle = \langle a | b \rangle = \delta_{ab}$$

As a result, B'_v and B_v are equally good basis sets for the Hilbert space V.

The inverse of the unitary operator $\hat{u}$, $\hat{u}^{-1} = \hat{u}^+$ can be written in matrix notation as

$$\underline{u}^+ = \underline{u}^{T*} \quad \text{or} \quad (u^+)_{ab} = u^*_{ba} \quad \text{or sometimes} \quad u^+_{ab} = u^*_{ba}$$

Example 4.8.1

If $\hat{u} = \sum_{ab} u_{ab}|a\rangle\langle b|$ then $\hat{u}^+$ can be calculated as

$$\hat{u}^+ = \sum_{ab} (u_{ab}|a\rangle\langle b|)^+ = \sum_{ab} (u_{ab})^+ |b\rangle\langle a| = \sum_{ab} u^*_{ab}|b\rangle\langle a|$$

Now notice that u_{ab} represents a single complex number and not the entire matrix so that the dagger can be replaced by the complex conjugate without interchanging the indices.

Example 4.8.2

Show for the previous example that $u^+u = 1$

$$\hat{u}^+\hat{u} = \left(\sum_{\alpha\beta} u^*_{\alpha\beta}|\beta\rangle\langle\alpha|\right)\left(\sum_{ab} u_{ab}|a\rangle\langle b|\right) = \sum_{\substack{ab \\ \alpha\beta}} u^*_{\alpha\beta}u_{ab}|\beta\rangle\langle b| \quad \delta_{a\alpha} = \sum_{\substack{ab \\ \beta}} u^*_{\alpha\beta}u_{ab}|\beta\rangle\langle b|$$

We need to work with the product of the unitary matrices.

$$\sum_a u^*_{\alpha\beta}u_{ab} = \sum_a (u^+)_{\beta a}u_{ab} = (u^+u)_{\beta b} = \delta_{\beta b}$$

Notice that we switched the indices when we calculated Hermitian adjoint of the matrix since we are referring to the entire matrix. Substituting this result for the unitary matrices gives us

$$\hat{u}^+\hat{u} = \sum_{\beta b} \delta_{\beta b}|\beta\rangle\langle b| = \sum_b |b\rangle\langle b| = 1$$

4.8.3   Visualizing Unitary Transformations

Unitary transformations change one basis set into another basis set.

$$B_v = \{|a\rangle\} \to B'_v = \{\hat{u}|a\rangle = |a'\rangle\}$$

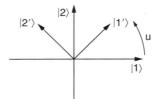

FIGURE 4.8.2
The unitary operator is determined by the mapping of the basis vectors.

Figure 4.8.2 shows the effect of the unitary transformation

$$\hat{u}|1\rangle = |1'\rangle \quad \hat{u}|2\rangle = |2'\rangle$$

The operator is defined by its affect on the basis vectors. The two objects $|1'\rangle\langle1|$ and $|2'\rangle\langle2|$, which are "basis vectors" for the vector space of operators $\{\hat{u}\}$, perform the following mappings

$$|1'\rangle\langle1| \quad \text{maps} \quad |1\rangle \to |1'\rangle \quad \text{since} \quad [|1'\rangle\langle1|]|1\rangle = |1'\rangle\langle1 \mid 1\rangle = |1'\rangle$$

$$|2'\rangle\langle2| \quad \text{maps} \quad |2\rangle \to |2'\rangle \quad [|2'\rangle\langle2|]|2\rangle = |2'\rangle\langle2 \mid 2\rangle = |2'\rangle$$

Putting both pieces together gives us a very convenient form for the operator

$$\hat{u} = |1'\rangle\langle1| + |2'\rangle\langle2|$$

The operator can be written just by placing vectors next to each other. The operator $\hat{u}$ can be left in the form

$$\hat{u} = \sum_a |a'\rangle\langle a|$$

to handle "rotations" in all directions. Of course, to use $\hat{u}$ for actual calculations, either $|a'\rangle$ must be expressed as a sum over $|a\rangle$ or vice versa.

4.8.4   Similarity Transformations

Assume there exists a linear operator $\hat{O}$ that maps the vector space into itself $\hat{O} : V \to V$. Assume the vectors $|v\rangle$ and $|w\rangle$ (not necessarily basis vectors) satisfy an equation of the form $\hat{O}|v\rangle = |w\rangle$. Now suppose that we transform both sides by the unitary transformation $\hat{u}$ and then use the definition of unitary $\hat{u}^+\hat{u} = 1$ to find

$$\hat{u}\hat{O}|v\rangle = \hat{u}|w\rangle \quad \to \quad \hat{u}\hat{O}\hat{u}^+\hat{u}|v\rangle = \hat{u}|w\rangle$$

Defining $\hat{O}' = \hat{u}\hat{O}\hat{u}^+$ and $|v'\rangle = \hat{u}|v\rangle$ and $|w'\rangle = \hat{u}|w\rangle$ provides

$$\hat{O}'|v'\rangle = |w'\rangle$$

which has the same form as the original equation. The difference is that the operator $\hat{O}$ is now expressed in the "rotated basis set" as

$$\hat{O}' = \hat{u}\hat{O}\hat{u}^+ \tag{4.8.7}$$

Changing basis vectors also changes the representation of the operator $\hat{O}$.

Transformations as those found in Equation (4.8.7) are "similarity" transformations. More generally, we write the similarity transformation as

$$\hat{O}' = \hat{S}\hat{O}\hat{S}^{-1} \qquad (4.8.8)$$

for a the general linear transformation $\hat{S}$. Equation 4.8.8 is equivalent to Equation (4.8.7) because $\hat{u}$ is unitary $\hat{u}^{-1} = \hat{u}^{+}$.

The similarity transformation can also be seen to have the form $\hat{O}' = \hat{u}\hat{O}\hat{u}^{+}$ by using the transformation $\hat{u}$ directly on the vectors in the basis vector expansion. For convenience, assume $\hat{O}: V \to V$ with $V = Sp\{|a\rangle\}$. Replacing $|a\rangle$ with $\hat{u}|a\rangle$ and $|b\rangle$ with $\hat{u}|b\rangle$ produces

$$\hat{O} = \sum_{ab} O_{ab}|a\rangle\langle b| \quad \to \quad \hat{O}' = \sum_{ab} O_{ab}\left(\hat{u}|a\rangle\right)\left(\hat{u}|b\rangle\right)^{+} = \sum_{ab} O_{ab}\hat{u}|a\rangle\langle b|\hat{u}^{+} = \hat{u}\hat{O}\hat{u}^{+}$$

which is the same result as before.

A string of operators can be rewritten using unitary transformation $\hat{u}$

$$\left(\hat{O}_1\hat{O}_2 + 5\hat{O}_3\hat{O}_4^3\right)|v\rangle = |w\rangle \quad \to \quad \left(\hat{O}_1'\hat{O}_2' + 5\hat{O}_3'\hat{O}_4'3\right)|v'\rangle = |w'\rangle$$

For example, $\hat{O}_4^3$ can be transformed by repeatedly inserting a "1" and applying $1 = \hat{u}^{+}\hat{u}$ as follows

$$\hat{u}\left(\hat{O}_4^3\right)\hat{u}^{+} = \hat{u}\left(\hat{O}_4\hat{O}_4\hat{O}_4\right)\hat{u}^{+} = \hat{u}\left(\hat{O}_41\hat{O}_41\hat{O}_4\right)\hat{u}^{+} = \hat{u}\hat{O}_4\hat{u}^{+}\hat{u}\hat{O}_4\hat{u}^{+}\hat{u}\hat{O}_4\hat{u}^{+} = \hat{O}_4'\hat{O}_4'\hat{O}_4' = \left(\hat{O}_4'\right)^3$$

Example 4.8.3

Write $\langle v'|\hat{T}'|w'\rangle$ in terms of the objects $|v\rangle$, $\hat{T}$, $|w\rangle$ where $|v'\rangle = \hat{u}|v\rangle$ and $\hat{T}' = \hat{u}\hat{T}\hat{u}^{+}$ and $|w'\rangle = \hat{u}|w\rangle$. This is done as follows

$$\langle v'|\hat{T}'|w'\rangle = \langle v|\hat{u}^{+}\left(\hat{u}\hat{T}\hat{u}^{+}\right)\hat{u}|w\rangle = \langle v|\hat{T}|w\rangle$$

again $\hat{O}' = \hat{u}\hat{O}\hat{u}^{+}$ is the representation of the operator $\hat{O}$ using the new basis set $B_v' = \{\hat{u}|a\rangle\}$.

4.8.5 Trace and Determinant

The trace is important for calculating averages. Similarity transformations leave the trace and determinant unchanged. That is, trace and determinant operations are invariant with respect to similarity transformations. Consider

$$\hat{A}' = \hat{u}\hat{A}\hat{u}^{+} \quad \text{and} \quad \hat{u}: V \to V$$

The cyclic property of the trace and the fact that $\hat{u}$ is a unitary operator provides

$$Tr(\hat{A}') = Tr(\hat{u}\hat{A}\hat{u}^{+}) = Tr(\hat{A}\hat{u}^{+}\hat{u}) = Tr(\hat{A})$$

The same calculation can be performed for the determinant

$$\text{Det}(\hat{A}') = \text{Det}(\hat{u}\hat{A}\hat{u}^{+}) = \text{Det}(\hat{u})\text{Det}(\hat{A})\text{Det}(\hat{u}^{+}) = \text{Det}(\hat{A})\text{Det}(\hat{u}\hat{u}^{+}) = \text{Det}(\hat{A})$$

4.9 Hermitian Operators and the Eigenvector Equation

The adjoint, self-adjoint and Hermitian operators play a central role in the study of quantum mechanics and the Sturm–Liouville problem for solving partial differential equations (refer to books on Boundary Value Problems).

In quantum mechanics, Hermitian operators represent physically observable quantities such as energy $\hat{H}$, momentum $\hat{p}$, and electric field. The "observable" refers to a quality of a particle that can be observed in the laboratory. In order to translate the physical world into mathematics, we represent the observables with Hermitian operators. Hermitian operators have eigenvectors that form a basis set $\{|n\rangle\}$ for the vector space. Physical systems require the completeness of the basis set in order to accommodate every possible physical situation. The "completeness" of a basis set is related to a "completeness" in nature. The eigenvalues are real. Physical systems need the real eigenvalues so that the results of measurement will yield real results.

4.9.1 Adjoint, Self-Adjoint and Hermitian Operators

Let $\hat{T} : V \to V$ be a linear transformation (Figure 4.9.1) defined on a Hilbert space $V = Sp\{|n\rangle : n = 1, 2, \ldots\}$. Let $|f\rangle, |g\rangle$ be two elements in the Hilbert space. We define the adjoint operator $\hat{T}^{+}$ to be the operator that satisfies

$$\left\langle g \mid \hat{T}f \right\rangle = \left\langle \hat{T}^{+}g \mid f \right\rangle \tag{4.9.1}$$

An operator $\hat{T}$ is self-adjoint or Hermitian if $\hat{T}^{+} = \hat{T}$.

Previous sections define the adjoint $\hat{T}^{+}$ as connected with the dual vectors space. We can demonstrate Equation 4.9.1 using the previous definition of the adjoint. Using the notation $\hat{T}|f\rangle = |\hat{T}f\rangle$, we find

$$\left\langle g \mid \hat{T}f \right\rangle = \langle g|\hat{T}|f\rangle = \left[\hat{T}^{+}|g\rangle \right]^{+} |f\rangle = \left[|\hat{T}^{+}g\rangle \right]^{+} |f\rangle = \left\langle \hat{T}^{+}g \mid f \right\rangle$$

Example 4.9.1

If $\hat{T} = \partial/\partial x$ then find $\hat{T}^{+}$ for the following Hilbert space

$$HS = \left\{ f : \frac{\partial f(x)}{\partial x} \ \text{exists and} \ f \to 0 \ \text{as} \ x \to \pm\infty \right\}$$

Solution

We want $\hat{T}^{+}$ such that $\langle f|\hat{T}g\rangle = \langle \hat{T}^{+}f|g\rangle$. Start with the quantity on the left

$$\left\langle f \mid \hat{T}g \right\rangle = \int_{-\infty}^{\infty} dx\, f^{*}(x)\hat{T}g(x) = \int_{-\infty}^{\infty} dx\, f^{*}(x)\frac{\partial}{\partial x}g(x)$$

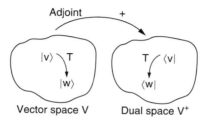

FIGURE 4.9.1
The vector and dual space.

The procedure usually starts with integration by parts:

$$\langle f | \hat{T}g \rangle = f^*(x)g(x)]_{-\infty}^{\infty} - \int_{-\infty}^{\infty} dx \, \frac{\partial f^*(x)}{\partial x} g(x)$$

In most cases, the surface term produces zero. Notice the Hermitian property of the operators depends on the properties of the Hilbert space. In the present case, the Hilbert space is defined such that $f^*(\infty)g(\infty) - f^*(-\infty)g(-\infty) = 0$; most physically sensible functions drop to zero for very large distances. Next move the minus sign and partial derivative under the complex conjugate to find

$$\langle f | \hat{T}g \rangle = \int_{-\infty}^{\infty} dx \left[-\frac{\partial f(x)}{\partial x} \right]^* g(x) = \langle \hat{T}^+ f | g \rangle$$

Note everything inside the bra $\langle \ |$ must be placed under the complex conjugate $(\ )^*$ in the integral. The operator $\hat{T}^+$ must therefore be $\hat{T}^+ = -\partial/\partial x$ or equivalently

Example 4.9.2
Find the adjoint operator $\hat{p}^+ = ((\hbar/i)(\partial/\partial_x))^+$ for the same set of functions as for Example 4.9.1 where $i = \sqrt{-1}$ and $\hbar$ represents a constant.

Solution

$$\hat{p}^+ = \left(\frac{\hbar}{i} \frac{\partial}{\partial x} \right)^+ = \left(\frac{\hbar}{i} \right)^+ \left(\frac{\partial}{\partial x} \right)^+ = \left(-\frac{\hbar}{i} \right) \left(-\frac{\partial}{\partial x} \right) = \frac{\hbar}{i} \frac{\partial}{\partial x} = \hat{p}$$

where the third term comes from Example 4.9.1. The operator $\hat{p}$ must be Hermitian and therefore corresponds to a physical observable.

As an important note, the boundary term $f^*(x) g(x)|_a^b$ (from the partial integration in the inner product) is always arranged to be zero. This depends on the definition of the Hilbert space. A number of different Hilbert spaces can produce a zero surface term. For example, if the function space is defined for $x \in [a, b]$, then the following conditions will work

1. $f(a) = f(b) = 0 \qquad \forall f \in V$
2. $f(a) = f(b)$ (without being equal to zero) $\forall f \in V$

Notice the property of an operator being Hermitian cannot be entirely separated from the properties of the Hilbert space since the surface terms must be zero.

4.9.2 Adjoint and Self-Adjoint Matrices

First, we derive the form of the adjoint matrix using the basis expansion of an operator. In the following, let $|m\rangle$ and $|n\rangle$ be basis vectors. Take the adjoint of the basis expansion

$$\hat{T} = \sum_{mn} T_{mn}|m\rangle\langle n| \quad \text{to get} \quad \hat{T}^+ = \sum_{mn} T_{mn}^*|n\rangle\langle m|$$

so now

$$\langle i|\hat{T}^+|j\rangle = \sum_{mn} T_{mn}^*\langle i|n\rangle\langle m|j\rangle = \sum_{mn} T_{mn}^* \delta_{in}\delta_{mj} = T_{ji}^*$$

The adjoint matrix requires a complex conjugate and has the indices reversed.

$$\left(T^+\right)_{ij} = T_{ji}^* \tag{4.9.2}$$

Now we show how the adjoint comes from the basic definition of the adjoint operator in Equation (4.9.1), specifically $\langle w|\hat{T}v\rangle = \langle \hat{T}^+w|v\rangle$. To work with $\langle w|\hat{T}v\rangle$, we need to use matrix notation for the inner product between two vectors $|w\rangle$ and $|y\rangle$

$$\langle w|y\rangle = \sum_a w_a^* y_a = (\underline{w}^*)^T \underline{y} = \underline{w}^+\underline{y} \tag{4.9.3}$$

The term $\langle w|\hat{T}v\rangle$ can be transformed into $\langle \hat{T}^+w|v\rangle$.

$$\left\langle w\left|\hat{T}v\right.\right\rangle = \sum_{ab} w_a^* T_{ab} v_b = \sum_{ab} (\underline{T}^T)_{ba} w_a^* v_b = \sum_{ab} \left[(\underline{T}^*{}^T)_{ba} w_a\right]^* v_b = [\underline{T}^{*T}\underline{w}]^+ \underline{v} = \left\langle \hat{T}^+w\left| v\right.\right\rangle$$

where the "+" in the last step comes from requiring that the column vector $\underline{y}^* = (\underline{T}^{*T}\underline{w})^*$ become a row vector to multiply into the column vector $\underline{v}$. The adjoint must therefore be $T^+ = T^{*T}$.

Finally, a specific form for a Hermitian matrix can be determined. A matrix is Hermitian provided $T = T^+$. For example, a 2×2 matrix is Hermitian if

$$\underline{T} = \underline{T}^+ \quad \text{so that} \quad \underline{T} = \begin{bmatrix} a & b \\ c & d \end{bmatrix} = \begin{bmatrix} a^* & c^* \\ b^* & d^* \end{bmatrix} = T^+$$

For $\underline{T}$ to be Hermitian, require $a = a^*$, $d = d^*$, so that "a, b" are both real and $b = c^*$. The self-adjoint form of the matrix $\underline{T}$ is then

$$\underline{T} = \begin{bmatrix} a & b \\ b* & d \end{bmatrix}$$

where both "a, d" are real.

4.9.3 Eigenvectors and Eigenvalues for Hermitian Operators

We now show some important theorems. The first theorem shows that Hermitian operators produce real eigenvalues. The importance of this theorem issues from representing all physically observable quantities by Hermitian operators. The result of making a measurement of the observable must produce a real number. For example, for a particle in an eigenstate $|n\rangle$ of the Hermitian energy operator $\hat{H}$ (i.e., the Hamiltonian), the result for measuring the energy $\hat{H}|n\rangle = E_n|n\rangle$ produces the real energy E_n. The particle has energy E_n when it occupies state $|n\rangle$. Energy can never be complex (except possibly for some mathematical constructs).

The second theorem shows that the eigenvectors of a Hermitian operator form a basis (we do not prove completeness). This basically says that for every observable in nature, there must always be a Hilbert space large enough to describe all possible results of measuring that observable. The state of the particle or system can be Fourier decomposed into the basis vectors.

Before discussing theorems, a few words should be said about notation conventions and about degenerate eigenvalues. We will assume that for each eigenvalue E_n there exists a single corresponding eigenfunction $|\phi_n\rangle$. We customarily label the eigenfunction by either the eigenvalue or by the eigenvalue number as

$$|\phi_n\rangle = |E_n\rangle = |n\rangle$$

Usually, the eigenvalues are listed in order of increasing value

$$E1 < E2 < \ldots$$

The condition of nondegenerate eigenvalues means that for a given eigenvalue, there exists only one eigenvector. The eigenvalues are "degenerate" if for a given eigenvalue, there are multiple eigenvectors.

non-degenerate	degenerate
$E_1 \leftrightarrow |E_1\rangle$	$E_1 \leftrightarrow |E_1\rangle$
$\vdots$	$E_2 \leftrightarrow |E_21\rangle, |E_22\rangle$
$E_n \leftrightarrow |E_n\rangle$	$E_3 \leftrightarrow |E_3\rangle$

The degenerate eigenvectors (which means both states have the same "energy" E_n) actually span a subspace of the full vector space. For example in the above table, the vectors $|E_21\rangle, |E_22\rangle$ corresponding to the eigenvalue E_2 form a two-dimensional subspace. Mathematically we can associate E_2 with any vector in the subspace spanned by $\{|E_2, 1\rangle, |E_2, 2\rangle\}$; however, it's best to choose one vector in the subspace such that it is orthogonal to the others in the set $\{|E_1\rangle, |E_3\rangle \ldots\}$. The Graham–Schmidt orthonormalization procedure helps here. After making the choice, we end up with a nondegenerate case: $|E_1\rangle, |E_2\rangle, |E_3\rangle, \ldots$.

THEOREM 4.9.1 Self-Adjoint Operators $\hat{H}$ have real eigenvalues

Proof: Assume the set $\{|n\rangle\}$ contains the eigenvectors corresponding to the eigenvalues $\{E_n\}$ so that the eigenvector equation can be written as $\hat{H}|n\rangle = E_n|n\rangle$. Consider

$$\langle n|\hat{H}|n\rangle = \langle n|E_n|n\rangle = E_n\langle n \mid n\rangle = E_n \tag{4.9.4}$$

This last equation provides the following string of equalities

$$E_n^* = \langle n|\hat{H}|n\rangle^* = \langle n|\hat{H}|n\rangle^+ = \langle n|\hat{H}^+|n\rangle = \langle n|\hat{H}|n\rangle = E_n$$

where the third term uses the fact that $* \rightarrow +$ for the complex number $\langle n|\hat{H}|n \rangle$, the fourth term reverse all factors and the fifth term uses the $\hat{H}^+ = \hat{H}$. Therefore we find $E_n = E_n^*$, which means that E_n must be real.

THEOREM 4.9.2 Orthogonal Eigenvectors
If $\hat{H}$ is Hermitian then the eigenvectors corresponding to different eigenvalues are orthogonal.

Proof: Assume $E_m \neq E_n$ and start with two separate eigenvalue equations

$$\hat{H}|E_m\rangle = E_m|E_m\rangle \qquad\qquad \hat{H}|E_n\rangle = E_n|E_n\rangle$$

$$\text{operate with } \langle E_n| \qquad\qquad\qquad \text{operate with } \langle E_m|$$

$$\langle E_n|\hat{H}|E_m\rangle = E_m\langle E_n \mid E_m\rangle \qquad \langle E_m|\hat{H}|E_n\rangle = E_n\langle E_m \mid E_n\rangle$$

$$\text{Take adjoint of both sides}$$

$$\langle E_n|\hat{H}|E_m\rangle = E_n\langle E_n \mid E_m\rangle$$

where the right hand column made use of the Hermiticity of the operator $\hat{H}$ and the reality of the eigenvalues E_n. Now subtract the results of the two columns to find

$$0 = (E_m - E_n)\langle E_n \mid E_m\rangle$$

We assumed that $E_m - E_n \neq 0$ and therefore $\langle E_n|E_m\rangle = 0$ as required to prove the theorem.

As a result of the last two theorems, the eigenvectors form a complete orthonormal set

$$B = \{|E_n\rangle = |n\rangle\} \qquad\qquad\qquad (4.9.5)$$

Next, examine what happens when two Hermitian operators $\hat{A}, \hat{B}$ commute. Each individual Hermitian operator must have a complete set of eigenvectors which means that each Hermitian operator generates a basis set for the vector space. The commutator $[\hat{A}, \hat{B}] = \hat{A}\hat{B} - \hat{B}\hat{A}$ indicates whether or not the operators commute. The next theorem shows that if the operators commute $[\hat{A}, \hat{B}] = 0$ then the operators $\hat{A}$ and $\hat{B}$ produce the same basis set for the vector space. The vectors space can be either a single space V or a direct product space $V \times W$.

THEOREM 4.9.3 A Single Basis Set for Commuting Hermitian Operators
Let $\hat{A}, \hat{B}$ be Hermitian operators that commute $[\hat{A}, \hat{B}] = 0$ then there exist eigenvectors $|\xi\rangle$ such that $\hat{A}|\xi\rangle = a_\xi|\xi\rangle$ and $\hat{B}|\xi\rangle = b_\xi|\xi\rangle$

Proof: Assume that A has a complete set of eigenvectors. Let $|\xi\rangle$ be the eigenvectors of $\hat{A}$ such that

$$\hat{A}|\xi\rangle = a_\xi|\xi\rangle \qquad\qquad\qquad (4.9.6)$$

Further assume that for each a_ξ there exists only one eigenvector $|\xi\rangle$. Consider

$$\hat{B}\hat{A}|\xi\rangle = \hat{B}a_\xi|\xi\rangle \qquad\qquad\qquad (4.9.7)$$

But $\hat{A}\hat{B} = \hat{B}\hat{A}$ since $[\hat{A}, \hat{B}] = 0$ and so the left-hand side of this last equation becomes

$$\hat{A}\left(\hat{B}|\xi\rangle\right) = \hat{A}\,\hat{B}|\xi\rangle = \hat{B}\hat{A}|\xi\rangle = \hat{B}a_\xi|\xi\rangle = a_\xi\left(\hat{B}|\xi\rangle\right) \qquad (4.9.8)$$

which requires $\hat{B}|\xi\rangle$ *to be an eigenvector of the operator* $\hat{A}$ *corresponding to the eigenvalue* a_{ξ}. *But there can only be one eigenvector for each eigenvalue so* $|\xi\rangle \sim \hat{B}|\xi\rangle$. *Rearranging this expression and inserting a constant of proportionality* b_{ξ} *we find* $\hat{B}|\xi\rangle = b_{\xi}|\xi\rangle$. *This is an eigenvector equation for the operator B; the eigenvalue is* b_{ξ}.

THEOREM 4.9.4 *Common Eigenvectors and Commuting Operators*

As an inverse to Theorem 4.9.3, if the operators $\hat{A}, \hat{B}$ *have a complete set of eigenvectors in common then* $[A,B] = 0$.

Proof: First, for convenience, let's represent the common basis set by $|\xi\rangle = |a, b\rangle$ *so that*

$$\hat{A}|a, b\rangle = a|a, b\rangle \quad and \quad \hat{B}|a, b\rangle = b|a, b\rangle$$

Let $|v\rangle$ *be an element of the direct product space of the eigenvectors for the operators* $\hat{A}, \hat{B}$ *so that it can be expanded as*

$$|v\rangle = \sum_{ab} \beta_{ab}|a\,b\rangle$$

then

$$\hat{A}\hat{B}|v\rangle = \sum_{ab} \beta_{ab}\hat{A}\hat{B}|ab\rangle = \sum_{ab} \beta_{ab}\hat{A}b|ab\rangle \sum_{ab} \beta_{ab}ba|ab\rangle$$

$$= \sum_{ab} \beta_{ab}a\hat{B}|ab\rangle = \sum_{ab} \beta_{ab}\hat{B}a|ab\rangle \sum_{ab} \beta_{ab}\hat{B}\hat{A}|ab\rangle$$

$$= \hat{B}\hat{A}|v\rangle$$

This is true for all vectors in the vector space and so $\hat{A}\hat{B} = \hat{B}\hat{A}$.

4.9.4 The Heisenberg Uncertainty Relation

If two operators $\hat{A}, \hat{B}$ commute then there exists a simultaneous set of basis functions $|a, b\rangle = |a\rangle|b\rangle$ such that

$$\hat{A}|a, b\rangle = a|a, b\rangle \quad and \quad \hat{B}|a, b\rangle = b|a, b\rangle$$

and vice versa. We can show that if two operators do not commute then there exists a Heisenberg uncertainty relation between them. The Heisenberg uncertainty relation shows the standard deviation for measurements of $\hat{A}$ and $\hat{B}$ can never be simultaneously zero (refer to the section on the relation between quantum theory and linear algebra for more detail). The standard deviation σ_A of an Hermitian operator $\hat{A}$ for the vector $|\psi\rangle$ (not necessarily a basis vector) is defined to be

$$\sigma_A^2 = \left\langle \left(\hat{A} - \langle\hat{A}\rangle\right)^2 \right\rangle = \left\langle \hat{A}^2 \right\rangle - \left\langle \hat{A} \right\rangle^2 = \langle\psi|\hat{A}^2|\psi\rangle - \langle\psi|\hat{A}|\psi\rangle^2$$

where $\langle\psi|\hat{A}|\psi\rangle$ represents the average of the operator. A non-Hermitian operator would require an adjoint operator.

We now show that two noncommuting Hermitian operators must always produce an uncertainty relation.

THEOREM 4.9.5 *If two operators* $\hat{A}$, $\hat{B}$ *are Hermitian and satisfy the commutation relation* $[\hat{A}, \hat{B}] = i\,\hat{C}$ *then the observed values "a, b" of the operators must satisfy a Heisenberg uncertainty relation of the form* $\sigma_a \sigma_b \geq 1/2 \,|\langle \hat{C} \rangle\,|$.

Proof: Consider the real, positive number defined by

$$\xi = \left\langle \left(\hat{A} + i\lambda\hat{B}\right)\psi \;\middle|\; \left(\hat{A} + i\lambda\hat{B}\right)\psi \right\rangle$$

which we know to be a real and positive since the inner product provides the length of the vector. The vector, in this case, is defined by

$$\left|\left(\hat{A} + i\lambda\hat{B}\right)\psi\right\rangle = \left(\hat{A} + i\lambda\hat{B}\right)|\psi\rangle$$

We assume that λ *is a real parameter. Now working with the number* ξ *and using the definition of adjoint, namely*

$$\left\langle \hat{O}f \,\middle|\, g \right\rangle = \left\langle f \,\middle|\, \hat{O}^+ g \right\rangle,$$

we find

$$\xi = \left\langle \psi \;\middle|\; \left(\hat{A} + i\lambda\hat{B}\right)^+ \left(\hat{A} + i\lambda\hat{B}\right)\psi \right\rangle = \langle\psi|\left(\hat{A} + i\lambda\hat{B}\right)^+\left(\hat{A} + i\lambda\hat{B}\right)|\psi\rangle$$

$$= \langle\psi|\left(\hat{A}^+ - i\lambda\hat{B}^+\right)\left(\hat{A} + i\lambda\hat{B}\right)|\psi\rangle = \langle\psi|\left(\hat{A} - i\lambda\hat{B}\right)\left(\hat{A} + i\lambda\hat{B}\right)|\psi\rangle$$

where the last step uses the Hermiticity of the operators $\hat{A}$, $\hat{B}$. *Multiply the operator terms in the bracket expression and suppress the reference to the wave function (for convenience) to obtain*

$$\xi = \left\langle \hat{A}^2 \right\rangle - \lambda\langle \hat{C} \rangle + \lambda^2 \left\langle \hat{B}^2 \right\rangle \geq 0$$

which must hold for all values of the parameter λ. *The minimum value of the positive real number* ξ *is found by differentiating with respect to the parameter* λ.

$$\frac{\partial\xi}{\partial\lambda} = 0 \rightarrow \lambda = \frac{\langle \hat{C} \rangle}{2\langle \hat{B}^2 \rangle}$$

The minimum value of the positive real number ξ *must be*

$$\xi_{min} = \langle \hat{A}^2 \rangle - \frac{1}{4}\frac{\langle \hat{C} \rangle^2}{\langle \hat{B}^2 \rangle} \geq 0$$

Multiplying through by $\langle \hat{B}^2 \rangle$ *to get*

$$\left\langle \hat{A}^2 \right\rangle\left\langle \hat{B}^2 \right\rangle \geq \frac{1}{4}\left\langle \hat{C} \right\rangle^2 \tag{4.9.9}$$

We could have assumed the quantities $\langle \hat{A} \rangle = \langle \hat{B} \rangle = 0$ *and we would have been finished at this point. However, the commutator* $[\hat{A}, \hat{B}] = i\hat{C}$ *holds for the two Hermitian operators defined by*

$$\hat{A} \rightarrow \hat{A} - \langle \hat{A} \rangle \quad \hat{B} \rightarrow \hat{B} - \langle \hat{B} \rangle$$

As a result, Equation (4.9.9) becomes

$$\left\langle \left(\hat{A} - \langle \hat{A} \rangle \right)^2 \right\rangle \left\langle \left(\hat{B} - \langle \hat{B} \rangle \right)^2 \right\rangle \geq \frac{1}{4} \langle \hat{C} \rangle^2$$

However, the terms in the angular brackets are related to the standard deviations σ_a, σ_b *respectively. We obtained the proof to the theorem by taking the square root of the previous expression*

$$\sigma_a \sigma_b \geq \frac{1}{2} \left| \langle \hat{C} \rangle \right| \tag{4.9.10}$$

Notice that this Heisenberg uncertainty relation involves the absolute value of the expectation value of the operator C. By its definition, the operator C must be Hermitian and its expectation value must be real.

Example 4.9.3

Find the σ_x, σ_p for operators satisfying $[\hat{x}, \hat{p}] = i\hbar$ where $i = \sqrt{-1}$ and $\hbar$ represents a constant.

Solution

The operator $\hat{C} = \hbar$ and so the results of Theorem (4.9.5) provides

$$\sigma_x \sigma_p \geq \frac{1}{2} \left| \langle \hat{C} \rangle \right| = \frac{\hbar}{2}$$

4.10 A Relation Between Unitary and Hermitian Operators

As previously discussed, a Hermitian operator $\hat{H} : V \rightarrow V$ has the property that $\hat{H} = \hat{H}^+$. This section shows how unitary operators can be expressed in the form $\hat{u} = e^{i\hat{H}}$ where $\hat{H}$ is a Hermitian operator.

We can show that the operator $\hat{u} = e^{i\hat{H}}$ is unitary by showing $\hat{u}^+ \hat{u} = 1$

$$\hat{u}^+ \hat{u} = \left(e^{i\hat{H}} \right)^+ \left(e^{i\hat{H}} \right) = e^{-i\hat{H}^+} e^{i\hat{H}} = e^{-i\hat{H}} e^{i\hat{H}} = e^0 = 1$$

This is a one line proof, but a few steps need to be explained as follows:

1. A function of an operator $f(\hat{A})$ must be interpreted as a Taylor expansion. Therefore, we define the *exponential of an operator* to be shorthand notation for a

Taylor series expansion in that operator. Recall that the Taylor series expansion of an exponential has the form

$$e^{ax} = \sum_{n=0}^{\infty} \frac{1}{n!} \frac{\partial^n e^{ax}}{\partial x^n}\bigg|_{x=0} x^n = 1 + \frac{\partial}{\partial x}(e^{ax})_{x=0}\, x + \cdots = 1 + ax + \frac{a^2}{2}x^2 + \cdots$$

In analogy, the exponential of an operator $\hat{H}$ (or equivalently of a matrix $\underline{H}$) can be written as

$$e^{i\underline{H}t} = \underline{1} + (i\underline{H})t + \frac{(i\underline{H})^2}{2}t^2 + \cdots$$

2. We wrote $e^{-i\hat{H}}e^{i\hat{H}} = e^{i(\hat{H}-\hat{H})} = e^0 = 1$. As shown in Section 4.6, $e^{\hat{A}}e^{\hat{B}} = e^{\hat{A}+\hat{B}}$ when $[\hat{A}, \hat{B}] = 0$. This condition is satisfied because $[\hat{H}, \hat{H}] = \hat{H}\hat{H} - \hat{H}\hat{H} = 0$.

Example 4.10.1
Find the unitary matrix corresponding to $e^{i\underline{H}}$ where

$$\underline{H} = \begin{bmatrix} 0.1 & 0 \\ 0 & 0.2 \end{bmatrix}$$

Solution
First note that the matrix H is Hermitian $\underline{H} = \underline{H}^+$

$$\underline{u} = e^{i\underline{H}} = \exp\left\{ i \begin{pmatrix} 0.1 & 0 \\ 0 & 0.2 \end{pmatrix} \right\} = \begin{bmatrix} 1 & 0 \\ 0 & 1 \end{bmatrix}$$

$$+ i \begin{bmatrix} 0.1 & 0 \\ 0 & 0.2 \end{bmatrix} + \frac{i^2}{2!} \begin{bmatrix} 0.1 & 0 \\ 0 & 0.2 \end{bmatrix}^2 + \ldots = \begin{bmatrix} e^{i0.1} & 0 \\ 0 & e^{i0.2} \end{bmatrix}$$

4.11 Translation Operators

Common mathematical operations such as rotating or translating coordinates are handled by operators in the quantum theory. Previous sections in this chapter show that states transform by the application of a single unitary operator whereas "operators" transform through a similarity transformation. The translation through the spatial coordinate x provides a standard example. Every operation in physical space has a corresponding operation in the Hilbert space.

4.11.1 The Exponential Form of the Translation Operator

Let $\hat{x}$ and $\hat{p}$ be the position operator and an operator defined in terms of a derivative

$$\hat{p} = \frac{1}{i}\frac{\partial}{\partial x}$$

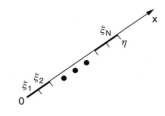

FIGURE 4.11.1
The total translation is divided into smaller translations.

which is the "position" representation of $\hat{p}$ and $i = \sqrt{-1}$. The position representation of $\hat{x}$ is x. The operator $\hat{p}$ is Hermitian (note that $\hat{p}$ is the momentum operator from quantum theory except the $\hbar$ has been left out of the definition given above). The coordinate kets satisfy $\hat{x}|x\rangle = x|x\rangle$ and the operators satisfy $[\hat{x}, \hat{p}] = [\hat{x}\hat{p} - \hat{p}\hat{x}] = i$ as can be easily verified

$$[\hat{x}, \hat{p}]f(x) = [\hat{x}\hat{p} - \hat{p}\hat{x}]f(x) = x\frac{1}{i}\frac{\partial}{\partial x}f - \frac{1}{i}\frac{\partial}{\partial x}(xf) = x\frac{1}{i}\frac{\partial}{\partial x}f - \frac{x}{i}\frac{\partial f}{\partial x} - \frac{1}{i}f = if$$

comparing both sides, we see that the *operator* equation $[\hat{x}, \hat{p}] = i$ holds. The commutator being nonzero defines the so-called conjugate variables. The translation operator uses products of the conjugate variables. The operator $\hat{p}$ is sometimes called the generator of translations. The Hamiltonian is the generator of translations in time.

This topic shows that the exponential $\hat{T}(\eta) = e^{-i\eta\hat{p}}$ translates the coordinate system according to

$$\hat{T}(\eta)f(x) = e^{-i\eta\hat{p}}f(x) = f(x - \eta)$$

where $\hat{p} = (1/i)(\partial/\partial x)$. The proof starts (Figure 4.11.1) by working with a small displacement ξ_k and calculating the Taylor expansion about the point "x"

$$f(x - \xi_k) \cong f(x) - \frac{\partial f(x)}{\partial x}\xi_k + \ldots = \left(1 - \xi_k\frac{\partial}{\partial x} + \ldots\right)f(x)$$

Substituting the operator for the derivative

$$\hat{p} = \frac{1}{i}\frac{\partial}{\partial x}$$

gives

$$f(x - \xi_k) = \left(1 - \xi_k\frac{\partial}{\partial x} + \ldots\right)f(x) = (1 - i\xi_k\hat{p} + \ldots)f(x) = \exp(-i\xi_k\hat{p})f(x)$$

Now, by repeated application of the infinitesimal translation operator, we can build up the entire length η

$$f(x + \eta) = \prod_k \exp(-i\xi_k\hat{p})f(x) = \exp\left(-\sum_k i\xi_k\hat{p}\right)f(x) = \exp(-i\eta\hat{p})f(x)$$

So the exponential with the operator $\hat{p}$ provides a translation according to

$$\hat{T}(\eta)f(x) = e^{-i\eta\hat{p}}f(x) = f(x - \eta)$$

Note that the translation operator is unitary $\hat{T}^+ = \hat{T}^{-1}$ for η real since $\hat{p}$ is Hermitian. It is easy to show $\hat{T}^+(-\eta) = \hat{T}(\eta)$. The operator $\hat{p}$ is the generator of translations.

In the quantum theory, the momentum conjugate to the displacement direction generates the translation according to

$$\hat{T}(\eta)f(x) = e^{-i\eta\hat{p}/\hbar}f(x) = f(x - \eta) \quad \text{where} \quad \hat{p} = \frac{\hbar}{i}\frac{\partial}{\partial x}$$

Notice the extra fractor of $\hbar$.

4.11.2   Translation of the Position Operator

This topic show that

$$\hat{T}^+(\eta)\,\hat{x}\,\hat{T}(\eta) = \hat{x} - \eta$$

where $\hat{T}(\eta) = e^{-i\eta\hat{p}}$. This is easy to show using the operator expansion theorem in Section 4.6

$$e^{\eta\hat{A}}\hat{B}e^{-\eta\hat{A}} = \hat{B} + \frac{\eta}{1!}\left[\hat{A}, \hat{B}\right] + \frac{\eta^2}{2!}\left[\hat{A}, \left[\hat{A}, \hat{B}\right]\right] + \cdots$$

Using $\hat{A} = i\hat{p}$ and the commutation relations $[\hat{x}, \hat{p}] = i$, we find

$$e^{i\eta\hat{p}}\hat{x}e^{-i\eta\hat{p}} = \hat{x} + \frac{\eta}{1!}\left[i\hat{p}, \hat{x}\right] + \frac{\eta^2}{2!}\left[i\hat{p}, \left[i\hat{p}, \hat{x}\right]\right] + \cdots = \hat{x} - \eta$$

4.11.3   Translation of the Position-Coordinate Ket

The position-coordinate ket $|x\rangle$ is an eigenvector of the position operator $\hat{x}$

$$\hat{x}|x\rangle = x|x\rangle$$

What position-coordinate ket $|\phi\rangle$ is an eigenvector of the translated operator

$$\hat{T}^+(\eta)\,\hat{x}\,\hat{T}(\eta) = \hat{x} - \eta$$

that is, what is the state $|\phi\rangle = \hat{T}^+(\eta)|x\rangle$? The eigenvector equation for the translated operator $\hat{x}_T = \hat{T}^+\hat{x}\hat{T}$ is

$$\hat{x}_T|\phi\rangle = \hat{T}^+(\eta)\hat{x}\hat{T}(\eta)|\phi\rangle = \left[\hat{T}^+(\eta)\hat{x}\hat{T}(\eta)\right]\hat{T}^+(\eta)|x\rangle = \hat{T}^+(\eta)\hat{x}|x\rangle = x\hat{T}^+(\eta)|x\rangle = x|\phi\rangle$$

However, we know the translated operator is $\hat{x}_T = \hat{x} - \eta$ and therefore the previous equation provides

$$x|\phi\rangle = \hat{x}_T|\phi\rangle = (\hat{x} - \eta)|\phi\rangle = (\phi - \eta)|\phi\rangle$$

Comparing both sides, we see $\phi = x + \eta$ which therefore shows that the translated position vector is

$$|\phi\rangle = \hat{T}^+(\eta)|x\rangle = |x + \eta\rangle$$

4.11.4 Example Using the Dirac Delta Function

Show that

$$|\phi\rangle = \hat{T}^{+}(\eta)\,|x'\rangle = |x' + \eta\rangle$$

using the fact that the position-ket represents the Dirac delta function in Hilbert space

$$|x'\rangle \equiv |\delta(\bullet - x')\rangle$$

where "$\bullet$" represents the missing variable. If "x" is a coordinate on the x-axis then

$$\langle x \mid x'\rangle \equiv \int_{-\infty}^{\infty} d\varsigma\, \delta(\varsigma - x)\,\delta(\varsigma - x') = \delta(x - x')$$

Applying the translation operator in the x-representation

$$\langle x|\hat{T}(\eta)|x'\rangle = e^{-i\eta\hat{p}_x}\langle x \mid x'\rangle = e^{-i\eta\hat{p}_x}\delta(x - x') = \delta(x - \eta - x') = \langle x \mid x' + \eta\rangle$$

Evidently

$$\hat{T}(\eta)\,|x'\rangle = |x' + \eta\rangle$$

4.12   Functions in Rotated Coordinates

This section shows how the form of a function changes under rotations. It then demonstrates a rotation operator.

4.12.1   Rotating Functions

If we know a function $f(x,y)$ in one set of coordinates (x,y) then what is the function $f'(x',y')$ for coordinates (x',y') that are rotated through an angle θ with respect to the first set (x,y).

Consider a point in space ξ as indicated in the picture. The single point can be described by the primed or unprimed coordinate system. The key fact is that the equations linking the two coordinate systems describe the single point ξ. The equations for coordinate rotations are

$$\underline{r}' = \underline{R}\,\underline{r} \tag{4.12.1}$$

where

$$\underline{r}' = \begin{pmatrix} x' \\ y' \end{pmatrix} \qquad \underline{R} = \begin{pmatrix} \cos\theta & \sin\theta \\ -\sin\theta & \cos\theta \end{pmatrix} \qquad \underline{r} = \begin{pmatrix} x \\ y \end{pmatrix} \tag{4.12.2}$$

$\underline{r}'$ and $\underline{r}$ represent the single point ξ. Notice the matrix differs by a minus sign from that discussed in Section 4.8.1 since Figure 4.12.1 relates one set of coordinates to another whereas Equation (4.12.1) rotates vectors.

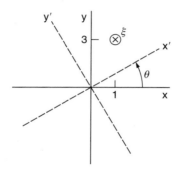

FIGURE 4.12.1
Rotated coordinates.

A value "z" associated with the point ξ is the same value regardless of the reference frame. Therefore, we require

$$z = f'(x',y') = f(x,y) \tag{4.12.3}$$

since (x',y') and (x,y) specify the same point ξ. We can write the last equation using Equation 4.12.1 as

$$f'(x',y') = f(x,y) = f(\underline{R}^{-1}\underline{r}) \tag{4.12.4}$$

where for the depicted 2-D rotation

$$\underline{R}^{-1} = \begin{pmatrix} \cos\theta & -\sin\theta \\ \sin\theta & \cos\theta \end{pmatrix}$$

Example 4.12.1
Suppose the value associated with the point $\underline{r} = \begin{pmatrix} 1 \\ 3 \end{pmatrix}$ is 10, that is $f(1,3) = 10$ what is $f'(x' = 3, y' = -1)$ for $\theta = \pi/2$?

Solution
Using Equation (4.12.4), we find

$$f'(3,-1) = f[\underline{R}^{-1}\underline{r}] = f\left[\begin{pmatrix} \cos\theta & -\sin\theta \\ \sin\theta & \cos\theta \end{pmatrix} \begin{pmatrix} 3 \\ -1 \end{pmatrix} \right] = f\left[\begin{pmatrix} 0 & -1 \\ 1 & 0 \end{pmatrix} \begin{pmatrix} 3 \\ -1 \end{pmatrix} \right] = f(1,3) = 10$$

4.12.2 The Rotation Operator

The unitary operator

$$\hat{R} = e^{i\vec{\alpha}\cdot\vec{L}/\hbar} \tag{4.12.5}$$

maps a function into another that corresponds to rotated position coordinates. Here, $\hat{L} = L_x\tilde{x} + L_y\tilde{y} + L_z\tilde{z}$ is the generator of rotations (later called the angular momentum operator) and $\vec{\alpha} = \alpha_x\tilde{x} + \alpha_y\tilde{y} + \alpha_z\tilde{z}$ gives a rotation angle. For example, α_z is the rotation angle around the $\tilde{z}$ axis. In the 3-D case, $|\vec{\alpha}|$ is the rotation angle about the unit axis $\vec{\alpha}/|\vec{\alpha}|$.

Consider the simple case of a rotation about the $\tilde{z}$ axis.

$$\hat{R} = e^{i\theta_0 \hat{L}_z/\hbar}$$

where the operator L_z has the form $L_z = i\hbar\, \partial/\partial\theta$. The nonzero commutator $[\theta, \hat{L}_z] = \hbar/i$ indicates the rotation operator uses products of conjugate variables similar to the translation operator. The operator $\hat{L}_z$ is sometimes termed the generator of rotations. Consider a function $\psi(r, \theta) \equiv \psi(\theta)$ and calculate a *new* function corresponding to the old one evaluated at $\theta \to \theta + \varepsilon$. The Taylor expansion gives

$$\psi'(\theta) = \psi(\theta + \varepsilon) = \psi(\theta) + \frac{\varepsilon}{1!}\frac{\partial}{\partial\theta}\psi(\theta)$$

$$+ \frac{\varepsilon^2}{2!}\frac{\partial^2}{\partial\theta^2}\psi(\theta) + \ldots = \left[\hat{1} + \frac{\varepsilon}{1!}\frac{\partial}{\partial\theta} + \frac{\varepsilon^2}{2!}\frac{\partial^2}{\partial\theta^2} + \ldots\right]\psi(\theta) = e^{\varepsilon\partial_\theta}\psi(\theta)$$

where $\partial_\theta = \partial/\partial\theta$. We can rearrange the exponential in terms of the z-component of the angular momentum $L_z = \frac{\hbar}{i}\frac{\partial}{\partial\theta}$ to find $\hat{R}(\varepsilon) = e^{\varepsilon\partial_\theta} = e^{i\varepsilon L_z/\hbar}$. Repeatedly applying the operator produces the rotation

$$\hat{R}(\theta_0) = e^{i\theta_0 L_z/\hbar} \quad \text{and} \quad \psi'(\theta) = \psi(\theta + \theta_0) = \hat{R}(\theta_0)\,\psi(\theta) \qquad (4.12.6)$$

Figure 4.12.2 shows that the rotation moves the function in the direction of a negative angle or rotates the coordinates in the positive direction. If we replace $\theta_0 \to -\theta_0$ then the rotation would be in the opposite sense.

We can easily show the generator of rotation $L_z = (\hbar/i)\partial_\theta$ can be replaced by $L_z = xp_y - yp_x = (x\partial_y - y\partial_x)\hbar/i$. The two sets of coordinates are related by

$$x = r\cos\theta \quad \text{and} \quad y = r\sin\theta$$

Therefore

$$\frac{\partial}{\partial\theta}\psi(x, y) = \frac{\partial\psi}{\partial x}\frac{\partial x}{\partial\theta} + \frac{\partial\psi}{\partial y}\frac{\partial y}{\partial\theta} = -r\sin\theta\frac{\partial\psi}{\partial x} + r\cos\theta\frac{\partial\psi}{\partial y} = \left(x\frac{\partial}{\partial y} - y\frac{\partial}{\partial x}\right)\psi = \frac{i}{\hbar}L_z\psi$$

as required.

The position operator can be written in rotated form. Denote the position operator by $\hat{r} = \hat{x}\tilde{x} + \hat{x}\tilde{y} + \hat{z}\tilde{z}$ where $\tilde{x}, \tilde{y}, \tilde{z}$ represent the usual Euclidean unit vectors. The position operator provides the relation $\hat{r}|\vec{r}_0\rangle = \vec{r}_0|\vec{r}_0\rangle$. Now consider a rotation of a function. The relation between the new and old functions gives $\langle\vec{r}|\psi'\rangle = \langle\vec{r}|\hat{R}|\psi\rangle \equiv \langle\vec{r}'|\psi\rangle$. We therefore

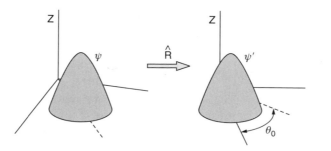

FIGURE 4.12.2
Rotating the function through an angle.

conclude that $|\vec{r}'\rangle = \hat{R}^+|\vec{r}\rangle$. For example, the coordinate ket might represent the wave function for a particle localized at the particular point $\vec{r}$. We see that the operator rotates the location in the positive angle direction. We can also see that the position operator must satisfy the relation

$$\hat{r}|\vec{r}'\rangle = \vec{r}'|\vec{r}'\rangle \quad \rightarrow \quad \hat{r}\hat{R}^+|\vec{r}\rangle = \vec{r}'\hat{R}^+|\vec{r}\rangle \quad \rightarrow \quad \hat{R}\hat{r}\hat{R}^+|\vec{r}\rangle = \vec{r}'|\vec{r}\rangle \quad \rightarrow \quad \hat{r}' = \hat{R}\hat{r}\hat{R}^+$$

which gives the rotated form of the position operator. We can also show

$$\langle \vec{r} \mid \psi'\rangle = \langle \vec{r}' \mid \psi\rangle \quad \rightarrow \quad \psi'(\vec{r}) = \psi(\vec{r}') \quad \rightarrow \quad \hat{R}\psi(\vec{r}) = \psi(\vec{r}' = R^{-1}\vec{r})$$

where $\hat{R}$ is the corresponding operator for Euclidean vectors. This shows that for every operation in coordinate space, there must correspond an operation in Hilbert space. The angle represents the coordinate space while the angular momentum represents the Hilbert space operation.

4.13 Dyadic Notation

This section develops the dyadic notation for the second rank tensor. We will see that it is equivalent to writing a 2-D matrix. Studies in solid state often use dyadic quantities to describe the effective mass of an electron or hole. For example, formulas relating the acceleration of a particle $\vec{a}$ to the applied force $\vec{F}$ have the form

$$\vec{F} = \ddot{m} \cdot \vec{a} \tag{4.13.1}$$

where the dyadic quantity $\ddot{m}$ represents the effective mass. This equation says that an applied force can produce an acceleration in a direction other than parallel to the force.
  A dyad can be written in terms of components for example

$$\ddot{A} = \sum_{ij} A_{ij}\hat{e}_i\hat{e}_j \tag{4.13.2}$$

where the unit vector $\hat{e}_i$ can be one of the basis vectors $\{\hat{x}, \hat{y}, \hat{z}\}$ for a 3-D space, and the $\hat{e}_i\hat{e}_j$ symbol places the unit vectors next to each other without an operator separating them.

Example 4.13.1
Find $A \cdot \vec{v}$ for $\ddot{A} = 1\hat{e}_1\hat{e}_1 + 2\hat{e}_3\hat{e}_2 + 3\hat{e}_2\hat{e}_3$ and $\vec{v} = 4\hat{e}_1 + 5\hat{e}_2 + 6\hat{e}_3$

Solution

$$\ddot{A} \cdot \vec{v} = (1\hat{e}_1\hat{e}_1 + 2\hat{e}_3\hat{e}_2 + 3\hat{e}_2\hat{e}_3) \cdot (4\hat{e}_1 + 5\hat{e}_2 + 6\hat{e}_3) = 4\hat{e}_1 + 10\hat{e}_3 + 18\hat{e}_2 = 4\hat{x} + 18\hat{y} + 10\hat{z}$$

The coefficients in Equation (4.13.2) can be arranged in a matrix. This means that a 3×3 matrix provides an alternate representation of the second rank tensor and the dyad. The matrix elements can easily be seen to be

$$\hat{e}_a \cdot \ddot{A} \cdot \hat{e}_b = \sum_{ij} A_{ij}\, \hat{e}_a \cdot \hat{e}_i\hat{e}_j \cdot \hat{e}_b = \sum_{ij} A_{ij}\, \delta_{ai}\delta_{jb} = A_{ab} \tag{4.13.3}$$

The procedure should remind you of Dirac notation for the matrix discussed in Chapter 5.

The unit dyad can be written as

$$\overset{\leftrightarrow}{1} = \sum_i \hat{e}_i \hat{e}_i \tag{4.13.4}$$

Applying the definition of the matrix elements in Equation 4.13.3 shows the unit dyad produces the unit matrix.

Example 4.13.2

Show that if $\overset{\leftrightarrow}{I} = \overset{\leftrightarrow}{A}$ then $A_{ab} = \delta_{ab}$

Solution

Operate with $\hat{e}_a$ on the left and $\hat{e}_b$ on the right to find

$$\hat{e}_a \cdot \overset{\leftrightarrow}{1} \cdot \hat{e}_b = \hat{e}_a \cdot \overset{\leftrightarrow}{A} \cdot \hat{e}_b = \hat{e}_a \cdot \left(\sum_{ij} A_{ij} \hat{e}_i \hat{e}_j \right) \cdot \hat{e}_b = \sum_{ij} A_{ij} \delta_{ai} \delta_{jb} = A_{ab}$$

Now let's discuss the inverse of a dyad. Suppose

$$\overset{\leftrightarrow}{1} = \overset{\leftrightarrow}{A} \cdot \overset{\leftrightarrow}{B} \tag{4.13.5}$$

then we can show that $\overset{\leftrightarrow}{B} = \overset{\leftrightarrow}{A}{}^{-1}$ where $\overset{\leftrightarrow}{A} = \sum_{ii'} A_{ii'} \hat{e}_i \hat{e}_{i'}$ and $\overset{\leftrightarrow}{B} = \sum_{jj'} B_{jj'} \hat{e}_j \hat{e}_{j'}$. Operating on the left of Equation (4.13.5) with $\hat{e}_a$ and on the right by $\hat{e}_b$ produces

$$\delta_{ab} = \hat{e}_a \cdot \left(\sum_{ii'} A_{ii'} \hat{e}_i \hat{e}_{i'} \cdot \sum_{jj'} B_{jj'} \hat{e}_j \hat{e}_{j'} \right) \cdot \hat{e}_b = \sum_{ii'jj'} A_{ii'} B_{jj'} \hat{e}_a \cdot \hat{e}_i \hat{e}_{i'} \cdot \hat{e}_j \hat{e}_{j'} \cdot \hat{e}_b$$

The dot products may produce Kronecker delta functions

$$\delta_{ab} = \sum_{ii'jj'} A_{ii'} B_{jj'} \delta_{ai} \delta_{i'j} \delta_{j'b} = \sum_j A_{aj} B_{jb}$$

which shows the *matrices* $\underline{A}$ and $\underline{B}$ must be inverses.

4.14 Minkowski Space

The study of the matter–light interaction often starts with the Lagrangian or Hamiltonian formulation. The tensor notation commonly found with studies of special relativity provides a compact, simplifying notation in many cases of interest. However, to be useful for special relativity, the notation must also accurately account for the pseudo inner product for Minkowski space necessary to make the speed of light independent of the observer. Refer to the companion volume for an introduction to the special relativity.

Minkowski space has four dimensions with coordinates (x_0, x_1, x_2, x_3) where for special relativity, the first coordinate is related to the time t. Rather than defining the inner product as $\langle v|w \rangle = \sum_n v_n w_n$, the inner product has the form

$$\langle v \mid w \rangle = v_0 w_0 - (v_1 w_1 + v_2 w_2 + v_3 w_3) \tag{4.14.1}$$

Based on this definition, the inner product for Minkowski space does not satisfy all the properties of the inner product. In particular, the pseudo inner product in Equation 4.14.1 does not require the vectors v and w to be zero when the inner product has the value of zero. The theory of relativity uses two types of notation. In the first, Minkowski 4-vectors use an imaginary number "i" to make the "inner product" appear similar to Euclidean inner products. In the second, a "metric" matrix is defined along with specialized notation. Additionally, a constant multiplies the time coordinate t in order to give it the same units as the spatial coordinates.

One variant of the 4-vector notation uses an imaginary "i" with the time coordinate $x_\mu = (ict, x, y, z) = (ict, \vec{r})$. The constant c, the speed of light, converts the time t into a distance. The pseudo inner product of the vector with itself then has the form

$$x_\mu x_\mu \equiv \sum_{\mu=1}^{4} x_\mu x_\mu = (ict, \vec{r}) \cdot (ict, \vec{r}) = -c^2 t^2 + x^2 + y^2 + z^2 \tag{4.14.2}$$

The imaginary number $i = \sqrt{-1}$ makes the calculation of length look like Pythagoras's theorem but produces the same result as for the pseudo inner product in Equation (4.14.1). Notice the "Einstein repeated summation convention" where repeated indices indicate a summation. The indices appear as subscripts. Notice this pseudo inner product does not require x_μ to be zero when $x_\mu x_\mu = 0$.

As an alternate notation, the imaginary number can be removed by using a "metric" matrix. As is conventional, we use natural units with the speed of light $c = 1$ and $\hbar = 1$ for convenience. The various constants can be reinserted if desired.

We represent the basic 4-vector with the index in the upper position. For example, we can represent the space–time 4-vector in component form as

$$x^\mu = (t, x, y, z) = (t, \vec{r}) \tag{4.14.3}$$

where time t comprises the $\mu = 0$ component. Notice the conventional order of the components. The position of the index is significant. To take a pseudo inner product, we could try writing $x^\mu x^\mu = t^2 + x^2 + \ldots$ where we have used a repeated index convention. However, the result needs an extra minus sign. Instead, if we write

$$x_\mu = (t, -\vec{r}) \tag{4.14.4}$$

then the summation becomes $x_\mu x^\mu = (t, -\vec{r}) \cdot (t, \vec{r}) = t^2 - r^2$ where the "extra" minus sign appears. Again the position of the index is important. Apparently, lowering an index places a minus sign on the spatial part of the 4-vector.

A *metric* (matrix) provides a better method of tracking the minus signs. Consider the following metric

$$g^{\mu\nu} = \begin{pmatrix} 1 & 0 & 0 & 0 \\ 0 & -1 & 0 & 0 \\ 0 & 0 & -1 & 0 \\ 0 & 0 & 0 & -1 \end{pmatrix} = g_{\mu\nu} \tag{4.14.5}$$

Ordinary matrix multiplication then produces

$$x_\mu = g_{\mu\nu}x^\nu \tag{4.14.6a}$$

Notice the form of this result and the fact that we sum over the ν index by the summation convention. We can also write

$$x^\mu = g^{\mu\nu}x_\nu \tag{4.14.6b}$$

Therefore to take a pseudo inner product, we write

$$x_\mu x^\mu = (g_{\mu\nu}x^\nu)x^\mu = (t, -\vec{r}) \cdot (t, \vec{r}) = t^2 - r^2 \tag{4.14.7}$$

The metric given here is the "West Coast" metric since it became most common on the west coast of the U.S. The east coast metric contains a minus sign on the time component and the rest have a "+" sign.
Derivatives naturally have lower indices.

$$\partial_\mu = (\partial_0, \partial_1, \partial_2, \partial_3) = \left(\frac{\partial}{\partial x^0}, \frac{\partial}{\partial x^1}, \frac{\partial}{\partial x^2}, \frac{\partial}{\partial x^3}\right) = \left(\frac{\partial}{\partial t}, \frac{\partial}{\partial x}, \frac{\partial}{\partial y}, \frac{\partial}{\partial z}\right) = \left(\frac{\partial}{\partial t}, \nabla\right) = \square \tag{4.14.8}$$

Notice the location of the indices. The upper-index case gives

$$\partial^\mu = g^{\mu\nu}\partial_\nu = (\partial_0, -\partial_1, -\partial_2, -\partial_3) = \left(\frac{\partial}{\partial t}, -\nabla\right) \tag{4.14.9}$$

Let's consider a few examples. The complex plane wave has the form

$$e^{i\left(\vec{k}\cdot\vec{r}-\omega t\right)} = e^{-i\left(\omega t-\vec{k}\cdot\vec{r}\right)} = e^{-ik_\mu x^\mu}$$

where $k^\mu = (\omega, \vec{k})$. Also notice that the wave equation

$$\left(\nabla^2 - \frac{\partial^2}{\partial t^2}\right)\psi = 0$$

can be written as

$$\partial_\mu \partial^\mu \psi = 0$$

just keep in mind the repeated index convention.
As a note, any valid theory must transform correctly. The inner product is relativistically correct since it is invariant with respect to Lorentz transformations.

4.15   Review Exercises

4.1 If $\vec{W} = (3+j)\hat{x} + (-1-j)\hat{y}$ with $j = \sqrt{-1}$ write $\langle W|$ in terms of $\langle 1|, \langle 2|$.

4.2 For the basis set $\{e^{in\pi x/L}/\sqrt{2L} : n = 0, \pm 1, \ldots\}$ write out the closure relation in terms of the Dirac delta function.

4.3 Use change of variables to find $\int_{-1}^{1} f(x)\,\delta(ax)\,dx$ where $a > 0$.

4.4 Use integration by parts to find $\int_{-1}^{1} dx\, f(x)\,\delta'(x)$ where $\delta'(x) = d/dx\delta(x)$.

4.5 Show that the set of Euclidean vectors $\{\vec{v} = a\tilde{x} + b\tilde{y} : a, b \in \mathscr{C}\}$ forms a vector space when the binary operation is ordinary vector addition. $\mathscr{C}$ denotes complex numbers and $\tilde{x}, \tilde{y}$ represent basis vectors.

4.6 Explain why the dot product satisfies the properties of the inner product.

4.7 Show that the set of 2-D Euclidean vectors terminating on the unit circle $\{\vec{v} : |\vec{v}| = 1\}$ do not form a vector space.

4.8 Find the sine series of $\cos(x)$ on the interval $(0, \pi)$.

4.9 Change the set of functions $\{1, x, x^2\}$ into a basis set on the interval $(-1, 1)$.

4.10 Starting with the vector space V, show that the dual space V^* must also be a vector space. That is, show that the vectors in V^* satisfy the properties required of a vector space.

4.11 Show that the adjoint operator induces an inner product on the dual space V^*. That is, show that we can define an inner product on V^*.

4.12 Find $\|\vec{v}\|^2$ when $|v\rangle = 2j|1\rangle + 3|2\rangle$ where $j = \sqrt{-1}$.

4.13 Find the Fourier transform of $\delta(x-1)$ and of $\frac{1}{2}\delta(x-1) + \frac{1}{2}\delta(x+1)$.

4.14 Show that the Fourier series in terms of complex exponentials

$$f(x) = \sum_{n=-\infty}^{\infty} D_n \frac{1}{\sqrt{2L}} \exp\left(i\frac{n\pi x}{L}\right)$$

must be equivalent to the Fourier series with the basis set

$$\left\{ \frac{1}{\sqrt{2L}}, \frac{1}{\sqrt{L}}\cos\left(\frac{n\pi x}{L}\right), \frac{1}{\sqrt{L}}\sin\left(\frac{n\pi x}{L}\right) : n = 1, 2, \ldots \right\}$$

Hint: Start with

$$f(x) = \frac{\alpha_0}{\sqrt{2L}} + \sum_{n=1}^{\infty} \alpha_n \frac{1}{\sqrt{L}}\cos\left(\frac{n\pi x}{L}\right) + \sum_{n=1}^{\infty} \beta_n \frac{1}{\sqrt{L}}\sin\left(\frac{n\pi x}{L}\right)$$

and rewrite the sine and cosine terms as complex exponentials. In the summation $\sum_{n=1}^{\infty} \frac{1}{\sqrt{L}}(\frac{\alpha_n + i\beta_n}{2})\exp(-i\frac{n\pi x}{L})$ replace "n" with "$-n$." Combine all terms under the summation and define new constants D_n. Relate these new coefficients to the old ones.

4.15 Show that $B_s = \{\psi_n(x) : n = 1, 2, 3 \ldots\} = \{\sqrt{\frac{2}{L}}\sin\left(\frac{n\pi x}{L}\right) n = 1, 2, 3 \ldots\}$ is orthonormal on $0 < x < L$.

4.16 Write the closure relation in the form of a Dirac delta function for the sine basis and the two forms of the Fourier series basis (Problem 4.14). Be sure to state the domain of integration correctly.

4.17 Show that the null space of a linear operator $\hat{T}$ defined by $N = \{|v\rangle : \hat{T}|v\rangle = 0\}$ forms a vector space.

4.18 Show that the inverse of a linear operator $\hat{T}$ does not exist when the null space $N = \{|v\rangle : \hat{T}|v\rangle = 0\}$ has more than one element.

4.19 Let $\hat{T} : V \to W$ be an "onto" linear operator. Let $V = Sp\{|\phi_i\rangle : i = 1, \ldots, n_v\}$ and $W = Sp\{|\psi_i\rangle : i = 1, \ldots, n_w\}$. Show that

$$\text{Dim}(V) = \text{Dim}(W) + \text{Dim}(N)$$

where $N = $ null space $N = \{|v\rangle : \hat{T}|v\rangle = 0\}$. Hint: Let $|1\rangle, \ldots, |n\rangle$ be the basis for N. Let $|1\rangle, \ldots, |n\rangle, |n+1\rangle, \ldots, |p\rangle$ be the basis for V. Use the definition of linearly independent. Note that $0 = \hat{T} \sum_{i=n+1}^{p} c_i |i\rangle$ requires $\sum_{i=n+1}^{p} c_i |i\rangle$ be in the null space. The null space has only $\vec{0}$ in common with $Sp\{|n+1\rangle, \ldots, |p\rangle\}$.

4.20 For vector spaces V and W and linear operator $\hat{T} : V \to W = \text{Range}(\hat{T})$, show that every vector $|w\rangle$ must have multiple preimages in V when the Null space $N = \{|v\rangle : \hat{T}|v\rangle = 0\}$ has multiple elements. Conclude the inverse of $\hat{T}$ does not exist. Hint: Suppose $|w\rangle \in W, |w\rangle \neq \vec{0}$ and $\hat{T}|v\rangle = |w\rangle$. Examine $N + \{|v\rangle\}$ where N represents the null space.

4.21 Let $\{|\phi_1\rangle, |\phi_2\rangle\}$ be a basis set. Write the following operator in matrix notation

$$\hat{L} = |\phi_1\rangle\langle\phi_1| + 2|\phi_1\rangle\langle\phi_2| + 3|\phi_2\rangle\langle\phi_2|$$

4.22 A Hilbert space V has basis $\{|\phi_1\rangle, |\phi_2\rangle\}$. Assume the linear operator $\hat{L} : V \to V$ has the matrix $\underline{L} = \begin{bmatrix} 0 & 1 \\ 2 & 3 \end{bmatrix}$. Write the operator in the form $\hat{L} = \sum_{ij} L_{ij} |\phi_i\rangle\langle\phi_j|$.

4.23 Write an operator $\hat{L} : V \to V$ in the form $\hat{L} = \sum L_{ab} |\phi_a\rangle\langle\phi_b|$ when $\hat{L}$ maps the basis set $\{|\phi_1\rangle, |\phi_2\rangle\}$ into the basis set $\{|\psi_1\rangle, |\psi_2\rangle\}$ according to the rule $\hat{L}|\phi_1\rangle = |\psi_1\rangle$ and $\hat{L}|\phi_2\rangle = |\psi_2\rangle$. Assume the two sets of basis vectors are related as follows

$$|\psi_1\rangle = \frac{1}{\sqrt{3}}|\phi_1\rangle + \sqrt{\frac{2}{3}}|\phi_2\rangle \quad \text{and} \quad |\psi_2\rangle = -\sqrt{\frac{2}{3}}|\phi_1\rangle + \frac{1}{\sqrt{3}}|\phi_2\rangle$$

4.24 Suppose $\hat{H} = \sum_n E_n |n\rangle\langle n|$ where $E_n \neq 0$ for all n. What value of c_n in $\hat{O} = \sum_n C_n |n\rangle\langle n|$ makes $\hat{O}$ the inverse of $\hat{H}$ so that $\hat{H}\hat{O} = 1 = \hat{O}\hat{H}$.

4.25 If $\hat{H} = 1|\phi_1\rangle\langle\phi_1| + 2|\phi_2\rangle\langle\phi_2|$ and $|\psi(0)\rangle = 0.86|\phi_1\rangle + 0.51|\phi_2\rangle$ is the wavefunction for an electron at $t = 0$. Find the average energy $\langle\psi(0)|\hat{H}|\psi(0)\rangle$.

4.26 Find the inverse of the following matrix using row operations

$$\underline{M} = \begin{bmatrix} 1 & 1 & 0 \\ 0 & 1 & 2 \\ 0 & 0 & 1 \end{bmatrix}$$

4.27 Show the following relations

$$\text{Det}(\hat{A}\hat{B}) = \text{Det}(\hat{A})\,\text{Det}(\hat{B}) \quad \text{and} \quad \text{Det}(\hat{A}\hat{B}\hat{C}) = \text{Det}(\hat{A})\,\text{Det}(\hat{B})\text{Det}(\hat{C})$$

You can use the first relation to prove the second one.

4.28 Show $\text{Det}(c\hat{A}) = c^N \text{Det}(\hat{A})$ using the completely antisymmetric tensor where $\hat{A} : V \to V$, $N = \text{Dim}(V)$ and c is a complex number.

4.29 Show the $\det(T)$ is independent of the particular basis chosen for the vector space. Hint: Use the unitary operator and a similarity transformation to change T, then use the results of previous problems.

4.30 Assume $\hat{A}, \hat{B}$ operate on a single vector space $V = Sp\{|1\rangle, |2\rangle, \ldots\}$. Show $Tr(\hat{A}\hat{B}) = Tr(\hat{B}\hat{A})$ by inserting the closure relation.

4.31 Show the relation $Tr(\hat{A}\hat{B}\hat{C}) = Tr(\hat{B}\hat{C}\hat{A}) = Tr(\hat{C}\hat{A}\hat{B})$ assuming $\hat{A}, \hat{B}, \hat{C}$ all operate on $V = Sp\{|1\rangle, |2\rangle, \ldots\}$ for simplicity.

4.32 Show the trace of the operator $\hat{T}$ is *independent* of the chosen basis set. Hint: Use a unitary operator to change basis and also use the closure relation.

4.33 Show that the set of linear operators $\{\hat{T} : V \to W\}$ mapping the vector space V into the vector space W forms a vector space.

4.34 Prove the required property $\langle \hat{A} | \alpha\hat{B} + \beta\hat{C} \rangle = \alpha\langle \hat{A} | \hat{B} \rangle + \beta\langle \hat{A} | \hat{C} \rangle$ for $\langle \hat{A} | \hat{B} \rangle = Tr\hat{A}^+\hat{B}$ to be an inner product. Use $L = \{\hat{T} : V \to V\}$.

4.35 Prove $\langle \hat{A} | \hat{A} \rangle = 0$ if and only if $\hat{A} = 0$ for $\hat{A} \in L = \{\hat{T} : V \to V\}$, the set of linear operators. Hint: Consider the expansion of an operator in a basis set.

4.36 (A) Find the "length" of a unitary operator $\hat{u} : V \to V$ where $\mathrm{Dim}(V) = N$. That is, calculate $\|\hat{u}\|^2 = \langle \hat{u} | \hat{u} \rangle = Tr(\hat{u}^+\hat{u})$. It's probably easiest to use matrices after taking the trace. (B) Find the length of an operator that doubles the length of every vector in an $N = 2$ vector space. (C) Find the length of the operator defined by $\hat{O}|v\rangle = c|v\rangle$.

4.37 Determine if the quantity $\langle \hat{L}_1 | \hat{L}_2 \rangle = Tr\{\hat{L}_1^+\hat{L}_2\}/\mathrm{Dim}(V)$ satisfies the requirements for an inner product where $L_1, L_2 : V \to V$.

4.38 Suppose $V = Sp\{|1\rangle, |2\rangle, \ldots, |n\rangle\}$ and $\hat{L} : V \to V$ according to

$$\hat{L}|1\rangle = |\phi_1\rangle \quad \text{and} \quad \hat{L}|2\rangle = |\phi_2\rangle$$

where $|\phi_1\rangle, |\phi_2\rangle$ are not necessarily orthogonal. Use the inner product $\langle \hat{L}_1 | \hat{L}_2 \rangle = Tr\{\hat{L}_1^+\hat{L}_2\}/\mathrm{Dim}(V)$ to show $\hat{L}$ has unit length so long as $|\phi_1\rangle, |\phi_2\rangle$ have unit length. Hint: First write $\hat{L} = |\phi_1\rangle\langle 1| + \ldots$, then calculate $\hat{L}^+\hat{L}$ having terms such as $|1\rangle\langle 1|\langle\phi_1|\phi_1\rangle + \ldots$, and then calculate the trace.

4.39 Prove properties 1–7 for the commutator given in Section 4.6.

4.40 If $\hat{A}^2 = \hat{A}$ and $[\hat{A}, \hat{B}] = 1$ then show $[e^{i\hat{A}x}, \hat{B}] = e^{ix} - 1$

4.41 Find $\sin \underline{A}$ where $\underline{A} = \begin{pmatrix} 1 & 0 \\ 0 & 2 \end{pmatrix}$. Hint: Use a Taylor expansion.

4.42 Find $\sin \underline{A}$ where $\underline{A} = \begin{pmatrix} 1 & 1 \\ 1 & 2 \end{pmatrix}$. Hint: Find a matrix $\underline{u}$ such that $\underline{u}\underline{A}\underline{u}^+ = \begin{pmatrix} \lambda_1 & 0 \\ 0 & \lambda_2 \end{pmatrix} = \underline{A}_D$ where λ_i represents the eigenvalues. Taylor expands $\sin \underline{A}$. Calculate $\hat{u}[\sin \underline{A}]\hat{u}^+$.

4.43 Consider a 3-D coordinate system. Write the matrix that rotates 45° about the x-axis.

4.44 Suppose an operator rotates vectors by $\theta = 30°$. Write the operator in the form $\sum_{a,b} c_{ab}|a\rangle\langle b|$ and write the matrix.

4.45 Consider a rotated basis set $|n'\rangle = \hat{u}|n\rangle$. Show that the closure relation in the primed system leads to the closure relation in the unprimed system.

$$1 = \sum |n'\rangle\langle n'| \quad \to \quad 1 = \sum |n\rangle\langle n|$$

4.46 Find a condition on "c" that makes the following matrix Hermitian

$$\begin{bmatrix} 1 & c \\ c & -1 \end{bmatrix}$$

4.47 Find a condition on "a" that makes the following operator Hermitian

$$\hat{L} = |1\rangle\langle 1| + aj|1\rangle\langle 2| + aj|2\rangle\langle 1| + |2\rangle\langle 2| \quad \text{where} \quad j = \sqrt{-1}$$

4.48 Show that the trace of a Hermitian operator $\hat{H}$ must be the sum of the Eigenvalues λ_i given by Trace $\hat{H} = \sum \lambda_i$.

Hint: Let $\{|n\rangle\}$ be the basis for the space V where $\hat{H} : V \to V$. Let $\hat{u}$ be the unitary operator that diagonalizes the operator.

$$\text{Tr } \hat{H} = \sum_n \langle \varphi_n | \hat{H} | \varphi_n \rangle = \sum_n \langle \varphi_n | \hat{u}^+ \hat{u} \hat{H} \hat{u}^+ \hat{u} | \varphi_n \rangle = \sum_n \langle \phi_n | \hat{H}_D | \phi_n \rangle = \text{Tr } \hat{H}_D$$

The eigenvalues must be on the diagonal of $\hat{H}_D$

$$\hat{H}_D | \phi_n \rangle = \lambda_n | \phi_n \rangle \quad \to \quad (\underline{H}_D)_{ab} = \langle \phi_a | \hat{H}_D | \phi_b \rangle = \lambda_b \delta_{ab}$$

4.49 Show the determinant of the operator in the previous problem must be the product of eigenvalues.

4.50 Use the definition of adjoint $\langle f | \hat{L} g \rangle = \langle \hat{L}^+ f | g \rangle$ for $\hat{L} = a \frac{d}{dx}$ to show that $\hat{L}^+ = -\hat{L}$ requires "a" to be purely real. Assume that the Hilbert space consists of functions $f(x)$ such that $f(\pm\infty) = 0$.

4.51 Use the definition of adjoint $\langle f | \hat{L} g \rangle = \langle \hat{L}^+ f | g \rangle$ for $\hat{L} = a \frac{d}{dx}$ to show that $\hat{L}^+ = \hat{L}$ requires "a" to be purely imaginary. Assume that the Hilbert space consists of functions $f(x)$ such that $f(\pm\infty) = 0$.

4.52 If $\hat{L} = \partial^2/\partial x^2$ then find $\hat{L}^+$ by partial integration. Assume a Hilbert space of differentiable functions $\{\psi(x)\}$ such that $\psi(x \to \pm\infty) = 0$.

4.53 Show $(\hat{A}\hat{B})^+ = \hat{B}^+ \hat{A}^+$ using $\langle f | \hat{T} g \rangle = \langle \hat{T}^+ f | g \rangle$.

4.54 Without multiplying the matrices, find the adjoint of the following matrix equation

$$\begin{bmatrix} a & b \\ c & d \end{bmatrix} \begin{bmatrix} e \\ f \end{bmatrix} = \begin{bmatrix} g \\ h \end{bmatrix}$$

4.55 Suppose $\hat{O} = \hat{O}^{(V)} \hat{O}^{(W)}$ where $V = Sp\{|\phi_a\rangle\}$ and $W = Sp\{|\psi_a\rangle\}$. Show

$$\underline{O}_{ab,cd} = \langle \phi_a | \hat{O}^{(V)} | \phi_c \rangle \langle \psi_b | \hat{O}^{(W)} | \psi_d \rangle$$

4.56 For the basis vector expansion of $|\Psi\rangle = \sum_{ab} \beta_{ab} |\phi_a \psi_b\rangle$ in the tensor product space $V \otimes W$ with $V = Sp\{|\phi_i\rangle\}$ and $W = Sp\{|\psi_j\rangle\}$, show the expansion coefficients must be $\beta_{ab} = \langle \phi_a \psi_b | \Psi \rangle$ and the closure relation has the form $\sum_{ab} |\phi_a \psi_b\rangle \langle \phi_a \psi_b| = \hat{1}$.

4.57 For a vector space V spanned by $\{|1\rangle, |2\rangle\}$ with $\hat{u}$ an orthogonal rotation by $45°$ and $\hat{T} = |1\rangle\langle 1| + 2|2\rangle\langle 2|$, find $\hat{T}$ in the new basis set. Hint: Find $\hat{u}$ by visual inspection and write in terms of the original basis.

4.58 Show $[\theta, \hat{L}_z] = \hbar/i$.

4.59 Prove the operator expansion theorem

$$\hat{O} = e^{x\hat{A}} \hat{B} e^{-x\hat{A}} = \hat{B} + x[\hat{A}, \hat{B}] + \frac{x^2}{2!}[\hat{A}, [\hat{A}, \hat{B}]] + \dots$$

by expanding the exponentials and collecting terms.

4.16   Further Reading

The following list contains references for background material.

Classics

1. Dirac P.A.M., *The Principles of Quantum Mechanics*, 4th ed., Oxford University Press, Oxford, 1978.
2. Von Neumann J., *Mathematical Foundations of Quantum Mechanics*, Princeton University Press, Princeton, 1996.

Introductory

3. Krause E.F., *Introduction to Linear Algebra*, Holt, Rinehart and Winston, New York, 1970.
4. Bronson R., *Matrix Methods—An Introduction*, Academic Press, New York, 1970.

Standard

5. Byron F.W., Fuller R.W., *Mathematics of Classical and Quantum Physics*, Dover Publications, New York, 1970.
6. von Neuman J., *Mathematical Foundations of Quantum Mechanics*, Princeton University Press, Princeton, 1996.
7. Schwinger J., *Quantum Kinematics and Dynamics*, W. A. Benjamin Inc., New York, 1970.

Involved

8. Loomis L.H., Sternberg S., *Advanced Calculus*, Addison-Wesley Publishing, Reading, MA, 1968.
9. *Green's Functions and Boundary Value Problems*, 2nd ed., I. Stakgold, John Wiley & Sons, New York, 1998.

5

Fundamentals of Dynamics

Quantum theory has formed a cornerstone for modern physics, engineering, and chemistry since the 1920s. It has found significant modern applications in engineering since the development of the semiconductor diode, transistor, and especially the laser in the 1960s. Not until the 1980s did the fabrication and materials growth technology become sufficiently developed to provide the ability to (1) produce quantum well devices (such as quantum well lasers) and (2) engineer the optical and electrical properties of materials (band-gap engineering). One purpose of this chapter is to summarize a small portion of modern quantum theory.

The first few sections of this chapter summarize Lagrange and Hamilton's approach to classical mechanics. These alternate formulations to Newton's formulation of classical mechanics allow us to use scalar quantities such as kinetic or potential energy to find the equations of motion. These alternate formulations are so powerful that they can be used to deduce Maxwell's and other continuous field equations. In fact, the quantum mechanical Hamiltonian comes from the classical one by substituting operators for the classical dynamical variables.

The chapter discusses the connection between linear algebra and quantum mechanics and reviews the basic theory. The discussion of the harmonic oscillator introduces the ladder operators and vacuum state, and prepares the way for the harmonic oscillator theory of the electromagnetic field encounter in quantum optics. The chapter includes quantum mechanical representation theory along with the time dependent perturbation theory. The density operator plays a central role in emission and absorption theory.

5.1 Introduction to Generalized Coordinates

The Lagrangian and Hamiltonian formulation of classical mechanics provide simple techniques for deriving equations of motion using energy relations. Rather than concerning ourselves with complicated vector relations in rectangular coordinates, these alternate formulations allow us to use the scalar quantities such as kinetic T and potential energy V in classical mechanics. The Hamiltonian generally represents the total energy of the system. It comes from the Lagrangian that satisfies a least action principle. The two functionals are related to each other by a Legandre transformation. They provide the gateway to quantizing systems of particles and fields (such as electromagnetic fields).

The Lagrangian L and Hamiltonian H are functionals of generalized coordinates. Generalized coordinates comprise any set of independent variables that describe the object (or objects) under scrutiny.

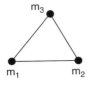

FIGURE 5.1.1
Three masses connected by rigid rods.

5.1.1 Constraints

Constraints represent *a priori* knowledge of a physical system. They reduce the total number of degrees of freedom available to the system. For example, Figure 5.1.1 shows a collection of masses interconnected by rigid (massless) rods. These rods constrain the distance between the masses and therefore reduce the number of degrees of freedom; however the whole system (of three masses) can translate or rotate. As another example, the walls of a container also impose constraints on a system. In this case, the constraints are important only when the molecules in the container make contact with the walls. For quantum theory, constraints are quite nonphysical since, in actuality, small particles experience forces and not constraints. For example, electrostatic forces (and not rigid rods) hold atoms in a lattice. Sometimes constraints appear in the quantum description to simplify problems. Evidently, constraints are mostly important for macroscopic classical systems.

5.1.2 Generalized Coordinates

Suppose a generalized set of coordinates $B_q = \{q_1, q_2, \ldots, q_k\}$ describes the position of N point particles. A single point particle has exactly three degrees of freedom corresponding to the three translational directions. Without constraints, N particles have $k = 3N$ degrees of freedom. Position vectors normally describe the location of the N particles

$$\vec{r}_1 = \vec{r}_1(q_1, \ldots, q_k, t)$$
$$\vdots$$
$$\vec{r}_N = \vec{r}_N(q_1, \ldots, q_k, t)$$

(5.1.1)

For example, the $\{q_i\}$ might be spherical coordinates. The q_i are independent of each other in this case. Constraints reduce the degrees of freedom so that $k < 3N$; that is, the constraints eliminate $3N - k$ degrees of freedom. As a note, we make use of the generalized coordinates especially for fields but not the constraints.

Example 5.1.1

A pulley system connecting two point particles

Assume a massless pulley as shown in Figure 5.1.2. Normally two point masses would have 6 degrees of freedom. Confining the masses to a 2-D plane reduces the degrees of freedom to 4. Allowing only vertical motion for the two masses reduces the degrees of freedom to 2. The string requires the masses to move together and reduces the number of degrees of freedom to 1. The motion of both masses can be described by the generalized coordinate $q_1 = q$. This single generalized coordinate describes the position vectors $\vec{r}_1, \vec{r}_2$ for the masses.

Configuration space consists of the collection of the "k" generalized coordinates $\{q_1, q_2, \ldots, q_k\}$ where each coordinate can take on a range of values. These generalized coordinates have special significance for the Lagrange formulation of dynamics. We can define generalized velocities by

$$\{\dot{q}_1, \dot{q}_2, \ldots, \dot{q}_k\}$$

(5.1.2)

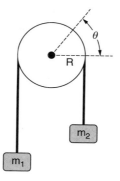

FIGURE 5.1.2
Two masses connected by a string passing over a pulley.

However, they are not independent of the generalized coordinates for the Lagrange formulation. That is, the variations δq, $\delta \dot{q}$ are not independent.

The generalized coordinates discussed so far constitute a discrete set whereby the coordinates are in one-to-one correspondence with a finite subset of the integers. The set can be infinite. A continuous set of coordinates would have elements in 1–1 correspondence with a dense subset of the *real* numbers. The distinction is important for a number of topics especially field theory.

Let's discuss a picture for the generalized coordinates and velocities especially important for field theories. We already know how to picture the position of particles in space for the case of x, y, z coordinates. So instead, let's take an example that illustrates the distinction between indices and generalized coordinates. Let's start with a collection of atoms arranged along a one-dimensional line oriented along the x-direction. Assume the number of atoms is k. As illustrated in the top portion of Figure 5.1.3, the atoms have equilibrium positions represented by the coordinates x_i. Given one atom for each equilibrium position x_i, the atoms can be labeled by either the respective equilibrium position x_i or by the number "i". The bottom portion of the figure shows the situation for the atoms displaced along the vertical direction. In this case, the generalized coordinates label the displacement from equilibrium. For the 1-D case shown, the generalized coordinates can be written equally well as either q_i or $q(x_i)$ so that $q_i = q(x_i) = q_{x_i}$. In this case, we think of x_i or i as indices to label a particular point in space or to label an atom. More generally for 3-D motion, each atom would have three generalized coordinates and three generalized velocities.

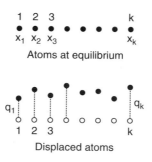

FIGURE 5.1.3
Example of generalized coordinates for atoms in a lattice.

Mathematically, these displacements q_i can be randomly assigned. It's only after applying the dynamics (Newton's laws, etc.) to the problem that the displacements become correlated. Mathematically, without dynamics (i.e., Newton's laws), atom #1 can be moved to position q_1 and atom #2 to position q_2 without there being any reason for choosing those two positions. The position of either atom can be independently assigned. This notion of independent translations leads to the alternate formulation of Newton's laws.

Let's briefly return to Figure 5.1.3 and discuss its importance to field theories. Let's focus on electromagnetics. When we write the electric field for example as $\vec{\mathscr{E}}(x, t)$, we think of x as an index labeling a particular point along a line in space. We think of $\vec{\mathscr{E}}$ as a displacement at point x. The displacement can vary with time. There must be three generalized coordinates at the point x. The three generalized coordinates are the three vector components of $\vec{\mathscr{E}}$. So $\vec{\mathscr{E}}$ really represents three displacements at the point x and not just one. In addition, we notice that the indices x form a continuum rather than the discrete set indicated in Figure 5.1.3.

5.1.3 Phase Space Coordinates

A system, which can consist of a single or multiple particles, evolves in time when it follows a curve in phase space as a function of time. Phase space consists of the generalized coordinates and conjugate momentum

$$\{q_1, q_2, \ldots, q_k, p_1, p_2, \ldots, p_k\} \tag{5.1.3}$$

all of which are assumed to be independent of one another. The momentum p_i is conjugate to the coordinate q_i because it describes the momentum of the particle corresponding to the direction q_i. Assigning particular values to the $2k$ coordinates in phase space specifies the "state of the system." The phase space coordinates are used primarily with the Hamiltonian of the system.

Example 5.1.3

The momentum P_x describes the momentum of a particle along the x direction.

Consider the pulley system shown in Figure 5.1.4. The momentum conjugate to the generalized coordinate $q = \theta$ is the total angular momentum along the axis of the pulley.

The Hamilton formulation of dynamics uses phase-space coordinates.

$$\{q_1, q_2, \ldots, q_k, p_1, p_2, \ldots, p_k\} \tag{5.1.4}$$

Each member of the set of the phase-space coordinates in Equation (5.1.4) has the same level of importance as any other member so that one cannot be more fundamental than another. For example, a point particle can be independently given position coordinates x, y, z and momentum coordinates $\{p_x, p_y, p_z\}$. This means that the particle can be assigned a random position and a random velocity. Given that the phase space coordinates are all independent, we can also vary the coordinates in an independent manner; that is, the variations δq, δp must be independent of one another. The term *configuration space* applies to the coordinates $\{q_1, q_2, \ldots, q_k\}$ and the term "phase space" applies to the full set of coordinates $\{q_1, q_2, \ldots, q_k, p_1, p_2, \ldots, p_k\}$. Essentially, in the absence of dynamics, position and momentum can be arbitrarily assigned to each particle.

5.2 Introduction to the Lagrangian and the Hamiltonian

The notion that nature follows a "law of least action" has a long history starting around 200 BC. The optical laws of reflection and refraction can be derived from the principle that light follows a path that minimizes the transit time. In the 1700s, the law was reformulated to require the dynamics of mechanical systems to minimize the action defined as Energy × Time. In the 1800s, Hamilton stated the most general form. A dynamical system will follow a path that minimizes the action defined as the time integral of the Lagrangian. A Legendre transformation of the Lagrangian then produces the total energy of the system in the form of the Hamiltonian. Today, the Lagrangian and Hamilton play central roles in quantum theory. The Schrodinger equation can be found from the classical Hamiltonian by replacing the classical dynamical variables with operators. The Feynman path integral provides a beautiful formulation of the quantum principle by incorporating the integral over all possible paths of the action. The form of the Lagrangian can be found from a variational method. This section derives the differential equation, Lagrange's equation, that provides the equations of motion for the generalized coordinates.

5.2.1 Lagrange's Equation from a Variational Principle

Hamilton's principle produces Lagrange's Equation for conservative systems. The method is particularly easy to generalize for systems consisting of continuous sets of coordinates (i.e., field theory). Of all the possible paths in configuration space that a system could follow between two fixed points $1 = (q_1^{(1)}, q_2^{(1)}, \ldots, q_k^{(1)})$ and $2 = (q_1^{(2)}, q_2^{(2)}, \ldots, q_k^{(2)})$, the path that it actually follows makes the following action integral an extremum (either minimum or maximum) as shown in Figure 5.2.1.

$$I = \int_1^2 dt\, L(q_1, q_2, \ldots, q_k, \dot{q}_1, \dot{q}_2, \ldots, \dot{q}_k, t) \qquad (5.2.5)$$

The Lagrangian "L" is a functional of the kinetic energy "T" and potential energy "V" according to $L = T - V$ for particles. The procedure assumes fixed endpoints but this can be generalized for variable endpoints. To minimize the notation, let $q_i, \dot{q}_i$ represent the entire collection of points in $\{q_1, q_2, \ldots, q_k, \dot{q}_1, \dot{q}_2, \ldots, \dot{q}_k\}$.

To find the extremum of the action integral

$$I = \int_1^2 dt\, L(q_i, \dot{q}_i, t)$$

define a new path in configuration space for each generalized coordinate q_i by

$$q_i'(t) = q_i(t) + \delta q_i$$

where the time "t" parameterizes the curve in configuration space. Assume q_i extremizes the integral I. We can find the functional form of each $q_i(t)$ by requiring the variation of the integral around q_i to vanish as follows.

$$0 = \delta I = \int_1^2 dt \sum_i \left[\frac{\partial L(q_i, \dot{q}_i, t)}{\partial q_i} \delta q_i + \frac{\partial L(q_i, \dot{q}_i, t)}{\partial \dot{q}_i} \delta \dot{q}_i \right] \qquad (5.2.6)$$

Partially integrate the second term using the fact that $\delta q_i(t_1) = 0 = \delta q_i(t_2)$ to find

$$0 = \delta I = \int_1^2 dt \sum_i \left[\frac{\partial L(q_i, \dot{q}_i, t)}{\partial q_i} - \frac{d}{dt} \frac{\partial L(q_i, \dot{q}_i, t)}{\partial \dot{q}_i} \right] \delta q_i$$

The small variations δq_i are assumed to be independent so that

$$\frac{\partial L}{\partial q_i} - \frac{d}{dt} \frac{\partial L}{\partial \dot{q}_i} = 0 \quad \text{for} \quad i = 1, 2, \ldots \tag{5.2.7}$$

where $L = T - V$.

The canonical momentum can be defined as

$$p_i = \frac{\partial L}{\partial \dot{q}_i} \tag{5.2.8}$$

p_i denotes the momentum conjugate to the coordinate q_i. The canonical momentum does not always agree with the typical momentum "mv" for a particle. The canonical momentum for an EM field interacting with a particle consists of the particle and field momentum.

Example 5.2.1

Consider a single particle of mass "m" constrained to move vertically along the "y" direction and acted upon by the gravitational force $F = -mg$

$$T = \frac{1}{2} m (\dot{y})^2 \qquad V = mgy \qquad L = T - V = \frac{1}{2} m (\dot{y})^2 - mgy$$

Lagrange's equation

$$\frac{\partial L}{\partial y} - \frac{d}{dt} \frac{\partial L}{\partial \dot{y}} = 0$$

gives Newton's second law for a gravitational force $-mg - m\ddot{y} = 0$ where the derivatives

$$\frac{\partial \dot{y}}{\partial y} = 0 = \frac{\partial y}{\partial \dot{y}}$$

since "y" and "$\dot{y}$" are taken to be independent. As a result, the equation of motion for the particle becomes $\ddot{y} = -g$ which gives the usual functional form of the height as $y = -\frac{g}{2} t^2 + v_o t + y_o$.

How can $y, \dot{y}$ be independent when they appear to be connected by $\dot{y} = dy/dt$? This relation assumes that the function y is already defined. Let's start with the step of defining the function y. At any value t, we can arbitrarily assign a value y and a value $\dot{y}$. The only requirement is that the function y must have fixed endpoints y_1 and y_2. These boundary conditions restrict only two points out of an uncountable infinite number. Figure 5.2.2 illustrates the concept. Notice that the value t can be assigned a large number of values of y and $\dot{y}$ without affecting the endpoints. Therefore, there can be

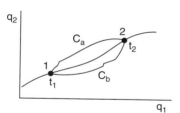

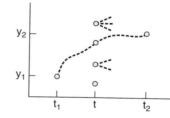

FIGURE 5.2.1
Three paths connecting fixed end points.

FIGURE 5.2.2
The function is determined by its value and slope at each point.

many curves connecting points $A = (t_1, y_1)$ and $B = (t_2, y_2)$. The equations $\dot{y} = dy/dt$ give a procedure for calculating the slope $\dot{y}$ only after we know the function y in some interval. For example, suppose we discuss the motion of a line of atoms so that the independent variables are $\{y, \dot{y}\}$ where $\dot{y}$ is the velocity. We can arbitrarily assign a displacement and a speed at each point x. Only after solving Newton's equations do we come to know how the speed and position at those points are inter-related.

Example 5.2.2

Find the equations of motion for the pulley system shown in Figure 5.2.3. Assume the pulley is massless, $m_2 > m_1$ and that $y_1(t) = 0$, $y_2(t) = h$. The kinetic energy is $T = \frac{1}{2} m_1 \dot{y}_1^2 + \frac{1}{2} m_2 \dot{y}_2^2$ and $V = m_1 g y_1 + m_2 g y_2$. The remaining 2 degrees of freedom y_1, y_2 can be reduced to one since $y_2 = h - y_1$.
  We therefore have

$$T = \frac{1}{2}(m_1 + m_2)\dot{y}_1^2 \quad V = m_1 g y_1 + m_2 g (h - y_1)$$

Lagrange's equation

$$\frac{\partial L}{\partial y_1} - \frac{d}{dt} \frac{\partial L}{\partial \dot{y}_1} = 0 \quad \text{produces} \quad \ddot{y}_1 = \frac{(m_1 - m_2)g}{(m_1 + m_2)}$$

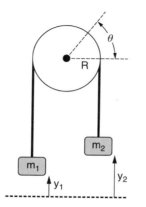

FIGURE 5.2.3
Pulley system.

5.2.2 The Hamiltonian

The Hamiltonian represents the total energy of a system. The quantum theory derives its operator Hamiltonian from the classical one by substituting operators for the classical dynamical variables.

Consider a closed, conservative system so that the Lagrangian L does not explicitly depend on time. The total energy and the total number of particles remain constant (in time) for a closed system. We define a conservative system to be one for which all of the forces can be derived from a potential. We do not consider any equations of constraint for quantum mechanics and field theory. Differentiating the Lagrangian provides

$$\frac{dL}{dt} = \sum_i \left(\frac{\partial L}{\partial q_i} \frac{dq_i}{dt} + \frac{\partial L}{\partial \dot{q}_i} \frac{d\dot{q}_i}{dt} \right) + \frac{\partial L}{\partial t} \tag{5.2.9}$$

The last term is zero by assumption $\partial L / \partial t = 0$. Substitute Lagrange's equation

$$\frac{\partial L}{\partial q_i} = \frac{d}{dt} \frac{\partial L}{\partial \dot{q}_i}$$

to find

$$\frac{dL}{dt} = \sum_i \left[\left(\frac{d}{dt} \frac{\partial L}{\partial \dot{q}_i} \right) \dot{q}_i + \frac{\partial L}{\partial \dot{q}_i} \frac{d\dot{q}_i}{dt} \right] = \sum_i \frac{d}{dt} \left(\frac{\partial L}{\partial \dot{q}_i} \dot{q}_i \right) \tag{5.2.10}$$

Recall the definition for the conjugate momentum

$$p_i = \frac{\partial L}{\partial \dot{q}_i} \tag{5.2.11}$$

Equation (5.2.10) becomes

$$\frac{d}{dt} \left[\sum_i \dot{q}_i p_i - L \right] = 0$$

The Hamiltonian "H" is defined to be

$$H = \sum_i \dot{q}_i p_i - L \tag{5.2.12}$$

which is the total energy of the system in this case. More thorough treatments show the Hamiltonian has the form $H = T + V$ for mechanical systems. *Important point:* We consider H to be a function of q_i, p_i whereas we consider L to be a function of $q_i, \dot{q}_i$.

5.2.3 Hamilton's Canonical Equations

The Hamiltonian leads to Hamilton's canonical equations

$$\dot{q}_j = \frac{\partial H}{\partial p_j} \qquad \dot{p}_j = -\frac{\partial H}{\partial q_j} \tag{5.2.13}$$

These equations allow us to find equations of motion from the Hamiltonian. Subsequent sections show that the quantum theory requires q_j and p_j to be operators satisfying commutation relations. The classical equivalent of the commutation relations appears in the next section on the Poisson brackets.

Hamilton's canonical equations (5.2.13) can now be demonstrated. Starting with Equation (5.2.12) we can write

$$\frac{\partial H}{\partial p_j} = \frac{\partial}{\partial p_j}\left[\sum_i \dot{q}_i p_i - L\right] = \dot{q}_j - \frac{\partial L}{\partial p_j} \tag{5.2.14}$$

Next noting that L depends on $q_i, \dot{q}_i$ and not p_i, we find

$$\frac{\partial H}{\partial p_j} = \dot{q}_j \tag{5.2.15}$$

which proves the first of Hamilton's equations. We can demonstrate the second of Hamilton's equations by using Lagrange's equation and the canonical momentum

$$\frac{\partial L}{\partial q_j} = \frac{d}{dt}\frac{\partial L}{\partial \dot{q}_j} \qquad p_j = \frac{\partial L}{\partial \dot{q}_j} \tag{5.2.16}$$

We find

$$\frac{\partial H}{\partial q_j} = \frac{\partial}{\partial q_j}\left[\sum_i \dot{q}_i p_i - L\right] = 0 - \frac{\partial L}{\partial q_j} = -\frac{d}{dt}\frac{\partial L}{\partial \dot{q}_j} = -\frac{d}{dt}p_j = -\dot{p}_j$$

Example 5.2.3

Find H and $\dot{q}_i, \dot{p}_i$ for a particle of mass m at a height y in a gravitational field.

Solution: The Lagrangian has the form

$$L = T - V = \frac{1}{2}m(\dot{y})^2 - mgy$$

The Hamiltonian H can be written as a function of the coordinate and its conjugate momentum. The relation for the canonical momentum for the Lagrangian

$$p = \frac{\partial L}{\partial \dot{y}} = m\dot{y}$$

allows "H" to be written as

$$H = \dot{y}p - L = \frac{p}{m}p - \left[\frac{1}{2}m\left(\frac{p}{m}\right)^2 - mgy\right] = \frac{p^2}{2m} + mgy$$

and then

$$\dot{y} = \frac{\partial H}{\partial p} = \frac{p}{m} \qquad \dot{p} = -\frac{\partial H}{\partial y} = mg$$

Example 5.2.4
For the pulley system in Example 5.2.2, find the Hamiltonian and Newton's equations of motion. Assume the pulley is massless.

Solution: The kinetic and potential energy can be written as

$$T = \frac{1}{2}(m_1 + m_2)\dot{y}_1^2 \quad V = m_1 g y_1 + m_2 g(h - y_1)$$

The Hamiltonian must be a function of momentum and not velocity. The Lagrangian gives the canonical momentum

$$p_1 = \frac{\partial L}{\partial \dot{y}_1} = \frac{\partial}{\partial \dot{y}_1} \frac{1}{2}(m_1 + m_2)\dot{y}_1^2 = M\dot{y}_1$$

where $M = m_1 + m_2$. Notice that p_1 is not the usual vector sum of the individual momenta. The kinetic energy can be rewritten as

$$T = \frac{1}{2}(m_1 + m_2)\dot{y}_1^2 = \frac{p_1^2}{2M}$$

The Hamiltonian can be written as

$$H = \dot{q}_1 p_1 - L = \frac{p_1}{M}p_1 - (T - V) = \frac{p_1^2}{M} - \frac{p_1^2}{2M} + m_1 g y_1 + m_2 g(h - y_1)$$

$$= \frac{p_1^2}{2M} + g y_1(m_1 - m_2) + m_2 g h$$

The Hamiltonian gives the time rate of change of momentum as

$$\dot{p}_1 = -\frac{\partial H}{\partial q_1} = -g(m_1 - m_2)$$

This last equation can be recognized as Newton's second law, which can be rewritten as a second-order differential equation if desired.

5.3 Classical Commutation Relations

The Hamiltonian is the primary quantity of interest for quantum theory. The specification of a quantum mechanical Hamiltonian follows several steps:

1. Determine the classical Hamiltonian.
2. Substitute operators for the classical dynamical variables (e.g., p's and q's).
3. Specify the commutation relations between those operators.

The commutation relations in quantum mechanics somewhat resemble the Poisson brackets in classical mechanics. The commutation relations and Poisson brackets determine the evolution of the dynamical variables. In the quantum theory, operators replace the classical dynamical variables. In fact, the Heisenberg quantum picture has the greatest resemblance to classical mechanics because the operators carry the system dynamics. In quantum theory, the commutation relations give time derivatives of operators. A commutator is defined by $\left[\hat{A}, \hat{B}\right] = \hat{A}\hat{B} - \hat{B}\hat{A}$ where $\hat{A}, \hat{B}$ are operators. The Poisson bracket as the classical version of the commutator uses partial derivatives whereas the quantum mechanical commutator does not.

Definition: Let $A = A(q_i, p_i)$, $B = B(q_i, p_i)$ be two differentiable functions of the generalized coordinates and momentum. We define the Poisson brackets by

$$[A, B] = \sum_i \left[\frac{\partial A}{\partial q_i} \frac{\partial B}{\partial p_i} - \frac{\partial B}{\partial q_i} \frac{\partial A}{\partial p_i} \right]$$

Sometimes we subscript the brackets with p, q

$$[A, B] = [A, B]_{q, p}$$

to indicate Poisson brackets. Using the definition of Poisson brackets, some basic properties can be proved.

1. Let A, B be functions of the phase space coordinates q, p and let c be a number; then

$$[A, A] = 0 \quad [A, B] = -[B, A] \quad [A, c] = 0$$

2. Let A, B, C be differentiable functions of the phase space coordinates q, p; then

$$[A + B, C] = [A, C] + [B, C] \quad [AB, C] = A[B, C] + [A, C]B$$

3. The time evolution of the dynamical variable A (for example) can be calculated by

$$\frac{dA}{dt} = [A, H] + \frac{\partial A}{\partial t}$$

Proof:

$$\frac{dA}{dt} = \sum_i \left[\frac{\partial A}{\partial q_i} \frac{dq_i}{dt} + \frac{\partial A}{\partial p_i} \frac{dp_i}{dt} \right] + \frac{\partial A}{\partial t}$$

We include the partial with respect to time in case the function A explicitly depends on time.

Substituting the two relations for the rate of change of position and momentum

$$\frac{dq_i}{dt} = \frac{\partial H}{\partial p_i} \qquad \frac{dp_i}{dt} = -\frac{\partial H}{\partial q_i}$$

the Poisson brackets become

$$\frac{dA}{dt} = \sum_i \left[\frac{\partial A}{\partial q_i} \frac{\partial H}{\partial p_i} - \frac{\partial A}{\partial p_i} \frac{\partial H}{\partial q_i} \right] + \frac{\partial A}{\partial t} = [A, H] + \frac{\partial A}{\partial t}$$

Although the order of multiplication $AH = HA$ does not matter in classical theory, the order must be maintained in quantum theory. In quantum theory, the order of two operators can only be switched by using the commutation relations.

4.
$$\dot{q}_m = [q_m, H] \qquad \dot{p}_m = [p_m, H]$$

Proof: Consider the first one for example

$$[q_m, H] = \sum_i \left[\frac{\partial q_m}{\partial q_i} \frac{\partial H}{\partial p_i} - \frac{\partial q_m}{\partial p_i} \frac{\partial H}{\partial q_i} \right] = \sum_i \left[\delta_{im} \frac{\partial H}{\partial p_i} - 0 \frac{\partial H}{\partial q_i} \right] = \frac{\partial H}{\partial p_m} = \dot{q}_m$$

5.
$$[q_i, q_j] = 0 [p_i, p_j] = 0 \quad [q_i, p_j] = \delta_{ij}$$

5.4 Classical Field Theory

So far we have discussed the classical Lagrangian and Hamiltonian for discrete sets of generalized coordinates and their conjugate momentum. Now we turn our attention to systems with an uncountably infinite number of coordinates. The section first discusses the relation between discrete and continuous system, and then shows how the Lagrangian for sets of discrete coordinates leads to the Lagrangian for the continuous set of coordinates. This latter Lagrangian begins the study of classical field theories since it can produce the Maxwell's equations, the Schrodinger equation, and it begins the quantum field theory for particles and the quantum electrodynamics. The present section demonstrates the Lagrangian for the wave motion in a continuous media that has applications to phonon fields and provides an example for the later field theory of electromagnetic fields.

5.4.1 Concepts for the Lagrangian and Hamiltonian Density

For systems with a continuous set of generalized coordinates, Lagrange and Hamilton's formulation of dynamics must be generalized. First, we discuss the generalized coordinates and velocities. Second, we show how a continuous system can be viewed as a discrete one with a countable number of generalized coordinates. Third, we derive the generalized momentum for the Hamiltonian density. We end with a summary. The following topics apply the procedure to wave motion in a continuous medium. The following topics apply the procedure to wave motion in a continuous medium.

For the continuous coordinate case, we posit the following imagery. Suppose the indices x, y, z in $\vec{r} = x\tilde{x} + y\tilde{y} + z\tilde{z}$ label points in space. The value of a function $\eta(\vec{r}, t) = \eta(x, y, z, t)$ serves as a generalized coordinate indexed by the point $\vec{r}$. Figure 5.4.1 shows some of the generalized coordinates along the z-direction. The lower left side

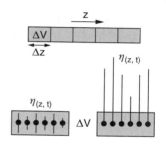

FIGURE 5.4.1

Top portion shows space divided into cells. Bottom portion shows two types of continuous coordinates. Left side shows a field and the right sides shows displacement of small masses.

shows a small volume of space with a field (EM in origin). The field has a different value for each point. The lower right side shows another example for the generalized coordinates. Here η represents the displacement of small masses. The generalized velocities are given by $\dot{\eta}$.

Now let's discuss how the continuous coordinates $\eta(\vec{r}, t)$ compare with the discrete ones q_i. The top portion of Figure 5.4.1 shows all of space divided into many cells of volume ΔV_i. In *each* cell, the field $\eta(z, t)$ takes on many similar values. We can define the *discrete* generalized coordinates by the average

$$q_i(t) = \frac{1}{\Delta V_i} \int_{\Delta V_i} dV \, \eta(\vec{r}, t) \qquad (5.4.1)$$

The q_i represent the average value of the continuous coordinate in the given cell. Making ΔV_i small enough means that the η under the integral is approximately constant so that

$$q_i(t) = \frac{1}{\Delta V_i} \int_{\Delta V_i} dV \, \eta(\vec{r}, t) \quad \rightarrow \quad \eta(\vec{r}, t) \qquad (5.4.2)$$

Notice that the small volume ΔV_i must be associated with the points x, y, z in space and not with the "tops" of $\eta(\vec{r}, t)$. In the next topic, we will show displaced small boxes but these will be different boxes. These boxes will refer to actual chunks of mass displaced from equilibrium. The procedure given in the present topic uses the small cells in Figure 5.4.1 to show how the continuous and discrete Lagrangians can be interrelated.

Next we compare the Lagrangians for the two systems. For continuous sets of coordinates, people usually work with the Lagrange density $\mathscr{L}$ defined through

$$L = \int_V dV \, \mathscr{L} \qquad (5.4.3)$$

where the Lagrange density has units of "energy per volume." The Lagrange density has the form

$$\mathscr{L} = \mathscr{L}(\eta, \dot{\eta}, \partial_i \eta) \qquad (5.4.4)$$

where $i = 1, 2, 3$ refers to derivatives with respect to x, y, z, respectively. The Lagrange density refers to a single point in space (or possibly two arbitrarily close points due to the derivatives).

On the other hand, suppose we divide all space into cells of volume ΔV_i with $q_i, \dot{q}_i$ being the generalized coordinate and velocity in cell #i, respectively. The full Lagrangian must have the form

$$L = L(q_i, \dot{q}_i, q_{i\pm1}) \qquad (5.4.5)$$

where the $q_{i\pm1}$ allows for derivatives. Especially note that all coordinates, including time, $i = 1, 2, 3, 4$, appear in the full Lagrangian. Now to make the connection with the Lagrange density, apply the cellular space to the full Lagrangian in Equation (5.4.5). Dividing up the volume V into cells so that $V = \sum_i \Delta V_i$ we can write

$$L(q_i, \dot{q}_i, q_{i\pm1}) = \int_V dV \ \mathscr{L}(q_i, \dot{q}_i, q_{i\pm1}) = \sum_i \int_{\Delta V_i} dV \ \mathscr{L}(q_i, \dot{q}_i, q_{i\pm1}) \qquad (5.4.6)$$

The definition of an average from calculus provides

$$\bar{\mathscr{L}}_i = \frac{1}{\Delta V_i} \int_{\Delta V_i} dV \ \mathscr{L} \quad \text{so that} \quad L = \sum_i \int_{\Delta V_i} dV \ \mathscr{L} = \sum_i \Delta V_i \bar{L}_i(q_i, \dot{q}_i, q_{i\pm1}) \qquad (5.4.7)$$

where now each ΔV_i has one q_i and one $\dot{q}_i$ associated with it on account of Equation (5.4.1). We can see that the two forms (Equations (5.4.7) and (5.4.4)) of the Lagrangian agree by using Equation (5.4.2) when we take the limit $\Delta V_i \to 0$

$$L = \sum_i \Delta V_i \bar{\mathscr{L}}_i(q_i, \dot{q}_i, q_{i\pm1}) \ \to \ \int dV \ \mathscr{L}(\eta, \dot{\eta}, \partial_i \eta) \qquad (5.4.8)$$

where the average on the Lagrangian density has been removed because the cell volume shrinks to a single point. This last equation shows how discrete coordinates and the corresponding Lagrangian produce the continuous coordinates and the Lagrangian density.

Finally, we compare the full Hamiltonian with the Hamiltonian density. The full Hamiltonian can be written as

$$H = H(q_i, p_i) = \sum_i p_i \dot{q}_i - L = \sum_i p_i \dot{q}_i - \sum_i \Delta V_i \ \bar{\mathscr{L}}_i \qquad (5.4.9)$$

We can calculate p_j by the usual method

$$p_j = \frac{\partial L}{\partial \dot{q}_j} = \frac{\partial}{\partial \dot{q}_j} \sum_i \Delta V_i \ \bar{\mathscr{L}}_i = \sum_i \Delta V_i \ \frac{\partial \bar{\mathscr{L}}_i}{\partial \dot{q}_j} = \Delta V_j \ \frac{\partial \bar{\mathscr{L}}_j}{\partial \dot{q}_j} \qquad (5.4.10)$$

where the summation in the last term disappears because we assume $\bar{\mathscr{L}}_i$ depends only on $\dot{q}_j$ (along with q_j) and the relation $d\dot{q}_i/d\dot{q}_j = \delta_{ij}$ holds. Notice how the momentum depends on the volume of the small box whereas the relation $q_j \to \eta$ does not. We write the momentum relation for a small mass whereby the momentum must be proportional to the mass and hence the volume $p_j \sim (\Delta m)\dot{q}_j \sim (\Delta V)\rho\dot{q}_j \sim (\Delta V)\pi_j$ where ρ represents the mass density. Therefore, the momentum density can be defined as

$$\Delta V_j \pi_j = p_j = \frac{\partial L}{\partial \dot{q}_j} = \Delta V_j \ \to \ \frac{\partial \bar{\mathscr{L}}_j}{\partial \dot{q}_j} \pi_j = \frac{\partial \bar{\mathscr{L}}_j}{\partial \dot{q}_j} \ \xrightarrow[\Delta V_i \to 0]{} \ \pi(\vec{r}, t) = \frac{\partial \mathscr{L}(\eta, \ldots)}{\partial \dot{\eta}} \qquad (5.4.11)$$

The full Hamiltonian can be written as a Hamiltonian density

$$H = \int dV \ \mathscr{H} \qquad (5.4.12)$$

We can write

$$\int d^3x \, \mathcal{H} = H = \sum_i p_i \dot{q}_i - L = \sum_i \Delta V_i \, \pi_i \dot{q}_i - \sum_i \Delta V_i \, \bar{\mathcal{L}}_i \; \rightarrow \; \int d^3x \left[\pi(\vec{r}, t) \, \dot{\eta}(\vec{r}, t) - \mathcal{L} \right]$$

and identify the Hamiltonian density as

$$\mathcal{H} = \pi(\vec{r}, t) \, \dot{\eta}(\vec{r}, t) - \mathcal{L} \qquad (5.4.13)$$

Summary of results

Lagrange density: $\qquad \mathcal{L} = \mathcal{L}(\eta, \dot{\eta}, \partial_i \eta)$

Lagrangian: $\qquad L = \int_V dV \, \mathcal{L}$

Hamiltonian density: $\quad \mathcal{H} = \pi(\vec{r}, t) \, \dot{\eta}(\vec{r}, t) - \mathcal{L}$

Hamiltonian: $\qquad H = \int dV \, \mathcal{H}$

Momentum density: $\quad \pi(\vec{r}, t) = \frac{\partial \mathcal{L}(\eta, \ldots)}{\partial \dot{\eta}}$

Hamilton's Canonical Equations:

$$\dot{\eta} = \frac{\partial \mathcal{H}}{\partial \pi} \qquad \dot{\pi} = -\frac{\partial \mathcal{H}}{\partial \eta}$$

5.4.2 The Lagrange Density for 1-D Wave Motion

Now we develop the Lagrangian for 1-D wave motion in a continuous medium. As discussed in the previous topic, we imagine each point in space to be labeled by indices x, y, z according to $\vec{r} = x\tilde{x} + y\tilde{y} + z\tilde{z}$. The value of a function $\eta(\vec{r}, t) = \eta(x, y, z, t)$ serves as a generalized coordinate indexed by the point $\vec{r}$. Figure 5.4.2 shows transverse wave motion along the z-axis with η giving the displacement. The generalized velocity at the point x, y, z can be written as $\dot{\eta}$. Two important notes are in order. First, note that x, y, z do not depend on time since they are treated as indices. Second, the small boxes appearing in Figure 5.4.2 represent small chunks of matter that the wave displaces from equilibrium. The coordinate q_i denotes the average displacement of the scalar field h for the small chunk.

The description of wave motion requires a partial differential equation involving partial derivatives. We require the partial derivatives to appear in the argument of the Lagrangian. These spatial derivatives take the form $\partial_i \eta$ where i refers to one of the indices x, y, z. For example, $i = 3$ gives $\partial_3 \eta = \partial \eta / \partial z$. For the purpose of the Lagrangian, the spatial derivatives must be independent of each other and of the coordinates.

$$\frac{\partial(\partial_i \eta)}{\partial(\partial_j \eta)} = \delta_{ij} \qquad \frac{\partial(\partial_i \eta)}{\partial \eta} = 0 \qquad \frac{\partial \eta}{\partial(\partial_i \eta)} = 0$$

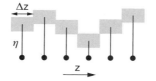

FIGURE 5.4.2

Displacement of masses at various points along the z-axis.

The Lagrangian can be written as

$$L = L(\eta, \dot{\eta}, \partial_i \eta) = L(\eta, \dot{\eta}, \partial_1 \eta, \partial_2 \eta, \partial_3 \eta) \tag{5.4.14}$$

For the transverse wave motion, the partial derivatives actually enter the Lagrangian as a result of the generalized forces acting on each element of volume.

We need to minimize the action

$$I = \int_{t_1}^{t_2} dt\, L \tag{5.4.15}$$

However, for continuous systems (i.e., systems with continuous sets of generalized coordinates), it is customary to work with the Lagrange density defined by

$$I = \int_{t_1}^{t_2} dt\, L = \int_{t_1}^{t_2} \int_{\vec{r}_1}^{\vec{r}_2} dt\, d^3x\, \mathscr{L}(\eta, \dot{\eta}, \partial_i \eta) \tag{5.4.16}$$

The Lagrange density $\mathscr{L}$ has units of energy per volume. To find the minimum action, we must vary the integral I so that $\delta I = 0$. In the process, a partial integration produces a "surface term." We must assume two boundary conditions: one for the time integral and one for the spatial integral. For the time integral, the set of displacements η must be fixed at times t_1, t_2 so that $\delta\eta(t_1) = 0 = \delta\eta(t_2)$. For the spatial integrals, we assume either periodic boundary conditions or fixed-endpoint conditions so that the surface term vanishes.

Now let's find the extremum of the action in Equation (5.4.16)

$$0 = \delta I = \int_{t_1}^{t_2} \int_{\vec{r}_1}^{\vec{r}_2} dt\, d^3x\, \delta\mathscr{L}(\eta, \dot{\eta}, \partial_i \eta) = \int_{t_1}^{t_2} \int_{\vec{r}_1}^{\vec{r}_2} dt\, d^3x \left[\frac{\partial \mathscr{L}}{\partial \eta} \delta\eta + \frac{\partial \mathscr{L}}{\partial \dot{\eta}} \delta\dot{\eta} + \frac{\partial \mathscr{L}}{\partial(\partial_i \eta)} \delta(\partial_i \eta) \right]$$

where we use the Einstein convention for repeated indices in a product, namely $A_i B_i = \sum_i A_i B_i$. Interchanging the differentiation with the variation produces

$$0 = \delta I = \int_{t_1}^{t_2} \int_{\vec{r}_1}^{\vec{r}_2} dt\, d^3x \left[\frac{\partial \mathscr{L}}{\partial \eta} \delta\eta + \frac{\partial \mathscr{L}}{\partial \dot{\eta}} \frac{\partial}{\partial t} \delta\eta + \frac{\partial \mathscr{L}}{\partial(\partial_i \eta)} \partial_i \delta\eta \right]$$

Integrating by parts and using the fact that both the temporal and spatial surface terms do not contribute, we find

$$\int_{t_1}^{t_2} \int_{\vec{r}_1}^{\vec{r}_2} dt\, d^3x \left[\frac{\partial \mathscr{L}}{\partial \eta} - \frac{\partial}{\partial t} \frac{\partial \mathscr{L}}{\partial \dot{\eta}} - \partial_i \frac{\partial \mathscr{L}}{\partial(\partial_i \eta)} \right] \delta\eta = 0$$

Given that the variation at each point is independent of every other, we find Lagrange's equations for the continuous media

$$\frac{\partial \mathscr{L}}{\partial \eta} - \frac{\partial}{\partial t} \frac{\partial \mathscr{L}}{\partial \dot{\eta}} - \partial_i \frac{\partial \mathscr{L}}{\partial(\partial_i \eta)} = 0 \tag{5.4.17}$$

where the repeated index convention must be enforced on the last term. Notice that the first two terms look very similar to the usual Lagrange equation for the discrete set of generalized coordinates. If desired, we can also include generalized forces in the formalism so that the motion of the waves can be "driven" by an outside force.

Example 5.4.1

Suppose the Lagrange density has the form $\mathscr{L} = \frac{\rho}{2}\dot{\eta}^2 + \frac{\beta}{2}(\partial_z\eta)^2$ for 1-D motion progating along the z-direction, where ρ, β resemble the mass density and spring constant (Young's modulus) for the material, and $\eta = \eta(z,t)$.

Solution: Lagrange's equation has the following terms

$$\frac{\partial\mathscr{L}}{\partial\eta} = 0 \qquad \frac{\partial}{\partial t}\frac{\partial\mathscr{L}}{\partial\dot{\eta}} = \rho\ddot{\eta} \qquad \frac{\partial}{\partial z}\frac{\partial\mathscr{L}}{\partial(\partial_z\eta)} = \beta\frac{\partial^2\eta}{\partial z^2}$$

Equation (5.4.17) then gives

$$\frac{\partial^2\eta}{\partial z^2} + \frac{\rho}{\beta}\ddot{\eta} = 0 \quad \text{with speed} \quad v = \sqrt{\beta/\rho}$$

5.5 Schrodinger Equation from a Lagrangian

The quantum theory relies primarily on the Schrodinger wave equation to describe the dynamics of quantum particles. The present section shows one method by which the Lagrangian formulation leads to the Schrodinger wave equation. The companion volume on quantum and solid state shows the beautiful connection with the Feynman path integral. Subsequent sections in the present volume explore the meaning of the Hamiltonian and the Schrodinger wave equation in more detail.

As a mathematical exercise, we start with the Lagrange *density*

$$\mathscr{L} = i\hbar\psi^*\dot{\psi} - \frac{\hbar^2}{2m}\nabla\psi^* \cdot \nabla\psi - V(r)\,\psi^*\psi \tag{5.5.1a}$$

or equivalently

$$\mathscr{L} = i\hbar\psi^*\dot{\psi} - \frac{\hbar^2}{2m}\sum_j \partial_j\psi^*\partial_j\psi - V(r)\,\psi^*\psi \tag{5.5.1b}$$

where $j = x, y, z$ the Lagrangian is

$$L = \int d^3x\,\mathscr{L} \tag{5.5.1c}$$

The Lagrange density is a functional of the independent coordinates ψ, ψ^* and their derivatives $\partial_j\psi, \partial_j\psi^*$ where $j = x, y, z$.

The variation of L leads to the Euler–Lagrange equations of the form

$$\frac{\partial \mathscr{L}}{\partial \phi} - \sum_a \partial_a \frac{\partial \mathscr{L}}{\partial(\partial_a \phi)} = 0 \qquad (5.5.2a)$$

where $a = x, y, z, t$ and $\phi = \psi$ or ψ^*. Setting $\phi = \psi^*$ provides

$$\frac{\partial \mathscr{L}}{\partial \psi^*} - \sum_a \partial_a \frac{\partial \mathscr{L}}{\partial(\partial_a \psi^*)} = 0 \qquad (5.5.2b)$$

Evaluating the first term produces

$$\frac{\partial \mathscr{L}}{\partial \psi^*} = \frac{\partial}{\partial \psi^*}\left[i\hbar\psi^*\dot{\psi} - \frac{\hbar^2}{2m}\sum_j \partial_j\psi^* \partial_j\psi - V(r)\,\psi^*\psi \right] = i\hbar\dot{\psi} - V(r)\,\psi$$

The argument of the second term in Equation (5.5.2b) produces

$$\frac{\partial \mathscr{L}}{\partial(\partial_a \psi^*)} = \frac{\partial}{\partial(\partial_a \psi^*)}\left\{ i\hbar\psi^*\partial_t\psi - \frac{\hbar^2}{2m}\sum_j \partial_j\psi^* \partial_j\psi - V(r)\,\psi^*\psi \right\} = \left\{ \begin{array}{ll} 0 & a = t \\ -\frac{\hbar^2}{2m}\partial_j\psi & a = j \end{array} \right.$$

Equation (5.5.2b) becomes

$$i\hbar\dot{\psi} - V(r)\,\psi + \frac{\hbar^2}{2m}\sum_j \partial_j\partial_j\psi = 0$$

Therefore, we find the Schrodinger wave equation

$$-\frac{\hbar^2}{2m}\nabla^2\psi + V(r)\,\psi = i\hbar\dot{\psi} \qquad (5.5.3)$$

We can find the classical Hamiltonian density (energy per unit volume)

$$\mathscr{H} = \pi\dot{\psi} - \mathscr{L} \qquad (5.5.4a)$$

where p is the momentum conjugate to ψ and the total energy is

$$H = \int d^3x\,\mathscr{H} \qquad (5.5.4b)$$

The conjugate momentum is defined by

$$\pi = \frac{\partial \mathscr{L}}{\partial \dot{\psi}} \qquad (5.5.5)$$

For the Lagrange density in Equation (5.5.1), we find

$$\pi = \frac{\partial \mathscr{L}}{\partial \dot{\psi}} = \frac{\partial}{\partial \dot{\psi}}\left\{ i\hbar\psi^*\dot{\psi} - \frac{\hbar^2}{2m}\sum_j \partial_j\psi^* \partial_j\psi - V(r)\,\psi^*\psi \right\} = i\hbar\psi^*$$

The classical Hamiltonian density becomes

$$\mathcal{H} = \pi\dot{\psi} - \mathcal{L} = i\hbar\psi^*\dot{\psi} - \left\{ i\hbar\psi^*\dot{\psi} - \frac{\hbar^2}{2m}\nabla\psi^* \cdot \nabla\psi - V(r)\,\psi^*\psi \right\}$$

$$= \frac{\hbar^2}{2m}\nabla\psi^* \cdot \nabla\psi + V(r)\,\psi^*\psi$$

Often times the Lagrange density is stated as

$$\mathcal{L} = i\hbar\psi^*\dot{\psi} + \frac{\hbar^2}{2m}\psi^*\nabla^2\psi - V(r)\,\psi^*\psi = \psi^*\left(i\hbar\partial_t + \frac{\hbar^2}{2m}\nabla^2 - V \right)\psi \qquad (5.5.6)$$

This last equation comes from Equations (5.5.1) by partially integrating and assuming the surface terms are zero. The Hamiltonian density then has the form

$$\mathcal{H} = \pi\dot{\psi} - \mathcal{L} = \psi^*\left(-\frac{\hbar^2}{2m}\nabla^2 + V \right)\psi \qquad (5.5.7)$$

In terms of the quantum theory, the classical Hamiltonian is most related to the average energy

$$H = \int d^3x\, \psi^*\left(-\frac{\hbar^2}{2m}\nabla^2 + V \right)\psi = \langle\psi|H_{sch}|\psi\rangle \qquad (5.5.8a)$$

where

$$H_{sch} = -\frac{\hbar^2}{2m}\nabla^2 + V \qquad (5.5.8b)$$

5.6 Linear Algebra and the Quantum Theory

The mathematical objects in the quantum theory must accurately model the physical world—linear algebra is a natural language. The theory must represent properties of particles and systems, predict the evolution of the system, and provide the ability to make and interpret observations. Quantum theory began in an effort to describe microscopic (atomic) systems when classical theory gave erred predictions. However, classical and quantum mechanical descriptions must agree for macroscopic systems—the correspondence principle.

Vectors in a Hilbert space represent specific properties of a particle or system. Every physically possible state of the system must be represented by one of the vectors. A single particle must correspond to a single vector (possibly time dependent). Hermitian operators represent physically observable quantities such as energy, momentum, and electric field. These operators provide values for the quantities when they act upon a vector in a Hilbert space. The discussion will show how the theory distinguishes measurement operators from Hermitian operators.

The Feynman path integral and principle of least action (through the Lagrangian) lead to the Schrodinger equation, which describes the system dynamics. The method essentially reduces to using a classical Hamiltonian and replacing the dynamical variables with operators. The operators must satisfy commutation relations somewhat similar to the Poisson brackets for classical mechanics.

We need to address the issue of how the particle dynamics (equations of motion) arise. In the classical sciences and engineering, dynamical variables such as position and momentum can depend on time. The Heisenberg representation in quantum theory gives the time dependence to the Hermitian operator version of the dynamical variables. In this description, the operators carry the dynamics of the system while the wave functions remain independent of time. The vectors/wave functions in Hilbert space appears as a type of "lattice" (or stage) for observation. The result of an observation depends on the time of making the observation through the operators. The Schrodinger representation of the quantum theory provides an interpretation most closely related to classical optics and electromagnetic theory. The wave functions depend on time but the operators do not. This is very similar to saying that the electric field (as the wave function) depends on time because the traveling wave, for example, has the form $e^{ikx-i\omega t}$. We will encounter an intermediate case, the interaction representation, where the operators carry trivial time dependence and the wave functions retain the time response to a "forcing function." All three representations contain identical information.

In this section, we address the following questions:

1. How do basis vectors differ from other vectors?
2. What physical meaning should be ascribed to the superposition of wave functions?
3. How should we interpret the expansion coefficients of a general vector in a Hilbert space?
4. How can we picture a time dependent wave function?
5. What does the collapse of the wave function mean and how does it relate to reality?
6. What does it mean to say "observables" cannot be "simultaneously and precisely" known? The results are summarized in Table 5.6.1.

5.6.1 Observables and Hermitian Operators

Every system must be capable of interacting with the physical world. In the laboratory, the systems come under scrutiny of other probing systems such as our own physical senses or the equipment in the laboratory. An observable, such as energy or momentum, is a quantity that can be observed or measured in the laboratory and can take on only real values. These values can be either discrete or continuous. For example, confined electrons have discrete energy values whereas the position of an electron can have a continuous range.

Suppose measurements of a particular property such as energy H of a system always produces the set of real values $\{E_1, E_2, \ldots\}$ and the particle is always found in one of the corresponding states $\{|E_1\rangle, |E_2\rangle, \ldots\}$. Based on these values and vectors, we define an energy operator (Hamiltonian $\hat{\mathscr{H}}$)

$$\hat{\mathscr{H}} = \sum_n E_n |E_n\rangle \langle E_n| \tag{5.6.1}$$

TABLE 5.6.1

Physical World, Linear Algebra and Quantum Theory

Physical World	Mathematics
Observables: Properties that can be measured in a laboratory	Hermitian operators $\hat{\mathcal{H}}$
Specific particle/system properties	Wave functions $|\psi\rangle$
Fundamental motions/states of existence	Basis/eigenvectors $|h\rangle$ of $\hat{\mathcal{H}}$
Value of observable in fundamental motion	$\hat{\mathcal{H}}|h\rangle = h|h\rangle$
Laboratory measured values, states	Sets $\{h\}$ and $\{|h\rangle\}$
Particle/system has characteristics of all fundamental motions	Superposed wave function $|\psi\rangle = \sum_h \beta_h |h\rangle$
Average behavior of a particle	$\langle\psi|\hat{\mathcal{H}}|\psi\rangle$
Probability of finding value or fundamental motion	Probability amplitude of finding "h" or $|h\rangle$ is $\langle h \lceil \psi\rangle = \beta_h$. Probabiltiy $= |\beta_h|^2$
Dynamics of system	Time dependence of operators or vectors— Schrodinger's equation
Measure state of particle/system	Collapse of $|\psi\rangle$ to basis vector $|h\rangle$. Random collapse does not have an equation of motion
Simultaneous measurements of two or more observables	*Commuting operators:* repeated measurements produce identical values *Noncommuting operators:* repeated measurements produce a range of values
Complete description of a particle/system	Largest possible set of commuting Hermitian operators

Applying the Hamiltonian to one of the states produces

$$\hat{\mathcal{H}}|E_n\rangle = E_n|E_n\rangle \qquad (5.6.2)$$

We naturally interpret the operation as measuring the value of $\hat{\mathcal{H}}$ for a system in the state $|E_n\rangle$. Notice that the operator in Equation (5.6.1) must be Hermitian since $\hat{\mathcal{H}}^+ = \hat{\mathcal{H}}$. By assumption, the eigenvalues are real. The number of eigenvectors equals the number of possible states for the system and forms a complete set. For these reasons, quantum theory represents observables by Hermitian operators.

The process of "making a measurement" cannot be fully modeled by the eigenvalue equation (5.6.2). The operators in the theory operate on vectors in a Hilbert space. A general vector can be written as a superposition of the eigenvectors of $\hat{\mathcal{H}}$ and therefore do not have just a single value for the measurement of $\hat{\mathcal{H}}$. A physical measurement of $\hat{\mathcal{H}}$ causes the wave function to collapse to a random basis vector, which does not follow from the dynamics and does not appear in the effect of the Hermitian operator.

5.6.2 The Eigenstates

The eigenvectors of a Hermitian operator, which correspond to an observable, represent the most fundamental states for the particle or system. Every possible fundamental motion of a particle must be observable (i.e., measurable). For example, the various orbitals in an atom correspond to the eigenvectors. This requires each fundamental physical state of a system or particle to be represented as a basis vector. The basis set must be complete so that all fundamental motions can be detected and represented in the theory. As mentioned in the previous topic, if measurements of particle energy $\hat{\mathscr{H}}$ produce the values $\{E_1, E_2, \ldots, E_n, \ldots\}$ then we can represent the *resulting* states by the eigenvectors $\{|E_1\rangle, |E_2\rangle, \ldots, |E_n\rangle, \ldots\}$ where $\hat{\mathscr{H}}\,|E_n\rangle = E_n|E_n\rangle$. These states must be the most basic states; they form the basis states. Any other state of the system must be a linear combination of these basis states having the form $|\psi\rangle = \sum \beta_n |E_n\rangle$.

The idea of "state" occurs in many branches of science and engineering. A particle or system can usually be described by a collection of parameters. We define a state of the particle or system to be a specific set of values for the parameters.

For classical mechanics, the position and momentum describe the motion of a point particle. Therefore the three position and three momentum components completely specify the state of motion for a single point particle. There are three degrees of freedom as discussed in previous sections. For optics, the polarization, wavelength, and the propagation vector specify the basic states (i.e., modes). Notice that we do not include the amplitude in the list because we can add any number of photons to the mode (i.e., produce any amplitude we choose) without changing the basic shape. The optical modes are eigenvectors of the time-independent Maxwell wave equation. We know these basic modes will be essentially sinusoidal functions for a Fabry-Perot cavity. They produce traveling plane waves for free space.

Example 5.6.1

Polarization in Optics

A single photon travels along the z-axis as shown in Figure 5.6.1. The photon has components of polarization along the x-axis and along the y-axis, for example, according to

$$\vec{s} = \frac{1}{\sqrt{2}}\tilde{x} + \frac{1}{\sqrt{2}}\tilde{y}$$

The electric field is parallel to the polarization $\vec{s}$. We view the single photon as simultaneously polarized along $\tilde{x}$ and along $\tilde{y}$. Suppose we place a polarizer in the path of the photon with its axis along the x-axis. There exists a 50% chance that the photon will be found polarized along the x-axis. The *polarization* state of the incident photon must be the superposition of two basis states $\tilde{x}, \tilde{y}$. We view the single incident photon as being *simultaneously in both polarization states*. The act of observing the photon causes the wave function to collapse to either the $\tilde{x}$ state or to the $\tilde{y}$ state. The polarizer absorbs the photon if the photon wave function collapses to the $\tilde{y}$-polarization. The polarizer allows the photon to pass if the photon wave function collapses to the $\tilde{x}$-polarization. For a single photon, the photon will be either transmitted or it will not; there can be no intermediate case.

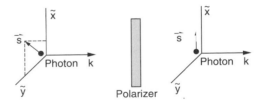

FIGURE 5.6.1
Polarization.

5.6.3   The Meaning of Superposition of Basis States and the Probability Interpretation

A quantum particle can "occupy" a state

$$|v\rangle = \sum_n \beta_n(t)\, |\phi_n\rangle \tag{5.6.3}$$

where basis set $\{|\phi_n\rangle\}$ represents the collection of fundamental physical states. The most convenient basis set consists of the eigenvectors of an operator of special interest to us, such as for example, the energy of the particle (i.e., the Hamiltonian $\hat{\mathscr{H}}$). We therefore choose the basis set to be the eigenvectors of the energy operator.

$$\hat{\mathscr{H}}|\phi_n\rangle = E_n|\phi_n\rangle$$

The superposed wave function $|v\rangle$ refers to a particle (or system) having attributes from all of the states in the superposition. The particle simultaneously exists in all of the basic states making up the superposition. In Figure 5.6.2 for example, an observation of the energy of the particle in the state $|v\rangle$ with the energy basis set will find it with energy E_1 or E_2 or E_3. Before the measurement, we view the particle as having some mixture of all three energies in a type of average. The measurement forces the electron to decide on the actual energy.

Not just any superposition wave function can be used for the quantum theory. All quantum mechanical wave functions must be normalized to have unit length $\langle v \mid v \rangle = 1$ including the basis functions satisfying $\langle \phi_m \mid \phi_n \rangle = \delta_{mn}$. All of the vectors are normalized to one in order to interpret the components as a probability (next topic).

Therefore, the functions appropriate for the quantum theory define a surface for which all of its points are exactly 1 unit away from the origin. For the 3-D case, the surface makes a unit sphere. The set of wave functions does not form a vector space since the zero vectors cannot be in the set.

5.6.4   Probability Interpretation

Perhaps most important, the quantum theory interprets the expansion coefficients β_n in the superposition $|v\rangle = \sum_n \beta_n|n\rangle = \sum_n |n\rangle\langle n \mid v\rangle$ as a probability amplitude.

$$\text{Probability amplitude} = \beta_n = \langle n \mid v \rangle \tag{5.6.4}$$

To be more specific, assume we make a measurement of the energy of the particle. The quantized system allows the particle to occupy a discrete number of *fundamental*

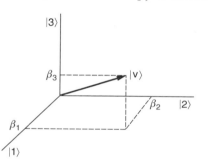

FIGURE 5.6.2
The vector is a linear combination of basis vectors.

states $|\phi_1\rangle$, $|\phi_2\rangle$, $|\phi_3\rangle$, ... with respective energies E_1, E_2, ... A measurement of the energy can only yield one of the numbers E_n and the particle must be found in one of the fundamental states $|\phi_n\rangle$. The probability that the particle is found in state $|n\rangle = |\phi_n\rangle$ is given by

$$P(n) = \left|\beta_n\right|^2 = \left|\langle n \mid v \rangle\right|^2 \qquad (5.6.5)$$

Keep in mind that a probability function must satisfy certain conditions including

$$P(n) \geq 0 \quad \text{and} \quad \sum_n P(n) = 1 \qquad (5.6.6)$$

Let's check that Equation (5.6.5) satisfies these last two properties. It satisfies the first property $P(n) \geq 0$ since the length of a vector must always be greater than or equal to zero. Let's check the second property. Consider

$$1 = \langle v \mid v \rangle = \sum_m \sum_n \beta_m^* \beta_n \langle \phi_m \mid \phi_n \rangle = \sum_n \left|\beta_n\right|^2 = \sum_n P(n) \qquad (5.6.7)$$

So the normalization condition for the wave function requires the summation of all probabilities to equal unity even though each individual β_n might change with time.

We can handle continuous coordinates in a similar fashion except use integrals and Dirac delta functions rather than the discrete summations and Kronecker delta functions. Projecting the wave function $|\psi\rangle$ onto the spatial-coordinate basis set $\{|x\rangle\}$ provides a probability amplitude as the component $\psi(x) = \langle x \mid \psi \rangle$.

$$|\psi\rangle = \int dx \, |x\rangle\langle x \mid \psi\rangle = \int dx \, |x\rangle \, \psi(x)$$

These wave functions $\psi(x)$ usually come from the Schrodinger equation. The square of the probability amplitude $\langle x \mid \psi \rangle = \psi(x)$ provides the probability density $\rho(x) = \psi^*(x)\,\psi(x)$ (probability per unit length); it describes the probability of finding the particle at "point x' (refer to Appendix 4 for a review of probability theory). We require that all quantum mechanically acceptable wave functions have unit length so that

$$1 = \langle \psi \mid \psi \rangle = \langle \psi \mid \hat{1} \mid \psi \rangle = \int_{\text{all } x} dx \langle \psi \mid x \rangle\langle x \mid \psi \rangle = \int_{\text{all } x} dx \, \psi^*(x)\,\psi(x)$$

For three spatial dimensions, $\rho(\vec{r})\,dV = \psi^*(\vec{r})\,\psi(\vec{r})\,dV$ represents the probability of finding a particle in the infinitesimal volume dV centered at the position $\vec{r}$

$$\text{PROB}(a \leq x \leq b, c \leq y \leq d, e \leq z \leq f) = \int_a^b \int_c^d \int_e^f dV \, \rho(x, y, z)$$

Several types of reasoning on probability are quite common for the quantum theory. Unlike classical probability theory, we cannot simply add and multiply probabilities. In quantum theory, the probability *amplitudes* "add" and "multiply." Consider a succession of events occurring at the space-time points $\{(x_0, t_0), (x_1, t_1), (x_2, t_2), \ldots\}$ on the history path in Figure 5.6.3. The probability amplitude $\psi(x, t)$ of the succession of events all on the same history path consists of the product $\psi(x, t) = \prod_i \psi_i(x_i, t_i)$. Without

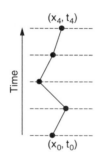

FIGURE 5.6.3
A succession of events on a single history path.

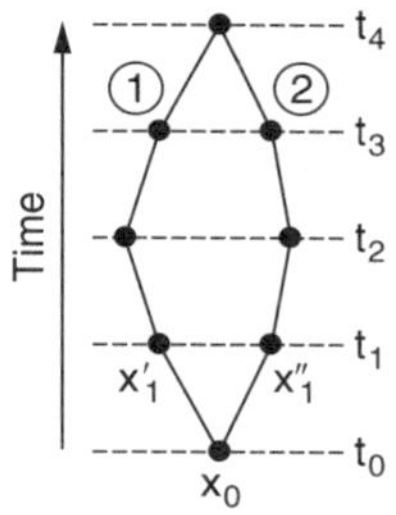

FIGURE 5.6.4
Parallel history paths.

superposition, the probability for successive events (the square of the amplitude) reduces to the product of the probabilities as found in classical probability theory. Superposition requires the phase of the amplitude to be taken into account similar to that of the electromagnetic field before calculating the total power.

For the case of two independent events such as two occurring at the same time, the probability amplitudes add (Figure 5.6.4) $\psi(x, t) = \psi_1(x_1', t_1) + \psi_2(x_2'', t_1)$ where all wave functions depend on (x, t) at the destination point (really need a propagator).

A measurement of an observable $\hat{A}$ destroys the phase relation between the components of the wave function $|\psi\rangle = \sum_n \beta_n |a_n\rangle$, forces the system to collapse to one of the eigenstates $\{|a_1\rangle, |a_2\rangle, \ldots\}$, and produces exactly one of the eigenvalues $\{a_1, a_2, \ldots\}$ for the results. The classical probability of finding the particle in state a_i or a_j can be written as

$$P(a_i \text{ or } a_j) = P(a_i) + P(a_j) - P(a_i \text{ and } a_j).$$

Since the wave function collapses to either a_i or a_j but not both, the two events (i.e., the result of the measurement) must be mutually exclusive in this case so that $P(a_i \text{ and } a_j) = 0$ and

$$P(a_i \text{ or } a_j) = |\beta_i|^2 + |\beta_j|^2.$$

When people look for the results of measurements on a quantum system, even though there exists an infinite number of wave functions $|\psi\rangle$, they consider only the basis states and eigenvalues.

5.6.5 The Average and Variance

We use the quantum mechanical probability density in a slightly different manner than the classical ones. Consider a particle (or system) in state

$$|\psi\rangle = \sum_n \beta_n |a_n\rangle \tag{5.6.8}$$

where $\{a_1, a_2, \ldots\}$ and $\{|a_1\rangle, |a_2\rangle, \ldots\}$ are the eigenvalues and eigenvectors for the observable $\hat{A}$. The quantum mechanical average value of $\hat{A}$ can be written as $\langle \psi | \hat{A} | \psi \rangle$.

We can project the wave function onto either the eigenvector basis set or the coordinate basis set. Consider the eigenvectors first. Using the expansion 5.6.8 we find

$$\langle \psi | \hat{A} | \psi \rangle = \sum_n a_n |\beta_n|^2 \tag{5.6.9}$$

This expression agrees with the classical probability expression for averages $E(A) = \sum_n a_n P_n$ where $E(A)$ represents the expectation value and where $\hat{A}$ takes the form of a random variable. In fact, the range of $\hat{A}$ can be viewed as the outcome space $\{a_1, a_2, \ldots\}$. Projecting into coordinate space, the average can be written as

$$\langle\psi|\hat{A}|\psi\rangle = \langle\psi|\left\{\int dx\, |x\rangle\langle x|\right\}\hat{A}|\psi\rangle = \int dx\, \psi^*(x)\,\hat{A}\,\psi(x) \tag{5.6.10a}$$

Notice that we must maintain the order of operators and vectors.

We define the variance of a Hermitian operator by

$$\sigma_{\hat{O}}^2 = E\left(\hat{O} - \langle\hat{O}\rangle\right)^2 = E\left(\hat{O}^2 - 2\hat{O}\langle\hat{O}\rangle + \langle\hat{O}\rangle^2\right) = E\left(\hat{O}^2\right) - \langle\hat{O}\rangle^2 = \langle\hat{O}^2\rangle - \langle\hat{O}\rangle^2 \tag{5.6.10b}$$

The standard deviation becomes

$$\sigma = \sqrt{\langle\hat{O}^2\rangle - \langle\hat{O}\rangle^2} \tag{5.6.10c}$$

A "Sharp value" refers to the case $\hat{\sigma}_{\hat{O}}^2 = 0$ such as for eigenstates of $\hat{O}$.

Three comments need to be made. First, to compute the expectation value or the variance, the wave function must be known. The components of the wave function give the probability amplitude. This is equivalent to knowing the probability function in classical probability theory. Second, from an ensemble point of view, the expectation of an operator really provides the average of an observable when making multiple observations on the *same* state. The quantity $\langle\hat{O}\rangle \equiv \langle\psi|\hat{O}|\psi\rangle$ provides the average of the observable $\hat{O}$ in the single state $|\psi\rangle$.

As a third comment, non-Hermitian operators do not necessarily have a unique definition for the variance. Consider a variance defined similar to a classical variance $\text{Var}(O) = \langle(O - \bar{O})^*(O - \bar{O})\rangle$. For simplicity, set $\bar{O} = 0$ so that $\text{Var}(O) = \langle O^*O\rangle$. Replacing O with $\hat{O}$ and O^* with $\hat{O}^+$ produces the three possibilities of $\langle\hat{O}^+\hat{O}\rangle$, $\langle\hat{O}\,\hat{O}^+\rangle$ and $\langle\frac{1}{2}\hat{O}^+\hat{O} + \frac{1}{2}\hat{O}\,\hat{O}^+\rangle$ out of an infinite number. The adjoint can be dropped for Hermitian operators and all possibilities reduce to the one listed in Equation (5.6.10c).

Example 5.6.2

The Infinitely Deep Square Well

Find the expectation value of the position "x" for an electron in state "n" where the basis functions are

$$\left\{\phi(x) = \sqrt{\frac{2}{L}}\,\text{Sin}\left(\frac{n\pi x}{L}\right)\right\}$$

Solution

$$\langle x\rangle = \langle n|x|n\rangle = \int_0^L dx\, u_n^*\, x\, u_n = \frac{2}{L}\int_0^L dx\, x\,\sin^2\left(\frac{n\pi x}{L}\right) = \frac{L}{2}$$

5.6.6 Motion of the Wave Function

As discussed in the next section, the Schrodinger wave equation provides the dynamics of particle through the wave function $|\Psi\rangle$.

$$\hat{\mathscr{H}}|\Psi\rangle = i\hbar\frac{\partial}{\partial t}|\Psi\rangle \tag{5.6.11}$$

Solving the Schrodinger equation by the method of orthonormal expansions provides the energy basis functions $\{|1\rangle = |\phi_1\rangle, |2\rangle = |\phi_2\rangle, \ldots\}$. It also gives the time dependence of $|\Psi\rangle$ which appears in the coefficients β in the basis vector expansion

$$|\Psi(t)\rangle = \sum_n \beta_n(t)|n\rangle$$

The wave function $|\Psi\rangle$ moves in Hilbert space since the coefficients β_n depend on time. Notice that the wave function stays within the given Hilbert space and never moves out of it! This is a result of the fact that the eigenvectors form a complete set.

A formal solution to Equation (5.6.11) can be found when the Hamiltonian does not depend on time

$$|\Psi(t)\rangle = e^{\frac{\hat{\mathscr{H}}(t-t_0)}{i\hbar}}|\Psi(t_0)\rangle \tag{5.6.12}$$

where $|\Psi(t_0)\rangle$ is the initial wave function. The operator

$$\hat{u}(t, t_0) = e^{\frac{\hat{\mathscr{H}}(t-t_0)}{i\hbar}} \tag{5.6.13}$$

moves the wave function $|\psi\rangle = |\psi(t)\rangle$ in time as shown in Figure 5.6.5. Also, because all quantum mechanical wave functions have unit length and never anything else, the operator "$\hat{u}$" must be unitary!

In general, operators that move the wave function in Hilbert space make the coefficients depend on time and therefore also the probabilities $P(n) = |\langle n|v(t)\rangle|^2 = |\beta_n(t)|^2$. If the total Hamiltonian does not depend on time and therefore, β's depend on time only through a trivial phase factor of the form $e^{i\omega t}$, then the probabilities $P(n) = |\beta_n|^2$ do not depend on time.

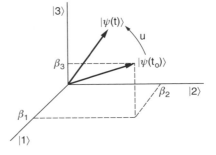

FIGURE 5.6.5
The evolution operator causes the wave function to move in Hilbert space. The unitary operator depends on the Hamiltonian. Therefore it is really the Hamiltonian that cause the wave function to move.

5.6.7 Collapse of the Wave Function

The collapse of the wave function is one of the most exciting aspects of quantum theory (certainly one of the most imaginative). The collapse deals with how a superposed wave function behaves while making a measurement of an observable. The collapse is random and outside the normal evolution of the wave function; a dynamical equation does not govern the collapse.

First we introduce the collapse of the wave function. Suppose we are most interested in the energy of the system (although any Hermitian operator will work) and that the energy has quantized values $\{E_1, E_2, \ldots\}$ where $\mathcal{H}|\phi_n\rangle = E_n|\phi_n\rangle$. Further assume that an electron resides in a superposed state

$$|\psi\rangle = \sum_n \beta_n |\phi_n\rangle \qquad (5.6.14)$$

Making a measurement of the energy produces a single energy value E_n (for example). To obtain the single value E_n, the particle must be in the single state $|\phi_n\rangle$. We therefore realize that making a measurement of the energy somehow changes the wave function from $|\psi\rangle$ to $|\phi_n\rangle$. The probability of the wave function $|\psi\rangle$ collapsing to the basis vectors $|\phi_n\rangle$ must be $P(n) = |\beta_n|^2$.

Let us discuss how we mathematically represent the process of measuring an observable. So far, we claim to model the measurement process by applying a Hermitian operator to a state. However, we've shown the process only for eigenstates

$$\hat{\mathcal{H}}|\phi_n\rangle = E_n|\phi_n\rangle \qquad (5.6.15)$$

In fact, the interpretation of Equation (5.6.15) does not match the processes of "measuring an observable" since we expect the results to be a number such as E_n and not the vector $E_n|\phi_n\rangle$.

How should we interpret the case when measuring an observable for a superposed wave function such as in Equation (5.6.14)? If we apply $\hat{\mathcal{H}}$ to the vector $|\psi\rangle$ we find

$$\hat{\mathcal{H}}|\psi\rangle = \sum_n \beta_n(t)\hat{\mathcal{H}}|\phi_n\rangle = \sum_n \beta_n(t)E_n|\phi_n\rangle \qquad (5.6.16)$$

This last equation attempts to measure the energy of a particle in state $|\psi\rangle$ at time t. While mathematically correct, this last equation does not accurately model the "act of observing!" Observing the superposition wave function must disturb it and cause it to collapse to one of the eigenstates! The process of observing a particle must therefore involve a projection operator! The collapse must introduce a time dependence beyond that in the coefficient $\beta_n(t)$. The interaction between the external measurement agent and the system introduces uncontrollable changes of the wave function in time. Once the wave function collapses to one of the basis states, a randomizing process must be applied to the system for the wave function to move away from that basis state.

Let us show how the "observation act" might be modeled. Suppose that the observation causes the wave function to collapse to state $|n\rangle$. The mathematical model for the "act of observing" the energy state should include a projection operator $\hat{P}_n = \frac{1}{\beta_n}\langle\phi_n|$ where $\hat{P}_n$ includes a normalization constant of $1/\beta_n$ for convenience (the symbol "P" should not be confused with the momentum operator and probability). The operator corresponding to the "act of observing" should be written as $\hat{P}_n \hat{\mathscr{H}}$. The results of the observation becomes

$$\hat{P}_n \hat{\mathscr{H}} |\psi\rangle = \sum_m \beta_m(t) \frac{1}{\beta_n} \langle \phi_n | \hat{\mathscr{H}} | \phi_m \rangle = E_n$$

However, we don't know *a priori* into which state the wave function will collapse. We can only give the probability of it collapsing into a particular state. The probability of it collapsing into state $|n\rangle$ must be $|\beta_n|^2 = \beta_n^* \beta_n = |\langle \phi_n | \psi \rangle|^2$. Quantities such as $\langle \psi | \hat{\mathscr{H}} | \psi \rangle$ give a single quantity E that represents an average over the potential collapse into any of the eigenstates. This means $E = \text{Ave}(E_n) = \text{Ave } \hat{P}_n \hat{\mathscr{H}} |\psi\rangle$.

So far in the discussion, we make a distinction between an undisturbed and a disturbed wave function. For the undisturbed wave function, the components in a generalized summation

$$|\psi\rangle = \sum_n \beta_n(t) |\phi_n\rangle \tag{5.6.17}$$

maintain their phase relation as the system evolves in time. In this case, the components $\beta_n(t)$ satisfy a differential equation (which implies the components must be continuous).

The undisturbed wave function follows the dynamics embedded in Schrodinger's equation. The general wave function satisfies

$$\hat{\mathscr{H}} |\psi\rangle = \sum_n \beta_n(t) \hat{\mathscr{H}} |\phi_n\rangle = \sum_n \beta_n(t) E_n |\phi_n\rangle \tag{5.6.18}$$

The collection of eigenvalues E_n make up the spectrum of the operator $\hat{\mathscr{H}}$. The coefficient β_n is the probability amplitude for the particle to be found in state ϕ_n with energy E_n.

The collapse of the wave function has several possible interpretations. For the first interpretation, people sometimes view the wave function as a mathematical construct describing the probability amplitude. They assume that the particle occupies a particular state although they don't know which one. They make a measurement to determine the state the particle (or system) actually occupies. Before a measurement, they have limited information of the system. They know the probability

$$P(n) = |\beta_n|^2$$

that the particle occupies a given fundamental state (basis vector). Therefore, they know a wave function by the superposition of $\beta_n |\phi_n\rangle$. Making a measurement naturally changes the wave function because they then have more information on the actual state of the particle. After the measurement, they know for certain that the electron must be in state "*i*" for example. Therefore, they know $\beta_i = 1$ while all the other β must be zero. In effect, the wave function collapses from ψ to ϕ_i. With this first view, they ascribe any wave motion of the electron to the probability amplitude while implicitly assuming that the electron occupies a single state and behaves as a point particle. Making a measurement removes their uncertainty. In this view, the collapse refers to probability and nothing more. However, apparently nature does not operate this way as seen from Bell's theorem.

As a second interpretation and probably the most profound, the collapse of the wave function can be viewed as more related to physical phenomena. The Copenhagen interpretation (refer to Max Jammer's book) of a quantum particle in a superposed state

$$|\psi\rangle = \sum_n \beta_n(t)\,|\phi_n\rangle \tag{5.6.19}$$

describes the particle as *simultaneously* existing in all of the fundamental states $|\phi_n\rangle$. Somehow the particle simultaneously has all of the attributes of all of the fundamental states. A measurement of the particle forces it to "decide" on one particular state. This second point of view produces one of the most profound theorems of modern times—Bell's theorem. Let's take an example connected with the EPR paradox (the Einstein–Podolski–Rosen paradox).

Suppose a system of atoms can emit two correlated photons (entangled) in opposite directions. We require the polarization of one to be tied with the polarization of the other. For example, suppose every time that we measure the polarization of photon A, we find photon B to have the same polarization. However, let's assume that each photon can be transversely polarized to the direction of motion according to

$$|\psi^a\rangle = \beta_1^a|1\rangle + \beta_2^a|2\rangle \tag{5.6.20}$$

where $|1\rangle, |2\rangle$ represent the x and y directions, and "a" represents particle A or B. This last equation represents a wave moving along the z-direction but polarized partly along the x-direction and partly along the y-direction. We regard each photon as simultaneously existing in both polarized states $|1\rangle, |2\rangle$. If a measurement is made on photon A, and its wave function collapses to state $|1\rangle$, then the wave function for photon B simultaneously collapses to state $|1\rangle$. The collapse occurs even though the photons might be separate by several light years! Apparently the collapse of one can influence the other at speeds faster than light! Most commercial bookstores carry a number of "easy to read" books on endeavors to make communicators using the effect.

5.6.8 Noncommuting Operators and the Heisenberg Uncertainty Relation

This topic provides an intuitive view of how the Heisenberg uncertainty relation arises from two non-commuting Hermitian operators $\hat{A}, \hat{B}$ corresponding to two observables. Figure 5.6.6 indicates that measuring $\hat{A}$ collapses the wavefunction $|\psi\rangle$ into one of many fundamental states. Suppose the wave function collapses to the state $|a\rangle$. Repeated measurements of observable A produces the sequence a, a, a and so on. The dispersion (standard deviation) for the sequence must be zero. We see that once the wave function collapses, the operator $\hat{A}$ cannot change the state since it produces the same state $\hat{A}|a\rangle = a|a\rangle$. Similar comments apply to $\hat{B}$. Now we can see what happens when two operators do not influence each others eigenstates.

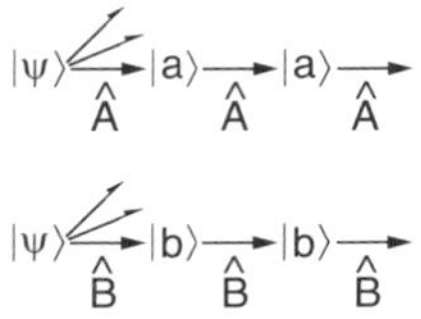

Let's suppose the two observables $\hat{A}, \hat{B}$ can be measured at the same time without dispersion; this means we can repeatedly measure $\hat{A}, \hat{B}$ and find the same result each time. We will use the shortcut phrase of "simultaneous observables." Let's assume that $|\phi\rangle$ characterizes the state of a particle such that $\hat{B}|\phi\rangle = b|\phi\rangle$ and $\hat{A}|\phi\rangle = a|\phi\rangle$. We can first apply $\hat{A}$ without affecting the results for $\hat{B}$. Applying $\hat{A}$ gives $\hat{A}|\phi\rangle = a|\phi\rangle$ and then applying $\hat{B}$ gives $\hat{B}\{\hat{A}|\phi\rangle\} = \hat{B}\{a|\phi\rangle\} = b\{a|\phi\rangle\} = b\{\hat{A}|\phi\rangle\}$. The result of observing $\hat{B}$

FIGURE 5.6.6
Repeatedly applying an operator to a state gives the same number.

must still be "*b*." Therefore $\hat{A}$ does not affect the state of the particle and therefore does not disturb a measurement of $\hat{B}$. As a matter of generalizing the discussion, consider the following string of equalities.

$$\hat{A}\hat{B}|\phi\rangle = b\hat{A}|\phi\rangle = ab|\phi\rangle = a\hat{B}|\phi\rangle = \hat{B}a|\phi\rangle = \hat{B}\hat{A}|\phi\rangle \tag{5.6.21}$$

This relation must hold for every vector in the space since it holds for each basis vector. We can conclude

$$\hat{A}\hat{B} = \hat{B}\hat{A} \;\rightarrow\; 0 = \hat{A}\hat{B} - \hat{B}\hat{A} \equiv \left[\hat{A}, \hat{B}\right] \tag{5.6.22}$$

Therefore simultaneous observables must correspond to operators that commute (refer to Section 4.9).

In this discussion, we say that we first apply $\hat{B}$ and then apply $\hat{A}$ according to their order in the product $\hat{A}\hat{B}$ or we might imagine using a time index. For example

$$\hat{A}(t_2)\hat{B}(t_1)|\psi\rangle \quad t_2 > t_1$$

In our case, the $\hat{A}, \hat{B}$ do not depend on time so that $t_2 \rightarrow t_1$. We might think of the order $\hat{A}\hat{B}|\psi\rangle$ as a remnant of mathematical notation (involving t). Physically it doesn't matter if we write $\hat{A}\hat{B}$ or $\hat{B}\hat{A}$ because we require them to be measured at the same time. We expect to find the *same answer* if the operators correspond to simultaneous observables. Therefore we expect $\hat{A}\hat{B} = \hat{B}\hat{A}$ for simultaneous observables.

Now let's consider the situation where two operators $\hat{A}, \hat{B}$ interfere with the measurement of each other. Suppose $\hat{B}$ disturbs the eigenvector of $\hat{A}$ where the eigenvectors of $\hat{A}$ satisfy

$$\hat{A}|\phi_1\rangle = a_1|\phi_1\rangle \quad \hat{A}|\phi_2\rangle = a_2|\phi_2\rangle \tag{5.6.23}$$

Suppose that $\hat{B}$ disturbs the eigenstates of $\hat{A}$ according to

$$\hat{B}|\phi_1\rangle = |v\rangle \tag{5.6.24}$$

which appears in Figure 5.6.7. Assume $|v\rangle$ has the expansion

$$|v\rangle = \beta_1|\phi_1\rangle + \beta_2|\phi_2\rangle \tag{5.6.25}$$

Now we can see that the order of applying the operators makes a difference. If we apply first $\hat{A}$ then $\hat{B}$, we find

$$\hat{B}\hat{A}|\phi_1\rangle = \hat{B}a_1|\phi_1\rangle = a_1|v\rangle \tag{5.6.26}$$

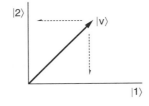

FIGURE 5.6.7
The vector collapses to either of two eigenvectors of *A*.

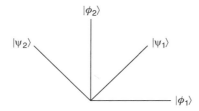

FIGURE 5.6.8
The two basis sets.

The reverse order $\hat{A}\hat{B}$ produces different behavior.

$$\hat{A}\hat{B}|\phi_1\rangle = \hat{A}|v\rangle = \hat{A}\{\beta_1|\phi_1\rangle + \beta_2|\phi_2\rangle\} = \beta_1 a_1|\phi_1\rangle + \beta_2 a_2|\phi_2\rangle \qquad (5.6.27)$$

The results of the two orderings do not agree. We therefore surmise

$$\hat{A}\hat{B} \neq \hat{B}\hat{A}$$

Therefore, operators that interfere with each other do not commute. Further, the collapse of the wave function $|v\rangle$ under the action of $\hat{A}$ can produce either $|\phi_1\rangle$ or $|\phi_2\rangle$ so that the standard deviation for the measurements of $\hat{A}$ can no longer be zero.

We now provide a "cartoon" view of how the non-commutivity of two observables gives rise to the Heisenberg uncertainty relation. Assume a 2-D Hilbert space with two different basis sets $\{|\phi_1\rangle, |\phi_2\rangle\}$ and $\{|\psi_1\rangle, |\psi_2\rangle\}$ where $\hat{A}|\phi_n\rangle = a_n|\phi_n\rangle$ and $\hat{B}|\psi_n\rangle = b_n|\psi_n\rangle$. The relation between the basis vectors appears in Figure 5.6.8. We make repeated measurements of $\hat{B}\hat{A}$. Suppose we start with the wave function $|\phi_1\rangle$ and measure $\hat{A}$; we find the result a_1. Next, let's measure $\hat{B}$. There's a 50% chance that $|\phi_1\rangle$ will collapse to $|\psi_1\rangle$ and a 50% chance it will collapse to $|\psi_2\rangle$. Let's assume it collapses to $|\psi_1\rangle$ and we find the value b_1. Next we measure $\hat{A}$ and find that $|\psi_1\rangle$ collapses to $|\phi_2\rangle$ and we observe value a_2 and so on. Suppose we find the following results for the measurements.

$$a_1 \quad b_1 \quad a_2 \quad b_1 \quad a_2 \quad b_2 \quad a_1 \quad b_1 \quad a_1 \quad b_2$$

Next let's sort this into two sets for the two operators

$$A \;\rightarrow\; a_1 \quad a_2 \quad a_2 \quad a_1 \quad a_1$$
$$B \;\rightarrow\; b_1 \quad b_1 \quad b_2 \quad b_1 \quad b_2$$

We therefore see that both A and B must have a nonzero standard deviation. Section 4.9 shows how the observables must satisfy a relation of the form $\sigma_A \sigma_B \geq$ constant $\neq 0$. We find a nonzero standard deviation when we measure two noncommuting observables and the wave function collapses to different basis vectors. Had we repeatedly measured A, we would have found $a_1 \quad a_1 \quad a_1 \quad a_1$ which has zero standard deviation.

5.6.9 Complete Sets of Observables

As previously discussed, we define the state of a particle or a system by specifying the values for a set of observables

$$\left\{\hat{O}_1, \hat{O}_2, \ldots\right\}$$

such as $\hat{O}_1 =$ energy, $\hat{O}_2 =$ angular momentum, and so on. We know that each Hermitian operator induces a basis set. The direct product space has a basis set of the form $|o_1, o_2, \ldots\rangle = |o_1\rangle|o_2\rangle\ldots$ where the eigenvalue "o_n" occurs in the eigenvalue relation $\hat{O}_n|o_1\ldots o_n\ldots\rangle = o_n|o_1\ldots o_n\ldots\rangle$. These operators all share a common basis set. Knowing the particle occupies the state $|o_1, o_2, \ldots\rangle$ means that we exactly know the outcome of measuring the observables $\{\hat{O}_1, \hat{O}_2, \ldots\}$. How do we know which observables to include in the set? Naturally we include observables of interest to us. We make the set as large as possible without including Hermitian operators that don't commute.

In quantum theory, we specify the basic states (i.e., basis states) of a particle or system by listing the observable properties. The particle might have a certain energy, momentum, angular momentum, polarization, etc. Knowing the value of all observable properties is equivalent to knowing the basis states of the particle or system. Each physical "observable" corresponds to a Hermitian operator $\hat{O}_i$ which induces a preferred basis set for the respective Hilbert space V_i (i.e., the eigenvectors of the operator comprises the "preferred" basis set). The multiplicity of possible observables means that a single particle can "reside" in many Hilbert spaces at the same time since there can be a Hilbert space V_i for each operator $\hat{O}_i$. The particle can therefore reside in the direct product space (see Chapters 2 and 3) given by

$$V = V_1 \otimes V_2 \otimes \cdots$$

where V_1 might describe the energy, V_2 might describe the momentum and so on. The basis set for the direct product space consists of the combination of the basis vectors for the individual spaces such as

$$|\Psi\rangle = |\phi, \eta, \ldots\rangle = |\phi\rangle|\eta\rangle\ldots$$

where we assume, for example, that the space spanned by $\{|\phi\rangle\}$ refers to the energy content and $\{|\eta\rangle\}$ refers to momentum, etc.

The basis states can be most conveniently labeled by the eigenvalues of the commuting Hermitian operators. For example, $|E_i, p_j\rangle$ represents the state of the particle with energy E_i and momentum p_j assuming of course that the Hamiltonian and momentum commute. These two operators might represent all we care to know about the system.

5.7 Basic Operators of Quantum Mechanics

This section reviews the basic quantities in the quantum theory and useable forms of some observables such as energy and momentum. We develop the Schrodinger wave equation.

5.7.1 Summary of Elementary Facts

Electrons, holes, photons, and phonons can be pictured as particles or waves. Momentum and energy usually apply to particles while wavelength and frequency apply to waves. The momentum and energy relations provide a bridge between the two pictures

$$p = \hbar k \quad E = \hbar\omega \tag{5.7.1a}$$

where $\hbar = h/2\pi$ and "h" is Planck's constant. For both massive and massless particles, the wave vector and angular frequency can be written as

$$k = \frac{2\pi}{\lambda} \qquad \omega = 2\pi\nu \tag{5.7.1b}$$

where λ and ν represent the wavelength and frequency (Hz). For massive particles, the momentum $p = mv$ can be related to the wavelength by

$$\lambda = \frac{h}{mv}$$

for mass m and velocity v.

5.7.2 Operators, Eigenvectors and Eigenvalues

"Hermitian operators" $\hat{O}$ represent observables, which are physically measurable quantities such as the momentum of a particle, electric field, and position. If "ϕ" is an eigenvector (basis vector), then the eigenvector equation $\hat{O}\phi = o\phi$ provides the result of the observation where "o," a *real* constant, represents the results of a measurement. If for example, "$\hat{O}$" represents the momentum operator, then "o" must be the momentum of the particle when the particle occupies state "ϕ." We can write an eigenfunction equation for every observable. The result of every physical observation must always be an eigenvalue. Quantum mechanics does not allow us to simultaneously and precisely know the values of all observables.

5.7.3 The Momentum Operator

The mathematical theory of quantum mechanics admits many different forms for the operators. The "spatial-coordinate representation" relates the momentum to the spatial gradient. To find an operator representing the momentum, consider the plane wave $\Psi = Ae^{i\vec{k}\cdot\vec{r} - i\omega t}$. The gradient gives

$$\nabla\Psi = i\vec{k}\Psi = i\frac{\vec{P}}{\hbar}\Psi$$

where $\vec{P} = \hbar\vec{k}$ is the momentum defined at the beginning of this section. We assume this form holds for all eigenvectors of the momentum operator. Therefore, comparing both sides of the last equation, it seems reasonable to identify the momentum operator with the spatial derivative

$$\hat{P} = \frac{\hbar}{i}\nabla = \frac{\hbar}{i}\left[\hat{x}\frac{\partial}{\partial x} + \hat{y}\frac{\partial}{\partial y} + \hat{z}\frac{\partial}{\partial z}\right] \tag{5.7.2}$$

The momentum operator has both a vector and operator character. The operator character comes from the derivatives in the gradient and the vector character comes from the unit vectors appearing in the gradient. We identify the individual components of the momentum as

$$\hat{P}_x = \frac{\hbar}{i}\frac{\partial}{\partial x} \qquad \hat{P}_y = \frac{\hbar}{i}\frac{\partial}{\partial y} \qquad \hat{P}_z = \frac{\hbar}{i}\frac{\partial}{\partial z}$$

Sometimes it's more convenient to work with alternate notation

$$x_m = \begin{cases} x & m = 1 \\ y & m = 2 \\ z & m = 3 \end{cases} \qquad \hat{P}_m = \begin{cases} \hat{P}_x & m = 1 \\ \hat{P}_y & m = 2 \\ \hat{P}_z & m = 3 \end{cases}$$

The position and momentum do not commute.

$$\left[x_m, \hat{P}_n \right] = i\hbar \delta_{mn}$$

In general, conjugate variables (i.e., $m = n$) refer to the same degree of freedom and do not commute.

5.7.4 Developing the Hamiltonian Operator and the Schrodinger Wave Equation

We can observe the total energy of a particle or a system (the word system usually denotes a collection of particles—not necessarily all of the same type). We know that there exists a Hermitian operator $\mathcal{H}$ representing the total energy. Earlier sections in this book on classical mechanics develop the special mathematical properties of the classical Hamiltonian and associated Lagrangian. Quantum theory models the act of observing the energy of a particle by an eigenvalue equation

$$\mathcal{H}|\Phi\rangle = E|\Phi\rangle \quad \text{or} \quad \mathcal{H}\Phi = E\Phi \tag{5.7.3}$$

where $|\Phi\rangle$ is the wave function (more accurately, a basis function) for the particle. The eigenvector equation cannot easily be solved without more details on the form of the operator. In general, we need a wave equation in order to find the wave motion associated with the probability of the quantum particles.

We now determine another form for the energy operator using a plane wave representation for the wave function of a particle. Even though we use a specific wave function, we require the partial differential equation to hold in general, even for arbitrary wave functions. A plane wave traveling along the $+z$ direction with phase velocity $v = \omega/k$ has the form $\Phi = Ae^{ikz-i\omega t}$. Differentiating with respect to time and using $E = \hbar\omega$ gives us

$$\frac{\partial \Phi}{\partial t} = -i\omega\Phi = -i\frac{E}{\hbar}\Phi \;\rightarrow\; i\hbar\frac{\partial \Phi}{\partial t} = E\Phi \tag{5.7.4}$$

We assume Equation (5.7.4) holds for all vectors Ψ in the Hilbert space. Comparing Equations (5.7.4) and (5.7.3), we are encouraged to write

$$\mathcal{H}\Psi = i\hbar\frac{\partial \Psi}{\partial t} \tag{5.7.5}$$

The Schrodinger wave equation (SWE) in Equation (5.7.5) provides the dynamics for the motion of the quantum particles. The motion in the SWE can refer to a variety of motions including the motion of a particle through space or the evolution of the spin of a particle. Any wave function solving Equation (5.7.5) can be Fourier expanded in a basis set of plane waves. Equation (5.7.5) has only a first derivative in time contrary to the usual form of a classical wave equation (the wave equation for electromagnetics for example). The reason is that, for the probability interpretation of the wave function to hold, and for conservation of particle number (i.e., an equation of continuity for probability), the second derivative in time must be replaced by a first derivative and complex numbers must be introduced.

We must specify the exact form of the energy operator in terms of other quantities related to the energy of the system. For a single particle, we know that the total energy can be related to the kinetic and potential energy. We must keep in mind throughout this procedure that $\hat{\mathscr{H}}$ is an operator; any expression for $\hat{\mathscr{H}}$ must therefore contain operators.

The usual procedure for finding the quantum mechanical Hamiltonian starts by writing the classical Hamiltonian (i.e., energy) and then substituting operators for the dynamical variables (i.e., observables). The operators are then required to satisfy commutation relations which accounts for the fact that the corresponding observables might or might not be simultaneously observable (i.e., the Heisenberg uncertainty relations must be satisfied).

The classical Hamiltonian for a particle with potential energy $V(\vec{r})$ can be written as

$$\mathscr{H} = \text{ke} + \text{pe} = \frac{p^2}{2m} + V(\vec{r})$$

The quantum mechanical Hamiltonian can be found by replacing all dynamical variables, which consist of $\vec{r}$ and $\vec{p}$ in this case, with the equivalent operator. We will work in the spatial-coordinate representation so that we denote the position vector by $\vec{r}$ and we use Equation (5.7.2) for the momentum. The quantum mechanical Hamiltonian can be written as

$$\hat{\mathscr{H}} = \frac{\hat{P}^2}{2m} + V(\vec{r}) = \frac{1}{2m}\left(\frac{\hbar}{i}\nabla\right)\cdot\left(\frac{\hbar}{i}\nabla\right) + V(\vec{r}) = -\frac{\hbar^2}{2m}\nabla^2 + V(\vec{r})$$

If we cannot simultaneously and precisely measure both momentum and position in the Hamiltonian, how can the energy ever have an exact value? We resolve this apparent contradiction by noting that the Hamiltonian is well defined for an *energy* eigenfunction basis set even though momentum and position cannot be simultaneously exactly known.

As a note, the basis vectors by themselves do not solve the Schrodinger equation. Instead, the functions of the form $e^{Et/i\hbar}|E\rangle$ and their superposition do solve the Schrodinger equation.

5.7.5 Infinitely Deep Quantum Well

We solve Schrodinger's equation for an electron confined to an infinitely deep well of width L in free space. The particle requires an infinitely large amount of energy to escape from the well.

$$V(x) = \begin{cases} 0 & x \in (0, L) \\ \infty & \text{elsewhere} \end{cases}$$

The boundary value problem consists of a partial differential equation and boundary conditions

$$-\frac{\hbar^2}{2m}\frac{\partial^2}{\partial x^2}\Psi = i\hbar\frac{\partial\Psi}{\partial t} \quad \Psi(0,t) = \Psi(L,t) = 0 \tag{5.7.6}$$

where "*m*" is the mass of an electron. There should also be an initial condition (*IC*) for the time derivative; it should have the form $\Psi(x,0) = f(x)$. The initial condition specifies the initial probability for each of the basis states. We are most interested in the basis states for now.

We use the technique for the separation of variables. Set $\Psi(x,t) = X(x)T(t)$, substitute into the partial differential equation, then divide both sides by Ψ, and finally use *E* as the separation constant to obtain

$$\frac{1}{u}\left(-\frac{\hbar^2}{2m}\frac{\partial^2}{\partial x^2}X\right) = E = i\hbar\frac{1}{T}\frac{\partial T}{\partial t} \tag{5.7.7}$$

We now have two equations

$$-\frac{\hbar^2}{2m}\frac{\partial^2 X}{\partial x^2} = EX \quad i\hbar\frac{\partial T}{\partial t} = ET \tag{5.7.8}$$

The last equation provides

$$T(t) = \exp\left(\frac{E}{i\hbar}t\right) = \exp(-i\omega t) \tag{5.7.9}$$

where $E = \hbar\omega$. Separation of variables also provides boundary conditions for $X(x)$. We find

$$\Psi(0,t) = 0 = \Psi(L,t) \;\rightarrow\; X(0)T(t) = 0 = X(L)T(t) \;\rightarrow\; X(0) = 0 = X(L) \tag{5.7.10}$$

The first of Equations (5.7.8) along with boundary conditions from Equations (5.7.10) constitute the Sturm–Liouville problem that produces the basis set $\{X_n(x)\}$. We must solve an eigenvalue equation $\hat{H} X(x) = E X(x)$

$$-\frac{\hbar^2}{2m}\frac{\partial^2 X}{\partial x^2} = EX \quad X(0) = 0 = X(L) \tag{5.7.11}$$

Three ranges for the separation constant $E < 0$, $E = 0$, $E > 0$ must be considered because the sign of *E* determines the character of the solution. All cases must be considered because the solution wave function becomes a summation over all eigenfunctions with the eigenvalues as the index. The $E < 0$, $E = 0$ cases lead to trivial solutions and not eigenfunctions.

The case of $E > 0$ produces the only eigenfunctions $X(x)$ having the form

$$X(x) = A'e^{-ikx} + B'e^{ikx} = A\cos(kx) + B\sin(kx) \quad \text{with} \quad k = \sqrt{2mE/\hbar^2} \tag{5.7.12}$$

The equation for *k* comes from substituting *X* into Equation (5.7.11). We have 3 unknowns *A*, *B*, *k* and only two boundary conditions in Equations (5.7.6). Clearly, we will not find values for all three parameters. The boundary conditions lead to multiple discrete values for *k* and hence for the energy *E*.

Let us determine the parameters *A*, *B*, *k* as much as possible. The boundary conditions $X(0) = 0$ and $X(L) =$ requires $X(x) = B\sin(kx)$ and $\sin(kL) = 0$, respectively. The last one can only happen when $k = n\pi/L$ for $n = 1, 2, 3, \ldots$ and therefore the wavelength must be given by $\lambda = 2\pi/k = 2L/n$, which requires multiples of half wavelengths to fit in the width of the well. The functions $X_n(x) = B\sin(n\pi x/L)$ are the eigenfunctions. The basis set comes from normalizing the eigenfunctions. We require $\langle X_n \mid X_n \rangle = 1$ so that $B = \sqrt{2/L}$ and the basis set must be

$$\left\{ X_n(x) = \sqrt{\frac{2}{L}} \sin\left(\frac{n\pi}{L} x\right) \right\} \tag{5.7.13}$$

These are also called *stationary solutions* because they do not depend on time. Stationary solutions satisfy the *time-independent* Schrodinger wave equation $\hat{H} X_n(x) = E_n X_n(x)$.

   A solution of the partial differential equation corresponding to an allowed energy E_n must be

$$\Psi_n = X_n T_n = B \sin\left(\frac{n\pi}{L} x\right) e^{-itE_n/\hbar} \tag{5.7.14}$$

The allowed energies E_n can be found by combining $k = \sqrt{2mE/\hbar^2}$ and $k = n\pi/L$

$$E_n = \frac{\hbar^2 k_n^2}{2m} = \frac{\hbar^2 \pi^2}{2mL^2} n^2 \tag{5.7.15}$$

The full wave function must be a linear combination of these fundamental solutions

$$\Psi(x, t) = \sum_E \beta_E \Psi_E = \sum_n \beta_n(t) \, X_n(t) \tag{5.7.16}$$

which has the form of the summation over basis vectors with time dependent components. The components of the vector must be $\beta_n(t) = \beta_n(0) \, e^{-itE_n/\hbar}$ where $\beta_n(0)$ are constants.

Example 5.7.1

Suppose a student places an electron in the infinitely deep well at $t = 0$ according to the prescription

$$\Psi(x, 0) = \frac{1}{\sqrt{2}} X_1 + \frac{1}{\sqrt{2}} X_2 \quad \text{or} \quad |\Psi(0)\rangle = \frac{1}{\sqrt{2}} |1\rangle + \frac{1}{\sqrt{2}} |2\rangle \tag{5.7.17}$$

The function $\Psi(x, 0)$ provides the initial condition. Find the full wave function.

Solution: The full wave function appears in Equation (5.7.16)

$$|\Psi(t)\rangle = \sum_n \beta_n e^{-itE_n/\hbar} |n\rangle \tag{5.7.18}$$

We need the coefficients β_n which come from the wave function evaluated at some fixed time such as $t = 0$. The expansion coefficients must have the form

$$\beta_n = \langle n \mid \Psi(0)\rangle = \langle n| \left\{ \frac{1}{\sqrt{2}} |1\rangle + \frac{1}{\sqrt{2}} |2\rangle \right\} = \frac{1}{\sqrt{2}} \delta_{1n} + \frac{1}{\sqrt{2}} \delta_{2n}$$

and the full wave function becomes

$$\Psi(x,t) = \sum_n \beta_n \sqrt{\frac{2}{L}} \sin\left(\frac{n\pi}{L}x\right) e^{-itE_n/\hbar} = \sum_n \left[\frac{1}{\sqrt{2}}\delta_{1n} + \frac{1}{\sqrt{2}}\delta_{2n}\right]\sqrt{\frac{2}{L}}\sin\left(\frac{n\pi}{L}x\right)e^{-itE_n/\hbar}$$

which reduces to

$$\Psi(x,t) = \frac{1}{\sqrt{L}}\sin\left(\frac{\pi}{L}x\right)e^{-itE_1/\hbar} + \frac{1}{\sqrt{L}}\sin\left(\frac{2\pi}{L}x\right)e^{-itE_2/\hbar} = \frac{1}{\sqrt{2}}X_1 e^{-itE_1/\hbar} + \frac{1}{\sqrt{2}}X_2 e^{-itE_2/\hbar}$$

where Equation 5.7.15 gives

$$E_n = \frac{\hbar^2 k_n^2}{2m} = \frac{\hbar^2 \pi^2}{2mL^2}n^2$$

5.8 The Harmonic Oscillator

The Schrodinger wave equation (SWE) describes the time evolution of the wave function. The Hamiltonian for the harmonic oscillator describes a particle of mass m in a quadratic potential. Displacing the mass from equilibrium produces a linear restoring force. We focus on the 1-D oscillator since a 3-D oscillator can be decomposed into three 1-D oscillators. Any coupling between the three 1-D oscillators can be included in the Hamiltonian later if desired.

The harmonic oscillator has important applications. Many systems have nonlinear potential functions. Expanding these nonlinear potentials in a Taylor series often produces a quadratic term as the lowest order approximation. As an example, the periodic motion of atoms about their equilibrium position can be modeled with the quadratic potential. We know this motion must be related to phonons moving through the material. We will see that the zero point motion of the atom can be described by the quantum mechanical vacuum state.

Quantum optics provides a somewhat surprising application for the quadratic potential. The electromagnetic fields can be modeled by quadratic kinetic and potential terms. Of course, these do not refer to an electron in an electrostatic potential. Nor do they refer to the position or momemtum of photons. Instead, they refer to the form assumed by the fields in the Hamiltonian. The quantized form of the electromagnetic fields can be immediately written by comparison with the wave functions for the electron in the quadratic potential.

5.8.1 Introduction to the Classical and Quantum Harmonic Oscillators

For a harmonic oscillator, the quadratic potential (Figure 5.8.1) produces a linear restoring force

$$V = \begin{cases} \dfrac{1}{2}kx^2 & \text{1-D} \\[2ex] \dfrac{1}{2}kr^2 & \text{3-D} \end{cases}$$

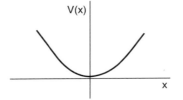

FIGURE 5.8.1
The quadratic potential.

where $r^2 = x^2 + y^2 + z^2$, and the equilibrium position occurs at the origin $x = 0$, the "spring constant" must be positive $k > 0$ and it describes the curvature of the potential (i.e., magnitude of the force).

The classical Hamiltonian has the form

$$H_c = \frac{p^2}{2m} + \frac{1}{2}kx^2 \tag{5.8.1}$$

where we consider the dynamic variables x, p to be independent of one another. Newton's second law can be demonstrated using Hamilton's canonical equation (refer to Section 5.2).

$$\dot{p} = -\frac{\partial H_c}{\partial x} = -kx = F$$

The Lagrangian shows that the momentum p must be related to the velocity by $p = mv = m\dot{x}$.

We want to compare and contrast solutions $x(t)$ to the classical and quantum harmonic oscillators. The classical Hamiltonian (the total energy) can be rewritten using Equation (5.8.1) and $p = m\dot{x}$

$$\frac{m}{2}\left(\frac{d\,x(t)}{dt}\right)^2 + \frac{1}{2}k\,(x(t))^2 = E \tag{5.8.2}$$

where E represents the total energy of the oscillator and $x(t)$ represents the position of the electron parameterized by the time t. The solution has the form

$$x(t) = A\sin(\omega_o t) \tag{5.8.3a}$$

The formula $\omega_o^2 = k/m$ relates the angular frequency of oscillation ω_o to the "spring constant" k. Substituting Equation (5.8.3a) into Equation (5.8.2) provides

$$A = \sqrt{\frac{2E}{k}} = \sqrt{\frac{2E}{m\omega_o^2}} \tag{5.8.3b}$$

The amplitude A represents the points on the potential plot $V(x)$ where the kinetic energy becomes zero (see Figure 5.8.2)

$$E = \frac{1}{2}kx^2\bigg|_{x=A} \quad \rightarrow \quad A = \sqrt{\frac{2E}{k}}$$

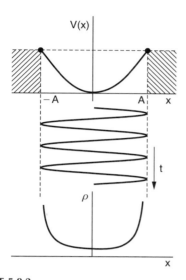

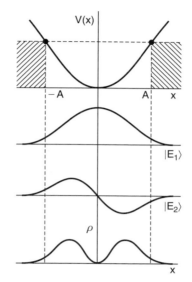

FIGURE 5.8.2

Motion of a harmonic oscillator. The probability density shows the most likely position of finding the mass m is at the turning points where the oscillator momentarily comes to rest.

FIGURE 5.8.3

The first two quantum mechanical solutions to the harmonic oscillator. The probability density for finding the particle at point x does not resemble the classical one.

Classically, the particle can only be found in the region $x \in [-A, A]$ and never outside that region. The probability density ρ for finding the particle at a point x appears similar to a delta function near the endpoints of the motion; this behavior occurs because the particle slows down near those points and spends more time there.

Several differences exist between the classical and quantum mechanical harmonic oscillators. Figure 5.8.3 shows the quantum mechanical solution to Schrodinger's equation with the quadratic potential. Unlike the classical particle, the quantum particle can be found in the classically forbidden region. The figure shows how the wave function exponentially decays in these classically forbidden regions. Classically, the particle doesn't have enough energy to enter the forbidden region. The basis functions have the form

$$\phi_n(x) = \left(\frac{\alpha}{\pi^{1/2} n! \, 2^n}\right)^{\frac{1}{2}} H_n(\alpha x) \exp\left(-\frac{\alpha^2 x^2}{2}\right) \tag{5.8.4}$$

where $\alpha^4 = (m\omega_o/\hbar)^2$. The exponential part of the solution ensures the wave function decreases in the classically forbidden region. The Hermite polynomials H_n primarily control the behavior in the classically allowed region near the center. They can be conveniently generated by differentiating an exponential according to

$$H_n(\xi) = (-1)^n \exp(\xi^2) \frac{d^n}{d\xi^n} \exp(-\xi^2) \tag{5.8.5a}$$

where $\xi = \alpha x$. The first three Hermite polynomials are

$$H_o(\xi) = 1, \quad H_1(\xi) = 2\xi, \quad H_2(\xi) = 4\xi^2 - 2 \tag{5.8.5b}$$

Continuing with Figure 5.8.3, perhaps most striking of all, the probability density function for the quantum particle decays to zero *near* the endpoints of motion and reaches its peak value (or values) near the center of the classical region $[-A, A]$. However, the classical probability of finding the classical particle assumes its minimum value near the origin.

Here's another difference between the classical and harmonic oscillator solutions. The classical oscillator energy can be increased by applying a driving force and increasing the oscillation amplitude $E = A^2 m\omega_o^2/2$. The angular oscillation frequency $\omega_o = \sqrt{k/m}$ remains constant for a fixed spring constant k. The energy of the quantum oscillator increases by also absorbing energy

$$E_n = \hbar\omega_n = \hbar\omega_o\left(n + \frac{1}{2}\right) \quad n = 0, 1, 2, \ldots \tag{5.8.6}$$

The integer "n" can be interpreted as either the "basis function number" or as the number of quanta stored in the motion. Contrary to the classical case, the angular frequency $\omega_n = w_o(n + \frac{1}{2})$ of the quantum oscillator changes even though the value ω_o remains fixed. The angular frequency does not refer to the rate at which the quantum particle bounces from side to side. We view the quantum particle as a stationary wave function. Larger numbers of quanta "n" result in larger "displacements" from equilibrium meaning the probability density has more peaks that move closer to the classically forbidden region.

We find similar plots for quantized EM waves. The energy of an EM oscillator (the EM waves) can be changed by changing the angular frequency (or wavelength) or by changing the amplitude (i.e., the number of quanta in the mode). We will see that the "position x" and "momentum p" become the "in-phase" and "out-of-phase" electric fields. Therefore, the wave functions in the EM case describe the probability of finding a particular value of the electric field.

5.8.2 The Hamiltonian for the Quantum Harmonic Oscillator

The quantum mechanical Hamiltonians come from the classical ones by replacing the dynamical variables x, p with the corresponding operators $\hat{x}, \hat{p}$ in $H_c = p^2/2m + kx^2/2$ to find

$$\hat{\mathscr{H}}\,|\Psi(t)\rangle = i\hbar\frac{\partial}{\partial t}\,|\Psi(t)\rangle \qquad \left(\frac{\hat{p}^2}{2m} + \frac{1}{2}k\hat{x}^2\right)|\Psi(t)\rangle = i\hbar\frac{\partial}{\partial t}\,|\Psi(t)\rangle \tag{5.8.7}$$

Operating with the "coordinate" projection operator $\langle x|$ produces $\hat{x} \to x$ and $\hat{p} \to (\hbar/i)(\partial/\partial x)$ (refer to Appendix 6) to obtain the Schrodinger equation

$$\hat{\mathscr{H}}\,\Psi(x, t) = i\hbar\frac{\partial\Psi(x, t)}{\partial t} \quad \text{or} \quad \left(\frac{-\hbar^2}{2m}\frac{\partial^2}{\partial x^2} + \frac{1}{2}kx^2\right)\Psi(x, t) = i\hbar\frac{\partial}{\partial t}\Psi(x, t) \tag{5.8.8}$$

The boundary conditions for the Schrodinger wave equation for the harmonic oscillator require the wave function to approach zero as "x" goes to infinity

$$\Psi(x \to \pm\infty, t) \to 0 \tag{5.8.9}$$

There are two methods for solving the Schrodinger equation for the harmonic oscillator. The first method uses a power series solution, which becomes very algebraically involved. The solution starts by separating variables in Equation (5.8.8) and using a power series to find the solutions to the Sturm–Liouville problem (the eigenvector problem). The second method uses the linear algebra of raising and lowering operators. We present the method of raising and lowering operators (commonly referred to as the algebraic approach). We will find the stationary solutions given in Equation (5.8.4) and the energy eigenvalues in Equation (5.8.6).

5.8.3 Introduction to the Operator Solution of the Harmonic Oscillator

The operator approach (i.e., algebraic approach) to solving Schrodinger's equation for the harmonic oscillator is simpler than the power series approach. In addition, it provides a great deal of insight into the mathematical structure of the quantum theory. The algebraic approach uses "raising $\hat{a}^+$ and lowering $\hat{a}$ operators" (i.e., ladder operators, or sometimes called promotion and demotion operators). Later chapters demonstrate the similarity between the ladder operators and the "creation/annihilation" operators most commonly found in advanced studies of quantum theory.

We will rewrite the Hamiltonian in terms of the raising and lowering operators in the form of the number operator $\hat{N} = \hat{a}^+ \hat{a}$. The raising and lowering operators map one basis vector into another one according to

$$\hat{a}^+|n\rangle = \sqrt{n+1}|n+1\rangle \quad \hat{a}|n\rangle = \sqrt{n}|n-1\rangle \tag{5.8.10}$$

as suggested by Figure 5.8.4. The lowering operator produces zero when operating on the vacuum state $\hat{a}|0\rangle = 0$. The number operator has two interpretations for the harmonic oscillator. First we will show the energy eigenvectors are also eigenvectors for the number operator according to $\hat{N}|n\rangle = n|n\rangle$. The number operator therefore tells us the number of the eigenstate occupied by a particle.

The number operator also tells us the number of energy quanta in the system as its second interpretation. We can say that a particle occupying one of the energy basis states $|n\rangle \in B_V = \{|0\rangle = |E_0\rangle, \; |1\rangle = |E_1\rangle, \; \ldots\}$ has n quanta of energy according to $E_n = \hbar\omega_o$ $(n + 1/2)$. Therefore the vacuum state $|0\rangle$ corresponds to a particle state without any quanta of energy $n = 0$. Interestingly, there exists energy in the vacuum state $E_0 = \hbar\omega_o/2$. Atoms executing zero-point motion (i.e., $T = 0$ K) in a solid, for example, are exhibiting vacuum energy. The atoms continue to move even though all of the extractable energy has been removed (i.e., $n = 0$). Absolute zero can never be achieved since it is a classical concept corresponding to stationary atoms. Studies in quantum optics indicate that the electric field also experiences vacuum fluctuations; these fluctuations produce spontaneous emission from an ensemble of excited atoms.

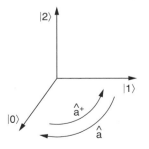

FIGURE 5.8.4
Raising and lowering operators move the harmonic oscillator from one state to another.

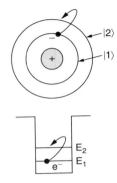

FIGURE 5.8.5
Physical examples showing the effect of a raising operators defined for an atom (top) and square well (bottom) rather than for the harmonic oscillator.

In the next few topics, we wish to find the energy eigenvectors $B_V = \{|0\rangle = |E_0\rangle, |1\rangle = |E_1\rangle, \ldots\}$ and eigenvalues for the harmonic oscillator. We assume non-degenerate eigenvalues E_n, which means that for each energy E_n, there corresponds exactly one eigenstate ϕ_n satisfying $\hat{H}|\phi_n\rangle = E_n|\phi_n\rangle$. We further assume an order for the energy levels $E_0 < E_1 < E_2 < \cdots$. The operator approach must reproduce the results found with the power series approach.

We first show how the Hamiltonian incorporates the raising–lowering operators (see, for example, Figure 5.8.5). We briefly discuss the mathematical description of the ladder operators and demonstrate the origin of their normalization constant. We then easily solve for the energy eigenvalues and eigenvectors.

5.8.4 Ladder Operators in the Hamiltonian

The Hamiltonian for the harmonic oscillator is

$$\hat{H} = \frac{\hat{p}^2}{2m} + \frac{m\omega_0\hat{x}^2}{2} \tag{5.8.11}$$

We define the lowering $\hat{a}$ and the raising $\hat{a}^+$ operators in terms of the position $\hat{x}$ and momentum operators $\hat{p}$.

$$\hat{a} = \frac{m\omega_0}{\sqrt{2m\hbar\omega_0}}\hat{x} + \frac{i\hat{p}}{\sqrt{2m\hbar\omega_0}} \tag{5.8.12a}$$

$$\hat{a}^+ = \frac{m\omega_0}{\sqrt{2m\hbar\omega_0}}\hat{x} - \frac{i\hat{p}}{\sqrt{2m\hbar\omega_0}} \tag{5.8.12b}$$

The raising operator in Equation (5.8.12b) comes from taking the adjoint of the lowering operator in Equation (5.8.12a) and using the fact that both $\hat{x}, \hat{p}$ must be Hermitian since they correspond to observables. Notice that the raising and lowering operators are not Hermitian $\hat{a} \neq \hat{a}^+$. These two equations for the lowering and raising operators can be solved for the position and momentum operators to find

$$\hat{x} = \sqrt{\frac{\hbar}{2m\omega_0}}(\hat{a} + \hat{a}^+) \quad \hat{p} = -i\sqrt{\frac{m\omega_0\hbar}{2}}(\hat{a} - \hat{a}^+) \tag{5.8.13}$$

We need the Hamiltonian written in terms of the ladder operators. We must first determine the commutation relations. We can demonstrate that the raising operator commutes with itself as does the lowering operator while the raising operator does not commute with the lowering operator

$$[\hat{a}, \hat{a}] = 0 = [\hat{a}^+, \hat{a}^+] \quad [\hat{a}, \hat{a}^+] = 1 \tag{5.8.14}$$

These last two relations can be proven using the commutation relations between the position and momentum operators

$$[\hat{x}, \hat{x}] = 0 = [\hat{p}, \hat{p}] \quad [\hat{x}, \hat{p}] = i\hbar \tag{5.8.15}$$

We prove $[\hat{a}, \hat{a}^+] = 1$ by first substituting Equations (5.8.12).

$$[\hat{a}, \hat{a}^+] = \left[\frac{m\omega_o}{\sqrt{2m\hbar\omega_o}}\hat{x} + \frac{i\hat{p}}{\sqrt{2m\hbar\omega_o}}, \frac{m\omega_o}{\sqrt{2m\hbar\omega_o}}\hat{x} - \frac{i\hat{p}}{\sqrt{2m\hbar\omega_o}} \right]$$

Distributing the terms provides

$$[\hat{a}, \hat{a}^+] = \left(\frac{m\omega_o}{\sqrt{2m\hbar\omega_o}} \right)^2 [\hat{x}, \hat{x}] + \frac{[\hat{p}, \hat{p}]}{2m\hbar\omega_o} + \frac{i}{2\hbar}[\hat{p}, \hat{x}] - \frac{i}{2\hbar}[\hat{x}, \hat{p}]$$

Substituting the commutation relations from Equation (5.8.15), we find the desired results

$$[\hat{a}, \hat{a}^+] = 0 + 0 + \frac{i}{2\hbar}(-i\hbar) - \frac{i}{2\hbar}(i\hbar) = 1$$

In the case of an ensemble of independent harmonic oscillators, each one has its own degrees of freedom $\hat{x}_i, \hat{p}_i$ that obey their own commutation relations.

$$[\hat{x}_i, \hat{x}_j] = 0 = [\hat{p}_i, \hat{p}_j] \quad [\hat{x}_i, \hat{p}_j] = i\hbar\delta_{ij}$$

As a result, there will be raising and lowering operators for each oscillator

$$[\hat{a}_i, \hat{a}_j] = 0 = [\hat{a}_i^+, \hat{a}_j^+] \quad [\hat{a}_i, \hat{a}_j^+] = \delta_{ij}$$

Using the definitions of the position and momentum operators, the Hamiltonian for the single harmonic oscillator can be rewritten by substituting relations (5.8.27).

$$\hat{\mathcal{H}} = \frac{\hat{p}^2}{2m} + \frac{1}{2}m\omega_o^2\hat{x}^2 = \frac{1}{2m}\left[-i\sqrt{\frac{m\omega_o\hbar}{2}}(\hat{a} - \hat{a}^+) \right]^2 + \frac{1}{2}m\omega_o^2\left[\sqrt{\frac{\hbar}{2m\omega_o}}(\hat{a} + \hat{a}^+) \right]^2 \tag{5.8.16a}$$

Squaring the constants provides

$$\hat{\mathcal{H}} = -\frac{\hbar\omega_o}{4}(\hat{a} - \hat{a}^+)^2 + \frac{\hbar\omega_o}{4}(\hat{a} + \hat{a}^+)^2$$

Squaring the operators and taking care not to commute them gives us

$$\hat{\mathscr{H}} = \frac{\hbar\omega_o}{4}\left\{-\hat{a}^2 + \hat{a}\hat{a}^+ + \hat{a}^+\hat{a} - \hat{a}^{+2} + \hat{a}^2 + \hat{a}\hat{a}^+ + \hat{a}^+\hat{a} + \hat{a}^{+2}\right\}$$

Combining the squared terms

$$\hat{\mathscr{H}} = \frac{\hbar\omega_o}{2}\left\{\hat{a}\hat{a}^+ + \hat{a}^+\hat{a}\right\} \tag{5.8.16b}$$

We must always use commutation relation to change the order of operators. Finally, by using the commutation relation $[\hat{a}, \hat{a}^+] = 1 \rightarrow \hat{a}\hat{a}^+ = 1 + \hat{a}^+\hat{a}$, the Hamiltonian becomes

$$\hat{\mathscr{H}} = \frac{\hbar\omega_o}{2}\left\{\hat{a}\hat{a}^+ + \hat{a}^+\hat{a}\right\} = \frac{\hbar\omega_o}{2}\left\{2\hat{a}^+\hat{a} + 1\right\}$$

As a result, the Hamiltonian for the single harmonic oscillator can be written as

$$\hat{\mathscr{H}} = \hbar\omega_o\left(\hat{a}^+\hat{a} + \frac{1}{2}\right) \tag{5.8.17a}$$

We can define the number operator $\hat{N} = \hat{a}^+\hat{a}$ and rewrite Equation (5.8.17a) as

$$\hat{\mathscr{H}} = \hbar\omega_o\left(\hat{N} + \frac{1}{2}\right) \tag{5.8.17b}$$

5.8.5 Properties of the Raising and Lowering Operators

Next, we demonstrate the relations

$$\hat{a}^+|n\rangle = \sqrt{n+1}|n+1\rangle \qquad \hat{a}|n\rangle = \sqrt{n}\,|n-1\rangle$$

by first showing $|n-1\rangle \sim \hat{a}|n\rangle$ and $|n+1\rangle \sim \hat{a}^+|n\rangle$ are eigenvectors of the number operator $\hat{N}$ corresponding to the eigenvalues $n-1$ and $n+1$, respectively. We next find the constants of proportionality. We will need two commutation relations. Using $[\hat{A}\hat{B}, \hat{C}] = \hat{A}[\hat{B}, \hat{C}] + [\hat{A}, \hat{C}]\hat{B}$ and $[\hat{A}, \hat{B}] = -[\hat{B}, \hat{A}]$ and Equation (5.8.14), we find

$$\left[\hat{N}, \hat{a}\right] = [\hat{a}^+\hat{a}, \hat{a}] = [\hat{a}^+, \hat{a}]\hat{a} = -\hat{a} \qquad \left[\hat{N}, \hat{a}^+\right] = [\hat{a}^+\hat{a}, \hat{a}^+] = \hat{a}^+[\hat{a}, \hat{a}^+] = \hat{a}^+ \tag{5.8.18}$$

We now show $\hat{N} = \hat{a}^+\hat{a}$ and $\hat{\mathscr{H}} = \hbar\omega_o(\hat{N} + 1/2)$ have eigenvectors $|n-1\rangle \sim \hat{a}|n\rangle$ and $|n+1\rangle \sim \hat{a}^+|n\rangle$. Suppose $|n\rangle$ represents one eigenvector then

$$\hat{N}[\hat{a}|n\rangle] = \hat{N}\hat{a}|n\rangle = \left\{[N, \hat{a}] + \hat{a}\hat{N}\right\}|n\rangle = \left\{-\hat{a} + \hat{a}\hat{N}\right\}|n\rangle = \left\{-\hat{a} + \hat{a}\,n\right\}|n\rangle = (n-1)[\hat{a}|n\rangle]$$

Therefore $\hat{a}|n\rangle$ must be an eigenvector of $\hat{N}$ with eigenvalue $(n-1)$. We can similarly show that $\hat{N}[\hat{a}^+|n\rangle] = (n+1)[\hat{a}^+|n\rangle]$ (see the chapter review exercises). Therefore, we conclude $\hat{a}^+|n\rangle = C_n|n+1\rangle$ and $\hat{a}\,|n\rangle = D_n|n-1\rangle$ since the eigenvalues are not degenerate where C_n and D_n denote constants of proportionality.

The eigenvalues of $\hat{N} = \hat{a}^+\hat{a}$ and $\hat{\mathscr{H}} = \hbar\omega_o(\hat{N} + \frac{1}{2})$ must be real because $\hat{N} = \hat{a}^+\hat{a}$ is Hermitian according to $\hat{N}^+ = (\hat{a}^+\hat{a})^+ = \hat{a}^+\hat{a} = \hat{N}$. Further the eigenvalues n must be greater than or equal to zero since the length of a vector must always be positive $n = \langle n|\hat{N}|n\rangle = \langle n|\hat{a}^+\hat{a}|n\rangle = \|\hat{a}|n\rangle\|^2 \geq 0$. We can also show that only integers represent the eigenvalues n.

Next, we find the normalization constants C_n and D_n occurring in the relations.

$$\hat{a}^+|n\rangle = C_n|n+1\rangle \qquad \hat{a}|n\rangle = D_n|n-1\rangle$$

Let's work with the lowering operator. To find D_n, consider the string of equalities

$$D_n^*D_n\langle n-1 \mid n-1\rangle = [D_n|n-1\rangle]^+[D_n|n-1\rangle] = \big[\hat{a}|n\rangle\big]^+\big[\hat{a}|n\rangle\big]$$

$$= \langle n|a^+a|n\rangle = \langle n|\hat{N}|n\rangle = \langle n|n|n\rangle = n\langle n \mid n\rangle$$

Now use the fact that all eigenvectors are normalized to one so that

$$\langle n-1 \mid n-1\rangle = 1 = \langle n \mid n\rangle$$

Therefore, the coefficient D_n must be

$$|D_n|^2 = n \;\rightarrow\; D_n = \sqrt{n}$$

where a phase factor has been ignored. Similarly, an expression for C_n can be developed

$$C_n^*C_n\langle n+1 \mid n+1\rangle = [C_n|n+1\rangle]^+[C_n|n+1\rangle] = \big[\hat{a}^+|n\rangle\big]^+\big[\hat{a}^+|n\rangle\big]$$

$$= \langle n|a\, a^+|n\rangle = \langle n|\, a^+a+1|n\rangle = \langle n|\hat{N}+1|n\rangle = \langle n|n+1|n\rangle = (n+1)\langle n \mid n\rangle$$

where a commutator has been used in the fifth term. Once again using the eigenvector normalization conditions and comparing both sides of the last equation

$$|C_n|^2 = n+1 \;\rightarrow\; C_n = \sqrt{n+1}$$

as expected. We therefore have the required relations.

$$\hat{a}^+|n\rangle = \sqrt{n+1}|n+1\rangle \qquad \hat{a}\,|n\rangle = \sqrt{n}|n-1\rangle \tag{5.8.19}$$

The set of eigenvectors $|0\rangle, |1\rangle, \ldots$ can be obtained by repeatedly using the relation $\hat{a}^+|n\rangle = \sqrt{n+1}|n+1\rangle$ as

$$|1\rangle = \frac{\hat{a}^+}{\sqrt{1}}|0\rangle, \;\; |2\rangle = \frac{\hat{a}^+}{\sqrt{2}}|1\rangle = \frac{(\hat{a}^+)^2}{\sqrt{2}\sqrt{1}}|0\rangle\, , \;\ldots, \; |n\rangle = \frac{(\hat{a}^+)^n}{\sqrt{n!}}|0\rangle, \;\ldots \tag{5.8.20}$$

Some Commutation Relations

1. $[\hat{\mathscr{H}}, a] = \hbar\omega_o[\hat{a}^+\hat{a}, \hat{a}] = \hbar\omega_o\hat{a}^+[\hat{a}, \hat{a}] + \hbar\omega_o[\hat{a}^+, \hat{a}]\hat{a} = -\hbar\omega_o\hat{a}$
2. $[\hat{\mathscr{H}}, \hat{a}^+] = \hbar\omega_o\hat{a}^+[\hat{a}, \hat{a}^+] = \hbar\omega_o\hat{a}^+$
3. $[\hat{N}, \hat{a}] = -\hat{a}[\hat{N}, \hat{a}^+] = \hat{a}^+$

5.8.6 The Energy Eigenvalues

The Hamiltonian for the harmonic oscillator can be written in terms of the ladder operators as given in Section 5.8.4, Equation (5.8.17b).

$$\hat{\mathscr{H}} = \hbar\omega_o \left(\hat{N} + \frac{1}{2} \right) \tag{5.8.21}$$

We already know the eigenvalues of the number operator to be $\hat{N}|n\rangle = n|n\rangle$. The allowed energy values can be found as follows

$$\hat{\mathscr{H}}|n\rangle = \hbar\omega_o \left(\hat{N} + \frac{1}{2} \right)|n\rangle = \hbar\omega_o \left(n + \frac{1}{2} \right)|n\rangle$$

Therefore the energy values must be

$$E_n = \hbar\omega_o \left(n + \frac{1}{2} \right) \tag{5.8.22}$$

5.8.7 The Energy Eigenfunctions

We know the energy eigenvectors can be listed in the sequence

$$|1\rangle = \frac{\hat{a}^+}{\sqrt{1}}|0\rangle, \ |2\rangle = \frac{\hat{a}^+}{\sqrt{2}}|1\rangle = \frac{(\hat{a}^+)^2}{\sqrt{2}\sqrt{1}}|0\rangle, \ \ldots, \ |n\rangle = \frac{(\hat{a}^+)^n}{\sqrt{n!}}|0\rangle, \ \ldots \tag{5.8.23}$$

from Equation (5.8.20). However, we would like to know the functional form of these functions. There exists a simple method for finding the energy eigenfunctions for the harmonic oscillator using the ladder operators. Starting with

$$0 = \hat{a}|0\rangle$$

operate on both sides using the bra operator $\langle x|$ and insert the definition for the lowering operator

$$0 = \langle x|\hat{a}|0\rangle = \langle x| \frac{m\omega_o \hat{x}}{\sqrt{2\hbar m\omega_o}} + \frac{i\hat{p}}{\sqrt{2\hbar m\omega_o}}|0\rangle = \langle x| \frac{m\omega_o \hat{x}}{\sqrt{2\hbar m\omega_o}}|0\rangle + \langle x| \frac{i\hat{p}}{\sqrt{2\hbar m\omega_o}}|0\rangle \tag{5.8.24}$$

Factor out the constants from the brackets and use the relations

$$\langle x|\hat{x}|0\rangle = x\langle x|0\rangle = x\,\phi_0(x) \quad \langle x|\hat{p}|0\rangle = \frac{\hbar}{i}\frac{\partial}{\partial x}\langle x|0\rangle = \frac{\hbar}{i}\frac{\partial}{\partial x}\phi_0(x)$$

where $\langle x \mid 0 \rangle = \phi_0(x)$ is the first energy eigenfunction in the set of eigenfunctions given by

$$\left\{ \phi_0(x), \phi_1(x), \ldots \right\}$$

Equation (5.8.24) now provides

$$0 = \langle x|a|0\rangle = \frac{m\omega_o x}{\sqrt{2\hbar m\omega_o}}\phi_0(x) + \frac{\hbar}{\sqrt{2\hbar m\omega_o}}\frac{\partial}{\partial x}\phi_0(x)$$

which is a simple first-order differential equation

$$\frac{d\phi_0}{dx} + \frac{m\omega_o}{\hbar}x\phi_o = 0$$

We can easily find the solution

$$\phi_0(x) = \phi_0(0)\,\exp\left(-\frac{m\omega_o}{2\hbar}x^2\right)$$

which represents the first energy eigenfunction. The normalization constant $\phi_0(0)$ is found by requiring the wave function to have unit length $1 = \langle\phi_0(x)\,|\,.\,\phi_0(x)\rangle$ which gives

$$\phi_0(x) = \left(\frac{m\omega_o}{\pi\hbar}\right)^{1/4}\exp\left(-\frac{m\omega_o}{2\hbar}x^2\right)$$

Now the other eigenfunctions can be found from $\phi_1(x)$ using the raising operator

$$\phi_1(x) = \langle x\,|\,1\rangle = \langle x|\frac{a^+}{\sqrt{1}}|0\rangle = \langle x|\frac{m\omega_o\hat{x}}{\sqrt{2\hbar m\omega_o}} - \frac{i\hat{p}}{\sqrt{2\hbar m\omega_o}}|0\rangle$$

where the constants can be factored out and the coordinate representation can be substituted for the operators to get

$$\phi_1(x) = \frac{m\omega_o x}{\sqrt{2\hbar m\omega_o}}\langle x|0\rangle - \frac{\hbar}{\sqrt{2\hbar m\omega_o}}\frac{\partial}{\partial x}\langle x|0\rangle = \frac{m\omega_o x}{\sqrt{2\hbar m\omega_o}}\phi_0(x) - \frac{\hbar}{\sqrt{2\hbar m\omega_o}}\frac{\partial}{\partial x}\phi_0(x)$$

Notice that we do not need to solve a differential equation to find the eigenfunctions $\phi_1, \phi_2, \ldots$ in the basis set. Differentiating $\phi_0(x)$ provides

$$\frac{\partial\phi_o}{\partial x} = \frac{\partial}{\partial x}\left(\frac{m\omega_o}{\pi\hbar}\right)^{1/4}\exp\left(-\frac{m\omega_o}{2\hbar}x^2\right) = -\frac{m\omega_o x}{\hbar}\phi_0(x)$$

Consequently the $n=1$ energy eigenfunction becomes

$$\phi_1(x) = \frac{m\omega_o x}{\sqrt{2\hbar m\omega_o}}\phi_0(x) + \frac{\hbar}{\sqrt{2\hbar m\omega_o}}\frac{m\omega_o x}{\hbar}\phi_0(x) = \frac{2\sqrt{m\omega_o}}{\sqrt{2\hbar}}x\,\phi_0(x)$$

$$= \frac{2\sqrt{m\omega_o}}{\sqrt{2\hbar}}\left(\frac{m\omega_o}{\pi\hbar}\right)^{1/4}x\,\exp\left(-\frac{m\omega_o x^2}{2\hbar}\right)$$

The $n=2$ energy eigenfunctions can be found repeating the procedure using

$$\phi_2(x) = \frac{\hat{a}^+}{\sqrt{2}}\phi_1(x)$$

Notice that the above procedure only requires the relation between the ladder operators and the momentum/position operators. At this point, the energy eigenvalues can be found using the time independent Schrodinger equation.

Special Integrals

The raising and lowering operators can be used to show the following integrals.

1. $\int_{-\infty}^{\infty} dx\, \phi_n(x) \frac{d}{dx} \phi_m(x) = \alpha[\delta_{m,n+1}\sqrt{(n+1/2)} - \delta_{m,n-1}\sqrt{n/2}]$ since $(d/dx) = (i/\hbar)\hat{p}$

2. $\int_{-\infty}^{\infty} dx\, \phi_n(x)\, x\, \phi_m(x) = \sqrt{(\hbar/2m\omega_o)}\left[\delta_{m,n+1}\sqrt{n+1} + \delta_{m,n-1}\sqrt{n}\right]$ where Problem 5.24 was combined with integral (1) above and with $E_{n+1} - E_n = \hbar\omega_o$

3. $\int_{-\infty}^{\infty} dx\, \phi_n(x)\, x^2\, \phi_m(x) = \delta_{m,n}(n+1/2\alpha^2) + \delta_{m,n\pm2}\sqrt{((n+1)(n+2)/2\alpha^2)}$. The closure relation can be used to prove this last one.

5.9 Quantum Mechanical Representations

A representation (or picture) in quantum theory refers to the manner in which the theory models the time evolution of fundamental dynamical quantities. For example, we might be interested in the time-dependence of energy, momentum, position, or angular momentum. In classical mechanics, these dynamical variables evolve in time according to an equation of motion such as Newton's equation. The values of the dynamical variables describe the state of the classical system. Usually the word "dynamics" refers to the motion of an object. In quantum theory, the dynamical variables correspond to Hermitian operators. However the quantum theory handles the time dependence in at least three ways. The Heisenberg picture assigns the time dependence to the Hermitian operators; this representation most closely mimics the classical approach. The Schrodinger picture assigns the time dependence to the wave functions; this representation most resembles that for classical wave motion. The interaction picture combines the best of both representations; the wave functions move in Hilbert space only due to driving forces not included in the Hamiltonian.

This section discusses three representations used in quantum theory in the first topic. The remainder of the section explores the mathematical descriptions of the particle as a result of using a particular representation.

5.9.1 Discussion of the Schrodinger, Heisenberg and Interaction Representations

Quantum theory generally employs the Schrodinger, Heisenberg, and interaction representations. For the Schrodinger representation, the wave functions (i.e., vectors in Hilbert space) carry the dynamics of the particle or system (not the basis states though). The states depend on time according to

$$|\Psi(t)\rangle = \sum_E \beta_E(t)|\phi_E\rangle = \sum_E \beta_E(t)|E\rangle$$

The wave function resides in a Hilbert space defined by the basis vectors. This wave function moves in the space and its components therefore change with time. The wave functions in optics (i.e., the electric field) most closely resemble those for the quantum mechanical Schrodinger representation. In optics we know the energy density and power flow (etc.) once we know the motion of the electric field. As part of the definition of the Schrodinger picture, we require the operators (especially those corresponding to

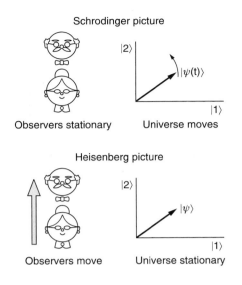

FIGURE 5.9.1
Cartoon representation of the Schrodinger and Heisenberg pictures.

observables) to be explicitly independent of time. For example, in the coordinate-representation of the Schrodinger picture, we know the momentum to be given by

$$\hat{P} = \frac{\hbar}{i}\nabla$$

It does not depend on time. In fact, we surmised this form of the momentum by working with the time dependent wave function $e^{ikx-i\omega t}$ (a wave function in the Schrodinger picture). The top portion of Figure 5.9.1 attempts to describe the situation. We model the motion (i.e., dynamics) of a physical system (denoted "universe") by the motion of the wave function in Hilbert space. The figure shows that the detection equipment—eyes in this case—does not change with time so that the manner of making an observation does not depend on time. The "act of observing" a particular quantity does not depend on when we make the observation. This point of view seems very natural since we assume that any change in a physical quantity must be due to changes in the physical systems and not our detection apparatus (eyes).

The *Heisenberg representation* assigns all of the time dependence to the operators and none to the wave functions. This representation resembles classical mechanics where the dynamical variables, such as momentum, depend on time. The wave functions in this representation do not depend on time. The wave functions in the Heisenberg representation consist of the superposition of the basis vectors of the form

$$|\Psi\rangle = \sum_E \beta_E |\phi_E\rangle = \sum_E \beta_E |E\rangle$$

where β_E does not depend on time. In some sense, the wave functions (i.e., basis) form the lattice-work of a stage that defines the specific system that can be observed. The operators contain all the dynamics but they need to have the wave functions to give information on the specific system. The bottom portion of Figure 5.9.1 attempts to illustrate this paradigm by having the observers move rather than the system. Observations made in the two portions of the figure must agree.

This brings out another point for comparing and contrasting operators and vectors in the quantum theory. Regardless of the representation, an operator must contain all possible outcomes to an observation or operation. We can understand this point of view using the basis vector expansion of an operator found in Chapter 4. For example using the energy basis set, the Hamiltonian

$$\hat{\mathscr{H}} = \sum_{\text{all } E} E|\phi_E\rangle \langle\phi_E| = \sum_{\text{all } E} E|E\rangle \langle E|$$

consists of all possible results of the observation because of the sum over all the energy eigenvalues E. However, the *wave functions* are written as a specific sum over the basis set; only a certain combination of basis vectors appears in the sum. For example, the wave function

$$|\Psi\rangle = \sum_{\text{some } E} \beta_E|\phi_E\rangle = \sum_{\text{some } E} \beta_E|E\rangle$$

contains information on only specific eigenvalues E. Even if it contains all eigenvalues, the sum refers to only one certain mixture (i.e., one vector in the Hilbert space) because of the specific values of β chosen. In summary, operators contain all possible results of a measurement while vectors represent specific instances of the system in question.

The *interaction representation* assigns some time dependence to the operators and some to the wave functions. We will find this representation especially suited for an "open" system. First, consider a "closed system" for which the number of particles and the total energy contained within the system remains constant. Basically we assume that we have solved Schrodinger's equation for this simple closed system. The time evolution of the system trivially involves only factors of the form $e^{-iEt/\hbar}$. We assign this trivial time dependence to the *operators*. With only the simple closed system present, the wave functions remain stationary in the vector space as defined by the time independent basis set. Essentially this much corresponds to the Heisenberg representation. Now, if we include extra forces (above and beyond those included for the trivial solution) then any additional motion induced in the system appears in the wave function. For example, we might have a chunk of semiconductor material for which we can find the solution to Schrodinger's equation for the holes and electrons. For the Heisenberg representation, we remove the time dependence from the wave function and assign it to the operators. The system consists of the chunk of material. Now second, consider the open system consisting of the semiconductor absorbing light. The original Hamiltonian for the closed system does not include this matter–light interaction so that the absorbed light will cause effects not taken into account by the original Hamiltonian. We assign this additional time dependence to the wave function. Of course we also work with the new Hamiltonian but it and all other operators are assigned the trivial time dependence. In this way, the wave functions move in Hilbert space only due to the additional forces not accounted for by the original closed system.

5.9.2 The Schrodinger Representation

We have previously shown how the time dependent wave function satisfies Schrodinger's equation.

$$\hat{\mathscr{H}}|\psi(t)\rangle = i\hbar \frac{\partial}{\partial t}|\psi(t)\rangle \tag{5.9.1}$$

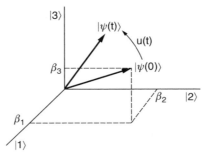

FIGURE 5.9.2
The wave function moves through Hilbert space in the Schrodinger picture.

The wave function moves in Hilbert space as shown in Figure 5.9.2. The components β depend on time but not the basis vectors. The unitary evolution operator moves the initial wave function forward in time according to

$$\hat{u}(t, t_o) \left| \psi(t_o) \right\rangle = \left| \psi(t) \right\rangle \tag{5.9.2}$$

without changing the normalization of the function. The evolution operator actually depends on the difference in time and can be written as $\hat{u}(t, t_o) = \hat{u}(t - t_o)$.

For either open or closed systems, we define the evolution operator

$$\hat{\mathscr{H}} \, \hat{u}(t, t_o) \left| \psi(t_o) \right\rangle = i\hbar \frac{\partial}{\partial t} \hat{u}(t, t_o) \left| \psi(t_o) \right\rangle \quad \text{or} \quad \hat{\mathscr{H}} \, \hat{u}(t, t_o) = i\hbar \frac{\partial}{\partial t} \hat{u}(t, t_o) \tag{5.9.3}$$

by substituting Equation (5.9.2) into Equation (5.9.1). Equation (5.9.2) gives the initial condition of $\hat{u}(t_o, t_o) = 1$.

Consider a closed system. For simplicity, set the initial time to zero $t_o = 0$. Schrodinger's equation can be formally integrated when the Hamiltonian does not depend on time (i.e., a close system). Rearranging Equation (5.9.1) provides

$$\frac{\partial}{\partial t} \left| \psi(t) \right\rangle = \frac{\hat{\mathscr{H}}}{i\hbar} \left| \psi(t) \right\rangle$$

Consider the Hamiltonian operator to be similar to a constant and solve the simple differential equation to obtain

$$\left| \psi(t) \right\rangle = \exp\left(\frac{\hat{\mathscr{H}} \, t}{i\hbar} \right) \left| \psi(0) \right\rangle = \hat{u}(t) \left| \psi(0) \right\rangle \tag{5.9.4}$$

As discussed in Chapter 4, the operator $\hat{u}(t)$ is unitary (i.e., $\hat{u}^{-1} = \hat{u}^+$) since the Hamiltonian $\hat{\mathscr{H}}$ is Hermitian. For the energy basis set $\{\phi_n(x)\}$, the time dependence of the wave function must be

$$\Psi(x, t) = \sum_n \beta_n(t)\phi_n(x) = \sum_n \beta_n(0) \exp\left(\frac{\hat{\mathscr{H}} \, t}{i\hbar} \right)\phi_n(x) = \sum_n \beta_n(0)\exp\left(\frac{E_n t}{i\hbar} \right)\phi_n(x)$$

The evolution operator will play a pivotal role for the Heisenberg representation.

5.9.3 Rate of Change of the Average of an Operator in the Schrodinger Picture

In this topic, we discuss how an observed *value* (not the operator!) evolves in time for the Schrodinger representation. The next topic on Ehrenfest's theorem then shows how Schrodinger's quantum mechanics reproduces results for classical mechanics. We expect the classical analog of a quantum mechanical system to involve an average over the quantum mechanical microscopic quantities. We expect to recover Newton's second law by calculating the rate of change of the expectation value of the quantum mechanical momentum operator. We therefore start the discussion by considering the rate of change of the expectation value of an operator using the Schrodinger picture.

Let $\hat{A} = \hat{A}(\vec{r}, t)$ be an operator in the Schrodinger picture where usually the operator does not explicitly depend on time. Suppose further that the wave vector $|\psi(t)\rangle$ is a solution to Schrodinger's equation. The time rate of change of the expectation value of the operator can be calculated

$$\frac{d}{dt}\langle\hat{A}\rangle = \frac{d}{dt}\langle\psi|\hat{A}|\psi\rangle = \left\langle\frac{\partial\psi}{\partial t}\Big|\hat{A}\Big|\psi\right\rangle + \left\langle\psi\Big|\frac{\partial\hat{A}}{\partial t}\Big|\psi\right\rangle + \left\langle\psi\Big|\hat{A}\Big|\frac{\partial\psi}{\partial t}\right\rangle$$

The derivatives moves into the bra since it symbolizes an integral with respect to spatial coordinates. Now use Schrodinger's equation for the time derivatives of the wave functions to obtain

$$\frac{d}{dt}\langle\hat{A}\rangle = \left\langle\frac{\hat{\mathscr{H}}}{i\hbar}\psi\Big|\hat{A}\Big|\psi\right\rangle + \left\langle\psi\Big|\frac{\partial\hat{A}}{\partial t}\Big|\psi\right\rangle + \left\langle\psi\Big|\hat{A}\Big|\frac{\hat{H}}{i\hbar}\psi\right\rangle$$

Evaluating the left-most inner product by using the definition of the adjoint

$$\left\langle\frac{\hat{\mathscr{H}}}{i\hbar}\psi\right| = \left[\Big|\frac{\hat{\mathscr{H}}}{i\hbar}\psi\Big\rangle\right]^{+} = \left[\frac{\hat{\mathscr{H}}}{i\hbar}|\psi\rangle\right]^{+} = \langle\psi|\frac{\hat{\mathscr{H}}^{+}}{(-i)\hbar} = \langle\psi|\frac{\hat{\mathscr{H}}}{(-i)\hbar}$$

The rate of change of the expectation value of the operator can now be rewritten as

$$\frac{d}{dt}\langle\hat{A}\rangle = \langle\psi|\frac{\hat{\mathscr{H}}}{(-i)\hbar}\hat{A}|\psi\rangle + \langle\psi|\frac{\partial\hat{A}}{\partial t}|\psi\rangle + \langle\psi|\hat{A}\frac{\hat{\mathscr{H}}}{i\hbar}|\psi\rangle$$

Collecting terms provides

$$\frac{d}{dt}\langle\hat{A}\rangle = \frac{i}{\hbar}\langle[\hat{\mathscr{H}},\hat{A}]\rangle + \left\langle\frac{\partial\hat{A}}{\partial t}\right\rangle \tag{5.9.5}$$

Usually the expectation value of the time derivatives of the operator (last term) is zero for the Schrodinger picture.

Example 5.9.1

For the infinitely deep potential well, calculate the rate of change of the momentum for electron.

Solution: The operator in Equation (5.9.5) becomes $\hat{A} = \hat{p}$. The Hamiltonian is given by $\hat{\mathscr{H}} = \hat{p}^2/2m$. It is easy to calculate that $[\hat{\mathscr{H}}, \hat{p}] = 0$. We assume $\partial\hat{p}/\partial t = 0$ as usual for the Schrodinger representation. Therefore, the rate of change of the expected value of momentum must be $d\langle\hat{p}\rangle/dt = 0$.

5.9.4 Ehrenfest's Theorem for the Schrodinger Representation

Now we discuss Ehrenfest's theorem showing that Schrodinger's quantum mechanics leads to Newton's second law. We first show that because a quantum particle can be considered as smeared-out over a volume of space (at least in the sense of statistics), the classical dynamical variable corresponds to the quantum mechanical average of the operator.

Consider an example for the force exerted on a body to see why quantum mechanics averages operators in the Schrodinger representation. Figure 5.9.3 shows the probability density for the location of a quantum mechanical particle. We might imagine the particle of mass m as "smeared-out" over the region. Suppose $\vec{\mathscr{F}}$ represents the force per unit mass. The total classical force must be $\vec{F} = \sum_i \vec{\mathscr{F}}_i \, \Delta m_i$ where the mass $m = \sum_i \Delta m_i$ might not be uniformly distributed across the region of space. The figure shows more mass near the center and less at the "boundaries." The amount of mass in a given region must be proportional to the probability density p_r of finding the electron in a small region Δx. For the one-dimensional case, we write $\Delta m_i \sim p_r \, dx \sim \psi^* \psi \, \Delta x$. We can therefore write the total force as

$$\vec{F} = \sum_i \vec{\mathscr{F}}_i \, \Delta m_i \sim \sum_i \psi^*(x_i)\vec{\mathscr{F}}_i(x_i)\psi(x_i)\Delta x \rightarrow \int \psi^*(x)\vec{\mathscr{F}}(x)\psi(x)\,dx = \langle \mathscr{F} \rangle$$

Therefore, because the quantum mechanical particle effectively occupies a large volume of space, classical quantities like force and interaction energy do not occur at one specific point; instead they occur over the region of space. We expect the quantum mechanical operator to be averaged over a region of space to produce the corresponding classical quantity. Furthermore, this shows that the time-dependence of the wave function translates to a time dependence of the classical quantity through the averaging procedure.

We now show *Ehrenfest's theorem*, which relates the classical force to the rate of change of the expected value of momentum for a single particle

$$\vec{F}_{\text{class}} = \frac{d\langle \hat{p} \rangle}{dt} \quad \text{with} \quad \frac{\partial \hat{p}}{\partial t} = 0$$

for the Schrodinger picture. The time rate of change of the expected value of an operator is obtained from Equation (5.9.5).

$$\frac{d\langle \hat{p} \rangle}{dt} = \frac{i}{\hbar}\left\langle \left[\hat{\mathscr{H}},\hat{p}\right]\right\rangle = \frac{i}{\hbar}\left\langle\left[\frac{\hat{p}^2}{2m} + V(\vec{r}),\hat{p}\right]\right\rangle = \frac{i}{\hbar}\left\langle\left[\frac{\hat{p}^2}{2m},\hat{p}\right]\right\rangle + \frac{i}{\hbar}\langle[V(\vec{r}),\hat{p}]\rangle$$

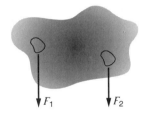

FIGURE 5.9.3
A quantum mechanical object described by a wave function.

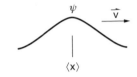

FIGURE 5.9.4
If a wave function depends on time then averages using that wave function must also depend on time.

where we have used the commutator identity $[\hat{A} + \hat{B}, \hat{C}] = [\hat{A}, \hat{C}] + [\hat{B}, \hat{C}]$. Then since $[\hat{p}^2/2m, \hat{p}] = 0$ we must have $\langle [\hat{p}^2/2m, \hat{p}] \rangle = 0$. Finally, we need to evaluate the commutator between the potential energy and the momentum

$$[V(\vec{r}), \hat{p}]\, f = \left[V(\vec{r}), \frac{\hbar}{i}\nabla\right] f = V\frac{\hbar}{i}\nabla f - \frac{\hbar}{i}\nabla(Vf) = V\frac{\hbar}{i}\nabla f - \frac{\hbar}{i}V\nabla f - \left(\frac{\hbar}{i}\nabla V\right)f = i\hbar(\nabla V)f$$

where we use an arbitrary function "f" because the commutator is an operator. As a result, we can conclude the operator relation $[V(\vec{r}), \hat{p}] = i\hbar\nabla V$. Putting all the steps together we arrive at Ehrenfest's theorem:

$$\frac{d\langle \hat{p} \rangle}{dt} = \frac{i}{\hbar}\left\langle \left[\hat{\mathscr{H}}, \hat{p}\right]\right\rangle = \frac{i}{\hbar}\langle i\hbar\nabla V \rangle = \langle -\nabla V \rangle = \langle F \rangle$$

Figure 5.9.4 shows a wave packet traveling to the right with speed "v." The wave function clearly depends on time because it moves. The expectation value of the position operator $\hat{x}$ gives the position of the center of the wave packet. Now because the wave packet moves, the expectation value of the position operator must depend on time $\langle \hat{x} \rangle = \bar{x}(t)$. We therefore find the average of an operator depends on time (through the average) even though the operator itself remains independent of time.

5.9.5 The Heisenberg Representation

The Heisenberg representation assigns the dynamics to the operators. None of the wave functions depend on time so that none of the dynamics appears in the wave functions. We can find the time dependent operators from those in the Schrodinger picture. The simplest procedure requires all expectation values to be invariant with respect to the particular picture. Suppose we represent the state of the system by the ket $|\psi_s(t)\rangle$ in the Schrodinger picture (where "s" denotes Schrodinger). The expectation value of an operator $\hat{O}_s$ can be written as

$$\langle \psi_s(t)|\hat{O}_s|\psi_s(t)\rangle = \langle \psi_h|\hat{u}^+\hat{O}_s\hat{u}|\psi_h\rangle \tag{5.9.6}$$

where $\hat{u}$ represents a unitary operator. For convenience, we set the origin of time to $t = 0$ rather than an arbitrary time t_o. That is, we define the Heisenberg wave function to be $|\psi_h\rangle = |\psi_s(0)\rangle$. Therefore, in order for the expectation value to be independent of picture, we define the time dependent Heisenberg operator to be

$$\hat{O}_h(t) = \hat{u}^+\hat{O}_s\hat{u} \tag{5.9.7}$$

We found the unitary evolution operator $\hat{u}$ for *closed* systems in the Schrodinger picture; a perturbation approach can be used for open systems. Recall the evolution operator has the form

$$\hat{u}(t) = \exp\left(\frac{\hat{\mathscr{H}}_s t}{i\hbar}\right) \tag{5.9.8}$$

where $\hat{H}_s$ denotes the Schrodinger Hamiltonian. We do not need to subscript the Hamiltonian with an "s" in this case because, as will be seen in the second example below, it has the same form in either the Schrodinger or Heisenberg representation. We can show that commutator expressions in the Schrodinger picture produce similar results in the Heisenberg picture (refer to the chapter exercises).

Example 5.9.2

Find the Heisenberg representation of the momentum operator $\hat{p}$ for the infinitely deep square well without an external interaction.

Solution: The Heisenberg momentum operator must be given by

$$\hat{p}_h = \hat{u}^+\hat{p}\hat{u} = \exp\left(-\frac{\hat{\mathcal{H}}}{i\hbar}t\right)\hat{p}\,\exp\left(\frac{\hat{\mathcal{H}}}{i\hbar}t\right)$$

Inside the well, the Schrodinger Hamiltonian has the form $\hat{\mathcal{H}} = \hat{p}^2/2m$. Now, since the momentum operator commutes with the Hamiltonian $[\hat{p}, \hat{\mathcal{H}}] = [\hat{p}, \hat{p}^2/2m] = 0$ then any function of the Hamiltonian must also commute with momentum

$$\left[\hat{p},\ \exp\left(\frac{\hat{\mathcal{H}}\,t}{i\hbar}\right)\right] = 0$$

as can be easily verified by Taylor expanding the exponential. Therefore, the Heisenberg representation of the momentum operator can be written as

$$\hat{p}_h = \hat{u}^+\hat{p}\hat{u} = \exp\left(-\frac{\hat{\mathcal{H}}}{i\hbar}t\right)\exp\left(\frac{\hat{\mathcal{H}}}{i\hbar}t\right)\hat{p} = \hat{p}$$

In the simple case of an infinitely deep well, we see that the Heisenberg and Schrodinger representations are the same for the momentum operator. Especially notice that the unitary operator $\hat{u}$ is written in terms of Schrodinger quantities.

Example 5.9.3

What is the Heisenberg representation of the Schrodinger Hamiltonian without an external interaction?

Solution: The Schrodinger and Heisenberg representations have identical Hamiltonians since $[\hat{u}, \hat{\mathcal{H}}_s] = 0$

$$\hat{\mathcal{H}}_h = \hat{u}^+\hat{\mathcal{H}}_s\hat{u} = \exp\left(-\frac{\hat{\mathcal{H}}_s}{i\hbar}t\right)\hat{\mathcal{H}}_s\exp\left(\frac{\hat{\mathcal{H}}_s}{i\hbar}t\right) = \hat{\mathcal{H}}_s$$

5.9.6 The Heisenberg Equation

Next, we show the principal method of calculating the time evolution of the Heisenberg operators. As demonstrated in the present topic, the dynamics of the Heisenberg operators can be found using the Heisenberg equation given by

$$\frac{d\hat{O}_h}{dt} = \frac{i}{\hbar}\left[\hat{\mathcal{H}}_h, \hat{O}_h\right] + \left(\frac{\partial}{\partial t}\hat{O}_s\right)_h \tag{5.9.9}$$

Often the last term is defined as $(\partial \hat{O}_s/\partial t)_h \equiv \partial \hat{O}_h/\partial t$. The Hamiltonian generates displacements in time. The commutator for the operators takes the place of the Schrodinger equation for the wave functions. This last equation has a form somewhat similar to that for the Schrodinger picture in Equation (5.9.5). For the Heisenberg representation, we do not need to calculate an expectation value. We will see how the operators in the Heisenberg representation obey equations of motion very similar to the dynamical variables in classical mechanics.

Equation (5.9.9) holds for either an open or closed system as we now show. Starting with Equation (5.9.7), $\hat{O}_h(t) = \hat{u}^+\hat{O}_s\hat{u}$, we find

$$\frac{d\hat{O}_h}{dt} = \frac{d}{dt}\hat{u}^+\hat{O}_s\hat{u} = \left(\frac{d}{dt}\hat{u}^+\right)\hat{O}_s\hat{u} + \hat{u}^+\left(\frac{d\hat{O}_s}{dt}\right)\hat{u} + \hat{u}^+\hat{O}_s\left(\frac{d}{dt}\hat{u}\right) \tag{5.9.10}$$

Equation (5.9.3), $\hat{\mathcal{H}}_s\hat{u}(t) = i\hbar\partial_t\hat{u}(t)$ for $t_o = 0$, provides $\partial_t\hat{u}^+ = i\hat{u}^+\hat{\mathcal{H}}_s/\hbar$ by taking the adjoint of both sides. Therefore, Equation (5.9.10) becomes

$$\frac{d\hat{O}_h}{dt} = \left(\frac{i}{\hbar}\hat{u}^+\hat{\mathcal{H}}_s\right)\hat{O}_s\hat{u} + \hat{u}^+\left(\frac{d\hat{O}_s}{dt}\right)\hat{u} + \hat{u}^+\hat{O}_s\left(-\frac{i}{\hbar}\hat{\mathcal{H}}_s\hat{u}\right)$$

Finally, substituting $\hat{u}\,\hat{u}^+ = 1$ between the Hamiltonian and the operator $\hat{O}$ provides

$$\frac{d\hat{O}_h}{dt} = \left(\frac{i}{\hbar}\hat{u}^+\hat{\mathcal{H}}_s\hat{u}\right)\left(\hat{u}^+\hat{O}_s\hat{u}\right) + \hat{u}^+\left(\frac{d\hat{O}_s}{dt}\right)\hat{u} + \hat{u}^+\hat{O}_s\hat{u}\left(-\frac{i}{\hbar}\hat{u}^+\hat{\mathcal{H}}_s\hat{u}\right) = \frac{i}{\hbar}\left[\hat{\mathcal{H}}_h, \hat{O}_h\right] + \left(\frac{d\hat{O}_s}{dt}\right)_h$$

as required.

Example 5.9.4
Show Equation (5.9.9) for the closed system using Equation (5.9.8) where $\hat{\mathcal{H}}_s = \hat{\mathcal{H}} = \hat{\mathcal{H}}_h$.

Solution

$$\frac{d\hat{O}_h}{dt} = \frac{d}{dt}\left(\hat{u}^+\hat{O}_s\hat{u}\right) = \frac{d}{dt}\left\{\exp\left(-\frac{\hat{\mathcal{H}}}{i\hbar}t\right)\hat{O}_s\exp\left(+\frac{\hat{\mathcal{H}}}{i\hbar}t\right)\right\}$$

$$= \left\{-\frac{\hat{\mathcal{H}}}{i\hbar}\exp\left(-\frac{\hat{\mathcal{H}}}{i\hbar}t\right)\right\}\hat{O}_s\exp\left(+\frac{\hat{\mathcal{H}}}{i\hbar}t\right) + \exp\left(-\frac{\hat{\mathcal{H}}}{i\hbar}t\right)\left\{\frac{\partial}{\partial t}\hat{O}_s\right\}\exp\left(+\frac{\hat{\mathcal{H}}}{i\hbar}t\right)$$

$$+ \exp\left(-\frac{\hat{\mathcal{H}}}{i\hbar}t\right)\hat{O}_s\left\{\frac{\hat{\mathcal{H}}}{i\hbar}\exp\left(+\frac{\hat{\mathcal{H}}}{i\hbar}t\right)\right\}$$

Using the definition of a Heisenberg operator (5.9.10) and combining terms produces

$$\frac{d\hat{O}_h}{dt} = -\frac{\hat{\mathscr{H}}}{i\hbar}\hat{O}_h + \hat{O}_h\frac{\hat{\mathscr{H}}}{i\hbar} + \left(\frac{\partial}{\partial t}\hat{O}_s\right)_h$$

where the time derivative of the Schrodinger operator is usually 0. Forming the commutator provides the required results in Equation (5.9.9).

5.9.7 Newton's Second Law from the Heisenberg Representation

We can easily recover Newton's second law of motion from the Heisenberg representation starting with the one-dimensional *Schrodinger* Hamiltonian, for example.

$$\hat{\mathscr{H}}_s = \frac{\hat{p}^2}{2m} + V(x)$$

This Hamiltonian represents a closed system so the demonstration can follow either of two routes. We use the general definition of the evolution operator in Equations (5.9.2) and (5.9.3), and leave the corresponding demonstration for the closed evolution operator to the chapter exercises. Let $\hat{p}_h = \hat{p}_h(t)$ be the Heisenberg momentum operator. We wish to calculate its rate of change using Equation (5.9.9).

$$\frac{d\hat{p}_h}{dt} = \frac{i}{\hbar}\left[\hat{\mathscr{H}}_h, \hat{p}_h\right] = \frac{i}{\hbar}\left[\hat{u}^+\hat{\mathscr{H}}_s\hat{u}, \hat{u}^+\hat{p}\hat{u}\right] = \frac{i}{\hbar}\hat{u}^+\left[\hat{\mathscr{H}}_s, \hat{p}\right]\hat{u} \tag{5.9.11}$$

since $\hat{u}^+\hat{u} = 1 = \hat{u}\,\hat{u}^+$. Substituting for the Hamiltonian we find

$$\frac{d\hat{p}_h}{dt} = \frac{i}{\hbar}\hat{u}^+\left[\frac{\hat{p}^2}{2m} + V(x), \hat{p}\right]\hat{u} = \frac{i}{\hbar}\hat{u}^+[V(x), \hat{p}]\hat{u} = \frac{i}{\hbar}\hat{u}^+\left[V(x), \frac{\hbar}{i}\frac{\partial}{\partial x}\right]\hat{u}$$

$$= \hat{u}^+\left(-\frac{\partial V}{\partial x}\right)\hat{u} = \hat{u}^+F\hat{u} = F_h$$

This last result is Newton's second law! We see that the Heisenberg operators most naturally take the place of the classical dynamical variables.

5.9.8 The Interaction Representation

As previously mentioned, the interaction representation combines portions of the Schrodinger and Heisenberg representations. Both the operators and wave functions depend on time. We identify the dynamics embedded in the wave function as due to the interaction between the system and an external agent. Therefore, the wave functions move in Hilbert space in response to the "extra" potentials imposed on the system. The operators carry the dynamics of the closed system.

Suppose the Hamiltonian for the system has the form

$$\hat{\mathscr{H}} = \hat{\mathscr{H}}_o + \hat{V} \tag{5.9.12}$$

where the closed-system Hamiltonian $\hat{\mathscr{H}}_o$ must be independent of time. Consider Schrodinger's equation in operator form

$$\hat{\mathscr{H}}|\Psi_s(t)\rangle = i\hbar\frac{\partial}{\partial t}|\Psi_s(t)\rangle \quad \text{or} \quad \left(\hat{\mathscr{H}}_o + \hat{V}\right)|\Psi_s(t)\rangle = i\hbar\frac{\partial}{\partial t}|\Psi_s(t)\rangle \tag{5.9.13}$$

Define the interaction wave function through the relation

$$|\Psi_s(t)\rangle = \hat{u}|\Psi_I(t)\rangle \tag{5.9.14}$$

using the unitary evolution operator previously defined for the closed system

$$\hat{u}(t) = \exp\left(\frac{\hat{\mathscr{H}}_o}{i\hbar}t\right) \tag{5.9.15}$$

The subscripts "*s*" and "*I*" stand for Schrodinger and Interaction, respectively. The inverse unitary operator $\hat{u}^+$ essentially removes the time dependence from the wave function $|\Psi_s(t)\rangle$ attributable to the Hamiltonian $\hat{\mathscr{H}}_o$. However, the wave function retains some time dependence due to the added potential $\hat{V}$ occurring in the full Hamiltonian $\hat{\mathscr{H}}$.

We can write the Schrodinger equation using the interaction representation. Substituting Equation (5.9.14) into Equation (5.9.13) produces

$$\left(\hat{\mathscr{H}}_o + \hat{V}\right)\hat{u}(t)|\Psi_I(t)\rangle = i\hbar\frac{\partial}{\partial t}\hat{u}(t)|\Psi_I(t)\rangle \tag{5.9.16}$$

Now, differentiate both terms on the right-hand side of Equation (5.9.16)

$$\left(\hat{\mathscr{H}}_o + \hat{V}\right)\hat{u}(t)|\Psi_I(t)\rangle = i\hbar\frac{\partial}{\partial t}\text{Exp}\left(\frac{\hat{\mathscr{H}}_o}{i\hbar}t\right)|\Psi_I(t)\rangle$$

$$= i\hbar\left\{\frac{\hat{\mathscr{H}}_o}{i\hbar}\exp\left(\frac{\hat{\mathscr{H}}_o}{i\hbar}t\right)|\Psi_I(t)\rangle + \exp\left(\frac{\hat{\mathscr{H}}_o}{i\hbar}t\right)\frac{\partial}{\partial t}|\Psi_I(t)\rangle\right\}$$

$$= \hat{\mathscr{H}}_o\,\hat{u}|\Psi_I(t)\rangle + \hat{u}\frac{\partial}{\partial t}|\Psi_I(t)\rangle$$

Canceling the terms involving $\hat{\mathscr{H}}_o$ from both sides produces

$$\hat{V}\hat{u}(t)|\Psi_I(t)\rangle = i\hbar\hat{u}\frac{\partial}{\partial t}|\Psi_I(t)\rangle$$

Operating on both sides with the adjoint of the evolution operator and defining the interaction potential as $\hat{V}_I = \hat{u}^+\hat{V}\hat{u}$ yields

$$\hat{u}^+\hat{V}\hat{u}\,|\Psi_I(t)\rangle = i\hbar\frac{\partial}{\partial t}|\Psi_I(t)\rangle \quad \text{or} \quad \hat{V}_I|\Psi_I(t)\rangle = i\hbar\frac{\partial}{\partial t}|\Psi_I(t)\rangle \tag{5.9.17}$$

As a result, the wave function satisfies a Schrodinger-like equation with the interaction potential $\hat{V}_I$ in the interaction representation taking the place of the Hamiltonian.

The next section on time dependent perturbation theory will demonstrate a unitary evolution operator $\hat{U}(t)$ that moves the interaction wave function through Hilbert space according to $|\Psi_I(t)\rangle = \hat{U}|\Psi_I(0)\rangle$. The operator $\hat{U}(t)$ should not be confused with the operator $\hat{u}$ that changes the Schrodinger wave function into the interaction one, $|\psi_s\rangle = \hat{u}|\psi_h\rangle$. The operator $\hat{U}(t)$ has the form $\hat{U} = \hat{T}e^{(1/i\hbar)\int_{t_o}^{t} dt_1 \hat{V}_I(t_1)}$ for an interaction that starts at t_o. The operator $\hat{T}$ denotes the time ordered product.

5.10 Time Dependent Perturbation Theory

Electromagnetic energy interacting with an atomic system can produce transitions between the energy levels. A Hamiltonian $\hat{\mathscr{H}}_o$ describes the atomic system and provides the energy basis states and the energy levels. The interaction potential $\hat{V}(t)$ (i.e., the perturbation) depends on time. The theory assumes the perturbation does not change the basis states or the energy levels, but rather induces transitions between these fixed levels. The perturbation rotates the particle (electron or hole) wave function through Hilbert space so that the probability of the particle occupying one energy level or another changes with time. Therefore, the goal of the time-dependent perturbation theory consists of finding the time dependence of the wave function components.

Subsequent sections apply the time dependent perturbation theory to an electromagnetic wave interacting with an atom or an ensemble of atoms. Fermi's golden rule describes the matter–light interaction in this semiclassical approach, which uses the non-operator form of the EM field. The quantized version will be given in a later chapter.

5.10.1 Physical Concept

The Hamiltonian

$$\hat{\mathscr{H}} = \hat{\mathscr{H}}_o + \hat{V}(t) \tag{5.10.1}$$

describes an atomic system subjected to a perturbation. The Hamiltonian $\hat{\mathscr{H}}_o$ refers to the atom and determines the energy basis states $\{|n\rangle = |E_n\rangle\}$ so that $\hat{\mathscr{H}}_o|n\rangle = E_n|n\rangle$. The interaction potential $\hat{V}(t)$ describes the interaction of an external agent with the atomic system.

Consider an electromagnetic field incident on the atomic system as indicated in Figure 5.10.1 for the initial time $t = 0$. Assume the atomic system consists of a quantum well with an electron in the first level as indicated by the dot in the figure. The atomic system can absorb a photon from the field and promote the electron from the first to the second level. The right-hand portion of Figure 5.10.1 shows the same information as

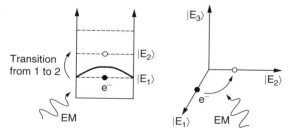

FIGURE 5.10.1

An electron absorbs a photon and makes a transition from the lowest level to next highest one.

the electron transitions from energy basis vector $|E_1\rangle$ to the basis vector $|E_2\rangle$ when the atom absorbs a quantum of energy. This transition of the electron from one basis vector to another should remind the reader of the effect of the ladder operators.

The transition of the electron from one state to another requires the electron occupation probability to change with time. Suppose the wave function for the electron has the form

$$|\psi(t)\rangle = \sum_n \beta_n(t)\,|n\rangle \tag{5.10.2}$$

In the case without any perturbation, the wave function evolves according to

$$|\psi(t)\rangle = e^{\hat{\mathscr{H}}_o t/(i\hbar)} \sum_n \beta_n(0)|n\rangle = \sum_n \beta_n(0)\,e^{E_n t/(i\hbar)}|n\rangle \quad \text{(no pert)} \tag{5.10.3}$$

where $\beta_n(t) = \beta_n(0)\,e^{E_n t/(i\hbar)}$. In this "no perturbation" case, the probability of finding the electron in a particular state n at time t, denoted by $P(n, t)$, does not change from its initial value at $t = 0$, denoted by $P(n, t = 0)$, since

$$P(n, t) = |\beta_n(t)|^2 = |\beta_n(0)\,e^{E_n t/(i\hbar)}|^2 = |\beta_n(0)|^2 = P(n, t = 0) \quad \text{(no pert)} \tag{5.10.4}$$

This behavior occurs because the Hamiltonian describes a "closed system" that does not interact with external agents. The eigenvectors are exact solutions to full Hamiltonian $\hat{\mathscr{H}}_o$ in this case. The exact Hamiltonian introduces only the trivial factor $e^{E_n t/(i\hbar)}$ into the motion of the wave function through Hilbert space.

What about the case of an atomic system interacting with the external agent? Now we see that Equation (5.10.3) cannot accurately describe the situation because of Equation (5.10.4). The perturbation $\hat{V}(t)$ must produce an expansion coefficient with more than just the trivial factor. We will see below that the wave function must have the form

$$|\psi(t)\rangle = \sum_n a_n(t)\,e^{E_n t/(i\hbar)}|n\rangle \tag{5.10.5}$$

in the Schrodinger picture where the trivial factor $e^{E_n t/(i\hbar)}$ comes from $\hat{\mathscr{H}}_o$ and the time dependent term $a_n(t)$ comes from the perturbation $\hat{V}(t)$. Essentially working in the Schrodinger picture produces the trivial factor $e^{E_n t/(i\hbar)}$ in the wave function. Using the interaction representation produces only the nontrivial time dependence in the wave function. *If the electron starts in state $|i\rangle$ at time $t = 0$* then the probability of finding it in state n after a time t must be

$$P(n, t) = |a_n(t)\,e^{E_n t/(i\hbar)}|^2 = |a_n(t)|^2 \tag{5.10.6}$$

At time $t = 0$, all of the a's must be zero except a_i because the electron starts in the initial state i. Also, $a_i(0) = 1$ because the probabilities sum to one. For later times t, any increase in a_n for $n \neq i$ must be attributed to increasing probability of finding the particle in state n. So, if the particle starts in state $|i\rangle$ then $a_n(t)$ gives the probability amplitude of a transition from state $|i\rangle$ to state $|n\rangle$ after a time t.

An example helps illustrate how motion of the wave function in Hilbert space correlates with the transition probability. Consider the three vector diagrams in Figure 5.10.2. At time $t = 0$, the wave function $|\psi(t)\rangle$ coincides with the $|1\rangle$ axis. The probability amplitude at $t = 0$ must be $\beta_n(0) = a_n(0) = \delta_{ni}$ and therefore the probability values must be $\text{Prob}(n = 1, t = 0) = 1$ and $\text{Prob}(n \neq 1, t = 0) = 0$. Therefore the particle definitely

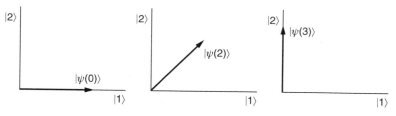

FIGURE 5.10.2
The probability of the electron occupying the second state increases with time.

occupies the first energy eigenstate at $t=0$. The second plot in Figure 5.10.2 at $t=2$, shows the electron partly occupies both the first and second eigenstates. There exists a nonzero probability of finding it in either basis state. According to the figure,

$$\text{Prob}(n=1,t=2) = \text{Prob}(n=2,t=2) = 0.5$$

The third plot in Figure 5.10.2 at time $t=3$ shows the electron must be in state $|2\rangle$ alone since the wavefunction $|\psi(3)\rangle$ coincides with basis vector $|2\rangle$. At $t=3$, the probability of finding the electron in state $|2\rangle$ must be

$$\text{Prob}(n=2,t=3) = |\beta_2|^2 = 1$$

Notice how the probability of finding the particle in state $|1\rangle$ decreases with time, while the probability of finding the particle in state $|2\rangle$ increases.

Unlike the unperturbed system, multiple measurements of the energy of the electron does not always return the same value. The reason concerns the fact that the eigenstates of $\hat{\mathcal{H}}_o$ do not describe the full system. In particular, it does not describe the external agent (light field) nor the interaction between the light field and the atomic system. The external agent, the electromagnetic field, disturbs the state of the particle between successive measurements. The basis function for the atomic system alone does not include one for the optical field. However, given the basis set for the full Hamiltonian $\hat{\mathcal{H}} = \hat{\mathcal{H}}_o + \hat{V} + \hat{\mathcal{H}}_{em} + \cdots$ then a measurement of $\hat{\mathcal{H}}$ must cause the full wave function to collapse to one of the full basis vectors from which it does not move.

Several points should be kept in mind while reading through the next topic. First, the procedure uses the Schrodinger representation but does not replace β_n with $a_n e^{E_n t/(i\hbar)}$. Instead, the procedure directly finds β_n, which then turns out to have the form $a_n e^{E_n t/(i\hbar)}$. Second, these components β_n have exact expressions until we make an approximation of the form $\beta(t) = \beta^{(0)}(t) + \beta^{(1)}(t) + \cdots$ (similar to the Taylor expansion). Third, assume the particle starts in state $|i\rangle$ so that $\beta_n(0) = \beta_n^{(0)}(0) = \delta_{ni}$ and $\beta_n^{(j)}(0) = 0 j \geq 1$. Fourth, the transition matrix elements $V_{fi} = \langle f|\hat{V}|i\rangle$ determine the final states f that can be reached from the initial states i. Stated equivalently, these selection rules determine the allowed transitions.

5.10.2 Time Dependent Perturbation Theory Formalism in the Schrodinger Picture

The perturbed Hamiltonian $\hat{\mathcal{H}} = \hat{\mathcal{H}}_o + \hat{V}(x,t)$ consists of the closed Hamiltonian $\hat{\mathcal{H}}_o$ for the system and the perturbation $\hat{V}(t)$. Schrodinger's equation becomes

$$\hat{\mathcal{H}}|\Psi(t)\rangle = i\hbar \frac{\partial}{\partial t}|\Psi(t)\rangle \rightarrow \left(\hat{\mathcal{H}}_o + \hat{V}\right)|\Psi(t)\rangle = i\hbar \frac{\partial}{\partial t}|\Psi(t)\rangle \qquad (5.10.7)$$

The unperturbed Hamiltonian $\hat{\mathcal{H}}_o$ produces the energy basis set $\{u_n = |n\rangle\}$ so that

$$\hat{H}_o|n\rangle = E_n|n\rangle.$$

We assume that the Hamiltonian $\hat{\mathcal{H}}$ has the same basis set $\{u_n = |n\rangle\}$ as $\hat{\mathcal{H}}_o$. The boundary conditions on the system determine the basis set and the eigenvalues. This step relegates the perturbation to causing transitions between the basis vectors.

As usual, we write the solution to the Schrodinger wave equation

$$\hat{\mathcal{H}}\big|\Psi(t)\big\rangle = i\hbar\frac{\partial}{\partial t}\big|\Psi(t)\big\rangle \tag{5.10.8}$$

as

$$\big|\Psi(t)\big\rangle = \sum_n \beta_n(t)\,|n\rangle \tag{5.10.9}$$

Recall that the wave vector $\big|\Psi(t)\big\rangle$ moves in Hilbert space in response to the Hamiltonian $\hat{\mathcal{H}}$ as indicated in Figure 5.10.3. The components $\beta_n(t)$ must be related to the probability of finding the electron in the state $|n\rangle$. As an *important point*, we assume that the particle starts in state $|i\rangle$ at time $t=0$ (where $i=1, 2, \ldots$ and should not be confused with the complex number $i = \sqrt{-1}$). To find the components $\beta_n(t)$, start by substituting $\big|\Psi(t)\big\rangle$ (Equation (5.10.9)) into Schrodinger's equation (5.10.8).

$$\left(\hat{\mathcal{H}}_o + \hat{\mathcal{V}}\right)\big|\Psi(t)\big\rangle = i\hbar\frac{\partial}{\partial t}\big|\Psi(t)\big\rangle \;\rightarrow\; \left(\hat{\mathcal{H}}_o + \hat{\mathcal{V}}\right)\sum_n \beta_n(t)\,|n\rangle = i\hbar\frac{\partial}{\partial t}\sum_n \beta_n(t)\,|n\rangle -$$

Move the unperturbed Hamiltonian and the potential inside the summation to find

$$\sum_n \beta_n(t)\left(E_n + \hat{\mathcal{V}}\right)|n\rangle = i\hbar\sum_n \dot{\beta}_n(t)\,|n\rangle$$

where the dot over the symbol β indicates the time derivative. Operate on both sides of the equation with $\langle m|$ to find

$$\sum_n \beta_n(t)\left(E_n\langle m\mid n\rangle + \langle m|\hat{\mathcal{V}}|n\rangle\right) = i\hbar\sum_n \dot{\beta}_n(t)\langle m\mid n\rangle$$

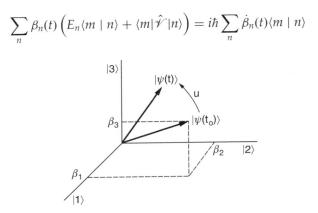

FIGURE 5.10.3
The Hamiltonian causes the wave functions to move in Hilbert space.

The orthonormality of the basis vectors $\langle m \mid n \rangle = \delta_{mn}$ transforms the previous equation to

$$E_m \beta_m(t) + \sum_n \beta_n(t) \langle m | \hat{\mathscr{V}}(x,t) | n \rangle = i\hbar \dot{\beta}_m(t)$$

which can be rewritten as

$$\dot{\beta}_m(t) - \frac{E_m}{i\hbar} \beta_m(t) = \frac{1}{i\hbar} \sum_n \beta_n(t) \, \mathscr{V}_{mn}(t) \tag{5.10.10}$$

where the matrix elements can be written as

$$\mathscr{V}_{mn}(t) = \langle m | \hat{\mathscr{V}}(x,t) | n \rangle = \int dx \, u_m^* \, \hat{\mathscr{V}}(x,t) \, u_n$$

for the basis set consisting of functions of "x."

We must solve Equation 5.10.10 for the components $\beta_n(t)$; this can most easily be handled by using an integrating factor $\mu_m(t)$. Rather than actually solve for the integrating factor, we will just state the results (see Appendix 1)

$$\mu_m(t) = \exp\left(-\frac{E_m}{i\hbar} t\right) \tag{5.10.11}$$

Multiplying the integrating factor on both sides of Equation (5.10.10), we can write

$$\mu_m \dot{\beta}_m - \frac{E_m}{i\hbar} \mu_m \beta_m = \frac{1}{i\hbar} \sum_n \mu_m \beta_n(t) \mathscr{V}_{mn} \tag{5.10.12}$$

Noting that

$$\frac{d}{dt}(\mu_m \beta_m) = \dot{\mu}_m \beta_m + \mu_m \dot{\beta}_m \quad \text{and} \quad \dot{\mu}_m = -\frac{E_m}{i\hbar} \exp\left(-\frac{E_m t}{i\hbar}\right) = -\frac{E_m}{i\hbar} \mu_m$$

Equation 5.10.12 becomes

$$\frac{d}{dt}[\mu_m(t)\beta_m(t)] = \frac{1}{i\hbar} \mu_m(t) \sum_n \beta_n(t) \, \mathscr{V}_{mn}(t) \tag{5.10.13}$$

We need to solve this last equation for the components $\beta_n(t)$ in the first and last terms. Assume that the perturbation starts at $t = 0$ and integrate both sides with respect to time.

$$\mu_m(t)\beta_m(t) = \mu_m(0)\beta_m(0) + \frac{1}{i\hbar} \int_0^t d\tau \, \mu_m(\tau) \sum_n \beta_n(\tau) \, \mathscr{V}_{mn}(\tau) \tag{5.10.14}$$

Substituting for $\mu_m(t)$, noting from Equation (5.10.11) that $\mu_m(0) = 1$, and using the fact that the particle starts in state $|i\rangle$ so that

$$\beta_n(0) = \delta_{ni} \tag{5.10.15}$$

we find

$$\beta_m(t) = \mu_m^{-1}(t)\delta_{mi} + \frac{\mu_m^{-1}(t)}{i\hbar} \sum_n \int_0^t d\tau \mu_m(\tau)\beta_n(\tau) \mathscr{V}_{mn}(\tau) \tag{5.10.16}$$

To this point, the solution is exact.

Now we make the approximation by writing the components $\beta_n(t)$ as a summation

$$\beta_n(t) = \beta_n^{(0)}(t) + \beta_n^{(1)}(t) + \cdots$$

where the superscripts provide the order of the approximation. Substituting the approximation for the components $\beta_n(t)$ into Equation (5.10.10) provides

$$\beta_m^{(0)}(t) + \beta_m^{(1)}(t) + \cdots = \mu_m^{-1}(t)\delta_{mi} + \frac{\mu_m^{-1}(t)}{i\hbar} \sum_n \int_0^t d\tau \mu_m(\tau)\left[\beta_n^{(0)}(\tau) + \beta_n^{(1)}(\tau) + \cdots\right]\mathscr{V}_{mn}(\tau)$$

Note that the approximation term $\beta_n^{(0)} V_{mn}$ has order "(1)" even though $\beta_n^{(0)}$ has order "(0)" since we consider the interaction potential V_{mn} to be small (i.e., it has order "(1)"). Equating corresponding orders of approximation in the previous equation provides

$$\beta_m^{(0)}(t) = \mu_m^{-1}(t)\delta_{mi} \tag{5.10.17}$$

$$\beta_m^{(1)}(t) = \frac{\mu_m^{-1}(t)}{i\hbar} \sum_n \int_0^t d\tau \, \mu_m(\tau)\beta_n^{(0)}(\tau)\mathscr{V}_{mn}(\tau) \tag{5.10.18}$$

and so on. Notice how Equation (5.10.18) invokes itself in Equation (5.10.17) in the integral. So once we solve for the zeroth-order approximation for the component, we can immediately find the first-order approximation. Higher-order terms work the same way. This last equation gives the lowest-order correction to the probability amplitude.

The Kronecker delta function in Equation (5.10.17) suggests considering two separate cases when finding the probability amplitude correction $\beta_m^{(1)}(t)$. The first case for $m = i$ corresponds to finding the probability amplitude for the particle remaining in the initial state. The second case $m \neq i$ produces the probability amplitude for the particle making a transition to state m.

Case $m = i$

We calculate the probability amplitude $\beta_i(t)$ for the particle to remain in the initial state. The lowest-order approximation gives (using Equations (5.10.17) and (5.10.11))

$$\beta_n^{(0)}(t) = \delta_{ni}\mu_n^{-1}(t) = \delta_{ni}\exp\left(\frac{E_n}{i\hbar}t\right) \tag{5.10.19}$$

Substituting Equation (5.10.19) into Equation (5.10.18) with $m = i$, we find

$$\beta_i^{(1)}(t) = \frac{\mu_i^{-1}(t)}{i\hbar} \sum_n \int_0^t d\tau \, \mu_i(\tau)\beta_n^{(0)}(\tau)\mathscr{V}_{in}(\tau) = \frac{\mu_i^{-1}(t)}{i\hbar}\int_0^t d\tau \, \mu_i(\tau)\exp\left(\frac{E_i}{i\hbar}\tau\right)\mathscr{V}_{ii}(\tau)$$

Substituting Equation (5.10.11) for the remaining integrating factors in the previous equation we find

$$\beta_i^{(1)}(t) = \frac{1}{i\hbar}\exp\left(\frac{E_i}{i\hbar}t\right)\int_0^t d\tau \, \mathscr{V}_{ii}(\tau)$$

So therefore the approximate value for $\beta_i(t)$ must be

$$\beta_i(t) = \beta_i^{(0)}(t) + \beta_i^{(1)}(t) + \cdots = \exp\left(\frac{E_i}{i\hbar}t\right) + \frac{1}{i\hbar}\exp\left(\frac{E_i}{i\hbar}t\right)\int_0^t d\tau\, \mathscr{V}_{ii}(\tau) + \cdots \qquad (5.1.20)$$

Case $m \neq i$

We find the component $\beta_m(t)$ corresponding to a final state $|m\rangle$ different from the initial state $|i\rangle$. The lowest-order approximation $\beta_m^{(0)}$ for $m \neq i$ must be

$$\beta_m^{(0)}(t) = 0$$

The procedure finds the probability amplitude for a particle to make a transition from the initial state $|i\rangle$ to a different final state $|m\rangle$.

We start with Equation (5.10.18).

$$\beta_m^{(1)}(t) = \frac{\mu_m^{-1}(t)}{i\hbar}\sum_n \int_0^t d\tau\, \mu_m(\tau)\beta_n^{(0)}(\tau)\mathscr{V}_{mn}(\tau) = \frac{\mu_m^{-1}(t)}{i\hbar}\sum_n \int_0^t d\tau\, \mu_m(\tau)\,\delta_{ni}\mu_i^{-1}(\tau)\,\mathscr{V}_{mn}(\tau)$$

Substitute Equation (5.10.11) for the integrating factors to find

$$\beta_m^{(1)}(t) = \frac{1}{i\hbar}\exp\left(\frac{E_m}{i\hbar}t\right)\int_0^t d\tau\,\exp\left(-\frac{E_m - E_i}{i\hbar}\tau\right)\mathscr{V}_{mi}(\tau)$$

We often write the difference in energy as $E_m - E_i = E_{mi}$ and also

$$\omega_{mi} = \omega_m - \omega_i = \frac{E_m - E_i}{\hbar} = \frac{E_{mi}}{\hbar} \qquad (5.10.21)$$

The reader must keep track of the distinction between matrix elements and this new notation for differences between quantities ... matrix elements refer to operators. Using this notation

$$\beta_m^{(1)}(t) = \frac{1}{i\hbar}\exp\left(\frac{E_m}{i\hbar}t\right)\int_0^t d\tau\,\exp\left(-\frac{E_{mi}}{i\hbar}\tau\right)\mathscr{V}_{mi}(\tau) \qquad (5.10.22)$$

Therefore, the components $\beta_m(t)$ for $m \neq i$ are approximately given by

$$\beta_m(t) = \beta_m^{(0)}(t) + \beta_m^{(1)}(t) + \cdots = 0 + \frac{1}{i\hbar}\exp\left(\frac{E_m}{i\hbar}t\right)\int_0^t d\tau\,\exp\left(-\frac{E_{mi}}{i\hbar}\tau\right)\mathscr{V}_{mi}(\tau) + \cdots$$

$$(5.10.23)$$

In summary, the expansion coefficients in

$$|\Psi(t)\rangle = \sum_n \beta_n(t)\,|n\rangle \qquad (5.10.24a)$$

are given by Equations (5.10.23) and (5.10.21)

$$\beta_m(t) = \delta_{mi}\exp\left(\frac{E_i}{i\hbar}t\right) + \frac{1}{i\hbar}\exp\left(\frac{E_m}{i\hbar}t\right)\int_0^t d\tau\,\exp\left(-\frac{E_{mi}}{i\hbar}\tau\right)\mathscr{V}_{mi}(\tau) + \cdots \qquad (5.10.24b)$$

5.10.3   Time Dependent Perturbation Theory in the Interaction Representation

The interaction representation for quantum mechanics is especially suited for time dependent perturbation theory. Once again, the Hamiltonian $\mathcal{H} = \mathcal{H}_o + \hat{V}(x, t)$ consists of the atomic Hamiltonian $\mathcal{H}_o$ and the interaction potential $\hat{V}(t)$ due to an external agent. The atomic Hamiltonian has the basis set $\{|n\rangle\}$ satisfying $\mathcal{H}_o|n\rangle = E_n|n\rangle$. Both the operators and the wave functions depend on time in the interaction representation. The wave functions move through Hilbert space only in response to the interaction potential $\hat{V}(t)$. A unitary operator $\hat{u} = \exp[\mathcal{H}_o t/(i\hbar)]$ removes the trivial motion from the wave function and places it in the operators; consequently, the operators depend on time. Without any potential $\hat{V}(t)$, the wave functions remain stationary and the operators remain trivially time dependent; that is, the interaction picture reduces to the Heisenberg picture. The motion of the wave function in Hilbert space reflects the dynamics embedded in the interaction potential.

The evolution operator removes the trivial time dependence from the wave function

$$\hat{u}(t) = \exp\left(\frac{\mathcal{H}_o}{i\hbar}t\right) \quad \text{with} \quad \mathcal{H} = \mathcal{H}_o + \hat{V}(t) \tag{5.10.25}$$

The interaction potential in the interaction picture has the form $\hat{V}_I = \hat{u}^+ \hat{V} \hat{u}$ and produces the interaction wave function $|\Psi_I\rangle$ given by

$$|\Psi_s\rangle = \hat{u}\,|\Psi_I\rangle \tag{5.10.26}$$

The wave function $|\Psi_s\rangle$ is the usual Schrodinger wave function embodying the dynamics of the full Hamiltonian $\mathcal{H}$. The equation of motion for the interaction wave function can be written as (Section 5.9)

$$\hat{V}_I|\Psi_I(t)\rangle = i\hbar\frac{\partial}{\partial t}|\Psi_I(t)\rangle \quad \text{or} \quad \frac{\partial}{\partial t}|\Psi_I(t)\rangle = \frac{1}{i\hbar}\hat{V}_I\,|\Psi_I(t)\rangle \tag{5.10.27}$$

We wish to find an expression for the wave function in the interaction representation. First, formally integrate Equation (5.10.27)

$$|\Psi_I(t)\rangle = |\Psi_I(0)\rangle + \frac{1}{i\hbar}\int_0^t d\tau\,\hat{V}_I(\tau)\,|\Psi_I(\tau)\rangle \tag{5.10.28}$$

where we have assumed that the interaction starts at $t = 0$. We can write another equation (see below) by substituting Equation (5.10.28) into itself, which assumes that the interaction wave functions only slightly move in Hilbert space for small interaction potentials.

0th Order Approximation

The lowest-order approximation can be found by noting small interaction potentials $\hat{V}(x, t)$ which lead to small changes in the wave function with time. Neglecting the small integral term in Equation (5.10.4) produces the lowest-order approximation

$$|\Psi_I(t)\rangle \cong |\Psi_I(0)\rangle = |\Psi_s(0)\rangle \tag{5.10.29}$$

where the second equality comes from the fact that $\hat{u}(0) = \hat{1}$ in Equation (5.10.1). This last equation says that to lowest order, the interaction-picture wave function remains stationary in Hilbert space. Therefore to lowest order, the probabilities calculated by projecting the wave function $|\Psi_I(t)\rangle$ onto the basis vectors remain independent of time. The trivial terms $e^{-iEt/\hbar}$ that occur in changing back from the interaction to Schrodinger picture do not have any effect on the probability of finding a particle in a given basis state.

Higher Order Approximation

We obtain subsequent approximations by substituting the wave functions into the integral. The total first-order approximation can be found by substituting Equation (5.10.29) into Equation (5.10.28).

$$|\Psi_I(t)\rangle = |\Psi_I(0)\rangle + \frac{1}{i\hbar} \int_0^t dt_1 \, \hat{\mathscr{V}}_I(t_1) |\Psi_I(0)\rangle \tag{5.10.30}$$

The total second-order approximation can be found by substituting Equation (5.10.30) into Equation (5.10.28) to obtain

$$|\Psi_I(t)\rangle = \left\{ 1 + \frac{1}{i\hbar} \int_0^t dt_1 \, \hat{\mathscr{V}}_I(t_1) + \left(\frac{1}{i\hbar}\right)^2 \int_0^t dt_1 \int_0^{t_1} dt_2 \, \hat{\mathscr{V}}_I(t_1)\hat{\mathscr{V}}_I(t_2) \right\} |\Psi_I(0)\rangle \tag{5.10.31}$$

We can continue this process to find any order of approximation.

5.10.4 An Evolution Operator in the Interaction Representation

We can find an unitary operator that moves the interaction wave function forward in time. Equation (5.10.31) essentially gives the evolution operator $\hat{U}$ defined by

$$|\psi_I(t)\rangle = \hat{U}(t) |\psi_I(0)\rangle \tag{5.10.32}$$

not to be confused with the operator $\hat{u}$ that maps between the Schrodinger and interaction pictures. Equation (5.10.31) approximates $\hat{U}$ by

$$\hat{U} = \left\{ 1 + \frac{1}{i\hbar} \int_0^t dt_1 \, \hat{\mathscr{V}}_I(t_1) + \left(\frac{1}{i\hbar}\right)^2 \int_0^t dt_1 \int_0^{t_1} dt_2 \, \hat{\mathscr{V}}_I(t_1)\hat{\mathscr{V}}_I(t_2) \right\} \tag{5.10.33}$$

which is somewhat reminiscent of writing the operator as an exponential. For example, if the interaction potential were independent of time (but it is not) then the operator would reduce to

$$\hat{U}^+ \asymp 1 + \frac{\hat{\mathscr{V}}_I t}{i\hbar} + \left(\frac{\hat{\mathscr{V}}_I t}{i\hbar}\right)^2 + \cdots = \exp\left(\frac{\hat{\mathscr{V}}_I t}{i\hbar}\right)$$

In order to see how this operator can be related to an exponential, we must digress and discuss the time ordered product.

We define the time ordered product $\hat{T}$ as follows:

$$\hat{T}\left\{\hat{\mathscr{V}}(t_1)\,\hat{\mathscr{V}}(t_2)\,\hat{\mathscr{V}}(t_3)\right\} = \hat{\mathscr{V}}(t_1)\,\hat{\mathscr{V}}(t_3)\,\hat{\mathscr{V}}(t_2) \quad \text{when} \quad t_1 > t_3 > t_2 \tag{5.10.34}$$

The time ordered product can also be defined in terms of a step function.

$$\Theta(t) = \begin{cases} 1 & t > 0 \\ 1/2 & t = 0 \\ 0 & t < 0 \end{cases} \tag{5.10.35}$$

Note the $1/2$ for $t = 0$. The third term in Equation (5.10.33) has two operators and notice that the integration limits require $t_1 > t_2$. We will want to change the limits on both integrals to cover the interval $(0, t)$. Therefore we must keep track of the time ordering. The time ordered product of two operators can be written in terms of the step function as

$$\hat{T}\hat{\mathscr{V}}(t_1)\,\hat{\mathscr{V}}(t_2) = \Theta(t_1 - t_2)\hat{\mathscr{V}}(t_1)\,\hat{\mathscr{V}}(t_2) + \Theta(t_2 - t_1)\hat{\mathscr{V}}(t_2)\,\hat{\mathscr{V}}(t_1) \tag{5.10.36}$$

$$\frac{1}{2!}\int_0^t dt_1 \int_0^t dt_2\, \hat{T}\hat{\mathscr{V}}_I(t_1)\hat{\mathscr{V}}_I(t_2) = \frac{1}{2}\int_0^t dt_1 \int_0^t dt_2\, \Theta(t_1 - t_2)\hat{\mathscr{V}}_I(t_1)\hat{\mathscr{V}}_I(t_2)$$
$$+ \frac{1}{2}\int_0^t dt_1 \int_0^t dt_2\, \Theta(t_2 - t_1)\hat{\mathscr{V}}_I(t_2)\hat{\mathscr{V}}_I(t_1)$$

Interchanging the dummy variables t_1, t_2 in the last integral shows that it's the same as the middle integral. Therefore, by the properties of the step function we find

$$\frac{1}{2!}\int_0^t dt_1 \int_0^t dt_2\, \hat{T}\, \hat{\mathscr{V}}_I(t_1)\hat{\mathscr{V}}_I(t_2) = \int_0^t dt_1 \int_0^{t_1} dt_2\, \hat{\mathscr{V}}_I(t_1)\hat{\mathscr{V}}_I(t_2) \tag{5.10.37}$$

which agrees with the second integral in Equation (5.10.33).

We are now in a position to write an operator that evolves the wave function in time for the interaction representation. Substituting Equation (5.10.33) into Equation (5.10.32) yields

$$|\Psi_I(t)\rangle = \hat{T}\left\{\hat{1} + \frac{1}{i\hbar}\int_0^t dt_1\,\hat{\mathscr{V}}_I(t_1) + \left(\frac{1}{i\hbar}\right)^2 \frac{1}{2!}\int_0^t dt_1 \int_0^t dt_2\,\hat{\mathscr{V}}_I(t_1)\hat{\mathscr{V}}_I(t_2) + \cdots\right\}|\Psi_I(0)\rangle \tag{5.10.38}$$

The term in brackets can be written as an exponential

$$\hat{T}\left\{\hat{1} + \frac{1}{i\hbar}\int_0^t dt_1\,\hat{\mathscr{V}}_I(t_1) + \left(\frac{1}{i\hbar}\right)^2 \frac{1}{2!}\int_0^t dt_1 \int_0^t dt_2\,\hat{\mathscr{V}}_I(t_1)\hat{\mathscr{V}}_I(t_2) + \cdots\right\} = \hat{T}e^{\frac{1}{i\hbar}\int_0^t dt_1\,\hat{\mathscr{V}}_I(t_1)}$$

5.11 Density Operator

The density operator and its associated equation of motion provide an alternate formulation for a quantum mechanical system. The density operator combines the probability functions of quantum and statistical mechanics into one mathematical object. The quantum mechanical part of the density operator uses the usual quantum mechanical wave function to account for the inherent particle probabilities. The statistical mechanics portion accounts for possible multiple wave functions attributable to random external influences. Typically, statistical mechanics deals with ensembles of many particles and only describes the dynamics of the system through statements of probability.

5.11.1 Introduction to the Density Operator

We usually assume we know the initial wave function of a particle or system. Consider the example wave function depicted in Figure 5.11.1 where the initial wave function consists of *two exactly specified basis functions with two exactly specified components*. Suppose the initial wave function can be written

$$|\psi(0)\rangle = 0.9|u_1\rangle + 0.43|u_2\rangle$$

As shown in Figure 5.11.2, the quantum mechanical probability of finding the electron in the first eigenstate must be

$$|\langle u_1 \mid \psi(0)\rangle|^2 = (0.9)^2 = 81\%$$

Similarly, the quantum mechanical probability that the electron occupies the second eigenstate must be

$$|\langle u_2 \mid \psi(0)\rangle|^2 = (0.43)^2 = 19\%$$

We know the values of these probabilities with certainty since we know the decomposition of the initial wave function $|\psi(0)\rangle$ and the coefficients (0.9 and 0.43) with 100% certainty. We assume that the wave function $|\psi\rangle$ satisfies the time dependent Schrodinger wave equation while the basis states satisfy the time-independent Schrodinger wave equation

$$\hat{H}|\psi\rangle = i\hbar\partial_t|\psi\rangle \qquad \hat{H}|u_n\rangle = E_n|u_n\rangle$$

What if we don't exactly know the initial preparation of the system? For example, we might be working with an infinitely deep well. Suppose we try to prepare a number of identical systems. Suppose we make four such systems with parameters as close as possible to each other. Figure 5.11.3 shows the ensemble of systems all having the same width L. Unlike the present case with only four systems, we usually (conceptually) make

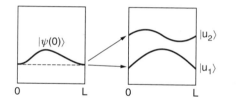

FIGURE 5.11.1
The initial wave function consists of exactly two basis functions.

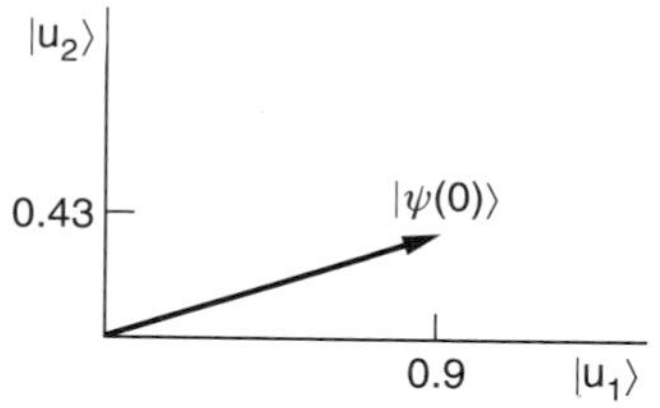

FIGURE 5.11.2
The components of the wave function.

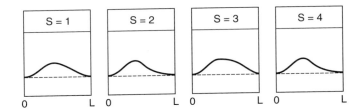

FIGURE 5.11.3
An ensemble of four systems.

an infinite number of systems to form an ensemble. Figure 5.11.3 shows that we were not able to prepare identical wave functions $|\psi\rangle$. Denote the wave function for system S by $|\psi_s\rangle$. Then the wave function $|\psi_s\rangle$ for each system must have different coefficients, as for example,

$$|\psi_1\rangle = 0.98\,|u_1\rangle + 0.19\,|u_2\rangle$$

$$|\psi_2\rangle = 0.90\,|u_1\rangle + 0.43\,|u_2\rangle$$

$$|\psi_3\rangle = 0.95\,|u_1\rangle + 0.31\,|u_2\rangle \qquad (5.11.1)$$

$$|\psi_4\rangle = 0.90\,|u_1\rangle + 0.43\,|u_1\rangle$$

The four wave functions appear in Figure 5.11.4. Notice how system $S=2$ and system $S=4$ both have the same wave function.

What actual wave function $|\psi\rangle$ describes the system? Answer: An *actual* $|\psi\rangle$ does not exist; we can only talk about an average wave function. In fact, if we had prepared many such systems, we would only be able to specify the probability that the system has a certain wave function. For example, for the four systems described above, the probability of each type of wave function must be given by

$$P(S=2) = \frac{1}{2} \qquad P(S=1) = \frac{1}{4} \qquad P(S=3) = \frac{1}{4}$$

For convenience, systems $S=2$ and $S=4$ have both been symbolized by $S=2$ since they have identical wave functions. Perhaps this would be clearer by writing

$$P\{0.90|u_1\rangle + 0.43|u_2\rangle\} = \frac{1}{2}, \ P\{0.98|u_1\rangle + 0.19|u_2\rangle\} = \frac{1}{4}, \ P\{0.95|u_1\rangle + 0.31|u_2\rangle\} = \frac{1}{4}$$

We can now represent the four systems by three vectors in our Hilbert space rather than four so long as we also account for the probability.

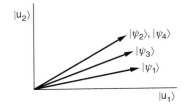

FIGURE 5.11.4
The different initial wave functions for the infinitely deep well.

Now let's do something a little unusual. Suppose we try to define an *average wave function* to represent a typical system (think of the example with the four infinitely deep wells)

$$\text{Ave}\{|\psi\rangle\} = \sum_s P_S |\psi_S\rangle$$

Recall, the classical average of a quantity "x_i" or "x" can be written as $\langle x_i \rangle = \sum_i x_i P_i$ and $\langle x \rangle = \int dx\, x\, P(x)$ for the discrete and continuous cases respectively (see Appendix 4). The average wave function would represent an average system in the ensemble. We look at the entire ensemble of systems (there might be an infinite number of copies) and say that the wave function $\text{Ave}\{|\psi\rangle\}$ behaves like the average for all those systems. The wave function $\text{Ave}|\psi\rangle$ would represent the quantum mechanical stochastic processes while the probabilities P_S represent the macroscopic probabilities. No one actually uses this average wave function. The sum of the squares of the components of $\text{Ave}\{|\psi\rangle\}$ do not necessarily add to one since the probabilities P_i are squared (see the chapter review exercises).

Now here comes the really unusual part where we define an average probability. If we exactly know the wave function, then we can exactly calculate probabilities using the quantum mechanical probability density $\psi^*(x)\,\psi(x)$ (it's a little odd to be combining the words "exact" and "probability"). Now let's extend this idea of probability using our ensemble of systems. We change notation and let P_ψ be the probability of finding one of the systems to have a wave function of $|\psi\rangle$. We define an average probability density function according to

$$\text{average}\,(\psi^*\psi) = \sum_\psi P_\psi \left(\psi^*(x)\,\psi(x) \right) \tag{5.11.2}$$

This formula contains both the quantum mechanical probability density $\psi^*\psi$ and the macroscopic probability P_ψ. We could use the "s" subscripts on P_s so long as we include only one type of wave function for each s. Equation (5.11.2) assumes a discrete number of possible wave functions $|\psi_S\rangle$. However, the situation might arise with so many wave functions that they essentially form a continuum in Hilbert space (i.e., "s" must be a continuously varying parameter). In such a case, we talk about the classical probability density ρ_S which gives the probability per unit interval s of finding a particular wave function.

$$\text{average}\,(\psi^*\psi) = \int dS\, \rho_S \left(\psi_S^*(x)\,\psi_S(x) \right)$$

The probability ρ_S is very similar to the density of states seen in later chapters; rather than a subscript of "S," we would have a subscript of energy and units of "number of states per unit energy per unit volume." We continue with Equation (5.11.2) since it contains all the essential ingredients.

Rearranging Equation (5.11.2), we obtain a "way to think of the average." First switch the order of the wave function and its conjugate.

$$\text{average}\,(\psi^*\psi) = \sum_\psi P_\psi\, \psi^*(x)\,\psi(x) = \sum_\psi P_\psi\, \psi(x)\,\psi^*(x)$$

Next write the wave functions in Dirac notation and factor out the basis kets $|x\rangle$

$$\text{average }(\psi^*\psi) = \sum_\psi P_\psi \langle x \mid \psi\rangle\langle\psi \mid x\rangle = \langle x| \left\{ \sum_\psi P_\psi \; |\psi\rangle\langle\psi| \right\} |x\rangle$$

We define the density operator to be

$$\hat{\rho} = \sum_\psi P_\psi \, |\psi\rangle\langle\psi| \qquad\qquad (5.11.3)$$

Example 5.11.1

Find the initial density operator $\hat{\rho}(0)$ for the wave functions given in the table. We assume four two-level atoms.

Initial Wave Function $|\psi_S(0)\rangle$	Probability P_s
$|\psi_1\rangle = 0.98 \, |u_1\rangle + 0.19 \, |u_2\rangle$	1/4
$|\psi_2\rangle = 0.90 \, |u_1\rangle + 0.43 \, |u_2\rangle$	1/2
$|\psi_3\rangle = 0.95 \, |u_1\rangle + 0.31 \, |u_2\rangle$	1/4

The initial density operator must be given by $\hat{\rho}(0) = \sum_{S=1}^{3} P_S|\psi_S(0)\rangle\langle\psi_S(0)|$. Substituting the probabilities and initial wave functions, we find

$$\hat{\rho}_S(0) = P_1|\psi_1(0)\rangle\langle\psi_1(0)| + P_2|\psi_2(0)\rangle\langle\psi_2(0)| + P_3|\psi_3(0)\rangle\langle\psi_3(0)|$$

$$= \frac{1}{4}[0.98|u_1\rangle + 0.19|u_2\rangle] \, [0.98\langle u_1| + 0.19\langle u_2| \,]$$

$$+ \frac{1}{2}[0.90|u_1\rangle + 0.43|u_2\rangle] \, [0.90\langle u_1| + 0.43\langle u_2| \,]$$

$$+ \frac{1}{4}[0.95|u_1\rangle + 0.31|u_2\rangle] \, [0.95\langle u_1| + 0.31\langle u_2| \,]$$

Collecting terms

$$\hat{\rho}(0) = 0.86 \, |u_1\rangle\langle u_1| + 0.307 \, |u_1\rangle\langle u_2| + 0.307 \, |u_2\rangle\langle u_1| + 0.14 \, |u_2\rangle\langle u_2|$$

Example 5.11.2

Assume that the probability of any wave function is zero except for the particular wave function $|\psi_o\rangle$. Find the density operator in both the discrete and continuous cases.

Solution: For the discrete case, the probability can be written as $P_\psi = \delta_{\psi,\psi_o}$ and the density operator becomes

$$\hat{\rho} = \sum_\psi P_\psi|\psi\rangle\langle\psi| = \sum_\psi \delta_{\psi,\psi_o}|\psi\rangle\langle\psi| = |\psi_o\rangle\langle\psi_o|$$

For the continuous case, the probability density can be written as $\rho_\psi = \delta(\psi - \psi_o)$ and the density operator becomes

$$\hat{\rho} = \int d\psi \, \rho_\psi \, |\psi\rangle\langle\psi| = \int d\psi \, \delta(\psi - \psi_o) \, |\psi\rangle\langle\psi| = |\psi_o\rangle\langle\psi_o|$$

5.11.2 The Density Operator and the Basis Expansion

The density operator can be written in a basis vector expansion. The density operator $\hat{\rho}$ has a range and domain within a single vector space. Suppose the set of basis vectors $\{|m\rangle = u_m\}$ spans the vector space of interest. People most commonly use the energy eigenfunctions as the basis set. Using the basis function expansion of an operator as described in Chapter 3, the density operator can be written as

$$\hat{\rho} = \sum_{mn} \rho_{mn} |m\rangle\langle n| \tag{5.11.4}$$

where $\langle n| = |n\rangle^+$. Recall that ρ_{mn} must be the matrix elements of the operator $\hat{\rho}$. We term the collection of coefficients $[\rho_{mn}]$ the "density matrix." Recall from Chapter 3

$$\langle a|\hat{\rho}|b\rangle = \sum_{mn} \rho_{mn}\langle a \mid m\rangle\langle n \mid b\rangle = \sum_{mn} \rho_{mn}\delta_{am}\delta_{bn} = \rho_{ab}$$

where $|a\rangle, |b\rangle$ are basis vectors. This topic shows how the density operator can be expanded in a basis and provides an interpretation of the matrix elements.

The density operator provides two types of average. The first type consists of the quantum mechanical average and the second consists of the ensemble average. For the ensemble average, we imagine a large number of systems prepared as nearly the same as possible. We imagine a collection of wave functions $\{|\psi_S(t)\rangle\}$ with one for each different system S. Again, we imagine that P_s denotes the probability of finding a particular wave function $|\psi_S(t)\rangle$. Assume that all of the wave functions of the systems can be described by vector spaces spanned by the set $\{|m\rangle = u_m\}$ as shown in Figure 5.11.5. Assume the same basis functions for each system. Each wave function $|\psi_S(t)\rangle$ can be expanded in the complete orthonormal basis set for each system.

$$|\psi_S(t)\rangle = \sum_m \beta_m^{(S)}(t) \, |m\rangle \tag{5.11.5}$$

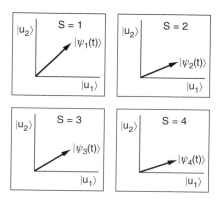

FIGURE 5.11.5
Four systems with the same basis functions.

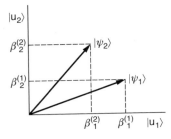

FIGURE 5.11.6
Two realizations of a system have different wave functions and therefore different components.

The superscript (S) on each expansion coefficient refers to a different system. However, a single set of basis vectors applies to all of the systems S in the ensemble of systems. Therefore, if two systems (a) and (b) have different wave functions, then the coefficients must be different $\beta_m^{(a)} \neq \beta_m^{(b)}$ (see Figure 5.11.6).

Using the definition of the density operator, we can write

$$\hat{\rho}(t) = \sum_S P_S |\psi_S(t)\rangle\langle\psi_S(t)| \tag{5.11.6}$$

Notice that the density operator in the Schrodinger picture can depend on time since the wave functions depend on time. Using the definition of adjoint

$$\langle\psi_S(t)| = |\psi_S(t)\rangle^+ = \left[\sum_n \beta_n^{(S)}|n\rangle\right]^+ = \sum_n \beta_n^{(S)^*}\langle n| \tag{5.11.7}$$

Substituting Equation (5.11.5) and (5.11.7) into Equation (5.11.6), we obtain

$$\hat{\rho}(t) = \sum_{mn}\sum_S P_S\, \beta_m^{(S)}\beta_n^{(S)^*}\, |m\rangle\,\langle n|$$

Now, compare this last expression with Equation (5.11.4) to see that the matrix of the density operator (i.e., the density matrix) must be

$$\rho_{mn} = \langle m|\hat{\rho}|n\rangle = \sum_S P_S \beta_m^{(S)}\beta_n^{(S)^*} = \left\langle \beta_m^{(S)}\beta_n^{(S)^*}\right\rangle = \overline{\beta_m^{(S)}\beta_n^{(S)^*}} \tag{5.11.8}$$

where the "e" subscript indicates the ensemble average.

Whereas the density *operator* $\hat{\rho}$ gives the *ensemble* average of the wavefuntion projection operator $\overline{|\psi\rangle\langle\psi|} = \langle|\psi\rangle\langle\psi|\rangle_e$ the density *matrix* element ρ_{mn} provides the ensemble average of the wave function coefficients $\rho_{mn} = \overline{\beta_m^{(S)}\beta_n^{(S)^*}} = \langle\beta_m^{(S)}\beta_n^{(S)^*}\rangle_e$. The averages must be taken over all of the systems S in the ensemble.

The whole point of the density operator is to simultaneously provide two averages. We use the quantum mechanical average to find quantities such as average position, momentum, energy, or electric field using only the quantum mechanical state of a given system. The ensemble average takes into account non-quantum mechanical influences such as variation in container size or slight differences in environment that can be represented by a probability P_s. Notice in the definition of density operator

$$\hat{\rho}(t) = \sum_S P_S |\psi_S(t)\rangle\langle\psi_S(t)| \tag{5.11.9}$$

that if one of the systems occurs at the exclusion of all others (say $S = 1$) so that

$$\hat{\rho}(t) = \left|\psi_1(t)\right\rangle\!\left\langle\psi_1(t)\right| = \left|\psi(t)\right\rangle\!\left\langle\psi(t)\right| \tag{5.11.10}$$

then the density operator only provides quantum mechanical averages. In such a case, the wave functions for all the systems in the ensemble have the same form since macroscopic conditions do not differently affect any of the systems. Density operators as in Equation (5.11.10) without a statistical mixture will be called "pure" states. Sometimes people refer to a density operator of the form $|\psi(t)\rangle\langle\psi(t)|$ as a "state" or a "wave function" because it consists solely of the wave function $|\psi(t)\rangle$. We will see later that in the case of Equation (5.11.10), the density operator and the wave function provide equivalent descriptions of the single quantum mechanical system and both obey a Schrodinger equation.

Now let's examine the conceptual meaning of the matrix elements $\rho_{mn} = \overline{\beta_m^{(S)}\beta_n^{(S)*}}$ in Equation (5.11.8). The diagonal matrix elements $\rho_{nn} = \overline{\beta_n^{(S)}\beta_n^{(S)*}} = \overline{P(n)}$ provide the average probability of finding the system in eigenstate n. In other words, even though the diagonal elements have the ensemble average, we still "think" of them as $\rho_{nn} \sim |\beta_n|^2 \sim P(n)$ where $P(n)$ represents the usual quantum mechanical probability. For an ensemble of systems with different wave functions $|\psi^{(s)}\rangle$, we must average the quantum probability over the various systems.

The off-diagonal elements of the density operator appear to be similar to the probability amplitude that a particle simultaneously exists in two states. For simplicity, assume the ensemble has only one type of wave function given by the superposition $|\psi\rangle = \sum_n \beta_n |u_n\rangle$ so that $\langle u_m \mid \psi \rangle = \sum_n \beta_n \langle u_m \mid u_n \rangle = \beta_m$. The off-diagonal elements have the form

$$\rho_{ab} = \langle u_a|\hat{\rho}|u_b\rangle = \langle u_a \mid \psi\rangle\langle\psi \mid u_b\rangle = \langle u_a \mid \psi\rangle\langle u_b \mid \psi\rangle^+ = \beta_a\beta_b^*$$

Recall that the classical probability of finding a particle in both states can be written as

$$P(a \text{ and } b) = P(a)P(b)$$

for independent events. But $P(a) = |\beta_a|^2$ and $P(b) = |\beta_b|^2$ so, combining the last several expressions provides

$$\rho_{ab} = \langle u_a|\hat{\rho}|u_b\rangle = \beta_a\beta_b^* \sim \sqrt{P(a \text{ and } b)}$$

Apparently, the off-diagonal elements of the density operator must be related to the probability of simultaneously finding the particle in both states "a" and "b." This should remind the reader of a transition from one state to another. In fact, we will see that the off-diagonal elements can be related to the susceptibility, which is related to the dipole moment and the gain or loss.

Example 5.11.3

For Example 5.11.1, find the density matrix.

Solution: The density matrix can be written as

$$\underline{\rho} = \begin{bmatrix} 0.86 & 0.307 \\ 0.307 & 0.14 \end{bmatrix}$$

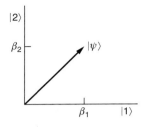

FIGURE 5.11.7
Wave function and components.

for the basis set $\{|u_1\rangle, |u_2\rangle\}$. Notice how the coefficients of the first and last term add to one ... this is not an accident. The diagonal elements of the density matrix correspond to the probability that a particle will be found in the level $|u_1\rangle, |u_2\rangle$.

Example 5.11.4

Find the coordinate and energy basis set representation for the density operator under the following conditions. Assume the density operator can be written as $\hat{\rho} = |\psi\rangle\langle\psi|$. Assume the energy basis set can be written as $\{|u_a\rangle\}$ so that $|\psi\rangle = \sum_n \beta_n |u_n\rangle$. What is the probability of finding the particle in state $|a\rangle = |u_a\rangle$?

Solution: First, the expectation of the density operator in the coordinate representation.

$$\langle x|\hat{\rho}|x\rangle = \langle x\,|\,\psi\rangle\langle\psi\,|\,x\rangle = \psi^*(x)\,\psi(x)$$

Second, the expectation of the density operator using a vector basis (Figure 5.11.7) produces the probability of finding the particle in the corresponding state (i.e., diagonal matrix elements give the probability of occupying a state).

$$\langle u_a|\hat{\rho}|u_a\rangle = \langle u_a\,|\,\psi\rangle\langle\psi\,|\,u_a\rangle = \langle u_a\,|\,\psi\rangle\langle u_a\,|\,\psi\rangle^+ = |\beta_a|^2$$

Third, the probability of finding the particle in state $|a\rangle$ is

$$P(a) = |\beta_a|^2 = \langle u_a|\hat{\rho}|u_a\rangle = \rho_{aa}$$

as seen in the last equation. Therefore, the diagonal elements provide the probability of finding the electron in the corresponding state.

Example 5.11.5

Show the diagonal terms of the denisity matrix add to 1. Assume the wave function $|\psi^{(s)}\rangle = \sum_n \beta_n^{(s)}|n\rangle$ describes system "s" and the density operator has the form $\hat{\rho} = \sum_s P_s |\psi^{(s)}\rangle\langle\psi^{(s)}|$.

Solution: The matrix element of the density operator can be written as

$$\rho_{aa} = \langle a|\hat{\rho}|a\rangle = \langle a|\left\{\sum_s P_s|\psi^{(s)}\rangle\langle\psi^{(s)}|\right\}|a\rangle = \sum_s P_s\langle a\,|\,\psi^{(s)}\rangle\langle\psi^{(s)}\,|\,a\rangle = \sum_s P_s|\beta_a^{(s)}|^2$$

Now summing over the diagonal elements (i.e., equivalent to taking the trace)

$$Tr(\hat{\rho}) = \sum_a \rho_{aa} = \sum_a \sum_s P_s|\beta_a^{(s)}|^2 = \sum_s P_s \sum_a |\beta_a^{(s)}|^2 = \sum_s P_s\,1 = \sum_s P_s$$

where the second to last result follows since the components for each individual wavefunction "s" must add to one. Finally, the sum of the probabilities P_s must add to one to get

$$Tr(\hat{\rho}) = \sum_s P_s = 1$$

This shows that the probability of finding the particle in any of the states must sum to one.

5.11.3 Ensemble and Quantum Mechanical Averages

For semiconductor lasers, the density operator most importantly provides averages of operators. We know averages of operators correspond to classically observed quantities. We will find the average of an operator has the form

$$\langle\langle \hat{O} \rangle\rangle = Tr\left(\hat{\rho}\hat{O}\right) \tag{5.11.11}$$

where for now the double brackets reminds us that the density operator involves two probabilities and therefore two types of average. This equation contains both the quantum mechanical and ensemble average. "Tr" means to take the trace. We will see that the average of the dipole moment leads to the polarization and susceptibility, which leads to the complex wave vector k_n, which then leads to the gain.

We define the quantum mechanical "q" and ensemble "e" averages for an operator $\hat{O}$ as follows:

Quantum Mechanical	Ensemble
$\langle \hat{O} \rangle_q = \langle \psi \vert \hat{O} \vert \psi \rangle$	$\langle \hat{O} \rangle_e = \sum_s P_s \, \hat{O}_s$

where $\vert\psi\rangle$ denotes a typical quantum mechanical wave function. In what follows, we take the operator in the *ensemble* average to be just a number that depends on the particular system S (for example, it might be the system temperature that varies from one system to the next).

Now we will show that the ensemble and quantum mechanical average of an operator $\hat{O}$ can be calculated using $\langle\langle \hat{O} \rangle\rangle = Tr(\hat{\rho}\hat{O})$. Recall the definition of trace,

$$Tr\left(\hat{\rho}\hat{O}\right) = \sum_n \langle n \vert \hat{\rho}\hat{O} \vert n\rangle \tag{5.11.12}$$

Although the trace does not depend on the particular basis set, equations of motion use the energy basis $\{\vert n\rangle = \vert u_n\rangle\}$ where $\hat{\mathscr{H}}\vert n\rangle = E_n \vert n\rangle$.

First let's find the quantum mechanical average of an operator for the specific system S starting with

$$\left\langle \hat{O} \right\rangle_q^{(S)} = \langle \psi_S \vert \hat{O} \vert \psi_S \rangle \quad \text{with} \quad \vert\psi_S(t)\rangle = \sum_n \beta_n^{(S)}(t) \vert u_n\rangle \tag{5.11.13}$$

provides the wave function for the system S. Substituting the wave function (5.11.12) into the operator expression (5.11.11) provides

$$\left\langle \hat{O} \right\rangle_q^{(S)} = \sum_n \beta_n^{*(S)} \langle u_n \vert \hat{O} \sum_m \beta_m^{(S)}(t) \vert u_m\rangle = \sum_{nm} \beta_n^{*(S)} \beta_m^{(S)} \langle u_n \vert \hat{O} \vert u_m\rangle = \sum_{mn} \beta_m^{(S)} \beta_n^{*(S)} O_{nm} \tag{5.11.14}$$

There is one such average for each different system S since there is a different wave function for each different system. For a given system S, this last expression gives the quantum mechanical average of the operator for that one system.

As a last step, take the ensemble average of Equation (5.11.14) using P_s as the probability.

$$\langle\langle\hat{O}\rangle\rangle = \left\langle \langle\hat{O}\rangle_q^{(S)}\right\rangle_e = \sum_S P_S \langle\hat{O}\rangle_q^{(S)} = \sum_S P_S \sum_{mn} \beta_m^{(S)} \beta_n^{*(S)} O_{nm}$$

Rearranging the summation and noting $Tr\left(\underline{\rho}\underline{O}\right) = \sum_{mn} \rho_{mn} O_{nm}$ provides the desired results.

$$\langle\langle\hat{O}\rangle\rangle = \sum_{mn}\left(\sum_S P_S \beta_m^{(S)} \beta_n^{*(S)}\right) O_{nm} = \sum_{mn} \overline{\beta_m^{(S)}\beta_n^{*(S)}} O_{nm} = \sum_{mn} \rho_{mn} O_{nm} = Tr\left(\hat{\rho}\hat{O}\right)$$

Example 5.11.6

Find the average of an operator for a pure state with $\hat{\rho} = |\psi(t)\rangle\,\langle\psi(t)|$

Solution: Equation (5.11.12) provides

$$\langle\hat{O}\rangle = Tr\left(\hat{\rho}\hat{O}\right) = \sum_n \langle u_n \mid \psi(t)\rangle\langle\psi(t)|\hat{O}|u_n\rangle = \sum_n \langle\psi(t)|\hat{O}|u_n\rangle\langle u_n \mid \psi(t)\rangle = \langle\psi(t)|\hat{O}|\psi(t)\rangle$$

where the first summation uses the definition of trace and the last step used the closure relation for the states $|u_n\rangle$. For the pure state, we see that the trace formula reduces to the ordinary quantum mechanical average of $\langle\hat{O}\rangle = \langle\psi(t)|\hat{O}|\psi(t)\rangle$.

Example 5.11.7 The Two Averages

The electron gun in a television picture tube has a filament to produce electrons and a high voltage electrode to accelerate them toward the phosphorus screen (see top portion of Figure 5.11.8). Suppose the high voltage section is slightly defective and produces small random voltage fluctuations. We therefore expect the momentum $p = \hbar k$ of the electrons to slightly vary similar to the bottom portion of Figure 5.11.8. Assume each individual electron is in a plane wave state $\psi^{(k)}(x, t) = \frac{1}{\sqrt{V}}e^{ikx - i\omega t}$ where the superscript "(k)" indicates the various systems rather than "(s)." Find the average momentum.

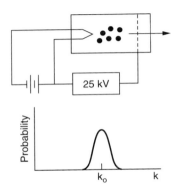

FIGURE 5.11.8
The electron gun (top) produces a slight variation in wave vector k (bottom).

Solution: The quantum mechanical average can be found

$$\left\langle \psi^{(k)} \middle| \hat{p} \middle| \psi^{(k)} \right\rangle_q = \left\langle \psi^{(k)} \middle| \frac{\hbar}{i} \frac{\partial}{\partial x} \middle| \psi^{(k)} \right\rangle_q$$

Substituting for the wave function, we find

$$\left\langle \psi^{(k)} \middle| \hat{p} \middle| \psi^{(k)} \right\rangle_q = \frac{1}{V} \int_V dV \, e^{-ikx+i\omega t} \frac{\hbar}{i} \frac{\partial}{\partial x} e^{ikx-i\omega t} = \hbar k$$

where we assume that the wave function is normalized to the volume V. We still need to average over the various electrons (i.e., the systems or k values) leaving the electron gun. The bottom portion of Figure 5.11.8 shows the k-vectors have a Gaussian distribution. Therefore, the average momentum must be $\langle \langle \hat{p} \rangle_q \rangle_e = \hbar k_o$.

Example 5.11.8

Let $\hat{H}$ be the Hamiltonian for a two-level system with energy eigenvectors $\{ |u_1\rangle, |u_2\rangle \}$ so that $\hat{H}|u_1\rangle = E_1|u_1\rangle$ and $\hat{H}|u_2\rangle = E_2|u_2\rangle$. What is the matrix of $\hat{H}$ with respect to the basis vectors $\{ |u_1\rangle, |u_2\rangle \}$?

Solution: The matrix elements of $\hat{H}$ can be written as $H_{ab} = \langle u_a|\hat{H}|u_b\rangle = E_b \delta_{ab}$ which can be written as

$$\underline{H} = \begin{bmatrix} E_1 & 0 \\ 0 & E_2 \end{bmatrix}$$

Example 5.11.9

What is the ensemble-averaged energy $\langle \hat{H} \rangle \equiv \langle \langle \hat{H} \rangle \rangle$? Assume all of the information remains the same as for Examples 5.11.8, 5.11.1, and 5.11.3.

Solution: We want to evaluate the average given by

$$\left\langle \hat{H} \right\rangle = Tr\left(\hat{\rho}\hat{H} \right)$$

We can insert basis vectors as required by the trace and then insert the closure relation between the two operators. We would then end up with the formula identical to taking the trace of the product of two matrices.

$$Tr\left(\hat{\rho}\hat{H} \right) = Tr\left(\underline{\rho}\underline{H} \right) = Tr \begin{bmatrix} 0.86 & 0.307 \\ 0.307 & 0.14 \end{bmatrix} \begin{bmatrix} E_1 & 0 \\ 0 & E_2 \end{bmatrix} = Tr \begin{bmatrix} 0.86E_1 & 0.307E_2 \\ 0.307E_1 & 0.14E_2 \end{bmatrix}$$

Of course, in switching from operators to matrices, we have used the isomorphism between operators and matrices. Operations using the operators must be equivalent to operations using the corresponding matrices. Summing the diagonal elements provides the trace of a matrix and we find

$$\left\langle \hat{H} \right\rangle = Tr\left(\hat{\rho}\hat{H} \right) = 0.86E_1 + 0.14E_2$$

So the average is no longer equal to the eigenvalue E_1 or E_2! The average energy represents a combination of the energies dictated by both the quantum mechanical and ensemble probabilities.

Example 5.11.10

What is the probability that an electron will be found in the state $|u_1\rangle$? Assume all of the information remains the same as for Examples 5.11.9, 5.11.8, 5.11.3, and 5.11.1

Solution: We assume the density matrix

$$\underline{\rho} = \begin{bmatrix} 0.86 & 0.307 \\ 0.307 & 0.14 \end{bmatrix}$$

The answer is Probability of state #1 $= \langle u_1 | \hat{\rho} | u_1 \rangle = \rho_{11} = 0.86$. In fact, we can find the probability of the first state being occupied directly from the definition of the density operator

$$\langle 1 | \hat{\rho} | 1 \rangle = \langle 1 | \left[\sum_S P_S | \psi_S \rangle \langle \psi_S | \right] | 1 \rangle = \sum_S P_S \langle 1 | \psi_S \rangle \langle \psi_S | 1 \rangle = \sum_S P_S \beta_1^{(S)} \beta_1^{(S)*} = \overline{\beta_1 \beta_1^*}$$

5.11.4 Loss of Coherence

In some cases, the physical system introduces uncontrollable phase shifts in the various components of the wave functions. Suppose the wave functions have the form

$$\left| \psi^{(\phi_1, \phi_2, \dots)} \right\rangle = \sum_n \beta_n^{(\phi_n)} | n \rangle \tag{5.11.15a}$$

where the phases $(\phi_1, \phi_2, \dots)$ label the wave function and assume a continuous range of values. The components have the form

$$\beta_n^{(\phi_n)} = \left| \beta_n \right| e^{i\phi_n} \tag{5.11.15b}$$

Let $P_\phi(\phi_1, \phi_2, \dots) = P(\phi_1) P(\phi_2) \dots$ be the probability for $|\psi^{(\phi_1, \phi_2, \dots)}\rangle$. The density operator assumes the form

$$\hat{\rho} = \int d\phi_1 d\phi_2 \dots P(\phi_1, \phi_2, \dots) \left| \psi^{(\phi_1, \phi_2, \dots)} \right\rangle \left\langle \psi^{(\phi_1, \phi_2, \dots)} \right| \tag{5.11.16}$$

Now we can demonstrate the loss of coherence. Expanding the terms in Equation (5.11.16) using Equations (5.11.15) produces

$$\hat{\rho} = \int d\phi_1 d\phi_2 \dots P(\phi_1) P(\phi_2) \dots \sum_{m,n} \left| \beta_m \right| \left| \beta_n \right| e^{i(\phi_m - \phi_n)} | m \rangle \langle n | \tag{5.11.17}$$

The exponential terms drop out for $m = n$. The integral over the probability density can be reduced using the property $\int d\phi_a P(\phi_a) = 1$.

$$\hat{\rho} = \sum_m \left| \beta_m \right|^2 | m \rangle \langle m | + \sum_{m \neq n} \left| \beta_m \right| \left| \beta_n \right| | m \rangle \langle n | \int d\phi_m P(\phi_m) \, e^{i\phi_m} \int d\phi_n P(\phi_n) \, e^{-i\phi_n} \tag{5.11.18}$$

Assume a uniform distribution $P(\phi) = 1/2\pi$ on $(0, 2\pi)$. The integrals produce

$$\int_0^{2\pi} d\phi_m \, P(\phi_m) \, e^{i\phi_m} = 0$$

and the density operator in Equation (5.11.18) becomes diagonal

$$\hat{\rho} = \sum_m |\beta_m|^2 |m\rangle\langle m| \tag{5.11.19}$$

Some mechanisms produce a loss of coherence. For example, making a measurement causes the wave functions to collapse to a single state. The wave functions become $|m\rangle$ with quantum mechanical probability $|\beta_m|^2$ so that the density operator appears as in Equation (5.11.19). Often the macroscopic and quantum probabilities are combined into a single number p_m and the density operator becomes

$$\hat{\rho} = \sum_m p_m |m\rangle\langle m| \tag{5.11.20}$$

Notice that the density matrix $\hat{\rho} = |\psi\rangle\langle\psi|$ for a pure state can always be reduced to a single entry by choosing a basis with $|\psi\rangle$ as one of the basis vectors. The mixed state in Equation (5.11.9) cannot be reduced from its diagonal form.

Example 5.11.11

Suppose a system contains N independent two-level atoms (per unit volume). Each atom corresponds to one of the systems that make up the ensemble. Given the density matrix ρ_{mn}, find the number of two-level atoms in level #1 and level #2.

Solution: The number of atoms in state $|a\rangle$ must be given by

$$N_a = (\text{total number})\,(\text{Prob of state } a) = N\rho_{aa} \tag{5.11.21}$$

Example 5.11.12

Suppose there are $N=5$ atoms as shown in Figure 5.11.9. Let the energy basis set be $\{|1\rangle = |u_1\rangle, |2\rangle = |u_2\rangle\}$. Assume a measurement determines the number of atoms in each level. Find the density matrix based on the figure.

Solution: Notice that the diagonal density-matrix elements can be calculated if we assume that the wave functions $|\psi_S\rangle$ can only be either $|u_1\rangle$ or $|u_2\rangle$. The density operator has the form

$$\hat{\rho} = \sum_{S=1}^{2} P_S |\psi_S\rangle\langle\psi_S| = P_1|u_1\rangle\langle u_1| + P_2|u_2\rangle\langle u_2|$$

or, equivalently, the matrix must be

$$\rho_{aa} = \langle u_a|\hat{\rho}|u_a\rangle \;\to\; \underline{\rho} = \begin{bmatrix} P_1 & 0 \\ 0 & P_2 \end{bmatrix}$$

Figure 5.11.9 clearly shows that $\text{Prob}(1) = P_1 = 3/5$ and $\text{Prob}(2) = P_2 = 2/5$. Therefore, the probability of an electron occupying level #1 must be $\rho_{11} = 2/5$ and the probability of an electron occupying level #2 must be $\rho_{22} = 3/5$.

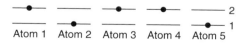

FIGURE 5.11.9
Ensemble of atoms in various states.

Example 5.11.13

What if we had defined the occupation number operator $\hat{n}$ to be

$$\hat{n}|1\rangle = 1|1\rangle \qquad \hat{n}|2\rangle = 2|2\rangle$$

Calculate the expectation value of $\hat{n}$ using the trace formula for the density operator.

Solution

$$\langle \hat{n} \rangle = Tr(\hat{\rho}\hat{n}) = Tr\begin{bmatrix} 2/5 & 0 \\ 0 & 3/5 \end{bmatrix}\begin{bmatrix} 1 & 0 \\ 0 & 2 \end{bmatrix} = \frac{8}{5}$$

This just says that the average state is somewhere between "1" and "2." We can check this result by looking at the figure. The average state should be

$$1 \cdot \text{Prob}(1) + 2 \cdot \text{Prob}(2) = 1\frac{2}{5} + 2\frac{3}{5} = \frac{8}{5}$$

as found with the density matrix.

5.11.5   Some Properties

1. If $P_\psi = 1$ so that $\hat{\rho} = |\psi\rangle\langle\psi|$ represents a pure state, then $\hat{\rho}\hat{\rho} = |\psi\rangle\langle\psi|\psi\rangle\langle\psi| = |\psi\rangle\langle\psi| = \hat{\rho}$.
   In this case, the operator $\hat{\rho}$ satisfies the property required for idempotent operators. The only possible eigenvalues for this particular density operator are 0 and 1.

   $$\hat{\rho}|v\rangle = v|v\rangle \;\rightarrow\; \hat{\rho}\hat{\rho}|v\rangle = v|v\rangle \;\rightarrow\; v^2|v\rangle = v|v\rangle \;\rightarrow\; v^2 = v \;\rightarrow\; v = 0, 1$$

2. All density operators are Hermitian

   $$\hat{\rho}^+ = \left\{\sum_\psi P_\psi|\psi\rangle\langle\psi|\right\}^+ = \sum_\psi P_\psi\{|\psi\rangle\langle\psi|\}^+ = \sum_\psi P_\psi|\psi\rangle\langle\psi| = \hat{\rho}$$

   since the probability must be a real number.

3. Diagonal elements of the density matrix give the probability that a system will be found in a specific eigenstate. The diagonal elements take into account both ensemble and quantum mechanical probabilities. Let $\{|a\rangle\}$ be a complete set of states (basis states) and let the wave function for each system have the form

   $$|\psi(t)\rangle = \sum_a \beta_a^{(\psi)}(t)|a\rangle$$

   The diagonal elements of the density matrix must be

   $$\rho_{aa} = \langle a|\hat{\rho}|a\rangle = \langle a|\left\{\sum_\psi P_\psi|\psi\rangle\langle\psi|\right\}|a\rangle = \sum_\psi P_\psi\langle a|\psi\rangle\langle\psi|a\rangle$$

   $$= \sum_\psi P_\psi\beta_a^{*(\psi)}\beta_a^{(\psi)} = \overline{|\beta_a|^2} = \overline{\text{prob}(a)}$$

4. The sum of the diagonal elements must be unity

$$Tr(\hat{\rho}) = \sum_n \rho_{nn} = 1$$

since the matrix diagonal contains all of the system probabilities.

5.12 Review Exercises

5.1 Using the Poisson Brackets, show

$$[A, A] = 0 \quad [A, B] = -[B, A] \quad [A, c] = 0$$

for A, B a function of phase space coordinates q, p and "c" a number.

5.2 Using the Poisson Brackets, show

$$[A + B, C] = [A, C] + [B, C] \quad [AB, C] = A[B, C] + [A, C]B$$

where A, B, C denote differentiable functions of the phase space coordinates q, p.

5.3 Using the Poisson Brackets, show

$$[q_i, q_j] = 0 [p_i, p_j] = 0 \quad [q_i, p_j] = \delta_{ij}$$

5.4 Explain why the following relation must hold for δx_i independent of each other.

$$\sum_{i=1}^{N} f(x_i)\delta x_i = 0 \rightarrow f(x_i) = 0$$

This is similar to a step in the procedure to derive Lagrange's equation. Hint: Consider a matrix solution. Keep in mind that δx_1, for example, can have any number of values such as 0.1, 0.001, etc.

5.5 Assume periodic boundary conditions. Show how

$$0 = \delta I = \int_{t_1}^{t_2} \int_{\vec{r}_1}^{\vec{r}_2} dt\, d^3x \left[\frac{\partial \mathcal{L}}{\partial \eta} \delta\eta + \frac{\partial \mathcal{L}}{\partial \dot{\eta}} \frac{\partial}{\partial t} \delta\eta + \frac{\partial \mathcal{L}}{\partial(\partial_i \eta)} \partial_i \delta\eta \right]$$

leads to

$$\int_{t_1}^{t_2} \int_{\vec{r}_1}^{\vec{r}_2} dt\, d^3x \left[\frac{\partial \mathcal{L}}{\partial \eta} - \frac{\partial}{\partial t} \frac{\partial \mathcal{L}}{\partial \dot{\eta}} - \partial_i \frac{\partial \mathcal{L}}{\partial(\partial_i \eta)} \right] \delta\eta = 0$$

Explain and show any necessary conditions of the limits of the spatial integral.

5.6 Suppose the Lagrange density has the form $\mathcal{L} = \frac{\rho}{2}\dot{\eta}^2 + \frac{\beta}{2}[(\partial_x \eta)^2 + (\partial_y \eta)^2]$ for 1-D motion, where ρ, β resemble the mass density and spring constant (Young's modulus) for the material, and $\eta = \eta(x, y, t)$. Find the equation of motion for η.

5.7 If $\mathcal{L} = \frac{\rho}{2}\dot{\eta}^2 + \frac{\beta}{2}(\nabla\eta)^2$ where $(\nabla\eta)^2 = \nabla\eta \cdot \nabla\eta$ and $\eta = \eta(x, y, z)$ then find the equation of motion for η.

5.8 Starting with $\mathcal{L} = i\hbar\psi^*\dot{\psi} - (\hbar^2/2m)\nabla\psi^* \cdot \nabla\psi - V(r)\,\psi^*\psi$, show the alternate form of the Lagrange density by partial integration.

$$\mathcal{L} = i\hbar\psi^*\dot{\psi} + \frac{\hbar^2}{2m}\psi^*\,\nabla^2\psi - V(r)\,\psi^*\,\psi = \psi^*\left(i\hbar\partial_t + \frac{\hbar^2}{2m}\nabla^2 - V\right)\psi$$

5.9 Show Hamiltonian

$$\mathcal{H} = \pi\dot{\psi} - \mathcal{L} = \psi^*\left(-\frac{\hbar^2}{2m}\nabla^2 + V\right)\psi$$

based on the Lagrange density

$$\mathcal{L} = \psi^*\left(i\hbar\partial_t + \frac{\hbar^2}{2m}\nabla^2 - V\right)\psi$$

5.10 For Hermitian operators, show that the following definitions of variance all reduce to the same thing.

$$\left\langle\left(\hat{O} - \bar{O}\right)^+\left(\hat{O} - \bar{O}\right)\right\rangle$$

$$\left\langle\left(\hat{O} - \bar{O}\right)\left(\hat{O} - \bar{O}\right)^+\right\rangle$$

$$\left\langle\frac{1}{2}\left(\hat{O} - \bar{O}\right)^+\left(\hat{O} - \bar{O}\right) + \frac{1}{2}\left(\hat{O} - \bar{O}\right)\left(\hat{O} - \bar{O}\right)^+\right\rangle$$

5.11 Find the standard deviation for an operator $\hat{O}$ in one of its eigenstates $|n\rangle$.

5.12 Show a particle does not move from an energy eigenstate once its wavefunction collapses to the eigenstate. Hint: consider the evolution operator.

5.13 Find the classical Hamiltonian for the harmonic oscillator starting with the classical Lagrangian $L = T - V$, finding momentum from L and then applying the Legendre transformation.

5.14 Show that momentum must be conserved if the Lagrangian does not depend on position. Repeat the demonstration using the Hamiltonian.

5.15 Assume the pulley has mass M and radius R and that it supports two masses as in Figure P4.1. The kinetic energy of the pulley is given by $T_p = \frac{1}{2}I\dot{\theta}^2$ where I is the moment of inertial given by $I = \int dm\, R^2$.

  1. Find I.

  2. Write the total kinetic and potential energy in terms of θ and $\dot{\theta}$.

  3. Use the Lagrangian to find the equation of motion and solve it.

5.16 Using Figure P5.15 and the results of Problem 5.15

  1. Write the Hamiltonian in terms of the angle.

  2. Find the equations of motion.

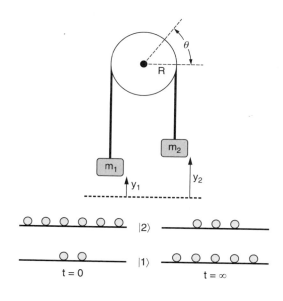

FIGURE P5.15
Pulley system.

5.17 Normalize the following functions (i.e., find A) to make them a probability density:

1. $y = A\,e^{ax}$ for $a < 0, x \in (0, \infty)$
2. $y = A\delta(x - 1) + (1 - A)\delta(x - 2)\ x \in (0, 3)$
3. Repeat part b for $x \in (0, 2)$
4. $y = A\sin(\pi x)\quad x \in (0, 1)$

5.18 For each of the density functions in Problem 5.17, find $\bar{x}$.

5.19 Fill in all the missing steps in Section 5.8.5. That is, solve the Schrodinger equation for the infinitely deep 1-D quantum well

$$V(x) = \begin{cases} 0 & x \in (0, L) \\ \infty & \text{elsewhere} \end{cases}$$

to show the energy eigenfunctions and eigenvalues have the form

$$\left\{ \phi(x) = \sqrt{\frac{2}{L}} \mathrm{Sin}\left(\frac{n\pi x}{L}\right) \right\}\quad E = \frac{n^2\pi^2\hbar^2 k^2}{2mL^2}$$

Hint: Separate variables in the Schrodinger equation, find the spatial functions and normalize.

5.20 Find the average momentum for the nth eigenstate for the infinitely deep quantum well given in Problem 5.19.

5.21 Suppose an engineer has a mechanism to place an electron in an initial state defined by

$$\Psi(x, 0) = \begin{cases} x & x \in (0, 1) \\ 2 - x & x \in (1, 2) \end{cases}$$

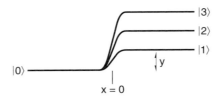

FIGURE P5.22
Electron wave divides among three paths on the right-hand side.

for an infinitely deep quantum well with width $L = 2$. The bottom of the well has potential $V = 0$.

1. At $t = 0$, what is the probability that the electron will be found in the $n = 2$ state?
2. What is the probability of finding $n = 2$ at time t?

5.22 An electron moves along a path located at a height $y = 0$ as shown in Figure P5.22. The path is along the x-direction as shown in the top figure. Near $x = 0$ the electron wave divides among three separate paths at heights $y = 1$, $y = 2$, $y = 3$. Suppose each path represents a possible state for the electron. Denote the states by $|0\rangle, |1\rangle, |2\rangle, |3\rangle$ so that the position operator $\hat{y}$ has the eigenvalue equations $\hat{y}|n\rangle = n|n\rangle$.
The set of $|n\rangle$ forms a discrete basis. Assume the full Hamiltonian has the form

$$\hat{H} = \frac{\hat{p}_x^2}{2m} + \hat{V} \quad \text{where} \quad \hat{V} = mg\hat{y}$$

Further assume $\hat{p}_x|n\rangle = p_n|n\rangle$ for $x \gg 0$ or $x \ll 0$.

1. Use the following probabilities (at time $t = 0$) for finding the particles on the paths $x \gg 0$

$$P_1 = \frac{1}{4} \quad P_2 = \frac{1}{2} \quad P_3 = \frac{1}{4}$$

to find suitable choices for the β_n in

$$|\psi(0)\rangle = \sum_{n=1}^{3} \beta_n|n\rangle$$

for the three paths $x \gg 0$. Neglect any phase factors.
2. Find the average $\langle \hat{V} \rangle = \langle \psi(0)|\hat{V}|\psi(0)\rangle$ for $x \gg 0$.
3. For $x \ll 0$, find $\langle \hat{H} \rangle$.
4. For $x \gg 0$, find $\langle \hat{H} \rangle$ in terms of n and p_n for $n = 1, 2, 3$.
5. Using the evolution operator $\hat{u}(t) = \exp \hat{H}t/(i\hbar)$, find $|\psi(t)\rangle$ for $x \ll 0$. Write the final answer in terms of n and p_n for $n = 1, 2, 3$.

5.23 Show $\hat{N}[\hat{a}^+|n\rangle] = (n+1)[\hat{a}^+|n\rangle]$ where $\hat{a}^+$ represents the harmonic oscillator raising operator and $\hat{N} = \hat{a}^+\hat{a}$.

5.24 Prove the classic integral relation

$$\frac{\hbar^2}{m} \int_{-\infty}^{\infty} dx \, u_b^* \frac{\partial u_a}{\partial x} = (E_a - E_b) \int_{-\infty}^{\infty} dx \, u_b^* x \, u_a$$

where $\hat{H}u_a = E_a u_a$, $\hat{H}u_b = E_b u_b$ and $\hat{H} = \frac{\hat{p}^2}{2m} + V(x)$. Use the following steps.

1. Show $[\hat{H}, \hat{x}] = -\frac{i\hbar}{m}\hat{p}$

2. Use the results of Part a to show

$$\frac{i\hbar}{m}\langle u_b | \hat{p} | u_a \rangle = -(E_b - E_a)\langle u_b | \hat{x} | u_a \rangle$$

   Show why $\langle u_b | \hat{H} = \langle u_b | E_b$.

3. Use the results of Part *b* to finally prove the relation stated at the start of this exercise.

5.25 Prove the special integrals at the end of Section 5.9 using ladder operators.

5.26 For the harmonic oscillator, calculate the second eigenfunction $u_2(x)$ using $\hat{a}^+$ and

$$u_1(x) = \left(\frac{\alpha}{2\sqrt{\pi}}\right)^{\frac{1}{2}} 2\alpha x e^{-\frac{\alpha^2 x^2}{2}} \quad \text{where} \quad \alpha^2 = \frac{m\omega_0}{\hbar}$$

5.27 Calculate $\langle \hat{p}^2/2m \rangle$ for a harmonic oscillator in the eigenstate $|u_n\rangle$.
Hint: Write the momentum operator in terms of the raising and lowering operators.

5.28 Find the Heisenberg representation of the momentum operator $\hat{p}$ for the infinitely deep square well when the bottom of the well has a constant potential of $V = c$.

5.29 Show $[\hat{x}_h, \hat{p}_h] = i\hbar$ for the Heisenberg representation using only the Schrodinger commutator $[\hat{x}_s, \hat{p}_s] = i\hbar$ and the fact that $\hat{u}$ is unitary.

5.30 Show

$$\left[V(x), \frac{\hbar}{i}\frac{\partial}{\partial x}\right] = \frac{\partial V}{\partial x}$$

5.31 Obtain Newton's second law $\hat{F}_h = d\hat{p}_h/dt$ using the evolution operator for a closed system and the rate of change of an operator in the Heisenberg representation.

5.32 Show $\dot{\hat{p}}_h = -k\hat{x}_h$ using the harmonic oscillator Hamiltonian $\hat{H} = (\hat{p}^2/2m) + (k/2)\hat{x}^2$ and the expression for the rate of change of Heisenberg operators.

5.33 Starting with the Heisenberg equation of motion $(d\hat{A}_h/dt) = (i/\hbar)[\hat{\mathcal{H}}_h, \hat{A}_h] + ((\partial/\partial t)\hat{A}_s)_h$ show the time average found for the Schrodinger representation.

$$\frac{d}{dt}\langle \hat{A} \rangle = \frac{i}{\hbar}\left\langle \left[\hat{\mathcal{H}}, \hat{A}\right]\right\rangle + \left\langle \frac{\partial \hat{A}}{\partial t}\right\rangle$$

Hint: Consider $\langle \psi(t_o) | (d\hat{A}_h/dt) | \psi(t_o) \rangle$.

5.34 Explain why the Heisenberg and Interaction representation become identical for closed systems.

5.35 Starting with $\dot{\beta}_m(t) - (E_m/i\hbar)\beta_m(t) = (1/i\hbar)\sum_n \beta_n(t)\mathcal{V}_{mn}(t)$ from Section 5.11, show the integrating factor must be $\mu_m(t) = \exp(-(E_m/i\hbar)t)$ and then show $\rightleftharpoons (d/dt)[\mu_m(t)\beta_m(t)] = (1/i\hbar)\mu_m(t)\sum_n \beta_n(t)\mathcal{V}_{mn}(t)$.

5.36 Show the first- and second-order terms in the interaction-representation perturbation theory

$$|\Psi_I(t)\rangle = \left\{1 + \frac{1}{i\hbar}\int_0^t dt_1 \hat{\mathcal{V}}_I(t_1)\right\}|\Psi_I(0)\rangle$$

reduces to the second order term for the Schrodinger representation

$$\beta_m(t) = \delta_{mi}\exp\left(\frac{E_i}{i\hbar}t\right) + \frac{1}{i\hbar}\exp\left(\frac{E_m}{i\hbar}t\right)\int_0^t d\tau\exp\left(-\frac{E_{mi}}{i\hbar}\tau\right)\mathscr{V}_{mi}(\tau)$$

Hint: Expand $|\Psi_I\rangle$ in basis $|n\rangle$, use the evolution operator, and project with $\langle m|$.

5.37 The chapter discusses time-dependent perturbation theory. Using the Schrodinger representation, derive the first-order correction to β using the coefficients a_n in

$$|\psi\rangle = \sum_n a_n(t)\,e^{\frac{E_n t}{i\hbar}}|n\rangle$$

where $\hat{H}|\psi\rangle = i\hbar\partial_t|\psi\rangle$ and $\hat{H}_o|n\rangle = E_n|n\rangle$ and $\hat{H} = \hat{H}_o + \hat{V}(t)$.

5.38 An engineering student prepares a two-level atomic system. The student doesn't know the exact wave function $|\psi\rangle$. After many attempts the student finds the following probability table.

$\lvert\psi\rangle$ at $t=0$	P_ψ	where
$0.98\lvert u_1\rangle + 0.19\lvert u_2\rangle$	$2/3$	$\hat{H}\lvert u_1\rangle = E_1\lvert u_1\rangle$
		$\hat{H}\lvert u_2\rangle = E_2\lvert u_2\rangle$
$0.90\lvert u_1\rangle + 0.43\lvert u_2\rangle$	$1/3$	

1. Write the density operator $\hat{\rho}(t=0)$ in a basis vector expansion.
2. What is the matrix of $\hat{\rho}(0)$?
3. What is the average energy $\langle\langle\hat{H}\rangle\rangle = \langle\hat{H}\rangle$?

5.39 Show the components of the average wave function $\text{Ave}\{|\psi\rangle\} = \sum P_s|\psi^{(s)}\rangle$ do not necessarily sum to one. Consider the simplest case: Assume that each wave function lives in a 2-D Hilbert space $|\psi^{(s)}\rangle = \beta_1^{(s)}|1\rangle + \beta_2^{(s)}|2\rangle$. Consider only two wave functions for $s=1, 2$. Assume all coefficients $\beta_n^{(s)}$ are real. To make the problem simpler, consider the case of $\beta_1^{(2)} = (1 + \varepsilon_1)\beta_1^{(1)}$ and $\beta_2^{(2)} = (1 + \varepsilon_2)\beta_2^{(1)}$.

1. *Show that the sum of the square of the components equals 1 if and only if $\varepsilon_1 = 0 = \varepsilon_2$.* Hint: Sum the squares of the coefficients of $\text{Ave}\{|\psi\rangle\}$ in the usual application of Pythagorean's theorem, collect the squared terms of P_1^2 and P_2^2, and add terms to 1 where appropriate. You should find a result similar to $1 + 2P_1P_2\{\beta_1^{(1)2}\varepsilon_1 + \beta_2^{(1)2}\varepsilon_2$, with ε_1 and $\varepsilon_2 \geq 0\}$.
2. Explain why the diagonal components of the density operator add to 1 but the sum of the square of the components of the average wave function do not.

5.40 For the wave function $|\psi^{(\phi)}\rangle = \beta_1|1\rangle + \beta_2 e^{i\phi}|2\rangle$ with the probability density $P(\phi) = 1/2\pi$ for $\phi \in (0, 2\pi)$, find the basis vector expansion of the density operator. Assume β_n are complex numbers.

5.41 Repeat Problem 5.40 for the case of $P(\phi) = \delta(\phi - 0)$.

5.42 The electron gun in a television picture tube has high voltage to accelerate electrons toward the phosphorus on the screen. Suppose the high-voltage section is slightly defective and produces small random voltage fluctuations. Suppose the wave vectors k are approximately uniformly and continuously distributed between k_1 and k_2. Assume each individual electron is in a plane wave state $\psi^{(k)}(x, t) = (1/\sqrt{V})\,e^{ikx - i\omega t}$.

1. Find the probability density $P(k)$ for the wave vector. Make sure it is correctly normalized so that its integral over k equals one.
2. Find the average $\bar{k}$.
3. Find $\langle\langle\hat{p}\rangle\rangle$.

$$\sum_{i=1}^{N} f(x_i)\,\delta x_i = 0 \quad \rightarrow f(x_i) = 0$$

$$0 = \delta I = \int_{t_1}^{t_2}\int_{\vec{r}_1}^{\vec{r}_2} dt\,d^3x\left[\frac{\partial L}{\partial \eta}\delta\eta + \frac{\partial L}{\partial\dot\eta}\frac{\partial}{\partial t}\delta\eta + \frac{\partial L}{\partial(\partial_i\eta)}\partial_i\delta\eta\right]$$

$$\int_{t_1}^{t_2}\int_{\vec{r}_1}^{\vec{r}_2} dt\,d^3x\left[\frac{\partial L}{\partial \eta} - \frac{\partial}{\partial t}\frac{\partial L}{\partial\dot\eta} - \partial_i\frac{\partial L}{\partial(\partial_i\eta)}\right]\delta\eta = 0$$

$$L = \frac{\rho}{2}\dot\eta^2 + \frac{\beta}{2}\left[(\partial_x\eta)^2 + (\partial_y\eta)^2\right]$$

$$\rho,\beta$$

$$\eta = \eta(x,y,t)$$

$$L = \frac{\rho}{2}\dot\eta^2 + \frac{\beta}{2}(\nabla\eta)^2 \quad (\nabla\eta)^2 = \nabla\eta\cdot\nabla\eta \quad \eta = \eta(x,y,z)\eta$$

$$L = i\hbar\psi^*\dot\psi - \frac{\hbar^2}{2m}\nabla\psi^*\cdot\nabla\psi - V(r)\,\psi^*\psi$$

$$L = i\hbar\psi^*\dot\psi + \frac{\hbar^2}{2m}\psi^*\nabla^2\psi - V(r)\,\psi^*\psi = \psi^*\left(i\hbar\partial_t + \frac{\hbar^2}{2m}\nabla^2 - V\right)\psi$$

$$H = \pi\dot\psi - L = \psi*\left(-\frac{\hbar^2}{2m}\nabla^2 + V\right)\psi$$

$$L = \psi^*\left(i\hbar\partial_t + \frac{\hbar^2}{2m}\nabla^2 - V\right)\psi$$

$$\left\langle(O-\bar{O})^+\left(\hat{O}-\bar{O}\right)\right\rangle \quad \left\langle\left(\hat{O}-\bar{O}\right)\left(\hat{O}-\bar{O}\right)^+\right\rangle$$

$$\left\langle\frac{1}{2}\left(\hat{O}-\bar{O}\right)^+\left(\hat{O}-\bar{O}\right) + \frac{1}{2}\left(\hat{O}-\bar{O}\right)\left(\hat{O}-\bar{O}\right)^+\right\rangle$$

$$\hat{O}\quad |n\rangle$$

$$L = T - V$$

$$T_p = \frac{1}{2}I\dot\theta^2$$

$$I = \int dm \, R^2$$

$$\theta \, \dot{\theta} \quad y = Ae^{ax} \, a < 0, \, x \in (0, \infty)$$

$$y = A\delta(x - 1) + (1 - A)\delta(x - 2)$$

$$x \in (0, 3)$$

$$x \in (1, 2)$$

$$y = A \sin(\pi x) \quad x \in (0, 1)$$

$$\bar{V}(x) = \begin{cases} 0 & x \in (0, L) \\ \infty & \text{elsewhere} \end{cases}$$

$$\left\{ \phi(x) = \sqrt{\frac{2}{L}} \sin\left(\frac{n\pi x}{L}\right) \right\} \quad E = \frac{n^2 \pi^2 \hbar^2 k^2}{2mL^2}$$

$$\Psi(x, 0) = \begin{cases} x & x \in (0, 1) \\ 2 - x & x \in (1, 2) \end{cases}$$

$$|0\rangle, |1\rangle, |2\rangle, |3\rangle$$

$$\hat{y} \, \hat{y}|n\rangle = n|n\rangle|n\rangle$$

$$\hat{H} = \frac{\hat{p}_x^2}{2m} + \hat{V} \quad \hat{V} = mg\hat{y}$$

$$\hat{p}_x|n\rangle = p_n|n\rangle$$

$$P_1 = \frac{1}{4}$$

$$P_1 = \frac{1}{4} \quad P_2 = \frac{1}{2} \quad P_1 = \frac{1}{4}$$

$$\beta_n \, |\psi(0)\rangle = \sum_{n=1}^{3} \beta_n|n\rangle$$

$$\langle \hat{V} \rangle = \langle \psi(0)|\hat{V}|\psi(0)\rangle$$

$$\langle \hat{H} \rangle$$

$$\langle \hat{H} \rangle$$

$$\hat{u}(t) = \exp \hat{H}t/(i\hbar) \quad |\psi(t)\rangle$$

$$\hat{N}[\hat{a}^+|n\rangle] = (n + 1)[\hat{a}^+|n\rangle] \quad \hat{a}^+$$

$$\hat{N} = \hat{a}^+\hat{a}$$

$$\frac{\hbar^2}{m}\int_{-\infty}^{\infty} dx\, u_b^*\frac{\partial u_a}{\partial x} = (E_a - E_b)\int_{-\infty}^{\infty} dx\, u_b^* x\, u_a \quad \hat{H}\,u_a = E_a u_a \quad \hat{H}\,u_b = E_b u_b \quad \hat{H} = \frac{\hat{p}^2}{2m} + V(x)$$

$$\left[\hat{H}, \hat{x}\right] = -\frac{i\hbar}{m}\hat{p}$$

$$\frac{i\hbar}{m}\langle u_b|\hat{p}|u_a\rangle = -(E_b - E_a)\langle u_b|\hat{x}|u_a\rangle$$

$$\langle u_b|\hat{H} = \langle u_b|E_b$$

$$u_2(x) \quad \hat{a}^+$$

$$u_1(x) = \left(\frac{\alpha}{2\sqrt{\pi}}\right)^{\frac{1}{2}} 2\alpha x e^{-\frac{\alpha^2 x^2}{2}} \quad \alpha^2 = \frac{m\omega_0}{\hbar}$$

$$\left\langle\frac{\hat{p}^2}{2m}\right\rangle \quad |u_n\rangle$$

$$\hat{p}[\hat{x}_h, \hat{p}_h] = i\hbar \quad [\hat{x}_s, \hat{p}_s] = i\hbar \quad \hat{u}$$

$$\left[V(x), \frac{\hbar}{i}\frac{\partial}{\partial x}\right] = -\frac{\partial V}{\partial x}$$

$$\hat{F}_h = d\hat{p}_h/dt$$

$$\dot{\hat{p}}_h = -k\hat{x}_h \quad \hat{H} = \frac{\hat{p}^2}{2m} + \frac{k}{2}\hat{x}^2$$

$$\frac{d\hat{A}_h}{dt} = \frac{i}{\hbar}\left[\hat{H}_h, \hat{A}_h\right] + \left(\frac{\partial}{\partial t}\hat{A}_s\right)_h$$

$$\frac{d}{dt}\langle\hat{A}\rangle = \frac{i}{\hbar}\langle\left[\hat{H}, \hat{A}\right]\rangle + \left\langle\frac{\partial\hat{A}}{\partial t}\right\rangle$$

$$\langle\psi(t_o)|\frac{d\hat{A}_h}{dt}|\psi(t_o)\rangle$$

$$\dot{\beta}_m(t) - \frac{E_m}{i\hbar}\beta_m(t) = \frac{1}{i\hbar}\sum_n \beta_n(t)\,V_{mn}(t)$$

$$\mu_m(t) = \exp\left(-\frac{E_m}{i\hbar}t\right) \quad \frac{d}{dt}[\mu_m(t)\beta_m(t)] = \frac{1}{i\hbar}\mu_m(t)\sum_n \beta_n(t)V_{mn}(t)$$

$$|\Psi_I(t)\rangle = \left\{1 + \frac{1}{i\hbar}\int_0^t dt_1 \hat{V}_I(t_1)\right\}|\Psi_I(0)\rangle$$

$$\beta_m(t) = \delta_{mi} \exp\left(\frac{E_i}{i\hbar}t\right) + \frac{1}{i\hbar}\exp\left(\frac{E_m}{i\hbar}t\right)\int_0^t d\tau \exp\left(-\frac{E_{mi}}{i\hbar}\tau\right)V_{mi}(\tau)$$

$$|\Psi_I \quad \rangle|n\rangle \quad \langle m|$$

$$\beta$$

$$|\psi\rangle = \sum_n a_n(t)\,e^{\frac{E_n t}{i\hbar}}|n\rangle \quad \hat{H}|\psi\rangle = i\hbar\partial_t|\psi\rangle \quad \hat{H}_o|n\rangle = E_n|n\rangle \quad \hat{H} = \hat{H}_o + \hat{V}(t)|\psi\rangle$$

$$\hat{\rho}(t=0)$$

$$\hat{\rho}(0)$$

$$\langle\langle\hat{H}\rangle\rangle = \langle\hat{H}\rangle$$

$$\text{Ave}\{|\psi\rangle\} = \sum_s P_s|\psi^{(s)}\rangle.$$

$$|\psi^{(s)}\rangle = \beta_1^{(s)}|1\rangle + \beta_2^{(s)}|2\rangle$$

$$\beta_n^{(s)}$$

$$\beta_1^{(2)} = (1+\varepsilon_1)\beta_1^{(1)}$$

$$\beta_2^{(2)} = (1+\varepsilon_2)\beta_2^{(1)}$$

$$\varepsilon_1 = 0 = \varepsilon_2$$

$$\text{Ave}\{|\psi\rangle\}$$

$$P_1^2 \quad P_2^2$$

$$1 + 2P_1P_2\left\{\beta_1^{(1)}2\varepsilon_1 + \beta_2^{(1)}2\varepsilon_2\right\}$$

$$|\psi^{(\phi)}\rangle = \beta_1|1\rangle + \beta_2 e^{i\phi}|2\rangle \quad P(\phi) = 1/2\pi$$

$$\phi \in (0, 2\pi) \quad \beta_n$$

$$P(\phi) = \delta(\phi - 0)$$

$$\psi^{(k)}(x,t) = \frac{1}{\sqrt{V}}e^{ikx - i\omega t}$$

$$P(k)$$

$$\bar{k}$$

$$\langle\langle\hat{p}\rangle\rangle$$

5.13 Further Reading

The following lists some well known references for the summary material in this chapter.

Mechanics

1. Marion J.B., *Classical Dynamics*, Academic Press, New York (1970).
2. Goldstein R., *Classical Mechanics*, Addison-Wesley Publishing, Reading, MA (1950).

Quantum Theory

3. Elbaz E., *Quantum, The Quantum Theory of Particles, Fields, and Cosmology*, Springer-Verlag, Berlin (1998).
4. Baym G., *Lectures on Quantum Mechanics*, Addison-Wesley Publishing, Reading, MA (1990).
5. Messiah A., *Quantum Mechanics*, Dover Publications, Mineola, NY (1999).

Density Operator

6. Blum K., *Density Matrix Theory and Applications*, 2nd ed., Plenum Press, New York (1996).

6

Light

The study of light through the centuries has produced a number of theories and experiments. Newton advanced the corpuscular theory, building on the early Greek view of matter composed of atoms. Then Young essentially proved the wave nature of light through the interference experiments. Maxwell provided the firm theoretical basis for the wave nature. During the late 1800s and early 1900s, further experiments showed predictions based on a continuous wave led to the well-known ultraviolet catastrophe. Plank and Einstein originated and developed the notion of the photon as an elementary quantum of propagating electromagnetic (EM) energy. The first half of the 20th century saw the development of the field quantization and the quantum electrodynamics (QED). Feynmann labeled QED as "the best theory we have" to describe the matter–light interaction. The second half of the 20th century witnessed the development of the coherent and squeezed optical states attributable to Glauber and Yuen. The study of light and electromagnetic fields has a long history. This chapter explores the nature of light in terms of quantum optics.

The properties of matter depend on the quantum mechanical state occupied by the constituent particles. Likewise, properties of light depend on the states for the photons. The present chapter begins with a discussion of the Classical Vector Potential and Gauge transformations. We discuss the solutions to the optical Schrodinger equation with special emphasis on the Fock, coherent, and squeezed states. The chapter introduces the Wigner function.

The material in the present chapter covers the "free field" case for light that applies to situations where the light does not interact with matter (except possibly for reflections). The Hamiltonian for the complete system consisting of both matter and light does not contain any interaction terms. The next chapter discusses the "matter–light interaction" case. The matter–light interaction gives rise to light emission and detection. The interaction of matter with the vacuum field produces the spontaneous emission necessary for both light emitting diodes and lasers.

6.1 A Brief Overview of the Quantum Theory of Electromagnetic Fields

As well known, atoms can emit light waves that are coherent with a driving optical field (stimulated emission) and they can also emit light on their own without a driving field (spontaneous emission). The spontaneous emission arises as a result of quantum vacuum fluctuations. We need to introduce quantum electrodynamics in order to discuss vacuum fluctuations.

Maxwell's electromagnetic (EM) equations can be solved to define a set of allowed EM modes. Sine and cosine functions represent the allowed modes for a cubic volume; for example, imagine sinusoidal standing waves in a Fabry–Perot cavity. The allowed wavelengths and polarization of light characterize these modes. The fields additionally have amplitude and phase; these attributes characterize only the field and not the mode. In QED, the electric field becomes an operator having the *form*

$$\hat{\mathscr{E}} \sim \hat{q}\sin(kz - \omega t) + \hat{p}\cos(kz - \omega t)$$

for a single mode. This can be recognized as an alternative to writing the field in terms of the amplitude and phase. The quadrature operators $\hat{q}, \hat{p}$ refer to amplitude and not to position and momentum. The operators do not commute and must operate on vectors in a Hilbert space that describe the amplitude and phase of the field. The various vectors in the amplitude space lead to the various EM fields with distinct properties. The states of light refer to the basis states of the amplitude space or to various combinations of the basis states. The Fock, coherent, and squeezed states represent three types of amplitude states.

The QED Fock state represents one of the most fundamental notions of Quantum ElectroDynamics (QED). A Fock state has a definite number of photons in the mode (this means that each mode has a definite average power) but completely random phase. The photons occupy the modes, which function as a type of framework or stage. In *classical* electrodynamics, a state without any photons corresponds to a mode without any amplitude. In QED, a state without any photons (the vacuum state) has an *average* electric field of zero, but nonzero variance (which is proportional to the square of the field). This means that the value of the electric field can fluctuate away from the average of zero. The nonzero variance refers to quantum fluctuations or noise; the vacuum state has the minimum quantum noise often termed vacuum fluctuations. Fock states make it easy to count photons but there exists a slight complication for engineering purposes! It turns out that *all* Fock states have zero average electric field because of the random phase. In addition, the noise associated with the Fock state must be larger than the minimum value set by the vacuum.

A *coherent* state has nonzero average electric field and fairly well-defined phase. The electric fields for these states can be pictured as sine and cosine waves; these states best describe laser emission. The coherent state actually consists of a linear combination of all Fock states. Coherent and Fock states can be seen to be quite different. One of the most important distinctions is that, for a coherent state with given amplitude, a Poisson probability distribution describes the number of photons n in the mode. A Fock state has an exact number of photons. The Poisson probability distribution links the standard deviation of photon number $\sigma = \sqrt{\langle n \rangle}$ with the average number of photons $\langle n \rangle$. For example, a beam with an average of $\langle n \rangle = 100$ photons has a standard deviation of $\sqrt{\langle n \rangle} = 10$ photons. One might reasonably expect the measured number of photons to range from 80 to 120 (almost 50% variation). The variation represents the shot noise. Now returning to the amplitude and phase, it just so happens that any coherent state has the same noise content as the vacuum state (regardless of the amplitude of the coherent state).

A *squeezed* vacuum state can be produced from the quantum vacuum state by reducing the noise (i.e., reducing the variance) in one set of parameters while adding it to another (i.e., "squeezing the noise out"). Squeezing the vacuum state is equivalent to squeezing the coherent state since the vacuum and coherent states have the same type and amount of noise. For example, noise can be removed from one quadrature term of the electric field

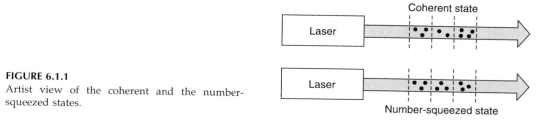

FIGURE 6.1.1

Artist view of the coherent and the number-squeezed states.

for "quadrature squeezing" but that removed noise reappears in the other quadrature term. Similarly, a "quiet" photon stream (i.e. a number squeezed state) obtains by removing noise from the photon-number but it reappears in the phase. "Sub-Poisson" statistics describe the quiet photon stream. Phase-squeezed states have less phase noise but more amplitude noise. Squeezed coherent states can be produced, detected, and used for low noise applications. Figure 6.1.1 shows examples of laser light moving past an observer. The top portion shows "coherent light" (i.e., light in a coherent state) where the number of photons in equal beam-lengths can vary from one length-interval to the next. The number of photons follows the Poisson probability distribution. The bottom portion of the figure shows a "number squeezed state" where the equal lengths have equal numbers of photons. Apparently, a number-squeezed state is related to a Fock state.

Spontaneous emission comprises another form of noise in the laser although we certainly should not term it as "noise" for a Light Emitting Diode (LED). We require spontaneous emission in a laser to start the laser oscillation but, in addition to producing larger than necessary threshold current, it also wastes energy. Interestingly, spontaneous emission is not solely a property of a collection of atoms, but arises from quantum vacuum fluctuations. The fluctuating electric field of the vacuum state initiates the spontaneous emission. Changing the number of vacuum modes coupled to the atomic ensemble can modify the rate of spontaneous emission—there exists one vacuum mode for each wavelength and polarization allowed by the boundary conditions on the enclosed volume. The field of cavity QED describes the theory and measurement of both spontaneous and stimulated emission for which these interesting cavity effects become important. These vacuum effects are essential for emitters (LEDs or lasers) that have physical sizes comparable to the wavelength of the emitted light (nanophotonics). Further, to characterize the effect of spontaneous emission on another laser or device, it is necessary to understand the effects of vacuum entropy.

Noise can be a problem because small (and low power) components do not deal with many particles (electrons, holes, and photons) at one time. For low particle numbers, as might be typical for small or low power components, the uncertainty (or standard deviation) in the signal can be roughly the same size as the magnitude of the quantity itself. Equivalently stated, small systems and signals have relatively large deviation of the number of particles carrying the signal compared with the average number. Ultimately, nanometer-scale devices (and for low power systems) have different types of noise with the quantum noise representing the commonly accepted lowest noise floor.

Noise can be more detrimental to an analog signal than a digital one. An analog signal usually caries information of a continuously varying parameter (such as distance, length, temperature, or music) and therefore, the noise determines the ultimate precision of the measurement or the quality of the impressed information. Noise as small as 0.1% can be significant for audio applications (for example). A digital system, however, must be capable of distinguishing between a logic "0" and "1." The signal strength must exceed a threshold value before the circuit recognizes the logic level. Many circuits and devices include a hysteresis effect to reduce the effect of noise. The "bit error rate" determines the accuracy of the digital system.

Noise problems can also appear in low power, high frequency RF or RADAR transmitters. These transmitters must operate at higher powers in order to keep the signal-to-noise ratios (S/N) as large as possible. For conventional electronic equipment (not just optical equipment) operating at modest powers of 10 W and 30 GHz, the quantum noise becomes a significant factor over a distance of 5 miles.

The theory of quantized fields mathematically unifies the pictures of light as particles and as waves. We know the photon as the basic quantum of light. The EM fields and the EM Hamiltonian are quantized similar to the electronic harmonic oscillator. The quantized electric field will be seen to consist of a wave portion (described by the complex traveling wave) and a particle portion consisting of creation and annihilation operators. Quantum field theory mathematically unifies the wave and particle pictures for all matter not just photons or electrons.

The previous few paragraphs point out the importance of quantum optics and some very interesting sources of noise in electromagnetic systems. Although quantum noise is interesting and important, other forms of noise such as RIN and thermal noise must be addressed.

6.2 The Classical Vector Potential and Gauges

In classical electrodynamics, we treat the magnetic and electric fields as physical quantities. A classical Hamiltonian (i.e., electromagnetic energy stored in free space) can be written in terms of these fields. However, the electric and magnetic fields can be derived from vector and scalar potential functions. The vector potential propagates as a wave and can be Fourier decomposed into plane waves. Replacing the Fourier amplitudes with operators quantizes the vector potential. Therefore the electromagnetic fields and Hamiltonian can also be quantized.

The procedure to find and quantize the vector potential uses the Coulomb gauge. Gauge transformations refer to certain changes that can be made to the vector and scalar potentials without affecting the mathematical expressions for the electric and magnetic fields. Consequently, both the fields and Maxwell's equations must be invariant with respect to gauge transformation.

There exists a number of different Gauge transformations with the Coulomb and Lorentz gauges being the most common. We use the *Coulomb* gauge to quantize the electromagnetic fields as a result of the fields having independent generalized coordinates. This gauge makes the vector potential a transverse field (the field is perpendicular to the direction of propagation). It also provides Poisson's equation for voltage; that is, the scalar potential provides the instantaneous voltage between any two spatially separated points. We do not use the Lorentz gauge that manifests the Lorentz invariance since it makes the fields more difficult to quantize.

Sometimes we consider the EM fields to be the "physical" objects and the vector potentials to be just mathematical constructions. However, the potentials produce real effects for device engineering. The Aharanov–Bohm devices provide an example where the vector potential (and not the fields) can be used for modulating currents.

The topics in this section (1) discuss the relation between the EM fields and the potential functions, (2) show that the resulting electric and magnetic fields satisfy

Maxwell's equations, (3) discuss the gauge transformation, and then (4) demonstrate the plane wave expansion of the vector potential.

6.2.1 Relation between the Electromagnetic Fields and the Potential Functions

This topic introduces the relations between the electromagnetic fields and the potentials. It shows that derived EM fields satisfy Maxwell's equations. The next topic specializes to a source-free region of space and then Section 6.2.3 introduces the gauge transformation and the importance of the Coulomb gauge.

The following two equations always relate the vector potential $\vec{A}(\vec{r}, t)$ and a scalar potential "Φ" to the magnetic and electric fields regardless of the gauge.

$$\vec{\mathscr{B}} = \nabla \times \vec{A}(\vec{r}, t) \quad \text{and} \quad \vec{\mathscr{E}} = -\frac{\partial \vec{A}(\vec{r}, t)}{\partial t} - \nabla \Phi \tag{6.2.1}$$

In this topic, we will make frequent use of the Coulomb gauge, which requires the vector potentials $\vec{A}(\vec{r}, t)$ to satisfy the Coulomb condition $\nabla \cdot \vec{A} = 0$. Section 6.2.3 discusses the origin and meaning of the Coulomb gauge. The present topic shows that the scalar potential Φ is the "electrostatic voltage due to charges" and that the vector potential (in the Coulomb gauge) can be pictured as a transverse traveling wave when the electric and magnetic fields are traveling waves.

First we show that the magnetic and electric fields derived from the potentials in the Coulomb gauge satisfy Maxwell's equations

$$\nabla \cdot \vec{\mathscr{B}} = 0 \qquad \nabla \times \vec{\mathscr{E}} = -\frac{\partial \vec{\mathscr{B}}}{\partial t}$$
$$\nabla \cdot \vec{\mathscr{D}} = \rho \qquad \nabla \times \vec{\mathscr{H}} = \vec{\mathscr{J}} + \frac{\partial \vec{\mathscr{D}}}{\partial t} \tag{6.2.2}$$

where $\vec{\mathscr{D}} = \varepsilon \vec{\mathscr{E}} = \varepsilon_0 \vec{\mathscr{E}} + \vec{\mathscr{P}} = \varepsilon_0 \vec{\mathscr{E}} + \varepsilon_0 \chi \vec{\mathscr{E}} = \varepsilon_0 (1 + \chi) \vec{\mathscr{E}}$, $\vec{\mathscr{B}} = \mu \vec{\mathscr{H}} = \mu_0 \vec{\mathscr{H}}$ and where ε_0, $\mathscr{P}$, χ, μ_0 denote the permittivity of free-space, polarization, susceptibility, and permeability of free-space. We assume an isotropic, homogeneous, nonmagnetic medium.

1. First we show that the magnetic field derived from the vector potential satisfies $\nabla \cdot \vec{\mathscr{B}} = 0$. The curl of the vector potential can be found using a determinant

$$\vec{\mathscr{B}} = \nabla \times \vec{A} = \begin{vmatrix} \tilde{x} & \tilde{y} & \tilde{z} \\ \partial_x & \partial_y & \partial_z \\ A_x & A_y & A_z \end{vmatrix}$$

and the divergence of the magnetic field gives the triple product

$$\nabla \cdot \vec{\mathscr{B}} = \nabla \cdot \nabla \times \vec{A} = \begin{vmatrix} \partial_x & \partial_y & \partial_z \\ \partial_x & \partial_y & \partial_z \\ A_x & A_y & A_z \end{vmatrix}$$

$$= \partial_x (\partial_y A_z - \partial_z A_y) - \partial_y (\partial_x A_z - \partial_z A_x) + \partial_z (\partial_x A_y - \partial_y A_x) = 0$$

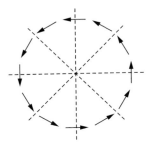

FIGURE 6.2.1
Divergence (along dotted line) of the curl must be zero.

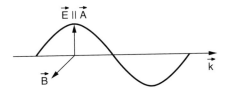

FIGURE 6.2.2
Vector potential propagates along the same direction as the EM wave.

This last result obviously holds since the curl of a vector field measures the amount of "rotation" of that vector field around a point (see the arrows in Figure 6.2.1); the direction of the curl vector points perpendicular to the plane of rotation. However, we might imagine that the divergence of a vector field measures the amount of the field that "diverges" away from a point (i.e., the change of the vector field along the dotted lines in the figure). Therefore, we expect the divergence of the curl to be zero.

Notice that we did not need the Coulomb condition for this result.

2. Next we demonstrate that the electric field and the magnetic induction E, B derived from the vector potential satisfy

$$\nabla \times \vec{\mathscr{E}} = -\frac{\partial \vec{B}}{\partial t}$$

Starting with $\nabla \times \vec{\mathscr{E}}$ and substituting for the electric field using $\vec{\mathscr{E}} = -\partial \vec{A}(\vec{r}, t)/\partial t - \nabla \Phi$ gives

$$\nabla \times \vec{\mathscr{E}} = \nabla \times \left(-\frac{\partial \vec{A}}{\partial t} - \nabla \Phi \right) = -\frac{\partial}{\partial t} \nabla \times \vec{A} - \nabla \times \nabla \Phi = -\frac{\partial}{\partial t} \vec{B}$$

which demonstrates the desired Maxwell equation. To arrive at the result, we have used $\vec{B} = \nabla \times \vec{A}(\vec{r}, t)$ and the fact that $\nabla \times \nabla \Phi$, the curl of the gradient, must be zero since

$$\nabla \times \nabla \Phi = \begin{vmatrix} \tilde{x} & \tilde{y} & \tilde{z} \\ \partial_x & \partial_y & \partial_z \\ \partial_x \Phi & \partial_y \Phi & \partial_z \Phi \end{vmatrix} = 0$$

The result can also be obtained by picturing the gradient as an outwardly pointing vector whereas the curl measures the amount of rotation. Again notice that we did not need the Coulomb condition and so this result holds for any vector and scalar potential.

3. Now we show that the electric field derived from the potentials satisfies the third of Maxwell's equations, namely

$$\nabla \cdot \vec{D} = \rho$$

In this case, we need the Coulomb gauge $\nabla \cdot \vec{A} = 0$ condition in order to show that the scalar potential Φ satisfies Poisson's equation. Substituting for the electric field we find

$$\frac{\rho}{\varepsilon_o} = \nabla \cdot \vec{\mathscr{E}} = \nabla \cdot \left(-\frac{\partial \vec{A}}{\partial t} - \nabla \Phi \right) = -\frac{\partial}{\partial t} \nabla \cdot \vec{A} - \nabla^2 \Phi = -\nabla^2 \Phi$$

We recognize this as Poisson's equation for the electrostatic potential Φ

$$\nabla^2 \Phi = -\frac{\rho}{\varepsilon_o}$$

The solution to Poisson's equation (neglecting a constant potential term)

$$\Phi(\vec{x}, t) = \frac{1}{4\pi\varepsilon_0} \int \frac{\rho(\vec{r}', t)}{|\vec{r} - \vec{r}'|} d^3 r'$$

gives the voltage at position $\vec{r}$ due to charge density $\rho(\vec{r}', t)$ located at position $\vec{r}'$. The integral in $\Phi(\vec{x}, t)$ has the effect of adding together the potential due to a point charge located at each point $\vec{r}'$—the integral represents the convolution. The function $G(\vec{r} - \vec{r}') = (-1)/(4\pi\varepsilon_o|\vec{r} - \vec{r}'|)$ is the Green function for Poisson's equation. It satisfies Poisson's equation for a unit point charge located at $\vec{r}'$ according to $\nabla^2 \Phi = -\delta(\vec{r} - \vec{r}')/\varepsilon_o$. We can easily show the Green function has the correct form by calculating the field, due to a positive point charge located at the origin $\vec{r}' = 0$ (see the chapter exercises).

The formula for $\Phi(\vec{x}, t)$ is identical to the one normally obtained for the electrostatic case. Interestingly, the instant the charge density appears at point $\vec{r}'$, it establishes a voltage at point $\vec{r}$. The solution would appear to violate a relativity principle prohibiting signals from propagating faster than the speed of light in vacuum c. The resolution to the apparent paradox resides with the fact that we are interested in the fields (rather than the potentials), integrals over 3-D space, and retarded and advanced Green functions—refer to O. L. Brill and B. Goodman, *Am. J. Phys.* 35, 832 (1967).

4. The final Maxwell equation provides a wave equation for the vector potential. Again we need the Coulomb gauge condition $\nabla \cdot \vec{A} = 0$. Starting with

$$\nabla \times \vec{B} = \varepsilon\mu \frac{\partial \vec{E}}{\partial t} + \mu \vec{J}$$

and using relations between fields and potentials

$$\vec{\mathcal{B}} = \nabla \times \vec{A}(\vec{r}, t) \quad \text{and} \quad \vec{E} = -\frac{\partial \vec{A}}{\partial t} - \nabla \Phi$$

provides

$$\nabla \times \nabla \times \vec{A} = \varepsilon \mu \frac{\partial}{\partial t}\left(-\frac{\partial \vec{A}}{\partial t} - \nabla \Phi\right) + \mu \vec{J} = -\varepsilon \mu \frac{\partial^2 \vec{A}}{\partial t^2} - \varepsilon \mu \frac{\partial}{\partial t}\nabla \Phi + \mu \vec{J} \tag{6.2.3}$$

The double cross product can be evaluated using the differential form of the "BAC–CAB" rule

$$\vec{A} \times \vec{B} \times \vec{C} = \vec{B}\left(\vec{A} \cdot \vec{C}\right) - \vec{C}\left(\vec{A} \cdot \vec{B}\right) \quad \text{so} \quad \nabla \times \nabla \times \vec{A} = \nabla\left(\nabla \cdot \vec{A}\right) - \nabla^2 \vec{A}$$

The first term on the right-hand side yields zero because of the Coulomb gauge condition $\nabla \cdot \vec{A} = 0$. Therefore, Equation (6.2.3) becomes a wave equation

$$\nabla^2 \vec{A} - \varepsilon \mu \frac{\partial^2 \vec{A}}{\partial t^2} = +\varepsilon \mu \frac{\partial}{\partial t}\nabla \Phi - \mu \vec{J} \tag{6.2.4}$$

with the speed of light in the medium $v = 1/\sqrt{\mu \varepsilon}$.

6.2.2 The Fields in a Source-Free Region of Space

In a source-free region of space, the charge and current density must be zero

$$\rho = 0 \quad \vec{\mathcal{J}} = 0$$

The magnetic and electric fields become

$$\vec{\mathcal{B}} = \nabla \times \vec{A}(\vec{r}, t) \quad \text{and} \quad \vec{\mathcal{E}} = -\frac{\partial \vec{A}(\vec{r}, t)}{\partial t}$$

assuming that the source-free voltage is zero (i.e., at least it should be independent of position) so that $\nabla \Phi = 0$. Step 3 in the previous topic shows that $\nabla \Phi = 0$ when $\rho = 0$ only for the Coulomb gauge. The wave equation for the vector potential given by Equation (6.2.4), with the source term

$$\varepsilon \mu \frac{\partial}{\partial t}\nabla \Phi - \mu \vec{J} = 0$$

becomes

$$\nabla^2 \vec{A} - \varepsilon \mu \frac{\partial^2 \vec{A}}{\partial t^2} = 0 \tag{6.2.5}$$

where, again, the speed of light in the medium is $v = 1/\sqrt{\mu \varepsilon}$.

Example 6.2.1

Find $\vec{E}$, $\vec{B}$, and an alternate expression for $\nabla \cdot \vec{A} = 0$ using the vector potential given by

$$\vec{A} = \frac{\vec{A}_0}{\omega} e^{i(\vec{k}\cdot\vec{r} - \omega t)}$$

where $\vec{k} = k\hat{z}$.

Solution: The electric field (Figure 6.2.2) can be written as

$$\vec{E} = -\frac{\partial \vec{A}}{\partial t} = i\vec{A}_0 e^{i(kz - \omega t)} \qquad (6.2.6)$$

The magnetic field can be written as

$$\vec{B} = \nabla \times \vec{A} = \frac{i\vec{k} \times \vec{A}_0}{\omega} e^{i(\vec{k}\cdot\vec{r} - \omega t)} = i\frac{k}{\omega}\hat{z} \times \vec{A}_0 e^{i(\vec{k}\cdot\vec{r} - \omega t)} = i\frac{\hat{z} \times \vec{A}_0}{c} e^{i(\vec{k}\cdot\vec{r} - \omega t)}$$

Finally, an alternate form for the gauge condition follows by substituting the vector potential into $\nabla \cdot \vec{A} = 0$

$$0 = \nabla \cdot \vec{A} = \frac{\vec{A}_0}{\omega} \nabla \cdot e^{i(\vec{k}\cdot\vec{r} - \omega t)} = \frac{i\vec{k} \cdot \vec{A}}{\omega}$$

to give

$$\vec{k} \cdot \vec{A} = 0 \qquad (6.2.7)$$

This last equation shows why people sometimes interchangeably use the terms "Coulomb gauge" and "transverse gauge." The direction of the vector $\vec{A}$ must be perpendicular to the propagation direction of the wave according to Equation (6.2.7). The direction of the vector $\vec{A}$ must be parallel to the electric field according to Equation (6.2.6).

6.2.3 Gauge Transformations

A gauge transformation changes the mathematical form of the vector and scalar potential but leaves the form of the electric and magnetic fields unaltered. Maxwell's equation must therefore be invariant with respect to gauge transformations. We can proceed using two methods. The first method uses 4-vector notation usually found with discussions of special relativity. The method simultaneously treats all components of the 4-vector potential. In the present topic, we discuss a classical second method that separates the components of the 4-vector potential into an ordinary 3-vector and a fourth potential term.

The vector potential can be written as

$$A^\mu = \left(A^0, \vec{A}\right) = \left(\Phi, \vec{A}\right)$$

where Φ is the electrostatic potential. The gauge transformation simultaneously changes all four terms. We can change the 4-vector potential by a 4-vector gradient of a scalar function without affecting the fields

$$A^\mu_{new} = A^\mu_{old} + \partial^\mu \Lambda = \left(\Phi_{old}, \vec{A}_{old}\right) + \left(\frac{\partial}{\partial t}\Lambda, -\nabla\Lambda\right)$$

where $\partial_\mu = ((\partial/\partial t), \nabla)$ and $\partial^\mu = ((\partial/\partial t), -\nabla)$. This equation can be written in the usual notation as

$$\Phi_{\text{new}} = \Phi_{\text{old}} + \frac{\partial}{\partial t}\Lambda \quad \vec{A}_{\text{new}} = \vec{A}_{\text{old}} - \nabla\Lambda \tag{6.2.8}$$

where the function $\Lambda = \Lambda(\vec{r}, t)$ is arbitrary.

We can see that the particular choice of gauge does not affect the expressions for the magnetic and electric fields. Calculating "new" fields from "new" potentials and then substituting the gauge transformation provides

$$\vec{\mathscr{E}}^{\text{new}} = -\frac{\partial}{\partial t}\vec{A}^{\text{new}} - \nabla\Phi^{\text{new}} = -\frac{\partial}{\partial t}\left(\vec{A}^{\text{old}} - \nabla\Lambda\right) - \nabla\left(\Phi^{\text{old}} + \frac{\partial\Lambda}{\partial t}\right) = -\frac{\partial\vec{A}^{\text{old}}}{\partial t} - \nabla\Phi^{\text{old}} = \vec{\mathscr{E}}^{\text{old}}$$

$$\vec{\mathscr{B}}^{\text{new}} = \nabla \times \vec{A}^{\text{new}} = \nabla \times \left(\vec{A}^{\text{old}} - \nabla\Lambda\right) = \nabla \times \vec{A}^{\text{old}} = \vec{\mathscr{B}}^{\text{old}}$$

We have used the fact that derivatives can be interchanged and that the curl of a gradient produces zero. These last two equations make it unnecessary to show that the fields derived from Equations (6.2.8) satisfy Maxwell's equations as we did in Section 6.2.1. We can equally well show the same results using the 4-vector notation. The field tensor $F^{\mu\nu} = \partial^\mu A^\nu - \partial^\nu A^\mu$ provides the electric and magnetic fields. Substituting the gauge transformation $A^\mu_{\text{new}} = A^\mu_{\text{old}} + \partial^\mu\Lambda$ produces

$$F^{\mu\nu}_{\text{new}} = \partial^\mu A^\nu_{\text{new}} - \partial^\nu A^\mu_{\text{new}} = \partial^\mu\left(A^\nu_{\text{old}} + \partial^\nu\Lambda\right) - \partial^\nu\left(A^\mu_{\text{old}} + \partial^\mu\Lambda\right)$$

$$= F^{\mu\nu}_{\text{old}} + \partial^\mu\partial^\nu\Lambda - \partial^\nu\partial^\mu\Lambda = F^{\mu\nu}_{\text{old}}$$

6.2.4 Coulomb Gauge

We start by showing the origin and significance of the Coulomb gauge. Recall that the vector potential must satisfy the Coulomb condition $\nabla \cdot \vec{A} = 0$ if we wish to operate in the Coulomb gauge. This means, starting with potentials $(\Phi, \vec{A})$, we must be able to find a scalar gauge function Λ to make the Coulomb condition true for the gauge-transformed vector potential $\nabla \cdot \vec{A}_{\text{new}} = 0$ even though it might not be true for the original vector potential. Taking the divergence of the second of Equations (6.2.8), we find

$$0 = \nabla \cdot \vec{A}_{\text{new}} = \nabla \cdot \vec{A}_{\text{old}} - \nabla^2\Lambda$$

Therefore, the Coulomb condition holds by suitable choice of the gauge function Λ.

We see that gauge transformations exist because of the arbitrariness in the definition of the potentials. For example, everyone is familiar with the fact that the zero of the electric potential V can be shifted without affecting the fields or the operation of electrical devices. We now start with $\vec{A}_{\text{old}}$ and demonstrate a scalar gauge function Λ that makes the Coulomb condition true for the gauge-transformed vector potential $\nabla \cdot \vec{A}_{\text{new}} = 0$. The new and old potentials must be related by Equations (6.2.8)

$$\Phi_{\text{new}} = \Phi_{\text{old}} + \frac{\partial}{\partial t}\Lambda \quad \vec{A}_{\text{new}} = \vec{A}_{\text{old}} - \nabla\Lambda \tag{6.2.9}$$

Starting with Gauss' law $\nabla \cdot \vec{E} = \rho/\varepsilon_0$ and substituting $\vec{E} = -\partial_t \vec{A}_{old}(\vec{r}, t) - \nabla\Phi_{old}$ from Equation (6.2.1) we find

$$\nabla \cdot \left[-\frac{\partial \vec{A}_{old}(\vec{r}, t)}{\partial t} - \nabla\Phi_{old} \right] = \frac{\rho}{\varepsilon_0} \tag{6.2.10}$$

Substitute the new vector potential to find

$$\nabla \cdot \left[-\frac{\partial}{\partial t} \left(\vec{A}_{new} + \nabla\Lambda \right) - \nabla\Phi_{old} \right] = \frac{\rho}{\varepsilon_0} \tag{6.2.11a}$$

From which we find

$$\frac{\partial}{\partial t} \left(\nabla \cdot \vec{A}_{new} + \nabla \cdot \nabla\Lambda \right) + \nabla \cdot \nabla\Phi_{old} = -\frac{\rho}{\varepsilon_0} \tag{6.2.11b}$$

We require the first term to be zero $\nabla \cdot \vec{A}_{new} = 0$ and find

$$\nabla^2 \left(\Phi_{old} + \dot{\Lambda} \right) = -\frac{\rho}{\varepsilon_0} \tag{6.2.12}$$

Therefore, the Coulomb condition $\nabla \cdot \vec{A}_{new} = 0$ requires

$$\Phi_{old} + \dot{\Lambda} = \Phi(\vec{x}, t) = \frac{1}{4\pi\varepsilon_0} \int \frac{\rho(\vec{x}', t)}{|\vec{x} - \vec{x}'|} d^3x' \tag{6.2.13}$$

as given in (3) of Section 6.2.1. This last equation tells us to choose a scalar function Λ that makes up for the difference $\dot{\Lambda} = \Phi(\vec{x}, t) - \Phi_{old}$. In other words, the Coulomb condition $\nabla \cdot \vec{A}_{new} = 0$ holds for the new vector potential $\vec{A}_{new}$ so long as we change the old scalar potential into the solution to Poisson's equation $-\varepsilon_0 \nabla^2\Phi = \rho$ from Gauss' law $\varepsilon_0 \nabla \cdot \vec{E} = \rho$. We make Φ_{new} the electrostatic potential (voltage). Equations (6.2.13) and (6.2.9) verify this.

$$\Phi_{new} = \Phi_{old} + \frac{\partial}{\partial t}\Lambda = \Phi(\vec{x}, t) = \frac{1}{4\pi\varepsilon_0} \int \frac{\rho(\vec{x}', t)}{|\vec{x} - \vec{x}'|} d^3x' \tag{6.2.14}$$

As a second result for the Coulomb potential, previous topics in this section show that the direction of the vector potential must be perpendicular to the wave vector $\vec{k}$ for plane waves.

$$\nabla \cdot \vec{A} = 0 \quad \rightarrow \quad \vec{k} \cdot \vec{A} = 0 \tag{6.2.15}$$

We use the Coulomb gauge for quantizing the electromagnetic field. We can rotate the 3-D coordinate system so that the z-axis is along the k-vector direction. The 4-vector potential can now be written as

$$\left(\Phi, \vec{A} \right) = \left(\Phi, A_x, A_y, 0 \right) \tag{6.2.16}$$

where $A_z = 0$ in view of condition (6.2.15). Two directions perpendicular to the wave vector provide two polarization modes. The two components of the vector potential A_x, A_y are independent dynamical variables for the traveling wave and they can be quantized. As mentioned previously, the scalar potential is the instantaneous Coulomb potential. Therefore, the 0th component of the 4-vector potential corresponds to the "instantaneously" propagating longitudinal component of the vector potential.

6.2.5   Lorentz Gauge

The Lorentz gauge uses potentials that satisfy the Lorentz condition

$$0 = \partial_\mu A^\mu = \frac{\partial A^0}{\partial t} + \nabla \cdot \vec{A} = \frac{\partial \Phi}{\partial t} + \nabla \cdot \vec{A} \qquad (6.2.17)$$

where we use the "repeated index" convention which means to sum over indices that occur twice in a product. Specifically, the Lorentz condition is

$$\frac{\partial \Phi}{\partial t} + \nabla \cdot \vec{A} = 0$$

We can find potentials

$$\Phi_{\text{new}} = \Phi_{\text{old}} + \frac{\partial}{\partial t} \Lambda \qquad \vec{A}_{\text{new}} = \vec{A}_{\text{old}} - \nabla \Lambda$$

to satisfy the Lorentz condition

$$\frac{\partial \Phi_{\text{new}}}{\partial t} + \nabla \cdot \vec{A}_{\text{new}} = 0$$

substituting to find

$$\nabla^2 \Lambda_{\text{old}} - \frac{\partial^2 \Lambda_{\text{old}}}{\partial t^2} = \nabla \cdot \vec{A}_{\text{old}} + \frac{\partial \Phi_{\text{old}}}{\partial t}$$

In particular, the Lorentz condition is

$$\nabla^2 \Lambda_{\text{old}} - \frac{\partial^2 \Lambda_{\text{old}}}{\partial t^2} = 0$$

This can be seen to hold for $\Lambda = e^{i\vec{k}\cdot\vec{r} - i\omega t}$. Therefore potentials satisfying the Lorentz gauge conditions exist. We do not use the Lorentz gauge for quantizing the EM field since not all components of A^μ are independent according to Equation (6.2.17).

6.3   The Plane Wave Expansion of the Vector Potential and the Fields

The quantum theory of the electromagnetic field starts by Fourier expanding the vector potential and then substituting operators for the amplitude terms. The quantum theory version of the vector potential (i.e., quantized vector potential) leads to that for the electric and magnetic fields since they can be derived from a vector potential according to

$$\vec{E} = -\frac{\partial \vec{A}(\vec{r}, t)}{\partial t} \quad \text{and} \quad \vec{B} = \nabla \times \vec{A}(\vec{r}, t) \tag{6.3.1}$$

in a source-free region of space with a zero scalar potential. We only need to discuss one vector field, namely $\vec{A}$, propagating through space as opposed to both $\vec{E}$ and $\vec{B}$. The direction of the vector potential $\vec{A}$ parallels the direction of the electric field $\vec{E}$. Therefore we only need to solve the wave equation (Equation (6.2.4)) in the Coulomb gauge

$$\nabla^2 \vec{A} - \frac{1}{c^2} \frac{\partial^2 \vec{A}}{\partial t^2} = 0 \tag{6.3.2}$$

to find the propagating electric and magnetic fields in a source-free region of space. For traveling waves, periodic boundary conditions are most convenient. The boundary value problem (wave equation and boundary conditions) produces a basis set of the form $B = \{\langle x \mid n \rangle = e^{ik_n x}/\sqrt{L} \quad n = 0, \pm 1, \pm 2 \ldots\}$. The general solution must be a (time-dependent) sum over the basis set according to

$$|A\rangle = \sum_n a_n(t)|n\rangle \tag{6.3.3}$$

A moments thought shows that the sum in Equation (6.3.3) must be the Fourier series when using the basis set B. Unlike the quantum theory with the single particle wavefunctions, the vector potential $|A\rangle$ does not need to be normalized to one. We will see the vector potential and EM fields involve two Hilbert spaces; one for the wave solution of the wave equation and another for the amplitude Hilbert space once operators replace the classical amplitudes.

Once having determined the solution in Equation (6.3.3), substituting operators for the Fourier coefficients "a" quantizes the vector potential. The quantization procedure uses the Coulomb gauge where the vector potential must satisfy $\nabla \cdot \vec{A} = 0$ so that the vector potential $\vec{A}$ for a traveling wave must be perpendicular to the wave-vector $\vec{k}$. Given the relation between the fields and the vector potential, we can also find the expression for the quantized electric and magnetic fields. Chapter 3 provides an expression for the electromagnetic energy stored in free space in terms of the electric and magnetic fields. Consequently, we can then write the quantum mechanical Hamiltonian for the electromagnetic field.

6.3.1 Boundary Conditions

We first review periodic boundary conditions and then the fixed end-point boundary conditions. Periodic boundary conditions require the wave to repeat itself over a distance L (for a 1-D propagating wave—see Figure 6.3.1). It does not require the wave to have a specific magnitude or phase at any particular point. This type of boundary condition most appropriately applies to traveling waves. The periodic boundary conditions require the Fourier expansion to be composed of periodic waves having wavelengths related to L by $\lambda = L/n$ where n represents an integer.

The Fourier series expansion of a function consists of the sum over sines and cosines. These functions must repeat themselves over some length scale L. The linear algebra

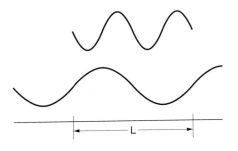

FIGURE 6.3.1
Periodic boundary conditions.

shows that the basis functions for a 1-D system have the form of $e^{ikx}/\sqrt{L}$. The length "L" represents the longest allowed wavelength and it's also the length of the interval of x-values $x \in \left(x_0 - \frac{L}{2}, x_0 + \frac{L}{2}\right)$. If we allow "$L$" to approach infinity, the 1-D basis set becomes uncountably infinite and the basis functions have the form of $e^{ikx}/\sqrt{2\pi}$; the Fourier series becomes the Fourier transform. Periodic boundary conditions produce periodic basis functions that span a Hilbert space of periodic wave functions $\psi(x + L) = \psi(x)$ as shown in Figure 6.3.1. Notice that the wavefunctions do not need to be zero at the boundaries.

For the basis set $\{e^{ik_n x}/\sqrt{L}\}$, the periodic boundary condition requires the wave vectors $\vec{k}$ to take on specific values. For the one-dimensional case, the allowed wavelengths must be

$$\lambda_n = \frac{L}{n} \quad n = 1, 2, 3 \ldots$$

and therefore the allowed wave vectors must be

$$k_n = \frac{2\pi}{\lambda_n} = \frac{2\pi n}{L} \quad n = \pm 1, \pm 2, \pm 3 \ldots \tag{6.3.4}$$

For the 3-D case, a function of three variables $\vec{r} = x\tilde{x} + y\tilde{y} + z\tilde{z}$ or (x,y,z) can be Fourier expanded using the complex exponential functions (assumed to be periodic with a unit cell of volume V)

$$\left\{ \phi_{\vec{k}}(\vec{r}) = \frac{e^{i\vec{k}\cdot\vec{r}}}{\sqrt{V}} \right\} \tag{6.3.5}$$

as a basis set where the volume V can be related to L by $V = L^3$. Each component of the wave vector $\vec{k}$ must satisfy an equation similar to Equations (6.3.4).

$$\vec{k} = \tilde{x}\left(\frac{2\pi m}{L}\right) + \tilde{y}\left(\frac{2\pi n}{L}\right) + \tilde{z}\left(\frac{2\pi p}{L}\right) \quad m, n, p = \pm 1, \pm 2, \pm 3 \ldots \tag{6.3.6}$$

Notice that we allow negative wave vectors (for waves propagating along negative directions). For light, the basic modes (i.e., the vectors in the basis set) are described by the allowed wavelengths (i.e., the allowed wave vectors or frequencies) and the polarization. For the most part, we ignore the polarization except possibly in final formulas.

Now consider the fixed-endpoint boundary conditions. This type of boundary condition requires the magnitude of the wave to have a specific magnitude at two separated points—for a 1-D problem, think of a string tied down at the ends. Figure 6.3.2 shows the typical fixed-endpoint type of boundary conditions; the function must be zero at the endpoints. This boundary condition determines the amplitude at two fixed points as well

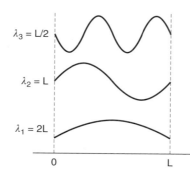

FIGURE 6.3.2
Fixed-endpoint boundary conditions.

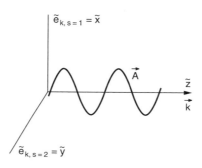

FIGURE 6.3.3
The vector potential is polarized in the x direction.

as the allowed wavelengths $\lambda = 2L/n$ appearing in the Fourier expansion. The wave vectors obtained from the fixed-endpoint boundary conditions

$$k_n = n\pi/L \quad n = 1, 2, 3 \ldots$$

differ by a factor of two from those allowed by the periodic boundary conditions.

6.3.2 The Plane Wave Expansion

The basis set for periodic boundary conditions (6.3.5) and the solution to the vector-potential wave equation (6.3.3) can be combined into the solution

$$\vec{A}(\vec{r}, t) = \sum_{\vec{k}} \sqrt{\frac{\hbar}{2\varepsilon_0 \omega_k}} \vec{A}_k(t) \frac{e^{i\vec{k}\cdot\vec{r}}}{\sqrt{V}} \tag{6.3.7}$$

for free space (we replace the free space permittivity ε_0 with ε for dielectric medium). The additional parameters $\sqrt{\hbar/(2\varepsilon_0 \omega_k)}$ provide the MKS units. Assume that the angular frequency of the electromagnetic wave must always be positive $\omega_k = \omega_{-k} > 0$.

We require real fields so that their quantum counterparts will be Hermitian and therefore observable. Using Equation (6.3.7) and the condition for real fields $\vec{A}^* = \vec{A}$, we obtain

$$\sum_{\vec{k}} \sqrt{\frac{\hbar}{2\varepsilon_0 \omega_k}} \vec{A}_k^*(t) \frac{e^{-i\vec{k}\cdot\vec{r}}}{\sqrt{V}} = \sum_{\vec{k}} \sqrt{\frac{\hbar}{2\varepsilon_0 \omega_k}} \vec{A}_k(t) \frac{e^{i\vec{k}\cdot\vec{r}}}{\sqrt{V}}$$

Replacing $\vec{k} \to -\vec{k}$ on the left-hand side and using $\omega_k = \omega_{-k}$ produces

$$\sum_{\vec{k}} \sqrt{\frac{\hbar}{2\varepsilon_0 \omega_k}} \vec{A}^*_{-k}(t) \frac{e^{+i\vec{k}\cdot\vec{r}}}{\sqrt{V}} = \sum_{\vec{k}} \sqrt{\frac{\hbar}{2\varepsilon_0 \omega_k}} \vec{A}_k(t) \frac{e^{i\vec{k}\cdot\vec{r}}}{\sqrt{V}}$$

Comparing both sides (i.e., using the orthonormality of the basis functions), we find that

$$\vec{A}_{-k}(t) = \vec{A}^*_k(t) \tag{6.3.8}$$

We next find a differential equation for the amplitudes $\vec{A}_k(t)$. The previous section shows that in the Coulomb gauge, the vector potential must satisfy the wave equation

$$\nabla^2 \vec{A} - \frac{1}{c^2} \frac{\partial^2 \vec{A}}{\partial t^2} = 0 \tag{6.3.9}$$

Substituting Equations (6.3.7) into Equation (6.3.9) requires us to substitute the results

$$\nabla^2 \vec{A}(\vec{r}, t) = \sum_{\vec{k}} \sqrt{\frac{\hbar}{2\varepsilon_0 \omega_k}} \vec{A}_k(t)(-k^2) \frac{e^{i\vec{k}\cdot\vec{r}}}{\sqrt{V}}$$

and

$$\frac{\partial^2}{\partial t^2} \vec{A}(\vec{r}, t) = \sum_{\vec{k}} \sqrt{\frac{\hbar}{2\varepsilon_0 \omega_k}} \frac{\partial^2 \vec{A}_k(t)}{\partial t^2} \frac{e^{i\vec{k}\cdot\vec{r}}}{\sqrt{V}}$$

So therefore, Equation (6.3.9) becomes

$$-k^2 \vec{A}_{\vec{k}} - \frac{1}{c^2} \frac{\partial^2 \vec{A}_{\vec{k}}}{\partial t^2} = 0 \quad \text{or} \quad \frac{\partial^2 \vec{A}_{\vec{k}}(t)}{\partial t^2} + \omega_k^2 \vec{A}_{\vec{k}}(t) = 0 \tag{6.3.10}$$

which uses the relation $c^2 = \omega^2/k^2$.

The amplitudes can now be determined. For simplicity, let's first treat the vector potential as a scalar function. The Fourier transformed wave equation (Equation (6.3.10)) has two solutions $e^{-i\omega_k t}$ and $e^{+i\omega_k t}$. The general solution must be

$$A_k = b_k e^{-i\omega_k t} + a_k e^{+i\omega_k t} \quad \omega_k > 0$$

Using the "reality" of the vector potential, or equivalently $A_{-k}(t) = A^*_k(t)$, we find

$$b^*_k e^{+i\omega_k t} + a^*_k e^{-i\omega_k t} = b_{-k} e^{-i\omega_k t} + a_{-k} e^{+i\omega_k t} \quad \text{so that} \quad b^*_k = a_{-k} \to a_k = b^*_{-k}$$

Therefore, the general solution to Equation (6.3.10) must be

$$A_k = b_k e^{-i\omega_k t} + b^*_{-k} e^{+i\omega_k t} \quad \omega_k > 0 \tag{6.3.11}$$

The vector nature of the vector potential cannot be ignored (Figure 6.3.3). Similar to electric and magnetic fields, which have *polarization* vectors to describe their "direction," the vector potential also has a polarization. Section 6.2 shows that the polarization of the electric field is parallel to the polarization of the vector potential (i.e. $\vec{A} \sim \vec{E}$). The Coulomb gauge supports two independent polarization modes for each wave vector $\vec{k}$; the polarization vectors must be transverse to the direction of motion (i.e., more specifically, perpendicular to the wave vector). We denote these as $\tilde{e}_{ks}$ where $s = 1,2$ represents the $\tilde{x}$ and $\tilde{y}$ directions for a wave propagating along the $\tilde{z}$ direction. Now we can summarize by stating the total number of EM modes. Each different $\vec{k}$ specifies a mode by virtue of its direction and magnitude (wavelength). The unit vectors $\tilde{e}_{ks}$ describe two modes. For now, we combine the k, s subscripts into the k subscript for simplicity. Writing Equation (6.3.11) in vector notation

$$\vec{A}_k = \vec{b}_k e^{-i\omega_k t} + \vec{b}_{-k}^* e^{+i\omega_k t} \quad \omega_k > 0$$

or, setting $\vec{b}_k = \tilde{e}_k b_k$ we obtain the solution of Equation (6.3.6).

$$\vec{A}_k = \tilde{e}_k b_k e^{-i\omega_k t} + \tilde{e}_{-k}^* b_{-k}^* e^{+i\omega_k t} \quad \omega_k > 0$$

The vector potential can be written as

$$\vec{A}(\vec{r}, t) = \sum_{\vec{k}} \sqrt{\frac{\hbar}{2\varepsilon_0 \omega_k}} \left(\tilde{e}_k b_k e^{-i\omega_k t} + \tilde{e}_{-k}^* b_{-k}^* e^{+i\omega_k t}\right) \frac{e^{i\vec{k}\cdot\vec{r}}}{\sqrt{V}}$$

This last equation can be rewritten by using the following observations:

1. The second summation is over all allowed wave vectors (i.e., all positive and negative components) so that, *for the second term*, we can make the replacements $-\vec{k} \to \vec{k}$ and $\sum_{-\vec{k}} \to \sum_{\vec{k}}$.
2. For this chapter, we assume a real polarization vector $\tilde{e}_k^* = \tilde{e}_k$.
3. The angular frequency must be positive $\omega_{-\vec{k}} = \omega_{\vec{k}}$.

The vector potential becomes

$$\vec{A}(\vec{r}, t) = \sum_{\vec{k}} \sqrt{\frac{\hbar}{2\varepsilon_0 \omega_k V}} \tilde{e}_k \left[\left(b_k e^{-i\omega_k t}\right) e^{i\vec{k}\cdot\vec{r}} + \left(b_k^* e^{+i\omega_k t}\right) e^{-i\vec{k}\cdot\vec{r}} \right] \quad (6.3.12a)$$

or, including the summation over the polarization,

$$\vec{A}(\vec{r}, t) = \sum_{k,s} \sqrt{\frac{\hbar}{2\varepsilon_0 \omega_k V}} \tilde{e}_{ks} \left[\left(b_{ks} e^{-i\omega_k t}\right) e^{i\vec{k}\cdot\vec{r}} + \left(b_{ks}^* e^{+i\omega_k t}\right) e^{-i\vec{k}\cdot\vec{r}} \right] \quad (6.3.12b)$$

We can simplify the equations by defining the functions

$$b_{ks}(t) = b_{ks} e^{-i\omega_k t} \quad \text{and} \quad b_{ks}^*(t) = b_{ks}^* e^{+i\omega_k t}$$

where

$$b_{ks} = b_{ks}(0) \quad \text{and} \quad b_{ks}^* = b_{ks}^*(0)$$

Section 6.4 shows that the coefficients b_{ks} and b_{ks}^* become the annihilation and creation operators in the quantum theory of EM fields. The annihilation operator removes a photon from a mode characterized by the wave vector $\vec{k}$ and the polarization $\tilde{e}_{ks}$.

6.3.3 The Fields

The electric and magnetic fields obtain from the relations between the fields and the potentials in a source-free region

$$\vec{\mathscr{E}} = -\frac{\partial \vec{A}(\vec{r}, t)}{\partial t} \quad \text{and} \quad \vec{\mathscr{B}} = \nabla \times \vec{A}(\vec{r}, t)$$

We find

$$\vec{E} = -\frac{\partial}{\partial t} \vec{A}(\vec{r}, t) = \frac{+i}{\sqrt{\varepsilon_0 V}} \sum_{\vec{k}} \sqrt{\frac{\hbar \omega_k}{2}} \left[b_k \, e^{i\vec{k}\cdot\vec{r} - i\omega_k t} - b_k^* \, e^{-i\vec{k}\cdot\vec{r} + i\omega_k t} \right] \tilde{e}_k \qquad (6.3.13a)$$

Similarly, we can calculate the magnetic field (see the chapter review exercises)

$$\vec{B} = \nabla \times \vec{A}(\vec{r}, t) = \frac{+i}{\sqrt{\varepsilon_0 V}} \sum_{\vec{k}} \sqrt{\frac{\hbar}{2\omega_k}} \left(\vec{k} \times \tilde{e}_k \right) \left[b_k \, e^{i\vec{k}\cdot\vec{r} - i\omega_k t} - b_k^* \, e^{-i\vec{k}\cdot\vec{r} + i\omega_k t} \right] \qquad (6.3.13b)$$

6.3.4 Spatial-Temporal Modes

The previous topic develops (Equations (6.3.13)) the vector potential solution to the wave equation

$$\nabla^2 \vec{A}_{\vec{k}} - \frac{1}{c^2} \frac{\partial^2 \vec{A}_{\vec{k}}}{\partial t^2} = 0$$

using spatial-temporal modes consisting of traveling plane waves

$$U(x, t) = e^{i\vec{k}\cdot\vec{r} - i\omega_k t}$$

The wave equation can have other solutions besides traveling waves; the type of solution depends on the boundary conditions. For example, a perfect no-loss Fabry–Perot cavity has standing sine wave solutions; the boundary conditions, in this case, require the fields to be zero at the boundaries. It is important to be able to identify creation and annihilation operators and quantize arbitrary EM fields. This topic shows how to write the vector potential in terms of other spatial-temporal modes $U(x, t)$. We will find the following forms for the vector potential

$$A(x,t) = \sum_k \sqrt{\frac{\hbar}{2\varepsilon_0 \omega_k}} \left[b_k \, U_k(x,t) + b_k^* \, U_k^*(x,t) \right]$$

$$A(x,t) = \sum_k \sqrt{\frac{\hbar}{2\varepsilon_0 \omega_k}} \left[b_k \, e^{-i\omega_k t} + b_k^* \, e^{+i\omega_k t} \right] \phi_k(x)$$

where $\phi_k(x)$ denotes a basis function and

$$b_{ks} = b_{ks}(0) \quad \text{and} \quad b_{ks}^* = b_{ks}^*(0)$$

The next section shows that the vector potential is quantized by substituting creation $\hat{b}^+$ and annihilation $\hat{b}$ operators for the amplitudes b^* and b, respectively. The creation operator $\hat{b}^+$ creates a photon in the mode U_k while the annihilation operator removes a photon from the mode U_k.

The general solution to the wave equation can be found by separating variables and applying boundary conditions. To solve the wave equation

$$\frac{\partial^2 A(x,t)}{\partial x^2} - \frac{1}{c^2} \frac{\partial^2 A(x,t)}{\partial t^2} = 0 \tag{6.3.14}$$

(where we ignore the vector nature of $\vec{A}$ for simplicity), we separate variables according to

$$A_k(x,t) = \phi_k(x) T_k(t) \tag{6.3.15}$$

Separating variables for three dimensions is similar. Substituting Equation (6.3.15) into the wave equation (6.3.14), separating variables, and taking $-\lambda_k$ as the separation constant (where $\lambda_k > 0$), we find

$$\frac{1}{\phi_k^2} \frac{\partial^2 \phi_k(x)}{\partial x^2} = -\lambda_k = \frac{1}{c^2} \frac{1}{T_k^2} \frac{\partial^2 T_k(t)}{\partial t^2}$$

The Sturm–Liouville problem for ϕ_k includes specific boundary conditions and provides the set of basis functions $\{\phi_k(x)\}$. The solution of the Sturm–Liouville problem includes the eigenvalues $\lambda_k = k^2$. The solution of the separated time equation

$$\frac{\partial^2 T_k(t)}{\partial t^2} = -\lambda_k c^2 T_k^2$$

is therefore found to be

$$T_k(t) = \sqrt{\frac{\hbar}{2\varepsilon_0 \omega_k}} \left[b_k e^{-i\omega_k t} + b_k^* e^{+i\omega_k t} \right]$$

where $\omega_k = ck$, $b_{ks} = b_{ks}(0)$ and $b_{ks}^* = b_{ks}^*(0)$. This last equation includes the normalization factor $\sqrt{\hbar/2\varepsilon_0 \omega_k}$. Therefore the general solution of the wave equation must be

$$A(x,t) = \sum_k \sqrt{\frac{\hbar}{2\varepsilon_0 \omega_k}} \left[b_k e^{-i\omega_k t} + b_k^* e^{+i\omega_k t} \right] \phi_k(x)$$

where $c = \omega_k/k$. For the vector potential to be real $(A = A^*)$, the eigenvectors ϕ_k must be real. Therefore, the general spatial-temporal mode is

$$U_k(x,t) = \phi_k(x)e^{-i\omega_k t}$$

so that

$$A(x,t) = \sum_k \sqrt{\frac{\hbar}{2\varepsilon_0 \omega_k}} \left[b_k U_k(x,t) + b_k^* U_k^*(x,t) \right]$$

It is perhaps more convenient to write this as

$$A(x,t) = \sum_k \sqrt{\frac{\hbar}{2\varepsilon_0 \omega_k}} \left[b_k(0)\, e^{-i\omega_k t} + b_k^*(0)\, e^{+i\omega_k t} \right] \phi_k(x)$$

We will see in the next topic that b_k and b_k^* become the annihilation $\hat{b}_k$ and the creation $\hat{b}_k^+$ operators, respectively, in the quantum theory of EM fields. The annihilation operator removes one photon from the mode ϕ_k while the creation operator adds one photon. Because photons are bosons, any number of them can occupy a single state; electrons and holes are Fermions and only one can occupy a given state at a given time.

   Including the polarization vector with the basis set, the vector potential

$$\vec{A}(x,t) = \sum_{ks} \sqrt{\frac{\hbar}{2\varepsilon_0 \omega_k}} \left[b_{ks}(0)\, e^{-i\omega_k t} + b_{ks}^*(0)\, e^{+i\omega_k t} \right] \vec{\phi}_{ks}(x)$$

provides the free-space electric and magnetic fields

$$\vec{E}(x,t) = -\frac{\partial}{\partial t} \vec{A}(x,t) = i \sum_{ks} \sqrt{\frac{\hbar}{2\varepsilon_0 \omega_k}} \omega_k \left[b_{ks}(0)\, e^{-i\omega_k t} - b_{ks}^*(0)\, e^{+i\omega_k t} \right] \vec{\phi}_{ks}(x)$$

$$\vec{B}(x,t) = \nabla \times \vec{A}(x,t) = \sum_{ks} \sqrt{\frac{\hbar}{2\varepsilon_0 \omega_k}} \left[b_{ks}(0)\, e^{-i\omega_k t} + b_{ks}^*(0)\, e^{+i\omega_k t} \right] \nabla \times \vec{\phi}_{ks}(x)$$

6.4 The Quantum Fields

We want a full quantum theory for the electromagnetic (EM) fields. The theory must incorporate both the particle and wave nature. Both of these can be included by quantizing the EM vector potential. The traveling wave part of the Fourier-expanded vector potential describes the wave properties of light. By converting the Fourier amplitudes (which have magnitude and phase) into operators, the particle aspects of light can be recovered. The operator form of the electric and magnetic fields can be substituted into the classical EM Hamiltonian to find the quantum one. The Hamiltonian for light will be similar to that for the electron harmonic oscillator. In fact, the EM quantum Hamiltonian has wavefunction solutions that describe the probability of finding an EM wave with a particular classical amplitude.

We should compare and contrast the classical and quantum pictures of the vector potential. The two pictures yield similar results for large numbers of photons but not for small numbers. The amplitudes in the classical vector potential represent numbers that can be exactly known. We might need measuring equipment to determine these amplitudes in the laboratory and we might need to do some statistical analysis to come close to the true or actual amplitude, but we don't need any additional mathematical construction to actually define the meaning of the amplitudes. A series of measurements of the amplitudes might lead to some variation in the observed values and so we must take averages over the series. This variation can be described by a probability distribution and leads to the notion of the ensemble and the probability P used in the density operator. In the classical theory, the parameters describing the electromagnetic field can be exactly known except possibly for experimental error in the measurements. The actual expression for the electric field depends on the results of the measurements.

For the quantum picture, the amplitudes must be changed into operators (with both magnitude and phase). The vector potential (and hence the electric and magnetic fields) also become operators. The creation and annihilation operators for a mode of the electric or magnetic field do not commute. Therefore, regardless of the experimental accuracy, the electric and magnetic field can never achieve a "true value." Every measurement of the field necessarily produces different results. To find average values of the fields, they must operate on a Hilbert space, which has vectors describing the possible states of the amplitude. We can call this the "amplitude Hilbert space." Now however, two types of probability enter into specifying the "classical amplitudes." First, because the amplitude operators do not all commute, there will be an uncertainty in measuring the fields. Second, the measurement apparatus can also introduce error into the specification of the fields. Therefore, we must describe the amplitude states representing the system by both classical and quantum probability distributions such as appear in the density operator. Finally, unlike the classical expression for the field, the operator form of the field does not depend on the results of any measurements. The state vectors in the Hilbert space reflect any physical attribute of the field.

The quantum theory of EM fields does not circumvent classical electromagnetics, but rather augments it. The quantum theory must still deal with modes and the corresponding electromagnetic wave functions typically found in the classical theory. However, the simple classical amplitudes become operators with commutation rules. These amplitude operators become more classical-like once they operate on a Hilbert space. In particular, the "classical value" can be obtained by finding the expectation value of the amplitude operator for a given state in Hilbert space. The particular state in Hilbert space defines the particular light beam of interest by defining the properties of the amplitude.

This section shows how the vector potential can be converted into an operator. The quantized fields operate on a vector in Hilbert space (usually a Fock state or a sum of Fock states). Once we know the quantized vector potential, we can find the quantized electric and magnetic fields using the relations from Section 6.2, namely $\vec{\mathscr{E}} = -\partial \vec{A}/\partial t$ and $\vec{\mathscr{B}} = \nabla \times \vec{A}$. The quantum Hamiltonian can be obtained by replacing the classical electric and magnetic fields with their operator counterparts. The electric and magnetic fields can be written in terms of either creation–annihilation or quadrature operators. We will see special uses for each type. The quadrature operators bring the quantum picture of the electric field the closest to the usual classical picture of the field as a sinusoidal wave. We will see that we cannot simultaneously and precisely know the amplitude and the phase of the field.

6.4.1 The Quantized Vector Potential

We start with the classical form of the vector potential

$$\vec{A}(\vec{r}, t) = \frac{1}{\sqrt{\varepsilon_0 V}} \sum_{\vec{k}} \sqrt{\frac{\hbar}{2\omega_k}} \, \tilde{e}_k \Big[\big(b_k e^{-i\omega_k t}\big) e^{i\vec{k}\cdot\vec{r}} + \big(b_k^* e^{+i\omega_k t}\big) e^{-i\vec{k}\cdot\vec{r}} \Big] \qquad (6.4.1)$$

with the classical Fourier components (from Section 6.3.2)

$$b_k(t) = b_k e^{-i\omega_k t} = b_k(0)\, e^{-i\omega_k t} \quad b_k^*(t) = b_k^* e^{+i\omega_k t} = b_k^*(0)\, e^{+i\omega_k t} \qquad (6.4.2)$$

The coefficient "b" provides the amplitude of a given optical mode "k." The derivation assumes a source-free region of space (i.e., no interaction potential).

Replacing the amplitudes with operators according to the prescription

$$b_k \rightarrow \hat{b}_k \quad \text{and} \quad b_k^* \rightarrow \hat{b}_k^+$$

produces the quantum version of the vector potential. We must later specify the Hilbert space from which the operators can assume a value. We can write the equations of motion for the operators using Equations (6.4.2).

$$\hat{b}_k(t) = \hat{b}_k \, e^{-i\omega_k t} = \hat{b}_k(0) e^{-i\omega_k t} \quad \hat{b}_k^+(t) = \hat{b}_k^+ \, e^{+i\omega_k t} = \hat{b}_k^+(0)\, e^{+i\omega_k t} \qquad (6.4.3)$$

In the quantum theory, these equations hold in the interaction representation (for any situation) or in the Heisenberg representation for a situation without sources. Example 6.5.1 develops the equations of motion in the Heisenberg representation for the case of free-fields (without matter to produce gain or absorption). These will have the same form as the equations of motion deduced using an interaction picture regardless of whether or not the wave interacts with matter. We can replace the classical amplitudes with creation and annihilation operators to produce the quantum version of the vector potential.

$$\hat{A}(\vec{r}, t) = \frac{1}{\sqrt{\varepsilon_0 V}} \sum_{\vec{k}} \sqrt{\frac{\hbar}{2\omega_k}} \, \tilde{e}_k \Big[\hat{b}_{\vec{k}}(t) e^{i\vec{k}\cdot\vec{r}} + \hat{b}_{\vec{k}}^+(t) e^{-i\vec{k}\cdot\vec{r}} \Big] \qquad (6.4.4)$$

The operator version of the vector potential contains all the possible creation and annihilation operators for the various modes. In a sense, the quantum EM field (and as we shall see later, the Hamiltonian) must contain all of the possibilities that can physically occur. The amplitude states in the amplitude Hilbert space contain the specific information on the system.

The creation and annihilation operators (the same ones as will be given in Section 6.5.4) must satisfy the equal-time commutation relations

$$\Big[\hat{b}_\xi(t), b_\eta(t)\Big] = 0 = \Big[\hat{b}_\xi^+(t), \hat{b}_\eta^+(t)\Big] \quad \text{for all} \quad \xi, \eta \qquad \Big[\hat{b}_\xi(t), \hat{b}_\eta^+(t)\Big] = \delta_{\xi\eta} \qquad (6.4.5)$$

We therefore can also write the commutation relation at the specific time $t = 0$ as

$$\left[\hat{b}_\xi, b_\eta\right] = 0 = \left[\hat{b}_\xi^+, \hat{b}_\eta^+\right] \text{ for all } \xi, \eta \qquad \left[\hat{b}_\xi, \hat{b}_\eta^+\right] = \delta_{\xi\eta} \qquad (6.4.6)$$

Some comments should be made regarding the commutation relations in Equations (6.4.5). First, both operators $\hat{b}_\xi(t), \hat{b}_\eta^+(t)$ must be evaluated at the same time t (i.e., equal times) thereby suggesting the name "equal-time commutator." We consider the modes $\xi \neq \eta$ to be independent; we can create a photon in state ξ independently of annihilating one in the state η. However for the mode $\xi = \eta$, we should anticipate a type of Heisenberg uncertainty relation since the corresponding operators do not commute as shown in the last commutation relation. However, the uncertainty relation requires Hermitian operators and not the nonHermitian creation–annihilation operators; we will later use the quadrature operators for this purpose. Second, we assume either periodic or fixed-endpoint boundary conditions, which lead to both the *discrete values* for the wave vector k and to the Kronecker delta function for the orthonormality relation.

In the Heisenberg representation, the time dependence of the creation and annihilation operators depends on the Hamiltonian. Example 6.5.1 in the next section produces the same results as given by Equations (6.4.3). The creation and annihilation operators will have a different time-dependence if we include the interaction between the fields and the matter. This occurs since matter can produce or absorb electromagnetic fields, which must necessarily change the operators describing the EM field. The Heisenberg representation makes the amplitudes depend on time in a manner very reminiscent of classical EM theory. For example, the amplitude of a wave decreases as it travels through an absorber. The states (yet to be discussed) in this case must be independent of time.

In contrast, the interaction representation always assigns the trivial time dependence to the operators; however, the states move due to the interaction potentials. The creation and annihilation operators always have the trivial time dependence given by Equations (6.4.6) regardless of whether or not the Hamiltonian has an interaction term. The interaction representation of the fields must always be the same as that in Heisenberg representation when an interaction potential does not appear. Therefore, when somebody writes Equation (6.4.1) without specifying the representation, it can be taken as in either (i) the interaction representation or (ii) the Heisenberg representation for the free-field case (no interaction term in the Hamiltonian).

The Schrodinger representation of the creation and annihilation operators must be independent of time. Setting $t = 0$ in Equations (6.4.3) provides the Schrodinger representation of the creation and annihilation operators. Equations (6.4.6) then give the commutation relations for these operators. All of the time dependence must reside in the Hilbert space vectors.

6.4.2 Quantizing the Electric and Magnetic Fields

Changing the Fourier amplitudes into operators quantizes the electromagnetic fields. As discussed in later sections, we can only specify values for the amplitudes (i.e., the fields) by providing a Hilbert space upon which the operators can act. The previous topic shows the quantum version of the vector potential for an electromagnetic wave. The *quantized electric field operator* (in the Coulomb gauge) is then found by differentiating the vector potential with respect to time.

Recall that the vector potential can be written as a Fourier Expansion

$$\hat{A}(\vec{r}, t) = \sum_{\vec{k}} \sqrt{\frac{\hbar}{2\varepsilon_0 \omega_k V}} \left[\hat{b}_{\vec{k}}(0)\, e^{i\vec{k}\cdot\vec{r} - i\omega_k t} + \hat{b}_{\vec{k}}^+(0)\, e^{-i\vec{k}\cdot\vec{r} + i\omega_k t} \right] \tilde{e}_k$$

Differentiating this vector potential with respect to time yields

$$\hat{E} = -\frac{\partial}{\partial t}\hat{A}(\vec{r},t) = \sum_{\vec{k}} i\sqrt{\frac{\hbar\omega_k}{2\varepsilon_0 V}}\left[\hat{b}_k(0)\,e^{i\vec{k}\cdot\vec{r}-i\omega_k t} - \hat{b}_k^+(0)\,e^{-i\vec{k}\cdot\vec{r}+i\omega_k t}\right]\tilde{e}_k \tag{6.4.7}$$

where $i = \sqrt{-1}$. Similar to Equation (4.2.13), we can calculate the *quantized magnetic field operator*

$$\hat{B} = \nabla\times\hat{A}(\vec{r},t) = \sum_{\vec{k}} i\sqrt{\frac{\hbar}{2\omega_k\varepsilon_0 V}}\left(\vec{k}\times\tilde{e}_k\right)\left[\hat{b}_k(0)\,e^{i\vec{k}\cdot\vec{r}-i\omega_k t} - \hat{b}_k^+(0)\,e^{-i\vec{k}\cdot\vec{r}+i\omega_k t}\right] \tag{6.4.8}$$

The polarization vectors $\tilde{e}_{k,s}$ give the electromagnetic field operators their vector character. Keep in mind that Equations (6.4.7) and (6.4.8) represent the quantized fields in free-space. To account for an increase in the electromagnetic wave as might occur when atoms produce stimulated emission, either the amplitude Hilbert space must contain time dependent vectors (Schrodinger Picture) or the creation–annihilation operators in the fields must be time dependent above and beyond the simple exponential time dependence already present. As written, Equations (6.4.7) and (6.4.8) do not contain a factor that can account for this increase. In this text, the *time independent* creation and annihilation operators will be denoted by "$\hat{b}^+$" and "$\hat{b}$" respectively. For Equations (6.4.7) and (6.4.8), $\hat{b}^+ = \hat{b}^+(0)$ and $\hat{b} = \hat{b}(0)$.

As mentioned in the introductory material, we know that electromagnetic energy consists of fundamental quanta—photons. This necessarily requires the EM Hamiltonian to be quantized. However, because we imagine the EM fields carry the energy across space, the existence of photons also requires the EM fields to be quantized. As another point, the reader certainly must be aware of the typical conceptual problems with picturing light as both particle and wave. The EM quantization procedure combines both pictures into one mathematical expression as in Equation (6.4.7). The traveling wave portion $e^{ikx-i\omega t}$ represents the wave nature of light whereas the creation and annihilation operators represent the particle nature of light. A thorough treatment of field quantization shows that all particles (not just photons) can be characterized by similar equations with both the wave and particle quantities (refer to the companion volume on second quantization for example). By the way, similar to Equations (6.4.7) and (6.4.8), the traveling wave portion can be replaced by other wave functions such as the sine and cosine for the Fabry–Perot cavity. Once again, we would see that the electric field operators contain the annihilation and creation operators for all of the possible modes of the system. The Hilbert space vectors (i.e., Fock states) describe the actual physical system (i.e., how many photons and what modes they occupy).

6.4.3 Other Basis Sets

Section 6.3 shows that the vector potential can be written in other basis sets besides the traveling waves. If the set $\{\phi_n(x)\}$ forms a basis set that satisfies the boundary conditions then the vector potential that satisfies the wave equation can be written as

$$\hat{A}(x,t) = \sum_{ks}\sqrt{\frac{\hbar}{2\varepsilon_0\omega_k}}\left[\hat{b}_{ks}(0)e^{-i\omega_k t} + \hat{b}_{ks}^+(0)e^{+i\omega_k t}\right]\vec{\phi}_{ks}(x)$$

where the polarization vector has been grouped with the basis functions and the creation and annihilation operators replace the expansion coefficients. As a result, the free-space electric and magnetic fields can be written as

$$\hat{E}(x,t) = -\frac{\partial}{\partial t}\hat{A}(x,t) = i\sum_{ks}\sqrt{\frac{\hbar}{2\varepsilon_0\omega_k}}\,\omega_k\Big[\hat{b}_{ks}(0)e^{-i\omega_k t} - \hat{b}_{ks}^+(0)e^{+i\omega_k t}\Big]\,\vec{\phi}_{ks}(x)$$

$$\hat{B}(x,t) = \nabla\times\hat{A}(x,t) = \sum_{ks}\sqrt{\frac{\hbar}{2\varepsilon_0\omega_k}}\Big[\hat{b}_{ks}(0)e^{-i\omega_k t} + \hat{b}_{ks}^+(0)e^{+i\omega_k t}\Big]\,\nabla\times\vec{\phi}_{ks}(x)$$

Example 6.4.1

Find the quantized electric field for the perfect Fabry–Perot cavity with the left mirror at $z=0$ and the right mirror at $z=L$.

Solution: The standing wave modes are

$$\vec{\phi}_n(z) = \tilde{x}\sqrt{\frac{2}{L}}\sin(k_n z)\quad\text{where}\quad k_n = \frac{\pi n}{L}\quad n = 1,2,3\ldots$$

The electric field is therefore given by $\hat{E} = -\partial_t\hat{A}$

$$\hat{E} = i\sum_k\sqrt{\frac{\hbar\omega_k}{2\varepsilon_0}}\Big[\hat{b}_k(0)\,e^{-i\omega_k t} - \hat{b}_k^+(0)\,e^{+i\omega_k t}\Big]\,\vec{\phi}_k(z) = i\tilde{x}\sum_k\sqrt{\frac{\hbar\omega_k}{\varepsilon_0 L}}\Big[\hat{b}_k\,e^{-i\omega_k t} - \hat{b}_k^+\,e^{+i\omega_k t}\Big]\sin(kz)$$

where the sum is over the allowed values of "k."

6.4.4 EM Fields with Quadrature Operators

Similar to the electron harmonic oscillator, the creation and annihilation operators can be related to quadrature operators $\hat{q}_k$ and $\hat{p}_k$ according to

$$\hat{b}_k = \frac{\omega_k\,\hat{q}_k}{\sqrt{2\hbar\omega_k}} + \frac{i\,\hat{p}_k}{\sqrt{2\hbar\omega_k}}\quad \hat{b}_k^+ = \frac{\omega_k\,\hat{q}_k}{\sqrt{2\hbar\omega_k}} - \frac{i\,\hat{p}_k}{\sqrt{2\hbar\omega_k}} \tag{6.4.9}$$

$$\hat{q}_k = \sqrt{\frac{\hbar}{2\omega_k}}\big(\hat{b}_k + \hat{b}_k^+\big)\quad \hat{p}_k = -i\sqrt{\frac{\hbar\omega_k}{2}}\big(\hat{b}_k - \hat{b}_k^+\big) \tag{6.4.10}$$

where $\hat{q}_k$ and $\hat{p}_k$ must be Hermitian operators. The subscripts "k" label the modes. We define the creation–annihilation operators in terms of these quadrature operators $\hat{q}_k$ and $\hat{p}_k$ similar to the ladder operators for the electron harmonic oscillator. Often the quadrature operators $\hat{q}_k$ and $\hat{p}_k$ are termed position and momentum operators because of their similarity to those used in the electron harmonic oscillator. *However, these position and momentum quadrature operators describe neither the spatial position $\vec{r}$ nor the photon momentum $\hbar\vec{k}$.* Knowing the commutator between the creation and annihilation operators $\hat{b}_k^+, \hat{b}_k$ allows us to deduce the commutation relations between the quadrature operators

$$\left[\hat{q}_k, \hat{q}_K\right] = 0 = \left[\hat{p}_k, \hat{p}_K\right] \quad \text{and} \quad \left[\hat{q}_k, \hat{p}_K\right] = i\hbar\delta_{k,K} \tag{6.4.11}$$

The fields can be written in terms of the position and momentum quadrature operators by substituting Equations (6.4.9) and (6.4.10) into Equations (6.4.6) through (6.4.8).

$$\vec{A}(\vec{r}, t) = \frac{1}{\sqrt{\varepsilon_0 V}} \sum_{\vec{k}} \tilde{e}_k \left[\hat{q}_k \cos\left(\vec{k}\cdot\vec{r} - \omega_k t\right) - \frac{\hat{p}_k}{\omega_k} \sin\left(\vec{k}\cdot\vec{r} - \omega_k t\right)\right] \tag{6.4.12a}$$

$$\vec{E}(\vec{r}, t) = \frac{-1}{\sqrt{\varepsilon_0 V}} \sum_{\vec{k}} \tilde{e}_k\, \omega_k \left[\hat{q}_k \sin\left(\vec{k}\cdot\vec{r} - \omega_k t\right) + \frac{\hat{p}_k}{\omega_k} \cos\left(\vec{k}\cdot\vec{r} - \omega_k t\right)\right] \tag{6.4.12b}$$

$$\vec{B}(\vec{r}, t) = \frac{-1}{\sqrt{\varepsilon_0 V}} \sum_{\vec{k}} \vec{k} \times \tilde{e}_k \left[\hat{q}_k \sin\left(\vec{k}\cdot\vec{r} - \omega_k t\right) + \frac{\hat{p}_k}{\omega_k} \cos\left(\vec{k}\cdot\vec{r} - \omega_k t\right)\right] \tag{6.4.12c}$$

Equations (6.4.12) give the meaning to the name quadrature operators since the q and p multiply sines and cosines, respectively. The reader should also recognize that these quadrature operators do not describe the polarization of the electromagnetic wave; however, different polarizations can correspond to different quadrature operators through the indices on the operators. Similarly, neither operator can be identified solely with just one of the fields (E or B). At $\vec{r} = 0$, notice that at $t = 0$ the electric field is directly proportional to the momentum operator $\hat{p}$

$$\vec{E}(0, 0) = \frac{-1}{\sqrt{\varepsilon_0 V}} \sum_{\vec{k}} \tilde{e}_k \hat{p}_k \tag{6.4.13a}$$

while at a later time the electric field is directly proportional to the position operator $\hat{q}$.

$$\vec{E}(\vec{r}, t') = \frac{-1}{\sqrt{\varepsilon_0 V}} \sum_{\vec{k}} \tilde{e}_k \omega_k \hat{q}_k \tag{6.4.13b}$$

Similarly, changing the point of observation $\vec{r}$ also changes the relation between the electric field and the operators. The magnitude of the field must be related to the sum of the square of the p's and q's.

6.4.5 An Alternate Set of Quadrature Operators

Quantum optics often defines an alternate set of quadrature operators $\hat{Q}, \hat{P}$ in order to make the electric field in Equation 6.4.12 appear more symmetrical and to provide a multiplying constant that has units of square root of energy. The normalized quadrature operators are defined by

$$\hat{Q}_{\vec{k}} = \hat{q}_{\vec{k}}\sqrt{\frac{\omega_{\vec{k}}}{\hbar}} = \frac{\hat{b}_{\vec{k}} + \hat{b}_{\vec{k}}^+}{\sqrt{2}} \qquad \hat{P}_{\vec{k}} = \hat{p}_{\vec{k}}\frac{1}{\sqrt{\hbar\omega_{\vec{k}}}} = -i\frac{\hat{b}_{\vec{k}} - \hat{b}_{\vec{k}}^+}{\sqrt{2}} \tag{6.4.14a}$$

These normalized quadrature operators can easily be shown to obey the following commutation relations.

$$\left[\hat{Q}_{\vec{k}}, \hat{Q}_{\vec{K}}\right] = 0 = \left[\hat{P}_{\vec{k}}, \hat{P}_{\vec{K}}\right] \quad \left[\hat{Q}_{\vec{k}}, \hat{P}_{\vec{K}}\right] = i\delta_{\vec{k}\vec{K}} \tag{6.4.14b}$$

As a result the electric field becomes

$$\vec{E}(\vec{r}, t) = -\sum_{\vec{k}} \tilde{e}_k \sqrt{\frac{\hbar\omega_k}{\varepsilon_0 V}} \; \hat{Q}_{\vec{k}} \sin\left(\vec{k}\cdot\vec{r} - \omega_k t\right) + \hat{P}_{\vec{k}} \cos\left(\vec{k}\cdot\vec{r} - \omega_k t\right)$$

These alternate quadrature components simplify the plots of the Wigner distribution.

6.4.6 Phase Rotation Operator for the Quantized Electric Field

Electric fields have an arbitrary origin of time, which is equivalent to setting the initial phase ϕ to an arbitrary value. Occasions arise when we would like an operator that "rotates" the electric field operator to an arbitrary phase θ such as occurs in

$$\hat{E} = i \sum_k \sqrt{\frac{\hbar\omega_k}{2\varepsilon_0 V}} \left[\hat{b}_k e^{i\vec{k}\cdot\vec{r} - i\omega_k t - i\theta_k} - \hat{b}_k^+ e^{-i\vec{k}\cdot\vec{r} + i\omega_k t + i\theta_k}\right] \tilde{e}_k$$

The subscript k on θ_k indicates that each mode can be independently rotated. Besides being interesting and important in its own right, the *single-mode* rotation operator

$$\hat{R}_k(\theta) = e^{-i\theta_k \hat{N}_k} = e^{-i\theta_k \hat{b}_k^+ \hat{b}_k}$$

allows us to interchange quadrature components in order to facilitate the discussion of the Wigner probability function. We can simultaneously rotate all of the modes by applying the rotation operator

$$R = \prod_k R_k$$

For now, we concentrate on rotating a single mode

$$\hat{E} = i \sqrt{\frac{\hbar\omega}{2\varepsilon_0 V}} \left[\hat{b}\, e^{i\vec{k}\cdot\vec{r} - i\omega t} - \hat{b}^+ e^{-i\vec{k}\cdot\vec{r} + i\omega t}\right]$$

Here the word "rotate" does not refer to the polarization, instead it describes the phase delay. As discussed later, the values of the electric field can be plotted on a 2-D phase–space graph with axes corresponding to the quadrature values p and q. The electric field rotates in this phase–space plot.

We will now show that

$$\hat{R}(\theta) = e^{-i\theta \hat{N}} = e^{-i\theta \hat{b}^+ \hat{b}} \tag{6.4.15}$$

defines a rotation operator. The operator $\hat{R}$ is obviously unitary with $\hat{R}^+(\theta) = \hat{R}^{-1}(\theta) = \hat{R}(-\theta)$ for the real phase parameter θ. The number operator $\hat{N} = \hat{b}^+\hat{b}$ being conjugate to the phase operator appears in the argument of the exponential. The number operator is the generator of phase rotations. We will find the rotated field by applying a similarity transformation

$$\hat{E}_R = \hat{R}^+\hat{E}\,\hat{R}$$

We can either rotate the state with something like $\hat{R}^+|\psi\rangle$ or rotate the operator using the similarity transformation but we should not do both! To apply the similarity transformation, we must know how the rotation affects the creation and annihilation operators.

First we will show two relations for the rotated annihilation and creation operators.

$$\hat{R}^+\hat{b}\,\hat{R} = e^{i\theta\hat{b}^+\hat{b}}\hat{b}\,e^{-i\theta\hat{b}^+\hat{b}} = \hat{b}\,e^{-i\theta} \quad \text{and} \quad \hat{R}^+\hat{b}^+\,\hat{R} = \hat{b}^+\,e^{i\theta} \qquad (6.4.16)$$

We need to use the operator expansion theorem from Chapter 4 which is

$$e^{x\hat{A}}\hat{B}e^{-x\hat{A}} = \hat{B} + \frac{x}{1!}[\hat{A},\hat{B}] + \frac{x^2}{2!}\left[\hat{A},\left[\hat{A},\hat{B}\right]\right] + \cdots$$

with $x = i\theta$, $\hat{A} = \hat{b}^+\hat{b}$ and $\hat{B} = \hat{b}$ we find

$$e^{i\theta\hat{b}^+\hat{b}}\hat{b}e^{-i\theta\hat{b}^+\hat{b}} = \hat{b} + \frac{i\theta}{1!}[\hat{b}^+\hat{b},\hat{b}] + \frac{(i\theta)^2}{2!}\left[\hat{b}^+\hat{b},\left[\hat{b}^+\hat{b},\hat{b}\right]\right] + \cdots = \hat{b} - i\theta\hat{b} + \cdots = \hat{b}\,e^{-i\theta}$$

where we used $[\hat{b},\hat{b}^+] = 1$ and $[\hat{A}\hat{B},\hat{C}] = \hat{A}[\hat{B},\hat{C}] + [\hat{A},\hat{C}]\hat{B}$. The second relation can be found by taking the adjoint of the first one.

Now we can show that the single-mode rotation operator rotates the phase of the electric field

$$\hat{E}_R = \hat{R}^+\hat{E}\hat{R} = i\sqrt{\frac{\hbar\omega}{2\varepsilon_0 V}}\left[\left(\hat{R}^+\hat{b}\hat{R}\right)e^{i\vec{k}\cdot\vec{r}-i\omega t} - \left(\hat{R}^+\hat{b}^+\hat{R}\right)e^{-i\vec{k}\cdot\vec{r}+i\omega t}\right]$$

$$= i\sqrt{\frac{\hbar\omega}{2\varepsilon_0 V}}\left[\left(\hat{b}e^{-i\theta}\right)e^{i\vec{k}\cdot\vec{r}-i\omega t} - \left(\hat{b}^+e^{i\theta}\right)e^{-i\vec{k}\cdot\vec{r}+i\omega t}\right] \qquad (6.4.17)$$

$$= i\sqrt{\frac{\hbar\omega}{2\varepsilon_0 V}}\left[\hat{b}e^{i\vec{k}\cdot\vec{r}-i\omega t-i\theta} - \hat{b}^+e^{-i\vec{k}\cdot\vec{r}+i\omega t+i\theta}\right]$$

as required. We will see that this rotation makes most sense for coherent states.

Next, we show that the rotation operator can be used to interchange the quadrature terms in Equation (6.4.12b), which we repeat in single mode form (for simplicity),

$$\hat{E} = \frac{-\omega}{\sqrt{\varepsilon_0 V}}\left[\hat{q}\sin\left(\vec{k}\cdot r - \omega t\right) + \frac{\hat{p}}{\omega}\cos\left(\vec{k}\cdot r - \omega t\right)\right]$$

We need to know how the rotation operator affects the quadrature operators. Equation (6.4.17) shows that we only need to make the replacement $\vec{k}\cdot\vec{r} - \omega t \rightarrow \vec{k}\cdot\vec{r} - \omega t - \theta$.

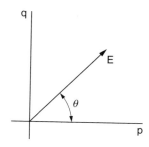

FIGURE 6.4.1
Phase angle.

Now if we set $\theta = (\pi/2)$ and use the relations $\cos(\phi - \pi/2) = \sin(\phi)$ and $\sin(\phi - \pi/2) = \cos(\phi)$, we obtain the rotated field

$$\hat{E} = \frac{-\omega}{\sqrt{\varepsilon_0 V}} \left[\hat{q} \cos\left(\vec{k} \cdot r - \omega t\right) + \frac{\hat{p}}{\omega} \sin\left(\vec{k} \cdot r - \omega t\right) \right]$$

Now the q and p operators correspond to the "x" and "y" axis respectively.

Figure 6.4.1 shows the general idea. A measurement of the electric field produces values for the quadrature operators $\hat{q}, \hat{p}$ although subsequent measurements produce different values since the operators don't commute. Suppose q and p represent the possible results of the measurements. Then the electric field might have the particular value shown in the figure. The phase rotation operator essentially changes the angle that the field makes with respect to the p axis.

6.4.7 Trouble with Amplitude and Phase Operators

Once given the amplitude space, we want to know the operators that give the amplitude and phase of the field. This has been an area of research and problems since the start of quantum electrodynamics. To see the simplest of problems, consider the electric field for a single mode using the quadrature operators (see Mandel and Wolf)

$$\hat{E}(z, t) = \frac{-\omega}{\sqrt{\varepsilon_0 V}} \left[\hat{q} \sin(kz - \omega t) + \frac{\hat{p}}{\omega} \cos(kz - \omega t) \right] \tag{6.4.18}$$

where

$$\hat{q}_k = \sqrt{\frac{\hbar}{2\omega_k}}\left(\hat{b}_k + \hat{b}_k^+\right) \quad \hat{p}_k = -i\sqrt{\frac{\hbar\omega_k}{2}}\left(\hat{b}_k - \hat{b}_k^+\right) \tag{6.4.19}$$

We might think to define a *classical* amplitude (for a single mode) by the sum of the squares of the coefficients of the sine and cosine terms similar to the classical case (see Problem 5.2). We find

$$\widehat{\mathrm{Ampl}}^2 = \left|\hat{E}\right|^2 \times \frac{\omega^2}{\varepsilon_0 V}\left[\hat{q}^2 + \left(\frac{\hat{p}}{\omega}\right)^2\right] = \frac{2\hbar^2\omega^2}{\varepsilon_0 V}\left(\hat{N} + \frac{1}{2}\right) \tag{6.4.20a}$$

where Equations (6.4.19) have been used. However, the $\frac{1}{2}$ should not be in the formula for a classical amplitude. If the readers carry through the calculation leading to the last term in Equation (6.4.20a), they will realize that the term arises due to the nonzero commutation relations. We could try using the normal-ordering symbol.

$$\widehat{Ampl}^2 = \left|\hat{E}\right|^2 = \frac{\omega^2}{\varepsilon_0 V} : \left[\hat{q}^2 + \left(\frac{\hat{p}}{\omega}\right)^2\right] : = \frac{2\hbar\omega}{\varepsilon_0 V}\hat{N} \qquad (6.4.20b)$$

The symbol :f: refers to the "normal" order, which here means to interchange the boson creation and annihilation operators without using the commutation relations (Fermion creation–annihilation operators must include an extra minus sign for each interchange). The normal order symbol has the following effect : $\hat{b}\hat{b}^+ : = \hat{b}^+\hat{b}$. If we agree to take expectation values of Equation (6.4.20b) and only afterwards take the square root, similar to the procedure for variance and standard deviation, then the equation provides reasonable results for the Fock and coherent states. However, if we first define $\widehat{Ampl} = \sqrt{\hat{N}2\hbar\omega/\varepsilon_0 V}$ then the term $\sqrt{\hat{N}}$ leads to problems for the coherent states since it produces $\langle\sqrt{n}\rangle$ rather than $\sqrt{\bar{n}}$.

We might try to write the amplitudes $\hat{b}, \hat{b}^+$ as a magnitude and phase. For example, it would be tempting to write $\hat{b}\hat{b}' \times e^{i\hat{\theta}}$ and assume the magnitude $\hat{b}'$ and phase $\hat{\theta}$ to be Hermitian operators. However, it can be shown that $e^{i\hat{\theta}}$ is not unitary and therefore $\hat{\theta}$ cannot be Hermitian. In order to discuss the phase, people often define the following Hermitian operators (refer to the Leonhardt book).

$$\widehat{Cos}\,\theta = \frac{1}{2}\left[e^{i\hat{\theta}} + e^{(i\hat{\theta})^+}\right] \quad \widehat{Sin}\,\theta = \frac{1}{2i}\left[e^{i\hat{\theta}} - e^{(i\hat{\theta})^+}\right] \qquad (6.4.21)$$

We also want to write a Heisenberg uncertainty relation of the form.

$$\Delta N \Delta\theta \geq 1/2 \qquad (6.4.22)$$

Comparing the evolution operator $\hat{u} = e^{\hat{H}t/i\hbar}$ with the rotation operator $\hat{R}^+(\theta) = e^{i\hat{N}\theta}$ which rotates a state through an angle θ, we might expect to transform the energy uncertainty relation $\Delta E \Delta t \geq \hbar/2$ into one for angle. Using the fact that $E = \hbar\omega(n + 1/2)$ and defining the phase angle in terms of the angular frequency as $\theta = \omega t$, we find

$$\Delta E \Delta t \geq \hbar/2 \quad \rightarrow \quad \hbar\omega(\Delta n)(\Delta\theta/\omega) \geq \hbar/2 \quad \rightarrow \quad \Delta n\,\Delta\theta \geq 1/2$$

In this case, the phase angle is just another definition for the time. The uncertainty relation tells us that we cannot simultaneously know the number of photons n and the phase of the wave. This is equivalent to the commutation relations for the quadrature operators.

6.4.8 The Operator for the Poynting Vector

The optical power flow is an important concept. Chapter 3 indicates that the Poynting vector for real fields can be written as

$$\vec{S} = \langle E \times H \rangle_{\substack{one \\ cycle}}$$

Now, substituting the quantum mechanical operators into

$$\hat{S} = \left\langle \hat{E} \times \hat{H} \right\rangle_{\substack{\text{one} \\ \text{cycle}}} \tag{6.4.23}$$

produces

$$\hat{S} = \sum_{\vec{k}} \frac{\hbar \omega_k c}{V} \left(\hat{N}_k + \frac{1}{2} \right) \tag{6.4.24}$$

where $\hat{S}$ has the units of power flow per unit area (see the chapter review exercises). The number operator $\hat{N}_k = \hat{b}_k^+ \hat{b}_k$, and the number of photons in the mode k, shows that the photons carry the energy where each photon has the energy $\hbar \omega_k$. The $\frac{1}{2}$ refers to the vacuum. The last equation has the units of energy density multiplied by velocity or Watts per area.

6.5 The Quantum Free-Field Hamilton and EM Fields

We have the necessary apparatus to quantize the free-field electromagnetic (EM) Hamiltonian. The typical method for transforming the classical Hamiltonian into the corresponding quantum mechanical one consists of replacing the classical dynamical variables with operators and requiring them to satisfy commutation rules. The development can proceed in two ways. The first method consists of substituting the classical vector potential into the Hamiltonian and changing the Fourier amplitudes into operators. This procedure does not complicate the derivation with the operator notation and issues of commutivity until the very end. It also conforms to the procedure outlined in the companion volume for quantizing a classical Hamiltonian. With the second method, the vector potential can be quantized and then substituted into the classical Hamiltonian. This method has the advantage of being conceptually simple and the most straightforward. We use the first method in this section.

Once having quantized the free-field Hamiltonian, we can find the equations of motion for the Fourier amplitudes and write Schrodinger's equation for the EM field. The wave functions have generalized coordinates as arguments. In this case, the wave function is the probability amplitude for finding a field to have specific amplitude. Actually, this development provides information on only one quadrature component. A full classical-like picture (amplitude and phase) must wait for the section on the Wigner distribution towards the end of the chapter.

6.5.1 The Classical Free-Field Hamiltonian

Section 3.5 shows that the divergence of the Poynting vector leads to an expression for the electromagnetic power flowing into/out of a volume (see also Chapter 7 on the Lagrangian and Hamiltonian for the electromagnetic fields). We identify the classical energy *density* in free space and the energy in a volume V as

$$\mathcal{H}_c = \frac{\varepsilon_0}{2} \vec{E} \cdot \vec{E} + \frac{1}{2\mu_0} \vec{B} \cdot \vec{B} \qquad H_c = \int_V dV \left(\frac{\varepsilon_0}{2} \vec{E} \cdot \vec{E} + \frac{1}{2\mu_0} \vec{B} \cdot \vec{B} \right) \tag{6.5.1}$$

where the subscript "*c*" refers to the classical case. Section 6.3 shows that the vector potential

$$\vec{A}(\vec{r},t) = \sum_{\vec{k}s} \sqrt{\frac{\hbar}{2\varepsilon_0 \omega_k V}} \tilde{e}_{ks} \left[b_{ks}(t)\, e^{i\vec{k}\cdot\vec{r}} + b_{ks}^*(t)\, e^{-i\vec{k}\cdot\vec{r}} \right]$$

for a free-space traveling wave leads to the classical electric field

$$\vec{E} = -\frac{\partial}{\partial t}\vec{A}(\vec{r},t) = \sum_{\vec{k}s} i\sqrt{\frac{\hbar\omega_k}{2\varepsilon_0 V}} \left[b_{ks}(t)\, e^{i\vec{k}\cdot\vec{r}} - b_{ks}^*(t)\, e^{-i\vec{k}\cdot\vec{r}} \right]\tilde{e}_{ks} \qquad (6.5.2)$$

and to the classical magnetic field

$$\vec{B} = \nabla\times\vec{A}(\vec{r},t) = \sum_{\vec{k}s} i\sqrt{\frac{\hbar}{2\varepsilon_0\omega_k V}} \left(\vec{k}\times\tilde{e}_{ks}\right)\left[b_{ks}(t)\, e^{i\vec{k}\cdot\vec{r}} - b_{ks}^*(t)\, e^{-i\vec{k}\cdot\vec{r}} \right] \qquad (6.5.3)$$

where the index "*s*" refers to the polarization of the mode and

$$b_{ks}(t) = b_{ks}(0)e^{-i\omega_k t} = b_{ks}e^{-i\omega_k t} \qquad b_{ks}^*(t) = b_{ks}^*(0)e^{i\omega_k t} = b_{ks}^*e^{i\omega_k t}$$

Keep in mind that the set

$$\left\{ u_{\vec{k}}(\vec{r}) = \frac{e^{i\vec{k}\cdot\vec{r}}}{\sqrt{V}} \right\}$$

consists of discrete basis vectors with the orthonormality relation of

$$\delta_{\vec{k}\vec{K}} = \left\langle u_{\vec{k}}(\vec{r})\,\middle|\,u_{\vec{K}}(\vec{r})\right\rangle = \int_V dV\, u_{\vec{k}}^*(\vec{r})\, u_{\vec{K}}(\vec{r}) = \int_V dV\,\frac{e^{i(\vec{K}-\vec{k})\cdot\vec{r}}}{V} \qquad (6.5.4)$$

The classical Hamiltonian can be written in terms of the Fourier amplitudes by substituting for the electric and magnetic fields in Equation (6.5.1). Here we rewrite the integral of the magnetic field since the procedure is slightly more complicated than the corresponding one for the electric field (see the chapter review exercises). For a while, we suppress the functional notation for the Fourier coefficients *b* and *b** to make the notation more compact.

$$\int_V dV\,\frac{1}{2\mu_0}\vec{B}\cdot\vec{B}$$

$$= \frac{-\hbar}{4\mu_0\varepsilon_0 V}\int_V dV \sum_{\vec{K}S}\sum_{\vec{k}s}\frac{1}{\sqrt{\omega_k\omega_K}}\left(\vec{k}\times\tilde{e}_{ks}\right)\left(\vec{K}\times\tilde{e}_{KS}\right)\left[b_{ks}e^{i\vec{k}\cdot\vec{r}} - b_{ks}^*e^{-i\vec{k}\cdot\vec{r}} \right]\left[b_{KS}e^{i\vec{K}\cdot\vec{r}} - b_{KS}^*e^{-i\vec{K}\cdot\vec{r}} \right]$$

$$= \frac{-\hbar}{4\mu_0\varepsilon_0 V}\int_V dV \sum_{\vec{K}S}\sum_{\vec{k}s}\frac{1}{\sqrt{\omega_k\omega_K}}\left(\vec{k}\times\tilde{e}_{ks}\right)\left(\vec{K}\times\tilde{e}_{KS}\right)\left[b_{ks}b_{KS}e^{i(\vec{k}+\vec{K})\cdot\vec{r}} + b_{ks}^*b_{KS}^*e^{-i(\vec{k}+\vec{K})\cdot\vec{r}} \right.$$

$$\left. - b_{ks}b_{KS}^*e^{i(\vec{k}-\vec{K})\cdot\vec{r}} - b_{ks}^*b_{KS}e^{-i(\vec{k}-\vec{K})\cdot\vec{r}} \right]$$

Using the orthonormality relation in Equation (6.5.4), this equation simplifies to

$$\int_V dV \frac{1}{2\mu_0} \vec{B} \cdot \vec{B} = \frac{-\hbar}{4\mu_0\varepsilon_0} \sum_{ksS} \frac{1}{\omega_k} \Big[\cdot \big(\vec{k} \times \tilde{e}_{ks}\big)\big(-\vec{k} \times \tilde{e}_{-kS}\big) b_{ks} b_{-\vec{k}S} + \big(\vec{k} \times \tilde{e}_{ks}\big)\big(-\vec{k} \times \tilde{e}_{-kS}\big) b_{ks}^* b_{-kS}^* S +$$

$$- \big(\vec{k} \times \tilde{e}_{ks}\big)\big(\vec{k} \times \tilde{e}_{kS}\big) b_{ks} b_{kS}^* - \big(\vec{k} \times \tilde{e}_{ks}\big)\big(\vec{k} \times \tilde{e}_{kS}\big) b_{ks}^* b_{kS} \Big]$$

where we have used the fact that $\omega_k = \omega_{-k}$. Next using the general vector relation

$$\big(\vec{A} \times \vec{B}\big) \cdot \big(\vec{C} \times \vec{D}\big) = \big(\vec{A} \cdot \vec{C}\big)\big(\vec{B} \cdot \vec{D}\big) - \big(\vec{A} \cdot \vec{D}\big)\big(\vec{B} \cdot \vec{C}\big) \tag{6.5.5}$$

and the fact that the polarization vectors satisfy an orthonormality relation

$$\tilde{e}_{ks} \cdot \tilde{e}_{kS} = \delta_{sS} \tag{6.5.6}$$

since different polarizations are orthogonal. We find

$$\big(\vec{k} \times \tilde{e}_{ks}\big)\big(\vec{k} \times \tilde{e}_{kS}\big) = k^2 \delta_{sS} \quad \text{and} \quad \big(\vec{k} \times \tilde{e}_{ks}\big)\big(-\vec{k} \times \tilde{e}_{-kS}\big) = -k^2 \delta_{sS}$$

The integral over the magnetic field becomes

$$\int_V dV \frac{1}{2\mu_0} \vec{B} \cdot \vec{B} = \frac{-\hbar}{4\mu_0\varepsilon_0} \sum_{\vec{k}s} \frac{1}{\omega_k} \Big[-k^2 b_{ks} b_{-\vec{k}s} - k^2 b_{ks}^* b_{-ks}^* - k^2 b_{ks} b_{ks}^* - k^2 b_{ks}^* b_{ks} \Big]$$

Making the substitution $(\mu_0\varepsilon_0)^{-1} = c^2 = (\omega_k/k)^2$, we find the result for the energy residing in the magnetic field

$$\int_V dV \frac{1}{2\mu_0} \vec{B} \cdot \vec{B} = \frac{-\hbar}{4} \sum_{\vec{k}s} \omega_k \Big[-b_{ks} b_{-\vec{k}s} - b_{ks}^* b_{-ks}^* - b_{ks} b_{ks}^* - b_{ks}^* b_{ks} \Big]$$

In a similar manner, but with a lot less trouble, we find the expression for the integral over the electric field to be

$$\int_V dV \frac{\varepsilon_0}{2} \vec{E} \cdot \vec{E} = \frac{-\hbar}{4} \sum_{\vec{k}s} \omega_k \Big[+b_{ks} b_{-\vec{k}s} + b_{ks}^* b_{-ks}^* - b_{ks} b_{ks}^* - b_{ks}^* b_{ks} \Big]$$

Therefore, the classical Hamiltonian in Equation (6.5.1) becomes

$$H_c = \int_V dV \left(\frac{\varepsilon_0}{2} \vec{E} \cdot \vec{E} + \frac{1}{2\mu_0} \vec{B} \cdot \vec{B} \right) = \frac{1}{2} \sum_{\vec{k}s} \hbar\omega_k \Big[b_{\vec{k}s}^*(t) b_{\vec{k}s}(t) + b_{\vec{k}s}(t) b_{\vec{k}s}^*(t) \Big] \tag{6.5.7}$$

where we have been careful not to commute the conjugate variables (i.e., b, b^*) since we know they will become creation and annihilation operators which do not commute.

6.5.2   The Quantum Mechanical Free-Field Hamiltonian

The classical Hamiltonian (total energy in volume V) in Equation (6.5.7) can be quantized by replacing the classical fields with operators

$$\hat{H} = \int_V dV \left(\frac{\varepsilon_0}{2} \hat{E}^2 + \frac{1}{2\mu_0} \hat{B}^2 \right)$$

This represents the total energy in a volume V. Rather than substituting the operators first as in this last equation, we have written a classical Hamiltonian in terms of the classical Fourier coefficients as in Equation (6.5.7). These classical Fourier amplitudes can be replaced with the corresponding creation and annihilation operators to quantize the Hamiltonian

$$\hat{H} = \frac{1}{2} \sum_{\vec{k}s} \hbar\omega_k \left[\hat{b}^+_{\vec{k}s}(t)\, \hat{b}_{\vec{k}s}(t) + \hat{b}_{\vec{k}s}(t)\, \hat{b}^+_{\vec{k}s}(t) \right] = \frac{1}{2} \sum_{\vec{k}s} \hbar\omega_k \left[\hat{b}^+_{\vec{k}s} \hat{b}_{\vec{k}s} + \hat{b}_{\vec{k}s} \hat{b}^+_{\vec{k}s} \right] \tag{6.5.8}$$

where the creation and annihilation operators depend on time according to

$$\hat{b}_{\vec{k}s}(t) = \hat{b}_{\vec{k}s}(0)\, e^{-i\omega_k t} = \hat{b}_{\vec{k}s}\, e^{-i\omega_k t} \quad \text{and} \quad \hat{b}^+_{\vec{k}s}(t) = \hat{b}^+_{\vec{k}s}(0)\, e^{+i\omega_k t} = \hat{b}^+_{\vec{k}s}\, e^{+i\omega_k t}$$

Notice that the time-dependence in the free-field Hamiltonian cancels out. The time-dependent annihilation and creation operators are operators in the interaction picture (refer to the next example) or equivalently, Heisenberg operators for a closed system. The required equal-time commutation relations

$$\left[\hat{b}_{\vec{k}s}(t), \hat{b}^+_{\vec{K}S}(t) \right] = \delta_{\vec{k}\vec{K}}\delta_{sS} \quad \left[\hat{b}_{\vec{k}s}(t), \hat{b}_{\vec{K}S}(t) \right] = 0 = \left[\hat{b}^+_{\vec{k}s}(t), \hat{b}^+_{\vec{K}S}(t) \right]$$

hold for all times including $t = 0$. Normal order for the creation and annihilation operators requires that the creation operators be positioned to the left of the annihilation operators. We therefore use the first commutation relation $\hat{b}_{\vec{k}s} \hat{b}^+_{\vec{k}s} = \hat{b}^+_{\vec{k}s} \hat{b}_{\vec{k}s} + 1$ to change the second term in Hamiltonian (6.5.8) so that

$$\hat{H} = \sum_{\vec{k}s} \hbar\omega_k \left(\hat{b}^+_{\vec{k}s} \hat{b}_{\vec{k}s} + \frac{1}{2} \right) = \sum_{\vec{k}s} \hbar\omega_k \left(\hat{N}_{\vec{k}s} + \frac{1}{2} \right) \tag{6.5.9}$$

To obtain the Hamiltonian in Equation (6.5.9), we converted the electromagnetic fields into operators by substituting the creation–annihilation $\hat{b}^+_{\vec{k}s}, \hat{b}_{\vec{k}s}$ operators for the amplitudes. Essentially the amplitude of a wave now must be specified by a Hilbert space of vectors. Different linear combinations in this amplitude space produce different amplitudes with different properties. To specify the total energy in the volume V, we must feed the Hamiltonian $\hat{H}$ a vector from the amplitude space. Normally, the number operator $\hat{N}_{\vec{k}s} = \hat{b}^+_{\vec{k}s} \hat{b}_{\vec{k}s}$ is interpreted as providing the total number of photons in the mode $\vec{k}, s$. Using the integral over the volume V as in Equation (6.5.7) suggests an interpretation as either (i) the number of photons in mode $\vec{k}, s$ for fixed endpoint boundary conditions or as (ii) the number of photons in volume V and mode $\vec{k}, s$ for periodic boundary conditions. The two interpretations become equivalent if $V \to \infty$. Interpretation (ii) would provide the number of photons per unit volume in mode $\vec{k}, s$. Recent work (refer to the book by Mandel and Wolf, Section 12.11) shows that photons cannot be localized to a finite region of space.

The Hamiltonian for light is similar to the Hamiltonian for the electron harmonic oscillator. The summation occurs in Equation (6.5.9) because there exists infinitely many light modes (i.e., wavelengths). We can add photons to any of these modes. There can be any number of photons in a mode. The harmonic oscillator Hamiltonian for the electron does not have the summation. The modes (basis set) accept only a single electron. In addition we use ladder operators to promote or demote electrons from one state to another. These ladder operators do not create free photons. Instead they add or subtract a quantum of energy to promote or demote the electron, respectively. The electromagnetic field appears as an ensemble of *independent* harmonic oscillators. The oscillators are called "independent" because Equation (6.5.9) doesn't have any cross-terms between modes. The number operator $\hat{N}_{\vec{ks}}$ provides the number of photons in a particular mode specified by the wave vector $\vec{k}$ and polarization "s."

Equation (6.5.9) for $\hat{H} = \sum_{\vec{ks}} \hbar \omega_k (\hat{N}_{\vec{ks}} + 1/2)$ contains a summation over the frequency for all of the possible modes, namely,

$$\frac{1}{2} \sum_{\vec{ks}} \hbar \omega_k \qquad (6.5.10)$$

The allowed frequencies can be infinitely large. Without photons $n_k = 0$, the energy represented by Equation (6.5.10) must be stored as a fluctuating electric field in the vacuum state. The summation in Equation (6.5.10) becomes infinite even for a finite volume V of integration initially used to calculate the energy! Physically, this implies a very large energy stored in the vacuum! In some cases, we can ignore the divergent term. For example, to calculate the rate of change of an operator $\hat{A}$ (Heisenberg picture for a closed system), we commute the operator with the Hamiltonian. If the operator involves a term like Equation (6.5.10) which is just a number (albeit an infinite one), then we see that the infinite term drops out.

$$\left[\hat{H}, \hat{A}\right] = \left[\sum_{\vec{ks}} \hbar \omega_k \hat{N}_{\vec{ks}}, \hat{A}\right] + \left[\sum_{\vec{ks}} \frac{\hbar \omega_k}{2}, \hat{A}\right] = \left[\sum_{\vec{ks}} \hbar \omega_k \hat{N}_{\vec{ks}}, \hat{A}\right]$$

Example 6.5.1

Calculate the time-dependence of the creation operator $b_K^+(t)$ in the Heisenberg picture for the free-fields.

Solution: Calculate the commutator of the creation operator with the Hamiltonian. Chapter 4 shows that the rate of change of the Heisenberg operator can be written as

$$\frac{d\hat{b}_K^+(t)}{dt} = \frac{i}{\hbar}\left[\hat{H}, \hat{b}_K^+(t)\right]$$

Substituting the Hamiltonian

$$\hat{H} = \sum_{\vec{k}} \hbar \omega_k \left(\hat{b}_{\vec{k}}^+(t)\hat{b}_{\vec{k}}(t) + \frac{1}{2}\right)$$

we find

$$\frac{d\hat{b}_{\bar{K}}^+(t)}{dt} = \frac{i}{\hbar}\left[\sum_{\bar{k}}\hbar\omega_k\left(\hat{b}_{\bar{k}}^+(t)\hat{b}_{\bar{k}}(t) + \frac{1}{2}\right), \hat{b}_{\bar{K}}^+(t)\right] = \frac{i}{\hbar}\sum_{\bar{k}}\hbar\omega_k\left\{\left[\hat{b}_{\bar{k}}^+(t)\hat{b}_{\bar{k}}(t), \hat{b}_{\bar{K}}^+(t)\right] + \left[\frac{1}{2}, \hat{b}_{\bar{K}}^+(t)\right]\right\}$$

The infinite vacuum sum produces the commutator at the end of this last equation. The commutator of a *c*-number with an operator produces a result of zero; consequently, the infinite divergence does not affect the calculated value. Using commutation rules, we can evaluate

$$\left[\hat{b}_{\bar{k}}^+\hat{b}_{\bar{k}}, \hat{b}_{\bar{K}}^+\right] = \left[\hat{b}_{\bar{k}}^+, \hat{b}_{\bar{K}}^+\right]\hat{b}_{\bar{k}} + \hat{b}_{\bar{k}}^+\left[\hat{b}_{\bar{k}}, \hat{b}_{\bar{K}}^+\right] = 0 + \hat{b}_{\bar{k}}^+\delta_{\bar{k}\bar{K}}$$

so that

$$\frac{d\hat{b}_{\bar{K}}^+(t)}{dt} = \frac{i}{\hbar}\sum_{\bar{k}}\hbar\omega_k\left[\hat{b}_{\bar{k}}^+(t)\hat{b}_{\bar{k}}(t), \hat{b}_{\bar{K}}^+(t)\right] = i\omega_K\hat{b}_{\bar{K}}^+(t)$$

This is a simple differential equation with a solution that agrees with our previous results

$$\hat{b}_{\bar{K}}^+(t) = \hat{b}_{\bar{K}}^+(0)e^{i\omega_K t} \tag{6.5.11a}$$

The complex conjugate provides the time dependence of the annihilation operator

$$\hat{b}_{\bar{K}}(t) = \hat{b}_{\bar{K}}(0)e^{-i\omega_K t} \tag{6.5.11b}$$

6.5.3 The EM Hamiltonian in Terms of the Quadrature Operators

Before specifying the Hilbert space, we consider an alternate form for the Hamiltonian that again shows its similarity to the Hamiltonian for a collection of independent harmonic oscillators. Later in this chapter, we will see that the quadrature-form of the EM Hamiltonian allows us to calculate the probability of finding a particular amplitude or phase for the EM field.

As previously mentioned, Equation (6.5.9)

$$\hat{H} = \sum_{\bar{k}s}\hbar\omega_k\left(\hat{b}_{\bar{k}s}^+\hat{b}_{\bar{k}s} + \frac{1}{2}\right) = \sum_{\bar{k}s}\hbar\omega_k\left(\hat{N}_{\bar{k}s} + \frac{1}{2}\right)$$

has the same form as that for a collection of independent harmonic oscillators. Similar to the harmonic oscillator, the creation and annihilation operators can be related to position-like $\hat{q}_k$ and momentum-like $\hat{p}_k$ quadrature operators according to

$$\hat{b}_{ks}(t) = \frac{\omega_k\hat{q}_{ks}(t)}{\sqrt{2\hbar\omega_k}} + \frac{i\hat{p}_{ks}(t)}{\sqrt{2\hbar\omega_k}} \quad \hat{b}_{ks}^+(t) = \frac{\omega_k\hat{q}_{ks}(t)}{\sqrt{2\hbar\omega_k}} - \frac{i\hat{p}_{ks}(t)}{\sqrt{2\hbar\omega_k}} \tag{6.5.12}$$

where $\hat{q}_{ks}(t)$ and $\hat{p}_{ks}(t)$ are taken to be Hermitian operators. Equations (6.5.12) hold for $t=0$ with the definitions $q_{ks} = q_{ks}(0)$ and $p_{ks} = p_{ks}(0)$. The subscripts "*k*" and "*s*" label the wavelength and polarization modes, respectively. We usually suppress the subscript "*s*."

The position $\hat{q}_k$ and momentum $\hat{p}_k$ quadrature operators are not related to the spatial position $\vec{r}$ nor to the photon momentum $\hbar\vec{k}$. The quadrature operators are related to the amplitude of the electric and magnetic fields. Solving Equations (6.5.12) for the position $\hat{q}_k$ and momentum $\hat{p}_k$ provides relations similar to those for the harmonic oscillator

$$\hat{q}_k(t) = \sqrt{\frac{\hbar}{2\omega_k}}\left[\hat{b}_k(t) + \hat{b}_k^+(t)\right] \quad \hat{p}_k(t) = -i\sqrt{\frac{\hbar\omega_k}{2}}\left[\hat{b}_k(t) - \hat{b}_k^+(t)\right] \tag{6.5.13}$$

Unlike that for the electron harmonic oscillator, the mass does not appear in these quadrature operators. The commutation relations for the creation and annihilation operators provide the commutation relations between the position and momentum quadrature operators as follows

$$\left[\hat{q}_i(t), \hat{q}_j(t)\right] = 0 = \left[\hat{p}_i(t), \hat{p}_j(t)\right] \quad \left[\hat{q}_i(t), \hat{p}_j(t)\right] = i\hbar\delta_{ij} \tag{6.5.14}$$

which hold for all times t including $t=0$.

The Hamiltonian and the fields can be written in terms of these position and momentum quadrature operators. Starting with the Hamiltonian

$$\hat{H} = \sum_{\vec{k}s} \hbar\omega_k\left(\hat{b}_{\vec{k}s}^+\hat{b}_{\vec{k}s} + \frac{1}{2}\right) = \sum_{\vec{k}s} \hbar\omega_k\left(\hat{N}_{\vec{k}s} + \frac{1}{2}\right)$$

Neglecting the polarization index and substituting Equations (6.5.12) for the creation and annihilation operators in the Hamiltonian provides

$$\hat{H} = \sum_{\vec{k}} \hbar\omega_k\left[\left(\frac{\omega_k\hat{q}_k}{\sqrt{2\hbar\omega_k}} + \frac{i\hat{p}_k}{\sqrt{2\hbar\omega_k}}\right)\left(\frac{\omega_k\hat{q}_k}{\sqrt{2\hbar\omega_k}} - \frac{i\hat{p}_k}{\sqrt{2\hbar\omega_k}}\right) + \frac{1}{2}\right]$$

Multiplying out the terms and taking care with noncommuting operators

$$\hat{H} = \sum_{\vec{k}} \hbar\omega_k\left[\frac{\omega_k^2\hat{q}_k^2}{2\hbar\omega_k} + \frac{\hat{p}_k^2}{2\hbar\omega_k} - i\frac{\omega_k}{2\hbar\omega_k}\left(\hat{q}_k\hat{p}_k - \hat{p}_k\hat{q}_k\right) + \frac{1}{2}\right]$$

Using the commutation relation $\left[\hat{q}_a, \hat{p}_b\right] = i\hbar\delta_{ab}$ and then simplifying gives

$$\hat{H} = \sum_{\vec{k}}\left(\frac{\hat{p}_{\vec{k}}^2}{2} + \frac{\omega_k^2}{2}q_{\vec{k}}^2\right) \tag{6.5.15}$$

The Hamiltonian consists of a sum of Hamiltonians for a collection of independent harmonic oscillators.

6.5.4 The Schrodinger Equation for the EM Field

We can find a Schrodinger wave equation for the wave function $\Psi(q_1, q_2, q_3, \ldots, q_N, t)$. The wave function gives the probability amplitude of finding mode #1,#2, ..., #N to have quadrature amplitude $q_1, q_2, \ldots, q_N$. For a single mode, the probability amplitude of finding the mode to have quadrature amplitude q is $\Psi(q,t)$ which can also be written in Dirac notation as $\Psi(q,t) = \langle q \mid \Psi(t)\rangle$. Similarly, by working with Fourier transforms, we can also find a wave function $\Psi(p_1, p_2, p_3, \ldots, p_N, t)$ that gives the probability amplitude of finding modes #1,#2,....,#N to have quadrature momentum $p_1, p_2, \ldots, p_N$. Again, for a single mode, we would find $\Psi(p,t)$. These wave functions must be the coordinate representations of vectors in the amplitude Hilbert space. We will discuss the amplitude Hilbert space starting in the next section concerning Fock states. For now, we realize that the operators can act on the wave functions Ψ.

Using the coordinate representation of the position and momentum quadrature operators, namely

$$\hat{q}_k \rightarrow q_k \quad \hat{p}_k \rightarrow \frac{\hbar}{i}\frac{\partial}{\partial q_k} \tag{6.5.16}$$

we can substitute them into the Hamiltonian (Equation (6.5.15))

$$\hat{H} = \sum_{\vec{k}} \left(\frac{\hat{p}_{\vec{k}}^2}{2} + \frac{\omega_{\vec{k}}^2}{2} q_{\vec{k}}^2 \right) \tag{6.5.17}$$

to obtain the coordinate representation of the Schrodinger equation

$$\sum_{\vec{k}} \left(-\frac{\hbar^2}{2}\frac{\partial^2}{\partial q_{\vec{k}}^2} + \frac{\omega_{\vec{k}}^2}{2} q_{\vec{k}}^2 \right) \Psi(q_1, q_2 \ldots, t) = i\hbar \frac{\partial}{\partial t} \Psi(q_1, q_2 \ldots, t) \tag{6.5.18}$$

We expect the solutions of this wave equation to be similar to that for the harmonic oscillator. We expect decaying exponentials multiplied by Hermite polynomials. For more information on solving the Schrodinger equation for the field amplitudes please refer to Section 5.7.

Let's continue the discussion on the meaning of the wavefunctions. Consider a single mode for simplicity. We can solve Schrodinger's equation twice to find the most probable values for q and p. Then we can find the most probable value for the electric field given by Equation (6.4.12b)

$$\hat{E}(z,t) = \frac{-\omega}{\sqrt{\varepsilon_0 V}} \left[\hat{q}\sin(kz - \omega t) + \frac{\hat{p}}{\omega}\cos(kz - \omega t) \right] \tag{6.5.19}$$

In fact, if we can find $\Psi(q,t)$ then we can find the average electric field by substituting the coordinate representation of the quadrature operators in Equation (6.5.19) and calculating

$$\langle \Psi|\hat{E}|\Psi\rangle = \frac{-\omega}{\sqrt{\varepsilon_0 V}} \left[\langle\Psi|\hat{q}|\Psi\rangle \sin(kz - \omega t) + \left\langle\Psi\left|\frac{\hat{p}}{\omega}\right|\Psi\right\rangle \cos(kz - \omega t) \right]$$

$$= \frac{-\omega}{\sqrt{\varepsilon_0 V}} \left[\sin(kz - \omega t) \int_{-\infty}^{\infty} dq \; \Psi^+(q,t) \; q \; \Psi(q,t) \right.$$

$$\left. + \cos(kz - \omega t) \int_{-\infty}^{\infty} dq \; \Psi^+(q,t) \; \frac{1}{\omega}\left(\frac{\hbar}{i}\frac{\partial}{\partial q}\right)\Psi(q,t) \right]$$

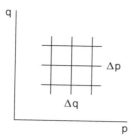

FIGURE 6.5.1
Dividing phase space into small areas $\Delta q \Delta p \sim \hbar/2$

Notice that the average occurs only over the amplitude operators.

A question naturally arises as to why the wave function Ψ does not have arguments including both q and p. The answer is that, because of the commutation relations in Equations (6.5.15), we cannot assign precise values to p and q at the same time at least in the quantum mechanical sense. Apparently, phase space consisting of all possible values of $\{q, p\}$ can only be defined by dividing the space into small squares of size $\Delta q \Delta p \sim \hbar/2$ as shown in Figure 6.5.1. The set of possible values $\{q, p\}$ then must label the individual rectangles.

It turns out that it is possible to define a joint probability distribution function for $\{q, p\}$ without reference to the subdivision rectangles. This so-called Wigner distribution brings the quantum picture of EM fields as close as possible to the classical picture. From the Wigner point of view, we can make a measurement of the field, but we must assign a probability to each value $\{q, p\}$ or equivalently, to each amplitude and phase. For example, coherent states have a Gaussian distribution for the quadrature amplitudes—the most likely set of quadrature amplitudes occurs at the center of the Gaussian distribution. This also sets the most likely amplitude and phase for the wave since the amplitude must be related to the sum of the squares of the quadratures (amplitude is similar to the hypotenuse of a triangle that has p and q as legs). There will be statistical variation between measurements. We will see later that the Wigner distribution combines the probabilities $|\Psi(q)|^2$ and $|\Psi(p)|^2$.

The difficulties come from the fact that the quadrature operators do not commute. We cannot find vectors that are simultaneously eigenvectors of both quadrature operators (the same is true for the creation and annihilation operators). This means we cannot find eigenvectors of the field. We cannot simultaneously and definitely know the two quadratures nor the magnitude and phase of the field. Other sections in this chapter discuss this more fully.

6.6 Introduction to Fock States

Previous sections have quantized the electromagnetic (EM) fields and the EM Hamiltonian by replacing classical dynamical variables with operators. In particular, the Fourier amplitudes become operators. The various vectors in the "amplitude" Hilbert space provide the various possible amplitudes and expectation values for the field operators. The operator expressions for the fields apply to a wide range of systems whereas the states in the amplitude Hilbert space provide the specifics of a particular system. We can represent a traveling light beam by a state in the amplitude space as a way of stating the power in the beam and other characteristics. In the Schrodinger and interaction representations, the state can evolve in time when material absorbs or produces light.

The present section begins the discussion of the Fock state as one type of amplitude state among other types including the coherent and squeezed states. We will see that Fock states specify the exact number of photons in the EM modes of a system; as a result of the Heisenberg uncertainty relation however, the the phases of those states must be completely unknown. They are the eigenstates of the EM Hamiltonian giving rise to the notion of the photon as an indivisible quantum of energy. This section shows Fock states have zero average electric field. Later sections in this chapter show that coherent states have classically sensible amplitudes and phases, and best describe the laser light. The coherent state describes the total amplitude and phase of the electric field.

6.6.1 Introduction to Fock States

The quantum fields and the Hamiltonian can be expressed by a traveling wave Fourier expansion with creation "$\hat{b}^+$" and annihilation "$\hat{b}$" operators for the Fourier amplitudes that satisfy commutation relations. These operators act on "amplitude space." The "Fock states" provide the first example of a basis set for this Hilbert space. The Fock states specify the exact number of photons (particles) in a given basic state of the system; the standard deviation of the number must be zero. The ket representing the Fock state consists of "place holders" for the number of photons in a given mode (basic state) $|n_1, n_2, \ldots\rangle$. Figure 6.6.1 shows buckets that can hold photons where the mode numbers label the buckets. For example, $m = 1$ might correspond to the longest wavelength mode in a Fabry–Perot resonator. The figure shows the system has two photons (for example) in the $m = 1$ mode, none in the $m = 2$ mode, and so on. In proper notation, the state would be represented by the ket $|2, 0, 1, \ldots\rangle$. The vacuum state, denoted by $|0, 0, 0, \ldots\rangle = |0\rangle$ represents a system without any photons in any of the modes. The Fock state lives in a direct product space so that it can be written as $|n_1, n_2, \ldots\rangle = |n_1\rangle |n_2\rangle \cdots$ with each ket representing a single mode. The Fock vectors for a system with only one mode characterized by the wavelength λ_1 have only one position. For example, $|n_1\rangle$ represents n_1 particles in the mode λ_1 and $|0\rangle$ represents the single mode vacuum state. The most important point of the Fock state is that it is an eigenstate of the number operator as we will see.

We should include the polarization in the description of the Fock state. The vector potential satisfies the Coulomb gauge condition $\nabla \cdot \vec{A} = 0$ and therefore, the polarization vector must be perpendicular to the direction of propagation. Using the relations for the fields $\vec{E} = -\dot{\vec{A}}$ and $\vec{B} = \nabla \times \vec{A}$, we see that these fields must also be perpendicular to the direction of propagation. Given that polarization refers to the direction of the electric field, we see that, as a transverse field, it can have two independent directions of polarization. These directions constitute the polarization modes. In general, we use two basic polarization directions $\tilde{e}_{ks}$ ($s = 1,2$) for each wave vector $\vec{k}$. If the wave propagates along the z-direction, then one polarization mode is along $\tilde{x}$, the $s = 1$ mode, and the other is along $\tilde{y}$, the $s = 2$ mode. Each index $\vec{k}$ value must be augmented with the polarization directions as indicated in Figure 6.6.2. Circular polarization unit vectors can also be used

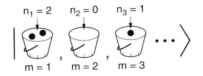

FIGURE 6.6.1
The Fock state describes the number of particles in the modes or states of the system. The diagram represents the ket $|201\ldots\rangle$.

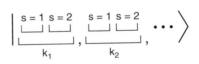

FIGURE 6.6.2
The modes must include polarization.

rather than the plane-wave polarization vectors used here.

As bosons characterized by integer spin $(0, 1, 2, \ldots)$, any number of photons (spin 1) can occupy a mode. For a given set of modes, each Fock state is a basis vector for the amplitude space. The set

$$\{|n_1, n_2, n_3, \ldots\rangle\}$$

represents the *complete* set of basis vectors where each n_i can range up to an *infinite* number of particles in the system. The orthonormality relation can be written as

$$\langle n_1, n_2, \ldots \mid m_1, m_2, \ldots \rangle = \delta_{n_1 m_1} \delta_{n_2 m_2} \ldots \tag{6.6.1}$$

and the closure relation as

$$\sum_{n_1, n_2 \ldots = 0}^{\infty} |n_1, n_2 \ldots \rangle \langle n_1, n_2 \ldots | = \hat{1} \tag{6.6.2}$$

A general vector in the Hilbert space must have the form

$$|\psi\rangle = \sum_{n_1, n_2 \ldots = 0}^{\infty} \beta_{n_1, n_2 \ldots} |n_1, n_2 \ldots \rangle \tag{6.6.3}$$

where quantum mechanical wave functions must be normalized to unity as usual. The component $\beta_{n_1, n_2 \ldots} = \langle n_1, n_2, \ldots \mid \psi \rangle$ represents the probability amplitude of finding n_1 photons in mode 1, n_2 photons in mode 2 (etc.) when the system has wave function $|\psi\rangle$.

A Fock state gives the exact number of photons in a mode but there isn't any information on the phase of the wave. The phase of the wave for the Fabry–Perot cavity does not refer to whether the wave looks like a sine or cosine. Rather the phase refers to the ϕ in $\sin(kx) \, e^{i\omega t + \phi}$ (or equivalently, the origin of time). The fact that the Fock state provides exact information on the photon number but none on the phase can be explained by a Heisenberg uncertainty relation between the particle number "n" and the phase "ϕ." The "n–ϕ" uncertainty relation has the form

$$\Delta n \, \Delta \phi \geq \frac{1}{2}$$

where, we know that Δ represents the standard deviation. Knowing the exact number of photons $\Delta n = 0$ then requires the phase to be completely random $\Delta \phi \sim \infty$.

Figure 6.6.3 suggests that although a mode might not have any photons, there still exists an electric field! The motion of the vacuum field is equivalent to the zero point motion of a molecule near absolute zero. For the optical case, although there isn't any available energy, there still exists a fluctuating electric field. If the vacuum field encounters excited atoms, it can produce spontaneous emission. The vacuum fields have a number of real-world effects. For example, vacuum fields can be shown to move two metal plates toward each other (the Casimir effect).

Fock states can also be constructed for fermions with half-integral spin, such as electrons with spin $\frac{1}{2}$; however, the Pauli exclusion principle limits the number per mode to at most 1. These properties originate in the commutation relations for the creation and annihilation operators.

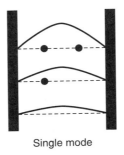

Single mode

FIGURE 6.6.3
A single mode with either 0,1, or 2 photons.

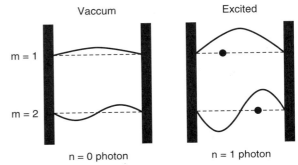

FIGURE 6.6.4
Both diagrams show the first two modes in a Fabry–Perot resonator. The left side shows an artist's view of the mode without any photons. The right side shows one photon in each mode. Adding a photon to a mode must increase the amplitude.

6.6.2 The Fabry–Perot Resonator as an Example

We consider a Fabry–Perot cavity as an example to introduce the Fock state and show its relation to the stored energy. The calculations for energy do not include the vacuum energy.

The Fabry–Perot cavity appears in Figure 6.6.4 with the $m=1$ and $m=2$ optical modes (the sine waves represent the electric field). There exist more than two optical modes but we have not drawn them. Notice how the mirrors (drawn as black boxes) provide "boundary conditions" and give rise to a discrete spectrum for the wavelength λ_m, which characterize the allowed modes (eigenfunctions).

$$\lambda_m = 2L, L, \frac{2L}{3} \cdots \frac{2L}{m} \quad m = 1, 2, 3 \ldots$$

The mode number "m" must be nonzero for this example. The energy of a *single* photon can be written as

$$E_m = \frac{hc}{\lambda_m} = \left(\frac{hc}{2L}\right) m$$

where Planck's constant is $h = 6.63 \cdot 10^{-34}$ and the speed of light in vacuum is $c = 3 \cdot 10^8$ in MKS units. The eigenfunctions

$$\phi_m(x) = \sqrt{\frac{2}{L}} \sin\left(\frac{m\pi}{L} x\right)$$

represent the modes of the Fabry–Perot cavity and correspond to the energy E_m.

The Fock state, denoted by $|n_1, n_2, n_3 \ldots\rangle$, lives in a direct product space and represents photons in the Fabry–Perot cavity. The first position in the ket $|\;\rangle$ stands for a mode with wave vector $\vec{k}_1$ (i.e., wavelength $\lambda = \lambda_1$ for 1-D). The symbol n_1 gives the number of photons in mode number 1. Similarly, n_2 represents the number of photons in the mode with wave vector $\vec{k}_2$ and wavelength $\lambda_2 = (2L/n) = L$. Consider the case of two excited modes as shown on the right-hand side of Figure 6.6.4. The first two modes have one photon each so that $n_1 = n_2 = 1$. The state vector must be $|1, 1, 000\ldots\rangle$. We can easily find the total energy stored in the cavity using the energy stored in mode #m for each photon in the mode.

$$E_m = \frac{hc}{\lambda_m} = \left(\frac{hc}{2L}\right) m$$

If mode #m has "n_m" photons, then the energy stored in mode #m must be

$$E_{m,\,tot} = n_m \frac{hc}{\lambda_m} = n_m \left(\frac{hcm}{2L}\right)$$

The total energy stored in all of the modes, for the example in Figure 6.6.4, must be

$$E_{tot} = E_{1,\,tot} + E_{2,\,tot} = 1\left(\frac{hc}{2L}\right) + 1\left(\frac{hc2}{2L}\right) = \frac{3}{2}\frac{hc}{L}$$

Unfortunately, this formulation does not include a $\frac{1}{2}$ accounting for energy stored in vacuum fields. Fock states $|n_1, n_2, \ldots\rangle$ explicitly track of the number of particles in a mode. The position in the Fock vector corresponds to a given mode, which can include polarization and wavelength for photons.

6.6.3 Creation and Annihilation Operators

The creation and annihilation operators create and remove photons from a mode characterized by a given wave vector and given polarization. "Adding a photon" to a mode means "adding energy." However, the amplitude of the electric field is directly related to the energy in an electromagnetic wave. Therefore adding a photon must increase the amplitude similar to that shown in Figure 6.6.3.

The creation and annihilation operators have both $\vec{k}$ and s subscripts (keep in mind that $\vec{k}$ represents three indices $i = k_x, k_y, k_z$). We define creation operators $\hat{b}_{is}^+ = \hat{b}_{is}^+(0)$ and annihilation operators $\hat{b}_{is} = \hat{b}_{is}(0)$ as

$$\hat{b}_{is}^+ |n_{1s}, n_{2s}, \ldots, n_{is}, \ldots\rangle = \sqrt{n_{is} + 1}\, |n_{1s}, n_{2s}, \ldots, n_{is} + 1, \ldots\rangle \qquad (6.6.4a)$$

$$\hat{b}_{is} |n_{1s}, n_{2s}, \ldots, n_{is}, \ldots\rangle = \sqrt{n_{is}}\, |n_{1s}, n_{2s}, \ldots, n_{is} - 1, \ldots\rangle \qquad (6.6.4b)$$

Usually, we will suppress the polarization index (sometimes called the spin index) and just keep track of the mode by the wave vector index. So that

$$\hat{b}_i^+ |n_1, n_2, \ldots, n_i, \ldots\rangle = \sqrt{n_i + 1}\,|n_1, n_2, \ldots, n_i + 1, \ldots\rangle \qquad (6.6.4c)$$

$$\hat{b}_i |n_1, n_2, \ldots, n_i, \ldots\rangle = \sqrt{n_i}\,|n_1, n_2, \ldots, n_i - 1, \ldots\rangle \qquad (6.6.4d)$$

where $\hat{b}_i^+$ creates a particle in mode "i" and $\hat{b}_i$ removes a particle. If the initial state is the quantum mechanical vacuum then $\hat{b}_i|0, 0, \ldots, 0_{\#i}, \ldots\rangle = 0$. The creation–annihilation operators satisfy commutation relations

$$\left[\hat{b}_{\vec{k}s}, \hat{b}_{\vec{K}S}\right] = 0 = \left[\hat{b}_{\vec{k}s}^+, \hat{b}_{\vec{K}S}^+\right] \quad \text{and} \quad \left[\hat{b}_{\vec{k}s}, \hat{b}_{\vec{K}S}^+\right] = \delta_{\vec{k}\vec{K}}\delta_{sS} \tag{6.6.5}$$

We usually suppress the "s" index.

The mode-number operator

$$\hat{N}_k = \hat{b}_k^+ \hat{b}_k \tag{6.6.6}$$

provides the number of photons in the state "k." The Fock states are eigenstates of the number operator

$$\hat{N}_k|n_1, \ldots, n_k, \ldots\rangle = n_k|n_1, \ldots, n_k, \ldots\rangle \tag{6.6.7}$$

The total number of particles in a Fock state can be found by using the total-number operator

$$\hat{N} = \sum_i \hat{N}_i \tag{6.6.8}$$

so that

$$\hat{N}|n_1, \ldots, n_k, \ldots\rangle = \left(\sum_i \hat{N}_i\right)|n_1, \ldots, n_k, \ldots\rangle = \left(\sum_i n_i\right)|n_1, \ldots, n_k, \ldots\rangle$$

The number operators have a "sharp" value for the Fock states which means their standard deviation must be zero. The standard deviation is zero for any operator $\hat{O}$ evaluated in its eigenstate $|\phi\rangle$ (i.e., $\hat{O}|\phi\rangle = \phi |\phi\rangle$) as can be seen by calculating

$$\sigma_{\hat{O}}^2 = \langle\phi| \left(\hat{O} - \bar{O}\right)^2 |\phi\rangle = \langle\phi| \hat{O}^2|\phi\rangle - \bar{O}^2 = \phi^2\langle\phi \mid \phi\rangle - \phi^2 = 0$$

Physically "sharp values" means that repeated measurements produce only one value (i.e., the measurement does not interfere with the system).

6.6.4 Comparison between Creation–Annihilation and Ladder Operators

The raising $\hat{a}^+$ and lowering $\hat{a}$ operators (i.e., ladder operators such as for the electron hormonic oscillator) map between basis vectors $\{|n\rangle = \phi_n\}$ according to

$$\hat{a}^+|n\rangle = \sqrt{n + 1} |n + 1\rangle \quad \hat{a}|n\rangle = \sqrt{n}|n - 1\rangle$$

or equivalently

$$\hat{a}^+ = \sum_{n=0}^{\infty} \sqrt{n + 1}|n + 1\rangle\langle n| \quad \hat{a} = \sum_{j=1}^{\infty} \sqrt{n}|n - 1\rangle\langle n|$$

We imagine that the raising operator removes an electron from energy eigenstate $|n\rangle$ and places it in state $|n+1\rangle$. We can re-interpret the ket $|n\rangle$ as representing the number of available quanta in the oscillator; i.e., there exists a 1–1 correspondence between the energy eigenstate occupied by a particle and the number of available quanta in that state. The ladder operators map an energy eigenstate into another one in sequence, which must be equivalent to adding or subtracting a quantum of energy.

The same operation of moving a particle from one Fock state to another requires two operations. For example, to move a photon from state n to state $n+1$

$$\hat{b}_{n+1}^{+}\hat{b}_{n}\left|0,\ldots,\underbrace{1}_{n},\underbrace{0}_{n+1},\ldots\right\rangle = \left|0,\ldots,\underbrace{0}_{n},\underbrace{1}_{n+1},\ldots\right\rangle$$

Therefore the raising operator must be somewhat equivalent to the product of the creation and annihilation operator as $\hat{a}^{+} \sim \hat{b}_{n+1}^{+}\hat{b}_{n}$. We should expect something like this since $\hat{a}^{+} = \sum \sqrt{n+1}|n+1\rangle\langle n|$ and the bra $\langle n|$ acts like the annihilation operator $\hat{b}_{n}$ while $|n+1\rangle$ is somewhat equivalent to the creation operator $\hat{b}_{n+1}^{+}$.

Example 6.6.1
Show the average electric field must be zero for a single mode Fock state

Solution: The average electric field can be found by using its definition in terms of the creation and annihilation operators found in the previous section and using Equation (6.6.4)

$$\langle n|\hat{\mathscr{E}}|n\rangle = \langle n|\left\{i\sqrt{\frac{\hbar\omega_{k}}{2\varepsilon_{0}V}}\left(\hat{b}\,e^{ikz-i\omega t} - \hat{b}^{+}e^{-ikz+i\omega t}\right)\right\}|n\rangle$$

$$= i\sqrt{\frac{\hbar\omega_{k}}{2\varepsilon_{0}V}}\left(\langle n|\hat{b}|n\rangle\,e^{ikz-i\omega t} - \langle n|\hat{b}^{+}|n\rangle e^{-ikz+i\omega t}\right)$$

$$= i\sqrt{\frac{\hbar\omega_{k}}{2\varepsilon_{0}V}}\left(\sqrt{n}\langle n\mid n-1\rangle\,e^{ikz-i\omega t} - \sqrt{n+1}\langle n\mid n+1\rangle e^{-ikz+i\omega t}\right) = 0$$

6.6.5 Introduction to the Fermion Fock States

Fermion creation and annihilation operators can also represent the half-integral spin particles known as Fermions. Only a single Fermion can occupy a single state at any given time.

Fermions are particles, such as electrons and holes, that have half-integral spin $1/2$, $3/2$, and so on. The commutation relations for Fermions demonstrate the Pauli exclusion principle which mandates that only a single Fermion can occupy a single state at one time. The Fermion creation $\hat{f}_{k}^{+}$ and annihilation $\hat{f}_{k}$ operators obey *anticommutation* relations given by

$$\left[\hat{f}_{k},\hat{f}_{K}\right]_{+} = 0 = \left[\hat{f}_{k}^{+},\hat{f}_{K}^{+}\right]_{+} \quad \text{and} \quad \left[\hat{f}_{k},\hat{f}_{K}^{+}\right]_{+} = \delta_{kK}$$

where the relation

$$\left[\hat{A}, \hat{B}\right]_+ = \hat{A}\hat{B} + \hat{B}\hat{A}$$

defines the anticommutator. Notice the anticommutator uses a "+" sign which makes all the difference for the particle statistics.

Let's try to create two Fermions in a single state (neglecting all but one mode). $\hat{b}^+\hat{b}^+|0\rangle$. The anticommutation relation for the creation operator provides

$$0 = \left[\hat{b}^+, \hat{b}^+\right]_+ = \hat{b}^+\hat{b}^+ + \hat{b}^+\hat{b}^+ = 2\hat{b}^+\hat{b}^+$$

so that the two particle Fermion ket becomes

$$\hat{b}^+\hat{b}^+|0\rangle = \frac{1}{2}\left[\hat{b}^+, \hat{b}^+\right]_+|0\rangle = 0$$

The anticommutation relations for Fermions therefore lead to the Pauli Exclusion Principle.

6.7 Fock States as Eigenstates of the EM Hamiltonian

Having quantized the electromagnetic (EM) Hamiltonian by replacing the Fourier expansion coefficients with operators obeying commutation relations, we now proceed to examine the "amplitude Hilbert space." The quantum fields operate on the amplitude Hilbert space to provide amplitudes for the EM waves along with information on the statistics for the photon number, quadratures, and phase. Because the operators defining the quantum fields do not commute, we cannot repeatedly measure the electric field (for example) and expect to find the same value each time. However, we can find the quantum expectation values by forming matrix elements of the fields using the basis vectors in the amplitude space. These expectation values provide the classically expected values for the fields. The later portions of this chapter additionally explore the density operator expressions of the fields that give both the quantum and ensemble averages.

This section shows that the Fock states are eigenvectors of the energy operator but not of the fields (recall that the amplitude operators do not commute); repeated measurements of the energy produce identical results. First we start the section by finding the wave functions defining the Fock states; these wave functions come from projecting the Fock vector into coordinate space. In this case for EM waves, the coordinate space consists of the quadrature components "q." Although the purpose of q resembles that of x for the electron harmonic oscillator, the q quadrature describes the amplitude generalize coordinate and not the position or momentum of the photon. We will set up Schrodinger's equation for the EM field and solve for the EM wave functions. The last portion of the section returns to the familiar use of the Fock states as energy eigenfunctions.

The subsequent section discusses the meaning of the wave function and its probability interpretation. There we show the Heisenberg uncertainty relation for the quadrature operators, number-phase operators using both the number and coordinate representations of the Fock states. As a note, the problems with the electric field, namely that the field cannot be repeatedly measured without finding a range of values, should not be too surprising. The Hamiltonian is really the most basic quantity of interest for many systems. Classical physics and engineering defines the electric field as a force per charge or as related to potential energy through the voltage. The electric field must interact with charge to produce energy. The interaction energy provides a more fundamental quantity.

6.7.1 Coordinate Representation of Boson Wavefunctions

In this topic, we first develop the solution to the EM Schrodinger equation obtained by separating variables. Starting with the coordinate representation of the vacuum state, we use the creation and annihilation operators to find the eigenfunctions corresponding to an arbitrary number of photons.

Section 6.5 shows that the quantized Hamiltonian for light can be cast into two equally valid but interrelated forms

$$\hat{H} = \sum_k \left[\frac{p_k^2}{2} + \frac{\omega^2 q_k^2}{2} \right] \qquad \hat{H} = \sum_k \hbar \omega_k \left(\hat{N}_k + \frac{1}{2} \right) \tag{6.7.1}$$

We first examine the coordinate representation of the Hamiltonian because we can then find the coordinate representation of the Fock vectors.

In the coordinate representation, the Hamiltonian has the form

$$\hat{H} = \sum_k \left[\frac{\hbar^2}{2} \frac{\partial^2}{\partial q_k^2} + \frac{\omega^2 q_k^2}{2} \right] \tag{6.7.2}$$

where we identify the coordinate representation of the momentum operator in Equation (6.7.1) as

$$\hat{p}_k = \frac{\hbar}{i} \frac{\partial}{\partial q_k} \tag{6.7.3}$$

The wavefunction must depend on the independent coordinates q_k so that Schrodinger's equation for light can be written as

$$\sum_{k=1}^N \left[\frac{\hbar^2}{2} \frac{\partial^2}{\partial q_k^2} + \frac{\omega^2 q_k^2}{2} \right] \Psi(q_1, q_2, \ldots, q_N, t) = i\hbar \frac{\partial}{\partial t} \Psi(q_1, q_2, \ldots, q_N, t) \tag{6.7.4}$$

where N represents the number of modes (we are ignoring polarization and wave vector direction). We can separate variables by letting

$$\psi(q_1, q_2, \ldots, q_N, t) = u_{E_1}(q_1) \, u_{E_2}(q_2) \ldots u_{E_N}(q_N) \, T(t) \tag{6.7.5}$$

and each basis function $u_{E_k}(q_k)$ satisfies a time-independent Schrodinger's equation with a form similar to the harmonic oscillator

$$\left(\frac{\hbar^2}{2} \frac{\partial^2}{\partial q_k^2} + \frac{\omega^2 q_k^2}{2} \right) u_{E_k}(q_k) = E_k \, u_{E_k}(q_k) \tag{6.7.6}$$

FIGURE 6.7.1
Harmonic motion of a wave.

Equation (6.7.5) has the product of basis functions for a direct product space.

Before proceeding with the solution, we should develop an intuitive understanding of a wavefunction such as $u_{E_k}(q_k)$. Equation (6.7.6) does *not* reference the spatial position of the photon; it suggests that we focus on the amplitude of the oscillations of the electromagnetic field (similar comments apply to phonons and other quantized fields). The wave function $\psi(q)$ is *not* the probability amplitude of finding a particle at the *spatial position* "q." Figure 6.7.1 shows that the harmonic motion of the wave must be more related to the oscillation of the field about its "equilibrium." The coordinate q_k represents an electric field amplitude for the kth mode. For example, Equation (6.4.12b) shows this by considering a point in time and space such that "$kz - \omega t = \pi/2$"

$$\vec{E}(\vec{r}, t) = \frac{-1}{\sqrt{\varepsilon_0 V}} \sum_{\vec{k}} \tilde{e}_k \, \omega_k \left[\hat{q}_k \sin\left(\vec{k} \cdot \vec{r} - \omega_k t\right) + \frac{\hat{p}_k}{\omega_k} \cos\left(\vec{k} \cdot \vec{r} - \omega_k t\right) \right]$$

which gives

$$\vec{E} \sim \frac{-1}{\sqrt{\varepsilon_0 V}} \sum_{\vec{k}} \tilde{e}_k \, \omega_k \hat{q}_k$$

Therefore, the wave function $u_{E_k}(q_k)$ represents the probability amplitude for finding a particular electric field amplitude as represented by "q_k."

Returning to Equation (6.7.6), the Hamiltonian has a form similar to that for the electronic harmonic oscillator. The eigenvalues must be

$$E_k = \hbar\omega_k \left(n_k + \frac{1}{2} \right) \tag{6.7.7a}$$

Therefore, the total energy for N modes must be given by

$$E = \sum_{k=1}^{N} \hbar\omega_k \left(n_k + \frac{1}{2} \right) \tag{6.7.7b}$$

The eigenvectors consist of exponentials and Hermite polynomials. The eigenfunctions with the time-dependent phase factor can be written as

$$\psi(q_1, q_2, \ldots, q_N, t) = u_{E_1}(q_1) \, u_{E_2}(q_2) \ldots u_{E_N}(q_N) \, T(t) = u_{E_1}(q_1) \, u_{E_2}(q_2) \ldots u_{E_N}(q_N) \, e^{itE/\hbar}$$

A general wavefunction in the multidimensional Hilbert space can be written as

$$\psi(q_1, q_2, \ldots, q_N, t) = \sum_{E_1, E_2 \ldots E_N} \beta(E_1, E_2, \ldots, E_N, t) \, u_{E_1}(q_1) \, u_{E_2}(q_2) \ldots u_{E_N}(q_N)$$

where β includes the phase factor. The value of E_i must be related to the number of photons in mode #*i* because of Equation 6.7.7a. This equation can be rearranged to derive the usual representation of the Fock states given in the previous section. We pursue a solution only in the simple case of a single mode by using creation and annihilation operators.

We want to find the *single mode* wave function $\psi_n(q)$ *where "n" stands for the number of photons in the mode* and the wavefunction satisfies Schrodinger's equation for light; that is

$$\langle q | n \rangle = u_n(q) \tag{6.7.8}$$

and

$$\hat{H}\, u_n(q) = E_n u_n(q) \tag{6.7.9}$$

The number of photons must set the particular energy eigenvalue according to Equation (6.7.7a). This makes sense since the energy can be found by essentially counting the number of photons. The wave functions $u_n(q)$ can be found by applying the annihilation and creation operators similar to the procedure used for the harmonic oscillator with the ladder operators. First apply the destruction operator to the vacuum state

$$\hat{b}|0\rangle = 0 \tag{6.7.10}$$

Use the position and momentum representation given in Equations (6.4.10)

$$\left(\frac{\omega\,\hat{q}}{\sqrt{2\hbar\omega}} + \frac{i\,\hat{p}}{\sqrt{2\hbar\omega}} \right) |0\rangle = 0$$

where ω is the angular frequency for light in the mode. Operating with the coordinate space operator $\langle q |$ (i.e., projecting into coordinate space) provides

$$\langle q | \left(\frac{\omega\,\hat{q}}{\sqrt{2\hbar\omega}} + \frac{i\,\hat{p}}{\sqrt{2\hbar\omega}} \right) |0\rangle = 0$$

or, inserting the coordinate representation of the operators, we find

$$\left(\frac{\omega\,q}{\sqrt{2\hbar\omega}} + \frac{\hbar}{\sqrt{2\hbar\omega}} \frac{\partial}{\partial q} \right) \langle q \,|\, 0 \rangle = 0$$

This simple first-order differential equation can be solved for $\langle q|0\rangle = u_0(q)$ to find

$$u_0(q) = \left(\frac{\omega}{\pi\hbar} \right)^{1/4} \exp\left[-\frac{\omega q^2}{2\hbar} \right] \tag{6.7.11}$$

where the constant comes from the normalization condition. Equation (6.7.11) gives the probability amplitude of finding the electric field amplitude to have value "*q*" when the system occupies the vacuum state (i.e., a system without any photons).

Just like the electron harmonic oscillator, we can find all of the ensuing wavefunctions by applying the creation operator. For the first excitation of the mode (i.e., $n = 1$ corresponding to a single photon)

$$|1\rangle = \hat{b}^+ |0\rangle$$

Operating with the coordinate space projector and substituting the coordinate representation of the creation operator provides

$$u_1(q) = \left(\frac{\omega\, q}{\sqrt{2\hbar\omega}} - \frac{\hbar}{\sqrt{2\hbar\omega}} \frac{\partial}{\partial q} \right) u_0(q)$$

since $\langle q \mid 1 \rangle \equiv \langle q \mid u_1 \rangle = u_1(q)$. We find a result similar to the harmonic oscillator

$$u_1(q) = \sqrt{\frac{2\omega}{\hbar}} \left(\frac{\omega}{\pi\hbar} \right)^{1/4} q\ \exp\left(-\frac{\omega q^2}{2\hbar} \right) \tag{6.7.12}$$

6.7.2   Fock States as Energy Eigenstates

We have seen some of the differences between the particle and wave pictures in the previous topic. Specifying the number of particles is not equivalent to specifying the amplitude (including phase). The previous topic shows the functional representation of the Fock states that provide an amplitude interpretation for a wave. Now we discuss the particle nature of light, which refers to the photon as an elementary unit of energy (at a given frequency). For this, we need to show that Fock states must be eigenstates of the Hamiltonian.

The Hamiltonian for a system of free-space photons can be written as

$$\hat{H} = \sum_{\vec{k}s} \hbar\omega_k \left(\hat{b}^+_{\vec{k}s} \hat{b}_{\vec{k}s} + \frac{1}{2} \right) = \sum_{\vec{k}s} \hbar\omega_k \left(\hat{N}_{\vec{k}s} + \frac{1}{2} \right) \tag{6.7.13}$$

where the creation and annihilation operators depend on time according to

$$\hat{b}_{ks} = \hat{b}_{ks}(t) = \hat{b}_{ks}(0)\, e^{-i\omega_k t} \quad \text{and} \quad \hat{b}^+_{ks} = \hat{b}^+_{ks}(t) = \hat{b}^+_{ks}(0)\, e^{+i\omega_k t} \tag{6.7.14}$$

which satisfy the commutation relations

$$\left[\hat{b}_{\vec{k}s}, \hat{b}^+_{\vec{K}S} \right] = \delta_{\vec{k}\vec{K}} \delta_{sS} \quad \text{and} \quad \left[\hat{b}_{\vec{k}s}, \hat{b}_{\vec{K}S} \right] = 0 = \left[\hat{b}^+_{\vec{k}s}, \hat{b}^+_{\vec{K}S} \right] \tag{6.7.15}$$

In the following, we suppress the polarization index. The number operator

$$\hat{N}_{\vec{k}} = \hat{b}^+_{\vec{k}} \hat{b}_{\vec{k}}$$

gives the number of photons with a particular wave vector $\vec{k}$ and polarization $\tilde{e}_{\vec{k}}$. Fock states are eigenvectors of the number operator according to

$$\hat{N}_{\vec{k}}\big|n_1,\ldots,n_{\vec{k}},\ldots\big\rangle = \hat{b}_{\vec{k}}^{+}\hat{b}_{\vec{k}}\big|n_1,\ldots,n_{\vec{k}},\ldots\big\rangle = \hat{b}_{\vec{k}}^{+}\sqrt{n_{\vec{k}}}\big|n_1,\ldots,n_{\vec{k}}-1,\ldots\big\rangle = n_{\vec{k}}\big|n_1,\ldots,n_{\vec{k}},\ldots\big\rangle$$

Therefore, the Fock states must be eigenvectors of the quantum electromagnetic Hamiltonian

$$\hat{H}\,|n_1,\ldots,n_{\vec{k}},\ldots\rangle = \sum_{\vec{K}}\hbar\omega_K\left(\hat{N}_{\vec{K}}+\frac{1}{2}\right)|n_1,\ldots,n_{\vec{k}},\ldots\rangle = \sum_{\vec{K}}\hbar\omega_K\left(n_{\vec{K}}+\frac{1}{2}\right)|n_1,\ldots,n_{\vec{k}},\ldots\rangle$$

For each basis state $|n_1,\ldots,n_{\vec{k}},\ldots\rangle$, the energy eigenvalue must be

$$\sum_{\vec{k}}\hbar\omega_k\left(n_{\vec{k}}+\frac{1}{2}\right)$$

There exists a different eigenvalue for each set of occupation numbers $n_1, n_2 \ldots$.
    The energy stored in the Fock state $|n_1,\ldots,n_{\vec{k}},\ldots\rangle$ must be given by

$$E = \sum_{\vec{k}}\hbar\omega_k\left(n_{\vec{k}}+\frac{1}{2}\right)$$

For the vacuum state $|0\rangle = |0,0,0,\ldots\rangle$, the stored energy must be

$$E = \sum_{\vec{k}}\frac{1}{2}\hbar\omega_k$$

The energy stored in the vacuum is infinite but we don't have access to it since there are no available quanta of energy. This energy corresponds to randomly oscillating electromagnetic fields that permeate all space (the vacuum fields). These fields are responsible for initiating spontaneous emission from an ensemble of excited atoms.

Example 6.7.1

What is the energy eigenvalue corresponding to a single photon in the first mode of a Fabry–Perot cavity? Assume the distance L between the mirrors.

Solution: The applicable Fock state is $|1,0,0\ldots\rangle$ and so we find

$$\hat{H}\,|1,0,0\ldots\rangle = \sum_{\vec{K}}\hbar\omega_K\left(\hat{N}_{\vec{K}}+\frac{1}{2}\right)|1,0,0\ldots\rangle = \hbar\omega_1\left(n_1+\frac{1}{2}\right)|1,0,0\ldots\rangle = \frac{3}{2}\hbar\omega_1\,|1,0,0\ldots\rangle$$

We can substitute for the angular frequency by writing $\omega = ck$ where

$$k = \frac{2\pi}{\lambda} = \frac{2\pi}{2L} = \frac{\pi}{L}$$

for the first *mode*.
    So the total energy in the first mode is $E = \frac{3}{2}\hbar\omega_1 = \frac{3}{2}\hbar c\frac{\pi}{L}$

6.7.3   Schrodinger and Interaction Representation

Consider a multi-mode wave function $|\Psi(t)\rangle$ expanded in the Fock basis set

$$|\Psi(t)\rangle = \sum_{n_1,\ldots} \beta_{n_1,\ldots}(t) |n_1,\ldots\rangle \qquad (6.7.16)$$

that satisfies the Schrodinger wave equation

$$\hat{H}|\Psi(t)\rangle = i\hbar\partial_t|\Psi(t)\rangle \quad \text{or} \quad |\Psi(t)\rangle = \exp\left[\hat{H}t/i\hbar\right]|\Psi(0)\rangle = \hat{u}(t)|\Psi(0)\rangle \qquad (6.7.17)$$

where $\hat{u}$ represent the evolution operator and the Hamiltonian has the form

$$\hat{H} = \sum_k \hbar\omega_k\left(\hat{N}_k + 1/2\right) \qquad (6.7.18)$$

Therefore

$$\beta_{n_1,\ldots}(t) = \beta_{n_1,\ldots}(0)\exp\left[\frac{t}{i\hbar}\sum_k \hbar\omega_k(n_k + 1/2)\right] \qquad (6.7.19)$$

In the Schrodinger representation, the creation and annihilation operators must be independent of time according to

$$\hat{b}_k = \hat{b}_k(0) \quad \text{and} \quad \hat{b}_k^+ = \hat{b}_k^+(0) \qquad (6.7.20)$$

The interaction representation removes the trivial time dependence induced by $\hat{u}$ from the wave function

$$|\Psi_s(t)\rangle = \hat{u}(t)|\Psi_I\rangle \qquad (6.7.21)$$

where s and I represent the Schrodinger and interaction representations, respectively. Therefore, Equation (6.7.16) provides the interaction representation by making the replacement $\beta_{n_1,\ldots}(t) \to \beta_{n_1,\ldots}(0)$. Working with a single mode k, the interaction representation produces time-dependent creation and annihilation operators according to

$$\hat{b}_k(t) = \hat{u}^+\hat{b}_k\hat{u} = e^{-\hat{H}t/i\hbar}\hat{b}_k e^{\hat{H}t/i\hbar} = \hat{b}_k e^{-i\omega_k t} \qquad (6.7.22)$$

where the operator expansion theorem from Section 4.6 was used. The form of Equation (6.7.22) agrees with that found in Section 6.3.

6.8   Interpretation of Fock States

The Fock states are eigenstates of the number operator and the EM free-field Hamiltonian. The electric field averages to zero whereas its variance remains nonzero for every Fock state. The electric field can be expressed in terms of the noncommuting quadrature operators. These quadratures satisfy a Heisenberg uncertainty relation that limits our ability to determine the electric field from a classical point of view.

6.8.1 The Electric Field for the Fock State

The Fock state is not an eigenstate of the electric field as can easily be seen by calculating $\hat{E}|n\rangle$. Using the creation–annihilation form of the electric field from Section 6.4 for a single mode, we find

$$\hat{E}|n\rangle = i\sqrt{\frac{\hbar\omega}{2\varepsilon_0 V}}\left[\hat{b}e^{ikz-i\omega t} - \hat{b}^+e^{-ikz+i\omega t}\right]|n\rangle \tag{6.8.1a}$$

The annihilation and creation operators operating on the Fock state produce

$$\hat{E}|n\rangle = i\sqrt{\frac{\hbar\omega}{2\varepsilon_0 V}}\left[n|n-1\rangle e^{ikz-i\omega t} - (n+1)|n+1\rangle e^{-ikz+i\omega t}\right] \tag{6.8.1b}$$

The ket $|n\rangle$ cannot be factored from the expression to produce an eigenvector equation. We must expect such a result because the operators $\hat{b},\hat{b}^+$ don't commute. Using an operator analog of the classical expression for the magnitude of the electric field produces $\langle n|\widehat{Ampl}|n\rangle = \langle n|\sqrt{\frac{2\hbar\omega}{\varepsilon_0 V}}\hat{N}|n\rangle = \sqrt{\frac{n2\hbar\omega}{\varepsilon_0 V}}$. However the phase cannot be *a priori* known.

The expected results from a series of measurements of the electric field produces an average of zero for the Fock state $|n\rangle$. Equation (6.8.1a) gives us

$$\bar{E} = \langle n|\hat{E}|n\rangle = i\sqrt{\frac{\hbar\omega}{2\varepsilon_0 V}}\left[n\langle n\mid n-1\rangle e^{ikz-i\omega t} - (n+1)\langle n\mid n+1\rangle e^{-ikz+i\omega t}\right] = 0 \tag{6.8.2}$$

which makes use of the orthonormality of the Fock states $\langle n\mid m\rangle = \delta_{nm}$. The average electric field in the Fock state must be 0 because, even though it has a definite number of photons, it has a completely unspecified phase according to $\Delta N \Delta \phi \geq 1/2$. The idea is somewhat equivalent to integrating over the entire cycle of the sine wave. Only here, we don't know if the wave should be pictured as in Figure 6.8.1 or with the peaks and valleys reversed (180° phase shift). That is, we cannot specify the phase ϕ in $e^{i\omega t+\phi}$.

6.8.2 Interpretation of the Coordinate Representation of Fock States

Recall from the previous section that the Fock states must be eigenstates of the number operator and hence the Hamiltonian. Using the quadrature operator form of the Hamiltonian and the number representation of the Fock state, the eigenvector equation can be written

$$\hat{H}(\hat{q},\hat{p})|n\rangle = E_n|n\rangle \tag{6.8.3}$$

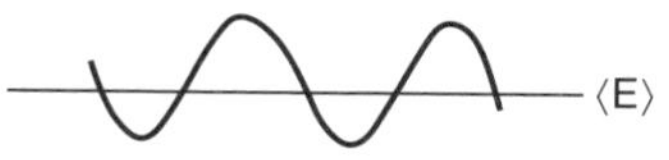

FIGURE 6.8.1
Representation of the average field.

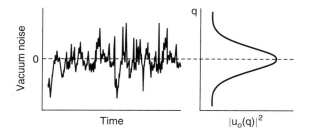

This time-independent Schrodinger equation can be written in the coordinate form

$$\left(-\frac{\hbar^2}{2}\frac{d^2}{dq^2}+\frac{\omega^2 q^2}{2}\right)u_n(q)=E_n u_n(q)\quad E_n=\hbar\omega\left(n+\frac{1}{2}\right) \tag{6.8.4}$$

where ω is the frequency of the mode and where $u_n(q)=\langle q \mid n\rangle$ is the coordinate representation for the Fock state with exactly n photons in the mode. As discussed in the previous section, we can either solve this second-order equation or use the creation–annihilation operators to find the solutions. The first two appear below.

$$u_0(q)=\left(\frac{\omega}{\pi\hbar}\right)^{1/4}\exp\left[-\frac{\omega q^2}{2\hbar}\right]\quad u_1(q)=\sqrt{\frac{2\omega}{\hbar}}\left(\frac{\omega}{\pi\hbar}\right)^{1/4}q\exp\left(-\frac{\omega q^2}{2\hbar}\right) \tag{6.8.5}$$

The first of Equations (6.8.5) provides the coordinate representation of the vacuum state (no photons). The eigenfunctions $u_n(q)$ represent the probability amplitude that a mode containing n photons will have the particular value of "q" for the quadrature amplitude. Repeated measurements of the field amplitude produce various values.

Figure 6.8.2 shows example measurements of the "electric field amplitude" q for the vacuum state $|0\rangle$. The probability density is plotted (sideways) next to the measured signal. Recall that the probability density is the modulus squared of the probability amplitude. The figure shows the greatest excursions in q from the average occur only a few times; the probability has the smallest value for these values of q. For a fixed number of photons (such as $n=0$) the electric field can be observed with a variety of q amplitudes (similar comments apply to the "p-quadratures"). Clearly, the "electric field amplitude q" must vary although the number of photons remains fixed from one measurement to the next.

6.8.3   Comparison between the Electron and EM Harmonic Oscillator

We can compare the results from the electron and EM harmonic oscillator. Figure 6.8.3 comes from the electronic harmonic oscillator with 10 quanta of energy, which means the electron occupies eigenstate $u_{10}(x)$. In this case, $|u_{10}(x)|^2$ represents the probability density of finding the electron at location x. The classical curve in the figure shows the classical probability density of finding the particle at the same location. In the limit of large numbers of quanta, the two curves more closely agree.

The EM quantum mechanical probability in Figure 6.8.3 shows that the quantum theory predicts the quadrature amplitude q can assume a range of possible values. Therefore, even though the system has a fixed total energy (fixed number of photons and fixed frequency), there must be a nonzero probability of finding the amplitude with any number of possible values. For the $n=0$ case (no photons), Equation (6.8.5) indicates

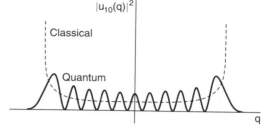

FIGURE 6.8.3
Comparing the quantum mechanical probability for a particular field amplitude (10 photons in the mode) with the classical counterpart.

a nonzero probability of finding the wave with nonzero amplitude! For very large numbers of photons, the quantum and classical theories become identical as though we can neglect the commutation relations. Keep in mind that there exists two quadrature operators and that the amplitude of the wave must account for both of them; this will become clear when we discuss the Wigner distribution.

6.8.4 An Uncertainty Relation between the Quadratures

Because the Hermitian quadrature operators $\hat{q}, \hat{p}$ appearing in the single-mode expression for the electric field (see Section 6.4)

$$\hat{E}(z,t) = \frac{-\omega}{\sqrt{\varepsilon_0 V}}\left[\hat{q}\sin(kz - \omega t) + \frac{\hat{p}}{\omega}\cos(kz - \omega t)\right] \tag{6.8.6}$$

do not commute $[\hat{q}, \hat{p}] = i\hbar$, they produce a Heisenberg uncertainty relation of the form

$$\Delta q\,\Delta p \geq \frac{\hbar}{2} \tag{6.8.7}$$

where Δq represents the standard deviation and therefore $\left(\Delta q\right)^2$ represents the variance.

$$\left(\Delta q\right)^2 = \left(\sigma_q\right)^2 = \left\langle\left(\hat{q} - \bar{q}\right)^2\right\rangle \tag{6.8.8}$$

The symbol $\bar{q}$ refers to the expected value $\langle\hat{q}\rangle$. The fact that the quadrature operators do not commute therefore indicates that multiple measurements of the same field will not produce the same identical results each time.

First, we indicate the calculation leading to Equation (6.8.7) for single mode Fock states $|n\rangle$. We need to calculate expressions of the form $\langle n|f(\hat{q}, \hat{p})|n\rangle$. The chapter review exercises ask for the same calculation but using the coordinate representation where

$$\langle n|f(\hat{q}, \hat{p})|n\rangle = \int dq\, u_n^*(q) f\left(q, \frac{\hbar}{i}\frac{\partial}{\partial q}\right) u_n(q)$$

The wave functions can be found using an equation similar to Equation (5.9.5).

In order to find the Heisenberg uncertain relation, we must calculate

$$\sigma_q^2 = \langle n|\hat{q}^2|n\rangle - \langle n|\hat{q}|n\rangle^2 \qquad \sigma_p^2 = \langle n|\hat{p}^2|n\rangle - \langle n|\hat{p}|n\rangle^2 \tag{6.8.9}$$

where the quadrature operators can be found from Equations 6.4.5

$$\hat{q} = \sqrt{\frac{\hbar}{2\omega}}\left(\hat{b} + \hat{b}^+\right) \qquad \hat{p} = -i\sqrt{\frac{\hbar\omega}{2}}\left(\hat{b} - \hat{b}^+\right) \tag{6.8.10}$$

Simple calculations using the creation–annihilation operator properties and commutators provide

$$\langle n|\hat{q}|n\rangle = \sqrt{\frac{\hbar}{2\omega}}\langle n|\left(\hat{b} + \hat{b}^+\right)|n\rangle = \sqrt{\frac{\hbar}{2\omega}}(\sqrt{n}\langle n \mid n - 1\rangle + \sqrt{n+1}\langle n \mid n+1\rangle) = 0 \tag{6.8.11a}$$

Next find the average of the square

$$\langle n|\hat{q}^2|n\rangle = \frac{\hbar}{2\omega}\langle n|(\hat{b} + \hat{b}^+)^2|n\rangle = \frac{\hbar}{2\omega}\langle n|\hat{b}^2 + \hat{b}^{+2} + \hat{b}\hat{b}^+ + \hat{b}^+\hat{b}|n\rangle = \frac{\hbar}{2\omega}\langle n|\hat{b}^2 + \hat{b}^{+2} + 2\hat{N} + 1|n\rangle$$

where $\hat{N} = \hat{b}^+\hat{b}$. Square terms such as $\langle n|\hat{b}^2|n\rangle = \sqrt{n}\langle n|\hat{b}|n-1\rangle = \sqrt{n(n-1)}\langle n \mid n-2\rangle = 0$ produce zero. This last expression therefore becomes

$$\langle n|\hat{q}^2|n\rangle = \frac{\hbar}{\omega}(n + 1/2) \tag{6.8.11b}$$

Combining Equations (6.8.11) produces the standard deviation of

$$\sigma_q^2 = \langle n|\hat{q}^2|n\rangle - \langle n|\hat{q}|n\rangle^2 = (\hbar/\omega)(n + 1/2) \tag{6.8.12a}$$

A similar procedure for σ_p produces the result (see the chapter review exercises)

$$\sigma_p^2 = \langle n|\hat{p}^2|n\rangle - \langle n|\hat{p}|n\rangle^2 = \hbar\omega(n + 1/2) \tag{6.8.12b}$$

Now we can demonstrate the Heisenberg uncertainty relation. Combining Equations (6.8.12) provides the relation.

$$\Delta q \Delta p = \sigma_q \sigma_p = \hbar(n + 1/2) \tag{6.8.13}$$

This last equation attains a minimum value for $n = 0$. Therefore we find the Heisenberg uncertainty relation

$$\Delta q \Delta p = \sigma_q \sigma_p \geq \hbar/2 \tag{6.8.14}$$

The equality holds for the vacuum state since $n = 0$. The vacuum state is a Gaussian and exhibits the minimum spread.

Next, we indicate why the inequality holds for Equation (6.8.13) when $n > 0$. Figure 6.8.4 compares the vacuum state with the $n = 1$ Fock state which has one photon. Each wave function gives an average of zero for the corresponding electric field. However, multiple measurements of the amplitude q can produce a range of values and not just zero. Therefore, the standard deviation cannot be zero for either case. Figure 6.8.4 shows the $n = 1$ state has most of its values away from $q = 0$ while the $n = 0$ state has most values near $q = 0$. The standard deviation must be larger for the $n = 1$ state. In fact, the standard deviation increases with n. Therefore, the value of $\Delta q \Delta p = \sigma_q \sigma_p$ must increase with n and the equality cannot hold. This uncertainty relation occurs because the quadrature operators do not commute $[\hat{q}_k, \hat{p}_k] = i\hbar$.

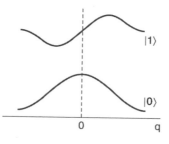

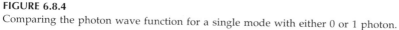

FIGURE 6.8.4
Comparing the photon wave function for a single mode with either 0 or 1 photon.

6.8.5 Fluctuations of the Electric and Magnetic Fields in Fock States

The previous topics have shown that the expected value of the electric field is zero for all Fock states. We have also seen that the quadratures can assume a range of values. Now we examine how this affects the fluctuations of the electric field. In other words, the electric field in the Fock state has an average of zero but that does not require every individual measurement to produce zero. In this topic, we calculate the standard deviation of the electric field for a Fock state. We expect multiple measurements of the electric field to give a range of values. For simplicity, let's consider a single mode traveling along the z-direction with frequency ω. The Fock state becomes

$$|n\rangle \equiv |n, 0, 0 \ldots\rangle$$

which describes the specifics of the system. The electric field operator is given by

$$\hat{E} = i\sqrt{\frac{\hbar\omega}{2\varepsilon_0 V}}\left[\hat{b}e^{ikz-i\omega t} - \hat{b}^+ e^{-ikz+i\omega t}\right] \tag{6.8.15}$$

The standard deviation must be the square root of the variance

$$\sigma_E^2 = \langle n|\left(\hat{E}^2 - \bar{E}^2\right)|n\rangle \tag{6.8.16}$$

The expectation value of the electric field $\bar{E}$ is given by Equation (6.8.6) as $\bar{E} = \langle n|\hat{E}|n\rangle = 0$. The definition of the standard deviation gives us

$$\sigma_E^2 = \langle n|\left(\hat{E}^2 - \bar{E}^2\right)|n\rangle = \langle n|\hat{E}^2|n\rangle \tag{6.8.17}$$

Using Equation (6.8.18), the square of the electric field operator becomes

$$\hat{E}^2 = -\frac{\hbar\omega}{2\varepsilon_0 V}\left[\hat{b}^2 e^{2ikz-2i\omega t} + \left(\hat{b}^+\right)^2 e^{-2ikz+2i\omega t} - \hat{b}\hat{b}^+ - \hat{b}^+\hat{b}\right]$$

Using the fact that $\langle n|\hat{b}^2|n\rangle = 0$ and $\langle n|\hat{b}^{+2}|n\rangle = 0$, we can write

$$\sigma_E^2 = \langle n|\hat{E}^2|n\rangle = -\frac{\hbar\omega}{2\varepsilon_0 V}\langle n|\left(-\hat{b}\hat{b}^+ - \hat{b}^+\hat{b}\right)|n\rangle$$

Now use the commutation relation $[\hat{b}, \hat{b}^+] = 1$ to substitute $\hat{b}\hat{b}^+ = 1 + \hat{b}^+\hat{b}$

$$\sigma_E^2 = \langle n|\hat{E}^2|n\rangle = -\frac{\hbar\omega}{2\varepsilon_0 V}\langle n|\left(-1 - 2\hat{b}^+\hat{b}\right)|n\rangle = \frac{\hbar\omega}{\varepsilon_0 V}\left(n + \frac{1}{2}\right) \qquad (6.8.18)$$

The last equation shows that even though the *average* electric field must be zero for Fock states, the variance can never be zero. Especially note that for the vacuum state with $n = 0$, the standard deviation must be

$$\sigma_E^2 = \langle n = 0|\hat{E}^2|n = 0\rangle = \frac{\hbar\omega}{2\varepsilon_0 V} \qquad (6.8.19)$$

Equation (6.8.19) shows that the electric field fluctuates away from the average value of zero for the vacuum state. The possibility of measuring nonzero values can also be seen in Figure 6.8.4. These vacuum field fluctuations initiate spontaneous emission from an ensemble of excited atoms. The fluctuations are equivalent to the zero point motion for the electron harmonic oscillator as discussed in Chapter 5.

6.9 Introduction to EM Coherent States

This section discusses coherent states of the electromagnetic (EM) field and contrasts them with Fock states. The formalism can be applied to optical and RF electromagnetic energy, phonons, and any other system that can be represented by a sum of "harmonic oscillators." As shown in previous sections, the quantized EM wave can be found by replacing classical *c*-number amplitudes with operators that must act on an "amplitude Hilbert space." These operators do not commute and they cannot be repeatedly and simultaneously measured without finding multiple values; this leads to a nonzero variance for the field. In the limit of large numbers of quanta, the noncommutivity of the operators has negligible affect and the quantum field becomes very similar to the classical one.

The manifestations of the quantum nature of EM waves depend greatly on the basis set employed for the amplitude space. The amplitude Hilbert space can have a number of different basis sets. The set of Fock vectors provides an example of the most fundamental basis set having definite numbers of photons but indefinite phase. The set of coherent states provides another example—the set actually has too many vectors to be a basis set (it's "over complete"). The coherent states appear as linear combinations of the Fock states. Strange, but true, the coherent states give finite uncertainty for the phase while increasing the uncertainty in the number of photons in the system (as a result of the summation over photon number n in the Fock states). Glauber and Yuen are the main early contributors to the study of coherent states although the work extends back to the time of Schrodinger.

This section introduces the coherent state and shows how translating the vacuum state can produce it. A subsequent section discusses the mathematical foundation of coherent states and the stochastic models.

6.9.1   The Electric Field in the Coherent State

The quantum picture of light described by a coherent state comes as close as nature allows to the classical picture of light as a sinusoidal wave with definite amplitude (magnitude and phase). For the coherent state, the *average* amplitude of the electric (or magnetic) field must be nonzero for the nonvacuum states unlike the zero averages found for Fock states. In addition, the coherent state has nonzero, finite variance for the amplitude and phase contrary to the infinite phase variance found for Fock states. However, the number of photons in a coherent beam cannot be fixed, but instead follows a Poisson distribution (Figure 6.9.1). The larger the number of photons, the more nearly the coherent state behaves similar to a classical state of light. One major distinction between the classical and coherent descriptions is that the coherent state requires uncertainty in the amplitude and phase of wave (i.e., noise).

We denote the coherent state for a single optical mode by $|\alpha\rangle$ where α is a complex number written as

$$\alpha = |\alpha|e^{i\phi} \tag{6.9.1}$$

The complex number α represents the *average* magnitude and phase of the electric field. Most importantly, we require the ket $|\alpha\rangle$ to be an eigenvector of the annihilation operator

$$\hat{b}|\alpha\rangle = \alpha|\alpha\rangle \tag{6.9.2a}$$

However, because the boson creation and annihilation operators do not commute, the coherent state cannot be an eigenstate of the creation operator. We can use the adjoint operator on Equation (6.9.2a) to write

$$\left[\hat{b}|\alpha\rangle = \alpha|\alpha\rangle\right]^{+} \quad \rightarrow \quad \langle\alpha|\hat{b}^{+} = \langle\alpha|\alpha^{*} \tag{6.9.2b}$$

The expressions for the EM fields contain the noncommuting creation–annihilation operators so that the coherent states cannot be eigenstates of those fields. Multiple measurements of the same field necessarily produce multiple complex amplitudes. The probability distribution associated with the coherent state describes the possible ranges of these values. The magnitude and phase of the classical wave must therefore be treated as random variables. The average of the magnitude and phase random variables produces the classical sinusoidal waves associated with EM phenomena. A single measurement of the complex amplitude for the EM coherent state can produce a result that differs from the average. These individual measurements produce waves with different magnitudes and phases. Figure 6.9.2 shows how the parameter α must be related to the average amplitude of the EM wave.

For a system such as a Fabry–Perot cavity or for a traveling wave with multiple modes, the coherent state can be written as

$$|\alpha_1, \alpha_2, \ldots\rangle = |\alpha_1\rangle|\alpha_2\rangle \cdots$$

Essentially this direct product state provides the complex amplitudes (magnitudes and phases) of the waves for each of the basic modes of the system. Each individual mode evolves independently of another unless there exists an explicit interaction between the

FIGURE 6.9.1
The number of photons in the coherent state follows a Poisson distribution.

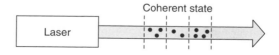

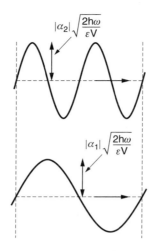

FIGURE 6.9.2
$|\alpha|$ describes the amplitude of the field.

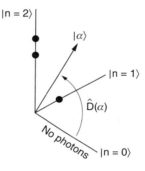

FIGURE 6.9.3
The displacement operator maps the vacuum state into the coherent state.

modes (mitigated by an interaction Hamiltonian). The multimode coherent state must be an eigenstate of an annihilation operator according to

$$\hat{b}_k|\alpha_1, \alpha_2, \ldots, \alpha_k, \ldots\rangle = \alpha_k|\alpha_1, \alpha_2, \ldots, \alpha_k, \ldots\rangle$$

The complex α_k represents the average wave amplitude and phase of mode #k.

The coherent state $|\alpha\rangle$ as a vector in the amplitude space must be composed of the Fock vectors $|n\rangle$ as shown in Figure 6.9.3 for a single mode. We will find the unitary displacement operator in the next section that maps the vacuum state into the coherent state according to $|\alpha\rangle = \hat{D}(\alpha)|0\rangle$. However, we will also discover that two coherent states *cannot* be orthogonal even though we can normalize them to 1.

$$\langle\alpha|\beta\rangle \neq 0 \quad \langle\alpha|\alpha\rangle = 1$$

The states are approximately orthogonal so long as α and β are sufficiently different.

The displacement of the vacuum state by displacement operator can be illustrated using the more physical quadrature representation. The coherent states can be obtained by moving (translating) the vacuum state $|0\rangle$ by the "distance" α in a Q-P plot (phase space) to find $|0 + \alpha\rangle = |\alpha\rangle$ as we will soon see. Apparently, we consider α to be the "distance" or "vector displacement" from an origin denoted by "0" in phase space. The coherent vacuum state is identical to the Fock vacuum state.

6.9.2 Average Electric Field in the Coherent State

To understand the relation between the EM field amplitude and the coherent state, consider the expression for the single mode quantized electric field found in Section 6.4 for traveling waves

$$\hat{E}_k = +i\sqrt{\frac{\hbar\omega_k}{2\varepsilon_0 V}}\left[\hat{b}_k e^{i\vec{k}\cdot\vec{r}-i\omega_k t} - \hat{b}_k^+ e^{-i\vec{k}\cdot\vec{r}+i\omega_k t}\right]\tilde{e}_k \qquad (6.9.3)$$

Now suppose we calculate the average electric field in the state $|\alpha_k\rangle$. Using relations (6.9.2), specifically $\hat{b}_k|\alpha_k\rangle = \alpha_k|\alpha_k\rangle$ and $\langle\alpha_k|\hat{b}_k^+ = \langle\alpha_k|\alpha_k^*$, the average becomes

$$\left\langle\hat{E}_k\right\rangle = \langle\alpha_k|\hat{E}_k|\alpha_k\rangle = +i\tilde{e}_k\sqrt{\frac{\hbar\omega_k}{2\varepsilon_0 V}}\left\{\langle\alpha_k|\hat{b}_k|\alpha_k\rangle e^{i\vec{k}\cdot\vec{r}-i\omega_k t} - \langle\alpha_k|\hat{b}_k^+|\alpha_k\rangle e^{-i\vec{k}\cdot\vec{r}+i\omega_k t}\right\}$$

$$= +i\tilde{e}_k\sqrt{\frac{\hbar\omega_k}{2\varepsilon_0 V}}\left\{\alpha_k\, e^{i\vec{k}\cdot\vec{r}-i\omega_k t} - \alpha_k^*\, e^{-i\vec{k}\cdot\vec{r}+i\omega_k t}\right\}$$

Using the expression found in Equation (6.9.1), specifically $\alpha_k = |\alpha_k|\,e^{i\phi_k}$, the average field can be rewritten as

$$\left\langle\hat{E}_k\right\rangle = \langle\alpha_k|\hat{E}_k|\alpha_k\rangle = +i\tilde{e}_k\sqrt{\frac{2\hbar\omega_k}{\varepsilon_0 V}}\left\{|\alpha_k|e^{i\vec{k}\cdot\vec{r}-i\omega_k t+i\phi_k} - |\alpha_k|e^{-i\vec{k}\cdot\vec{r}+i\omega_k t-i\phi_k}\right\}$$

Factor out the modulus and include the imaginary "*i*" with the exponentials to find

$$\left\langle\hat{E}_k\right\rangle = \langle\alpha_k|\hat{E}_k|\alpha_k\rangle = \tilde{e}_k\sqrt{\frac{2\hbar\omega_k}{\varepsilon_0 V}}\,|\alpha_k|\sin\left(\vec{k}\cdot\vec{r} - \omega_k t + \phi_k - \pi\right) \qquad (6.9.4)$$

The α that appears in the ket for the coherent state $|\alpha\rangle$ is the phasor (average) amplitude of the electric field (to within the normalization constant $\sqrt{2\hbar\omega_k/\varepsilon_0 V}$) as indicated in Figure 6.9.2. We can choose any desired phase for the average field just by adjusting ϕ_k, which is equivalent to rotating the ket $|\alpha\rangle$. However, the expectation value of the field (i.e., the classical field) has the same value whether we rotate the coherent-state ket or the electric field operator using the unitary rotation operator first introduced in Section 6.4.6.

6.9.3 Normalized Quadrature Operators and the Wigner Plot

The single-mode electric field operator $\hat{E}_k$ incorporates the quadrature operators

$$\hat{E}_k = +i\sqrt{\frac{\hbar\omega_k}{2\varepsilon_0 V}}\left[\hat{b}_k e^{i\vec{k}\cdot\vec{r}-i\omega_k t} - \hat{b}_k^+ e^{-i\vec{k}\cdot\vec{r}+i\omega_k t}\right]\tilde{e}_k \qquad (6.9.5)$$

Often for simplicity, new quadrature operators $\hat{Q}_k$ and $\hat{P}_k$ are defined in terms of the orginal ones $\hat{q}_k$ and $\hat{p}_k$ according to

$$\hat{Q}_{\vec{k}} = \hat{q}_{\vec{k}}\sqrt{\frac{\omega_{\vec{k}}}{\hbar}} = \frac{\hat{b}_{\vec{k}} + \hat{b}_{\vec{k}}^+}{\sqrt{2}} \qquad \hat{P}_k = \hat{p}_k\frac{1}{\sqrt{\hbar\omega_k}} = -i\frac{\hat{b}_{\vec{k}} - \hat{b}_{\vec{k}}^+}{\sqrt{2}} \qquad (6.9.6)$$

where, as discussed in Section 6.4, the original quadrature operators satisfy commutation relations of the form $[\hat{q}_K, \hat{p}_k] = i\hbar\delta_{k,K}$. Substituting the new quadrature operators into Equation 6.9.5 produces

$$\vec{E}_k(\vec{r}, t) = -\tilde{e}_k\sqrt{\frac{\hbar\omega_k}{\varepsilon_0 V}}\left[\hat{Q}_k \sin\left(\vec{k}\cdot\vec{r} - \omega_k t\right) + \hat{P}_k \cos\left(\vec{k}\cdot\vec{r} - \omega_k t\right)\right] \qquad (6.9.7)$$

so that $\hat{Q}_k$ and $\hat{P}_k$ appear as amplitudes without additional constants for $\hat{P}_k$ unlike the $\hat{q}_k$ and $\hat{p}_k$ in Equation 6.4.12. The new quadrature operators satisfy new commutation relations that can be obtained from the original ones

$$\left[\hat{Q}_k, \hat{P}_K\right] = \frac{\omega_k}{\hbar\omega_k}[\hat{q}_k, \hat{p}_K] = i\delta_{kK} \qquad \left[\hat{Q}_k, \hat{Q}_K\right] = 0 = \left[\hat{P}_k, \hat{P}_K\right] \qquad (6.9.8)$$

Clearly, the two Hermitian quadrature operators for the same mode do not commute $[\hat{Q}_k, \hat{P}_k] = i$ which yields an uncertainty relation as discussed in Chapter 4. Therefore, repeated measurements of the quadrature amplitudes $\hat{Q}_k$ and $\hat{P}_k$ yield a range of measured values $\{Q_k\}$ and $\{P_k\}$, respectively. These ranges of values can be described by a quasi-classical probability density (refer to the Wigner function).

6.9.4   Introduction to the Coherent State as a Displaced Vacuum in Phase Space

As we know from previous sections, the quantum electric field has operators in place of c-number Fourier amplitudes. Making measurements of the field necessarily means that all quadrature components must be measured. Because the operators do not commute, the result of a measurement can be positioned within a range of values for the quadratures. Let's consider a single mode k. We make measurements on the same field to find the range of possible values for the numbers Q_k, P_k. We want a pictorial representation of the range of possible values. The values can mostly be found inside a small circle enclosing a region of the Q–P plane. The displacement of this circle from the origin (the vacuum state—Figure 6.9.4) represents the complex amplitude of the field given by $|\alpha|e^{i\varphi}$ for $|\alpha\rangle$ in "amplitude space." This vector has both a length $|\alpha|$ (as measured from the origin—Figure 6.9.5) and an angle φ (as measured with respect to an axis). The displacement operator $D(\alpha)$ moves the circle from the origin through a distance $\sqrt{2}\,\alpha$ and produces the result of $|\alpha\rangle = \hat{D}(\alpha)|0\rangle$.

We can show that the statistical properties of the coherent state $|\alpha\rangle$ must be the same as that for the vacuum state $|0\rangle$ (except for the average value $|\alpha|$). The displacement operator moves the vacuum state away from the origin in phase space to produce the new state $|\alpha\rangle$ without changing the probability distribution (except for the mean value α) and therefore without changing the size of the small circle.

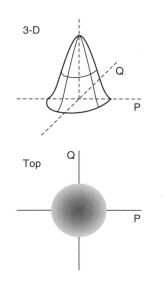

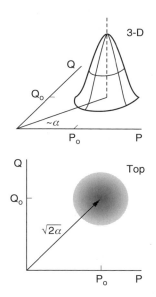

FIGURE 6.9.4

Quasi-classical probability distribution for the electric field in the vacuum state.

FIGURE 6.9.5

The coherent state is a displace vacuum.

In this topic, we first show the expected range of Q–P values in terms of a small area $\Delta Q_k \Delta P_k$. Using the results for the Fock vacuum (which is the same as the coherent state vacuum), we show that this small area must be a circle. Then we show how the position of the circle center must be related to the field amplitude $\alpha = |\alpha| e^{i\varphi}$. Finally, we discuss how measurements of the field produce results mostly found within the circle periphery.

We can see that the amplitude for the vacuum electric field must have an average value of zero, which means that any probability distribution representing the field must be centered about the origin (see Figure 6.9.4). The vacuum state (i.e., zero photon state) corresponds to a coherent state with zero average amplitude for the electric (and magnetic) field $\langle \hat{E}_k \rangle = \langle 0_k | \hat{E}_k | 0_k \rangle = 0$ as is easy to verify by setting $\alpha = 0$ in Equation (6.9.4). The variance of the measured field must be nonzero because the quadrature operators $\hat{Q}_k$ and $\hat{P}_k$ (or, equivalently, the creation $\hat{b}_k^+$ and annihilation $\hat{b}_k$ operators) appearing in the field operator $\hat{E}_k$ do not commute. Equation (6.8.19) with $n = 0$ in the previous section shows that the electric field (in the vacuum state) has a variance of

$$\sigma_E^2 = \langle 0 | \hat{E}_k | 0 \rangle = \frac{\hbar \omega_k}{2 \varepsilon_0 V}$$

This nonzero variance indicates that the quadratures also have nonzero variance. Section 6.7 solves Schrodinger's equation for the coordinate representation of the Fock state wavefunctions. The Q and P-space coordinate representation of the $n = 0$ Fock state $u_0(Q_k)$ and $u_0(P_k)$ respectively, as indicated in Equation (6.7.11), have Gaussian distributions (see the chapter review exercises). Therefore, the distribution of Q–P values must appear as a Gaussian along either the Q or the P axis. We therefore surmise the *joint* distribution for both Q and P must have a Gaussian shape $f(P, Q)$ as indicated in Figure 6.9.4. Given that most (but not all) of the *area* under a Gaussian $f(P)$ distribution must be contained within a *length* of twice the standard deviation, we can likewise define an *area* for which the Gaussian distribution $f(P, Q)$ has most (but not all) of its volume. Previous sections show that the vacuum state produces the minimum uncertainty Heisenberg relation

$$\Delta Q_k \, \Delta P_k = \left(\frac{\omega_k}{\sqrt{\hbar \omega_k}} \Delta q_k \right) \left(\frac{\Delta p_k}{\sqrt{\hbar \omega_k}} \right) = \frac{\omega_k}{\hbar \omega_k} \Delta q_k \Delta p_k = \frac{1}{2}$$

Therefore, for the Gaussian shown in Figure 6.9.4 representing the vacuum state distribution of the quadratures, the small circle with area approximately given by $\Delta q_k \Delta p_k = \hbar/2$ can be used to represent the most likely values of the quadratures. Repeated measurements of the quadratures produce a range of measured values Q_k and P_k (note the absence of the caret above the symbol) that, on average, must be located within the interior of the circle. For the vacuum state, these values have an *average* of zero. *Individual* measurements of the field amplitude do not necessarily produce zero. These occasionally measured nonzero values represent the vacuum fluctuations of the field.

Displacing the vacuum produces a coherent state as suggested by Figure 6.9.5. The parameter α in the coherent state $|\alpha\rangle$ is a complex number that gives the center of the distribution according to

$$\alpha_k = |\alpha_k| e^{i\phi} = \text{Re}(\alpha_k) + i \, \text{Im}(\alpha_k) = \frac{1}{\sqrt{2}} (Q_0 + i P_0) \qquad (6.9.9)$$

which is easy to verify by using Equation (6.9.6) with the definitions

$$\langle \alpha | \hat{Q} | \alpha \rangle = Q_0 \quad \langle \alpha | \hat{P} | \alpha \rangle = P_0$$

Therefore, the average electric field must be proportional to the hypotenuse from the origin to the point (P_0, Q_0) as can be seen from Equation (6.9.4)

$$|E| = \sqrt{\frac{\hbar \omega}{\varepsilon_0 V}} \sqrt{Q_0^2 + P_0^2} = \sqrt{\frac{2 \hbar \omega}{\varepsilon_0 V}} |\alpha|$$

The parameter α represents the average amplitude.

The "phase–space" plots (i.e., P–Q plots) in Figure 6.9.7 show why two coherent states can only be approximately orthogonal. The argument of the coherent-state ket (for example, α in $|\alpha\rangle$) represents the average amplitude. There can be significant overlap of the distributions for two neighboring states such as $|\alpha\rangle$ and $|\beta\rangle$. The integral over Q–P phase space for the inner product $\langle \alpha \mid \beta \rangle$ must be nonzero. However, two states $|\gamma\rangle$ and $|\delta\rangle$ widely separated in phase space essentially have zero overlap and the inner product must be approximately zero $\langle \gamma \mid \delta \rangle \cong 0$. Apparently, as long as the two circles do not touch, the two corresponding coherent states will be approximately orthogonal. This is easy to see since the distribution for γ is zero where as the distribution for δ is nonzero and vice versa so that the integral (for the inner product) is always zero.

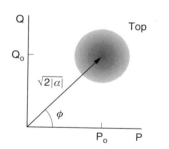

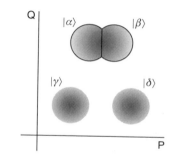

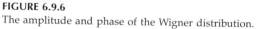

FIGURE 6.9.6
The amplitude and phase of the Wigner distribution.

FIGURE 6.9.7
The overlap of coherent states control the inner product.

6.9.5 Introduction to the Nature of Quantum Noise in the Coherent State

The term "noise" refers to the dispersion (i.e., standard deviation) in the electric field and quadrature terms. For the phase space plots, the area of the circle represents the noise. The amount of noise in the coherent state is exactly the same as the amount of noise in the vacuum because the distribution has been translated without a change of shape. Each time we make a measurement of the electric field in the coherent state $|\alpha\rangle$, we expect to find a different value of amplitude and phase, denoted by the phasor α'. Although the results of the measurements will be known, we cannot accurately predict those results before hand. Sometimes in classical EM theory we imagine there must exist an actual EM wave but the measurements only provide a range for the amplitude simply because of measurement error. With quantum fields, only the average field can be known. This is different from the classical case where we assume multiple measurements provide an average closer to the true value. In quantum theory, there does not exist the "true value." The value we will find upon measurement can only be known through a probability distribution—the Wigner distribution.

In quantum theory, the amplitude and phase of each measured α' can be found from the measured values Q and P similar to Equation (6.9.9)

$$\alpha'_k = |\alpha'_k|e^{i\phi'} = Re(\alpha'_k) + i\,\mathrm{Im}(\alpha'_k) = \frac{1}{\sqrt{2}}(Q' + iP') \tag{6.9.10}$$

This can be alternately expressed by saying the distance $\sqrt{2}\,|\alpha'_k|$ (which defines the amplitude of the *detected or measured* wave) and the phase ϕ' must be positioned within the circle representing the possible range of values (see Figure 6.9.6). Therefore, the interior of the circle gives the collection of vectors $\alpha' = |\alpha'|e^{i\phi'}$ or quadrature values Q', P' most likely to be found from any given measurement.

The Schrodinger representation of the coherent state $|\alpha\rangle$ has the form

$$|\alpha\rangle_s = e^{-i\omega t/2}|\alpha(t)\rangle$$

where $\alpha(t) = \alpha e^{-i\omega t}$ (refer to Section 6.10). The term $e^{\hbar\omega t/2i\hbar}$ is an unimportant phase factor. The magnitude $|\alpha|$ does not change with time but the phasor rotates at a rate ω.

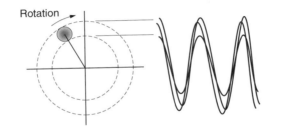

FIGURE 6.9.8
The moving Wigner plot provides a range of sine waves.

Figure 6.9.8 shows a small "uncertainty" circle in the QP plane. Every phasor terminating in the circle, which represents a possible outcome of a measurement, provides a different value for the magnitude and phase of a sine wave. The right portion of the figure shows the three possible results of a measurement. Each sine wave has a slightly different phase and amplitude but identical frequency ω. The measured electric field has the form

$$\vec{E}'(\vec{r},t) = -\tilde{e}_k \sqrt{\frac{\hbar\omega_k}{\varepsilon_0 V}} \left[Q' \sin\left(\vec{k}\cdot\vec{r} - \omega_k t\right) + P' \cos\left(\vec{k}\cdot\vec{r} - \omega_k t\right) \right]$$

where Q', P' represent the measured quadrature amplitudes. The measured values of the quadratures must depend on time because the average values of the quadratures depend on time. Figure 6.9.8 illustrates how the Gaussian distribution must rotate in a circle about the $\alpha = 0$ origin. The position of a point in the uncertainty circle corresponds to the particular amplitude and phase of the sinusoidal wave. The motion of the circle gives the sinusoidal shape to the wave. The area of the circle gives the range of possible values for the amplitude and phase.

6.9.6 Comments on the Theory

The structure of quantum theory regarding the relation between the operator and the state should be more evident now. The operators such as the Hamiltonian for the free field

$$\hat{H} = \sum_k \hbar\omega_k \left(\hat{b}_k^+ \hat{b}_k + \frac{1}{2} \right)$$

always appear the same (no real need to find a new formula) and contain all the possible outcomes in the summation. However, the states describe the specifics of the system. The operator such as $\hat{H}$ can be used with either Fock states or coherent states. The expectation value of $\hat{H}$ in the Fock state $|n_1, 0, \ldots\rangle$, for example, is

$$\langle n_1, 0, \ldots | \hat{H} | n_1, 0, \ldots \rangle = \hbar\omega_1 \left(n_1 + \frac{1}{2} \right)$$

whereas for the coherent state, using the same Hamiltonian, the expectation value is

$$\langle \alpha_1, 0, \ldots | \hat{H} | \alpha_1, 0, \ldots \rangle = \hbar \omega_1 \left(|\alpha|^2 + \frac{1}{2} \right)$$

Either way, it's the same formula for the operator.

6.10 Definition and Statistics of Coherent States

By definition, the number and annihilation operators have the Fock and coherent states, respectively, as eigenstates. The Fock states have a definite number of photons in each mode, but for the coherent states, the number of photons follows a Poisson distribution with nonzero variance. Because Fock states provide a basis set for amplitude space, the coherent states can be expressed as a sum over the Fock states. The average and standard deviation (and higher moments) characterize the probability distribution for the photon number in the coherent states.

In this section, we find the orthonormal expansion of the coherent states in terms of the Fock basis states. We next demonstrate the probability of finding a number of photons in a coherent state using the expansion coefficients.

6.10.1 The Coherent State in the Fock Basis Set

By definition, the annihilation operator has the coherent state as an eigenstate.

$$\hat{b}_k | \alpha_1, \ldots, \alpha_k, \ldots \rangle = \alpha_k | \alpha_1, \ldots, \alpha_k, \ldots \rangle \tag{6.10.1}$$

To within a normalization constant, the average electric field amplitude for mode "k" can be represented by the complex parameter $\alpha_k = |\alpha_k| e^{i\phi_k}$. Obviously, the coherent state vector $|\alpha_1, \ldots, \alpha_k, \ldots \rangle = |\alpha_1\rangle |\alpha_2\rangle \ldots$ lives in a direct product space. In what follows, for simplicity, we focus on a single mode $|\alpha_k\rangle$ and drop the subscript "k." The basic definition of the coherent state becomes

$$\hat{b} | \alpha \rangle = \alpha | \alpha \rangle \tag{6.10.2}$$

By applying the adjoint operator to both sides of Equation (6.10.2), the basic definition can be equivalently stated as

$$\langle \alpha | \hat{b}^+ = \langle \alpha | \alpha^*$$

The following discussion demonstrates the expansion of the coherent state in the Fock basis set

$$|\alpha\rangle = e^{-|\alpha|^2/2} \sum_{n=0}^{\infty} \frac{\alpha^n}{\sqrt{n!}} |n\rangle \tag{6.10.3}$$

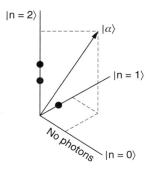

FIGURE 6.10.1
The coherent state as an element of Fock space by virtue of Eq. 6.10.10. The
solid circles indicate the number of photons residing in the corresponding
Fock state.

Recall that $\{|n_k\rangle\}$ spans a single-mode space; however, it can be part of a multimode
(i.e., direct product) space so that $|n_1, \ldots, n_k, \ldots\rangle = |n_1\rangle |n_2\rangle \ldots |n_k\rangle \ldots$. Although the
expansion appears to be complicated, the Fock states make it quite easy to use.

Interestingly, the expansion in Equation (6.10.3) and the resulting Poisson probability
distribution only require the eigenvalue equation (6.10.2) and the normalization
requirement $\langle \alpha \mid \alpha \rangle = 1$. We start with a linear combination of Fock states of the form

$$|\alpha\rangle = \sum_{n=0}^{\infty} C_n |n\rangle \tag{6.10.4}$$

Apply the annihilation operator to Equation (6.10.4) and require Equation (6.10.2) to hold

$$\alpha|\alpha\rangle = \hat{b}|\alpha\rangle = \sum_{n=0}^{\infty} C_n \hat{b}|n\rangle = \sum_{n=0}^{\infty} C_n \sqrt{n}|n - 1\rangle$$

Substitute Equation (6.10.4) for the left-most term to obtain

$$\sum_{n=0}^{\infty} \alpha C_n |n\rangle = \sum_{n=1}^{\infty} C_n \sqrt{n} \,|n - 1\rangle$$

where the second sum starts at $n = 1$ since $\sqrt{0} = 0$. A recursion relation can be found
for the expansion coefficients C_n. In the second summation, let $n - 1 \rightarrow n$ to find

$$\sum_{n=0}^{\infty} \alpha C_n |n\rangle = \sum_{n=0}^{\infty} C_{n+1} \sqrt{n + 1} \,|n\rangle$$

Comparing sides (or equivalently, operating with $\langle m|$ on both sides) provides
$C_{n+1} = C_n \alpha / \sqrt{n + 1}$. Assume C_0 is known. We find the following sequence

$$C_0 \quad C_1 = C_0 \frac{\alpha}{\sqrt{1}} \quad C_2 = C_1 \frac{\alpha}{\sqrt{2}} = C_0 \frac{\alpha^2}{\sqrt{1 \cdot 2}} \quad \ldots \quad C_n = C_0 \frac{\alpha^n}{\sqrt{n!}}$$

Now Equation (6.10.4) can be rewritten as

$$|\alpha\rangle = \sum_{n=0}^{\infty} C_0 \frac{\alpha^n}{\sqrt{n!}} |n\rangle \tag{6.10.5}$$

Normalizing the coherent state vector to 1 yields the constant C_0

$$1 = \langle \alpha \mid \alpha \rangle = \left(\sum_{m=0}^{\infty} C_0 \frac{\alpha^m}{\sqrt{m!}} \mid m \rangle \right)^+ \sum_{n=0}^{\infty} C_0 \frac{\alpha^n}{\sqrt{n!}} \mid n \rangle = \sum_{mn} C_0^* C_0 \frac{(\alpha^m)^*}{\sqrt{m!}} \frac{\alpha^n}{\sqrt{n!}} \langle m \mid n \rangle$$

Using the orthonormality relation for Fock states $\langle m \mid n \rangle = \delta_{mn}$ provides

$$1 = \langle \alpha \mid \alpha \rangle = |C_0|^2 \sum_n \frac{|\alpha|^{2n}}{n!} \qquad (6.10.6)$$

Comparing this last expression with the Taylor series expansion of $e^x = \sum_n x^n/n!$ gives

$$\sum_n \frac{|\alpha|^{2n}}{n!} = \exp |\alpha|^2 \qquad (6.10.7)$$

Substituting Equation (6.10.7) into (6.10.6) provides the constant C_0

$$1 = |C_0|^2 \exp |\alpha|^2 \quad \rightarrow \quad C_0 = e^{-|\alpha|^2/2} \qquad (6.10.8)$$

where the phase is set equal to unity. Finally, Equations (6.10.4) and (6.10.5) can be written as

$$|\alpha\rangle = e^{-|\alpha|^2/2} \sum_{n=0}^{\infty} \frac{\alpha^n}{\sqrt{n!}} \mid n \rangle \qquad (6.10.9)$$

6.10.2 The Poisson Distribution

This topic derives the Poisson probability distribution that characterizes the photon number in a coherent state. The coherent state exhibits shot noise.

What is the probability that a measurement of the number of photons for the coherent state $|\alpha\rangle$ will find "m" photons in the mode (of volume V)? The question can be answered by using the Fock basis $\{|n\rangle\}$ expansion (Figure 6.10.1) in Equation (6.10.9)

$$|\alpha\rangle = e^{-|\alpha|^2/2} \sum_{n=0}^{\infty} \frac{\alpha^n}{\sqrt{n!}} \mid n \rangle \qquad (6.10.10)$$

The probability *amplitude* for the coherent state having "m" photons can be found by projecting the coherent state $|\alpha\rangle$ onto the basis state $|m\rangle$ (which is the Fock state with the number of photons "m"). Therefore, the *probability* of finding m-photons in coherent state $|\alpha\rangle$ must be

$$P_\alpha(m) = |\langle m \mid \alpha \rangle|^2 \qquad (6.10.11)$$

Recall that the quantity $\langle m \mid \alpha \rangle$ is an expansion coefficient similar to those discussed in Chapter 4.

The probability $P_\alpha(m)$ can be found as follows. First operate on Equation 6.10.10 with $\langle m|$ to get

$$\langle m \mid \alpha \rangle = e^{-|\alpha|^2/2} \sum_{n=0}^{\infty} \frac{\alpha^n}{\sqrt{n!}} \langle m \mid n \rangle = e^{-|\alpha|^2/2} \frac{\alpha^m}{\sqrt{m!}}$$

since $\langle m \mid n \rangle = \delta_{mn}$. Substituting into Equation (6.10.11) provides

$$P_\alpha(m) = |\langle m \mid \alpha \rangle|^2 = e^{-|\alpha|^2} \frac{|\alpha|^{2m}}{m!} \tag{6.10.12}$$

This expression can be compared with the Poisson probability distribution usually written as

$$P(m) = \frac{\xi^m e^{-\xi}}{m!} \tag{6.10.13}$$

where $\xi = |\alpha|^2$ is the *average number of photons* in the coherent state $|\alpha\rangle$. There exists a possibility of having extremely large numbers of photons in the beam even for small field amplitudes α. As a note, Equations (6.10.12) and (6.10.13) represent probabilities and must sum to unity according to

$$\sum_m P_\alpha(m) = \sum_m |\langle m \mid \alpha \rangle|^2 = \sum_m e^{-|\alpha|^2} \frac{|\alpha|^{2m}}{m!} = 1 \tag{6.10.14}$$

6.10.3 The Average and Variance of the Photon Number

What is the average number of photons $\langle m \rangle$ in a field characterized by the coherent state $|\alpha\rangle$? The following discussion shows that the average number must be

$$\bar{n} = \xi = |\alpha|^2 \tag{6.10.15}$$

Figure 6.10.2 shows the discrete Poisson distribution for three coherent states. The curve for the $|\alpha = 0\rangle$ state has only one point for the coherent state since the state is also the Fock vacuum state $|0\rangle$. Notice the standard deviation (i.e., spread of the distribution) increases with the average number of photons in the mode. We can calculate the average occupation number by two methods. We next deal with the operator method and leave the series solution to the chapter review exercises.

Let $\hat{N} = \hat{b}^+\hat{b}$ be the number operator. The expected number of photons is then

$$\bar{n} \equiv \left\langle \hat{N} \right\rangle = \langle \alpha|\hat{b}^+\hat{b}|\alpha \rangle = \left[\hat{b}|\alpha\rangle\right]^+ \hat{b}|\alpha\rangle = [\alpha|\alpha\rangle]^+\alpha|\alpha\rangle = |\alpha|^2 \langle \alpha \mid \alpha \rangle = |\alpha|^2 \tag{6.10.16}$$

since $|\alpha\rangle$ is an eigenstate of the annihilation operator.

What is the standard deviation σ_N for the number of photons in an electromagnetic mode characterized by the coherent state $|\alpha\rangle$? Recall that the standard deviation σ_N can be found from the variance according to

$$\sigma_N^2 = \left\langle \hat{N}^2 \right\rangle - \left\langle \hat{N} \right\rangle^2 = \langle \alpha|\hat{N}^2|\alpha \rangle - \langle \alpha|\hat{N}|\alpha \rangle^2 \tag{6.10.17}$$

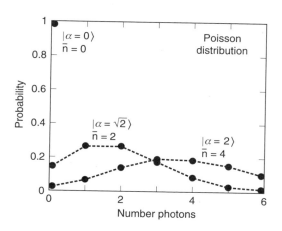

FIGURE 6.10.2
The Poisson distribution for averages of 0, 2, 4 photons in a mode.

Equation 6.10.16 provides the last term in Equation 6.10.17. Now calculate the first term.

$$\langle \alpha | \hat{N}^2 | \alpha \rangle = \langle \alpha | \hat{b}^+ \hat{b} \hat{b}^+ \hat{b} | \alpha \rangle$$

The middle two operators need to be commuted using the commutation relations $[\hat{b}, \hat{b}^+] = 1 \rightarrow \hat{b}\hat{b}^+ = \hat{b}^+\hat{b} + 1$ to get

$$\langle \alpha | \hat{N}^2 | \alpha \rangle = \langle \alpha | \hat{b}^+ \left(\hat{b}^+ \hat{b} + 1 \right) \hat{b} | \alpha \rangle = \langle \alpha | \hat{b}^+ \hat{b}^+ \hat{b}\hat{b} | \alpha \rangle + \langle \alpha | \hat{b}^+ \hat{b} | \alpha \rangle$$

Using relations of the form $\hat{b} | \alpha \rangle = \alpha | \alpha \rangle$ and $\hat{b}\hat{b} | \alpha \rangle = \alpha^2 | \alpha \rangle$ (and so on), we find

$$\langle \alpha | \hat{N}^2 | \alpha \rangle = \langle \alpha | \hat{b}^+ \hat{b}^+ \hat{b}\hat{b} | \alpha \rangle + \langle \alpha | \hat{b}^+ \hat{b} | \alpha \rangle = (\alpha^*)^2 \alpha^2 \langle \alpha | \alpha \rangle + \alpha^* \alpha \langle \alpha | \alpha \rangle$$

The coherent states are normalized to one, so that

$$\langle \alpha | \hat{N}^2 | \alpha \rangle = |\alpha|^4 + |\alpha|^2$$

Equation 6.10.17 provides the variance as

$$\sigma_N^2 = \left\langle \hat{N}^2 \right\rangle - \left\langle \hat{N} \right\rangle^2 = \langle \alpha | \hat{N}^2 | \alpha \rangle - \langle \alpha | \hat{N} | \alpha \rangle^2 = |\alpha|^4 + |\alpha|^2 - |\alpha|^4 = |\alpha|^2 = \bar{n}$$

so the standard deviation must be

$$\sigma_N = \sqrt{\bar{n}} \tag{6.10.18}$$

6.10.4 Signal-to-Noise Ratio

Fiber communication systems require semiconductor lasers with low signal-to-noise ratios (SNRs). At sufficiently high power, the lasers operate in a state that closely approximates the coherent state. The average number of photons in the beam represents the signal strength (see Equation 6.10.15) and the standard deviation provides a measure of the noise. The signal-to-noise ratio can be defined as (from Equations 6.10.15 and 6.10.18)

$$\text{SNR} = \frac{\bar{n}}{\sqrt{\bar{n}}} = \sqrt{\bar{n}} \tag{6.10.19}$$

Equation (6.10.19) shows that smaller numbers of photons produce smaller SNRs. This occurs because the unavoidable quantum noise inherent to the coherent state depends on the square root of the number of photons in the beam. For example, an optical beam with 100 photons has a standard deviation of 10 and a signal-to-noise ratio of 10. We can also see that a blue beam of light with intensity I will have a lower SNR than a red beam with the same power. This occurs because each photon in the blue beam has larger energy than each red photon, and therefore there must be fewer photons in the blue beam to make up the intensity I.

For systems composed of a small number of atoms, such as nanometer scale devices, low SNR can be a problem. For example, a device with dimensions smaller than $100 \times 100 \times 100$ angstrom might consist of $30 \times 30 \times 30$ atoms (using about 3 angstrom per atom). The total number of atoms is less than 27,000. Now if the collection is electrically pumped so that 10% are emitting light at any particular time then there are at most 2700 photons. The standard deviation in this case is about 50. The expected total variation of the signal is roughly twice the standard deviation or about 100. Therefore, the signal can be expected to vary by at least 4% due to inherent quantum noise. The percentage can be higher for systems with fewer atoms. For many analog applications, this is an unacceptably high noise level. Subsequent sections show that it might be possible to reduce the detected noise by working with "squeezed states."

6.10.5 Poisson Distribution from a Binomial Distribution

On many occasions, experiments make use of devices or processes that exhibit the binomial distribution in order to approximate a Poisson distribution. This often occurs for the testing of number squeezed light and for the discussion of noise (see Chapter 1). The binomial distribution describes the probability that exactly m out of n objects will be found when the probability of a single event (1 out of 1) has the probability p. The probability of the event not occurring must be $q = 1 - p$. The p and q are standard symbols and should not be confused with the quadratures. For example, consider Figure 6.10.3 where the reflectivity controls the probability p that a photon will pass through the partially reflective plate and $q = 1 - p$ that it will not pass through. The binomial probability has the form

$$P(m; n) = \binom{n}{m} p^m (1 - p)^{n-m} \qquad (6.10.20a)$$

The figure shows how the partially reflective plate introduces "partition" noise into the transmitted and reflected beams. The incident beam has perfectly arranged photons (conceptually at least) and the plate produces beams with some photons missing. This necessarily increases the variance.

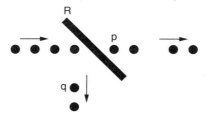

FIGURE 6.10.3
A partially reflective plate divides a stream of photons into two streams.

In the limit of large R, the transmitted beam follows a Poisson distribution. Let n become large and p become small such that the average number $np = \xi$ remains constant. Equation (6.10.20) can be rewritten as $P(m; n) = \frac{n!}{(n-m)!m!} \left(\frac{\xi}{n}\right)^m \left(1 - \frac{\xi}{n}\right)^n \left(1 - \frac{\xi}{n}\right)^{-m}$. Regrouping terms produces

$$P(m; n) = \frac{n!}{n^m(n-m)!} \left(1 - \frac{\xi}{n}\right)^{-m} \frac{\xi^m}{m!} \left(1 - \frac{\xi}{n}\right)^n \qquad (6.10.20b)$$

In the limit $n \to \infty$, the second term approaches 1 as $(1 - \xi/n)^{-m} \to 1$ and the last two terms become $(\xi^m/m!)(1 - \xi/n)^n \to e^{-\xi}\xi^m/m!$ as required for the Poisson distribution. The first term approaches 1 as can be seen using a form of the Sterling approximation $n! \cong \sqrt{2\pi}\, e^{-n}n^{n+1/2}$. For $n \gg m$ we have

$$\frac{n!}{n^m(n-m)!} \sim \frac{e^{-n}n^{n+1/2}}{n^m e^{-n+m}n^{n-m+1/2}} \sim \frac{e^{-n}n^{n+1/2}}{e^{-n}n^{n+1/2}} \cong 1$$

Therefore in the large n and small p limit, the binomial distribution produces the Poisson distribution described by

$$P(m) = e^{-\xi}\xi^m/m! \qquad (6.10.20c)$$

6.10.6   The Schrodinger Representation of the Coherent State

The unitary operator $\hat{u} = \exp \hat{H}t/i\hbar$ relates the interaction wave function to the Schrodinger wave function according to $|\Psi_s\rangle = \hat{u}|\Psi_I\rangle$. The coherent state $|\alpha\rangle$ without time dependence must be in the interaction picture. Therefore the Schrodinger representation of the single-mode coherent state must be

$$|\alpha_s\rangle = e^{\hat{H}t/i\hbar}|\alpha\rangle = e^{\hat{H}t/i\hbar}e^{-|\alpha|^2/2}\sum_{n=0}^{\infty}\frac{\alpha^n}{\sqrt{n!}}|n\rangle = e^{-|\alpha|^2/2}\sum_{n=0}^{\infty}\frac{\alpha^n}{\sqrt{n!}}e^{\hat{H}t/i\hbar}|n\rangle$$

where the single-mode Hamiltonian has the form $\hat{H} = \hbar\omega\left(\hat{N} + 1/2\right)$. The Schrodinger wave function becomes

$$|\alpha\rangle_s = e^{-|\alpha|^2/2}\sum_{n=0}^{\infty}\frac{\alpha^n}{\sqrt{n!}}e^{\hbar\omega(n+1/2)t/i\hbar}|n\rangle = e^{\hbar\omega t/2i\hbar}e^{-|\alpha|^2/2}\sum_{n=0}^{\infty}\frac{\left(\alpha e^{\hbar\omega t/i\hbar}\right)^n}{\sqrt{n!}}|n\rangle = e^{\hbar\omega t/2i\hbar}|\alpha(t)\rangle$$

where $\alpha(t) = \alpha e^{\hbar\omega t/i\hbar}$ and $e^{\hbar\omega t/2i\hbar}$ is an unimportant phase factor.

6.11   Coherent States as Displaced Vacuum States

Previous sections discuss the coherent state as an eigenvector of the annihilation operator and show how it produces classical-style fields and the Poisson distribution. Now we turn our attention to the displacement operator and show how any coherent state can be obtained by displacing the vacuum state. The displacement can be made in either the amplitude Hilbert space or in the physically intuitive (P, Q) phase space. For simplicity, we restrict the discussion to a single optical mode.

We can use the coordinate representation of the annihilation operator to find the coordinate representation of the coherent state. This coherent-state coordinate representation produces a minimum uncertainty Gaussian distribution identical to that for the vacuum state. Recall that there exist many types of probability in quantum theory. Whenever we project a ket onto a basis vector, the resulting inner product gives the probability amplitude of finding the system in the corresponding state. For example, projecting the coherent state onto a Fock state, which is an eigenstate of the number operator, provides the probability amplitude of finding the EM system with a given number of photons. As another example, projecting the coherent state into coordinate space (Q-space for example) gives the probability amplitude of finding the EM system with a given quadrature amplitude.

The Gaussian distribution for the coherent-state coordinate representation furnishes the probability of finding the EM wave with given quadrature amplitudes. Because the vacuum state can be translated in phase space to produce the coherent state, the Gaussian distribution for the vacuum state must be identical to the Gaussian distribution for the coherent state. The coordinate representation in either Q or P must have a Gaussian profile. This provides our first introduction to the idea of the Wigner distribution which is a function of both Q and P.

6.11.1 The Displacement Operator

We can find an expression for the displacement operator by starting with the definition of the coherent state as a sum over Fock basis states $\{|n\rangle = |n_k\rangle\}$ from Section 6.10

$$|\alpha\rangle = e^{-|\alpha|^2/2} \sum_{n=0}^{\infty} \frac{\alpha^n}{\sqrt{n!}} |n\rangle \tag{6.11.1}$$

where "n" denotes the photon occupation number. We need to write the Fock state $|n\rangle$ in terms of the vacuum state $|0\rangle$. The resulting relation between $|0\rangle$ and the coherent state $|\alpha\rangle$ must be the displacement operator. Start with the boson creation operator to relate the Fock state $|n\rangle$ to the vacuum $|0\rangle$

$$|1\rangle = \frac{\hat{b}^+}{\sqrt{1}}|0\rangle \quad |2\rangle = \frac{\hat{b}^+}{\sqrt{2}}|1\rangle = \frac{\left(\hat{b}^+\right)^2}{\sqrt{2\cdot 1}}|0\rangle \quad \cdots \quad |n\rangle = \frac{\left(\hat{b}^+\right)^n}{\sqrt{n!}}|0\rangle \tag{6.11.2}$$

Consequently, the coherent state in Equation (6.11.1) becomes

$$|\alpha\rangle = e^{-|\alpha|^2/2} \sum_{n=0}^{\infty} \frac{\alpha^n \left(\hat{b}^+\right)^n}{n!} |0\rangle$$

However, the summation can be rewritten as an exponential

$$|\alpha\rangle = e^{-|\alpha|^2/2} e^{\alpha \hat{b}^+} |0\rangle \tag{6.11.3}$$

Equation (6.11.3) shows explicitly that the coherent state $|\alpha\rangle$ is a displaced vacuum state. The "displacement operator" must be

$$\hat{D}(\alpha) = e^{-|\alpha|^2/2} e^{\alpha \hat{b}^+}$$

(6.11.4a)

which is unitary (as shown later). It is customary to make the displacement operator appear more symmetric in the argument of the second exponential. Notice that any exponential of the destruction operator maps the vacuum state into itself as can be seen by making a Taylor expansion

$$e^{-\alpha^* \hat{b}} |0\rangle = \left[1 - \alpha^* \hat{b} + \frac{\left(\alpha^* \hat{b}\right)^2}{2!} + \cdots \right] |0\rangle = |0\rangle$$

where $\hat{b}^n |0\rangle = 0$. Inserting this last expression between the exponential and the vacuum state in Equation (6.11.3) provides

$$|\alpha\rangle = e^{-|\alpha|^2/2} e^{\alpha \hat{b}^+} e^{-\alpha^* \hat{b}} |0\rangle$$

(6.11.4b)

We have chosen a specific exponential function of the annihilation operators for later convenience. The displacement operator must be

$$D(\alpha) = e^{-|\alpha|^2/2} e^{\alpha \hat{b}^+} e^{-\alpha^* \hat{b}}$$

(6.11.4c)

We still aren't finished with the form of the displacement operator. In some cases, we might want to combine the three exponentials in Equation (6.11.4c). Using the Campbell–Baker–Hausdorff equation

$$\exp\left(\hat{A} + \hat{B}\right) = \exp\hat{A} \, \exp\hat{B} \, \exp\left[-\frac{[\hat{A}, \hat{B}]}{2}\right] \quad \text{with} \quad \left[\hat{A}, \left[\hat{A}, \hat{B}\right]\right] = 0 = \left[\hat{B}, \left[\hat{A}, \hat{B}\right]\right]$$

Setting $\hat{A} = \alpha \hat{b}^+$ and $\hat{B} = -\alpha^* \hat{b}$, provides

$$D(\alpha) = e^{\alpha \hat{b}^+ - \alpha^* \hat{b}} |\alpha\rangle = e^{\alpha \hat{b}^+ - \alpha^* \hat{b}} |0\rangle$$

(6.11.5)

6.11.2 Properties of the Displacement Operator

1. The displacement operator is unitary with $\hat{D}^+(\alpha) = \hat{D}^{-1}(\alpha) = \hat{D}(-\alpha)$
   As discussed in Chapter 4, if an operator $\hat{O}$ is Hermitian then the operator $u = e^{i\hat{O}}$ must be unitary since

$$\hat{u}\hat{u}^+ = e^{i\hat{O}} \left(e^{i\hat{O}}\right)^+ = e^{i\hat{O}} e^{-i\hat{O}^+} = e^{i\hat{O}} e^{-i\hat{O}} = 1$$

For the displacement operator $D(\alpha) = e^{\alpha \hat{b}^+ - \alpha^* \hat{b}}$, we can define $i\hat{O} = \alpha \hat{b}^+ - \alpha^* \hat{b}$ so that $\hat{O} = -i(\alpha \hat{b}^+ - \alpha^* \hat{b})$ must be Hermitian.

The inverse of D must be

$$D(-\alpha) = e^{-\left(\alpha\hat{b}^+ - \alpha^*\hat{b}\right)}$$

since then $D(\alpha)D(-\alpha) = e^{\alpha\hat{b}^+ - \alpha^*\hat{b}}e^{-\left(\alpha\hat{b}^+ - \alpha^*\hat{b}\right)} = e^{\alpha\hat{b}^+ - \alpha^*\hat{b} - \left(\alpha\hat{b}^+ - \alpha^*\hat{b}\right)} = 1$ where the exponentials were combined because the arguments $i\hat{O} = \alpha\hat{b}^+ - \alpha^*\hat{b}$ and $-i\hat{O}$ commute $[i\hat{O}, -i\hat{O}] = 0$.

2. The displaced creation and annihilation operators can be found by a similarity transformation

$$D^+(\alpha)\,\hat{b}\,D(\alpha) = \hat{b} + \alpha \quad D^+(\alpha)\,\hat{b}^+D(\alpha) = \hat{b}^+ + \alpha^*$$

For example, consider the first relation. The operator expansion theorem from Chapter 4

$$e^{-\hat{A}}\hat{B}e^{\hat{A}} = \hat{B} - [\hat{A},\hat{B}] + \frac{1}{2!}\left[\hat{A},[\hat{A},\hat{B}]\right] + \cdots$$

with $\hat{A} = \alpha\hat{b}^+ - \alpha^*\hat{b}$ and $\hat{B} = \hat{b}$ provides

$$e^{-\left(\alpha\hat{b}^+ - \alpha^*\hat{b}\right)}\hat{b}\,e^{\alpha\hat{b}^+ - \alpha^*\hat{b}} = \hat{b} - \left[\alpha\hat{b}^+ - \alpha^*\hat{b},\hat{b}\right] + 0 + \cdots = \hat{b} + \alpha$$

3. The displacement operator acting on a nonzero coherent state produces another coherent state with the sum of two amplitudes and a complex phase factor.

4. The "phase space" representation of the displacement operator. The "phase space" operators $\hat{Q}, \hat{P}$ are defined through

$$\hat{b} = \frac{\omega\hat{q}}{\sqrt{2\hbar\omega}} + \frac{i\hat{p}}{\sqrt{2\hbar\omega}} = \frac{\hat{Q}}{\sqrt{2}} + \frac{i\hat{P}}{\sqrt{2}} \quad \text{and} \quad \hat{b}^+ = \frac{\omega\hat{q}}{\sqrt{2\hbar\omega}} - \frac{i\hat{p}}{\sqrt{2\hbar\omega}} = \frac{\hat{Q}}{\sqrt{2}} - \frac{i\hat{P}}{\sqrt{2}}$$

where

$$\left[\hat{Q},\hat{P}\right] = \left[\frac{\omega\hat{q}}{\sqrt{\hbar\omega}},\frac{\hat{p}}{\sqrt{\hbar\omega}}\right] = \frac{\omega}{\hbar\omega}[\hat{q},\hat{p}] = i$$

The values Q_0, P_0 define the center of the Wigner distribution

$$\alpha = \frac{1}{\sqrt{2}}\,[P_0 + iQ_0]$$

for the coherent state $|\alpha\rangle$. Therefore, the displacement operator can be written as

$$D(\alpha) = \exp\left(\alpha\hat{b}^+ - \alpha^*\hat{b}\right) = \exp\left[\frac{\alpha}{\sqrt{2}}\left(\hat{Q} - i\hat{P}\right) - \frac{\alpha^*}{\sqrt{2}}\left(\hat{Q} + i\hat{P}\right)\right] = \exp\left[iP_0\hat{Q} - iQ_0\hat{P}\right]$$

$$(6.11.6)$$

The derivation of the Wigner function makes use of Equation 6.11.6. An alternate form of this equation is useful for finding the coordinate representation of the coherent state (similar to $\psi(q)$ for Fock space). The Campbell–Baker–Hausdorff equation

$$\exp\left(\hat{A} + \hat{B}\right) = \exp\hat{A}\,\exp\hat{B}\,\exp\left[-\frac{[\hat{A},\hat{B}]}{2}\right] \quad \text{and} \quad \left[\hat{A},\left[\hat{A},\hat{B}\right]\right] = 0 = \left[\hat{B},\left[\hat{A},\hat{B}\right]\right]$$

with $\hat{A} = iP_0\hat{Q}$ and $\hat{B} = -iQ_0\hat{P}$ yields $[\hat{A},\hat{B}] = [iP_0\hat{Q},\,-iQ_0\hat{P}] = (i)(-i)P_0Q_0[\hat{Q},\hat{P}] = iP_0Q_0$ and

$$D(\alpha) = \exp\left[iP_0\hat{Q} - iQ_0\hat{P}\right] = \exp\left(iP_0\hat{Q}\right)\,\exp\left(-iQ_0\hat{P}\right)\,\exp\left(-\frac{i}{2}P_0Q_0\right) \qquad (6.11.7)$$

6.11.3 The Coordinate Representation of a Coherent State

Let $|\alpha\rangle$ be a single-mode coherent-state vector in an abstract Hilbert space. Rather than represent the coherent state as an abstract vector, we want to represent it as a function of the "position" coordinate, denoted by $U_\alpha(Q) = \langle Q \mid \alpha\rangle$. This is similar to the coordinate wave functions found for the Fock states in Section 6.7.1. Although the vector notation $|\alpha\rangle$ helps to show the vacuum displacement, it does not explicitly show the range of electric field amplitudes Q to be expected. The coordinate representation $U_\alpha(Q)$ helps to demonstrate that the "electric field amplitudes" Q must be normally distributed. In addition, the functions $U_\alpha(Q)$ pave the path for the Wigner distribution (refer to Figure 6.11.1 below). This section shows that the coherent fields must be normally distributed using two methods. The first method uses the basic definition of the coherent state as an eigenstate of the annihilation operator. The second method requires the displacement operator. Both use the coordinate representation.

Method 1   Coordinate Representation of the Coherent State Using the Annihilation Operator

We need to find $\langle Q \mid \alpha\rangle = U_\alpha(Q)$, the probability amplitude leading to the Gaussian distribution. To do this, we treat the coherent state $|\alpha\rangle$ as an eigenvector of the annihilation operator $\hat{b}$ so that $\hat{b}|\alpha\rangle = \alpha|\alpha\rangle$. Next, we write $\hat{b}$ in terms of the quadrature operators using their coordinate representation. The eigenvector equation becomes a first order differential equation for $U_\alpha(Q)$ which can easily be solved.

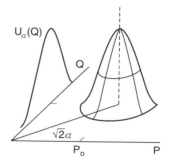

FIGURE 6.11.1
The Wigner distribution of the coherent state $|\alpha\rangle$ is shown as the 3-D relief plot. The coordinate representation $U_\alpha(Q)$ is the projection onto the plane containing the Q-axis.

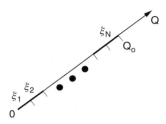

FIGURE 6.11.2
The displacement Q_o consists of smaller displacements ξ_i.

By definition, the arbitrary coherent state $|\alpha\rangle$ with the complex amplitude $\alpha = |\alpha|\, e^{i\phi}$ must be an eigenstate of the annihilation operator

$$\hat{b}|\alpha\rangle = \alpha|\alpha\rangle \qquad (6.11.8)$$

The annihilation operator can be written in terms of the quadrature operators

$$\hat{b} = \frac{\omega\hat{q}}{\sqrt{2\hbar\omega}} + \frac{i\hat{p}}{\sqrt{2\hbar\omega}} = \frac{\hat{Q}}{\sqrt{2}} + \frac{i\hat{P}}{\sqrt{2}} \qquad (6.11.9)$$

where

$$\hat{Q} = \frac{\omega\hat{q}}{\sqrt{\hbar\omega}} \qquad \hat{P} = \frac{\hat{p}}{\sqrt{\hbar\omega}} \qquad (6.11.10)$$

As a first comment, we can work with either the original quadrature operators $\hat{q}$–$\hat{p}$ or the new ones $\hat{Q}$–$\hat{P}$ used for our "phase–space" representations in the previous section. The new momentum-like operator has the "coordinate" representation

$$\hat{P} = \frac{\hat{p}}{\sqrt{\hbar\omega}} = \frac{1}{\sqrt{\hbar\omega}}\frac{\hbar}{i}\frac{\partial}{\partial q} = \frac{1}{\sqrt{\hbar\omega}}\frac{\hbar}{i}\frac{\partial Q}{\partial q}\frac{\partial}{\partial Q} = \frac{1}{i}\frac{\partial}{\partial Q} \qquad (6.11.11)$$

where we used the relation (6.11.11) in the form $Q = \omega q/\sqrt{\hbar\omega}$.

As a second comment, there exists a second method for demonstrating the coordinate representation of the new momentum-like operator $\hat{P}$ as given in Equation (6.11.11). Consider

$$\hat{P} = c\frac{\partial}{\partial Q}$$

where we wish to determine the constant "c." We require these new quadrature operators to satisfy familiar commutation relations.

$$\left[\hat{Q}, \hat{P}\right] = i \quad \rightarrow \quad \left[Q, c\frac{\partial}{\partial Q}\right] = i$$

Letting the second commutator operate on an arbitrary function $f(Q)$ shows that the only choice for "c" is $c = 1/i$ which agrees with the results for Equation (6.11.12) which is

$$\hat{P} = \frac{1}{i}\frac{\partial}{\partial Q}$$

Returning to the main discussion, we next rewrite the eigenvector equation (6.11.8) in the Q-coordinate representation because the P-quadrature involves a derivative which will produce a first order differential equation. Projecting both sides of Equation 6.11.8 onto the coordinate Q yields

$$\langle Q| \, \hat{b}\left(\hat{Q}, \hat{P}\right) |\alpha\rangle = \alpha \langle Q \mid \alpha\rangle \tag{6.11.12a}$$

where the notation $\hat{b}(\hat{Q}, \hat{P})$ serves as a reminder that the annihilation operator depends on the position and momentum operators. The coordinate representation $U_\alpha(Q) = \langle Q \mid \alpha\rangle$ of the coherent state $|\alpha\rangle$ provides the probability density $|U_\alpha(Q)|^2 = |\langle Q \mid \alpha\rangle|^2$ of finding the wave to have quadrature amplitude Q. Equation (6.11.12a) becomes

$$\hat{b}(Q) \, U_\alpha(Q) = \alpha \, U_\alpha(Q) \tag{6.11.12b}$$

where now the annihilation operator depends on Q and the derivative with respect to Q

$$\hat{b}(\hat{Q}, \hat{P}) = \frac{\hat{Q}}{\sqrt{2}} + \frac{i\hat{P}}{\sqrt{2}} \quad \rightarrow \quad \hat{b}(Q) = \frac{1}{\sqrt{2}}\left(Q + \frac{\partial}{\partial Q}\right)$$

Equation (6.11.12b) can be written as

$$\frac{1}{\sqrt{2}}\left(Q + \frac{\partial}{\partial Q}\right) U_\alpha(Q) = \alpha U_\alpha(Q)$$

This is a first-order, ordinary differential equation with the solution

$$U_\alpha(Q) = C_\alpha \exp\left[-\frac{\left(Q - \alpha\sqrt{2}\right)^2}{2}\right] \tag{6.11.3}$$

Normalizing the function U provides the constant C_α (by setting $\int_{-\infty}^{\infty} dQ \, U^* U = 1$).

$$C_\alpha = \frac{1}{\pi^{1/4}} \exp\left(\operatorname{Im} \alpha\right)^2 \tag{6.11.14}$$

Equation (6.11.13) shows that the electric field amplitude (represented by Q) must be normally distributed and centered at $\alpha\sqrt{2}$; that is, a Gaussian distribution can represent the probability density U^*U. Figure 6.11.1 shows that $U_\alpha(Q)$ is the projection of the Wigner distribution onto the plane containing the Q-axis. A momentum wave function $U_\alpha(P)$ would provide a similar distribution for the plane containing the P-axis. Therefore $U_\alpha(Q)$ and $U_\alpha(P)$ are the "shadows" from which to deduce the Wigner distribution.

Method 2 Coordinate Representation of the Coherent State Using the Displacement Operator

This demonstrates the coherent-state Gaussian distribution for the quadrature amplitudes by using the displacement operator to transform the vacuum state, which has a Gaussian distribution, into the nonvacuum coherent state. We first explicitly demonstrate how the displacement operator must produce a translation of the distribution in phase space (Q–P space). Then we show that it translates the vacuum-state probability distribution to the nonzero coherent-state distribution. This method emphasizes the notion of the displacement operator as a generator of translations in the abstract Hilbert space. The procedure is somewhat more complicated than the first one although it does not require the solution of a differential equation.

The displacement operator translates the vacuum state $|0\rangle$ to the coherent state $|\alpha\rangle$ according to

$$|\alpha\rangle = D(\alpha, \hat{Q}, \hat{P})\,|0\rangle \qquad (6.11.15)$$

where operators explicitly appear in the argument of D. Equation (6.11.7) provides the appropriate form for the displacement operator

$$D(\alpha, \hat{Q}, \hat{P}) = \exp\left[iP_0\hat{Q} - iQ_0\hat{P}\right] = \exp\left(iP_0\hat{Q}\right)\exp\left(-iQ_0\hat{P}\right)\exp\left(-\frac{i}{2}P_0Q_0\right) \qquad (6.11.16)$$

where $\alpha = (Q_0 + iP_0)/\sqrt{2}$. Projecting the Equations (6.11.15) and (6.11.16) onto the Q coordinates provides

$$U_\alpha(Q) = D(\alpha, Q)\,U_0(Q) \qquad (6.11.17)$$

In Equation (6.11.17), $U_\alpha(Q)$ is the Q-coordinate representation of the coherent state α, that is $\langle Q \mid \alpha \rangle = U_\alpha(Q)$. The operator $D(\alpha, Q)$ comes from Equation (6.11.11) using

$$\hat{Q} \to Q \quad \text{and} \quad \hat{P} \to \frac{1}{i}\frac{\partial}{\partial Q}$$

since the procedure is to be carried out in the Q-coordinate representation. The last term in Equation (6.11.16) is a complex constant. The middle term requires some discussion.

When Equation (6.11.16) is combined with Equation (6.11.17), one of the factors has the form

$$\text{factor} = \exp\left(-iQ_0\hat{P}\right)U_0(Q) \qquad (6.11.18)$$

To realize that the exponential represents a translation operator in Q-space (Figure 6.11.2), make a Taylor series expansion of the function $U_0(Q + \xi_k)$ about the point Q (where ξ_k is a small addition to Q). The expansion provides

$$U_0(Q + \xi_k) \cong U_0(Q) + \frac{\partial U_0(Q)}{\partial Q}\xi_k + \cdots = \left(1 + \xi_k\frac{\partial}{\partial Q} + \cdots\right)U_0(Q)$$

Replacing the derivative with

$$\hat{P} \leftrightarrow \frac{1}{i}\frac{\partial}{\partial Q}$$

gives

$$U_0(Q + \xi_k) = \left(1 + \xi_k\frac{\partial}{\partial Q} + \cdots\right)U_0(Q) = \left(1 + i\xi_k\hat{P} + \cdots\right)U_0(Q) = \exp\left(+i\xi_k\hat{P}\right)U_0(Q)$$

Now, by repeatedly applying the infinitesimal translation operator, we can build up the entire length Q_0

$$U_0(Q + Q_0) = \prod_k \exp\left(i\xi_k \hat{P}\right) U_0(Q) = \exp\left(\sum_k i\xi_k \hat{P}\right) U_0(Q) = \exp\left(iQ_0 \hat{P}\right) U_0(Q) \quad (6.11.19)$$

where the exponentials can be combined because the arguments commute. Equation 6.11.18 shows that $\exp(iQ_0\hat{P})$ represents a translation operator. Replacing Q_0 with $-Q_0$ shows how the "factor" in Equation (6.11.17) behaves

$$\text{factor} = \exp\left(-iQ_0\hat{P}\right) U_0(Q) = U_0(Q - Q_0)$$

Continuing to work with the combination of Equations (6.11.16) and (6.11.17) provides

$$U_\alpha(Q) = D(\alpha, Q)U_0(Q) = \exp(iP_0 Q) \, \exp\left(-iQ_0\hat{P}\right) \, \exp\left(-\frac{i}{2}P_0 Q_0\right) U_0(Q)$$

$$= \exp(iP_0 Q)\left\{\exp\left(-iQ_0\hat{P}\right) U_0(Q)\right\} \exp\left(-\frac{i}{2}P_0 Q_0\right)$$

$$= \exp(iP_0 Q) \, U_0(Q - Q_0) \, \exp\left(-\frac{i}{2}P_0 Q_0\right)$$

Substituting the expression for U_0

$$U_0(Q) = \frac{1}{\pi^{1/4}} \, \exp\left[-\frac{Q^2}{2}\right]$$

into the last equation provides

$$U_\alpha(Q) = \frac{1}{\pi^{1/4}} \, \exp\left[-\frac{(Q - Q_0)^2}{2} + iP_0 Q - \frac{iP_0 Q_0}{2}\right] \quad (6.11.20)$$

where the exponentials have been combined because the arguments commute. This last equation agrees with the *combination* of Equations (6.11.13) and (6.11.14) given by the first method.

As an important note, the coordinate space wavefunctions can never be specified as $U(P, Q)$ since the phase space operators $\hat{P}, \hat{Q}$ do not commute and cannot be consistently specified together. However, the Wigner function provides a semiclassical probability distribution for which it is possible to speak of the c-numbers P, Q together.

6.12   Quasi-Orthonormality, Closure and Trace for Coherent States

The annihilation operator has the coherent-state vector as an eigenvector. The average of the electric field operator for this state comes as close as possible to the classical paradigm of the field. The coherent state of light produces minimum uncertainty in amplitude and phase for the EM wave. The number of photons in the beam conforms to a Poisson

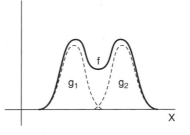

FIGURE 6.12.1
The function "f" is the sum of two different Gaussians g_1 and g_2.

probability distribution with nonzero variance since the state consists of a summation of Fock basis states each representing different numbers of photons.

The coherent state also represents a phase–space translated vacuum state. Both the coherent and vacuum states have coordinate representations, which provides the probability amplitude for the quadratures. Unlike the photon number with a Poisson probability distribution, the Q-quadrature amplitude follows a Gaussian distribution. In fact, the probability of finding the EM wave with quadrature amplitude P must likewise have a Gaussian distribution. These results for the two quadrature components foreshadow the description of EM states (not just coherent ones) by the Wigner distribution.

In the present section, we investigate the vector properties of the coherent-state vectors. In particular, we examine completeness, normalization, and orthogonality and closure.

6.12.1   The Set of Coherent-State Vectors

Each coherent state resides in an abstract Hilbert space since it must be a summation over the Fock basis set. However, neither the collection of Fock states nor coherent states form a vector space! As sets, they do not contain all of the vectors in the Hilbert space. For coherent states, the summation over Fock states produces the Gaussian-shaped coordinate representation. The sum of two coherent states g_1 and g_2 does not necessarily produce a third state "f" with a Gaussian distribution. In addition, the sum of two coherent states does not have unit length. Therefore the set of coherent states must violate the closure property for the definition of vector space as shown in Figure 6.12.1.

As we will see, the collection of coherent states can be treated similarly to basis vectors in that they span the Hilbert space. However, the set must be overly complete and the vectors cannot be independent. Naturally, the lack of independence precludes the vectors from being orthogonal.

6.12.2   Normalization

We wish to examine the orthonormality properties of the coherent states. First, consider normalization of the coherent state. Restricting the length to 1 allows for the probability amplitude interpretation for inner products.

Recall that the single mode coherent states can be defined as a summation of the Fock basis states according to

$$|\alpha\rangle = \mathrm{e}^{-|\alpha|^2/2} \sum_{n=0}^{\infty} \frac{\alpha^n}{\sqrt{n!}} |n\rangle \quad \text{or equivalently} \quad \langle\alpha| = \mathrm{e}^{-|\alpha|^2/2} \sum_{n=0}^{\infty} \frac{(\alpha^*)^n}{\sqrt{n!}} \langle n| \qquad (6.12.1)$$

Evidently, these states must have unit length according to

$$\langle \alpha \mid \alpha \rangle = e^{-|\alpha|^2} \sum_{mn=0}^{\infty} \frac{(\alpha^*)^m (\alpha)^n}{\sqrt{m!}\sqrt{n!}} \langle m \mid n \rangle = e^{-|\alpha|^2} \sum_{mn=0}^{\infty} \frac{(\alpha^*)^m (\alpha)^n}{\sqrt{m!}\sqrt{n!}} \delta_{mn} = e^{-|\alpha|^2} \sum_{n=0}^{\infty} \frac{|\alpha|^{2n}}{n!} = 1$$

since

$$\sum_{n=0}^{\infty} \frac{|\alpha|^{2n}}{n!} = e^{|\alpha|^2}$$

6.12.3   Quasi-Orthogonality

Next examine the orthogonality of two coherent states. Two nonzero vectors $|f\rangle, |g\rangle$ must be orthogonal when their inner product produces zero such

$$\langle f \mid g \rangle = \int dx\, f^*(x) g(x) = 0$$

The integral can be zero under two conditions. First the "shape" of the functions f and g might be such that the product fg is positive as much as it is negative over the range of interest. For example, f and g might be a sine and a cosine. Second, fg itself might be zero over the entire range of integration even though f and g are not zero everywhere. For example, this condition can be satisfied if $f=0$ for $x<x_0$ and $g=0$ for $x>x_0$.

We can intuitively see that the coherent states cannot ever be exactly orthogonal. Figure 6.12.2 shows a two-dimensional representation of a Wigner plot for four coherent states. The states $|\alpha\rangle$ and $|\beta\rangle$ can be represented as two overlapping Gaussians in the coordinate representation. We can write (recall the definition of closure for coordinate space in Chapter 4)

$$\langle \alpha | \beta \rangle = \langle \alpha | 1 | \beta \rangle = \langle \alpha | \left\{ \int dQ\, |Q\rangle\langle Q| \right\} | \beta \rangle = \int dQ\, \langle \alpha \mid Q \rangle\langle Q \mid \beta \rangle = \int dQ\, \alpha^*(Q)\beta(Q)$$

where previous sections define $\alpha(Q) = U_\alpha(Q)$ and $\beta(Q) = U_\beta(Q)$, which must be Gaussians (since they come from translated vacuum states). Here the notation $\alpha(Q)$ refers to a function centered a distance $\alpha\sqrt{2}$ from the origin. We see that the "shapes" of the functions $\alpha(Q)$ and $\beta(Q)$ do not produce a product function $\alpha^*\beta$ that has as many positive values as negative. As a matter of fact, the functions $\alpha(Q)$ and $\beta(Q)$ are never exactly zero over any finite region—they exponentially approach zero. Therefore, we

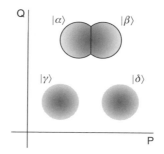

FIGURE 6.12.2
The overlap of coherent states controls the inner product.

expect the inner product between two coherent states to only approximate zero when the states have sufficient distance between them.

We can easily see two coherent states $|\alpha\rangle$ and $|\beta\rangle$ can only be approximately orthogonal. Writing the inner product of the two states using the expansions in the Fock basis sets provides

$$\langle \alpha | \beta \rangle = \exp\left[-\frac{|\alpha|^2+|\beta|^2}{2}\right] \sum_{n,m=0}^{\infty} \frac{(\alpha^*)^m}{\sqrt{m!}} \frac{\beta^n}{\sqrt{n!}} \langle m | n \rangle = \exp\left[-\frac{|\alpha|^2+|\beta|^2}{2}\right] \sum_{n,m=0}^{\infty} \frac{(\alpha^*\beta)^n}{n!}$$

where we used the orthonormality of the single-mode Fock states $\langle m \mid n \rangle = \delta_{mn}$. The summation gives an exponential.

$$\langle \alpha | \beta \rangle = \exp\left[-\frac{|\alpha|^2+|\beta|^2}{2}\right] \exp(\alpha^*\beta) = \exp\left[-|\alpha - \beta|^2\right] \qquad (6.12.2)$$

The overlap between the two states exponentially decreases as the separation between coherent states increases. This behavior is consistent with the fact that the Wigner probability distributions have Gaussian profiles. Equation 6.12.2 the inner product must be unity when $\alpha = \beta$.

Example 6.12.1

What is the inner product between the coherent state $|\alpha\rangle$ with an average of $\bar{n}_\alpha = 25$ photons and $|\beta\rangle$ having $\bar{n}_\beta = 16$. Assume that α, β are real.

Solution: Sections 6.10.2 and 6.10.3 show that $|\alpha|^2 = \bar{n}$. Ignoring the phase, the amplitudes can be written as $\alpha = \sqrt{\bar{n}_\alpha} = 5$ and $\beta = 4$. By ignoring the phase, we assume that the states both must be positioned on the right-hand side of the orgin; this represents the closest possible separation. Therefore $\langle \alpha \mid \beta \rangle \leq \exp[-(5-4)^2] = e^{-1} = 0.37$. If one of the states has nonzero phase, then the states must be further separated and the overlap becomes negligible.

6.12.4   Closure

The set of single-mode coherent states $\{|\alpha\rangle\}$ satisfy a closure relation of the form

$$\frac{1}{\pi} \int_{\substack{\alpha-\\ \text{plane}}} d^2\alpha \, |\alpha\rangle\langle\alpha| = 1 \qquad (6.12.3)$$

where α is a complex number $\alpha = re^{i\phi} = \alpha_x + i\alpha_y$ and the integral is over the entire α-plane with $d^2\alpha = d\alpha_x \, d\alpha_y$. The set of coherent states form an overcomplete quasi-basis set; we do not need all of the vectors in the set in order to span the vector space.

We start the proof of Equation (6.12.3) by substituting the Fock expansion for the coherent state

$$|\alpha\rangle = e^{-|\alpha|^2/2} \sum_{n=0}^{\infty} \frac{\alpha^n}{\sqrt{n!}} |n\rangle = e^{-r^2/2} \sum_{n=0}^{\infty} \frac{\alpha^n}{\sqrt{n!}} |n\rangle$$

into the left-hand side of Equation (6.12.3) to get

$$\frac{1}{\pi}\int_{\substack{\alpha-\\ \text{plane}}} d^2\alpha \,|\alpha\rangle\langle\alpha| = \frac{1}{\pi}\sum_{n,m}\frac{|n\rangle\langle m|}{\sqrt{n!m!}}\int d^2\alpha\, e^{-r^2/2}(\alpha^*)^n e^{-r^2/2}\alpha^m = \frac{1}{\pi}\sum_{n,m}\frac{|n\rangle\langle m|}{\sqrt{n!m!}}\int d^2\alpha\, e^{-r^2}\alpha^{*n}\alpha^m$$

Substituting the polar-coordinate element of area $d^2\alpha$ and writing α in polar form provides

$$\frac{1}{\pi}\int_{\substack{\alpha-\\ \text{plane}}} d^2\alpha \,|\alpha\rangle\langle\alpha| = \frac{1}{\pi}\sum_{n,m}\frac{|n\rangle\langle m|}{\sqrt{n!m!}}\int r\,dr\,d\phi\, e^{-r^2}r^{m+n}e^{i(m-n)\phi}$$

$$= \frac{1}{\pi}\sum_{n,m}\frac{|n\rangle\langle m|}{\sqrt{n!m!}}\int dr\,d\phi\, e^{-r^2}r^{m+n+1}e^{i(m-n)\phi}$$

The integral over the angle provides

$$\int_0^{2\pi} d\phi\, e^{i(m-n)\phi} = 2\pi\delta_{mn}$$

since for $m\neq n$ the range of integration includes multiple numbers of complete cycles. The closure integral becomes

$$\frac{1}{\pi}\int_{\substack{\alpha-\\ \text{plane}}} d^2\alpha \,|\alpha\rangle\langle\alpha| = 2\sum_{n,m}\frac{|n\rangle\langle m|}{\sqrt{n!m!}}\delta_{mn}\int dr\, e^{-r^2}r^{m+n+1} = 2\sum_{n}\frac{|n\rangle\langle n|}{n!}\int_0^\infty dr\, e^{-r^2}r^{2n+1}$$

Integral tables provide the last integral

$$\int_0^\infty dr\, e^{-r^2}r^{2n+1} = \frac{n!}{2}$$

Therefore, as required, the closure integral becomes

$$\frac{1}{\pi}\int_{\substack{\alpha-\\ \text{plane}}} d^2\alpha \,|\alpha\rangle\langle\alpha| = 2\sum_{n}\frac{|n\rangle\langle n|}{n!}\frac{n!}{2} = \sum_{n}|n\rangle\langle n| = 1$$

6.12.5 Coherent State Expansion of a Fock State

If coherent states form a quasi-basis set, then it must be possible to express other basis sets in terms of the coherent states. That is, we can express the single mode Fock states $\{|n\rangle\}$ as an expansion of coherent states $\{|\alpha\rangle\}$. This is accomplished by using the coherent-state closure relation

$$|n\rangle = 1\,|n\rangle = \left(\frac{1}{\pi}\int d^2\alpha\,|\alpha\rangle\langle\alpha|\right)|n\rangle = \frac{1}{\pi}\int d^2\alpha\,|\alpha\rangle\langle\alpha\mid n\rangle$$

Recall that the probability amplitude $\langle\alpha\mid n\rangle$ must be related to the Poisson probability distribution. Evaluating the inner product using the definition of coherent state gives

$$\langle\alpha\mid n\rangle = e^{-|\alpha|^2/2}\sum_{m=0}^\infty \frac{(\alpha^*)^m}{\sqrt{m!}}\langle m\mid n\rangle = e^{-|\alpha|^2/2}\sum_{m=0}^\infty \frac{(\alpha^*)^m}{\sqrt{m!}}\delta_{mn} = e^{-|\alpha|^2/2}\frac{(\alpha^*)^n}{\sqrt{n!}}$$

so the Fock state $|n\rangle$ becomes

$$|n\rangle = 1 \, |n\rangle = \frac{1}{\pi} \int d^2\alpha \, |\alpha\rangle \, \langle \alpha \mid n \rangle = \int d^2\alpha \, |\alpha\rangle \, e^{-|\alpha|^2/2} \frac{(\alpha^*)^n}{\pi\sqrt{n!}} \qquad (6.12.4)$$

Notice that every Fock basis vector can be written as a linear combination of the coherent states. Therefore coherent states span the same amplitude space as do the Fock vectors.

We can identify the matrix elements of the transformation in Equation (6.12.4). First, notice that the equation has an integral rather than a summation, which occurs because the parameters α are continuous. The transformation matrix elements must be

$$T_{n\alpha} = e^{-|\alpha|^2/2}(\alpha^*)^n/\pi\sqrt{n!}$$

6.12.6 Over-Completeness of Coherent States

The set of vectors $\{|\alpha\rangle\}$ is over-complete in the sense that each one can be expressed as a sum over the others. The situation can be compared with having three vectors span a 2-D vector space; obviously, we don't need one of them.

The fact that one coherent-state vector can be expressed as a sum of the others can be seen as follows:

$$|\alpha\rangle = 1|\alpha\rangle = \frac{1}{\pi} \int d^2\beta \, |\beta\rangle \langle \beta \mid \alpha \rangle$$

Using the inner product between two coherent states given in Equation 6.12.2 $\langle \beta \mid \alpha \rangle = \exp\left[-|\beta - \alpha|^2\right]$ provides

$$|\alpha\rangle = \frac{1}{\pi} \int d^2\beta \, |\beta\rangle \, \exp\left[-|\beta - \alpha|^2\right]$$

The more separated are the parameters α and β, the less the state $|\beta\rangle$ contributes to the state $|\alpha\rangle$. The integral is similar to a summation.

6.12.7 Trace of an Operator Using Coherent States

The trace formula involves the factor of "$1/\pi$" similar to the closure relation. Starting with the definition of trace using single mode Fock states, then inserting the coherent-state closure relation, and then removing the Fock states using the Fock state closure relation produces the following formula.

$$Tr\hat{O} = \sum_n \langle n|\hat{O}|n\rangle = \sum_n \langle n|1\,\hat{O}|n\rangle = \sum_n \langle n| \left\{ \frac{1}{\pi} \int d^2\alpha \, |\alpha\rangle \, \langle\alpha| \right\} \hat{O} \, |n\rangle = \frac{1}{\pi} \int d^2\alpha \sum_n \langle n \mid \alpha \rangle \, \langle \alpha|\hat{O}|n\rangle$$

Interchanging the order of the matrix elements in the last term gives

$$Tr\,\hat{O} = \frac{1}{\pi} \int d^2\alpha \sum_n \langle\alpha|\hat{O}|n\rangle\langle n \mid \alpha\rangle = \frac{1}{\pi} \int d^2\alpha \, \langle\alpha|\hat{O} \left\{ \sum_n |n\rangle\langle n| \right\} |\alpha\rangle = \frac{1}{\pi} \int d^2\alpha \, \langle\alpha| \, \hat{O} \, |\alpha\rangle$$

Therefore, the formula is similar to the Fock state trace formula except that an integral appears along with the factor of $1/\pi$.

$$Tr\,\hat{O} = \frac{1}{\pi}\int d^2\alpha\,\langle\alpha|\,\hat{O}\,|\alpha\rangle$$

6.13 Field Fluctuations in the Coherent State

This section shows that electromagnetic fields exhibit minimum uncertainty for the coherent states. The variance of the electromagnetic field (at any point in space-time) measures the uncertainty. As will be seen, the electric field has smaller variance for coherent states than for Fock states. However, the Hamiltonian has smaller variance for the Fock states. The difference between the two cases has to do with the fact that the coherent states must be eigenstates of the annihilation operator (which appears in the field expression) whereas the Fock states must be eigenstates of the Hamiltonian (since the Hamiltonian depends on the number operator).

For this section, recall that the (single mode) electric field can be written in either of the two equivalent forms as

$$\hat{E}_k = +i\sqrt{\frac{\hbar\omega}{2\varepsilon_0 V}}\left[\hat{b}\,e^{i\vec{k}\cdot\vec{r}-i\omega t} - \hat{b}^+\,e^{-i\vec{k}\cdot\vec{r}+i\omega t}\right] \tag{6.13.1a}$$

or

$$\vec{E}_k(\vec{r},t) = -\sqrt{\frac{\hbar\omega}{\varepsilon_0 V}}\left[\hat{Q}\sin\left(\vec{k}\cdot\vec{r}-\omega t\right) + \hat{P}\cos\left(\vec{k}\cdot\vec{r}-\omega t\right)\right] \tag{6.13.1b}$$

where the mode subscripts and polarization vector are suppressed, and ω, $\vec{k}$ represent the angular frequency and wave vector of the traveling waves. The creation $\hat{b}^+$ and annihilation $\hat{b}$ operators can be related to the "position" $\hat{Q}$ and "momentum" $\hat{P}$ operators according to $\hat{Q} = [\hat{b}+\hat{b}^+]/\sqrt{2}$ and $\hat{P} = -i[\hat{b}-\hat{b}^+]/\sqrt{2}$. The operators satisfy the following commutation relations:

$$\left[\hat{b},\hat{b}^+\right] = 1 \quad \left[\hat{b},\hat{b}\right] = 0 = \left[\hat{b},\hat{b}^+\right] \quad \left[\hat{Q},\hat{P}\right] = i \quad \left[\hat{Q},\hat{Q}\right] = 0 = \left[\hat{P},\hat{P}\right]$$

As discussed in the Section 6.11, the coherent states have Gaussian probability distribution functions for the quadratures.

The quadrature operators for the electric field satisfy commutation relations. Section 4.9 provides the relation $\sigma_a\sigma_b \geq \frac{1}{2}\,|\langle\hat{C}\rangle|$ when $\hat{A}$, $\hat{B}$ satisfy $[\hat{A},\hat{B}] = i\hat{C}$. In this case for Equation (6.13.1b), the commutator provides $[\hat{Q},\hat{P}] = i$ so that $\hat{C} = 1$ and therefore $\Delta Q\,\Delta P \geq 1/2$. The form of the electric field operator $\hat{E}$ requires this uncertainty relation without regard for the amplitude states. However, the specific form of the amplitude state determines whether the uncertainty is equal to or larger than $\frac{1}{2}$ and whether the Q or P quadrature produces the smallest dispersion. The production of the EM wave by matter determines the amplitude state of the wave. Although the field operator requires the uncertainty relation, the production of the wave by matter ultimately determines the specific nature of the wave. We will see how matter can produce coherent and squeezed states of light.

Coherent states with Gaussian wave functions produce the "minimum-area" uncertainty relation.

$$\Delta Q \, \Delta P = \frac{1}{2} \tag{6.13.2}$$

The term "area" is used because $\Delta Q \, \Delta P$ is an area in phase space. The term "minimum" indicates the use of the "=" sign. For any other wave function, the area must be larger than $\frac{1}{2}$.

In the ensuing topics, we determine the uncertainty relations for the quadrature components in the Fock and coherent states. As will be shown, coherent states produce minimum-area uncertainty relations and minimum variance electromagnetic fields.

6.13.1 The Quadrature Uncertainty Relation for Coherent States

We now demonstrate the minimum-area uncertainty relation for coherent states. We must evaluate the standard deviation for the quadrature operators $\hat{Q}$ and $\hat{P}$. The variance of $\hat{Q}$ can be found as follows:

$$\sigma_Q^2 = \langle \alpha | \hat{Q}^2 | \alpha \rangle - \langle \alpha | \hat{Q} | \alpha \rangle^2$$

First calculate the average $\langle \alpha | \hat{Q} | \alpha \rangle$ by using the creation–annihilation operators for the "position" operator $\hat{Q}$ and also the definition of the coherent state $\hat{b} | \alpha \rangle = \alpha | \alpha \rangle$ or equivalently $\langle \alpha | \hat{b}^+ = \langle \alpha | \alpha^*$. The expectation value produces

$$\langle \alpha | \hat{Q} | \alpha \rangle^2 = \frac{1}{2} \langle \alpha | \left[\hat{b} + \hat{b}^+ \right] | \alpha \rangle^2 = \frac{1}{2} (\alpha + \alpha^*)^2 = \frac{\alpha^2 + \alpha*^2}{2} + |\alpha|^2$$

Next calculate $\langle \alpha | \hat{Q}^2 | \alpha \rangle$ using the commutation relation $\left[\hat{b}, \hat{b}^+ \right] = 1$

$$\langle \alpha | \hat{Q}^2 | \alpha \rangle = \frac{1}{2} \langle \alpha | \left[\hat{b} + \hat{b}^+ \right]^2 | \alpha \rangle = \frac{1}{2} \langle \alpha | \left[\hat{b}^2 + \left(\hat{b}^+ \right)^2 + \hat{b} \hat{b}^+ + \hat{b}^+ \hat{b} \right] | \alpha \rangle$$

$$= \frac{1}{2} (\alpha^2 + \alpha*^2) + \frac{1}{2} \langle \alpha | \left(\hat{b} \hat{b}^+ + \hat{b}^+ \hat{b} \right) | \alpha \rangle = \frac{1}{2} (\alpha^2 + \alpha*^2) + \frac{1}{2} \langle \alpha | \left(2 \hat{b}^+ \hat{b} + 1 \right) | \alpha \rangle$$

$$= \frac{\alpha^2 + \alpha*^2}{2} + |\alpha|^2 + \frac{1}{2}$$

Therefore, the variance of $\hat{Q}$ becomes

$$(\Delta Q)^2 = \sigma_Q^2 = \langle \alpha | \hat{Q}^2 | \alpha \rangle - \langle \alpha | \hat{Q} | \alpha \rangle^2 = \frac{1}{2}$$

Similarly, we can show that the variance of the "momentum" operator in the coherent state must be

$$(\Delta P)^2 = \sigma_P^2 = \frac{1}{2}$$

Regardless of the value of α, the uncertainty relation for the coherent state must be

$$\Delta Q \, \Delta P = \frac{1}{2} \tag{6.13.3}$$

Alternatively, the uncertainty relation for coherent states (6.13.3) could be calculated using vacuum expectation values. The reason, as previously pointed out, is that displacing the vacuum state produces the coherent state; the noise remains unaffected by the displacement operation. This statement becomes obvious by setting $\alpha = 0$ in the derivation of Equation (6.13.3), which is independent of α.

6.13.2 Comparison of Variance for Coherent and Fock States

First we compare the $\Delta P \Delta Q$ uncertainty relation for the coherent and Fock states. The variance in P for the Fock states can be calculated as

$$\sigma_P^2 = \langle n|\hat{P}^2|n\rangle - \langle n|\hat{P}|n\rangle^2 = \langle n|\hat{P}^2|n\rangle = \langle n|\frac{-i}{\sqrt{2}}[\hat{b} - \hat{b}^+]^2\,|n\rangle = -\frac{1}{2}\langle n|\,(\hat{b}^2 + \hat{b}^{+2} - \hat{b}^+\hat{b} - \hat{b}\hat{b}^+)\,|n\rangle$$

Noting $\langle n|\hat{b}^2|n\rangle \sim \langle n\,|\,n+2\rangle = 0$ with similar results for the creation operator and using the commutation relations provides $\sigma_P^2 = \langle n|(\hat{b}^+\hat{b} + 1/2)|n\rangle = n + 1/2$ where we have used the number operator $\hat{N} = \hat{b}^+\hat{b}$. A similar result holds for the "position" operator $\sigma_Q^2 = n + 1/2$. Therefore, the Q–P uncertainty relation for a nonvacuum Fock state $|n\rangle$ becomes

$$\Delta Q\,\Delta P = \sigma_Q \sigma_P = n + \frac{1}{2} \geq \frac{1}{2}$$

Using Equation (6.13.3), we find the uncertainty in the field for the Fock state must always be larger than (except $n=0$) than that for the coherent state.

$$(\Delta Q\,\Delta P)_{\text{Fock}} > (\Delta Q\,\Delta P)_{\text{coherent}} \qquad n > 0 \text{ (nonvacuum)}$$

$$(\Delta Q\,\Delta P)_{\text{Fock}} = (\Delta Q\,\Delta P)_{\text{coherent}} = \frac{1}{2} \qquad n = 0 \text{ (vacuum)}$$

In fact, the difference between the two types of states increases with n. The coherent states represent the minimum uncertainty states. The noise for the electric field in a coherent state is sometimes called the "standard quantum limit" (SQL). Essentially it is the lowest possible noise level. Squeezing techniques can reduce the noise in one quadrature; however, the noise in the other increases (similarly for "number" and "phase"). This behavior occurs because the uncertainty relation $\Delta Q\,\Delta P \geq 1/2$ must still hold.

Next we compare the variance in the electric field for the Fock and coherent states. The variance of the electric field can be evaluated for a Fock state $|n\rangle$ (see Section 6.8)

$$\sigma_E^2 = \frac{\hbar\omega}{\varepsilon_0 V}\left(n + \frac{1}{2}\right) \qquad \text{Fock} \tag{6.13.4a}$$

The variance for the electric field in the coherent state can be written as

$$\sigma_E^2 = \langle \alpha|\hat{E}^2|\alpha\rangle - \langle \alpha|\hat{E}|\alpha\rangle^2$$

$$= -\frac{\hbar\omega}{2\varepsilon_0 V}\left\{ \langle \alpha|\left[\hat{b}\,e^{i\vec{k}\cdot\vec{r}-i\omega t} - \hat{b}^+\,e^{-i\vec{k}\cdot\vec{r}+i\omega t}\right]^2 |\alpha\rangle - \langle \alpha|\left[\hat{b}\,e^{i\vec{k}\cdot\vec{r}-i\omega t} - \hat{b}^+\,e^{-i\vec{k}\cdot\vec{r}+i\omega t}\right]|\alpha\rangle^2 \right\}$$

Being careful to use the commutation relations, the variance for the electric field evaluated in a coherent state becomes

$$\sigma_E^2 = \frac{\hbar\omega}{2\varepsilon_0 V} \tag{6.13.4b}$$

Notice that the value α does not appear in the variance. Equations (6.13.4) show the coherent state provides the minimum achievable dispersion for the electric field.

Therefore, the uncertainty in the electric field must always be larger for the Fock state than for the coherent state (except for the vacuum state).

Finally, we compare the variance of the energy as described by the Hamiltonian $\hat{H} = \hbar\omega(\hat{b}^+\hat{b} + 1/2) = \hbar\omega(\hat{N} + 1/2)$. Given that Fock states must be eigenstates of the Hamiltonian, the variance of the energy must be zero

$$\sigma_H = 0 \quad \text{Fock} \tag{6.13.5}$$

The variance of the energy in the coherent state can be calculated as

$$\sigma_H^2 = \langle\alpha|\hat{H}^2|\alpha\rangle - \langle\alpha|\hat{H}|\alpha\rangle^2 = (\hbar\omega)^2\langle\alpha|\left(\hat{b}^+\hat{b}\,\hat{b}^+\hat{b} + \hat{b}^+\hat{b} + \frac{1}{4}\right)|\alpha\rangle - (\hbar\omega)^2\langle\alpha|\left(\hat{b}^+\hat{b} + \frac{1}{2}\right)|\alpha\rangle^2$$

Using the commutator $\left[\hat{b}, \hat{b}^+\right] = 1$ to reverse the order of $\hat{b}\hat{b}^+$ and noting $\hat{b}|\alpha\rangle = \alpha|\alpha\rangle$ and $\langle\alpha|\hat{b}^* = \langle\alpha|\alpha^*$ provides

$$\sigma_H^2 = (\hbar\omega)^2\left\{|\alpha|^4 + 2|\alpha|^2 + \frac{1}{4}\right\} - (\hbar\omega)^2\left(|\alpha|^4 + |\alpha|^2 + \frac{1}{4}\right) = (\hbar\omega)^2|\alpha|^2$$

Recall from Equation (6.10.15) that the average number of photons in the coherent state $|\alpha\rangle$ is given by $\bar{n}_\alpha = |\alpha|^2$. Therefore, the energy in the coherent state is not definite since repeated measurements of the energy on the same coherent state produces a range of values characterized by the standard deviation

$$\sigma_H = \hbar\omega\sqrt{\bar{n}} \quad \text{Coherent} \tag{6.13.6}$$

The uncertainty in the energy must be due to the fluctuations in the electric field. The power in the electromagnetic beam must fluctuate. In the vacuum state $\alpha = 0$, there isn't any uncertainty in the energy since the vacuum is also a Fock state.

The uncertainty in energy and power is always larger for the coherent states. The number-squeezed state is the closest relative to the Fock state. All of the number-noise (i.e., amplitude noise) would need to be squeezed out of the coherent state to transform it into a Fock state. This shows the reason for the *number-phase* uncertainty relation. Removing noise (i.e., reducing the standard deviation) from the amplitude necessarily requires the uncertainty in the phase to increase in order to maintain the Q–P uncertainty relations. Removing all of the number-noise causes the phase to be completely unspecified.

6.14 Introduction to Squeezed States

Previous sections define the Fock states $|n\rangle$ as eigenstates of the number operator (and Hamiltonian) $\hat{N}|n\rangle = n|n\rangle$ where $\hat{N} = \hat{b}^+\hat{b}$. They represent EM waves with definite number of photons but indefinite phase—the photon-number probability density must be a Dirac delta function. A product of Hermite polynomials and decaying exponential functions describe the Fock states in the Q-quadrature representation. The pure Fock states have not been produced experimentally in the laboratory.

The coherent states $|\alpha\rangle$ are defined as eigenvectors of the annihilation operator $\hat{b}|\alpha\rangle = \alpha|\alpha\rangle$). These states can be written as summations of the Fock basis sets or as the displacement of the vacuum state. The outcome of a measurement for the amplitude and phase (or equivalently, the quadratures) can assume a value in a range of values. The amplitude of the EM wave does not have an *a priori* well-defined value. The photon number must follow a Poisson distribution. The coherent states provide the most well defined magnitude and phase. They come closest to the classical notion of a wave. For coherent states $|\alpha\rangle$, the classical amplitude (within a multiplicative constant) can be defined by the complex parameter $\alpha = |\alpha|e^{i\phi}$. The electric field can be written in terms of quadratures

$$\hat{E}(\vec{r}, t) = -\sqrt{\frac{\hbar\omega}{2\varepsilon_0 V}}\left[\hat{Q}\sin\left(\vec{k}\cdot\vec{r} - \omega t\right) + \hat{P}\cos\left(\vec{k}\cdot\vec{r} - \omega t\right)\right]$$

so that the classical values of the quadrature (i.e., averages) indicate the center of the Wigner probability plot and can be related to the parameter α by

$$\alpha = \frac{1}{\sqrt{2}}\left(\bar{Q} + i\bar{P}\right) \qquad (6.14.1)$$

where $\bar{Q} = \langle\alpha|\hat{Q}|\alpha\rangle = \sqrt{2}\text{Re}(\alpha)$ and $\bar{P} = \langle\alpha|\hat{P}|\alpha\rangle = \sqrt{2}\text{Im}(\alpha)$. In the coordinate representation, the range of quadrature values must be guided by a normal distribution. We will see in subsequent sections that the normal distributions for the Q and P quadratures can be combined into a quasi-classical probability distribution—the Wigner distribution.

We can define the "squeezed EM states" in terms of "squeezed" annihilation operators. We will find that squeezed states can be characterized by reduced noise (i.e., standard deviation) in one parameter but with added noise in the conjugate parameter. Figure 6.14.1 shows various types of squeezing as represented by a Wigner probability distribution plot. The amplitude-squeezed light, for example, has decreased magnitude variance and increased phase variance; the reverse is true for phase-squeezed light. For P-quadrature squeezed light, the possible range of "P," denoted by ΔP, decreases while the range of "Q," denoted by ΔQ, increases; however, the product $\Delta Q \Delta P = 1/2$ remains unaltered. A measurement of the electric field for squeezed light can assume a magnitude and phase out of the range of values characterized by the ovals in the figure.

The squeezed states must be part of the amplitude Hilbert space. Figure 6.14.2 illustrates a quick method for relating the "oval shapes" in phase space to the amplitude and phase of a sine wave. Rather than associate the time dependence with the quantum EM field, we can associate it with the squeezed state vector (as we will see later). The state then rotates about the origin of phase space as shown in Figure 6.14.2. The top portion of the figure shows phase squeezing. The angle must be confined to a very narrow region at any time, but not so for the amplitude of the vector from the origin to the center of the distribution. Notice at t_1, how the length of the oval defines a range of values for the amplitude of the left-hand sine waves. Also notice how the sine waves line up one under the other. The bottom portion shows amplitude squeezing. The length of the vector from the phase–space origin must be fairly well defined but not the angle. In this case, at time t_1 (oval at top), the tops of the sine waves approximately coincide with the horizontal line. However, the tops can be horizontally separated from one another according to the width of the oval. We will see later that the vacuum state can also be squeezed; the results appear similar to those in the figure except that the average amplitude must be zero. All squeezed electromagnetic waves can be related to coherent states and, in particular, to squeezed vacuum states.

There exists two sets of mathematical operations that produce squeezed states. Let $\hat{S}(\eta)$ denote an operator that "squeezes" a coherent state and let $\hat{D}(\alpha)$ be the displacement

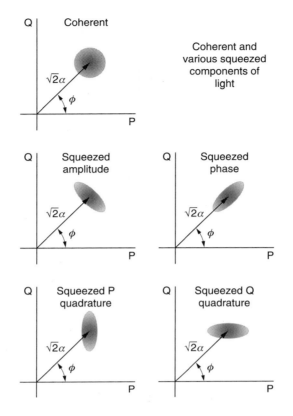

FIGURE 6.14.1
Various types of squeezed states as represented by the corresponding Wigner distribution. The coherent state has a value for the squeezing parameter of zero.

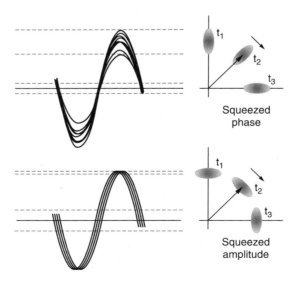

FIGURE 6.14.2
The top portion represents squeezed phase and the bottom represents squeezed amplitude.

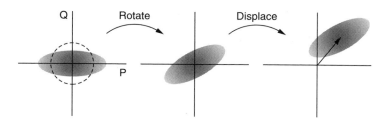

FIGURE 6.14.3
The vacuum is squeezed and then displaced to produce the squeezed coherent state.

operator that defines a coherent state $|\alpha\rangle = \hat{D}(\alpha)|0\rangle$ from the vacuum. The complex parameter $\eta = r\,e^{i\theta}$ uses $r > 0$ to describe the degree of squeezing and the angle θ to describe the angle of the "long axis" of the distribution. The first set of operations consists of first squeezing the vacuum

$$\hat{S}(\eta)|0\rangle = |0, \eta\rangle \tag{6.14.2}$$

and then displacing the "new vacuum" through the phase–space distance α

$$\hat{D}(\alpha)\hat{S}(\eta)|0\rangle = |\alpha, \eta\rangle \tag{6.14.3}$$

The word "new vacuum" appears in quotes because it is not actually a physical vacuum; it is more similar to a multi-photon state. Figure 6.14.3 shows the sequence of operations required by Equation (6.14.3). The second set of operations consists of first displacing the vacuum and then squeezing the resulting coherent state (the reverse of Equation (6.14.3)). We do not consider this second type of squeezed state further because the final result does not have as simple an interpretation as the first type.

The next sections provide the mathematical detail on squeezed states. The discussion starts with Q-squeezed vacuums using $\theta = 0$ in the squeeze parameter $\eta = r\,e^{i\theta}$ so that "P" becomes the "long" axis. The length of the oval along the P-axis must be a *factor* of $e^{r}/\sqrt{2}$ longer than the diameter of the circle representing the vacuum state $|0\rangle$; the Q-axis must be "shorter" by the factor $e^{-r}/\sqrt{2}$. The Q-squeezed vacuum can be rotated to any desired angle. The rotated, squeezed vacuum can be displaced to a new location by using the displacement operator $D(\alpha)$. The modulus of the parameter $\eta = r\,e^{-i\theta}$ determines the amount of squeeze and the angle determines the axis of squeezing (i.e., amplitude or phase squeezed, etc.).

6.15 The Squeezing Operator and Squeezed States

The squeezing operator $\hat{S}(\eta)$ transforms a coherent state into a squeezed state. The complex parameter η determines the type of squeezed state. The four common types include quadrature, amplitude, phase, and number squeezed. Amplitude squeezed states exhibit reduced number fluctuations only over a limited range of the squeezing parameter. In the limit of infinite squeezing, the amplitude squeezed state does not approach a Fock state. The probability distribution of the number-squeezed state (termed sub-Poisson) can be characterized by a standard deviation smaller than that for the coherent state. An anti-squeezed number state (phase squeezed) obeys super-Poisson statistics.

This section explores the squeezing operator $\hat{S}(\eta)$ along with some of its elementary properties. As mentioned in the previous section, the squeezed state can be defined to be the eigenvector of a "squeezed annihilation operator." However, to define the "squeezed annihilation operator," we must first know the squeezing operator $\hat{S}(\eta)$. After stating a definition for the squeezing operator, we then define the squeezed vacuum state and then the squeezed (nonzero) coherent state. Having expressions for these states can be useful for picturing them in phase space. However, calculations proceed using operators and commutators. Therefore, we must show how the squeezing operator affects the three basic types of operators, explored so far in this chapter, namely creation–annihilation operators, the quadrature operators and the displacement operators. We can then show how the complex squeezing parameter $\eta = r\,e^{i\theta}$ orients the "variance oval" (i.e., the region where the amplitude and phase of the electric field can most likely be found when measured). Finally, we show the coordinate representation of the squeezed states in phase space.

6.15.1 Definition of the Squeezing Operator

The squeezing operator is defined by

$$\hat{S}(\eta) = \exp\left(\frac{\eta^*}{2}\hat{b}^2 - \frac{\eta}{2}\hat{b}^{+2}\right) \qquad (6.15.1)$$

We define the squeezing parameter $\eta = r e^{-i\theta}$ with the minus sign. The operator $\hat{S}$ must be a unitary operator since $\hat{S}^+(\eta) = \hat{S}(-\eta) = \hat{S}^{-1}(\eta)$. We can still find useful relations that eventually provide (i) the variance of squeezed operators (such as electric field and energy), (ii) the coordinate representations of squeezed states, and (iii) the photon statistics for squeezed states. A subsequent section shows that homodyne sensor systems can be used to detect and measure the amount of squeezing.

6.15.2 Definition of the Squeezed State

As mentioned in the previous section, the simplest (but not the only) squeezed coherent state is obtained by first squeezing the vacuum and then displacing the result

$$|\alpha, \eta\rangle = \hat{D}(\alpha)\hat{S}(\eta)|0\rangle \qquad (6.15.2)$$

where recall from Section 6.11, displacing the vacuum state $|0\rangle$ produces the coherent state $|\alpha\rangle = \hat{D}(\alpha)|0\rangle$. Note that the order of parameters in $|\alpha, \eta\rangle$ matches the order of the operators in Equation (6.15.2). Although not yet mathematically clear, the squeezed state in Equation (6.15.2) can be represented by the sequence shown in Figure 6.15.1. Eventually, one would like to calculate transition rates using squeezed coherent states as might be important for communication devices.

6.15.3 The Squeezed Creation and Annihilation Operators

Squeezing a state provides convenient pictures while squeezing an operator provides convenient mathematics. We might want to know the expected result of measurement of an operator $\hat{O}$ (such as the electric field) when the system occupies a squeezed state

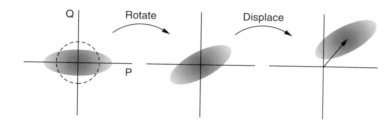

FIGURE 6.15.1
The vacuum is squeezed and then displaced to produce the squeezed coherent state.

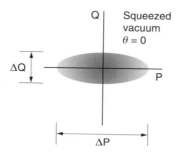

FIGURE 6.15.2
The Q-quadrature is squeezed.

$|\alpha, \eta\rangle = \hat{D}(\alpha)\hat{S}(\eta)|0\rangle$ as given by Equation (6.15.2). The expectation value of the operator $\hat{O}$ then takes the form

$$\langle \alpha, \eta|\hat{O}|\alpha, \eta\rangle = \langle 0|\hat{S}^+\hat{D}^+\hat{O}\hat{D}\,\hat{S}|0\rangle \qquad (6.15.3)$$

We must either know the state $|\alpha, \eta\rangle$ or the new operator $\hat{O}'' = \hat{S}^+\hat{D}^+\hat{O}\hat{D}\,\hat{S}$ or we can calculate a combination

$$\langle \alpha, \eta|\hat{O}|\alpha, \eta\rangle = \langle 0|\left(\hat{S}^+\hat{D}^+\hat{O}\hat{D}\hat{S}\right)|0\rangle = \left(\langle 0|\hat{S}^+\right)\hat{D}^+\hat{O}\hat{D}\left(\hat{S}|0\rangle\right)$$

If the original operator $\hat{O}$ can be written as a functional of the creation and annihilation operators $\hat{O} = \hat{O}(\hat{b}, \hat{b}^+)$, then we can calculate both $\hat{O}' = \hat{D}^+\hat{O}\hat{D}$ and $\hat{O}'' = \hat{S}^+\hat{D}^+\hat{O}\hat{D}\,\hat{S}$. Therefore, the transformation of any operator $\hat{O}$ under $\hat{S}$ is known so long as $\hat{O} = \hat{O}(\hat{b}, \hat{b}^+)$ and the transformations of $\hat{b}$ and $\hat{b}^+$ under $\hat{S}$ are known.

First let's examine the effects of the displacement operator. If the operator $\hat{O}$ can be written as a functional of the creation–annihilation operators $\hat{O} = \hat{O}\left(\hat{b}, \hat{b}^+\right)$ then the operator product $\hat{D}^+\hat{O}\hat{D}$, in Equation (6.15.3), can be easily calculated since previous sections show

$$\hat{D}^+(\alpha)\,\hat{b}\,\hat{D}(\alpha) = \hat{b} + \alpha \quad \text{and} \quad \hat{D}^+(\alpha)\hat{b}^+\hat{D}(\alpha) = \hat{b}^+ + \alpha^* \qquad (6.15.4a)$$

Products of the form $\hat{b}^+\hat{b}$ become

$$\hat{D}^+\left(\hat{b}^+\hat{b}\right)\hat{D} = \left(\hat{D}^+\,\hat{b}^+\hat{D}\right)\left(\hat{D}^+\,\hat{b}\,\hat{D}\right) = \left(\hat{b}^+ + \alpha^*\right)\left(\hat{b} + \alpha\right) \qquad (6.15.4b)$$

and so on.

We must determine the effect of the squeezing operators on the annihilation–creation operators. The first order of business consists of showing

$$\hat{S}^{+}\hat{b}\hat{S} = \hat{b}\cosh(r) - \hat{b}^{+}e^{i\theta}\sinh(r) \quad \hat{S}^{+}\hat{b}^{+}\hat{S} = \hat{b}^{+}\cosh(r) - \hat{b}\,e^{-i\theta}\sinh(r) \tag{6.15.5}$$

where $\eta = r\,e^{-i\theta}$ and $\hat{b}$, $\hat{b}^{+}$ are the annihilation and creation operators, respectively. We show the first of Equations (6.15.5) by applying the operator expansion theorem from Chapter 4, namely

$$e^{-x\hat{A}}\hat{B}e^{x\hat{A}} = \hat{B} - \frac{x}{1!}[\hat{A},\hat{B}] + \frac{x^{2}}{2!}\left[\hat{A},\left[\hat{A},\hat{B}\right]\right] + \cdots \tag{6.15.6}$$

For the squeezing operator in Equation (6.15.1), specifically $\hat{S}(\eta) = \exp(\eta^{*}\hat{b}^{2}/2 - \eta\hat{b}^{+2}/2)$, we set $\hat{A} = \frac{1}{2}(\eta^{*}\hat{b}^{2} - \eta\hat{b}^{+2})$ and $x = 1$, and substitute into the expansion formula (Equation 6.15.6)

$$\hat{S}^{+}\hat{b}\,\hat{S} = e^{-\frac{1}{2}\left(\eta^{*}\hat{b}^{2}-\eta\hat{b}^{+2}\right)}\hat{b}e^{\frac{1}{2}\left(\eta^{*}\hat{b}^{2}-\eta\hat{b}^{+2}\right)}$$

$$\hat{b} - \left[\frac{1}{2}\left(\eta^{*}\hat{b}^{2} - \eta\hat{b}^{+2}\right),\hat{b}\right] + \frac{1}{2!}\left[\frac{1}{2}\left(\eta^{*}\hat{b}^{2} - \eta\hat{b}^{+2}\right),\left[\frac{1}{2}\left(\eta^{*}\hat{b}^{2} - \eta\hat{b}^{+2}\right),\hat{b}\right]\right] + \cdots$$

The commutators in this expression can be evaluated using $[\hat{b},\hat{b}^{+}] = 1$ to produce

$$\hat{S}^{+}\hat{b}\,\hat{S} = \hat{b} - \eta^{*}\hat{b}^{+} + \frac{1}{2!}|\eta|^{2}\hat{b} - \frac{1}{3!}|\eta|^{2}\eta^{*}\hat{b}^{+} + \cdots = \hat{b}\left(1 + \frac{1}{2!}r^{2} + \frac{1}{4!}r^{4} + \cdots\right)$$
$$- \hat{b}^{+}e^{i\theta}\left(r + \frac{1}{3!}r^{3} + \cdots\right)$$

We therefore find

$$\hat{S}^{+}\hat{b}\,\hat{S} = \hat{b}\cosh(r) - \hat{b}^{+}e^{i\theta}\sinh(r) \tag{6.15.7}$$

which proves the first relation where $r = |\eta|$ is the modulus of the squeezing parameter $\eta = r\,e^{-i\theta}$. Taking the adjoint proves the second relation.

6.15.4 The Squeezed EM Quadrature Operators

The electromagnetic quadrature operators provide a first example for squeezing functionals of the creation and annihilation operators. The "position and momentum" operators are defined by

$$\hat{Q} = \frac{1}{\sqrt{2}}\left(\hat{b} + \hat{b}^{+}\right) \quad \hat{P} = \frac{1}{i\sqrt{2}}\left(\hat{b} - \hat{b}^{+}\right) \tag{6.15.8}$$

These operators prove to be important for plots of the Wigner probability distribution. Recall from the introduction to squeezed states (e.g., Figure 6.14.3) that the standard deviation of these operators determines, in an easy way, the direction of squeezing.

Using Equation 6.15.5, the squeezed operator $\hat{Q}_S = \hat{S}^+\hat{Q}\hat{S}$ can be evaluated as

$$\hat{S}^+\hat{Q}\,\hat{S} = \frac{1}{\sqrt{2}}\left[\hat{S}^+\hat{b}\hat{S} + \hat{S}^+\hat{b}^+\hat{S}\right] = \frac{\hat{b} + \hat{b}^+}{\sqrt{2}}\cosh(r) - \frac{\hat{b}e^{-i\theta} + \hat{b}^+e^{i\theta}}{\sqrt{2}}\sinh(r)$$

This last expression can be simplified to

$$\hat{S}^+\hat{Q}\,\hat{S} = \hat{Q}\cosh(r) - \hat{Q}_R\sinh(r) \tag{6.15.9a}$$

The "rotated" operator $\hat{Q}_R$ is shorthand notation for $\hat{Q}_R = (\hat{b}\,e^{-i\theta} + \hat{b}^+e^{i\theta}/\sqrt{2})$. The rotated operator can also be written as $\hat{Q}_R = \hat{R}^+\hat{Q}\,\hat{R}$ where $\hat{R} = e^{-i\hat{N}\theta} = e^{-i\hat{b}^+\hat{b}\theta}$ as discussed in Section 6.4.6.

Similarly, the squeezed $\hat{P}$ operator $\hat{P}_S = \hat{S}^+\hat{P}\hat{S}$ can be evaluated

$$\hat{S}^+\hat{P}\,\hat{S} = \frac{1}{i\sqrt{2}}\hat{S}^+\left(\hat{b} - \hat{b}^+\right)\hat{S} = \hat{P}\cosh(r) - \hat{P}_R\sinh(r) \tag{6.15.9b}$$

where $\hat{P}_R = (\hat{b}\,e^{-i\theta} - \hat{b}^+e^{i\theta})/(i\sqrt{2})$.

6.15.5 Variance of the EM Quadrature

Using the results of the previous topic, we can now demonstrate the origin of the "ovals" in the phase space plots (cf. Figure 6.15.1). As previously discussed, an "oval" represents the most likely region to find an amplitude vector when making a measurement; the coordinate-representation wavefunction approaches zero for regions outside the oval. The "ideally squeezed coherent states" come from first squeezing the vacuum state and then displacing it. The displacement operator does not change the "amount of squeezing" and therefore does not change the size of the oval. We can calculate the size of the oval by using the squeezed vacuum rather than the ideally-squeezed coherent state to simplify the calculation. The length and width of the ovals represent the standard deviation for two independent variables. We can find these sizes by calculating the standard deviation of the quadratures $\hat{Q}$ and $\hat{P}$ in the squeezed vacuum. We should realize that even though we remove the noise (i.e., reduce the standard deviation) from one parameter, it appears in the other which explains why the region has an oval shape. This says that the Heisenberg uncertainty relation continues to hold regardless of the value of the squeezing parameter η. We show these various aspects of the "ovals" in the following discussion. We limit our attention to the squeezed vacuum (i.e., $\alpha = 0$) because the size of the oval does not change when we translate the vacuum state.

We will need to calculate the variance of the quadrature operators $\hat{Q}, \hat{P}$. Recall that these operators appear in the expression for the single mode electric fields

$$\vec{E}_k(\vec{r}, t) = -\sqrt{\frac{\hbar\omega}{\varepsilon_0 V}}\left[\hat{Q}\sin\left(\vec{k}\cdot\vec{r} - \omega t\right) + \hat{P}\cos\left(\vec{k}\cdot\vec{r} - \omega t\right)\right] \tag{6.15.10a}$$

where the mode subscripts and polarization vector are suppressed, and ω, $\vec{k}$ are the angular frequency and wave vector of the traveling waves. The quadrature operators $\hat{Q}$ and $\hat{P}$ are defined by

$$\hat{Q} = \frac{1}{\sqrt{2}}\left[\hat{b} + \hat{b}^+\right] \quad \hat{P} = \frac{1}{i\sqrt{2}}\left[\hat{b} - \hat{b}^+\right] \tag{6.15.10b}$$

and these operators satisfy the commutation relations

$$\left[\hat{Q}, \hat{P}\right] = i \quad \left[\hat{Q}, \hat{Q}\right] = 0 = \left[\hat{P}, \hat{P}\right] \quad \left[\hat{b}, \hat{b}^{+}\right] = 1 \quad \left[\hat{b}, \hat{b}\right] = 0 = \left[\hat{b}, \hat{b}^{+}\right] \tag{6.15.10c}$$

The commutation relations imply that the two-quadrature terms in the electric field cannot be simultaneously and precisely known.

First calculate the variance of $\hat{Q}$ for the squeezed vacuum state $|0, \eta\rangle = \hat{S}(\eta)\,|0\rangle$ where $\eta = re^{-i\theta}$. The variance is

$$\sigma_Q^2 = \langle 0, \eta|\hat{Q}^2|0, \eta\rangle - \langle 0, \eta|\hat{Q}|0, \eta\rangle^2$$

The average of the "position" operator $\hat{Q}$ can be evaluated using Equations 6.15.5

$$\langle 0, \eta|\hat{Q}|0, \eta\rangle = \langle 0|\hat{S}^{+}\hat{Q}\hat{S}|0\rangle = \langle 0| \left\{ \frac{\hat{b} + \hat{b}^{+}}{\sqrt{2}} \cosh(r) - \frac{\hat{b}\,e^{-i\theta} + \hat{b}^{+}e^{i\theta}}{\sqrt{2}} \sinh(r) \right\} |0\rangle = 0$$

Next, evaluate $\langle 0, \eta|\,\hat{Q}^2|0, \eta\rangle$ using $\hat{S}\,\hat{S}^{+} = 1 = \hat{S}^{+}\hat{S}$

$$\begin{aligned}
\langle 0, \eta|\hat{Q}^2|0, \eta\rangle &= \langle 0|\hat{S}^{+}\hat{Q}^2\hat{S}|0\rangle = \langle 0|\,\hat{S}^{+}\hat{Q}\hat{S}\,\hat{S}^{+}\hat{Q}\hat{S}\,|0\rangle \\
&= \langle 0| \left[\hat{Q}\cosh(r) - \hat{Q}_R \sinh(r)\right] \left[\hat{Q}\cosh(r) - \hat{Q}_R\sinh(r)\right] |0\rangle \\
&= \langle 0|\hat{Q}^2|0\rangle\cosh^2(r) + \langle 0|\hat{Q}_R^2|0\rangle\sinh^2(r) - \langle 0|\left(\hat{Q}\hat{Q}_R + \hat{Q}_R\hat{Q}\right)|0\rangle\cosh(r)\sinh(r)
\end{aligned}$$

$$\tag{6.15.11}$$

Next, we must substitute expressions for $\hat{Q}$ and $\hat{Q}_R$ and use the commutation relations for the creation and annihilation operators. Evaluating each of the four terms separately in the previous equation, we find

$$\langle 0|\hat{Q}^2|0\rangle = \langle 0| \left(\frac{\hat{b} + \hat{b}^{+}}{\sqrt{2}}\right)^2 |0\rangle = \langle 0| \frac{\hat{b}^2 + \hat{b}^{+2} + \hat{b}\hat{b}^{+} + \hat{b}^{+}\hat{b}}{2} |0\rangle = \frac{1}{2}$$

$$\langle 0|\hat{Q}_R^2|0\rangle = \langle 0| \left(\frac{\hat{b}\,e^{-i\theta} + \hat{b}^{+}e^{i\theta}}{\sqrt{2}}\right)^2 |0\rangle = \frac{1}{2}$$

$$\langle 0|\hat{Q}\,\hat{Q}_R|0\rangle = \langle 0| \left(\frac{\hat{b} + \hat{b}^{+}}{\sqrt{2}}\right) \left(\frac{\hat{b}\,e^{-i\theta} + \hat{b}^{+}e^{i\theta}}{\sqrt{2}}\right) |0\rangle = \frac{1}{2}\langle 0|\hat{b}\,\hat{b}^{+}e^{i\theta} + \hat{b}^{+}\hat{b}\,e^{-i\theta}|0\rangle = \frac{e^{i\theta}}{2}$$

$$\langle 0|\hat{Q}_R\hat{Q}|0\rangle = \frac{1}{2}\langle 0|\hat{b}^{+}\hat{b}\,e^{i\theta} + \hat{b}\,\hat{b}^{+}\,e^{-i\theta}|0\rangle = \frac{e^{-i\theta}}{2}$$

Substituting all of these into the results for Equation (6.15.11) provides

$$\langle 0, \eta|\hat{Q}^2|0, \eta\rangle = \frac{1}{2}\left\{\cosh^2(r) + \sinh^2(r) - 2\cos(\theta)\sinh(r)\cosh(r)\right\}$$

Therefore the variance of $\hat{Q}$ in the state $|0, \eta\rangle$ must be

$$\sigma_Q^2 = \langle 0, \eta | \hat{Q}^2 | 0, \eta \rangle - \langle 0, \eta | \hat{Q} | 0, \eta \rangle^2 = \frac{1}{2} \left\{ \cosh^2(r) + \sinh^2(r) - 2\cos(\theta)\sinh(r)\cosh(r) \right\}$$

(6.15.12a)

where $\eta = re^{-i\theta}$. The variance of the $\hat{P}$ quadrature operator can be similarly demonstrated

$$\sigma_P^2 = \langle 0, \eta | \hat{P}^2 | 0, \eta \rangle = \frac{1}{2} \left\{ \cosh^2(r) + \sinh^2(r) + 2\cos(\theta)\sinh(r)\cosh(r) \right\}$$

(6.15.12b)

Now consider the special case of $\theta = 0$ (Figure 6.15.2). The variance of the EM quadratures operators (see Equation 6.15.11) become

$$\sigma_Q = \frac{e^{-r}}{\sqrt{2}} \quad \text{and} \quad \sigma_P = \frac{e^{r}}{\sqrt{2}}$$

(6.15.13)

The ovals appear in Figure 6.15.1 (these represent a projection of the squeezed-vacuum Wigner distribution into the 2-D plane). The greatest variance appears along the "P" axis while the least variance appears along the "Q" axis. Notice that the squeeze parameter "r" must always be positive.

The angle θ in the squeezing parameter $\eta = re^{-i\theta}$ controls the "direction" of squeezing. It rotates the "long" axis of squeezing by an angle $\theta/2$ as shown in Figure 6.15.3. This is easy to see from Equations (6.15.12) by letting $\theta = 180$. We see that $\cos\theta$ changes sign and rather than subtracting a term for Q, it adds a term to make the variance of Q larger.

An important point is that the Heisenberg uncertainty relation remains unchanged for squeezed versus nonsqueezed coherent states. For the squeezed vacuum, the Heisenberg uncertainty relation for the two quadrature components becomes

$$\sigma_Q \sigma_P = \frac{e^{-r}}{\sqrt{2}} \frac{e^{r}}{\sqrt{2}} = \frac{1}{2}$$

(6.15.14)

Although the noise is squeezed out of one quadrature, it reappears in the other. The noise in the total electric field and Hamiltonian never decreases. Squeezing the vacuum always increases the variance of the Hamiltonian and total electric field (both quadratures).

6.15.6 Coordinate Representation of Squeezed States

The wave function for the squeezed vacuum can be represented in either the Q or P coordinate representation. Consider the normal distributions for the vacuum and

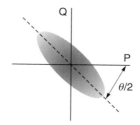

FIGURE 6.15.3
The angle θ in $\eta = re^{-i\theta}$ rotates the "squeezed direction" by $\theta/2$.

squeezed vacuum states shown in Figure 6.15.4. The distributions with dotted lines in the figure represent the coordinate projection of the coherent state. The distributions with the solid lines represent the projection of the squeezed vacuum onto the coordinates. As an example, consider the P coordinate. Notice that the Gaussian distribution for squeezed state is wider than that for the coherent-state vacuum. Also notice that the squeezed vacuum has a Gaussian distribution for either the "P" or the "Q" coordinates.

Leonhardt's book shows that the coordinate representations of the squeezed vacuum with the squeezing parameter $\eta = r e^{i\theta}$ (with $\theta = 0$) must be given by

$$\Psi(Q) = e^{r/2} U_0(e^r Q) \quad \text{and} \quad \Psi(P) = e^{-r/2} U_0(e^{-r} P) \tag{6.15.15}$$

where $U_0(Q)$ is the coordinate representation of the coherent-state vacuum as given in Section 6.11.3. The factors $e^{\mp r}$ compress or expand the scale of the axis which changes a circle into the oval. The multiplicative factors $e^{\pm r/2}$ normalize the wave functions. Leonhardt shows that Equations 6.15.15 lead to the correct form of the squeezing operator by differentiating with respect to "r" to get

$$\frac{\partial \Psi}{\partial r} = \frac{1}{2}\left(Q\frac{\partial}{\partial Q} + \frac{\partial}{\partial Q}Q\right)\psi = \frac{i}{2}\left(\hat{Q}\hat{P} + \hat{P}\hat{Q}\right)\psi$$

This differential equation can be solved by separating variables r, Ψ to find

$$\Psi(Q) = \exp\left\{\frac{ir}{2}\left(\hat{Q}\hat{P} + \hat{P}\hat{Q}\right)\right\}\Psi(0)$$

This last result is consistent with Equation (4.15.1) for $\theta = 0$.

$$\hat{S} = \exp\left\{\frac{ir}{2}\left(\hat{Q}\hat{P} + \hat{P}\hat{Q}\right)\right\} = \exp\left\{\frac{r}{2}\left(\hat{b}^2 - \hat{b}^{+2}\right)\right\}$$

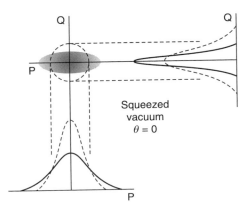

FIGURE 6.15.4
Comparison of a squeezed state (oval) and a coherent state (circle).

6.16   Some Statistics for Squeezed States

The ideally squeezed coherent state $|\alpha, \eta\rangle = D(\alpha)S(\eta)|0\rangle$ comes about by first squeezing the vacuum and then displacing it by the complex parameter α in amplitude space. Recall that there exists a number of different types of statistical distributions for quantum mechanical states; two of the most common include the normal distribution for the quadratures and another for the photon number. The parameter α characterizes the center of the quadrature distributions for both squeezed and unsqueezed coherent states. Translating a squeezed vacuum through the distance α does not affect the noise content of the state; this means that aside from the average, the other statistics remain unaffected by translations. Consequently, we should be able to calculate the variance of an operator in either the squeezed coherent state or the squeezed vacuum state and find the same result. We provide two examples of using squeezed states by calculating the average and variance of the electric field and the Hamiltonian. We also discuss the statistical distribution for the photon number and we'll see why squeezed states might more properly be termed "multi-photon states."

6.16.1   The Average Electric Field in a Squeezed Coherent State

The ideal squeezed coherent state is found by first squeezing the vacuum and then displacing the results (Figure 6.16.1). As will be shown, the average of the electric field operator remains unaffected by the squeezing (i.e., the average electric field in the ideal squeezed coherent state is identical to the average electric field in the coherent state). However, squeezing increases the variance of the *total* electric field compared with that for the coherent state alone. As will be seen, the displacement operator does not affect the amount of noise in the field.

The quantized electric field operator for a single mode can be written as

$$\hat{E} = +i\sqrt{\frac{\hbar\omega}{2\varepsilon_0 V}}\left[\hat{b}e^{ikx-i\omega t} - \hat{b}^+ e^{-ikx+i\omega t}\right] \tag{6.16.1}$$

The average electric field in the ideal squeezed coherent state $|\alpha, \eta\rangle = \hat{D}(\alpha)\,\hat{S}(\eta)|0\rangle$ becomes

$$\langle\alpha, \eta|\hat{E}|\alpha, \eta\rangle = \langle 0|\hat{S}^+\hat{D}^+\,\hat{E}\,\hat{D}\,\hat{S}|0\rangle$$

$$= +i\sqrt{\frac{\hbar\omega}{2\varepsilon_0 V}}\left\{\langle 0|\hat{S}^+\left(\hat{D}^+\hat{b}\,\hat{D}\right)\hat{S}|0\rangle e^{ikx-i\omega t} - \langle 0|\hat{S}^+\left(\hat{D}^+\hat{b}^+\,\hat{D}\right)\hat{S}|0\rangle e^{-ikx+i\omega t}\right\}$$

The displacement operator transforms the annihilation and creation operators according to

$$\hat{D}^+(\alpha)\,\hat{b}\,\hat{D}(\alpha) = \hat{b} + \alpha \quad\text{and}\quad \hat{D}^+(\alpha)\,\hat{b}^+\,\hat{D}(\alpha) = \hat{b}^+ + \alpha^* \tag{6.16.2}$$

so that

$$\langle\alpha, \eta|\hat{E}|\alpha, \eta\rangle = +i\sqrt{\frac{\hbar\omega}{2\varepsilon_0 V}}\left\{\langle 0|\hat{S}^+\left(\hat{b} + \alpha\right)\hat{S}|0\rangle e^{ikx-i\omega t} - \langle 0|\hat{S}^+\left(\hat{b}^+ + \alpha^*\right)\hat{S}|0\rangle e^{-ikx+i\omega t}\right\}$$

Using the fact that S must be unitary, substituting Equations 9.15.5 from the previous section, namely $\hat{S}^+ \hat{b} \hat{S} = \hat{b} \cosh(r) - \hat{b}^+ e^{i\theta} \sinh(r)$ and $\hat{S}^+ \hat{b}^+ \hat{S} = \hat{b}^+ \cosh(r) - \hat{b} e^{-i\theta} \sinh(r)$, and using $\langle 0|(\hat{b}\cosh(r))|0\rangle = 0$ (etc.), we find

$$\langle \alpha, \eta | \hat{E} | \alpha, \eta \rangle = +i\sqrt{\frac{\hbar\omega}{2\varepsilon_0 V}} \{\alpha e^{ikx-i\omega t} - \alpha^* e^{-ikx+i\omega t}\} = \langle \alpha | \hat{E} | \alpha \rangle \qquad (6.16.3)$$

This last equation shows that the average electric field is independent of the squeezing. Therefore the center of the Wigner distribution does not depend on the squeezing.

6.16.2 The Variance of the Electric Field in a Squeezed Coherent State

This section calculates the variance of the total electric field. Displacing the squeezed vacuum does not change the noise content. For simplicity, use Equation (6.16.1) with $x = 0$ and $t = 0$ to calculate the variance of the electric field

$$\hat{E} = +i\sqrt{\frac{\hbar\omega}{2\varepsilon_0 V}} \left[\hat{b} - \hat{b}^+\right]$$

The variance is given by

$$\sigma_E^2 = \langle \alpha, \eta | \hat{E}^2 | \alpha, \eta \rangle - \langle \alpha, \eta | \hat{E} | \alpha, \eta \rangle^2$$

We know the average from Equation (6.16.3). Calculating the term $\langle \alpha, \eta | \hat{E}^2 | \alpha, \eta \rangle$ as outlined in the chapter review exercises provides

$$\sigma_E^2\big|_{\text{sqz}} = \langle \alpha, \eta | \hat{E}^2 | \alpha, \eta \rangle - \langle \alpha, \eta | \hat{E} | \alpha, \eta \rangle^2 = \frac{\hbar\omega}{2\varepsilon_0 V} \{2\cos(\theta)\cosh(r)\sinh(r) + 1\} \qquad (6.18.8a)$$

where $r > 0$ (and so sinh must always be greater than zero). Equation (6.16.4) is independent of the displacement parameter α, as is necessary for the displacement operator not to influence the noise content.

The total-field variance in the ideal squeezed coherent state can be seen to be always larger than the variance of the field in the pure coherent state. The variance of the pure coherent state can be found by substituting $r = 0$ in Equation (6.16.4)

$$r = 0 \rightarrow \sigma_E^2\big|_{\text{coherent}} = \frac{\hbar\omega}{2\varepsilon_0 V}$$

Therefore

$$\sigma_E^2\big|_{\text{sqz}} \geq \sigma_E^2\big|_{\text{coherent}} \qquad (6.16.4b)$$

Systems can be designed to only detect or amplify the "quiet" quadrature component and ignore the noisy one. Similarly, squeezed number states can be detected by photodetectors, which ignore the noisy phase.

6.16.3 The Average of the Hamiltonian in a Squeezed Coherent State

We will see that the average energy (i.e., expectation value of the Hamiltonian) in a squeezed state must always be larger than the average energy in the corresponding coherent state. This occurs because the squeezed state carries the energy of the corresponding coherent state plus the energy required to squeeze the vacuum. In particular, we surmise that the squeezed vacuum has an average energy larger than the true vacuum. Squeezed states consist of pairs of photons rather than single ones. The squeezed-state variance for the Hamiltonian is always larger than the coherent-state variance.

The single mode electromagnetic Hamiltonian can be written as

$$\hat{H} = \hbar\omega \left(\hat{b}^+ \hat{b} + \frac{1}{2} \right) \tag{6.16.9}$$

where ω is the angular frequency of the mode. The expectation value of the energy in the ideal squeezed coherent state must be

$$\langle \alpha, \eta | \hat{H} | \alpha, \eta \rangle = \hbar\omega \langle \alpha, \eta | \left(\hat{b}^+ \hat{b} + \frac{1}{2} \right) | \alpha, \eta \rangle = \hbar\omega \langle \alpha, \eta | \hat{b}^+ \hat{b} | \alpha, \eta \rangle + \frac{1}{2} \hbar\omega$$

The average number in the first term on the right-hand side can be written as

$$\langle \alpha, \eta | \hat{b}^+ \hat{b} | \alpha, \eta \rangle = \langle 0 | \hat{S}^+ \hat{D}^+ \hat{b}^+ \hat{b} \, \hat{D} \, \hat{S} | 0 \rangle = \langle 0 | \hat{S}^+ \left(\hat{D}^+ \hat{b}^+ \hat{D} \right) \left(\hat{D}^+ \hat{b} \, \hat{D} \right) \hat{S} | 0 \rangle$$

which uses the unitary nature of the displacement operator "D." Substituting for the displaced creation and annihilation operators

$$\hat{D}^+(\alpha) \, \hat{b} \, \hat{D}(\alpha) = \hat{b} + \alpha \quad \text{and} \quad \hat{D}^+(\alpha) \, \hat{b}^+ \hat{D}(\alpha) = \hat{b}^+ + \alpha^* \tag{6.16.6}$$

the last expression becomes

$$\langle \alpha, \eta | \hat{b}^+ \hat{b} | \alpha, \eta \rangle = \langle 0 | \hat{S}^+ \hat{D}^+ \hat{b}^+ \hat{b} \, \hat{D} \, \hat{S} | 0 \rangle = \langle 0 | \hat{S}^+ \left(\hat{b}^+ + \alpha^* \right) \left(\hat{b} + \alpha \right) \hat{S} | 0 \rangle$$

$$= \langle 0 | \hat{S}^+ \left(\hat{b}^+ \hat{b} + \alpha \hat{b}^+ + \alpha^* \hat{b}^+ + |\alpha|^2 \right) \hat{S} | 0 \rangle \tag{6.16.7}$$

$$= \langle 0 | \hat{S}^+ \hat{b}^+ \hat{b} \, \hat{S} | 0 \rangle + \alpha \langle 0 | \hat{S}^+ \hat{b}^+ \hat{S} | 0 \rangle + \alpha^* \langle 0 | \hat{S}^+ \hat{b} \, \hat{S} | 0 \rangle + |\alpha|^2$$

Noting that the squeezed creation and annihilation operators consist of linear combinations of the creation and annihilation operators,

$$\hat{S}^+ \hat{b} \, \hat{S} = \hat{b} \cosh(r) - \hat{b}^+ e^{i\theta} \sinh(r) \quad \hat{S}^+ \hat{b}^+ \hat{S} = \hat{b}^+ \cosh(r) - \hat{b} e^{-i\theta} \sinh(r) \tag{6.16.8}$$

we realize that the vacuum expectation values of the squeezed operators must be zero. Therefore

$$\langle \alpha, \eta | \hat{b}^+ \hat{b} | \alpha, \eta \rangle = \langle 0 | \hat{S}^+ \hat{b}^+ \hat{b} \, \hat{S} | 0 \rangle + |\alpha|^2 = \langle 0 | \left(\hat{S}^+ \hat{b}^+ \hat{S} \right) \left(\hat{S}^+ \hat{b} \hat{S} \right) | 0 \rangle + |\alpha|^2$$

Substituting Equation 6.16.12 into the result of Equation 6.16.11, the average number becomes

$$\langle \alpha, \eta | \hat{b}^+ \hat{b} | \alpha, \eta \rangle = \langle 0 | \left(\hat{b}^+ \cosh(r) - \hat{b} e^{-i\theta} \sinh(r) \right) \left(\hat{b} \cosh(r) - \hat{b}^+ e^{i\theta} \sinh(r) \right) | 0 \rangle + |\alpha|^2$$

The terms $\langle 0 | \hat{b}^2 | 0 \rangle$, $\langle 0 | \hat{b}^{+2} | 0 \rangle$ give zero as do the terms $\langle 0 | \hat{b}^+ \hat{b} | 0 \rangle$. We are left with

$$\langle \alpha, \eta | \hat{b}^+ \hat{b} | \alpha, \eta \rangle = \langle 0 | \hat{b} \hat{b}^+ | 0 \rangle \sinh^2(r) + |\alpha|^2 = \sinh^2(r) + |\alpha|^2$$

Therefore the expectation value for the Hamiltonian in the ideal squeezed coherent state must be

$$\langle \alpha, \eta | \hat{H} | \alpha, \eta \rangle = \hbar \omega \langle \alpha, \eta | \hat{b}^+ \hat{b} | \alpha, \eta \rangle + \frac{1}{2} \hbar \omega = \hbar \omega \left\{ \sinh^2(r) + |\alpha|^2 + \frac{1}{2} \right\} \qquad (6.16.13)$$

Notice that $r = 0$ gives the average energy in the coherent state $|\alpha\rangle$. Squeezing the coherent state (i.e., $r > 0$) causes the average energy to increase. Equation (6.16.9) consists of three terms. The last term describes the vacuum energy, the middle term gives the energy stored in the coherent state (similar to the square of the electric field) and the first term describes the energy due to squeezing.

We can see that the average energy for the *squeezed vacuum* must be larger than the energy in the *vacuum* alone.

$$\langle 0, \eta | \hat{H} | 0, \eta \rangle = \hbar \omega \left\{ \sinh^2(r) + \frac{1}{2} \right\}$$

Similarly, the average energy in the *coherent* state $|\alpha\rangle$ can be written as

$$\langle \alpha | \hat{H} | \alpha \rangle = \hbar \omega \left\{ |\alpha|^2 + \frac{1}{2} \right\} \qquad (6.16.10)$$

The average energy in an ideal *squeezed coherent* state (Equation (6.16.9)) can be written in terms of the average energy in a coherent state as

$$\langle \alpha, \eta | \hat{H} | \alpha, \eta \rangle = \langle \alpha | \hat{H} | \alpha \rangle + \hbar \omega \sinh^2(r)$$

so that

$$\langle \alpha, \eta | \hat{H} | \alpha, \eta \rangle \geq \langle \alpha | \hat{H} | \alpha \rangle \qquad (6.16.11)$$

The energy must be larger for squeezed states because the energy expended to squeeze the state must be stored with the state. As will be seen in the next topic, the squeezed vacuum is actually a multi-photon state. In fact, the photons come along in pairs and are never found with an "odd" number.

6.16.4 Photon Statistics for the Squeezed State

The squeezed vacuum is a multi-photon state which consists of a linear combination of Fock states with an even number of photons; that is, the photons occur in pairs. The probability of finding an odd number of photons is zero. This behavior is a result of the fact that the creation and annihilation operators are squared in the argument of the exponential for the squeezing operator

$$\hat{S}(\eta) = \exp\left[\frac{\eta^*}{2}\hat{b}^2 - \frac{\eta}{2}\hat{b}^{+2}\right] \tag{6.16.16}$$

For example, consider the case of a "Q-squeezed" vacuum (i.e., $\theta = 0$ in the squeezing parameter $\eta = r\,e^{i\theta}$). Expanding the squeezing operator in powers of "r," provides

$$|0, r\rangle = \hat{S}(r)|0\rangle = \exp\left[\frac{r}{2}\left(\hat{b}^2 - \hat{b}^{+2}\right)\right]|0\rangle$$

$$\cong \left[1 + \left(\frac{r}{2}\right)\left(\hat{b}^2 - \hat{b}^{+2}\right) + \frac{1}{2!}\left(\frac{r}{2}\right)^2\left(\hat{b}^2 - \hat{b}^{+2}\right)^2 + \frac{1}{3!}\left(\frac{r}{2}\right)^3\left(\hat{b}^2 - \hat{b}^{+2}\right)^3 + \cdots\right]|0\rangle \tag{6.16.7}$$

The terms with the creation and annihilation operators can be expanded to provide

$$|0, r\rangle = \left[1 - \frac{1}{\sqrt{2}}\left(\frac{r}{2}\right)^2 + \cdots\right]|0\rangle + \left[-\sqrt{2}\left(\frac{r}{2}\right) + \left(\frac{r}{2}\right)^3\left(\frac{4 + 4!}{6\sqrt{2}}\right) + \cdots\right]|2\rangle + \left[\left(\frac{r}{2}\right)^2\frac{\sqrt{4!}}{\sqrt{2}} + \cdots\right]|4\rangle + \cdots \tag{6.16.18}$$

It is clear that all terms in Equation (6.16.13) (and hence Equation (6.16.14)) involve only even exponents of creation and annihilation operators. The squeezed vacuum must be a sum of even Fock states. The probability for finding "n" photons in the squeezed vacuum during a measurement must be

$$\text{prob}(n) = |\langle n \mid 0, r\rangle|^2 \tag{6.16.19}$$

The probability for odd "n" must be zero since only even Fock states appear in Equation 6.16.18. As a note for $r = 0$ (i.e., no squeezing), the series in Equation (6.16.14) reduces to the unsqueezed vacuum state. Figure 6.16.2 shows the probability of finding n photons for a vacuum with squeezing r. For $r = 0$, there would be a 100% probability of finding $n = 0$ photons for the unsqueezed vacuum.

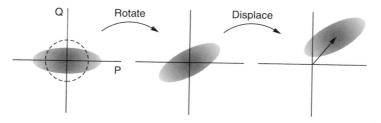

FIGURE 6.16.1
The vacuum is squeezed and then displaced to produce the squeezed coherent state.

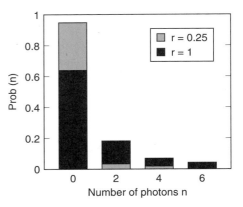

FIGURE 6.16.2
Probability distribution for finding "n" photons in the squeezed vacuum for two values of the squeezing parameter "r." A value of $r = 0.25$ produces an almost unsqueezed state.

Alternatively, Leonhardt calculates the probability by using the coordinate representation of the squeezed vacuum. The coordinate wavefunction is given in Section 6.15.6 as Equation (6.15.15)

$$\Psi(Q) = e^{r/2} U_0(e^r Q)$$

where U_0 is the unsqueezed vacuum wavefunction (a Gaussian). The probability amplitude is therefore

$$\langle n | \hat{S}(r) | 0 \rangle = \int_{-\infty}^{\infty} dQ \, \psi_n(Q) e^{r/2} U_0(e^r Q)$$

where $\psi_n(Q)$ is the coordinate representation of the Fock wavefunction for a state consisting of "n" photons. Leonhardt gives the formula

$$\text{Prob}(n) = \begin{cases} 0 & n = 1, 3, 5 \ldots \\ \binom{n}{n/2} \dfrac{1}{\cosh(r)} \left(\dfrac{\tan h(r)}{2} \right)^n & n = 0, 2, 4 \ldots \end{cases}$$

where

$$\binom{n}{m} = \frac{n!}{m! \, (n-m)!}$$

Figure 6.16.2 shows a plot of the probability of "n" photons for $r = 0.25$ and $r = 1$.

6.17 The Wigner Distribution

Previous sections discuss Fock, coherent, and squeezed states of the electromagnetic field. We expect to describe number-phase and the quadratures by probability distributions. Poisson-like distributions describe the photon number while normal distributions apply to the quadratures. We can define a Wigner distribution that treats the phase space coordinates (Q and P) as random variables. The Wigner distribution represents the closest quantum analog to the classical probability distribution. However, the Wigner distribution can become negative (a nonclassical probability property) for certain types of states (such as Fock states). For coherent and squeezed states, the Wigner distribution provides a near-classical picture. The so-called "*P*" and "*Q*" quasi-probability distributions can also be defined. The "*P*" distribution provides a mathematical object very similar to a density operator.

6.17.1   The Wigner Formula and an Example

The Wigner distribution is defined to be the joint quasi-probability density for the electromagnetic (EM) field using the Q and P coordinates. This section shows that the Wigner function can be written as

$$W(Q,P) = \frac{1}{2\pi} \int_{-\infty}^{\infty} dx \; \exp(iPx) \left\langle Q - \frac{x}{2} \middle| \hat{\rho} \middle| Q + \frac{x}{2} \right\rangle \qquad (6.17.1)$$

where $\hat{\rho}$ represents the density operator (from Chapter 5).

The Wigner function connects the quantum and classical probability theory through the use of marginal probabilities. For a classical probability density function $W(x,y)$, the probability of finding (for example) the values of $x \in (a,b)$ and $y \in (c,d)$ are given by

$$\text{Prob}\,(a < x < b,\ c < y < d) = \int_{a}^{b} \int_{c}^{d} dx\, dy\, W(x,y)$$

The probability that $x \in (a,b)$ regardless of the value of "*y*" is

$$\text{Prob}\,(a < x < b) = \text{prob}\,(a < x < b,\ -\infty < y < \infty) = \int_{a}^{b} \int_{-\infty}^{\infty} dx\, dy\, W(x,y)$$

Therefore, the probability density function for "*x*" can be identified as

$$W(x) \equiv \int_{-\infty}^{\infty} dy\, W(x,y)$$

Wigner defined his formula in such a way that simultaneous measurements of $\hat{P}, \hat{Q}$ do not need to be considered. The Wigner formulation requires the density functions for Q and P to be given by

$$W_Q(Q) \equiv \int_{-\infty}^{\infty} dP\, W(Q,P) = \langle Q|\hat{\rho}|Q\rangle \quad \text{and} \quad W_P(P) \equiv \int_{-\infty}^{\infty} dQ\, W(Q,P) = \langle P|\hat{\rho}|P\rangle$$

$$(6.17.2)$$

The subscripts P and Q appear in Equation (6.17.2) to remind the reader of the independent variable—something not always obvious from the formulas. The 1-D probability density functions can be related to the density operator by

$$W_Q(Q) = \langle Q|\hat{\rho}|Q\rangle \sim \langle Q\mid\psi\rangle\langle\psi\mid Q\rangle = |\psi(Q)|^2 \qquad (6.17.3)$$

and similarly for the density that depends on the "*P*" coordinate. This formula is reminiscent of the "shadow" plots such as in Figure 6.15.4.

What are Q and P? Recall that the single-mode electric field can be written as

$$\vec{E}(\vec{r}, t) = -\sqrt{\frac{\hbar\omega}{\varepsilon_0 V}}\left[\hat{Q}\sin\left(\vec{k}\cdot\vec{r} - \omega t\right) + \hat{P}\cos\left(\vec{k}\cdot\vec{r} - \omega t\right)\right] \qquad (6.17.4)$$

This last equation can be in either the interaction representation or the free-field Heisenberg representation. We know that measuring the field requires us to simultaneously find the values Q and P. Although we can make measurements and find specific values of Q and P, repeated measurements do not produce identical values. We think of Q and P as coordinates in a phase space plot. Given the results of a particular measurement, we can define the (normalized) electric field amplitude and the phase as

$$a = \frac{1}{\sqrt{2}}(Q + iP) \quad \tan\phi = P/Q \qquad (6.17.5)$$

which would be the annihilation operator if Q, P were operators. Notice that for the coherent state $|\alpha\rangle$, the value "a" is not necessarily the same as α. The value α represents an average amplitude according to $\alpha = \frac{1}{\sqrt{2}}(\bar{Q} + \bar{P})$ where $\bar{Q}, \bar{P}$ represent the average values of $\hat{Q}, \hat{P}$ in the coherent state whereas we might consider Q, P to be measured values. Each set of values Q, P leads to a different sine wave since Q, P must be related to an amplitude and phase for the EM field (see, for example, Figures 6.14.2 and 6.9.6).

Example 6.17.1
Find the Wigner density function for the vacuum state.

Solution: To fix our thoughts, we look for the Wigner function for the coherent state $|\alpha\rangle$ by setting $\hat{\rho} = |\alpha\rangle\langle\alpha|$ (the Wigner function for nonpure states can also be found). A term such as $\langle Q|\alpha\rangle$ in

$$\langle Q|\hat{\rho}|Q\rangle = \langle Q \mid \alpha\rangle\langle\alpha \mid Q\rangle = |\langle Q \mid \alpha\rangle|^2 \qquad (6.17.6)$$

is the coordinate representation of the wave function. Recall the coordinate wave function for the vacuum from Equations (6.11.14) and (6.11.15).

$$\langle Q \mid \alpha = 0\rangle = \psi_0(Q) = \frac{1}{\pi^{1/4}}e^{-Q^2/2} \qquad (6.17.7)$$

The Wigner function can therefore be written as

$$\begin{aligned}
W(Q, P) &= \frac{1}{2\pi}\int_{-\infty}^{\infty} dx \, \exp(iPx)\left\langle Q - \frac{x}{2}\middle|\hat{\rho}\middle|Q + \frac{x}{2}\right\rangle \\
&= \frac{1}{2\pi}\int_{-\infty}^{\infty} dx \, \exp(iPx)\left\langle Q - \frac{x}{2}\middle|\alpha = 0\right\rangle\left\langle\alpha = 0\middle|Q + \frac{x}{2}\right\rangle \qquad (6.17.8) \\
&= \frac{1}{2\pi}\int_{-\infty}^{\infty} dx \, \exp(iPx)\,\psi_0^*\left(Q + \frac{x}{2}\right)\psi_0\left(Q - \frac{x}{2}\right)
\end{aligned}$$

Notice the arguments $Q \pm x/2$ just tells us to shift the distribution to the right or to the left by an amount of $x/2$. Substituting for the wavefunctions in Equation 6.17.8 provides

$$W(Q, P) = \frac{1}{2\pi} \int_{-\infty}^{\infty} dx \; \exp(iPx) \frac{1}{\pi^{1/4}} \exp\left[-\frac{1}{2}\left(Q + \frac{x}{2}\right)^2\right] \frac{1}{\pi^{1/4}} \exp\left[-\frac{1}{2}\left(Q - \frac{x}{2}\right)^2\right]$$

$$= \frac{1}{\pi} \exp\left(-Q^2 - P^2\right)$$

(6.17.9)

The last line follows after some simplifying algebra and integration. Equation 6.17.9 verifies that the Wigner function is a Gaussian distribution centered on the phase–space origin for the vacuum. The displacement operator translates W to a new center characterizing $|\alpha\rangle$ without changing its shape. Therefore a Gaussian distribution must describe all coherent states.

6.17.2 Derivation of the Wigner Formula

The derivation presented here resembles the one in Leonhardt's book. The argument uses the notation Q', P' for the rotated coordinates which reduce to Q and P for a rotation of $0°$ (it's ok to think of Q, P instead). The following list provides the sequence of steps required for the derivation. The function W' is the Wigner function of Q', P'; it reduces to W when the coordinates are not rotated.

1. The classical marginal probability distribution is defined as

$$W_Q(Q') \equiv \int_{-\infty}^{\infty} dP' \, W'(Q', P')$$

(6.17.10)

which has the form of a "Radon" transformation. The function $W'(Q', P')$ is the quantity of interest; i.e., it is the Wigner function. The inverse transformation of this equation must be developed in order to isolate $W'(Q', P')$.

2. The "characteristic function" is the Fourier transform of the probability *density* $W'(Q', P')$ appearing in the integrand of Equation (6.17.10) in Step 1. The Fourier transform of the *marginal* probability will be found to be

$$\tilde{W}_Q(\xi, \theta) = \sqrt{2\pi} \, \tilde{W}(\xi \cos \theta, \xi \sin \theta) \equiv \sqrt{2\pi} \, \tilde{W}(u, v)$$

(6.17.11)

where "~" denotes Fourier transform, ξ is the Fourier transform variable conjugate to Q', θ is the rotation angle for the coordinates and $u = \xi \cos \theta$ and $v = \xi \sin \theta$. This last equation relates the Fourier transform of the marginal probability $\tilde{W}_Q$ to the Fourier transform of Wigner function $\tilde{W}$. Essentially, this step provides the desired inversion for the Radon transfomation—just solve for $\tilde{W}(u, v)$. We're not quite finished though since we have not yet included the quantum mechanical probability–Equation (6.17.11) represents the classical probability. The characteristic function $\tilde{W}(u, v)$ must be compared to a corresponding quantity from quantum theory.

3. The density operator $\hat{\rho}$ defines the quantum probability distribution through

$$W(Q') = \langle Q'|\hat{\rho}|Q'\rangle \tag{6.17.12}$$

and similar to Step 2, the Fourier transform provides *marginal* probability density

$$\tilde{W}_Q(\xi, \theta) = \frac{1}{\sqrt{2\pi}} Tr\left\{\hat{\rho} \exp\left(-iu\hat{Q} - iv\hat{P}\right)\right\} \tag{6.17.13}$$

where $\exp(-iu\hat{Q} - iv\hat{P})$ is known as the Weyl operator; it is structurally similar to the displacement operator.

4. Equating the results of Steps 2 and 3 provides an expression for the Fourier transform of the Wigner distribution

$$\tilde{W}(u, v) = \frac{1}{2\pi} Tr\left\{\hat{\rho} \exp\left(-iu\hat{Q} - iv\hat{P}\right)\right\} \tag{6.17.14}$$

The Wigner distribution can be found from this last expression by taking an inverse Fourier transform and using the wavefunction translation properties inherent to the Weyl operator.

The rest of the discussion executes the program plan in steps 1 through 4 above. First we discuss the coordinate rotations as step 0.

Step 0   The Rotations

This step defines some rotated quantities that will be required in the remainder of the section. In particular, we recall how coordinates transform and how functions of these coordinates transform. These will be used in Step 2 to write the Fourier transform of the marginal probability density. We examine how operators and eigenstates transform under coordinate rotations (Figure 6.17.1) for finding the quantum probability in Step 3.

The coordinates Q', P' can be defined by the rotation

$$\begin{pmatrix} Q' \\ P' \end{pmatrix} = \begin{pmatrix} \cos\theta & \sin\theta \\ -\sin\theta & \cos\theta \end{pmatrix} \begin{pmatrix} Q \\ P \end{pmatrix} \tag{6.17.15}$$

Recall that the coordinate transformation relates a function of the new coordinates to a function of the old coordinates (refer to Section 4.12.1) by

$$W'(Q', P') = W(Q, P) \tag{6.17.16}$$

The W and W' are the Wigner functions in the original and rotated coordinates. The coordinates Q and P must be the eigenvalues of the operators $\hat{Q}, \hat{P}$ (respectively) according to

$$\hat{Q}|Q\rangle = Q|Q\rangle \quad \text{and} \quad \hat{P}|P\rangle = P|P\rangle \tag{6.17.17}$$

Recall that the rotation operator $\hat{R}$, which rotates the quadrature operators $\hat{Q}, \hat{P}$, can be written as

$$\hat{R} = e^{-i\hat{N}\theta} \tag{6.17.18}$$

where $\hat{N}$ represents the number operator $\hat{N} = \hat{b}^+\hat{b}$, and $\hat{b}^+, \hat{b}$ represent the creation and annihilation operators respectively. The rotated "position" and "momentum" quadrature operators can be written as

$$\hat{Q}' = \hat{R}^+\hat{Q}\hat{R} \quad \text{and} \quad \hat{P}' = \hat{R}^+\hat{P}\hat{R} \tag{6.17.19}$$

For example, $\hat{Q}'$ is found to be

$$\hat{Q}' = \hat{R}^+\hat{Q}\hat{R} = e^{i\hat{N}\theta} \frac{1}{\sqrt{2}} \left(\hat{b} + \hat{b}^+ \right) e^{-i\hat{N}\theta} = \frac{1}{\sqrt{2}} \left(\hat{b}e^{-i\theta} + \hat{b}^+e^{i\theta} \right) = \hat{Q}\cos\theta + \hat{P}\sin\theta \tag{6.17.20}$$

Let the states be defined by

$$|Q,\theta\rangle \equiv |Q'\rangle = \hat{R}^+|Q\rangle \quad \text{and} \quad |P,\theta\rangle \equiv |P'\rangle = \hat{R}^+|P\rangle \tag{6.17.21}$$

(note the "$\hat{R}^+$"). The adjoint rotation operator $\hat{R}^+$, which is unitary, has the property that $\hat{R}^+(\theta) = \hat{R}(-\theta)$. The rotated coordinate kets must be eigenkets of the rotated operator

$$\hat{Q}'|Q'\rangle = \left(\hat{R}^+\hat{Q}\hat{R} \right)\hat{R}^+|Q\rangle = \hat{R}^+\hat{Q}|Q\rangle = \hat{R}^+Q|Q\rangle = Q\hat{R}^+|Q\rangle = Q|Q'\rangle \tag{6.17.22}$$

Note that Q remains the eigenvalue.

Step 1   The Marginal Probability Distribution

The classical marginal probability distribution is defined as

$$W_Q(Q') \equiv \int_{-\infty}^{\infty} dP'\, W'(Q', P') \tag{6.17.23}$$

To find the "inverse" of the transformation, one needs to work with the Fourier transformation.

Step 2   The Characteristic Function

The Fourier transform of

$$W_Q(Q') \equiv \int_{-\infty}^{\infty} dP'\, W'(Q', P') \tag{6.17.24}$$

is given by

$$\tilde{W}_Q(\xi, \theta) = \int_{-\infty}^{\infty} dQ'\, W_Q(Q') \frac{e^{-i\xi Q'}}{\sqrt{2\pi}} = \int_{-\infty}^{\infty} dQ' \frac{e^{-i\xi Q'}}{\sqrt{2\pi}} \int_{-\infty}^{\infty} dP'\, W'(Q', P')$$

$$= \iint_{\mathbb{R}^2} dQ'\, dP' \frac{e^{-i\xi Q'}}{\sqrt{2\pi}} W'(Q', P')$$

The Jacobian of the transformation provides $dQ\,dP = dQ'\,dP'$, the definition for the rotation of a function (Section 4.12.1) gives $W'(Q',P') = W(Q,P)$ and the coordinate rotation from Equation (6.17.15) yields $Q' = Q\cos\theta + P\sin\theta$. Therefore $\tilde{W}_Q(\xi,\theta)$ becomes

$$
\begin{aligned}
\tilde{W}_Q(\xi,\theta) &= \iint dQ dP \frac{e^{-i\xi(Q\cos\theta + P\sin\theta)}}{\sqrt{2\pi}} W(Q,P) \\
&= \sqrt{2\pi} \iint dQ dP\, W(Q,P) \frac{e^{-iQ(\xi\cos\theta) - iP(\xi\sin\theta)}}{2\pi}
\end{aligned}
\tag{6.17.25}
$$

The characteristic function is the Fourier transform of the probability density

$$
\tilde{W}(u,v) = \iint dQ\, dP\; W(Q,P) \frac{e^{-iQu - iPv}}{2\pi}
\tag{6.17.26}
$$

where the transform variables are related to polar coordinates by

$$
u = \xi\cos\theta \quad\text{and}\quad v = \xi\sin\theta
\tag{6.17.27}
$$

Therefore, Equation (6.17.25) relates the characteristic function (Fourier transform of W) to the Fourier transform of the marginal probability

$$
\tilde{W}_Q(\xi,\theta) = \sqrt{2\pi}\,\tilde{W}(u,v) = \sqrt{2\pi}\,\tilde{W}(\xi\cos\theta, \xi\sin\theta)
\tag{6.17.28}
$$

Step 3 Fourier Transform of the Quantum Probability Density Function

The quantum probability density (in the Q' coordinate) is given by

$$
W_Q(Q') = \langle Q'|\hat{\rho}|Q'\rangle = \langle Q|\hat{R}\hat{\rho}\hat{R}^+|Q\rangle
\tag{6.17.29}
$$

The Fourier transform of this function is

$$
\tilde{W}_Q(\xi,\theta) = \int_{-\infty}^{\infty} dQ\,\langle Q|\hat{R}\hat{\rho}\hat{R}^+|Q\rangle \frac{e^{-i\xi Q}}{\sqrt{2\pi}} = \int_{-\infty}^{\infty} dQ\,\langle Q|\hat{R}\hat{\rho}\hat{R}^+ \frac{e^{-i\xi Q}}{\sqrt{2\pi}}|Q\rangle
\tag{6.17.30}
$$

where the constant exponential term is moved inside the coordinate expectation value. Using $e^{-i\xi\hat{Q}}|Q\rangle = e^{-i\xi Q}|Q\rangle$ (as can be seen by Taylor expanding $e^{-i\xi\hat{Q}}$), we obtain

$$
\tilde{W}_Q(\xi,\theta) = \int_{-\infty}^{\infty} dQ\,\langle Q|\hat{R}\hat{\rho}\hat{R}^+ \frac{e^{-i\xi\hat{Q}}}{\sqrt{2\pi}}|Q\rangle = \mathrm{Tr}\left[\hat{R}\hat{\rho}\hat{R}^+ \frac{e^{-i\xi\hat{Q}}}{\sqrt{2\pi}}\right]
$$

where the coordinate basis is used for the trace. The order of the operators in the trace can be permuted (see Section 4.5) to get

$$
\tilde{W}_Q(\xi,\theta) = \int_{-\infty}^{\infty} dQ\,\langle Q|\hat{R}\hat{\rho}\hat{R}^+ \frac{e^{-i\xi\hat{Q}}}{\sqrt{2\pi}}|Q\rangle = \frac{1}{\sqrt{2\pi}}\mathrm{Tr}\left[\hat{\rho}\hat{R}^+ e^{-i\xi\hat{Q}}\hat{R}\right]
\tag{6.17.31}
$$

However, Step 0 shows that the rotated quadrature operator can be written as $\hat{Q}' = \hat{R}^+ \hat{Q} \hat{R} = \hat{Q} \cos \theta + \hat{P} \sin \theta$. Therefore, expanding the exponential in Equation (6.17.31), and using the unitary property of $\hat{R}$,

$$\hat{R}^+ e^{-i\xi\hat{Q}} \hat{R} = \hat{R}^+ \left(\sum_n \frac{1}{n!} (-i\,\xi)^n \hat{Q}^n \right) \hat{R} = \sum_n \frac{1}{n!} (-i\,\xi)^n \left(\hat{R}^+ \hat{Q} \hat{R} \right)^n = e^{-i\xi(\hat{Q} \cos \theta + \hat{P} \sin \theta)}$$

Substituting this into Equation (6.17.31) along with

$$u = \xi \cos \theta \quad \text{and} \quad v = \xi \sin \theta$$

provides

$$\tilde{W}_Q(\xi, \theta) = \frac{1}{\sqrt{2\pi}} \operatorname{Tr}\left[\hat{\rho} \hat{R}^+ e^{-i\xi\hat{Q}} \hat{R} \right] = \frac{1}{\sqrt{2\pi}} \operatorname{Tr}\left[\hat{\rho} e^{-i\hat{Q}u - i\hat{P}v} \right] \tag{6.17.31}$$

Step 4   Equate the Results of Steps 2 and 3

Equating the results of Steps 2 and 3 provides

$$\tilde{W}(u, v) = \frac{1}{2\pi} \operatorname{Tr}\left[\hat{\rho} e^{-i\hat{Q}u - i\hat{P}v} \right] \tag{6.17.33}$$

The Weyl operator can be rewritten using the Baker–Hausdorff formula as

$$e^{\hat{A}+\hat{B}} = e^{\hat{A}} e^{\hat{B}} e^{-[\hat{A}, \hat{B}]/2} \quad \text{so long as} \quad \left[\hat{A}, \left[\hat{A}, \hat{B}\right]\right] = 0 = \left[\hat{B}, \left[\hat{A}, \hat{B}\right]\right]$$

(refer to Chapter 4). The Weyl operator becomes $e^{-i\hat{Q}u - i\hat{P}v} = e^{-i\hat{Q}u} e^{-i\hat{P}v} e^{-[-i\hat{Q}u, -i\hat{P}v]/2} = e^{-i\hat{Q}u} e^{-i\hat{P}v} e^{+iuv/2}$ since $[\hat{Q}, \hat{P}] = i$. Equation (6.17.33) can be written

$$
\begin{aligned}
\tilde{W}(u, v) &= \frac{1}{2\pi} \operatorname{Tr}\left[\hat{\rho} e^{-i\hat{Q}u - i\hat{P}v} \right] = \frac{1}{2\pi} \int_{-\infty}^{\infty} dQ \, \langle Q | \hat{\rho} e^{-iu\hat{Q} - iv\hat{P}} | Q \rangle \\
&= \frac{e^{+iuv/2}}{2\pi} \int_{-\infty}^{\infty} dQ \, \langle Q | \hat{\rho} e^{-iu\hat{Q}} e^{-iv\hat{P}} | Q \rangle
\end{aligned}
\tag{6.17.34}
$$

Using the fact that $e^{-iv\hat{P}}$ is a translation operator in Q space according to Section 6.11, provides

$$e^{-iv\hat{P}} | Q \rangle = | Q + v \rangle \tag{6.17.35a}$$

so that

$$e^{-iu\hat{Q}} e^{-iv\hat{P}} | Q \rangle = e^{-iu\hat{Q}} | Q + v \rangle = e^{-iu(Q+v)} | Q + v \rangle \tag{6.17.35b}$$

Equation (6.17.34) becomes

$$\tilde{W}(u, v) = \frac{e^{iuv/2}}{2\pi} \int_{-\infty}^{\infty} dQ \, \langle Q | \hat{\rho} e^{-iu(Q+v)} | Q + v \rangle = \frac{e^{-iuv/2}}{2\pi} \int_{-\infty}^{\infty} dQ \, e^{-iuQ} \langle Q | \hat{\rho} | Q + v \rangle$$

Making the substitution $x = Q - \dfrac{v}{2}$ yields

$$\tilde{W}(u,v) = \frac{1}{2\pi} \int_{-\infty}^{\infty} dx\, e^{-iux} \left\langle x - \frac{v}{2} \middle| \hat{\rho} \middle| x + \frac{v}{2} \right\rangle \tag{6.17.36}$$

Step 5   *Inverse Transform*
The inverse transform of Equation (6.17.8) is

$$W(Q,P) = \iint_{(-\infty,\infty)} du\, dv\, \tilde{W}(u,v) \frac{e^{iuQ+ivP}}{(\sqrt{2\pi})^2} = \frac{1}{2\pi} \iint dx\, dv \left\langle x - \frac{v}{2} \middle| \hat{\rho} \middle| x + \frac{v}{2} \right\rangle e^{ivP} \int_{-\infty}^{\infty} du\, \frac{e^{iu(Q-x)}}{2\pi}$$

$$\tag{6.17.37}$$

The definition for the Dirac delta function

$$\int_{-\infty}^{\infty} du\, \frac{e^{iu(Q-x)}}{2\pi} = \delta(Q - x) \tag{6.17.38}$$

gives

$$W(Q,P) = \frac{1}{2\pi} \iint dx\, dv \left\langle x - \frac{v}{2} \middle| \hat{\rho} \middle| x + \frac{v}{2} \right\rangle e^{ivP} \delta(Q - x)$$

Finally, we obtain Wigner's formula for a quasi-probability distribution

$$W(Q,P) = \frac{1}{2\pi} \int_{-\infty}^{\infty} dv\, \exp(ivP) \left\langle Q - \frac{v}{2} \middle| \hat{\rho} \middle| Q + \frac{v}{2} \right\rangle \tag{6.17.39}$$

where Q and P are coordinates instead of operators.

6.17.3   Example of the Wigner Function
Wigner's formula

$$W(Q,P) = \frac{1}{2\pi} \int_{-\infty}^{\infty} dx\, \exp(ixP) \left\langle Q - \frac{x}{2} \middle| \hat{\rho} \middle| Q + \frac{x}{2} \right\rangle$$

is easy to evaluate so long as the wave function $|\psi\rangle$ in $\hat{\rho} = |\psi\rangle\langle\psi|$ is known as a function of the "position" coordinate Q. The inner product then produces

$$\left\langle Q - \frac{x}{2} \middle| \hat{\rho} \middle| Q + \frac{x}{2} \right\rangle = \left\langle Q - \frac{x}{2} \middle| \psi \right\rangle \left\langle \psi \middle| Q + \frac{x}{2} \right\rangle = \psi\left(Q - \frac{x}{2}\right) \psi^*\left(Q + \frac{x}{2}\right)$$

Example 6.17.2   Wigner representation of the Fock state $|1\rangle$
The coordinate representation of the first Fock state can be written as

$$\langle Q \mid 1 \rangle = \psi_1(Q) = C_1 Q \exp\left(-\frac{Q^2}{2}\right)$$

from Equation (8.7.12). Therefore, set $\rho_1 = |1\rangle\langle 1|$ to get

$$\left\langle Q - \frac{x}{2}\Big|\hat{\rho}\Big|Q + \frac{x}{2}\right\rangle = \left\langle Q - \frac{x}{2}\Big|1\right\rangle\left\langle 1\Big|Q + \frac{x}{2}\right\rangle = \psi_1\left(Q - \frac{x}{2}\right)\psi_1^*\left(Q + \frac{x}{2}\right)$$

which gives

$$\left\langle Q - \frac{x}{2}\Big|\hat{\rho}\Big|Q + \frac{x}{2}\right\rangle = |C_1|^2\left(Q - \frac{x}{2}\right)\exp\left(-\frac{(Q-x/2)^2}{2}\right)\left(Q + \frac{x}{2}\right)\exp\left(-\frac{(Q+x/2)^2}{2}\right)$$

The Wigner function becomes

$$W(Q,P) = \frac{|C_1|^2}{2\pi}\int_{-\infty}^{\infty} dx\ \exp(ixP)\left(Q - \frac{x}{2}\right)\exp\left(-\frac{(Q-x/2)^2}{2}\right)\left(Q + \frac{x}{2}\right)\exp\left(-\frac{(Q+x/2)^2}{2}\right)$$

which can be integrated. The function can be evaluated by (1) multiplying terms, (2) completing the square $\mathrm{Exp}\{ixP - x^2/4\} = \exp(-P^2)\exp\{-(x - 2iP)^2/4\}$, (3) using integrals of the form $\int_{-\infty}^{\infty} dx\,[\sqrt{2\pi}\sigma]^{-1}\exp[-(x - x_o)^2/2\sigma^2] = 1$ and $\int_{-\infty}^{\infty} dx\,x^2\exp[-y^2/2] = \sqrt{\pi}$.

$$W(P,Q) = \frac{\sqrt{2}|c_1|^2}{\sqrt{2\pi}}e^{-(P^2+Q^2)}(P^2 + Q^2) - \frac{|c_1|^2}{2\sqrt{2\pi}}e^{-(P^2+Q^2)}$$

Notice the circular symmetry in P, Q. The probability density becomes negative, for example, near $P = Q = 0$. A classical probability can never behave this way. This indicates that the Fock states cannot be classical states. Figure 6.17.2 shows the Wigner distribution $W/|c_1|^2$. The higher order Fock states have larger numbers of oscillations along the radial direction.

6.18 Measuring the Noise in Squeezed States

In general, we want to make use of the low noise light in optical systems such as those for communications. Previous sections describe several types of squeezed states including amplitude/phase squeezed and quadrature squeezed. Over a limited range of the squeezing parameter, the amplitude squeezed state exhibits reduced number fluctuation (number squeezed). The number squeezed state can be readily employed in optical communications using a simple photodetector since it responds to the number of photons in the beam. So long as the photodetector has high efficiency, the noise of the photocurrent will closely match the noise in the number-squeezed light beam. Given that the quality of an optical link partly depends on the noise content of the light, it seems reasonable to investigate methods of measuring the noise. Noise in number-squeezed light can be measured by connecting a high efficiency photodetector to a RF spectrum analyzer. In general though, homodyne detection represents the most common method for measuring the electromagnetic noise in a squeezed state. There exists both balanced and unbalanced detectors; for RF waves, these systems date back to World War II. This section discusses a simple homodyne detection system.

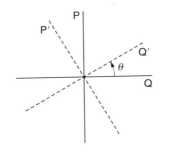

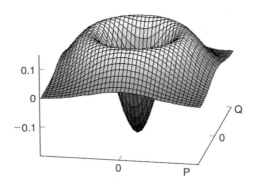

FIGURE 6.17.1
Definition of the rotated coordinates.

FIGURE 6.17.2
Wigner Distribution for the $n = 1$ Fock state.

Figure 6.18.1 shows a block diagram for producing and detecting optical electromagnetic waves in the squeezed state. A single photodetector could be used but it would detect the total light emission rather than the quadrature component. A laser generates a large amplitude optical signal (which can be represented by a classical wave). A system, denoted by "Sqz," converts the laser signal to squeezed light. The squeezed light is used in any desired manner after it leaves the "squeezer." For example, we might want to know how a device (i.e., the device under test (DUT) in the figure) affects the noise in the optical beam. A beam splitter taps-off a reference signal from the laser beam; we assume the reference signal can be represented by a classical wave. The reference signal maintains a fixed phase relation with the squeezed light. The homodyne detection system consists of perfect 50–50 beam splitter (no surface reflections and no absorption), two high-efficiency photodetectors and a summer with one input inverted. By adjusting the phase θ of the reference signal, it is possible to measure the quadrature component $\hat{Q}_\theta = \hat{R}^+ \hat{Q} \hat{R}$ where the rotation operator is $\hat{R}(\theta) = e^{-i\hat{N}\theta}$ and $\hat{N} = \hat{b}^+ \hat{b}$. Recall that $\hat{Q}_\theta$ can be either the "in-phase" component $\hat{P}$ or the "out-of-phase" component $\hat{Q}$ of the electric field. Therefore by varying the phase of the reference signal, properties of squeezed and anti-squeezed light can be examined.

Next, focus on the homodyne detection system shown in Figure 6.18.2. Let's represent the reference signal by a classical amplitude α_{LO} (i.e., the mode has a large amplitude). The symbols $\hat{a}, \hat{a}_1, \hat{a}_2$ refer to the operator amplitudes of the light waves. The following discussion shows that the difference in the number of photons arriving at the two photodetectors must be given by

$$\hat{n}_{21} = \sqrt{2}|\alpha_{LO}|\hat{Q}_\theta \qquad (6.18.1)$$

An increase of the signal from the local oscillator LO (i.e., the reference) also increases the size of the detected signal n_{21}; consequently the system provides an amplified version of the quadrature signal Q_θ. The detection system responds only to the frequency of the local oscillator and the desired quadrature component can be selected by using the appropriate value for θ. Besides detecting the quadrature component, the variance of the photon-number difference provides the variance of the selected quadrature component, which gives a measure of the amount of noise present.

The detectors respond to the power in each beam, which provides a measure of the number of photons in each beam. The number of photons in output beam #1 and beam

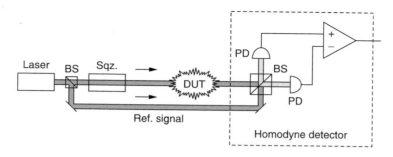

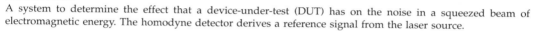

FIGURE 6.18.1

A system to determine the effect that a device-under-test (DUT) has on the noise in a squeezed beam of electromagnetic energy. The homodyne detector derives a reference signal from the laser source.

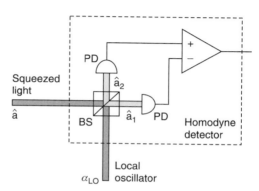

FIGURE 6.18.2

The homodyne detection system combines squeezed light and the large-amplitude local-oscillator signal at the beam splitter (mixing).

#2, respectively, can be written as

$$\hat{n}_1 = \hat{a}_1^+ \hat{a}_1 \qquad \hat{n}_2 = \hat{a}_2^+ \hat{a}_2 \tag{6.18.2a}$$

where we define the amplitudes by

$$\hat{a}_1 = \frac{1}{\sqrt{2}}(\hat{a} - \alpha_{LO}) \qquad \hat{a}_2 = \frac{1}{\sqrt{2}}(\hat{a} + \alpha_{LO}) \tag{6.18.2b}$$

where the amplitude $\hat{a}_1$ includes a $180°$ phase shift since the classical beam reflects from the internal interface of the beam splitter. We do not include similar phase shifts for the quantum amplitude $\hat{a}$ since we assume that the states in amplitude space will take care of any phase shifts. This really means that different portions of the input beam might be described by different states in amplitude space. The difference in the number of photons incident on the detectors is

$$\hat{n}_{21} = \hat{n}_2 - \hat{n}_1 = \hat{a}_2^+ \hat{a}_2 - \hat{a}_1^+ \hat{a}_1$$

$$= \frac{1}{\sqrt{2}}\left(\hat{a}^+ + \alpha_{LO}^*\right) \frac{1}{\sqrt{2}}(\hat{a} + \alpha_{LO}) - \frac{1}{\sqrt{2}}\left(\hat{a}^+ - \alpha_{LO}^*\right) \frac{1}{\sqrt{2}}(\hat{a} - \alpha_{LO}) \tag{6.18.3}$$

$$= \alpha_{LO}\hat{a}^+ + \alpha_{LO}^*\hat{a}$$

Using the definition for the creation and annihilation operators in terms of the "position and momentum" operators,

$$\hat{a} = \frac{1}{\sqrt{2}}\left(\hat{Q} + i\hat{P}\right) \quad \hat{a}^{+} = \frac{1}{\sqrt{2}}\left(\hat{Q} - i\hat{P}\right) \tag{6.18.4}$$

the difference in photon number becomes

$$\hat{n}_{21} = \alpha_{LO}\hat{a}^{+} + \alpha_{LO}^{*}\hat{a} = \frac{\alpha_{LO}^{*} + \alpha_{LO}}{\sqrt{2}}\hat{Q} + i\frac{\alpha_{LO}^{*} - \alpha_{LO}}{\sqrt{2}}\hat{P} = \hat{Q}\sqrt{2}\text{Re}(\alpha_{LO}) + \hat{P}\sqrt{2}\text{Im}(\alpha_{LO})$$

$$= \sqrt{2}|\alpha_{LO}|\left(\hat{Q}\cos\theta + \hat{P}\sin\theta\right) = \sqrt{2}|\alpha_{LO}|\hat{Q}_{\theta}$$

which proves Equation (6.18.1).

6.19   Review Exercises

6.1 Demonstrate the relations

$$\nabla \cdot e^{i\left(\vec{k}\cdot\vec{r} - \omega t\right)} = i k e^{i\left(\vec{k}\cdot\vec{r} - \omega t\right)} \quad \text{and} \quad \nabla \times \vec{A} = \frac{i\vec{k} \times \vec{A}_0}{\omega}e^{i\left(\vec{k}\cdot\vec{r} - \omega t\right)}$$

Assume the spatial variation of $\vec{A}_0$ is much smaller than for the high frequency optical term.

6.2 For a point particle at the point $\vec{x}_0$, find the potential $V = (1/4\pi\varepsilon_0)\int d^3x'(\rho(\vec{x}',t)/|\vec{x}-\vec{x}'|)$.

6.3 Show $G(\vec{r}) = (-1/4\pi\varepsilon_0|\vec{r}|)$ represents the potential due to a unit point charge at $\vec{r}' = 0$. Hint: Start with Maxwell's equation for the electric field in terms of a delta function charge density. After integrating, use the integral relation between voltage and field.

6.4 Show the function $\phi_k(z) = e^{ikz}/\sqrt{L}$ satisfying periodic boundary conditions over the length L has the correct orthonormal relation for a basis set $\langle \phi_k \mid \phi_K \rangle = \delta_{kK}$.

6.5 Show Equations (6.3.13)

$$\vec{E} = -\frac{\partial}{\partial t}\vec{A}(\vec{r},t) = \frac{+i}{\sqrt{\varepsilon_0 V}}\sum_{\vec{k}}\sqrt{\frac{\hbar\omega_k}{2}}\left[b_k\,e^{i\vec{k}\cdot\vec{r} - i\omega_k t} - b_k^{*}\,e^{-i\vec{k}\cdot\vec{r} + i\omega_k t}\right]\tilde{e}_k$$

$$\vec{B} = \nabla \times \vec{A}(\vec{r},t) = \frac{+i}{\sqrt{\varepsilon_0 V}}\sum_{\vec{k}}\sqrt{\frac{\hbar}{2\omega_k}}\left(\vec{k} \times \tilde{e}_k\right)\left[b_k\,e^{i\vec{k}\cdot\vec{r} - i\omega_k t} - b_k^{*}\,e^{-i\vec{k}\cdot\vec{r} + i\omega_k t}\right]$$

For $\vec{B}$, consider the formula $\nabla \times f\vec{v} = (\nabla f) \times \vec{v}$ for $\vec{v}$ a constant vector.

6.6 Derive Equations (6.4.12)

$$\vec{A}(\vec{r},t) = \frac{1}{\sqrt{\varepsilon_0 V}} \sum_{\vec{k}} \tilde{e}_k \left[\hat{q}_k \cos\left(\vec{k}\cdot\vec{r} - \omega_k t\right) - \frac{\hat{p}_k}{\omega_k} \sin\left(\vec{k}\cdot\vec{r} - \omega_k t\right) \right]$$

$$\vec{E}(\vec{r},t) = \frac{-1}{\sqrt{\varepsilon_0 V}} \sum_{\vec{k}} \tilde{e}_k \, \omega_k \left[\hat{q}_k \sin\left(\vec{k}\cdot\vec{r} - \omega_k t\right) + \frac{\hat{p}_k}{\omega_k} \cos\left(\vec{k}\cdot\vec{r} - \omega_k t\right) \right]$$

$$\vec{B}(\vec{r},t) = \frac{-1}{\sqrt{\varepsilon_0 V}} \sum_{\vec{k}} \vec{k} \times \tilde{e}_k \left[\hat{q}_k \sin\left(\vec{k}\cdot\vec{r} - \omega_k t\right) + \frac{\hat{p}_k}{\omega_k} \cos\left(\vec{k}\cdot\vec{r} - \omega_k t\right) \right]$$

starting with Equations (6.4.7) and (6.4.8).

6.7 Show the phase delay (rotation) operator $\hat{R}(\theta) = e^{-i\theta \hat{N}} = e^{-i\theta \hat{b}^+ \hat{b}}$ is unitary.

6.8 Show $\hat{R}^+ \hat{b}^+ \hat{R} = \hat{b}^+ e^{i\theta}$ using the operator expansion theorem

$$e^{x\hat{A}} \hat{B} e^{-x\hat{A}} = \hat{B} + \frac{x}{1!}[\hat{A}, \hat{B}] + \frac{x^2}{2!}\left[\hat{A}, \left[\hat{A}, \hat{B}\right]\right] + \ldots$$

6.9 Show the phase-rotated field must be

$$\hat{E} = i \sum_k \sqrt{\frac{\hbar\omega_k}{2\varepsilon_0 V}} \left[\hat{b}_k e^{i\vec{k}\cdot\vec{r} - i\omega_k t - i\theta_k} - \hat{b}_k^+ e^{-i\vec{k}\cdot\vec{r} + i\omega_k t + i\theta_k} \right] \tilde{e}_k$$

when the rotation operator is $\hat{R} = \prod_k \hat{R}_k$ and $\hat{R}_k(\theta) = e^{-i\theta_k \hat{N}_k} = e^{-i\theta_k \hat{b}_k^+ \hat{b}_k}$.

6.10 Find $\hat{q}_R = \hat{R}^+ \hat{q}\,\hat{R}$ and $\hat{p}_R = \hat{R}^+ \hat{p}\,\hat{R}$ where $\hat{R}(\theta) = e^{-i\theta \hat{N}} = e^{-i\theta \hat{b}^+ \hat{b}}$ and

$$\hat{q} = \sqrt{\frac{\hbar}{2\omega}}\left(\hat{b} + \hat{b}^+\right) \quad \hat{p} = -i\sqrt{\frac{\hbar\omega}{2}}\left(\hat{b} - \hat{b}^+\right)$$

6.11 Show the quadrature operators satisfy

$$\left[\hat{q}_k, \hat{q}_K\right] = 0 = \left[\hat{p}_k, \hat{p}_K\right] \quad \text{and} \quad \left[\hat{q}_k, \hat{p}_K\right] = i\hbar\delta_{k,K}$$

based on the commutation relations for the annihilation and creation operators.

6.12 Work out the right side of the amplitude operator by observing commutation relations.

$$\widehat{\text{Ampl}^2} = \left|\hat{E}\right|^2 \frac{\omega^2}{\varepsilon_0 V}\left[\hat{q}^2 + \left(\frac{\hat{p}}{\omega}\right)^2\right] = \frac{2\hbar^2\omega^2}{\varepsilon_0 V}\left(\hat{N} + \frac{1}{2}\right)$$

6.13 Find $\langle n|\hat{\mathscr{E}}^2|n\rangle$ for a single mode Fock State $|n\rangle$.

6.14 Starting with the expression $\hat{S} = \langle \hat{E} \times \hat{H} \rangle_{\substack{\text{one}\\\text{cycle}}}$, substitute the quantum fields and show

$$\hat{S} = \sum_{\vec{k}} \frac{\hbar\omega_k c}{V}\left(\hat{N}_k + \frac{1}{2}\right)$$

6.15 Find the average and variance of the Poynting vector for the Fock state $|n\rangle$ and coherent state $|\alpha\rangle$.

6.16 As in Section 6.5, show the relation

$$\int_V dV \frac{\varepsilon_0}{2} \vec{E} \cdot \vec{E} = \frac{-\hbar}{4} \sum_{\vec{ks}} \omega_k \left[+b_{ks}b_{-\vec{ks}} + b_{ks}^* b_{-ks}^* - b_{ks}b_{ks}^* - b_{ks}^* b_{ks} \right]$$

6.17 Show the number operator $\hat{N} = \sum_i \hat{N}_i$ has a sharp value for Fock states.

6.18 Consider a single mode. Suppose a system has the following wave function consisting of the sum of two Fock states.

$$|\psi\rangle = \frac{1}{\sqrt{3}} |n=1\rangle + \sqrt{\frac{2}{3}} |n=2\rangle$$

1. What is the probability of finding two photons?
2. Show the average electric field $\langle \psi | \hat{\mathscr{E}} | \psi \rangle$ for the single mode in part "*a*" has the form

$$\frac{4}{3} \sqrt{\frac{\hbar\omega}{2\varepsilon_0 V}} \sin(kz - \omega t + \pi)$$

6.19 Consider a single optical mode. Using the Fock basis set, show that the operator $\hat{b}^+$ does not have any right-hand eigenvectors: $\hat{b}^+ |v\rangle \ne \lambda |v\rangle$.

6.20 Show $\hat{b}_{\vec{K}}(t) = \hat{b}_{\vec{K}}(0)e^{-i\omega_K t}$ by calculating $[\hat{H}, \hat{b}_K(t)]$ in the Heisenberg representation.

6.21 Repeat Problem 6.20 but use the explicit form for the interaction representation $\hat{b}_k(t) = \hat{u}^+ \hat{b}_k \hat{u} = e^{-\hat{H}t/i\hbar} \hat{b}_k e^{\hat{H}t/i\hbar}$ and the operator expansion theorem found in Section 4.6. Explain why the result agrees with that from Problem 6.20 even though one refers to the Heisenberg representation and the other refers to the interaction representation.

6.22 Find the coordinate representation for a single Fock mode with $n=2$ photons using creation operators.

6.23 What is the approximate probability of finding the amplitude q in the range Δq centered on $q=0$ for the $n=0$ and $n=1$ photon cases for the Fock states.

6.24 Find $\langle n|\hat{S}|n\rangle$ for a single mode Fock state $|n\rangle$.

6.25 Show $\sigma_p^2 = \langle n|\hat{p}^2|n\rangle - \langle n|\hat{p}|n\rangle^2 = \hbar\omega(n + 1/2)$ using the number form of the Fock state.

6.26 Show $\Delta q \, \Delta p \geq \frac{\hbar}{2}$ for the vacuum state using the coordinate representation of the Fock states and operators. That is, use expressions of the form

$$\overline{f(q)} = \langle u_0(q)|f(q)|u_0(q)\rangle = \int_{-\infty}^{\infty} dq u_0^*(q) \, f(q) \, u_0(q)$$

$$\overline{f(\hat{p})} = \langle u_0(q)|f(\hat{p})|u_0(q)\rangle = \int_{-\infty}^{\infty} dq u_0^*(q) \frac{\hbar}{i} \frac{\partial}{\partial q} u_0(q)$$

and make direct substitutions for the vacuum wave functions.

6.27 For an electric field given by $A \sin(kz - \omega t) + B \cos(kz - \omega t)$ find the electric field in the form $C \sin(kz - \omega t + \phi)$. Identify C in terms of A and B and also identify ϕ. Assume A and B are real numbers.

6.28 Think of an example on how the coherent state $|\alpha\rangle$ can depend on position x according to $|\alpha(x)\rangle$.

6.29 Find the complex value of α (magnitude and phase) for the coherent state $|\alpha\rangle$ assuming the state represents the classical electric field $E = \eta \sin\left(kz - \omega t + \frac{\pi}{2}\right)$.

6.30 A mirror deflects a coherent beam at a $90°$ angle. Explain why we need two modes to describe the situation. Hint: remember that k vectors label the modes.

6.31 A student makes a homojunction laser and finds the emitted light can best be described by a coherent state.

1. If the student repeatedly measures the single-mode electric field

$$\hat{E} = i\sqrt{\frac{\hbar\omega}{2\varepsilon_0 V}} \left[\hat{b}\, e^{i\vec{k}\cdot\vec{r} - i\omega t} - \hat{b}^\dagger e^{-i\vec{k}\cdot\vec{r} + i\omega t} \right]$$

she finds the average

$$E = \sqrt{\frac{\hbar\omega}{2\varepsilon_0 V}}\sqrt{8} \sin\left(\vec{k}\cdot\vec{r} - \omega t - \pi\right)$$

6.32 What is the value of α?

1. What is the probability of finding 0 photons when a measurement is made?

6.33 Starting with the Hamiltonian for $\hat{H} = (\hat{p}^2/2) + (\omega^2/2)\hat{q}^2$, find the probability amplitude for the momentum p, specifically $u_0(p)$. Use the following steps.

1. If $\hat{q} = c\,\partial/\partial p$ show that $c = i\hbar$ using commutation relations.
2. Rewrite the Hamiltonian so that $\hat{H}' = \hat{H}/\omega^2$ and define $\xi = 1/\omega$. Note the roles of $\hat{q}$ and $\hat{p}$ have reversed. What must be the eigenvalues of $\hat{H}'$ and therefore $\hat{H}$?
3. Starting with the vacuum state $u_0(q)$, write the vacuum state $u_0(p)$.

6.34 Write the single mode Hamiltonian in terms of the normalized quadrature operators $\hat{P}, \hat{Q}$.

6.35 Starting with the solution for the vacuum state $u_0(q)$ and the normalization integral, find the correctly normalized wave function $u_0(Q)$.

6.36 Find the average number of photons $\langle m \rangle = |\alpha|^2$ in a coherent state $|\alpha\rangle$ by directly summing the series $\langle m \rangle = \sum_{m=0}^{\infty} m P_\alpha(m)$ where $P_\alpha(m)$ represent the Poisson distribution.

6.37 Find the average of the Poynting vector $\langle \alpha | \hat{S} | \alpha \rangle$ and the standard deviation $\sigma_{\hat{S}}$ for the coherent state $|\alpha\rangle$. Find the signal-to-noise ratio $SNR = \langle \hat{S} \rangle / \sigma_{\hat{S}}$. Write all answers in terms of the average number of photons in the mode $\bar{n}$. Explain any differences between your results and that found in Problem 2.22.

6.38 Blue and infrared beams of light have wavelengths λ_B, λ_R and powers, P_B, P_R. Using similar reasoning to Problem 6.37, show the ratio of the noise must be

$$\frac{\text{Noise Blue}}{\text{Noise Red}} = \frac{\sigma_B}{\sigma_R} = \sqrt{\frac{\lambda_R}{\lambda_B}\frac{P_B}{P_R}}$$

assuming $\bar{n} \gg 1/2$. Provide a physical explanation of why for equal powers, the blue light has the higher noise content. What difference does the factor of $\frac{1}{2}$ make and what does it mean?

6.39 Show $D^+(\alpha)\,\hat{b}^+D(\alpha) = \hat{b}^+ + \alpha^*$ by using the adjoint and also by using the operator expansion theorem.

6.40 Show that the displacement operator acting on a nonzero coherent state produces another coherent state with the sum of two amplitudes and a complex phase factor.

$$\hat{D}(\alpha)\hat{D}(\beta)|0\rangle \sim |\alpha + \beta\rangle$$

Hint: Combine exponentials using the Campbell–Baker–Hausdorff theorem.

6.41 Suppose two coherent states $|\alpha\rangle$ and $|\beta\rangle$ each have an average of 25 photons.

1. If Phase(α) = Phase(β), find the inner product between them.

2. If Phase(α) = Phase(β) + $180°$, find the inner product between them.

3. Find a finite nonzero vector orthogonal to the coherent state $|\alpha\rangle = e^{-|\alpha|^2/2}\sum_{n=0}^{\infty}(\alpha^n/\sqrt{n!})|n\rangle$. Can the vector be normalized and still remain orthogonal? Explain in detail whether or not the vector is a coherent state.

6.42 Explain why $\int dQ\,|Q\rangle\langle Q| = 1$ but $\int d\alpha\,|\alpha\rangle\langle\alpha| = \pi$ even though $\{|Q\rangle\}$ and $\{|\alpha\rangle\}$ have uncountably infinite numbers.

6.43 Show $(\Delta P)^2 = \sigma_P^2 = 1/2$ for the coherent state $|\alpha\rangle$.

6.44 Show the variance of the electric field $\sigma_E^2 = \frac{\hbar\omega}{2\varepsilon_0 V}$ for the coherent state.

6.45 Show $\hat{S}^+\hat{b}^+\hat{S} = \hat{b}^+\cosh(r) - \hat{b}\,e^{-i\theta}\sinh(r)$ using the operator expansion theorem.

6.46 Show $\hat{S}^+\hat{P}\hat{S} = \frac{1}{i\sqrt{2}}\hat{S}^+(\hat{b} - \hat{b}^+)\hat{S} = \hat{P}\cosh(r) - \hat{P}_R\sinh(r)$ where $\hat{P}_R = \frac{\hat{b}\,e^{-i\theta} - \hat{b}^+e^{i\theta}}{i\sqrt{2}}$

6.47 Show $\sigma_P^2 = \langle 0, \eta|\hat{P}^2|0, \eta\rangle = \frac{1}{2}\{\cosh^2(r) + \sinh^2(r) + 2\cos(\theta)\sinh(r)\cosh(r)\}$ for the squeezed state $|0, \eta\rangle$.

6.48 Starting with $\sigma_E^2 = \langle\alpha, \eta|\hat{E}^2|\alpha, \eta\rangle - \langle\alpha, \eta|\hat{E}|\alpha, \eta\rangle^2$, show

$$\sigma_E^2\big|_{\mathrm{sqz}} = \langle\alpha, \eta|\hat{E}^2|\alpha, \eta\rangle - \langle\alpha, \eta|\hat{E}|\alpha, \eta\rangle^2 = \frac{\hbar\omega}{2\varepsilon_0 V}\{2\cos(\theta)\cosh(r)\sinh(r) + 1\}$$

You will need to use the unitary properties of $\hat{D}$ and $\hat{S}$ for terms such as for $\hat{D}^+\hat{b}^2\hat{D} = (\hat{D}^+\hat{b}\hat{D})(\hat{D}^+\hat{b}\hat{D}) = (\hat{b} + \alpha)(\hat{b} + \alpha)$ and $\hat{S}^+\hat{b}\hat{b}^+\hat{S} = \hat{S}^+\hat{b}\hat{S}\,\hat{S}^+\hat{b}^+\hat{S} = \hat{b}_s\hat{b}_s^+$. Also note certain "squared" terms produce zero such as $\langle 0|\hat{b}^+\hat{b}|0\rangle = 0$. You will need the elementary relation $\cosh^2 r - \sinh^2 r = 1$ and the fact that $\langle 0|\hat{b}\hat{b}^+|0\rangle = 1$.

6.49 What can you say about the vacuum expectation values of odd powers of the form b_s^X, $b_s^X b_s^Y b_s^Z$ and so on, where X, Y, and Z represent "+" or "−" for creation and annihilation operators, respectively.

6.50 Find the variance of the number operator for the following state: Q-squeeze the vacuum and displace it to $\bar{Q}$ on the Q axis.

6.51 For the vacuum Wigner function, show

$$W(Q, P) = \frac{1}{2\pi^{3/2}}\int_{-\infty}^{\infty} dx\,\exp(iPx)\exp\left[-\frac{1}{2}\left(Q + \frac{x}{2}\right)^2\right]\exp\left[-\frac{1}{2}\left(Q - \frac{x}{2}\right)^2\right]$$

$$= \frac{1}{\pi}\exp(-Q^2 - P^2)$$

6.52 Starting with the definition of trace for a discrete basis $\text{Tr}(\hat{A}) = \sum_n \langle n | \hat{A} | n \rangle$, derive the formula for the coordinate representation $\text{Tr}(\hat{A}) = \int dx \, \langle x | \hat{A} | x \rangle$.

6.53 Derive the Wigner distribution for the $n=1$ Fock state

$$W(P, Q) = \frac{\sqrt{2}|c_1|^2}{\sqrt{2\pi}} e^{-(P^2+Q^2)}(P^2 + Q^2) - \frac{|c_1|^2}{2\sqrt{2\pi}} e^{-(P^2+Q^2)}$$

6.54 Derive the Wigner distribution for the $n=2$ Fock state.

6.55 Show $\hat{D}^+(\alpha)\hat{Q}\hat{D}(\alpha) = \hat{Q} + \bar{Q}$ and $\hat{D}^+(\alpha)\hat{P}\hat{D}(\alpha) = \hat{P} + \bar{P}$ and its significance in terms of the Wigner plot.

6.56 Read and write a summary of the following publication.

- Johnston, Photon states made easy: a computational approach to quantum radiation theory, *Am. J. Phys.* 64, 245 (1996).

6.57 Number squeezed light has been widely reported from lasers and LEDs. Read the following journal papers and summarize your findings. List experimental conditions.

- Richardson *et al.*, Squeezed photon-number noise and sub-Poissonian electrical partition noise in a semiconductor laser, *Phys. Rev. Lett.*, 66, 2867 (1991).
- Richardson *et al.*, Nonclassical light from a semiconductor laser operating at 4 K, *Phys. Rev. Lett.* 64, 400 (1990).
- Teich, *J. Opt. Soc. Am.* B2, 275(85).
- Machida *et al.*, Observation of amplitude squeezing in a constant-current-driven semiconductor laser, *Phys. Rev. Lett.* 58, 1000 (1987).
- Kitching and Yariv, Room temperature generation of amplitude squeezed light from a semiconductor laser with weak optical feedback, *Phys. Rev. Lett.* 74, 3372 (1995).
- Freeman *et al.*, Wavelength-tunable amplitude-squeezed light from a room temperature quantum-well laser, *Opt. Lett.* 18, 2141 (1993).

6.58 Greater number squeezing can be achieved using multiple optical sources. Read the following publication and summarize your findings.

- Sumitomo *et al.*, Wideband deep penetration of photon-number fluctuations into the quantum regime in series-coupled light-emitting diodes, *Opt. Lett.* 24, 40 (1999).

6.59 Number squeezed light can improve communications and the bit error rate. Read the following publications and summarize your findings.

- Saleh and Teich, Information transmission with photon-number-squeezed light, *Proc. IEEE*, 80, No. 3, 451 (1992).
- Mortensen, Amplitude-squeezed light promises quiet devices, *Laser Focus World*, November, p. 32 (1997).

6.60 The transverse-junction-stripe laser might be better for producing number squeezed light. Read the following journal publication and summarize your findings.

- Lathi *et al.*, Transverse-junction-stripe GaAs-AlGaAs lasers for squeezed light generation,' *J. Quant. Electr.* 35, 387 (1997).

6.61 A number of methods have been used to generate number squeezed light. Read the following explanations for its production and summarize.

- Abe *et al.*, Observation of the collective Coulomb blockade effect in a constant-current-driven high-speed light-emitting diode, *J. Opt. Soc. Am. B* 14, 1295 (1997).
- Kakimoto *et al.*, Laser diodes in photon number squeezed state, *J. Quant. Electr.* 33, 824 (1997).
- Kim *et al.*, Macroscopic Coulomb-blockade effect in a constant-current-driven light-emitting diode, *Phys. Rev. B* 52, 2008 (1995).
- Imamoglu *et al.*, Noise suppression in semiconductor p-i-n junctions: transition from macroscopic squeezing to mesoscopic Coulomb blockade of electron emission processes, *Phys. Rev. Lett.*, 70, 3327 (1993).

6.62 Summarize the points made in the following publication.

- Yuen and Chan, Noise in homodyne and heterodyne detection, *Opt. Lett.* 8, 177 (1983).

6.63 Lasers exhibit spontaneous emission noise. Some methods have been suggested to reduce the spontaneous emission. Read the following articles and summarize.

- Feld and An, The Single-Atom Laser, *Scientific American*, July 1998, p. 57.
- Yablonovitch, Inhibited spontaneous emission in solid-state physics and electronics, *Phys. Rev. Lett.* 58, 2059 (1987).
- Yablonovitch, Inhibited and enhanced spontaneous emission from optically thin AlGaAs/GaAs double heterostructures, *Phys. Rev. Lett.* 61, 2546 (1988).

6.20 Further Reading

Many excellent texts on Quantum Optics and Quantum Electrodynamics (QED) are available. They contain a wealth of references to journal publications.

Vector Potential and Gauges

1. Sakurai J.J., *Advance Quantum Mechanics*, Addison-Wesley Publishing, Reading, MA, 1980.
2. Jackson J.D., *Classical Electrodynamics*, 2nd ed., John Wiley & Sons, New York, 1975.
3. Greiner W., Reinhardt J., *Field Quantization*, Springer, Berlin, 1993.

Quantum Optics

4. Mandel L. and Wolf E., *Optical Coherence and Quantum Optics*, Cambridge University Press, Cambridge, 1995.
5. Leonhardt U., *Measuring the Quantum State of Light*, Cambridge University Press, Cambridge, 1997.
6. Carmichael H.J., *Statistical Methods in Quantum Optics 1, Master Equations and Fokker–Planck Equations*, Springer, Berlin, 1999.

7. Kim J., Somani S., Yamamoto Y., *Nonclassical Light from Semiconductor Lasers and LEDs*, Springer, Berlin, 2001.
8. Meystre P., Sargent III M., *Elements of Quantum Optics*, 3rd ed., Springer, Berlin, 1999.
9. Walls D.F., Milburn G.J., *Quantum Optics*, Springer, Berlin, 1995.
10. Haus H.A., *Electromagnetic Noise and Quantum Optical Measurments*, Springer, Berlin, 2000.
11. Bachor H.A., *A Guide to Experiments in Quantum Optics*, John Wiley & Sons, New York, 1998.
12. Schleich W.P., *Quantum Optics in Phase Space*, John Wiley & Sons, New York, 2001.

7

Matter–Light Interaction

Lasers, LEDs and photodetectors all share the same basic physical principles. The rate equations provide a fundamental description of optical emission and absorption. However, as in Chapter 2, the gain and spontaneous recombination terms often appear as phenomenological terms. Maxwell's equations and the Poynting vector explain emission, absorption and transport in terms of the classical theory of the microscopic dipoles. While quite successful, the description does not account for the quantum nature of matter and light, and does not explain basic phenomena such as the spontaneous emission.

We now explore the matter-light interaction culminating in a quantum description of gain and the rate equations. The study begins with the time-dependent perturbation theory and the semiclassical approach to Fermi's golden rule for optical transitions. Fermi's golden rule relates the transition rate to the optical power, frequency and dipole moment. The time-dependent perturbation theory for electromagnetic interactions uses a Hamiltonian $\hat{\mathscr{H}} = \hat{\mathscr{H}}_o + \hat{V}$ with $\hat{\mathscr{H}}_o$ representing the atomic system such as an atom or a quantum well for example. The interaction Hamiltonian $\hat{V}$ describes the effect of the electromagnetic wave interacting with the atomic system. The theory assumes that $\hat{V}$ does not change the energy basis states obtained from $\hat{\mathscr{H}}_o$ but rather induces particle transitions between them. The semiclassical approach does not use the quantized form of the electromagnetic field and, for this reason, the full Hamiltonian $\hat{\mathscr{H}}$ does not include the free-field Hamiltonian $\hat{\mathscr{H}}_{em}$. The semiclassical approach does not account for spectral broadening and gain saturation due to the interaction of the radiating system with its environment. Sections 7.4 through 7.6 discuss Hamilton's and Lagrange's classical formulation for the EM interaction and shows how the vector potential modifies Schrodinger's equation and produces the dipole moment.

The density operator describes the quantum and classical state of the particle and field system. This operator lives in the tensor product space consisting of (at minimum) the product of the spaces for the matter and field. The number of spaces can increase depending on the number of degrees of freedom. For example, the wave functions and density operator describing the state of the reservoirs reside in their own Hilbert space. The combination of the matter, fields and reservoirs constitutes a complete system. The density operator appears in both the semiclassical and the full quantum theory.

The motion of the density operator in its direct product space describes the system transitions. An atom, for example, might have an electron in the first excited state but, through the matter–field interaction, it transitions to the lower state, emits a photon and excites one of the optical modes. A first order partial differential equation, known as the Liouville equation or the master equation, describes the system dynamics. Rather than working with complicated environmental Hamiltonians, the Liouville equation uses phenomenological terms for the environmental relaxation effects while retaining the semiclassical theory for the matter–field interaction. The relaxation terms do not maintain phase coherence in the wavefunction and produce spectral broadening (homogeneous broadening) and naturally lead to gain saturation.

The remainder of chapter discusses the Jaynes-Cummings model and begins to explore the fully quantized model for the electromagnetic interaction. This material treats the interactions in terms of reservoirs and derives the Liouville equation.

7.1 Introduction to the Quantum Mechanical Dipole Moment

Matter and fields can interact by a variety of mechanisms. Electric dipoles emit and absorb fields and therefore produce the refractive index, gain and absorption. The classical and quantum mechanical descriptions of the matter–field interaction therefore, both incorporate the dipole at a fundamental level. In the classical theory, absorption occurs when an incident wave induces a dipole moment along the direction of the wave polarization and then the surrounding medium dissipates the energy. Emission occurs when an excited dipole synchronously radiates energy with its motion. Maxwell's wave equation incorporates the dipoles in terms of the susceptibility. The quantum theory of the matter–field interaction uses the operator form of the dipole. Subsequent sections on Fermi's golden rule and the density matrix formalism use the dipole moment as part of the interaction Hamiltonian.

The present section compares the classical and quantum descriptions of the electric dipole. It introduces how the quantum theory describes transitions by matrix elements of the dipole in the energy basis. Those matrix elements connecting distinct states can produce transitions. Finally, the section provides a picture of an oscillating charge in a quadratic potential by using the coherent state of the electron.

7.1.1 Comparison of the Classical and Quantum Mechanical Dipole

The classical electric dipole moment $\vec{\mu}$ describes two charges separated by a distance R

$$\vec{\mu} = q\vec{R} \tag{7.1.1}$$

where $q < 0$ for an electron as described in Chapter 3 and shown in Figure 7.1.1. Chapter 3 discusses optics in terms of the induced or oscillating electric dipole moment. A classically static dipole does not move nor does the separation between the charges change. This type of permanent dipole does not affect the usual optical properties of a material.

The dipole can also be described in terms of a charge distribution according to

$$\vec{\mu} = \int d^3r\, \vec{r}\, \rho(\vec{r}) \tag{7.1.2}$$

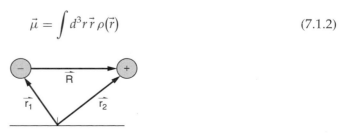

FIGURE 7.1.1
An example of the classical dipole with two point charges.

For discrete charges as shown in Figure 7.1.1, we can recover equation 7.1.1 by using the Dirac delta function for the distribution

$$\rho = \sum_i q_i \delta(\vec{r} - \vec{r}_i) \tag{7.1.3}$$

and then substituting into Equation 7.1.2

$$\vec{\mu} = \int d^3r\, \vec{r}\, \rho(\vec{r}) = q_1\vec{r}_1 + q_2\vec{r}_2 = e(\vec{r}_2 - \vec{r}_1) = e\vec{R} \tag{7.1.4}$$

where $e > 0$ represents the elementary charge. The classical dipole moment can be related to the quantum mechanical one as follows. Suppose that a massive positive charge (such as an atomic core) is located at the origin so that $\vec{r}_2 = 0$. Similar to the procedure used in Section 5.9, we view the position of the electron as smeared out in a manner consistent with its wave function (at least in the statistical sense).

$$\rho = -e\, \psi^* \psi \tag{7.1.5}$$

so that the integral in Equation 7.1.3 can be written as

$$\vec{\mu} = \int d^3r\, \psi^*(-e\vec{r})\, \psi = \langle \psi | q\vec{r} | \psi \rangle \tag{7.1.6}$$

where $q = -e$ for electrons.

In the quantum theory, the dipole $\vec{\mu} = q\vec{r}$ becomes the dipole operator in the coordinate representation. An electron in an energy eigenstate has an average dipole moment of

$$\mu_{nn} = \langle n | \mu | n \rangle \tag{7.1.7}$$

Wave functions with definite parity, such as the atomic S or P orbitals, do not have permanent dipoles. For example, the spherically symmetric electron wave function in Figure 7.1.2 does not produce a dipole moment. For the 1-D case, we can calculate the expected dipole moment as

$$\langle a | \hat{\mu}_x | a \rangle = q \langle u_a | x | u_a \rangle = q \int_{-\infty}^{\infty} dx\, \underbrace{u_a^*(x)\, x\, u_a(x)}_{\leftarrow odd \rightarrow} = 0$$

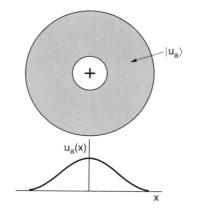

FIGURE 7.1.2
Spherically symmetric wave functions do not have a permanent dipole moment.

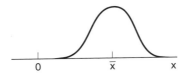

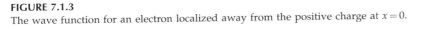

FIGURE 7.1.3
The wave function for an electron localized away from the positive charge at $x = 0$.

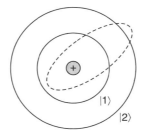

FIGURE 7.1.4
Cartoon view of an atomic dipole.

However, wave functions without symmetry can have a nonzero electric dipole moment. For example, consider the situation depicted in Figure 7.1.3 where the electron mostly lives on the right hand side and the probability distribution has the form of a Gaussian

$$\psi = \left(\frac{1}{\sqrt{2\pi}\,\sigma} \right)^{1/2} e^{-\frac{(x-\bar{x})^2}{4\sigma^2}} \quad \rightarrow \quad \psi^*\psi = \frac{1}{\sqrt{2\pi}\,\sigma} e^{-\frac{(x-\bar{x})^2}{2\sigma^2}} \tag{7.1.8}$$

where $\bar{x}, \sigma$ represent the center of the distribution and the approximate width (standard deviation) respectively. The dipole moment becomes $\langle \psi | qx | \psi \rangle = q\bar{x}$ by definition of the average position $\bar{x}$.

The average value of an operator provides the classically expected value. The average most generally uses the density operator as discussed in Section 5.11.

$$\langle \hat{\mu} \rangle = Tr(\hat{\rho}\hat{\mu}) \tag{7.1.9}$$

For a pure state $\hat{\rho} = |\psi\rangle\langle\psi|$, Equation 7.1.9 reduces to Equation 7.1.6. The polarization can be written as

$$\vec{\mathscr{P}} = \frac{\text{Total Dipole Moment}}{\text{Vol}} = N\langle\hat{\mu}\rangle = N\,Tr(\hat{\rho}\hat{\mu}) \tag{7.1.10}$$

where N represents the number of dipoles per unit volume. Assuming a 2 level atom and expanding the trace produces

$$\vec{\mathscr{P}} = N\,Tr(\hat{\rho}\hat{\mu}) = N\{\rho_{11}\mu_{11} + \rho_{12}\mu_{21} + \cdots\} = N\{\rho_{12}\mu_{21} + \rho_{21}\mu_{12}\} \tag{7.1.11}$$

where we assume states of definite parity do not have dipole moments $\mu_{11} = 0 = \mu_{22}$.

7.1.2 The Quantum Mechanical Dipole Moment

A perturbing potential can change or induce a dipole moment. Imagine an atom system with circular orbits (say levels 1 and 2) for example. We might imagine an electric field modifying the normally circular orbits into elliptical ones, as shown in Figure 7.1.4. The elliptical orbit might be considered some linear combination of the circular ones. The electron spends as some time in both the 1st and 2nd circular orbit. If the electric field fluctuates at the optical frequency and if we make a measurement of the electron energy, then we might find the electron in either the 1st or 2nd circular orbit. If the electron started in the 1st orbit, then the electron might transition to the second. The quantum theory accurately models this situation by assuming that the orbitals remain fixed but an EM field produces an electron wave function with a mixture of the two orbitals. The dipole moment develops between the two states $\langle 2|\hat{\mu}|1\rangle$.

Section 7.1.1 shows the expected dipole moment has the form

$$\langle \hat{\mu} \rangle = Tr(\hat{\rho}\hat{\mu}) = \rho_{12}\mu_{21} + \rho_{21}\mu_{12} \qquad (7.1.9)$$

for a 2 level atom where we assume states of definite parity do not have dipole moments $\mu_{11} = 0 = \mu_{22}$.

Consider the off-diagonal elements of the dipole matrix element. These induced dipole moments are given by

$$\langle a| \hat{\mu} |b\rangle = \mu_{ab} \quad \text{with} \quad a \neq b$$

and are related to the transition selection rules. The off-diagonal elements of the density matrix might be imagined as the "probability amplitude of finding a particle in two states" at the same time

$$\rho_{ab} = \langle a|\hat{\rho}|b\rangle \sim \langle a \mid \psi \rangle \langle \psi \mid b \rangle = \beta_a \beta_b^* \sim \sqrt{P(a)\,P(b)} \qquad (7.1.10)$$

where P denotes probability and we assume a pure state for simplicity and we denote the wave function by the usual form $|\psi\rangle = \sum \beta_n|n\rangle$. The off-diagonal elements describe the probability of the particle simultaneously existing in two states. Making a measurement of the energy causes the wave function to collapse to either one of the states. Therefore, any perturbation producing off-diagonal elements of the density operator can produce transitions between states. However, we know the induced dipole gives rise to transitions. Therefore we surmise that the off-diagonal elements of the density operator must be related to dipole matrix elements.

Let us now investigate the implications of the induced dipole moment for the polarization. As in 7.1.1, the dipole operator can be written as $\vec{\mu} = q\vec{r}$ or $\hat{\mu} = q\hat{x}$ for 1-D where $q = -e$ for an electron and $e > 0$. As before, the polarization has the form

$$\vec{\mathscr{P}} = \frac{\text{Total Dipole Moment}}{\text{Vol}} = N\langle \hat{\mu} \rangle = N\,Tr(\hat{\rho}\hat{\mu})$$

We assume that a perturbing EM field provides an interaction energy $\hat{V}$ so that

$$\hat{H} = \hat{H}_o + \hat{V} \quad \text{and} \quad \hat{H}|n\rangle = E_n|n\rangle \qquad (7.1.11)$$

where we assume a 2 level atom so that $n = 1,2$. We only need the energy eigenstates of the *unperturbed* Hamiltonian for this calculation. We use the isomorphism between operators and matrices.

$$\vec{\mathscr{P}} = N\,Tr(\hat{\rho}\hat{\mu}) = N\,Tr\,\underline{\rho}\underline{\mu} = N\sum_a \left(\underline{\rho}\underline{\mu}\right)_{aa} = N\{\rho_{12}\mu_{12} + \rho_{21}\mu_{21}\} \qquad (7.1.12)$$

since we assume the permanent dipole has a value of zero (i.e., $\mu_{11} = \mu_{22} = 0$). If we assume (as is usually the case) that

$$\mu = \mu_{12} = \mu_{21} = \text{real}$$

then the equation for the polarization becomes

$$\mathscr{P} = N\{\rho_{12}\mu_{12} + \rho_{21}\mu_{21}\} = N\mu(\rho_{12} + \rho_{21}) \tag{7.1.13}$$

So we see that the off-diagonal elements of the density matrix must be related to the polarization.

We can go a step further by using the fact that the density operator is Hermitian.

$$\hat{\rho} = \hat{\rho}^+ \rightarrow (\hat{\rho})_{12} = (\hat{\rho}^+)_{12} \rightarrow \rho_{12} = \rho_{21}^*$$

Now the polarization formula becomes

$$\mathscr{P} = N\mu\ (\rho_{12} + \rho_{21}) = N\mu(\rho_{12} + \rho_{12}^*) = 2N\mu\ \text{Re}(\rho_{12}) \tag{7.1.14}$$

As a result, the off-diagonal elements of the density operator must be related to polarization $\mathscr{P}$ and the induced dipole moment. The off-diagonal elements produce gain and absorption. This is quite different from the classical picture. In the quantum picture, the states are fixed but the electron wave function becomes a superposition of both which leads to the nonzero dipole moment. This point of view can be traced back to that for the time-dependent perturbation theory.

7.1.3 A Comment on Visualizing an Oscillating Electron in a Harmonic Potential

An electron in a coherent state in a quadratic potential can be pictured as a localized particle executing simple harmonic motion about the symmetry point of the potential. In the case of an electric dipole, we might imagine the force on the electron as proportional to its displacement from $x = 0$. The picture is not accurate for the Coulomb potential but would apply to a point of equilibrium for the electron or a dipole.

Suppose an electron executes simple harmonic motion about an equilibrium point $x = 0$. The Hamiltonian has the form given in Section 5.8

$$\hat{H} = \frac{\hat{p}^2}{2m} + \frac{k}{2}\hat{x}^2 \quad \text{with} \quad \omega = \sqrt{k/m} \tag{7.1.15}$$

Define the quadrature operators $\hat{Q}, \hat{P}$ as

$$\hat{Q} = \frac{\hat{a} + \hat{a}^+}{\sqrt{2}} \quad \hat{P} = -i\frac{\hat{a} - \hat{a}^+}{\sqrt{2}} \quad \text{or} \quad \hat{x} = \hat{Q}\sqrt{\frac{\hbar}{m\omega}} \quad \hat{p} = \hat{P}\sqrt{m\omega\hbar} \tag{7.1.16}$$

where $\hat{a}, \hat{a}^+$ represent the ladder operators and $[\hat{Q}, \hat{P}] = i$.

The electron coherent state $|\alpha\rangle$ can be defined similar to that for the photon by requiring $\hat{a}|\alpha\rangle = \alpha|\alpha\rangle$ where $\alpha = |\alpha|e^{i\phi}$. The coherent state must be a linear combination of the usual energy basis set so that (at $t = 0$)

$$|\alpha\rangle = e^{-|\alpha|^2/2} \sum_n \frac{\alpha^n}{\sqrt{n!}} |n\rangle \tag{7.1.17}$$

The evolution operator with $E_n = \hbar\omega(n + 1/2)$ produces the time-dependent wave function as

$$|\alpha(t)\rangle \equiv e^{\hat{H}t/i\hbar}|\alpha\rangle = e^{i\omega t/2}e^{-|\alpha(t)|^2/2} \sum_n \frac{\alpha(t)^n}{\sqrt{n!}} |n\rangle \tag{7.1.18}$$

where $\alpha(t) \equiv \alpha e^{-i\omega t} = |\alpha|e^{i\phi - i\omega t}$. The center position of the coherent state must oscillate according to

$$\bar{Q}(t) = \langle\alpha(t)|\hat{Q}|\alpha(t)\rangle = \sqrt{2}\,\mathrm{Re}[\alpha(t)] = \sqrt{2}\,|\alpha|\cos(\phi - \omega t) \tag{7.1.19}$$

This explicitly shows that the electron oscillates back and forth across the region of space with the quadratic potential. We finally show that the position of the electron must be normally distributed. So we can consider the electron to be localized similar to a point particle but with a size on the order of the standard deviation σ.

The coherent state can be written as a displacement operator

$$\hat{D}(\alpha) = e^{\alpha\hat{a}^+ - \alpha^*\hat{a}} \quad \text{or} \quad |\alpha\rangle = \hat{D}(\alpha)|0\rangle = e^{\alpha\hat{a}^+ - \alpha^*\hat{a}}|0\rangle \tag{7.1.20}$$

Converting the displacement operator to the position-momentum representation $\hat{D}(\alpha) = e^{i\bar{P}\hat{Q} - i\bar{Q}\hat{P}}$ produces the position-dependent wave function

$$\psi_\alpha(Q) = \langle Q \mid \alpha\rangle = \langle Q|\hat{D}(\alpha)|0\rangle = e^{i\bar{P}Q}e^{-i\bar{P}\bar{Q}/2}\langle Q|e^{-i\bar{Q}\hat{P}}|0\rangle = e^{i\bar{P}Q}e^{-i\bar{P}\bar{Q}/2}u_o(Q - \bar{Q}) \tag{7.1.21}$$

where $u_o(Q - \bar{Q})$ represents the vacuum state displaced to the location $\bar{Q} = \bar{Q}(t)$ and $\bar{P} = \langle\alpha|\hat{P}|\alpha\rangle$. Section 6.11 then provides

$$\psi = \frac{1}{\pi^{1/4}}e^{-(Q-\bar{Q})^2/2} \qquad \psi^*\psi = \frac{1}{\pi^{1/2}}e^{-(Q-\bar{Q})^2} \tag{7.1.22}$$

which shows the electron position is normally distributed with a mean that oscillations in time. Comparing this last expression with that in Equation 7.1.8 shows that $\sigma_Q = (1/\sqrt{2})$. A similar expression can be applied to the P coordinate to provide a Heisenberg relation $\sigma_Q\sigma_P = 1/2$.

The position x oscillates according to

$$x(t) = \langle\alpha(t)|\hat{x}|\alpha(t)\rangle = \sqrt{\frac{2E}{m\omega^2}}\cos(\phi - \omega t) \tag{7.1.23}$$

where

$$E = \langle\alpha|\hat{H}|\alpha\rangle = \langle\alpha|\hbar\omega(\hat{a}^+\hat{a} + 1/2)|\alpha\rangle = \hbar\omega(|\alpha|^2 + 1/2) \sim \hbar\omega|\alpha|^2 \tag{7.1.24}$$

for large $|\alpha|^2$. This last result agrees with Equation 5.9.3b.

7.2   Introduction to Optical Transitions

The time-dependent perturbation theory applied to the interaction of a classical electromagnetic (EM) wave with a quantized atomic system comprises the semiclassical description of the matter-light interaction. A coherent wave interacting with the atomic system produces absorption and emission. However, the semiclassical model does not describe the spontaneous emission (fluorescence) since it does not describe the vacuum fluctuations responsible for this emission. This section finds the probability of a transition while the next one discusses the rate of transition using Fermi's golden rule. The time-dependent perturbation theory and Fermi's golden rule must be augmented by the interaction between the environment and the ensemble of atoms in order to describe the spectral broadening mechanisms. Fermi's golden rule provides a delta function shaped spectral line and not the Gaussian or Lorenzian shape expected from interactions with the reservoirs. Subsequent sections incorporate the broadening mechanisms into the Liouville equation for the density operator.

7.2.1   The EM Interaction Potential

Suppose an electromagnetic wave washes over an atom with a single electron occupying an energy eigenstate. What is the probability that the electron will make an upward or a downward transition to a higher or lower energy level? Interestingly, the frequency of the electromagnetic wave necessary to induce a transition does not necessarily need to be in the optical range—it all depends on the type of "atom." Figure 7.2.1 shows the electromagnetic wave and the electron in the second energy level. If the atom (i.e., electron) absorbs energy from the wave (stimulated absorption) then the electron makes an upward transition. If the wave induces a downward transition (stimulated emission), then the atom releases energy to the bathing field. As mentioned previously, "semiclassical" theory describes the effects of a classical electromagnetic traveling wave. This form of interaction ignores the particle properties (i.e., discrete energy properties) of the electromagnetic wave. As discussed in the previous chapter, the coherent optical states most closely describe classical electromagnetic waves. The present section applies the time-dependent perturbation theory to the interaction of the classical wave to the atomic system.

Classically a material can produce or absorb light when an electromagnetic field interacts with dipoles within the material. The classical expression for the dipole interaction energy can be written as proportional to $\vec{p} \cdot \vec{\mathscr{E}}$ where $\vec{p}, \vec{\mathscr{E}}$ denote the dipole moment and electric field respectively. Quantum mechanically, we represent the interaction energy by operators such as $\hat{V} = \hat{\mu}\mathscr{E}$. The "dipole moment" operator $\hat{\mu}$, which is Hermitian, describes the strength of the interaction between the oscillating electric field

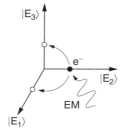

FIGURE 7.2.1
The EM wave can induce upward and downward transitions.

and the atom. People often write the interaction energy in the explicitly Hermitian form $\hat{V} = (1/2)\{\hat{\mu}\mathscr{E} + (\hat{\mu}\mathscr{E})^+\}$. With this form, a complex field $E = E_o e^{-i\omega t}$ does not affect the Hermiticity of the interaction energy. The two terms provide specific results. The first term $(\hat{\mu}\mathscr{E})$ corresponds to absorption when using the complex field and the second term $(\hat{\mu}\mathscr{E})^+$ corresponds to emission. Similar to the basis sets or the eigenvector expansion of a Hermitian operator, the interaction energy must contain all possible outcomes.

Let's assume that the unperturbed Hamiltonian $\hat{\mathscr{H}}_o$ describes an "atom" (located at the origin of the coordinate system) without any incident EM wave. The Hermitian interaction energy can be written as

$$\hat{\mathscr{V}}(x, t) = \begin{cases} \hat{\mu}(x)\dfrac{E_o}{2}e^{-i\omega t} + \left[\hat{\mu}(x)\dfrac{E_o}{2}e^{-i\omega t}\right]^+ & t \geq 0 \\ \\ 0 & t < 0 \end{cases} \tag{7.2.1}$$

which provides a small perturbation for the full Hamiltonian $\hat{\mathscr{H}} = \hat{\mathscr{H}}_o + \hat{\mathscr{V}}(x, t)$. Some books set $\hat{V} = \hat{\mu}E_o/2$. We assume that the angular frequency of the incident electromagnetic field is always positive i.e. $\omega > 0$. The matrix elements of the dipole operator $\hat{\mu}$ will be real constants of proportionality (refer to Section 7.1). We also assume a real amplitude E_o for the oscillating electric field. We see from Equation 7.2.1 that the interaction potential must be Hermitian $\hat{\mathscr{V}} = \hat{\mathscr{V}}^+$ and therefore it must be an observable. Equation 7.2.1 can be rewritten as

$$\hat{\mathscr{V}}(x, t) = \hat{\mu}(x)\frac{E_o}{2}\left(e^{-i\omega t} + e^{+i\omega t}\right) = \hat{\mu}(x)E_o \cos(\omega t) \tag{7.2.2}$$

We can see that the interaction potential must be Hermitian from this last expression by noting the dipole moment operator $\hat{\mu}$ must be Hermitian. The reader should realize that a phase factor could be added to the exponential term in the interaction energy to obtain a sine wave rather than the cosine wave. As it is appropriate for time-dependent perturbation theory, assume the set $\{|u_n\rangle = |n\rangle\}$ contains the energy eigenvectors for the unperturbed Hamiltonian.

7.2.2 The Integral for the Probability Amplitude

In Section 5.10, we show the wave function

$$|\Psi(t)\rangle = \sum_n \beta_n(t)\,|n\rangle \tag{7.2.3}$$

approximately satisfies Schrodinger's equation

$$\hat{\mathscr{H}}|\Psi(t)\rangle = i\hbar\frac{\partial}{\partial t}|\Psi(t)\rangle \tag{7.2.4}$$

provided

$$\beta_n(t) = \begin{cases} e^{-i\omega_i t} + \dfrac{1}{i\hbar}e^{-i\omega_i t}\displaystyle\int_o^t d\tau\,\mathscr{V}_{ii}(\tau) + \cdots & n = i \\ \\ \dfrac{1}{i\hbar}e^{-i\omega_n t}\displaystyle\int_o^t d\tau\,e^{i\omega_{ni}\tau}\mathscr{V}_{ni}(\tau) + \cdots & n \neq i \end{cases} \tag{7.2.5}$$

where $\omega_{ni} = (E_n/\hbar) - (E_i/\hbar)$ and the electron is assumed to start in state $|i\rangle$. For example, $|i\rangle = |2\rangle$ for Figure 7.2.1. Recall that the component $\beta_n(t)$ of the vector essentially describes the probability of finding the electron in state $|n\rangle$ after a time "t." Obviously therefore, the component $\beta_n(t)$ must be related to the probability (and the transition rate) of the electron making a transition from state $|i\rangle$ to state $|n\rangle$ since the electron started in state $|i\rangle$.

$$\text{Prob}(i \rightarrow n) = \left|\beta_n(t)\right|^2 \tag{7.2.6}$$

We can take the case of either $n=i$ or $n \neq i$. The case of $n=i$ provides the probability that the particle will *not* make a transition. Although interesting in itself, we have more interest in the case of $n \neq i$. We can find the rate of transition for the gain by taking the time derivative of the probability

$$R_{i \rightarrow n} = \frac{d}{dt} \text{Prob}(i \rightarrow n) \tag{7.2.7}$$

To find the probability and rate of transition (to first-order approximation) for the case of $n \neq i$, we must calculate the integral in

$$\beta_n(t) = \frac{1}{i\hbar} e^{-i\omega_n t} \int_o^t d\tau\, e^{i\omega_{ni}\tau}\, V_{ni}(x,\tau) = \frac{1}{i\hbar} e^{-i\omega_n t} \int_o^t d\tau\, e^{i\omega_{ni}\tau} \langle n|\hat{V}(x,\tau)|i\rangle \tag{7.2.8}$$

from Equation 5.1.23. Notice in the matrix element $\langle n|\hat{\mathcal{V}}|i\rangle$ how the perturbation induces a transition from *right* to *left*. The reader should keep in mind that ω represents the angular frequency of the electromagnetic wave whereas ω_{ni} denotes the angular frequency corresponding to the difference in energy. In chapter 1, we would have said that the atom requires light to have angular frequency $\omega_{ni} = (E_n - E_i)/\hbar$ in order for the atom to participate in stimulated absorption or emission. However, in this section we can have $\omega \neq \omega_{ni}$. The integral in Equation 7.2.8 can be evaluated by substituting Equation 7.2.1 to get

$$\beta_n(t) = \frac{1}{i\hbar} e^{-i\omega_n t} \int_o^t d\tau e^{i\omega_{ni}\tau} \langle n|\hat{V}(x,\tau)|i\rangle$$

$$= \frac{1}{i\hbar} e^{-i\omega_n t} \int_o^t d\tau e^{i\omega_{ni}\tau} \langle n| \left\{ \hat{\mu}\, \frac{E_o}{2} e^{-i\omega\tau} + \left[\hat{\mu}\, \frac{E_o}{2} e^{-i\omega\tau} \right]^+ \right\} |i\rangle$$

Now calculate the adjoint, distribute the projection operator and the ket through the braces, and use the definition

$$\langle n|\hat{\mu}|i\rangle = \mu_{ni} \tag{7.2.9}$$

to find

$$\beta_n(t) = \frac{1}{i\hbar} e^{-i\omega_n t} \int_o^t d\tau e^{i\omega_{ni}\tau} \left\{ \mu_{ni} \frac{E_o}{2} e^{-i\omega\tau} + \mu_{ni} \frac{E_o}{2} e^{+i\omega\tau} \right\}$$

Keep in mind that the matrix element μ_{ni} is just a constant of proportionality that describes the strength of the interaction between the impressed electromagnetic field and the atom. This nontrivial induced-dipole matrix element μ_{ni} provides the "transition selection rules" and will be explored in more detail later. Factoring out the constant values from the integral yields

$$\beta_n(t) = \frac{1}{i\hbar} e^{-i\omega_n t} \mu_{ni} \frac{E_o}{2} \int_o^t d\tau \left[e^{i(\omega_{ni}-\omega)\tau} + e^{i(\omega_{ni}+\omega)\tau} \rightleftharpoons \right] \qquad (7.2.10)$$

Performing the integration provides

$$\beta_n(t) = \frac{-1}{\hbar} e^{-i\omega_n t} \mu_{ni} \frac{E_o}{2} \left[\frac{e^{i(\omega_{ni}-\omega)t} - 1}{\omega_{ni} - \omega} + \frac{e^{i(\omega_{ni}+\omega)\tau} - 1}{(\omega_{ni} + \omega)} \right] \qquad (7.2.11)$$

Equation 7.2.11 contains terms for both absorption and emission of light. The denominators show that the first term dominates when $\omega \cong \omega_{ni}$ and the second term dominates when $\omega \cong -\omega_{ni}$. Recalling the definition

$$\omega_{ni} = \frac{E_n}{\hbar} - \frac{E_i}{\hbar} \qquad (7.2.12)$$

and the fact that the angular frequency of the incident light must always be positive $\omega > 0$, we see that the first term in Equation 7.2.11 corresponds to the absorption of light since

$$0 < \omega \cong \omega_{ni} = \frac{E_n}{\hbar} - \frac{E_i}{\hbar} \rightarrow E_n \geq E_i \qquad (7.2.13)$$

so that the energy of the final must be larger than the energy of the initial state (see Figure 7.2.2). The second term in Equation 7.2.11 corresponds to emission since

$$0 < \omega \cong -\omega_{ni} = \frac{E_i}{\hbar} - \frac{E_n}{\hbar} \rightarrow E_i \geq E_n \qquad (7.2.14)$$

so that the initial state, in this case, has a larger energy than the final state which can only happen when the atom emits a photon. The figure should remind the reader of the two-level atoms discussed in the first 2 chapters. Although we use the denominators of Equation 7.2.11 to determine which term corresponds to absorption and emission, another method consists of looking at the arguments of the exponential functions in Equation 7.2.10. We come back to the problem of calculating the probability of absorption and emission after a brief interlude for the monumentally important subject of the *rotating wave approximation*.

7.2.3 Rotating Wave Approximation

We wish to evaluate integrals such as

$$\beta_n(t) = \frac{1}{i\hbar} e^{-i\omega_n t} \mu_{ni} \frac{E_o}{2} \int_o^t d\tau \left[e^{i(\omega_{ni}-\omega)\tau} + e^{i(\omega_{ni}+\omega)\tau} \right] \qquad (7.2.15)$$

The exponentials have arguments that correspond to very high frequencies or very low frequencies. For example, when $\omega \cong \omega_{ni}$, we see that the first exponential has approximately constant value while the second one has frequency $\omega + \omega_{ni} \cong 2\omega_{ni}$.

There are two methods to evaluate integrals with "slow" and "fast" functions. 7.2.2 shows one method of evaluating the integral in Equation 7.2.15

$$\beta_n(t) = \frac{-1}{\hbar} e^{-i\omega_n t} \mu_{ni} \frac{E_o}{2} \left[\frac{e^{i(\omega_{ni}-\omega)t} - 1}{\omega_{ni} - \omega} + \frac{e^{i(\omega_{ni}+\omega)\tau} - 1}{(\omega_{ni} + \omega)} \right] \tag{7.2.16}$$

and neglecting terms based on the size of the denominator. When the angular frequency of the wave ω approximately matches the atomic resonant frequency $\omega \cong \omega_{ni}$ then the first term in Equation 7.2.16 dominates the second term. Of course, we could also have $\omega \cong -\omega_{ni}$, in which case the second term dominates by virtue of the denominator.

In the second method (refer to Appendix 7), the rotating wave approximation, averages a sinusioidal wave over many cycles and finds a result of zero. This method applies to integral of the form

$$\beta_n(t) = \frac{1}{i\hbar} e^{-i\omega_n t} \mu_{ni} \frac{E_o}{2} \int_o^t d\tau \left[e^{i(\omega_{ni}-\omega)\tau} + e^{i(\omega_{ni}+\omega)\tau} \right] \tag{7.2.17}$$

Using, for example, $\omega \cong \omega_{ni}$ requires $\exp\{i(\omega_{ni} - \omega)t\}$ to be approximately constant while $\exp\{i(\omega_{ni} + \omega)t\}$ must be a high-frequency sinusoidal wave. The integral looks very similar to an average from calculus given by

$$\langle f \rangle = \frac{1}{t} \int_o^t dt' \ f(t') \tag{7.2.18}$$

If over the interval $(0, t)$, the first integrand in Equation 7.2.17 doesn't change much, then its integral will be nonzero. On the other hand, the second term runs through many oscillations (rotating wave) and the average over the interval (Equation 7.2.18) yields zero.

7.2.4 Absorption

Now return to the calculation for the probability of a transition. We first consider the case for absorption where $\omega \cong \omega_{ni}$. We have Equation 7.2.11 as

$$\beta_n(t) = \frac{-1}{\hbar} e^{-i\omega_n t} \mu_{ni} \frac{E_o}{2} \left[\frac{e^{i(\omega_{ni}-\omega)t} - 1}{\omega_{ni} - \omega} + \frac{e^{i(\omega_{ni}+\omega)\tau} - 1}{(\omega_{ni} + \omega)} \right]$$

from Equation 7.2.10

$$\beta_n(t) = \frac{1}{i\hbar} e^{-i\omega_n t} \mu_{ni} \frac{E_o}{2} \int_o^t d\tau \left[e^{i(\omega_{ni}-\omega)\tau} + e^{i(\omega_{ni}+\omega)\tau} \right]$$

The rotating wave approximation allows us to drop the second term in Equations 7.2.10 and 7.2.11. Therefore, absorption produces the time-dependent probability amplitude

$$\beta_n(t) = \frac{-1}{\hbar} e^{-i\omega_n t} \mu_{ni} \left[\frac{E_o}{2} \frac{e^{i(\omega_{ni}-\omega)t} - 1}{\omega_{ni} - \omega} \right] \tag{7.2.19}$$

Recall that β_n represents the component of the wave function parallel to the $|n\rangle$ axis. The component β_n depends on time in a nontrivial manner and causes the wave function to move away from the ith axis and move closer to the nth axis.

We can find the transition probability for absorption

$$
\begin{aligned}
\text{Prob}(i \to n) = |\beta_n|^2 &= \left(\frac{\mu_{ni}E_o}{2\hbar}\right)^2 \frac{\left[e^{i(\omega_{ni}-\omega)t}-1\right]\left[e^{i(\omega_{ni}-\omega)t}-1\right]^*}{(\omega_{ni}-\omega)^2} \\
&= \left(\frac{\mu_{ni}E_o}{2\hbar}\right)^2 \frac{2 - e^{i(\omega_{ni}-\omega)t} - e^{-i(\omega_{ni}-\omega)t}}{(\omega_{ni}-\omega)^2}
\end{aligned}
\tag{7.2.20}
$$

Using the trigonometric identities,

$$
e^{i(\omega_{ni}-\omega)t} + e^{-i(\omega_{ni}-\omega)t} = 2\cos[(\omega_{ni}-\omega)t] \quad \text{and} \quad \cos(2\theta) = \cos^2(\theta) - \sin^2(\theta) = 1 - 2\sin^2(\theta)
$$

where $\theta = (\omega_{ni}-\omega)t/2$ in the last equation provides the probability of an upward transition

$$
\text{Prob}_{\text{abs}}(i \to n) = |\beta_n|^2 = \left(\frac{\mu_{ni}E_o}{\hbar}\right)^2 \frac{\sin^2\left[\frac{1}{2}(\omega_{ni}-\omega)t\right]}{(\omega_{ni}-\omega)^2}
\tag{7.2.21}
$$

Before discussing this last result, we consider the case for stimulated emission.

7.2.5 Emission

The case for emission obtains when $\omega \cong -\omega_{ni} > 0$. Equation 7.2.10

$$
\beta_n(t) = \frac{1}{i\hbar} e^{-i\omega_n t} \mu_{ni} \frac{E_o}{2} \int_0^t d\tau \left[e^{i(\omega_{ni}-\omega)\tau} + e^{i(\omega_{ni}+\omega)\tau}\right]
$$

gives Equation 7.2.11

$$
\beta_n(t) = \frac{-1}{\hbar} e^{-i\omega_n t} \mu_{ni} \left[\frac{E_o}{2}\frac{e^{i(\omega_{ni}-\omega)t}-1}{\omega_{ni}-\omega} + \frac{e^{i(\omega_{ni}+\omega)\tau}-1}{(\omega_{ni}+\omega)}\right]
$$

The rotating wave approximation allows us to drop the first term in Equations 7.2.10 and 7.2.11. Therefore, for emission, the component of the wave function parallel to the $|n\rangle$ axis (the probability amplitude) must be

$$
\beta_n(t) = \frac{-1}{\hbar} e^{-i\omega_n t} \mu_{ni} \left[\frac{E_o}{2}\frac{e^{i(\omega_{ni}+\omega)t}-1}{\omega_{ni}+\omega}\right]
$$

Following the same procedure as for absorption, we find

$$
\text{Prob}_{\text{emis}}(i \to n) = |\beta_n|^2 = \left(\frac{\mu_{ni}E_o}{\hbar}\right)^2 \frac{\sin^2\left[\frac{1}{2}(\omega_{ni}+\omega)t\right]}{(\omega_{ni}+\omega)^2}
\tag{7.2.22}
$$

The reader might be surprised to find the probability for absorption to be numerically the same as the probability for emission. This is easy to see from the last equation by setting

$$\omega_{ni} = \frac{E_n}{\hbar} - \frac{E_i}{\hbar} = -\left(\frac{E_i}{\hbar} - \frac{E_n}{\hbar}\right) = -\omega_{in}$$

to get

$$\text{Prob}_{\text{emis}}(i \to n) = |\beta_n|^2 = \left(\frac{\mu_{ni}E_o}{\hbar}\right)^2 \frac{\sin^2\left[\frac{1}{2}(-\omega_{in} + \omega)t\right]}{(-\omega_{in} + \omega)^2} = \text{Prob}_{\text{abs}}(i \to n) \qquad (7.2.23a)$$

from Equation 7.2.21. Because the probabilities are equal, we leave off the subscript for absorption and emission and write

$$\text{Prob}(i \to n) = |\beta_n|^2 = \left(\frac{\mu_{ni}E_o}{\hbar}\right)^2 \frac{\sin^2\left[\frac{1}{2}(\omega_{ni} - \omega)t\right]}{(\omega_{ni} - \omega)^2} \qquad (7.2.23b)$$

Notice however, an atom in its ground state cannot emit a photon and so the probability of emission must be zero.

We should make a few comments. First Equation 7.2.23a shows absorption and emission have the same probability

$$\text{Prob}_{\text{emis}}(i \to n) = \text{Prob}_{\text{abs}}(i \to n)$$

The transition must occur between the same two states (as shown in Figure 7.2.3). The dipole matrix element has the same value for either transition $i \to n$ or $n \to i$ since it is Hermitian (and assumed real) and therefore $\mu_{in} = \mu_{ni}$. We cannot expect the relation to hold in the case of three levels for example when $2 \to 1$ and $2 \to 3$. In this case, the dipole matrix element is not necessarily the same for the two transitions.

As a second comment, the reader should realize that the two-level lasers discussed in Chapters 1-2 have many atoms. Some of these atoms occupy the ground state and some

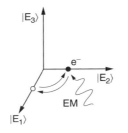

FIGURE 7.2.2
The sign of ω_{ni} indicates absorption or emission.

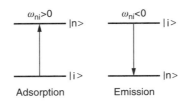

FIGURE 7.2.3
Absorption and emission of a quantum of energy between two states.

occupy the excited state. The two-level laser has an ensemble of atoms and not just one atom as discussed above. We will need the "density operator" discussed in Section 5.11.

7.2.6 Discussion of the Results

Figure 7.2.4 shows a plot of the probability as a function of angular frequency for two different times t_1 and t_2 ($t_2 > t_1$). Notice that the probability becomes narrower for larger times.

Let's discuss the case of stimulated emission with the proviso that the same considerations hold for the case of absorption. The highest probability for emission occurs when $\omega = \omega_{ni}$ as shown in the figure. We can find the peak probability from Equation 7.2.23b by Taylor expanding the sine term (assuming the argument is small) to get

$$\text{Peak Prob} = |\beta_n|^2 = \left(\frac{\mu_{ni}E_o}{\hbar}\right)^2 \frac{t^2}{4} \quad \omega = \omega_{ni} \tag{7.2.24}$$

which occurs when the frequency of the electromagnetic wave ω exactly matches the natural resonant frequency of the atom ω_{ni}. The width of the probability curve can be estimated by finding the point where it touches the horizontal axis. Setting the sine term in the Equation 7.2.23b to zero

$$\sin^2\left[\frac{1}{2}(\omega_{ni} - \omega)t\right] = 0$$

which occurs at $(\omega_{ni} - \omega)t = \pm\pi$, we find that the width is

$$W = \frac{4\pi}{t}$$

According to Figure 7.2.4, a frequency off-resonance can induce a transition.

Equations 7.2.23 and 7.2.24 show that stronger electric fields increase the rate of transition. Equation 7.2.23 shows that for small times (as is appropriate for the approximation of the probability amplitudes β_n), that the transition probability increases linearly with time. This might lead someone to anticipate that the transition requires some average time. If we know the probability as a function of time $P(t)$ then we can calculate an average time as

$$\bar{t} = \langle t \rangle = \int_0^\infty t\, P(t)\, dt$$

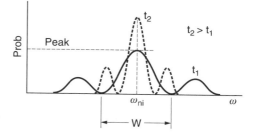

FIGURE 7.2.4
Plot of probability vs. driving frequency and parameterized by time.

We can similarly calculate the variance for the time for emission as

$$\sigma_t^2 = E(t - \bar{t})^2 = \langle t^2 \rangle - \langle t \rangle^2$$

We then see that the exact difference in energy E between the initial and final level is not exactly known by the Heisenberg uncertainty relation

$$\sigma_E \sigma_t \geq \frac{\hbar}{2}$$

7.3 Fermi's Golden Rule

Fermi's golden rule gives the *rate* of transition from a single state to a set of states, which can be described by the "density of state" function. The first topic in this section introduces the intuitive meaning of density of states and subsequent chapters discuss the extended and localized states for devices ranging from bulk to 1-D.

As shown in Figure 7.3.1, an electron makes a transition from an initial state $|i\rangle$ to one of the many final states $|n\rangle$. The probability of transition must be given by

$$\text{Total Prob} = \sum_n P(i \rightarrow n) \tag{7.3.1}$$

For a semiconductor, the final states closely approximate a continuum. In such a case, the probability $P(i \rightarrow n)$ should be interpreted as the probability of transition per final state and the summation should be changed to an integral over the final states.

The total probability in Equation 7.3.1 requires a sum over the integer corresponding to the final states $|n\rangle$. Apparently, we imagine the electron as lodging itself in one of the final energy basis states. However, we know that the final wave function might also be a linear combination of the energy basis states $|n\rangle$. In such a case, the electron simultaneously exists in two or more states $|n\rangle$ (consider two for simplicity). According to classical probability theory, we must subtract this probability from Equation 7.3.1 to find

$$\text{Prob(A or B)} = \text{Prob(A)} + \text{Prob(B)} - \text{Prob(A and B)}$$

However, we assume that a measurement of the energy of the electron has taken place, the wavefunction has collapsed, and that the electron resides in one of the energy basis states. Therefore the Prob(A or B) reduces to the sum of probabilities as in Equation 7.3.1. Fermi's golden rule therefore integrates over the range of final states to find the number of transitions occuring per unit time.

This section also shows how Fermi's golden rule can be used to demonstrate the semiconductor gain. A detailed treatment must wait for discussions on the denisty operator, the Bloch wave function and the reduced density of states.

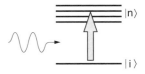

FIGURE 7.3.1
Schematic illustration of an electromagnetically induced transition from an initial state *i* to one of the final states *n*.

7.3.1 Definition of the Density of States

In this topic, we discuss the counting procedure for the energy density of states. The localized states provide the simplest starting point because we do not need the added complexity of finding the allowed wave vectors.

The energy density of states (DOS) function measures the number of energy states in each unit energy interval in each unit volume of the crystal

$$g(E) = \frac{\text{\# states}}{\text{Energy*XalVol}} \qquad (7.3.2)$$

We need to explore the reasons for dividing by the energy and the crystal volume.

Let's first discuss the reason for the "per unit energy." Suppose we have a system with the energy levels shown on the left side of Figure 7.3.2. Assume for now that the states reside in a unit volume of material (say $1\,\text{cm}^3$). The figure shows four energy states in the energy interval between 3 and $4\,\text{eV}$. The density of states at $E = 3.5$ must be

$$g(3.5) = \frac{\text{\# states}}{\text{Energy} \times \text{Vol}} = \frac{4}{1\,\text{eV} \times 1\,\text{cm}^3} = 4$$

Similarly, between four and five electron volts, we find two states and the density of states function has the value $g(4.5) = 2$ and so on. Essentially, we just add up the number of states with a given energy. The graph shows the number of states versus energy; for illustration, the graph has been flipped over on its side. Generally we use finer energy scales and the material has larger numbers of states (10^{17}) so that the graph generally appears much smoother than the one in Figure 7.3.2 since the energy levels essentially form a continuum. The "per unit energy" characterizes the type of state and the type of material.

The definition of density of states uses "per unit crystal volume" in order to remove geometrical considerations from the measure of the type of state. Obviously, if each unit volume has N_v states (electron traps for example) given by

$$N_v = \int_o^\infty dE g(E) = \int d(\text{energy}) \frac{\text{\# states}}{\text{Energy*vol}} = \frac{\text{\# states}}{\text{vol}} \qquad (7.3.3)$$

then the volume V must have $N = N_v V$ states. Changing the volume changes the total number. To obtain a measure of the "type of state," we need to remove the trivial dependence on crystal volume.

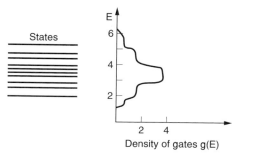

FIGURE 7.3.2
The density of states for the discrete levels shown on the left-hand side. The plot assumes the system has unit volume ($1\,\text{cm}^3$) and the levels have energy measured in *eV*.

What are the states? The states can be those in an atom. The states can also be traps that an electron momentarily occupies until being released back into the conduction band. The states might be recombination centers that electrons enter where they recombine with holes. Traps and recombination centers can be produced by defects in the crystal. Surface states occur on the surface of semiconductors as an inevitable consequence of the interrupted crystal structure. The density of defects can be low within the interior of the semiconductor and high near the surface; as a result, the density of states can depend on position. Later we discuss the extended states in a semiconductor. These all represent localized states that tend to confine the electron to a given region of space. Later sections discuss the more important extended states that represent plane waves (definitely not localized). These states come from periodic boundary conditions on the electron wave function and the resulting allowed k-values for the electron. These states correspond to the Bloch wave functions in a crystal. For now, our main interests centers on a method for picturing the states and the density of states.

Let's consider several examples for the density of localized states. Suppose a crystal has two discrete states (i.e. single states) in each unit volume of crystal. Figure 7.3.3 shows the two states on the left side of the graph. The density-of-state function consists of two Dirac delta functions of the form

$$g(E) = \delta(E - E_1) + \delta(E - E_2)$$

Integrating over energy gives the number of states in each unit volume
$N_v = \int_o^\infty dE g(E) = \int_o^\infty dE[\delta(E - E_1) + \delta(E - E_2)] = 2$ If the crystal has the size $1 \times 4\,\text{cm}^3$ then the total number of states in the entire crystal must be given by

$$N = \int_o^4 dV\, N_v = 8$$

as illustrated in Figure 7.3.4. Although this example shows a uniform distribution of states within the volume V, the number of states per unit volume N_v can depend on the position within the crystal. For example, the growth conditions of the crystal can vary or perhaps the surface becomes damaged after growth.

As a second example, consider localized states near the conduction band of a semiconductor as might occur for amorphous silicon. Figure 7.3.5 shows a sequence of graphs.

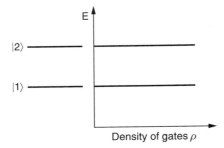

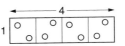

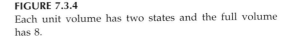

FIGURE 7.3.3
The density of states for two discrete states shown on the left side.

FIGURE 7.3.4
Each unit volume has two states and the full volume has 8.

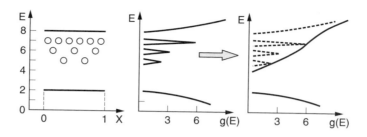

FIGURE 7.3.5
Transition from discrete localized states to the continuum.

The first graph shows the distribution of states versus the position "*x*" within the semiconductor. Notice that the states come closer together (in energy) near the conduction band edge. As a note, amorphous materials have mobility edges rather than band edges. The second graph shows the density of states function versus energy. A sharp Gaussian spike represents the number of states at each energy. At seven electron volts, the material has six states (traps) per unit length in the semiconductor as shown in the first graph. The second graph shows a spike at seven electron volts. Actual amorphous silicon has very large numbers of traps near the upper mobility edge and they form a continuum as represented in the third graph. This example shows how the density of states depends on position and how closely space discrete levels form a continuum.

7.3.2 Equations for Fermi's Golden Rule

Section 7.2 showed that the probability of a transition from an initial state $|i\rangle$ to a final state $|n\rangle$ can be written as

$$\text{Prob}(i \rightarrow n) = |\beta_n|^2 = \left(\frac{\mu_{ni} E_o}{\hbar}\right)^2 \frac{\sin^2[(1/2)(\omega_{ni} - \omega)t]}{(\omega_{ni} - \omega)^2} \tag{7.3.4}$$

with an applied electric field of

$$E(x, t) = E_o \cos(\omega t) \tag{7.3.5}$$

which leads to the perturbing interaction energy

$$\hat{V}(x, t) = \hat{\mu}(x) \frac{E_o}{2} \left(e^{-i\omega t} + e^{+i\omega t}\right) = \hat{\mu}(x) E_o \cos(\omega t) \tag{7.3.6}$$

The dipole moment operator $\hat{\mu}$ provides the matrix elements μ_{ni} that describe the interaction strength between the field and the atom. The dipole matrix element μ_{ni} can be zero for certain final states $|n\rangle$ and Equation 7.3.4 then shows that the transition from the initial to the proposed final state cannot occur. As in Section 7.2, the symbol ω_{ni} represents the difference in energy between the final state $|n\rangle$ and initial state $|i\rangle$

$$\omega_{ni} = \frac{E_n - E_i}{\hbar}$$

where ω_{ni} gives the angular frequency of emitted/absorbed light when the system makes a transition from state $|i\rangle$ to state $|n\rangle$. The incident electromagnetic field has angular frequency ω. Equation 7.3.4 gives the probability of transition for each *final* state $|n\rangle$ and each *initial* state $|i\rangle$. In this topic, we are interested in the density of final states but not in the density of initial states. We therefore take the units for Equation 7.3.4 as the *probability per final state*.

Equation 7.3.1 shows that the total probability of the electron leaving an initial state "i" must be related to the probability that it makes a transition into any number of final states. How can we change the formula if the final states have the same energy? As an answer, transition to final states all having the same energy must have equal probability as can be seen from Equation 7.3.2 (the same ω_{ni}). For N final states with the same energy, we then expect

$$\text{Total Prob} = \sum_n P(i \rightarrow n) = N P(i \rightarrow n)$$

What is the transition probability if some of the final states have energy E_1, some have energy E_2 and so on? Let $g(E_n)$ be the number of states at energy E_n (i.e., the density of states). Then we expect

$$\text{Total Prob} = \sum_n P(i \rightarrow n) = g(E_1)P(i \rightarrow 1) + g(E_2)P(i \rightarrow 2) + \cdots = \sum_n g(E_n)P(i \rightarrow n)$$

Therefore, for a unit volume of crystal, the *total* probability of transition P_V can be written as

$$P_V = \sum_E \left(\frac{\# \text{ states}}{\text{energy vol}} \right) \left(\frac{\text{prob}}{\text{state}} \right) \Delta E \rightarrow \int dE\, g(E)\, P(i \rightarrow n) \tag{7.3.7}$$

where $P(i \rightarrow n) = P(E_i \rightarrow E_n)$ is the probability of transition (per state) and the integral must be over the energy of the final states. Insert Equation 7.3.4 into Equation 7.3.7 to find

$$P_V = \int dE\, g(E) \left(\frac{\mu_{ni} E_o}{\hbar} \right)^2 \frac{\sin^2\left[\frac{1}{2}(\omega_{ni} - \omega)t \right]}{(\omega_{ni} - \omega)^2}$$

where the transition frequency

$$\omega_{ni} = (E_n - E_i)/\hbar = (E - E_i)/\hbar$$

includes the energy of final states E. It is more convenient to write the integral in terms of the *transition* energy

$$E_T = E - E_i = \hbar \omega_{ni}$$

E_T, which is the energy between the initial state and final states as shown in Figure 7.3.6. We find

$$P_V = \int dE_T\, g(E_i + E_T)\, (\mu_{ni} E_o)^2\, \frac{\sin^2\left[\frac{1}{2\hbar}(E_T - \hbar\omega)t \right]}{(E_T - \hbar\omega)^2} \tag{7.3.8}$$

The quantity $\hbar\omega$ represents the energy of the electromagnetic wave inducing the transition. The dipole matrix element μ_{ni} depends on the energy of the final state E

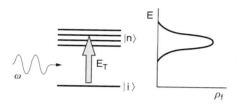

An electromagnetic wave induces a transition from state "*i*" to one of the final states.

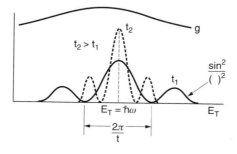

FIGURE 7.3.7
The "*S*" function becomes very narrow for larger times.

through the index "*n*." Therefore the dipole moment can be written as $\mu_{ni} = \mu(E)$ for fixed initial state *i*. In this section, we assume the dipole matrix element to be independent of the energy of the final state. Therefore, we take $\mu_{ni} = \mu$ to be a constant and remove it from the integral in Equation 7.3.8. This assumes that the final states all have the same transition characteristics; the interaction strength between the electromagnetic wave and the system (i.e., atom) remains the same for all possible final states under consideration.

Next, look at the last term in the integral in Equation 7.3.8

$$S = \frac{\sin^2\left[(\frac{1}{2\hbar})(E_T - \hbar\omega)t\right]}{(E_T - \hbar\omega)^2}$$

As discussed in 7.2.6, the function S become sharper as time increases.. For sufficiently large times t, the function S will become very sharp compared to the density of states ρ in Equation 7.3.8 as shown in the Figure 7.3.7. The S function essentially becomes the Dirac Delta function $S = \delta(E_T - \hbar\omega)$. The S function allows the density of states $\rho(E)$ to be removed from the integral with the substitution of $E_T = \hbar\omega$ in $\rho(E)$. Equation 7.3.8 becomes

$$P_V = (\mu E_o)^2 g(E = E_i + \hbar\omega) \int_{-\infty}^{\infty} dE_T \frac{\sin^2\left[(\frac{1}{2\hbar})(E_T - \hbar\omega)t\right]}{(E_T - \hbar\omega)^2}$$

Now evaluating the integral using a change of variable and checking the integration tables for $\int_{-\infty}^{\infty} dx\,(\sin^2 x)/x^2$, we find

$$P_V = (\mu E_o)^2 g(E = E_i + \hbar\omega) \frac{\pi t}{2\hbar}$$

which can also be written as

$$P_V = (\mu E_o)^2 g_f\big(E_f = E_i \pm \hbar\omega\big) \frac{\pi t}{2\hbar} \qquad (7.3.9)$$

where E_f and E_i are the energy of the final and initial states, respectively. Equation 7.3.9 includes the "+" for absorption and the "−" for emission. Equation 7.3.7 provides the probability (per initial state per unit volume) of the system absorbing energy from the electromagnetic waves and making a transition from E_i to E_f. Notice how the probability depends on the frequency of the EM wave through the density of states.

Fermi's Golden Rule gives the rate of stimulated emission and stimulated absorption from Equation 7.3.7. The rate of transition is found to be

$$R_{i \to f} = \frac{d}{dt} P_V = \frac{\pi}{2\hbar} (\mu \, E_o)^2 \, g_f \, (E_f = E_i \pm \hbar\omega) \tag{7.3.10}$$

Notice that the transition rate must be proportional to the optical power i.e.,

$$\text{Optical power} \propto E_o^2$$

Fewer available final states at energy E_f implies a lower transition rate because of the density of states that appears in Equation 7.3.10. We will find that lowering the number of final states *lowers* the rate of *spontaneous emission*, which can lower the laser threshold current.

For a single final level, the density of states function must be a Dirac Delta function centered at the energy E_f

$$R_{i \to f} = \frac{d}{dt} P_V = \frac{\pi}{2\hbar} (\mu E_o)^2 \delta(E_f = E_i + \hbar\omega)$$

The Dirac Delta function ensures that transition process conserves energy. We could integrate this last equation over energy to find a rate of transition.

7.3.3   Introduction to Laser Gain

At this point, we can begin to relate the laser gain to the rate of transition. Suppose a total of N atoms can occupy either state $|i\rangle$ or the state $|f\rangle$. We often refer to a collection of these atoms as an ensemble of two-level atoms. Suppose there are n_i atoms with electrons in state $|i\rangle$ and there are n_f in state $|f\rangle$. The total number of atoms can therefore be written as $N = n_i + n_f$. The net rate of stimulated emission must be

$$\text{Net Emission Rate} = \text{Emission--Absorption} = R_{f \to i} n_f - R_{i \to f} n_i = R_{f \to i}(n_f - n_i) \tag{7.3.11}$$

where the last step follows because the probability of an upward transition must be the same as the probability of the downward transition as discussed in Section 7.2. Using Equation 7.3.10, specifically

$$R_{i \to f} = \frac{d}{dt} P_V = \frac{\pi}{2\hbar} (\mu E_o)^2 g_f (E_f = E_i + \hbar\omega)$$

and defining g' by

$$g' = \frac{\pi}{2\hbar} \mu^2 g_f$$

Equation 7.3.11 becomes

$$\text{NER} = \text{Net Emission Rate} = g' \cdot (n_f - n_i) E_o^2$$

The "net emission rate" has units of photons per second and must be related to $d\gamma/dt$ from the rate equations in Chapter 2. Similarly, the factor E_o^2 is proportional to the optical power in the laser, which in turn, is related to photon density γ. Combining the constants of proportionality from $NER \sim d\gamma/dt$ and $E_o^2 \sim \gamma$ with the factor g' and calling the product g_o, we find

$$\frac{d\gamma}{dt} = g_o \cdot (n_f - n_i)\gamma$$

The author fervently hopes that the reader will recognize this formula as part of the laser rate equations first developed in Chapter 2.

7.4 Introduction to the Electromagnetic Lagrangian and Field Equations

The Lagrangian and Hamiltonian formulation of mechanics and electrodynamics have primary roles in both the classical and quantum theory of the matter–field interaction. Previous chapters show how the Hamiltonian appears in the Schrodinger wave equation and the quantum theory of fields. The Lagrangian and Hamiltonian formulation provide a foundation for the quantum mechanical form of the matter–field interaction most commonly associated with the dipole moment. The present section first introduces the Lagrangian density for the electromagnetic field, derives the Euler-Lagrange equations and then shows how Maxwell's equations can be reproduced from the formalism. The 4-vector notation ensures the expressions remain relativistically invariant. The next section derives the Hamiltonian for the matter, fields and their interactions based on the Lagrangian. The subsequent section then shows how the interaction Hamiltonian provides the dipole-field interaction term. As is common for the field theory, the section uses the Lorentz-Heaviside units for the electromagnetics and sets the speed of light in vacuum to $c=1$. In the end, the usual MKS units will be inserted.

7.4.1 A Summary of Results and Notation

Chapters 4 and 5 introduced 4-vector notation along with the Lagrangian and Hamiltonian densities, and equations of motion. The tables below provide a brief summary of some of the results.

Mechanics

Lagrange Density:	$\mathscr{L} = \mathscr{L}(\eta, \dot{\eta}, \partial_i\eta)$
Lagrangian:	$L = \int_V dV \, \mathscr{L}$
Hamitonian Density:	$\mathscr{H} = \pi(\vec{r}, t)\,\dot{\eta}(\vec{r}, t) - \mathscr{L}$
Hamiltonian:	$H = \int dV \, \mathscr{H}$
Momentum Density:	$\pi(\vec{r}, t) = \dfrac{\partial \mathscr{L}(\eta, \ldots)}{\partial \dot{\eta}}$
Hamilton's Canonical Equations:	$\dot{\eta} = \dfrac{\partial \mathscr{H}}{\partial \pi} \quad \dot{\pi} = -\dfrac{\partial \mathscr{H}}{\partial \eta}$

Mathematics

$$
\text{Metric:} \qquad g^{\mu\nu} = \begin{pmatrix} 1 & 0 & 0 & 0 \\ 0 & -1 & 0 & 0 \\ 0 & 0 & -1 & 0 \\ 0 & 0 & 0 & -1 \end{pmatrix} \mu, \nu = 0,1,2,3
$$

$$
\text{Vectors:} \qquad x^{\mu} = \left(t, \vec{r}\right) \text{ or } x_{\mu} = g_{\mu\nu}x^{\nu} = \left(t, -\vec{r}\right) i, j = 1,2,3
$$

$$
\text{Derivative:} \quad \partial_{\mu} = \frac{\partial}{\partial x^{\mu}} = \left(\frac{\partial}{\partial t}, \nabla\right) \text{ or } \partial^{\mu} = g^{\mu\nu}\partial_{\nu} = \left(\frac{\partial}{\partial t}, -\nabla\right)
$$

Electromagnetics

$$
\text{Field Tensor:} \qquad F^{\mu\nu} = \partial^{\mu}A^{\nu} - \partial^{\nu}A^{\mu} \text{ with } F^{\mu\nu} = -F^{\nu\mu}, F^{\mu\mu} = 0
$$

$$
\text{4-Vector Potential:} \quad A^{\mu} = \left(\Phi, \vec{A}\right) \text{ or } A_{\mu} = g_{\mu\nu}A^{\nu} = \left(\Phi, -\vec{A}\right)
$$

$$
\text{Current Density:} \quad j^{\mu} = \left(\rho, \vec{j}\right) \rho, \vec{j} = \text{charge and current density}
$$

7.4.2   The EM Lagrangian and Hamiltonian

The generalized coordinates for the electromagnetic field consist of the vector potential fields $A^{\mu} = (\Phi, \vec{A}) = (\Phi, A_x, A_y, A_z)$; there exists 4 functions for each point $\vec{r}$ in space. Before assigning the dynamical equations (Maxwell's or Newton's or Schrodinger's equations etc.), we can independently vary the magnitude of each field at each space time point $(t, \vec{r})$. The motions of the fields at neighboring points only become correlated after applying the equations of motion to the system.

We assume that the Lagrangian depends on the fields and the various possible derivatives.

$$
L = L\left(A_{\mu}, \partial_{\nu}A_{\mu}\right) \tag{7.4.1}
$$

By comparison, the Lagrangian for the simplest case of a point particle executing 1-D motion depends on x and $\dot{x}$. The terms in the set $\{A_{\mu}\}$ are independent of one another so that $(\partial A^{\mu}/\partial A^{\nu}) = \delta_{\mu\nu}$ where $\delta_{\mu\nu}$ denotes the Kronecker delta function. Similarly, the terms in $\{\partial_{\nu}A_{\mu}\}$ are independent so that $(\partial(\partial_{\nu}A_{\mu})/\partial(\partial_{\alpha}A_{\beta})) = \delta_{\alpha\nu}\delta_{\beta\mu}$. Finally, every term in one set $\{A_{\mu}\}$ is independent of every term in the other $\{\partial_{\nu}A_{\mu}\}$ so that partial derivatives between sets produce results of zero.

The Lagrange density L (units of energy per unit volume) can be defined by

$$
L = \int d^3x \, \mathcal{L} \tag{7.4.2}
$$

and so the action integral

$$
I = \int_{x_1^{\xi}}^{x_2^{\xi}} dt \, L\left(A_{\mu}, \partial_{\nu}A_{\mu}\right) = \int_{x_1^{\xi}}^{x_2^{\xi}} dt \, d^3x \, \mathcal{L}\left(A_{\mu}, \partial_{\nu}A_{\mu}\right) = \int_{x_1^{\xi}}^{x_2^{\xi}} d^4x \, \mathcal{L}\left(A_{\mu}, \partial_{\nu}A_{\mu}\right)
$$

must be minimized to find the equations of motion – Maxwell's equations in this case. The limits on the integral have the form $x^\xi = (t, \vec{r})$. We assume that fields obey either periodic or fixed-endpoint boundary conditions for the spatial part of the endpoints and that the fields at the endpoints are fixed in time so that their time-variation at the endpoints produce zero. The behavior of the integrand at the endpoints is arranged so as to eliminate surface terms after partial integration.

Varying the action integral produces

$$0 = \delta I = \int_{x_1^\xi}^{x_2^\xi} d^4x \, \delta \mathscr{L}(A_\mu, \partial_\nu A_\mu) = \int_{x_1^\xi}^{x_2^\xi} d^4x \left[\frac{\partial \mathscr{L}}{\partial A_\mu} \delta A_\mu + \frac{\partial \mathscr{L}}{\partial \partial_\nu A_\mu} \delta \partial_\nu A_\mu \right] \qquad (7.4.3)$$

Interchanging orders in the last term $\delta \partial_\nu A_\mu = \partial_\nu \delta A_\mu$ and partially integrating yields

$$0 = \frac{\partial \mathscr{L}}{\partial \partial_\nu A_\mu} \delta A_\mu \bigg|_{x_1^\xi}^{x_2^\xi} + \int_{x_1^\xi}^{x_2^\xi} d^4x \left[\frac{\partial \mathscr{L}}{\partial A_\mu} - \partial_\nu \frac{\partial \mathscr{L}}{\partial \partial_\nu A_\mu} \right] \delta A_\mu$$

where repeated indices in a product of terms must be summed. Assuming the surface term don't contribute and that the four variations in the integral represented by δA_μ are all arbitrary and independent, we find Lagrange's (i.e., Euler's) equations

$$\frac{\partial \mathscr{L}}{\partial A_\mu} - \partial_\nu \frac{\partial \mathscr{L}}{\partial \partial_\nu A_\mu} = 0 \qquad (7.4.4)$$

where we must sum over repeated indices.

7.4.3 4-Vector Form of Maxwell's Equations from the Lagrangian

In this section we show that the Lagrange density

$$\mathscr{L} = -\frac{1}{4} F_{\mu\nu} F^{\mu\nu} - j^\mu A_\mu \qquad (7.4.5)$$

leads to Maxwell's equations. The last term represents the matter-light interaction term. Keep in mind that repeated indices in a product must be summed as for example, we should write $F_{\mu\nu} F^{\mu\nu} \equiv \sum_{\mu,\nu=0}^{3} F_{\mu\nu} F^{\mu\nu}$. The antisymmetry property of the field tensor $F^{\mu\nu} = -F^{\nu\mu}$ follows from the basic definition

$$F^{\mu\nu} = \partial^\mu A^\nu - \partial^\nu A^\mu \qquad (7.4.6)$$

and provides $F^{\mu\mu} = 0$ with or without a summation on the index. As a final note, it is customary to use the microscopic form of Maxwell's equations. This means that we do not include the susceptibility and permeability for anything but the vacuum. The polarization must be stated in terms of the its component parts.

Having found Euler-Lagrange equations from the variational principle in 7.4.2

$$\frac{\partial \mathscr{L}}{\partial A_\mu} - \partial_\nu \frac{\partial \mathscr{L}}{\partial \partial_\nu A_\mu} = 0 \qquad (7.4.7)$$

we can now find the equations of motion for the EM field using Equations 7.4.5 and 7.4.7. First calculate $\partial \mathscr{L}/\partial A_\mu$

$$\frac{\partial \mathscr{L}}{\partial A_\alpha} = \frac{\partial}{\partial A_\alpha}\left[-\frac{1}{4}F_{\mu\nu}F^{\mu\nu} - j^\mu A_\mu\right] = \frac{\partial}{\partial A_\alpha}\left[-\frac{1}{4}(\partial_\mu A_\nu - \partial_\nu A_\mu)(\partial^\mu A^\nu - \partial^\nu A^\mu) - j^\mu A_\mu\right]$$

The first two terms in parenthesis do not contain A_α by themselves. Therefore, we find

$$\frac{\partial \mathscr{L}}{\partial A_\alpha} = -\frac{\partial}{\partial A_\alpha}j^\mu A_\mu = j^\mu \delta_{\mu\alpha} = j^\alpha \tag{7.4.8}$$

where the last result comes from the sum over the index μ.

Next calculate $\partial \mathscr{L}/\partial(\partial_\alpha A_\beta)$

$$\frac{\partial \mathscr{L}}{\partial \partial_\alpha A_\beta} = -\frac{1}{4}\frac{\partial}{\partial \partial_\alpha A_\beta}(\partial_\mu A_\nu - \partial_\nu A_\mu)(\partial^\mu A^\nu - \partial^\nu A^\mu) - \frac{\partial}{\partial \partial_\alpha A_\beta}j^\mu A_\mu$$

$$= -\frac{1}{4}\frac{\partial}{\partial \partial_\alpha A_\beta}(\partial_\mu A_\nu - \partial_\nu A_\mu)(\partial^\mu A^\nu - \partial^\nu A^\mu)$$

Notice that the derivative of $j^\mu A_\mu$ gave nothing because the coordinates must be independent of the generalized velocities. The last equation can be rewritten using the metric as

$$\frac{\partial \mathscr{L}}{\partial \partial_\alpha A_\beta} = -\frac{1}{4}\frac{\partial}{\partial \partial_\alpha A_\beta}(\partial_\mu A_\nu - \partial_\nu A_\mu)g^{\mu\mu'}g^{\nu\nu'}(\partial_{\mu'}A_{\nu'} - \partial_{\nu'}A_{\mu'})$$

The product rule then provides

$$\frac{\partial \mathscr{L}}{\partial \partial_\alpha A_\beta} = -\frac{g^{\mu\mu'}g^{\nu\nu'}}{4}\left[(\delta_{\alpha\mu}\delta_{\beta\nu} - \delta_{\alpha\nu}\delta_{\beta\mu})(\partial_{\mu'}A_{\nu'} - \partial_{\nu'}A_{\mu'}) + (\partial_\mu A_\nu - \partial_\nu A_\mu)(\delta_{\alpha\mu'}\delta_{\beta\nu'} - \delta_{\alpha\nu'}\delta_{\beta\mu'})\right]$$

Finally, using the metric and summing over repeated indices yields

$$\frac{\partial \mathscr{L}}{\partial \partial_\alpha A_\beta} = -\frac{1}{4}\left[(g^{\alpha\mu'}g^{\beta\nu'} - g^{\beta\mu'}g^{\alpha\nu'})(\partial_{\mu'}A_{\nu'} - \partial_{\nu'}A_{\mu'})\right] - \frac{1}{4}\left[(\partial^\mu A_\nu - \partial_\nu A_\mu)(g^{\mu\alpha}g^{\beta\nu'} - g^{\mu\beta}g^{\nu\alpha})\right]$$

$$= -\frac{1}{4}\left[2\partial^\alpha A^\beta - 2\partial^\beta A^\alpha\right] - \frac{1}{4}\left[2\partial^\alpha A^\beta - 2\partial^\beta A^\alpha\right] = \partial^\alpha A^\beta - \partial^\beta A^\alpha$$

Therefore, the last term in 7.4.2 becomes

$$\partial_\alpha \frac{\partial \mathscr{L}}{\partial \partial_\alpha A_\beta} = \partial_\alpha(\partial^\alpha A^\beta - \partial^\beta A^\alpha) = \partial_\alpha F^{\alpha\beta} \tag{7.4.9}$$

Combining 7.4.6, 7.4.5 and 7.4.2 provides Maxwell's equations in the form

$$\partial_\alpha F^{\alpha\beta} = j^\beta \tag{7.4.10}$$

with one equation for each $\beta = 0, \ldots, 3$.

7.4.4 3-Vector Form of Maxwell's Equations

Using the definition of the field tensor, and the relation between the field tensor and the currents,

$$F^{\mu\nu} = \partial^\mu A^\nu - \partial^\nu A^\mu \qquad \partial_\mu F^{\mu\nu} = j^\nu \tag{7.4.11}$$

we can give an explicit relation for the field tensor (matrix) and derive Maxwell's equations.

The Lagrangian density must be related to the energy density; this is especially true for the Hamiltonian. It is therefore interesting to note that we can start with the field *energy* and derive Maxwell's equations.

Gauss' Law

We first show how Equation 7.4.10, specifically $\partial_\mu F^{\mu\nu} = j^\nu$, leads to Gauss' law $\nabla \cdot \vec{\mathscr{E}} = \rho$ (in Lorentz-Heaviside units). We then use this result to construct the entries of the field tensor $F^{\mu\nu}$ for the first column and top row.

To find Gauss' law, start with $\nu = 0$ in Equation 7.4.10 to find

$$\partial_\mu F^{\mu 0} = j^0 \quad \rightarrow \quad \partial_\mu \partial^\mu A^0 - \partial_\mu \partial^0 A^\mu = j^0 \quad \rightarrow \quad \partial_\mu \partial^\mu \Phi - \partial_\mu \partial^0 A^\mu = \rho$$

Expand the individual terms

$$\partial_\mu \partial^\mu \Phi = \frac{\partial^2 \Phi}{\partial t^2} - \nabla \cdot \nabla \Phi \quad \text{and} \quad \partial_\mu \partial^0 A^\mu = \frac{\partial}{\partial t} \left(\frac{\partial \Phi}{\partial t} + \nabla \cdot \vec{A} \right)$$

and recombine into $\partial_\mu \partial^\mu \Phi - \partial_\mu \partial^0 A^\mu = \rho$ to get

$$\left[\frac{\partial^2 \Phi}{\partial t^2} - \nabla \cdot \nabla \Phi \right] - \left[\frac{\partial}{\partial t} \left(\frac{\partial \Phi}{\partial t} + \nabla \cdot \vec{A} \right) \right] = \rho \quad \rightarrow \quad \nabla \cdot \left[-\dot{\vec{A}} - \nabla \Phi \right] = \rho \quad \rightarrow \quad \nabla \cdot \vec{E} = \rho$$

where the last result comes from the definition of the vector potentials found in Section 6.2.

Gauss' law determines some of the tensor elements in *F*. Using $\nu = 0$ along with 7.4.8, we find

$$\partial_\mu F^{\mu 0} = j^0 = \rho \quad \rightarrow \quad \frac{\partial F^{00}}{\partial t} + \frac{\partial F^{10}}{\partial x} + \frac{\partial F^{20}}{\partial y} + \frac{\partial F^{30}}{\partial z} = \rho$$

with $F^{00} = 0$ since *F* is antisymmetric. We recognize this last equation as Gauss' law for the electric field $\vec{E}$. Therefore we now know

$$F^{\mu\nu} = \begin{pmatrix} 0 & -\mathscr{E}_x & -\mathscr{E}_y & -\mathscr{E}_z \\ \mathscr{E}_x & 0 & & \\ \mathscr{E}_y & & 0 & \\ \mathscr{E}_z & & & 0 \end{pmatrix}$$

where the top row comes from the antisymmetry property of *F*. We will determine the remaining entries in the next subtopic.

Ampere's Law

Let's start with Equation 7.4.10, namely $\partial_\mu F^{\mu\nu} = j^\nu$, and focus on $\nu = 2$. We find

$$\frac{\partial F^{02}}{\partial t} + \frac{\partial F^{12}}{\partial x} + \frac{\partial F^{22}}{\partial y} + \frac{\partial F^{32}}{\partial z} = j^2 \quad \rightarrow \quad -\frac{\partial \mathscr{E}_y}{\partial t} + \frac{\partial F^{12}}{\partial x} + \frac{\partial F^{32}}{\partial z} = j_y$$

where $F^{22} = 0$ due to the antisymmetry of F. After some thought, we recognize the F-terms as part of a cross product. Let's use the antisymmetry property of F and place the time derivative on the right hand side to rewrite the last equation as

$$\frac{\partial F^{12}}{\partial x} - \frac{\partial F^{23}}{\partial z} = j_y + \frac{\partial \mathscr{E}_y}{\partial t}$$

We recognize this as the y-component of the curl of the magnetic field. In particular, we identify $F^{12} = -\mathscr{B}_z$ and $F^{23} = -\mathscr{B}_x$. Similarly, we can show $F^{13} = +\mathscr{B}_y$. The field tensor now has the form

$$F^{\mu\nu} = \begin{pmatrix} 0 & -\mathscr{E}_x & -\mathscr{E}_y & -\mathscr{E}_z \\ \mathscr{E}_x & 0 & -\mathscr{B}_z & \mathscr{B}_y \\ \mathscr{E}_y & \mathscr{B}_z & 0 & -\mathscr{B}_x \\ \mathscr{E}_z & -\mathscr{B}_y & \mathscr{B}_x & 0 \end{pmatrix}$$

Alternatively, we can also show the y-component of the magnetic field working with the potentials. Consider $F^{13} = \partial^1 A^3 - \partial^3 A^1$, and notice that the partials have the index in the upper position which requires $\partial^\mu = (\partial_t, -\nabla)$. We therefore have

$$F^{13} = \partial^1 A^3 - \partial^3 A^1 = -\frac{\partial A_z}{\partial x} + \frac{\partial A_x}{\partial z} = B_y$$

All of the components can be treated similarly.

Magnetic Monopole Relation

The tensor formalism can be used to show that nature does not generally produce magnetic monopoles as described by the equation $\nabla \cdot \vec{\mathscr{B}} = 0$. Let i, j, k be restricted to the integers 1, 2, 3. We can easily see that

$$\partial_1 F^{23} + \partial_2 F^{31} + \partial_3 F^{12} = 0 \tag{7.4.12}$$

by substituting the potentials into $F^{\mu\nu} = \partial^\mu A^\nu - \partial^\nu A^\mu$ and then switching the order of differentiation—keep in mind that $F^{\mu\nu}$ has derivatives with the upper index, which requires an "extra" minus sign. By substituting the various terms for F into Equation 7.4.12, we find

$$\frac{\partial}{\partial x}(-\mathscr{B}_x) + \frac{\partial}{\partial y}(-\mathscr{B}_y) + \frac{\partial}{\partial z}(-\mathscr{B}_z) = 0 \quad \rightarrow \quad \nabla \cdot \vec{\mathscr{B}} = 0$$

Lenz's Law

Lenz's law relates the curl of the electric field to the rate of change of the magnetic field. We can use the same maneuver as for Equation 7.4.12 to show

$$\partial_0 F^{12} - \partial_1 F^{20} - \partial_2 F^{01} = 0 \tag{7.4.13}$$

To prove this last relation, remember the "extra" minus sign because the derivatives in F have the upper index. Now we can substitute the fields from the field tensor into the last equation to find

$$\partial_0 F^{12} - \partial_1 F^{20} - \partial_2 F^{01} = 0 \quad \rightarrow \quad \frac{\partial}{\partial t}(-\mathscr{B}_z) - \frac{\partial}{\partial x}(+\mathscr{E}_z) - \frac{\partial}{\partial y}(-\mathscr{E}_x) = 0$$

The other components can be similarly demonstrated to find

$$\nabla \times \vec{\mathscr{E}} = -\frac{\partial \vec{\mathscr{B}}}{\partial t}$$

7.5 The Classical Hamiltonian for Fields, Particles and Interactions

We now demonstrate the Hamiltonian that includes the free EM fields, the free particles and the interaction between the fields and the particles. We will see that the Hamiltonian in natural units ($c = 1, \hbar = 1$) has the form (for a single charge particle)

$$H = \underbrace{\frac{\left(\vec{p} - q\vec{A}\right)^2}{2m}}_{\substack{\text{Particle and} \\ \text{Particle-Field}}} + \underbrace{V(\vec{r})}_{\text{Atomic}} + \underbrace{H_{em}}_{\substack{\text{Free} \\ \text{Field}}} \cong \underbrace{\frac{p^2}{2m}}_{\substack{\text{Free} \\ \text{Particle}}} - \underbrace{\frac{q}{m}\vec{p} \cdot \vec{A}}_{\substack{\text{Matter-Field} \\ \text{Interaction}}} + \underbrace{V(\vec{r})}_{\text{Atomic}} + \underbrace{\frac{1}{2}\int dV \left(\mathscr{E}^2 + \mathscr{H}^2\right)}_{\text{Free Electromagnetic Fields}} \tag{7.5.1}$$

where q, m, $\vec{A}$, $\vec{\mathscr{E}}$, $\mathscr{H}$ denote the charge of the particle ($q < 0$ for electrons), mass m of the particle, the vector potential, and the electric and magnetic fields, respectively. The symbol $\vec{p}$ represents the canonical momentum comprised of the particle and field momenta. For small fields $\vec{A}$, the canonical momentum $\vec{p}$ reduces to the particle momentum $m\vec{v}$. Note that H represents the total energy of the combined system. In the low field limit, the same results can be found from Poynting vector calculations. The next section shows common form of $\hat{\mu} \cdot \vec{\mathscr{E}}$ for the matter–field interaction using Equation 7.5.1 as a starting point. As a matter of notation, we will use $\mathscr{H}$ as the Hamiltonian density and retain $\vec{\mathscr{H}}$ as the magnetic field.

The present section starts with the Lagrange density

$$\mathscr{L} = -\frac{1}{4}F_{\mu\nu}F^{\mu\nu} - j^{\mu}A_{\mu} \tag{7.5.2}$$

and derives the Hamiltonian density (energy density) and the Hamiltonian (energy) for the subsystem consisting of the fields and the interaction. It then uses the results to demonstrate the Hamiltonian for the combined system given by Equation 7.5.1. A list of the notational conventions are discussed in 7.4.1.

7.5.1   The EM Hamiltonian Density

Section 5.4 discusses the classical Lagrangian density and shows how the canonical momentum density "π" arises. Basically, the potentials (for the case at hand) comprise the generalized coordinates. On the other hand, the canonical momentum, obtained from the Lagrangian *density*, must have units of "per unit volume per second." The total canonical momentum must be the integral over volume of the momentum density.

Recall the definition of the Hamiltonian using the discrete coordinates q_x.

$$H = \sum_x p_x \dot{q}_x - L \tag{7.5.3}$$

where the canonical momentum p_x corresponds to the generalized coordinate q_x and can be found from the following relation.

$$p_x = \frac{\partial L}{\partial \dot{q}_x} \tag{7.5.4}$$

We can rewrite the momentum in terms of the momentum density π_x as

$$p_x = \pi_x \, d^3x \tag{7.5.5}$$

where d^3x is a small volume element. Also recall the definition of the Lagrange density $\mathscr{L}$ as

$$L = \int d^3x \, \mathscr{L} \tag{7.5.6}$$

Combining these last four equations to find

$$H = \sum_x p_x \dot{q}_x - L = \int d^3x \left(\pi \dot{q} - \mathscr{L} \right) \tag{7.5.7}$$

where the limit of small "x" produces the integral. The momentum density is given by

$$\pi = \frac{\partial \mathscr{L}}{\partial \dot{q}} \tag{7.5.8}$$

and the Hamiltonian density is defined to be

$$\mathscr{H} = \pi \dot{q} - \mathscr{L} \tag{7.5.9}$$

Keep in mind that there must be a different generalized coordinate and momentum for each position in space.

For the EM problem, we have four potentials assigned to each point in space. Therefore, the Hamiltonian density must have the form

$$\mathscr{H} = \sum_\mu \pi_\mu \dot{q}^\mu - \mathscr{L} \equiv \pi^\mu \dot{A}_\mu - \mathscr{L}\left(A_\mu, \partial_\nu A_\mu\right) \tag{7.5.10}$$

where we must remember to sum over repeated indices. The momentum density becomes

$$\pi^\mu = \frac{\partial \mathscr{L}}{\partial \dot{A}_\mu} \quad \text{where} \quad \dot{\partial A}_\mu = \partial_o A_\mu \tag{7.5.11}$$

To write the Hamiltonian, we only need to identify the canonical momentum densities using Equation 7.5.8, substitute into Equation 7.5.9, and make sure that the Hamiltonian density is written only in terms of the generalized coordinates and momentum (and not $\partial_\nu A_\mu$).

7.5.2 The Canonical Field Momentum

The momentum density can be calculated using Equations 7.5.10 and 7.5.2

$$\pi^\alpha = \frac{\partial \mathcal{L}}{\partial \dot{A}_\alpha} = \frac{\partial}{\partial \dot{A}_\alpha}\left(-\frac{1}{4}F_{\mu\nu}F^{\mu\nu} - j^\mu A_\mu\right) = -\frac{1}{4}\frac{\partial}{\partial \dot{A}_\alpha}F_{\mu\nu}F^{\mu\nu}$$

Using the metric to lower the indices on $F^{\mu\nu}$ provides

$$\pi^\alpha = -\frac{g^{\mu\mu'}g^{\nu\nu'}}{4}\frac{\partial}{\partial_0 A_\alpha}(\partial_\mu A_\nu - \partial_\nu A_\mu)(\partial_{\mu'}A_{\nu'} - \partial_{\nu'}A_{\mu'})$$

$$= -\frac{g^{\mu\mu'}g^{\nu\nu'}}{4}\left[(\delta_{0\mu}\delta_{\alpha\nu} - \delta_{0\nu}\delta_{\alpha\mu})(\partial_{\mu'}A_{\nu'} - \partial_{\nu'}A_{\mu'}) + (\partial_\mu A_\nu - \partial_\nu A_\mu)(\delta_{0\mu'}\delta_{\alpha\nu'} - \delta_{0\nu'}\delta_{\alpha\mu'})\right]$$

Using the Kronecker delta and eliminating the metrics by raising appropriate indices provides

$$\pi^\alpha = -\frac{1}{4}\left[(g^{0\mu'}g^{\alpha\nu'} - g^{\alpha\mu'}g^{0\nu'})(\partial_{\mu'}A_{\nu'} - \partial_{\nu'}A_{\mu'}) + (\partial_\mu A_\nu - \partial_\nu A_\mu)(g^{\mu 0}g^{\nu\alpha} - g^{\mu\alpha}g^{\nu 0})\right]$$

$$= -\frac{1}{4}\left[(\partial^0 A^\alpha - \partial^\alpha A^0) - (\partial^\alpha A^0 - \partial^0 A^\alpha) + (\partial^0 A^\alpha - \partial^\alpha A^0) - (\partial^\alpha A^0 - \partial^0 A^\alpha)\right]$$

Adding the terms in the last expression provides

$$\pi^\alpha = -\partial^0 A^\alpha + \partial^\alpha A^0 \tag{7.5.12}$$

Two cases of Equation 7.5.12 should be considered. First set $\alpha = 0$ to find

$$\pi^0 = -\partial^0 A^0 + \partial^0 A^0 = 0 \tag{7.5.13}$$

Next, set $\alpha = i$ where $i = 1, 2, 3$.

$$\pi^i = -\partial^0 A^i + \partial^i A^0 = -\dot{A}^i - \partial_i A^0 = -\dot{A}^i - \partial_i \Phi = -\dot{A}_i - (\nabla\Phi)_i \tag{7.5.14}$$

This last equation gives the ith component of the canonical momentum density. We can put this all together as follows

$$\pi^\alpha = \left\{\begin{matrix} 0 & \alpha = 0 \\ -\dot{\vec{A}} - \nabla\Phi & \alpha \neq 0 \end{matrix}\right\} = \left\{\begin{matrix} 0 & \alpha = 0 \\ \vec{\mathcal{E}} & \alpha \neq 0 \end{matrix}\right. \tag{7.5.15}$$

where we used the relation between the electric field and the potentials.

7.5.3 Evaluating the Hamiltonian Density

Now we calculate the Hamiltonian density using Equations 7.5.10 and 7.5.2

$$\mathcal{H} = \pi^{\mu} \dot{A}_{\mu} + \frac{1}{4} F_{\mu\nu} F^{\mu\nu} + j^{\mu} A_{\mu} \tag{7.5.16}$$

As shown in the previous section, the field tensor $F^{\mu\nu} = \partial^{\mu} A^{\nu} - \partial^{\nu} A^{\mu}$ has the following form.

$$F^{\mu\nu} = \begin{pmatrix} 0 & -\mathcal{E}_x & -\mathcal{E}_y & -\mathcal{E}_z \\ \mathcal{E}_x & 0 & -\mathcal{B}_z & \mathcal{B}_y \\ \mathcal{E}_y & \mathcal{B}_z & 0 & -\mathcal{B}_x \\ \mathcal{E}_z & -\mathcal{B}_y & \mathcal{B}_x & 0 \end{pmatrix} \quad F_{\mu\nu} = g_{\mu\alpha} F^{\alpha\beta} g_{\beta\nu} = \begin{pmatrix} 0 & \mathcal{E}_x & \mathcal{E}_y & \mathcal{E}_z \\ -\mathcal{E}_x & 0 & -\mathcal{B}_z & \mathcal{B}_y \\ -\mathcal{E}_y & \mathcal{B}_z & 0 & -\mathcal{B}_x \\ -\mathcal{E}_z & -\mathcal{B}_y & \mathcal{B}_x & 0 \end{pmatrix}$$

It's easy to calculate the first and last terms in Equation 7.5.16. The first term in Equation 7.5.16 gives

$$\pi^{\mu} \dot{A}_{\mu} = \pi^0 \dot{A}_o + \pi^i \dot{A}_i = 0 - \vec{\mathcal{E}} \cdot \dot{\vec{A}} = -\vec{\mathcal{E}} \cdot \dot{\vec{A}}$$

Next substitute for the time derivative using $\vec{\mathcal{E}} = -\dot{\vec{A}} - \nabla \Phi$

$$\pi^{\mu} \dot{A}_{\mu} = \vec{\mathcal{E}} \cdot \left(\vec{\mathcal{E}} + \nabla \Phi \right) = \mathcal{E}^2 + \vec{\mathcal{E}} \cdot \nabla \Phi \tag{7.5.17}$$

The last term in Equation 7.5.16 gives

$$j^{\mu} A_{\mu} = \rho \Phi - \vec{j} \cdot \vec{A} \tag{7.5.18}$$

Next, let's calculate the middle term in 7.5.15, namely $F_{\mu\nu} F^{\mu\nu}$. We can multiply corresponding entries in $F_{\mu\nu}$ and $F^{\mu\nu}$ and then add the results to find

$$F_{\mu\nu} F^{\mu\nu} = -2 \left(\mathcal{E}^2 - \mathcal{B}^2 \right) \tag{7.5.19}$$

Combining Equations 7.5.19, 7.5.18, 7.5.17 with the Equation 7.5.16 for the Hamiltonian density $\mathcal{H} = \pi^{\mu} \dot{A}_{\mu} + (1/4) F_{\mu\nu} F^{\mu\nu} + j^{\mu} A_{\mu}$ we find

$$\mathcal{H} = \frac{1}{2} \left(\mathcal{E}^2 + \mathcal{B}^2 \right) + \vec{\mathcal{E}} \cdot \nabla \Phi + \rho \Phi - \vec{j} \cdot \vec{A} \tag{7.5.20}$$

for the free fields and matter-light interaction. In the next section we add the terms for free particles and derive an expression usually used as the starting point for quantum systems interacting with light.

7.5.4 The Field and Interaction Hamiltonian

We can now write the Hamiltonian

$$H_{em} + H_{int} = \int dV \, \mathcal{H} = \int dV \left[\frac{1}{2} \left(\mathcal{E}^2 + \mathcal{B}^2 \right) + \vec{\mathcal{E}} \cdot \nabla \Phi + \rho \Phi - \vec{j} \cdot \vec{A} \right]$$

We can cancel the middle two terms by partially integrating as follows

$$\int dV \left[\vec{\mathscr{E}} \cdot \nabla\Phi + \rho\Phi\right] = \int dV \left[\left(-\nabla \cdot \vec{\mathscr{E}}\right)\Phi + \rho\Phi\right] = \int dV \left[-\rho\,\Phi + \rho\Phi\right] = 0$$

where we have assumed periodic or fixed-endpoint boundary conditions in order to neglect the surface term resulting from the integration. The EM and interaction Hamiltonian becomes

$$H_{em} + H_{int} = \int dV \left[\frac{1}{2}\left(\mathscr{E}^2 + \mathscr{B}^2\right) - \vec{j} \cdot \vec{A}\right]$$

In MKS units, the Hamiltonian can be written as

$$H_{em} + H_{int} = \int dV \left[\frac{\mu_o}{2}\vec{\mathscr{H}}^2 + \frac{\varepsilon_o}{2}\vec{\mathscr{E}}^2\right] + \int dV \left(-\vec{j} \cdot \vec{A}\right) \qquad (7.5.21)$$

Recall for free fields $\vec{\mathscr{B}} = \mu_o\vec{\mathscr{H}}$.

Finally, we can show the last term in Equation 7.5.21 can be reduced for point particles. The current density is

$$\vec{j} = q\rho\vec{v}$$

where q, ρ, $\vec{v}$ are the charge of the particle ($q < 0$ for electrons), the charge density and the velocity of the charge density respectively. For a point particle, we must have

$$\rho = \delta(x - x_o)$$

where x_o must be the position of the particle at the time of interest. Therefore, the Hamiltonian in Equation 7.5.21 becomes

$$H_{em} + H_{int} = \int dV \left[\frac{\mu_o}{2}\vec{\mathscr{H}}^2 + \frac{\varepsilon_o}{2}\vec{\mathscr{E}}^2\right] - q\vec{v} \cdot \vec{A} \qquad (7.5.22)$$

The last term becomes the starting point for showing the Hamiltonian for the matter-light interaction in the form of the dipole moment and also for demonstrating the conventional form of the Hamiltonian in the next topic. Notice that the last term comes straight from Equation 7.5.2 without any need for the fancy mathematical manipulations to find the EM Hamiltonian. We could have used the results from Equation 3.5.11 in Section 3.5 with an extra time derivative to convert from power to energy.

Hamiltonians use momentum rather than velocity. We need to convert the last term in Equation 7.5.22 into one involving momentum. In the low field limit, we make the identification $\vec{p} = m\vec{v}$ and find

$$H_{em} + H_{int} = \int dV \left[\frac{\mu_o}{2}\vec{\mathscr{H}}^2 + \frac{\varepsilon_o}{2}\vec{\mathscr{E}}^2\right] - \frac{q}{m}\vec{p} \cdot \vec{A} \qquad (7.5.23)$$

For larger fields, the momentum $\vec{p}$ must include both the particle and field momenta. This last equation represents the classical Hamiltonian for the fields and the interaction. It does not include a Hamiltonian for the particle and therefore does not provide any particle dynamics. Equation 7.5.23 treats the particle momentum as unchanging and independent of time. The next topic shows the Hamiltonian that includes the particle dynamics.

As a comment, if we had used "proper time" for calculating the extremum of the action, then the Lagrangian $\mathscr{L}$ and all of the calculations leading to the Hamiltonian would have observed the principles of relativity. We didn't go through any extra work simply because of the ease with which Maxwell's equations can be written in covariant form using the 4-vector notation. In fact, we used it as a compact notation without emphasizing the importance for special relativity. For simplicity, the next section abandons the formalism and assumes the charged particles travel much slower than the speed of light.

7.5.5   The Hamiltonian for Fields, Particles and their Interactions

The typical form of the classical Hamiltonian (energy and not energy density) for the entire system includes the particle, fields and their interactions according to

$$H = \frac{\left(\vec{p} - q\vec{A}\right)^2}{2m} + V(\vec{r}) + H_{em} \tag{7.5.24}$$

where V denotes the usual potential energy associated with the environment of the electron and H_{em} represents the Hamiltonian for the free EM fields. Squaring out the kinetic term provides

$$H = \frac{p^2}{2m} - \frac{q}{2m}\left(\vec{p}\cdot\vec{A} + \vec{A}\cdot\vec{p}\right) + \frac{q^2}{2m}A^2 + V(\vec{r}) + H_{em} \tag{7.5.25}$$

The second term can be reduced to agree with Equation 7.5.23

$$-\frac{q}{2m}\left(\vec{p}\cdot\vec{A} + \vec{A}\cdot\vec{p}\right) = -\frac{q}{m}\vec{p}\cdot\vec{A} \tag{7.5.26}$$

The matter-light interaction uses the quantum mechanical Hamiltonian for a charged particle where the dynamical variables become operators that don't necessarily commute. In fact, the original form of the second term in Equation 7.5.25 is most appropriate because, with operators, it can be seen to be Hermitian.

We now derive the most common form of the Hamiltonian given in Equation 7.5.24. We will need to generalize the Lagrangian density $\mathscr{L} = -(1/4)F_{\mu\nu}F^{\mu\nu} - j^\mu A_\mu$ in the previous section to include the particle dynamics. This will also generalize the previous form of the Hamiltonian $H_{em} + H_{int} = \int dV\,[(\mu_0/2)\vec{\mathscr{H}}^2 + (\varepsilon_0/2)\vec{\mathscr{E}}^2] - (q/m)\vec{p}\cdot\vec{A}$.

The total Lagrangian can be written as the summation of the terms for the free particle, the EM fields and the interaction between them

$$L = L_{\substack{\text{free} \\ \text{particle}}} + \int dV \sum_{\mu\nu}\left(-\frac{1}{4}F_{\mu\nu}F^{\mu\nu} - j^\mu A_\mu\right) \tag{7.5.27}$$

The kinetic energy for the free particle

$$L_{\substack{\text{free} \\ \text{particle}}} = \sum_i \frac{1}{2}m\dot{\eta}_i^2 \tag{7.5.28}$$

uses η_i to mean the ith position coordinate of the free particle (recall that for the EM fields, we treat x, y, z as indices). Also recall that the current-field interaction term can be rewritten as

$$-\int dV \, j^\mu A_\mu = -\int dV \left(\rho \Phi - \vec{j} \cdot \vec{A} \right) = q\vec{v} \cdot \vec{A} - \int dV \, \rho \Phi \tag{7.5.29}$$

where we used the relations $\rho = qn$, $\vec{j} = qn\vec{v}$, and $n(x) = \delta(x - x')$ for a single point particle. Combining Equations 7.5.28 and 7.5.29 into 7.5.27, we find

$$L = \sum_i \frac{1}{2} m\dot{\eta}_i^2 + \int dV \sum_{\mu\nu} \left(-\frac{1}{4} F_{\mu\nu} F^{\mu\nu} \right) - \int dV \, \rho \Phi + q\vec{v} \cdot \vec{A} \tag{7.5.30}$$

where $\vec{v} = \dot{\eta}_1 \hat{x} + \dot{\eta}_2 \hat{y} + \dot{\eta}_3 \hat{z}$.

Looking at Equation 7.5.30, we see that we can divide the total Lagrangian L into two parts

$$L_1 = \sum_i \frac{1}{2} m\dot{\eta}_i^2 + q\vec{v} \cdot \vec{A} \quad L_2 = \int dV \sum_{\mu\nu} \left(-\frac{1}{4} F_{\mu\nu} F^{\mu\nu} \right) - \int dV \, \rho \Phi \tag{7.5.31}$$

so that the coordinates for the point particle do not enter into L_2. The classical Hamiltonian can be written as $H = H_1 + H_2$ where the previous section shows

$$H_2 = \int dV \left[\frac{\mu_o}{2} \vec{\mathscr{H}}^2 + \frac{\varepsilon_o}{2} \vec{\mathscr{E}}^2 \right] \tag{7.5.32}$$

and where we now define

$$H_1 = \left(\sum_i p_i \dot{\eta}_i - L_1 \right) = \sum_i p_i \dot{\eta}_i - \sum_i \frac{1}{2} m\dot{\eta}_i^2 - q\vec{v} \cdot \vec{A} \tag{7.5.33}$$

so that the total Hamiltonian must have the form

$$H = H_1 + H_2 = \left(\sum_i p_i \dot{\eta}_i - L_1 \right) + H_2 = \left(\sum_i p_i \dot{\eta}_i - \sum_i \frac{1}{2} m\dot{\eta}_i^2 - q\vec{v} \cdot \vec{A} \right) + H_2 \tag{7.5.34}$$

The momentum p_i must be conjugate to the coordinate η_i. We take the electric field in this last equation to be the transverse component of the travelling wave. Any electrostatic potential can be included using the potential V. Equation 7.5.34 has Hamiltonians for the free particle, particle-field interaction, and the free EM fields.

We want to reduce this last equation to the form $H = (\vec{p} - q\vec{A})^2/2m + V(\vec{r}) + H_{em}$. We assume the symbol V represents an external potential originating from fields not included in the kinetic term. Because the coordinates $\dot{\eta}_i$ do not appear in $H_2 = H_{em}$ we can work with Equation 7.5.33 in isolation. We need to find the momentum p_i and the rewrite H_1 in terms of η_i and p_i. In the usual manner, we find

$$p_j = \frac{\partial L_1}{\partial \dot{\eta}_j} = \frac{\partial}{\partial \dot{\eta}_j} \left(\sum_i \frac{1}{2} m\dot{\eta}_i^2 + q\vec{v} \cdot \vec{A} \right) = \frac{\partial}{\partial \dot{\eta}_j} \sum_i \left(\frac{1}{2} m\dot{\eta}_i^2 + q\dot{\eta}_i \cdot A_i \right) = m\dot{\eta}_j + qA_j \tag{7.5.35a}$$

The momentum p_j conjugate to the coordinate η_j is not equal to mv_j as a person might expect. The total momentum includes both the particle and the field-like momentum. This observation explains the result expressed by the previous equation.

$$p_j = m\dot{\eta}_j + qA_j \tag{7.5.35b}$$

The Hamiltonian does not explicitly involve the position $\vec{r}$ and therefore momentum must be conserved according to Hamilton's canonical equation

$$\dot{p}_j = -\frac{\partial H}{\partial x_j} = 0 \qquad \rightarrow \qquad \vec{p} = \text{constant}$$

The particle can exchange momentum with the field and vice versa just so long as the total is conserved.

The Hamiltonian H_1 in Equation 7.5.33 can be rewritten

$$\begin{aligned}
H_1 &= \sum_i \left(p_i \dot{\eta}_i - \frac{1}{2} m\dot{\eta}_i^2 - q\dot{\eta}_i A_i \right) \\
&= \sum_i \left[p_i \frac{p_i - qA_i}{m} - \frac{1}{2} m\left(\frac{p_i - qA_i}{m}\right)^2 - q\left(\frac{p_i - qA_i}{m}\right) A_i \right]
\end{aligned} \tag{7.5.36a}$$

We can rearrange terms to find

$$H_1 = \sum_i \left[(p_i - qA_i)\left(\frac{p_i - qA_i}{m}\right) - \frac{1}{2} m\left(\frac{p_i - qA_i}{m}\right)^2 \right] = \sum_i \frac{(p_i - qA_i)^2}{2m}$$

We can include the extra potential $V(\vec{r})$ to find

$$H_1 = \sum_i \frac{(p_i - qA_i)^2}{2m} + V(\vec{r}) \tag{7.5.36b}$$

or in vector notation

$$H_1 = \frac{(\vec{p} - q\vec{A})^2}{2m} + V(\vec{r}) \tag{7.5.36c}$$

where as usual, the square means to take the dot product. Finally, including the electromagnetic Hamiltonian, we find

$$H = \frac{(\vec{p} - q\vec{A})^2}{2m} + V(\vec{r}) + \int dV \left[\frac{\mu_o}{2}\vec{\mathcal{H}}^2 + \frac{\varepsilon_o}{2}\vec{\mathcal{E}}^2\right] \tag{7.5.37}$$

7.5.6   Discussion

The Hamiltonian in Equation 7.5.37 describes the free particle, the interaction between the particle and the propagating fields, the freely propagating fields, and the interaction between the particle and an extra external potential *V*. Usually *V* describes the environment of the charge *q* (for example the potential for a finitely deep well).

The Hamiltonian in Equation 7.5.37 can be quantized. For EM field quantization, we replace the vector potential with a Fourier summation that has operators for amplitudes. Of course, this requires the electric and magnetic fields to be similarly expressed. The quantization procedure can be found in Chapter 6. The kinetic energy term can be quantized by making the identification $\vec{p} = (\hbar/i)\nabla$.

The matter-light interaction term in Equation 7.5.37 comes from squaring the first term

$$-\frac{q}{2m}\left(\vec{p}\cdot\vec{A} + \vec{A}\cdot\vec{p}\right) + \frac{q^2}{2m}A^2 \tag{7.5.38a}$$

Using the Poynting vector in Chapter 3, we write the interaction as

$$-\frac{q}{m}\vec{p}\cdot\vec{A} \tag{7.5.38b}$$

The first two terms in Equation 7.5.38a reduce to 7.5.38b when $\vec{p}, \vec{A}$ are ordinary vectors. Also, the result in 7.5.38a has the "extra" A^2 term. At first glance, the two equations appear identical except for this A^2 term. However, this isn't true because in Equation 7.5.38a, we have $\vec{p} = m\vec{v} + q\vec{A}$ from Equation 7.5.35b, while in Equation 7.5.38b we have $\vec{p} = m\vec{v}$. The quantity $m\vec{v}$ refers to the momentum of the particle alone while $\vec{p} = m\vec{v} + q\vec{A}$ refers to the particle and field momentum. Equation 7.5.36a shows that the A^2 enters from terms of the form

$$\sum_i (p_i - qA_i)A_i = \vec{p}\cdot\vec{A} - q\vec{A}\cdot\vec{A} = \vec{p}\cdot\vec{A} - qA^2 \tag{7.5.39}$$

For "*A*" sufficiently small, we can approximate

$$\vec{p} = m\vec{v} + q\vec{A} \approx m\vec{v} \tag{7.5.40}$$

In view of Equations 7.5.39 and 7.5.40, we therefore see that for sufficiently small $\vec{A}$, Equation 7.5.38a reduces to 7.5.38b and the definitions of the two momentum become synonymous.

As a final comment for quantizing the EM field as in Chapter 6, when the vector potential is quantized in terms of the creation $\hat{b}^+$ and annihilation $\hat{b}^-$ operators, the A^2 produces terms of the form $\hat{b}^+\hat{b}^+$, $\hat{b}^+\hat{b}^-$, $\hat{b}^-\hat{b}^-$

These represent two photon processes. The first term creates two photons (emission), the second term destroys one but creates another (absorption and emission), and the last one removes two photons (absorption). Two photon processes only become important at large light levels, which occurs for large "*A*." That is, we require a large number of photons for the two-photon processes to become significant. Neglecting the two photon processes must be somewhat equivalent to the approximation given in 7.5.43.

7.6   The Quantum Hamiltonian for the Matter–Light Interaction

The classical Hamiltonian for a particle interacting with a field has the form

$$H = \underbrace{\frac{\left(\vec{p} - q\vec{A}\right)^2}{2m}}_{\substack{\text{Particle and} \\ \text{Particle-Field}}} + \underbrace{V(\vec{r})}_{\text{Atomic}} + \underbrace{H_{em}}_{\substack{\text{Free} \\ \text{Field}}} \cong H_{\substack{\text{Free} \\ \text{Particle}}} + H_{\text{int}} + V(\vec{r}) + H_{em}$$

where q, m, $\vec{A}$, $\vec{\mathscr{E}}$, $\mathscr{H}$ denote the charge of the particle ($q < 0$ for electrons), mass m of the particle, the vector potential, and the electric and magnetic fields, respectively. The symbol $\vec{p}$ represents the canonical momentum comprised of the particle and field momenta. For small fields $\vec{A}$, the canonical momentum $\vec{p}$ reduces to the particle momentum $m\vec{v}$.

We now discuss the quantum mechanical form of the Hamiltonian. Typically, the quantum Hamiltonian is found by replacing all of the dynamical variable with operators in the classical Hamiltonian (with care taken to properly order the operators). Rather than starting with the Lagrangian as done in the previous sections, we use the results from the Poynting vector found in Chapter 3. The first topic in this section therefore contains a brief summary of the procedure without the "overhead" required by the Lagrangian approach. We focus on the particle-field interaction term and do not include the free-field Hamiltonian.

As a final topic, we show the two forms for the interaction Hamiltonian namely $\hat{\mu} \cdot \hat{\mathscr{E}}$ and $\hat{p} \cdot \hat{A}$. The first type $\hat{\mu} \cdot \hat{\mathscr{E}}$ with the dipole operator $\hat{\mu}$ has already been discussed in connection with Fermi's golden rule. We will also need it for the Liouville treatment of the matter-light interaction. The second form will be used in connection with some of the material in the next chapter for computing the gain or absorption of semiconductor devices. Either way, the two expressions for the interaction energy are equivalent to one another.

7.6.1   Discussion of the Classical Interaction Energy

The classical Hamiltonian in Section 3.5 can be quantized to provide the semiclassical Hamiltonian and the proper form for the matter-light interaction. This topic shows the matter-light interaction has the form

$$H_{A-L} = -\frac{q}{2m}\left(\vec{p} \cdot \vec{A} + \vec{A} \cdot \vec{p}\right)$$

where $\vec{A}$ denotes the vector potential and $\vec{p}$ represents a momentum.

For the development of the quantum theory of the electromagnetic fields in Chapter 6, we started with the classical expression for the energy stored in an EM field. We then quantized it by substituting operators for the Fourier amplitudes, and required the operators to satisfy commutation relations. The expression for the electromagnetic power flow (Poynting vector) found in Chapter 3 not only contains reference to the free field Hamiltonian but also the power due to the interaction between the fields and matter. The interaction term in Equation 3.5.11 has the form

$$P = \int_V dV\, \vec{\mathscr{E}} \cdot \vec{j} \tag{7.6.1}$$

where $\vec{\mathscr{E}}$ and $\vec{j}$ represent the electric field and current density (Amps per area). We look for energy rather than power and must remove a time derivative. Chapter 6 shows that the electric field can be related to the vector potential $\vec{A}$ by

$$\vec{E} = -\frac{\partial \vec{A}(\vec{r}, t)}{\partial t}$$

when there isn't any electrostatic potential. Assuming that the electric field produces only small fluctuations in the current density, we can write Equation 7.6.1 as

$$P = -\frac{\partial}{\partial t} \int_V dV \, \vec{A} \cdot \vec{j}$$

and therefore the interaction energy *density* (energy per volume) must be

$$\mathscr{H}_{A-L} = -\vec{A} \cdot \vec{j}$$

To make sure that the Hamiltonian remains Hermitian once we substitute operators, let's write

$$\mathscr{H}_{A-L} = -\frac{1}{2}\left[\vec{A} \cdot \vec{j} + \vec{j} \cdot \vec{A}\right] \tag{7.6.2}$$

because in the quantum theory, this interaction Hamiltonian will have the form

$$\mathscr{H}_{A-L} = -\frac{1}{2}\left[\vec{A} \cdot \vec{j} + \left(\vec{A} \cdot \vec{j}\right)^+\right]$$

which must be explicitly Hermitian $\mathscr{H}_{A-L}^+ = \mathscr{H}_{A-L}$. We will take care not to arbitrarily commute the dynamical variables $\vec{A}$ and $\vec{j}$ even though we can do so in the classical theory. As a note, we will see that, upon integration, Equation 7.6.2 takes the form $\vec{p} \cdot \vec{A} + \vec{A} \cdot \vec{p}$ and, using the Coulomb gauge, the expression reduces to $\vec{A} \cdot \vec{p}$. This is equivalent to retaining only the $\vec{A} \cdot \vec{j}$ term in Equation 7.6.2.

The current density can be written as

$$\vec{j} = qn\vec{v} \tag{7.6.3}$$

where n denotes the charge density (e.g., number of electrons per volume), $\vec{v}$ denotes the average speed of the charge, and $q < 0$ for electrons. Assume just one point charge in the volume. The charge density must be related to a Dirac delta function

$$n = \delta(\vec{x} - \vec{x}_o) \tag{7.6.4}$$

so that when we integrate overall space, we find exactly one electron. Substituting the current density into the interaction Hamiltonian

$$H_{A-L} = -\int_V dV \, \vec{A} \cdot \vec{j} \tag{7.6.5}$$

provides

$$H_{A-L} = -\int_V dV \, \vec{A} \cdot \vec{j} = -\int_V dV \, \vec{A} \cdot qn\vec{v} = -q\vec{v} \cdot \vec{A} \qquad (7.6.6)$$

We want to write the interaction Hamiltonian in terms of momentum rather than velocity (since that's how the Hamiltonian should be written). Here's where a discrepancy occurs between the classical Hamiltonian field theory (Section 7.5) and using the Poynting vector approach. Normally, we would say that the appropriate momentum for Equation 7.6.6 must be $\vec{p} = m\vec{v}$ so that the interaction Hamiltonian becomes

$$H_{A-L} = -\int_V dV \, \vec{A} \cdot \vec{j} = -\frac{q}{m}\vec{p} \cdot \vec{A} = -\frac{q}{2m}\left(\vec{p} \cdot \vec{A} + \vec{A} \cdot \vec{p}\right) \qquad (7.6.7)$$

However, the momentum conjugate to the charge position vector, according to the classical field theory approach in Equation 7.5.35b, must be

$$\vec{p} = m\vec{v} + q\vec{A} \qquad (7.6.8a)$$

The total momentum of the system must be the mechanical momentum of the electron plus a contibution related to the field. Basically, the Hamiltonian does not explicitly involve the particle position (lacks potential) and therefore momentum must be conserved according to Hamilton's canonical equation

$$\dot{p}_j = -\frac{\partial H}{\partial x_j} = 0 \qquad \rightarrow \qquad \vec{p} = \text{constant}$$

The particle can exchange momentum with the field and vice versa so long as the total remains constants. Now, if the momentum of the field remains negligibly small, then

$$\vec{p} = m\vec{v} + q\vec{A} \cong m\vec{v} \qquad (7.6.8b)$$

and Equation 7.6.6 will be a reasonable approximation. The next topic discusses the quantum mechanical form of the Hamiltonian that includes the interaction. It turns out, we never really need to identify the momentum $\vec{p}$ with the mechanical momentum of the electron alone (and it's incorrect to do so).

7.6.2 Schrodinger's Equation with the Matter–Light Interaction

The previous topic indicates the interaction Hamiltonian has the form

$$H_{A-L} = -\frac{q}{m}\vec{p} \cdot \vec{A} = -\frac{q}{2m}\left(\vec{p} \cdot \vec{A} + \vec{A} \cdot \vec{p}\right) \qquad (7.6.9)$$

where q, m, $\vec{p}$, $\vec{A}$ denote the charge ($q < 0$ for electrons), mass m, total momentum of particle and EM field (approximately particle momentum), and vector potential, respectively. Now if we include the usual Hamiltonian for a particle with kinetic and potential energy, we can write

$$H = \frac{p^2}{2m} + V + H_{A-L} = \frac{p^2}{2m} - \frac{q}{2m}\left(\vec{p} \cdot \vec{A} + \vec{A} \cdot \vec{p}\right) + V \qquad (7.6.10)$$

Equation 7.6.10 can be approximately written as

$$H = \frac{\left(\vec{p} - q\vec{A}\right)^2}{2m} + V \tag{7.6.11}$$

This last equation has exactly the correct form. Had we followed the classical field theory approach given in Sections 7.4 and 7.5, we would have found the extra term $q^2 A^2/2m$ appearing in Equation 7.6.9.10. Apparently, neglecting this extra term must be equivalent to assuming small field momentum so that $\vec{p} = m\vec{v} + q\vec{A} \cong m\vec{v}$. This "extra" term accounts for two photon processes.

In quantum theory, we know to replace the dynamical variables in Equation 7.6.10 with operators and require them to satisfy certain commutation relations. In the semiclassical approach, we replace the vector $\vec{p}$ with $(\hbar/i)\nabla$ but leave the vector potential as a classical vector. To include the quantum nature of the EM fields, we must also make the vector potential an operator.

For the semiclassical quantum description of a charged particle interacting with electromagnetic fields, we can write

$$\hat{H} = \frac{\left(\hat{p} - q\vec{A}\right)^2}{2m} + \hat{V} = \frac{1}{2m}\left(\frac{\hbar}{i}\nabla - q\vec{A}\right)^2 + \hat{V} \tag{7.6.12}$$

where Schrodinger's equation has the form

$$\hat{H}|\psi\rangle = i\hbar\frac{\partial}{\partial t}|\psi\rangle \tag{7.6.13}$$

The wave function $|\psi\rangle$ consists only of the electron basis vectors for the semiclassical approach. Once we quantize the fields however, then the wave function must include basis vectors of the Hilbert space for the EM wave.

7.6.3 The Origin of the Dipole Operator

Let us now demonstrate how the interaction potential depends on the dipole operator in $\hat{H}_{A-L} = \hat{V} = \hat{\mu} \cdot \hat{\mathscr{E}}$. We first used this expression in Sections 7.1–7.3 without proof. The interaction potential perturbs the atom and causes the electron to make transitions between atomic basis states. These transitions occur when the atom absorbs or emits photons. Having laid all the preliminary groundwork, let's now follow some of the development given in Yariv's Quantum Electronics and his references.

For quantum particles interacting with EM fields, we can write

$$\hat{H} = \frac{\left(\hat{p} - e\hat{A}\right)^2}{2m} + \hat{V}(x) \tag{7.6.14a}$$

where the free particle Hamiltonian (atomic Hamiltonian) is

$$\hat{H}_A = \frac{\hat{p}^2}{2m} + \hat{V}(x) \tag{7.6.14b}$$

and A denotes the vector potential. First, we will show the full Hamiltonian can be written as

$$\hat{H} = \hat{H}_A - \frac{e}{m}\hat{p} \cdot \hat{A}$$
(7.6.14c)

where the term $\hat{p} \cdot \hat{A} + \hat{A} \cdot \hat{p}$ in Equation 7.6.14a must be shown to reduce to $\hat{p} \cdot \hat{A}$ in Equation 7.6.13c.

The Hamiltonian in Equation 7.6.14a can be written as

$$\hat{H} = \frac{\left(\hat{p} - e\hat{A}\right)^2}{2m} + \hat{V} = \hat{H}_A - \frac{e\,\hat{p} \cdot \hat{A} + e\,\hat{A} \cdot \hat{p} - e^2\hat{A}^2}{2m}$$
(7.6.15)

where we must be careful not to commute the momentum and vector potential operators since $[\hat{A}, \hat{p}] \neq 0$. We focus on the term $\vec{p} \cdot \vec{A} - \vec{A} \cdot \vec{p}$. It can be cast into an alternate form by working in the coordinate representation $\vec{p} = (\hbar/i)\nabla$. Keeping in mind that $\vec{p} \cdot \vec{A} - \vec{A} \cdot \vec{p}$ must operate on functions ψ, we calculate $(\vec{p} \cdot \vec{A} - \vec{A} \cdot \vec{p})\psi$ to find

$$\vec{p} \cdot \vec{A} - \vec{A} \cdot \vec{p} = \frac{\hbar}{i}\nabla \cdot \vec{A}$$

However, working in the Coulomb gauge, we require the Coulomb gauge condition $\nabla \cdot \vec{A} = 0$ to hold. We therefore see that

$$\vec{p} \cdot \vec{A} = \vec{A} \cdot \vec{p}$$

in the Coulomb gauge. The Hamiltonian can now be written as

$$\hat{H} = \hat{H}_A - \frac{e}{m}\hat{p} \cdot \hat{A}$$
(7.6.16)

where we neglect the term $\hat{A}^2$ in Equation 7.6.14b since it refers to two-photon processes. For those who have read Chapter 6, this is easy to see by writing the vector potential in terms of creation and annihilation operators and considering Fock space matrix elements such as

$$\langle n + 2|\hat{A}^2|n\rangle \sim \langle n + 2|\left(\hat{b}^+\right)^2|n\rangle + \cdots$$

By the way, the creation operator $\hat{b}^+$ has the "b" to represent "boson" (the photon is a boson since it has integer spin). Equation 7.6.15 shows the matter-light interaction potential must be

$$\hat{H}_{A-L} = -\frac{e}{m}\hat{p} \cdot \hat{A}$$
(7.6.16)

We want to calculate transition matrix elements of the form $(\hat{H}_{A-L})_{fi} = \langle u_f, \gamma_f|\hat{H}_{A-L}|u_i, \gamma_i\rangle$ for the perturbation theory and Fermi's golden rule. In this last expression, the states $|u_i\rangle$ and $|u_f\rangle$ are eigenstates of the unperturbed Hamiltonian $\hat{H}_A$, and $|\gamma_i\rangle$ and $|\gamma_f\rangle$ are the initial and final basis states of the light-space.

We want to separate the matix element into the product of two matrix elements $\langle u_f | \hat{p} | u_i \rangle \cdot \langle \gamma_f | \hat{A} | \gamma_i \rangle$. We need two pieces of information. First, the dipole approximation allows us to separate out the matrix element for the vector potential $\langle \gamma_f | \hat{A} | \gamma_i \rangle$. However, $\hat{p}$ still has reference to the vector potential. Second, we make the low field approximation given in Equation 7.6.8a, specifically $\hat{p} = \hat{p}_e + q\hat{A} \cong \hat{p}_e$, where $\hat{p}_e$ refers to particle (i.e., electron) momentum. This allows us to write $\langle u_f | \hat{p} | u_i \rangle$ since then $\hat{p}$ does not contain any reference to the photon states. This is equivalent to dropping the $\hat{A}^2$ term. The matrix element becomes

$$\left(\hat{H}_{A-L} \right)_{fi} = \langle u_f, \gamma_f | \hat{H}_{A-L} | u_i, \gamma_i \rangle = -\frac{e}{m} \langle u_f | \hat{p} | u_i \rangle \cdot \langle \gamma_f | \hat{A} | \gamma_i \rangle \tag{7.6.18}$$

As a comment, many times the photon Fock states $|n\rangle$ are used for the states $|\gamma\rangle$ to show that photons can be removed (absorbed) from the field or added (emitted) into the field. Fock states $|n\rangle$ give the number of photons in the mode so that the creation and annihilation operators give $\hat{b}^+ |n\rangle = \sqrt{n+1} |n+1\rangle$ and $\hat{b}^- |n\rangle = \sqrt{n} |n-1\rangle$, respectively.

We can now finish showing the dipole form of the interaction energy $\hat{H}_{A-L} = \hat{\mu} \cdot \vec{\mathcal{E}}$. We start with the relation

$$\vec{p} = \frac{im}{\hbar} \left[\hat{H}_A, \hat{r} \right] \tag{7.6.19a}$$

(see Problem 7.14). The matrix element of the momentum operator in Equation 7.6.17 becomes

$$\langle u_f | \hat{p} | u_i \rangle = \frac{im}{\hbar} \langle u_f | \left[\hat{H}_A, \hat{r} \right] | u_i \rangle \tag{7.6.19b}$$

or, by expanding the commutator, and noting that the kets are eigenstates of the unperturbed Hamiltonian, we can write

$$\langle u_f | \hat{p} | u_i \rangle = \frac{im}{\hbar} (E_f - E_i) \langle u_f | \hat{r} | u_i \rangle \tag{7.6.20}$$

Equation 7.6.17 becomes

$$\left(\hat{H}_{A-L} \right)_{fi} = \langle u_f, \gamma_f | \hat{H}_{A-L} | u_i, \gamma_i \rangle = -\frac{i}{\hbar} (E_f - E_i) \langle u_f | e\hat{r} | u_i \rangle \cdot \langle \gamma_f | \hat{A} | \gamma_i \rangle \tag{7.6.21}$$

If the atom absorbs a photon then the final electron energy must be larger than the initial energy $E_f - E_i > 0$. The reverse must be true for the case of an atom emitting a photon.

Next we must incorporate the operator form of the vector potential in terms of creation and annihilation operators. The operator form of the vector potential as shown in 6.4.4 is

$$\hat{A}(\vec{r}, t) = \frac{1}{\sqrt{\varepsilon_o V}} \sum_{\vec{k}} \sqrt{\frac{\hbar}{2\omega_{\vec{k}}}} \, \tilde{e}_{\vec{k}} \left[\underbrace{\hat{b}_{\vec{k}} e^{i\vec{k}\cdot\vec{r} - i\omega_{\vec{k}} t}}_{\text{absorption}} + \underbrace{\hat{b}_{\vec{k}}^+ e^{-i\vec{k}\cdot\vec{r} + i\omega_{\vec{k}} t}}_{\text{emission}} \right] \tag{7.6.22}$$

where $\tilde{e}_{\vec{k}}$ gives the polarization unit vector of the plane wave. In principle, the summation should run over all indices "s" in $\tilde{e}_{\vec{k},s}$ (which can be one of the unit vectors $\tilde{x}, \tilde{y}$ when the wave travels along the z-direction). However, for linearly polarized light, we can always reduce the summation to Equation 7.6.22 by defining $\tilde{e}_{\vec{k}}$ to be along the direction of wave polarization rather than along the unit vectors $\tilde{x}, \tilde{y}$.

The matrix elements $\langle \gamma_f | \hat{A} | \gamma_i \rangle$ provide either emission or absorption as indicated in Equation 7.6.22. Using energy conservation, we realize that absorption of a photon requires $E_f - E_i = \hbar \omega_{\vec{k}}$, whereas emission of a photon requires $E_i - E_f = \hbar \omega_{\vec{k}}$. Therefore, the energy terms and matrix element in Equation 7.6.22 can be rewritten as

$$i \frac{E_f - E_i}{\hbar} \langle \gamma_f | \hat{A} | \gamma_i \rangle = \langle \gamma_f | -\frac{\partial}{\partial t} \hat{A} | \gamma_i \rangle \qquad (7.6.23)$$

Therefore, Equation 7.6.21 becomes

$$\left(\hat{H}_{A-L} \right)_{fi} = -\frac{i}{\hbar} (E_f - E_i) \langle u_f | e\hat{r} | u_i \rangle \cdot \langle \gamma_f | \hat{A} | \gamma_i \rangle = \langle u_f | -e\hat{r} | u_i \rangle \cdot \langle \gamma_f | -\frac{\partial}{\partial t} \hat{A} | \gamma_i \rangle \qquad (7.6.24)$$

We see that the interaction Hamiltonian can be equivalently written as

$$\hat{H}_{A-L} = \hat{\mu} \cdot \hat{\mathscr{e}} \qquad (7.6.25)$$

where $\hat{\mu} = -e\hat{r}$.

7.6.4 The Semiclassical Form of the Interaction Hamiltonian

Let us now demonstrate the semiclassical form of the interaction used in connection with Fermi's golden rule. Using the coherent state $|\alpha\rangle$ we define the semiclassical interaction Hamiltonian as

$$\hat{H}_{AL} = \langle \alpha | \hat{H}_{A-L} | \alpha \rangle = \hat{\mu} \cdot \langle \alpha | \hat{\mathscr{e}} | \alpha \rangle = \hat{\mu} \cdot \vec{\mathscr{e}} \qquad (7.6.26)$$

where the classical electric field can be found in Section 7.2.

7.7 Stimulated and Spontaneous Emission Using Fock States

Fermi's golden rule predicts the rate of stimulated emission for the semiclassical system consisting of the quantized atomic system with a classical electromagnetic wave. However, this semiclassical approach does not predict the rate of spontaneous emission nor does it include damping effects. This section modifies the Fermi formula to include the quantum nature of the EM fields (but it still does not include damping terms). The matrix elements of the interaction Hamiltonian in the Fock basis shows how the vacuum state and the inherent vacuum fluctuations give rise to spontaneous emission.

7.7.1 Restatement of Fermi's Golden Rule

Fermi's golden rule can be restated using Fock states and the quantized electromagnetic fields. Let $R_{i \to f}$ be the emission rate from a single initial state $|i\rangle$ to a single final state $|f\rangle$

$$R_{i \to f} = \frac{2\pi}{\hbar} \left| \hat{\mathscr{V}}_{fi} \right|^2 \delta(E_f - E_i) \tag{7.7.1}$$

where $\hat{\mathscr{V}}_{fi}$ denotes the matrix element of the interaction potential for the transition. Things are not as they appear however. The initial and final states must reside in a direct-product Hilbert space since the full system has an atom and the electromagnetic field as subsystems. Let's assume that the atom starts in an excited state $|2\rangle$, emits a photon with frequency ω_k, and ends up in the state $|1\rangle$. Initially, the light field consists of n_k photons in the single-mode Fock state $|n_k\rangle$ (of frequency ω_k) and then makes a transition to the state $|n_k + 1\rangle$ when the atom emits the photon. Figure 7.7.1 shows the situation. The initial and final states can be written as

$$|i\rangle = |2, n_k\rangle = |2\rangle|n_k\rangle \quad \text{and} \quad |f\rangle = |1, n_k + 1\rangle = |1\rangle|n_k + 1\rangle$$

Notice how separate and distinct Hilbert spaces represent the atom and photon field. Without an interaction Hamiltonian to link the two spaces, the atom would evolve according to its Hamiltonian, while the light evolves according to the free-field Hamiltonian. Only by including an interaction term $\hat{V}$ does the evolution of a state in one space affect the evolution of a state in the other. Also notice that the Dirac delta function in Equation 7.7.1 must include both the photon and the electron energy.

The matter-light interaction for a single mode "k" can be written

$$\hat{V} = \hat{\mu} \cdot \hat{E}_k \tag{7.7.2}$$

as discussed in the previous section. For traveling waves, the electric field operator has the form

$$\hat{E}_k(\vec{r}, t) = -i\tilde{e}_{k,s}\sqrt{\frac{\hbar \omega_k}{2\varepsilon_o V}}\left[\hat{b}_{ks}(t)e^{-i\vec{k}\cdot\vec{r}} - \hat{b}_{ks}^+(t)e^{+i\vec{k}\cdot\vec{r}}\right] \tag{7.7.3}$$

where $s = 1,2$ gives the state of polarization and where

$$\hat{b}_{ks}(t) = \hat{b}_{ks} e^{+i\omega_k t} \quad \text{and} \quad \hat{b}_{ks}^+(t) = \hat{b}_{ks}^+ e^{-i\omega_k t}$$

Recall, Fermi's golden rule explicitly accounts for the time dependence of the electric field; we included $e^{\pm i\omega t}$ in the expression and then integrated. In order not to include the

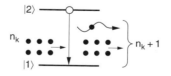

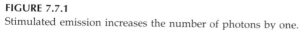

FIGURE 7.7.1
Stimulated emission increases the number of photons by one.

$e^{\pm i\omega t}$ a second time for this quantized field, we should set $t = 0$ in Equation 7.7.3 to eliminate the time parameter. Therefore, the electric field becomes

$$\hat{E}_k(\vec{r}) = -i\tilde{e}_{k,s}\sqrt{\frac{\hbar\omega_k}{2\varepsilon_0 V}}\left[\hat{b}_{ks}e^{-i\vec{k}\cdot\vec{r}} - \hat{b}_{ks}^{+}e^{+i\vec{k}\cdot\vec{r}}\right] \tag{7.7.4}$$

Section 7.3 stated Fermi's golden rule for either absorption or emission; however, both can be included by using Equation 7.7.4.

Fermi's golden rule becomes

$$R_{i\to f} = \frac{2\pi}{\hbar}\left|\hat{\mathcal{V}}_{fi}\right|^2 \delta(E_2 - E_1 - \hbar\omega_k)$$

$$= \frac{2\pi}{\hbar}\left|\langle 1, n_k + 1|\,\hat{\mu}\cdot\hat{E}_k(\vec{r})\,|2, n_k\rangle\right|^2 \delta(E_2 - E_1 - \hbar\omega_k) \tag{7.7.5}$$

To continue, we make the "dipole approximation." The approximation assumes the wavelength to be significantly larger than the size of the atom. The approximation remains valid for x-rays with atoms on the order of 10 nm or smaller.

7.7.2 The Dipole Approximation

Now we can apply the dipole approximation (Appendix 8) to Equation 7.7.5 by noting that the inner product

$$\langle 1, n_k + 1|\,\hat{\mu}\cdot\hat{E}_k(\vec{r})\,|2, n_k\rangle$$

has an integral over the spatial variables due to the atomic states $|1\rangle$, $|2\rangle$ where the dipole moment operator is $\hat{\mu} = -e\vec{r}$. Assume the wavelength of the electric field is much larger than the size of the dipole defined by the states $|1\rangle$ and $|2\rangle$ as shown in Figure 7.7.2. In such a case, the dipole approximation provides

$$\langle 1, n_k + 1|\,\hat{\mu}\cdot\hat{E}_k(\vec{r})\,|2, n_k\rangle = \langle n_k + 1|\ \langle 1|\hat{\mu}|2\rangle\,\cdot\hat{E}_k(\vec{r})\ |n_k\rangle$$

Now, because the dipole moment doesn't have any field operators (i.e., creation or annihilation operators) associated with it, we can move it outside the Fock-space inner product.

$$\langle 1, n_k + 1|\,\hat{\mu}\cdot\hat{E}_k(\vec{r})\,|2, n_k\rangle = \langle 1|\hat{\mu}|2\rangle\,\cdot\,\langle n_k + 1|\hat{E}_k(\vec{r})|n_k\rangle$$

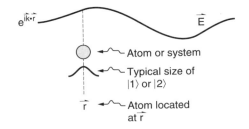

FIGURE 7.7.2

The electron wavefunction is nonzero over a region that is small compared with the wavlength of light.

For simplicity, we assume that the atom is located at $\vec{r} = 0$.

$$\langle 1, n_k + 1 | \hat{\mu} \cdot \hat{E}_k(\vec{r}) | 2, n_k \rangle \cong \langle 1 | \hat{\mu} | 2 \rangle \cdot \langle n_k + 1 | \hat{E}_k(0) | n_k \rangle$$

Notice the dot product between the two matrix elements since both the dipole moment operator and the electric field have a direction associated with them. Substituting this last result into Equation 7.7.5, we can write the rate of transition in the computationally simpler form

$$R_{i \to f} = \frac{2\pi}{\hbar} \left| \langle 1 | \hat{\mu} | 2 \rangle \cdot \langle n_k + 1 | \hat{E}_k(0) | n_k \rangle \right|^2 \delta(E_2 - E_1 - \hbar \omega_k) \tag{7.7.6}$$

7.7.3 Calculate Matrix Elements

Let us now continue to evaluate the rate of transition using Equation 7.7.6. Defining the expectation value of the dipole operator as

$$\vec{\mu}_{12} = \langle 1 | \hat{\mu} | 2 \rangle = \int dV \, u_1^*(\vec{r}) \, \hat{\mu}(\vec{r}) \, u_2(\vec{r})$$

Next we evaluate the expectation value of the electric field operator

$$\langle n_k + 1 | \hat{E}_k(0) | n_k \rangle$$

by substituting the operator form of the single-mode electric field from Equation 7.7.4.

$$\hat{E}_k(\vec{r}) = i \, \tilde{e}_{k,s} \sqrt{\frac{\hbar \omega_k}{2 \varepsilon_o V}} \left[\hat{b}_{ks} e^{-i\vec{k}\cdot\vec{r}} - \hat{b}_{ks}^+ e^{+i\vec{k}\cdot\vec{r}} \right]$$

The matrix element becomes

$$\langle n_k + 1 | \hat{E}_k(0) | n_k \rangle = \langle n_k + 1 | \left\{ i \tilde{e}_{k,s} \sqrt{\frac{\hbar \omega_k}{2 \varepsilon_o V}} \left[\hat{b}_{ks} - \hat{b}_{ks}^+ \right] \right\} | n_k \rangle$$

Distributing the linear operators to each term provides

$$\langle n_k + 1 | \hat{E}_k(0) | n_k \rangle = i \tilde{e}_{k,s} \sqrt{\frac{\hbar \omega_k}{2 \varepsilon_o V}} \left\{ \langle n_k + 1 | \hat{b}_{ks} | n_k \rangle - \langle n_k + 1 | \hat{b}_{ks}^+ | n_k \rangle \right\} \tag{7.7.7}$$

Using the definition of the annihilation operator and the orthonormality of the Fock states, the first term becomes

$$\langle n_k + 1 | \hat{b}_{ks} | n_k \rangle = \sqrt{n_k} \langle n_k + 1 \mid n_k - 1 \rangle = 0$$

The second term in Equation 7.7.7 is easily identified as the emission term since the creation operator adds a photon to the field. The emission term provides

$$\langle n_k + 1 | \hat{E}_k(0) | n_k \rangle = i \tilde{e}_{k,s} \sqrt{\frac{\hbar \omega_k}{2 \varepsilon_o V}} \langle n_k + 1 | \hat{b}_{ks}^+ | n_k \rangle$$

$$= i \tilde{e}_{k,s} \sqrt{\frac{\hbar \omega_k}{2 \varepsilon_o V}} \sqrt{n_k + 1} \tag{7.7.8}$$

Now we can write the rate of transition (i.e., Fermi's golden rule) as

$$R_{i \to f} = \frac{2\pi}{\hbar} \left| \vec{\mu}_{12} \cdot i\tilde{e}_{k,s}\sqrt{\frac{\hbar\omega_k}{2\varepsilon_o V}} \sqrt{n_k + 1}\, e^{-i\omega_k t} \right|^2 \delta(E_2 - E_1 - \hbar\omega_k)$$

Calculating the modulus provides

$$R_{i \to f} = \frac{2\pi}{\hbar} \left(\vec{\mu}_{12} \cdot \tilde{e}_{k,s} \right)^2 \frac{\hbar\omega_k}{2\varepsilon_o V} (n_k + 1)\ \delta(E_2 - E_1 - \hbar\omega_k) \qquad (7.7.9)$$

where we assume that the dipole moment and the polarization unit vector are real. Notice the dot product between the polarization and the induced dipole moment. If the dipole moment is perpendicular to the direction of the electric field there won't be any transition. This fact turns out to be important for certain quantum well devices. Equation 7.7.9 does not give the complete story because it does not account for damping and therefore cannot explain broadening mechanisms; this will be remedied in the next few sections.

7.7.4 Stimulated and Spontaneous Emission

Previous sections have discussed how the full electrodynamic system consists of several subsystems including the atom, the light field and the environment. The complete Hamiltonian must include these systems along with the interactions between them. We have limited our scope in the present section to the matter-light interaction. Also, we have not stated the model for the system (the Hamiltonian) in a manner that describes the possible broadening mechanisms due to the interaction; this will be remedied in the next few sections. Starting with Fermi's golden rule, we substitute the interaction Hamiltonian $\hat{\mu} \cdot \hat{\mathscr{E}}$ and find the matrix elements for the direct product space consisting of the atomic and photonic states. We find that the electron can make an upward or downward transition by subtracting or adding a quantum of energy to the photon field. The previous few topics focus on the downward transitions to highlight the stimulated and spontaneous emission processes.

Equation 7.7.9,

$$R_{i \to f} = \frac{2\pi}{\hbar} \left(\vec{\mu}_{12} \cdot \tilde{e}_{k,s} \right)^2 \frac{\hbar\omega_k}{2\varepsilon_o V} (n_k + 1)\ \delta(E_2 - E_1 - \hbar\omega_k)$$

specifically consists of the sum of two terms. Stimulated emission requires photons to induce the transition. Chapter 2 shows that larger numbers of photons produce greater rates of transition. Therefore we identify stimulated emission with the photon term n_k in Equation 7.7.9. The rate of stimulated emission must be

$$R_{i \to f}^{stim} = \frac{2\pi}{\hbar} \left(\vec{\mu}_{12} \cdot \tilde{e}_{k,s} \right)^2 \frac{\hbar\omega_k}{2\varepsilon_o V}\ n_k\ \delta(E_2 - E_1 - \hbar\omega_k) \qquad (7.7.10)$$

The rate of *spontaneous* emission must be independent of the number of photons. We therefore identify the rate of spontaneous emission with

$$R_{i \to f}^{spon} = \frac{2\pi}{n} \left(\vec{\mu}_{12} \cdot \tilde{e}_{k,s} \right)^2 \frac{n\omega_k}{2\varepsilon_o V}\ \delta(E_2 - E_1 - n\omega_k) \qquad (7.7.11)$$

where

$$R_{i \to f} = R_{i \to f}^{\text{stim}} + R_{i \to f}^{\text{spon}}$$

Equation 7.7.10 relates the rate of stimulated emission directly to the number of incident photons (per second); the photons induce the transitions within the active region of the laser. The laser rate equations in Chapter 2 show how the number of photons n_k in the laser must be related to rate of stimulated emission and to the optical loss mechanisms. In addition, we notice that n_k represents the number of photons in the mode characterized by the wave vector $\bar{k}$ (i.e., the wavelength). Although not evident from Equation 7.7.10, the gain (vs. wavelength) for a laser or optical amplifier has a finite spectral width which means that the total rate of stimulated emission can only be found by summing over the various wave vectors $\bar{k}$. We shall see more of this in the next chapter when we discuss the gain for semiconductor lasers.

Equation 7.7.11 shows that the spontaneous emission occurs without any incident photons. The fluctuations of the electric field in the vacuum state can be written as

$$\sigma_E^2 = \frac{\hbar \omega}{\varepsilon_o V} \left(n + \frac{1}{2} \right) \bigg|_{n=0} = \frac{\hbar \omega}{2 \varepsilon_o V}$$

The rate of spontaneous emission can then be written as

$$R_{i \to f}^{\text{spon}} = \frac{2\pi}{\hbar} \left(\vec{\mu}_{12} \cdot \tilde{e}_{k,s} \right)^2 \sigma_E^2 \, \delta(E_2 - E_1 - \hbar \omega_k)$$

where the polarization vector $\tilde{e}_{k,s}$ represents the polarization vector of the allowed modes of the system. These modes are always present even when they remain empty of photons. To find the total rate of spontaneous emission, we need to sum over all of the modes. Recent studies with lasers indicate that the rate of spontaneous emission can be reduced by removing available modes into which the atom can spontaneously emit. The boundary conditions determine the available modes. Photonic crystals provide one mechanism for reducing the number of optical modes available to the system. Incorporating these devices, the threshold current of the semiconductor laser can be reduced (recall, most of the threshold current is due to spontaneous emission). However, the laser still requires some spontaneous emission to initiate laser action. This spontaneous emission only needs to have the lasing frequency and a propagation vector parallel to that of the stimulated emission. This laser-initiating spontaneous emission accounts for less than 1 percent of the total spontaneous emission. In theory then, we should be able to reduce present laser threshold currents to the microamp scale.

7.8 Introduction to Matter and Light as Systems

Previous sections and chapters discuss the Hamiltonian for free particles and free fields. Particles interacting with a potential cannot be considered as "free." Likewise, EM fields interacting with matter cannot be considered "free fields." The interaction links the fields to the matter. The interaction causes matter to emit or absorb electromagnetic energy.

7.8.1 The Complete System

An atom (or collection of atoms) can interact with a variety of sources. The interaction between a collection of atoms and electromagnetic wave represents the most important interaction for emitters and detectors. But, for example, the atom can also interact with phonons within a crystal or with electrons. Figure 7.8.1 indicates the interaction with various sources. Each system shown in the figure has an associated Hamiltonian. Interactions link the various systems. The total Hamiltonian can therefore be written as

$$\mathscr{H} = H_{\text{atom}} + H_{\text{Light}} + H_{\text{Xal}} + H_{\text{Source}} + \mathscr{V}_{A-L} + \mathscr{V}_{A-X} + \mathscr{V}_{A-S}$$

where A, L, X and S denote the atom, light, crystal and source, respectively. Our discussion of perturbation theory and Fermi's golden rule to this point omits most of these terms and focuses on $H_{\text{atom}} + \mathscr{V}_{A-L}$ where the interaction potential links matter and light through the dipole moment and the electric field $\hat{\mathscr{V}}_{A-L} = \hat{\mu} \cdot \vec{\mathscr{E}}$. On the other hand, Chapter 6 shows how to write the Hamiltonian for the light system. The present chapter now begins to account for all of the systems and their effects upon the radiating atom.

For the moment, let's focus on the matter and light systems. In this case, the full Hamiltonian has the form

$$\hat{\mathscr{H}} = \hat{H}_{\text{atom}} + \hat{\mathscr{V}}_{A-L} + \hat{H}_{\text{Light}}$$

There must exist functions $|\psi\rangle$ such that

$$\hat{\mathscr{H}}|\psi\rangle = i\hbar \frac{\partial}{\partial t}|\psi\rangle$$

However, the $|\psi\rangle$ must simultaneously refer to two separate Hilbert Spaces—one for the atom and another for the light. Without the interaction term, a wavefunction $|\psi_{\text{atom}}\rangle$ and a wavefunction $|\psi_{\text{lite}}\rangle$, in separate spaces, would independently evolve. The evolution would be controlled by their separate respective Hamiltonians $\hat{H}_{\text{atom}}$ and $\hat{H}_{\text{Light}}$. The separate spaces can be pictured as somehow adjacent to each other as shown in Figure 7.8.2. The wavefunction $|\psi\rangle$ can be similarly represented by $|\psi\rangle = |\psi_{\text{atom}}\rangle |\psi_{\text{lite}}\rangle$. Because the two Hamiltonians (and other related operators) refer to separate spaces, we can write equations such as

$$\hat{H}_{\text{atom}}|\psi\rangle = \left\{ \hat{H}_{\text{atom}}|\psi_{\text{atom}}\rangle \right\}|\psi_{\text{lite}}\rangle \quad \text{and} \quad \hat{H}_{\text{Light}}|\psi\rangle = |\psi_{\text{atom}}\rangle \left\{ \hat{H}_{\text{Light}}|\psi_{\text{lite}}\rangle \right\}$$

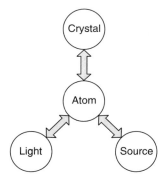

FIGURE 7.8.1

The atom interacts with the crystal environment (phonons), light and other sources such as energy pumps.

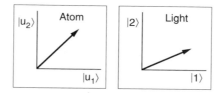

FIGURE 7.8.2
The atom and light Hilbert Spaces.

Notice how the operators only affect the vectors in their corresponding spaces.

You might recall from the study of linear algebra that $|\psi\rangle = |\psi_{\text{atom}}\rangle|\psi_{\text{lite}}\rangle$ denotes a vector in the direct product space. The basis vectors for the direct product space have the form $\{|u_a\rangle|\gamma\rangle\}$ where $\{|u_a\rangle\}$ and $\{|\gamma\rangle\}$ represent the atomic and light basis vectors, respectively. Also recall that the most general vector in the direct product space has the form

$$|\psi\rangle = \sum_{a,\gamma} \beta_{a,\gamma}|u_a\rangle|\gamma\rangle = \sum_{a,\gamma} \beta_{a,\gamma}|u_a\,\gamma\rangle$$

where $\beta_{a,\gamma}$ can depend on time. Note in particular, that this most general wave vector $|\psi\rangle$ cannot be decomposed as a vector in atom-space multiplying a vector in light-space. The single coefficient $\beta_{a,\gamma}$ entangles the two spaces (links them). We cannot write $|\psi\rangle \neq \sum_a \{\beta_a|u_a\rangle\} \sum_\gamma \{\beta_\gamma|\gamma\rangle\}$ without assuming that the vector $|\psi\rangle$ starts in a disentangled state and that the interaction potential must be zero. The $\beta_{a,\gamma}$ have the same probability interpretation as for the simpler Hilbert spaces. The probability of finding the system in atomic state $|u_b\rangle$ *and* light state $|\gamma'\rangle$ must be

$$\text{Prob}(b,\gamma') = |\beta_{b\gamma'}|^2 \quad \text{where} \quad \beta_{b\gamma'} = \langle u_b\,\gamma' \mid \psi\rangle$$

We can see that the motion of the wave functions in the two vector spaces must be completely independent if we can write $\beta_{a,\gamma} = \beta_a\beta_\gamma$. First the wave functions can be decoupled

$$\sum_{a\gamma} \beta_{a\gamma} \underbrace{|u_a\rangle}_{\text{atom}}\underbrace{|\gamma\rangle}_{\text{light}} = \sum_{a\gamma} \beta_a\beta_\gamma|u_a\rangle|\gamma\rangle = \sum_a \beta_a|u_a\rangle \sum_\gamma \beta_\gamma|\gamma\rangle = |\psi_a\rangle|\psi_\gamma\rangle$$

The Schrodinger equation can be written as

$$\left(\hat{\mathscr{H}}_a + \hat{\mathscr{H}}_\gamma\right)|\psi_a\rangle|\psi_\gamma\rangle = i\hbar\frac{\partial}{\partial t}|\psi_a\rangle|\psi_\gamma\rangle$$

Therefore, Schrodinger's equation can be divided into two parts

$$\hat{\mathscr{H}}_a|\psi_a\rangle = i\hbar\frac{\partial}{\partial t}|\psi_a\rangle \qquad \hat{\mathscr{H}}_\gamma|\psi_\gamma\rangle = i\hbar\frac{\partial}{\partial t}|\psi_\gamma\rangle$$

since the "a" and "γ" terms refer to distinct Hilbert spaces. Therefore the motion of the wave functions through each Hilbert space must be independent of one another.

The interaction Hamiltonians link the various spaces together. The motion of the vector in one affects the motion of the vector in the other one and vice versa. For example, suppose an atom starts in state $|u_1\rangle$ and light starts in a state with n photons $|n\rangle$. If the atom absorbs a photon, then the atom moves to state $|u_2\rangle$ and the light moves to state

$|n - 1\rangle$. As a result, the motion of the wave functions in the two spaces must be linked. The interaction Hamiltonian provides the link. We therefore expect the interaction Hamiltonian to involve operators for both spaces. For example, one term in the interaction Hamiltonian might be $\hat{f}_2^+ \hat{f}_1^- \hat{b}^-$. The photon annihilation operator $\hat{b}^-$ removes a photon from the Fock state $|n\rangle$, the electron annihilation operator $\hat{f}_1^-$ removes the electron from atomic state $|u_1\rangle$ while the electron creation operator $\hat{f}_2^+$ places it in state $|u_2\rangle$. The operator $\hat{f}_2^+ \hat{f}_1^- \hat{b}^-$ can therefore account for the combined motion $|u_1\rangle|n\rangle \rightarrow |u_2\rangle|n - 1\rangle$.

7.8.2 Introduction to Homogeneous Broadening

This topic introduces key points for homogeneous broadening and for atomic transitions. Consider two views.

1. Let's assume that a system has the *total* Hamiltonian $\hat{\mathscr{H}} = \hat{H}_{\text{atom}} + \hat{\mathscr{V}}_{A-L} + \hat{H}_{\text{Light}}$. This equation ignores any sources that originally produce the electromagnetic field along with any environmental effects experienced by the atom. We look for solutions to the time-independent Schrodinger wave Equation $\hat{\mathscr{H}}|\phi\rangle = E|\phi\rangle$. The basis set $|\phi\rangle$ applies to the direct product space (atom+light space together). Each $|\phi\rangle$ will be some irreducible combination of the basis vectors for each space. Here's the point. The basis vectors $|\phi\rangle$ are exact. If we make a measurement of the total energy, the system wavefunction $|\psi\rangle$ will collapse to one of the basis vectors $|\phi\rangle$. The system will stay in a state with the same total energy as the state $|\phi\rangle$ unless an extra perturbation acts on the system (such as a phonon interaction with the crystal).

2. The behavior of the combined system is totally different from what we find from Fermi's golden rule (or the density matrix equations in this chapter). Fermi's golden rule shows that an electron makes a transition from one basis state $|u_n\rangle$ to another. When we make a measurement of the energy, the electron wavefunction will collapse to an atomic basis state $|u_n\rangle$ but then because of the EM interaction, it will move away from this state. The difference between the two situations has to do with the fact that Fermi's golden rule does not account for the entire system. It only includes the Hamiltonian for the atom and the interaction and not the light system.

We can reconcile these two pictures by noting that the total system can only be described by the total Hamiltonian $\hat{\mathscr{H}} = \hat{H}_{\text{atom}} + \hat{\mathscr{V}}_{A-L} + \hat{H}_{\text{Light}}$ which gives the eigen vectors $|\phi\rangle$ for *both* systems (atom + light) together $|\phi\rangle = \sum_{a,\gamma} C_{a,\gamma}|u_a \gamma\rangle$. If we restrict our attention to the basis states $|u_a\rangle$, we would find the wavefunction moving from one to the other. At the same time, the light wavefunction would be continuously moving from one light basis state to another. The wave functions in the individual Hilbert spaces would be moving in such a way that the total energy remains constant. We will later see that we can mathematically reconcile the two pictures when using the density matrix by taking the trace over the entire set of basis vectors for one space or the other.

Now let's briefly discuss homogeneous broadening. Normally, we might expect the optical emission spectrum for a collection of electrons making transitions from atomic state $|u_2\rangle$ to $|u_1\rangle$ to be very sharp (a Dirac delta function). A plot of optical power versus frequency should have a single infinitely narrow line at a frequency $\omega_o = (E_2 - E_1)/\hbar$. But we actually observe a line somewhat broadened as shown in Figure 7.8.3. Again we can reconcile the discrepancy by realizing that the eigenfunctions $|u_n\rangle$ cannot be the exact eigenfunctions for the entire system. From the previous discussion, we realize that once

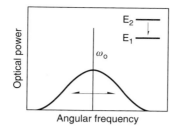

FIGURE 7.8.3

A broadened spectral line.

the whole system (atom + light) lodges itself in one of the exact basis states $|\phi\rangle$ with a specified total energy E, then the system cannot change energy unless some other energy perturbs the system. However, the system can still make random transitions between the various degenerate atomic basis states $|u_n\rangle$ so long as the total energy of the system remains fixed.

The transfer of energy to/from a system is responsible for homogeneous broadening. Given that a probability distribution describes the atomic transition process, the exact time required for the transition can not be a-priori known. This means that there exists an uncertainty Δt (a standard deviation) associated with the transition time. As a result, there must be an uncertainty in the transition energy ΔE because of the Heisenberg uncertainty relation $\Delta E \Delta t \geq \hbar/2$. The light spectrum therefore has a width on the order of $\Delta \omega = \Delta E / \hbar$. The broadening occurs anytime a system has the capability of making a transition from one eigenstate to another (with distinct energy). This can only happen when these eigenstates are not the eigenstates of the entire system. This means there is an external agent causing a perturbation.

What about spontaneous emission? Do we expect to see a range of frequency for spontaneously emitted light? Yes. Here the vacuum fluctuations of the electromagnetic field provide the external agent for the perturbation. We can therefore narrow the line widths and reduce the spontaneous emission by reducing the number of modes available to the electromagnetic field. Section 7.7 showed the rate of spontaneous emission and how it related to the vacuum fields.

7.9   Liouville Equation for the Density Operator

A collection of atoms can interact with other systems such as a light field and the environment. We imagine the collection of atoms embedded in an environment such as a crystal lattice (for semiconductor lasers) or a gas mixture (for gas lasers). Many physical processes influence the behavior of natural or man-made atoms. The complete Hamiltonian must describe not only the interaction between the collection of atoms and the electromagnetic field for optical emission and absorption but also the interaction between the atoms and the environment. These environmental interactions include collisions between the atoms and phonons (or carriers), the pumping mechanism, and the spontaneous recombination. Subsequent sections will show how reservoirs can be used to model the effect of the environment. The interactions between the atomic system and these other systems lead to atomic transitions and spectral broadening.

The density operator provides the most complete (and perhaps the simplest) method of describing the interactions between an ensemble of atoms and their environment.

The density operator $\hat{\rho}$ describes the occupation and transition probability. The diagonal elements of $\hat{\rho}$, namely ρ_{nn}, give the probability of finding the system in state "n." If the density operator depends on time, then the occupation probability can change with time. The rate of change of the diagonal terms of the density operator must be related to the transition rate. Therefore, we need an expression for the time rate of change of the density operator.

In this section, we use a Hamiltonian that includes all relevant interactions such as the matter-light interaction, electrical and optical pumping, and collisions. We will develop the Liouville equation for the density operator that includes phenomenological terms for the interaction between the atoms and the environment. The phenomenological terms describe steady-state conditions and give rise to the homogeneous broadening. The next section develops the steady-state density matrix in terms of the physical time-constants and pump currents. The subsequent section then demonstrates a solution to the Liouville equation for the density matrix.

7.9.1 The Liouville Equation Using the Full Hamiltonian

We define the total Hamiltonian as

$$\hat{\mathscr{H}} = \hat{H} + \hat{H}_{\text{env}} \tag{7.9.1}$$

where

$$\hat{H} = \hat{H}_o + \hat{\mathscr{V}} \tag{7.9.2}$$

includes only the stimulated emission and absorption. As for Fermi's golden rule, the Hamiltonian $\hat{H}_o$ describes the atom while the interaction energy $\hat{\mathscr{V}}$ describes the interaction between an applied electromagnetic wave and the atom. We expect laser gain to be related to $\hat{H}$. The environmental term $\hat{H}_{\text{env}}$ describes the effects of pumping currents, collisions, spontaneous emission and other terms not described by stimulated emission and absorption. The Hamiltonian $\hat{H}_{\text{env}}$ can be divided into separate terms such as

$$\hat{H}_{\text{env}} = \hat{H}_{\text{pump}} + \hat{H}_{\text{coll}} + \hat{H}_{\text{spont}} + \cdots \tag{7.9.3}$$

The next several topics treat the Hamiltonians embedded in Equation 7.9.3 on a phenomenological basis due to the complexity of the interactions they represent.

We look for a differential equation for the density operator. Part of the differential equation describes the stimulated emission processes and other terms describe the effect of the "environment" such as the pump. The density operator (and hence, the number of excited atoms or carriers) changes with time due to the stimulated processes and the interaction with the environment. These environmental interactions introduce certain time constants. For example, electrons can "relax" from the conduction band to the valence band by interacting with phonons (produced by a heated crystal lattice). The "relaxation" must be characterized by a time constant. The pump term provides another example. Carriers can be added to the conduction and valence bands at some rate by external sources. We use a time constant to describe the rate.

We start with the density operator

$$\hat{\rho}(t) = \sum_{\psi} P_{\psi} |\psi(t)\rangle\langle\psi(t)| \tag{7.9.4}$$

where the wave function satisfies Schrodingers equation for the complete Hamiltonian in Equation 7.9.1 according to $\hat{\mathcal{H}}|\psi\rangle = i\hbar\partial_t|\psi\rangle$. We assume the electromagnetic and environmental interactions produce transitions between the energy basis states of the atomic Hamiltonian $\hat{H}_o$. The Liouville equation for the density operator can be found by differentiating the density operator in Equation 7.9.4 with respect to time

$$\frac{\partial\hat{\rho}}{\partial t} = \sum_\psi P_\psi \frac{\partial|\psi\rangle}{\partial t}\langle\psi| + \sum_\psi P_\psi |\psi\rangle\frac{\partial\langle\psi|}{\partial t} = \sum_\psi P_\psi \frac{\partial|\psi\rangle}{\partial t}\langle\psi| + \sum_\psi P_\psi |\psi\rangle\left\langle\frac{\partial}{\partial t}\psi\right| \tag{7.9.5}$$

We moved the time derivative inside the last term by recalling the definition of the bra of a function

$$\frac{\partial}{\partial t}\langle\psi| = \frac{\partial}{\partial t}\int dV\, \psi^*(\vec{r},t) = \int dV \frac{\partial}{\partial t}\psi^*(\vec{r},t) = \int dV\left(\frac{\partial}{\partial t}\psi\right)^* = \left\langle\frac{\partial\psi}{\partial t}\right|$$

where we treat the integral as an operator. Next, the Schrodinger's equation produces

$$\hat{\mathcal{H}}|\psi\rangle = i\hbar\frac{\partial}{\partial t}|\psi\rangle \qquad \rightarrow \qquad \frac{\partial|\psi\rangle}{\partial t} = \frac{\hat{\mathcal{H}}}{i\hbar}|\psi\rangle$$

Using the definition of adjoint and the Hermiticity of the Hamiltonian, we find

$$\left\langle\frac{\partial\psi}{\partial t}\right| = \left\langle\frac{\hat{\mathcal{H}}\psi}{i\hbar}\right| = \left|\frac{\hat{\mathcal{H}}\psi}{i\hbar}\right\rangle^+ = \left(\frac{\hat{\mathcal{H}}}{i\hbar}|\psi\rangle\right)^+ = \langle\psi|\left(\frac{\hat{\mathcal{H}}}{i\hbar}\right)^+ = \langle\psi|\left(\frac{\hat{\mathcal{H}}}{-i\hbar}\right)$$

Inserting these last two results into Equation 7.9.5 provides

$$\frac{\partial\hat{\rho}}{\partial t} = \sum_\psi P_\psi \frac{\hat{\mathcal{H}}}{i\hbar}|\psi\rangle\langle\psi| - \sum_\psi P_\psi |\psi\rangle\langle\psi|\frac{\hat{\mathcal{H}}}{i\hbar}$$

$$= \frac{\hat{\mathcal{H}}}{i\hbar}\left\{\sum_\psi P_\psi |\psi\rangle\langle\psi|\right\} - \left\{\sum_\psi P_\psi |\psi\rangle\langle\psi|\right\}\frac{\hat{\mathcal{H}}}{i\hbar}$$

Recognizing the density operator as the terms in the braces, we can write this last expression as a commutator relation

$$\frac{\partial\hat{\rho}}{\partial t} = \frac{1}{i\hbar}\left[\hat{\mathcal{H}},\hat{\rho}\right]$$

Finally, inserting the definition for the Hamiltonian from Equation 7.9.1 produces Liouville's equation for the density operator

$$\frac{\partial\hat{\rho}}{\partial t} = \frac{1}{i\hbar}\left[\hat{H},\hat{\rho}\right] = \frac{1}{i\hbar}\left[\hat{\mathcal{H}} + \hat{H}_{\text{env}},\hat{\rho}\right] = \frac{1}{i\hbar}\left[\hat{H},\hat{\rho}\right] + \frac{1}{i\hbar}\left[\hat{H}_{\text{env}},\hat{\rho}\right] \tag{7.9.6}$$

The final term in this last equation shows how the environment produces a change in the density operator

$$\left(\frac{\partial\hat{\rho}}{\partial t}\right)_{\text{env}} = \frac{1}{i\hbar}\left[\hat{H}_{\text{env}}, \hat{\rho}\right]$$

Later discussion treats this term phenomenologically because of the complexity of the interactions between the atoms and the environment. Therefore, an alternate form for Liouville's equation for the density operator must be

$$\frac{\partial\hat{\rho}}{\partial t} = \frac{1}{i\hbar}\left[\hat{\mathscr{H}}, \hat{\rho}\right] = \frac{1}{i\hbar}\left[\hat{H} + \hat{H}_{\text{env}}, \hat{\rho}\right] = \frac{1}{i\hbar}\left[\hat{H}, \hat{\rho}\right] + \left(\frac{\partial\hat{\rho}}{\partial t}\right)_{\text{env}} \tag{7.9.7}$$

The reader should think of this equation as saying that the occupation probability changes (see the $\partial\hat{\rho}/\partial t$ term) due to the interaction of an electromagnetic field with the atom (the $(1/i\hbar)[\hat{H}, \hat{\rho}]$ term) and due to other sources such as electrical pump currents (the $(\partial\hat{\rho}/\partial t)_{\text{env}} = (1/i\hbar)[\hat{H}_{\text{env}}, \hat{\rho}]$ term). This last term is the carrier relaxation term that brings a system back to equilibrium once the electromagnetic perturbation is removed. Therefore, the last term in Equation 7.9.7 leads to relaxation times τ.

If we use the alternate form for $\hat{H}_{\text{env}} = \hat{H}_{\text{pump}} + \hat{H}_{\text{coll}} + \hat{H}_{\text{spont}} + \cdots$ in Equation 7.9.6, we find that the Liouville Equation has the form

$$\frac{\partial\hat{\rho}}{\partial t} = \frac{1}{i\hbar}\left[\hat{H}, \hat{\rho}\right] + \frac{1}{i\hbar}\left[\hat{H}_{\text{pump}}, \hat{\rho}\right] + \frac{1}{i\hbar}\left[\hat{H}_{\text{coll}}, \hat{\rho}\right] + \frac{1}{i\hbar}\left[\hat{H}_{\text{spont}}, \hat{\rho}\right] + \cdots \tag{7.9.8}$$

which can also be written as

$$\frac{\partial\hat{\rho}}{\partial t} = \frac{1}{i\hbar}\left[\hat{H}, \hat{\rho}\right] + \left(\frac{\partial\hat{\rho}}{\partial t}\right)_{\text{pump}} + \left(\frac{\partial\hat{\rho}}{\partial t}\right)_{\text{coll}} + \left(\frac{\partial\hat{\rho}}{\partial t}\right)_{\text{spont}} + \cdots \tag{7.9.9}$$

Now we see that there can be three or more time-constants associated with the three (or more) terms labeled as "pump," "coll" and "spont" in Equations 7.9.8 and 7.9.9.

Equation 7.9.9 begins to resemble the rate equation discussed in Chapter 2. Letting N_v to be the number of atoms per unit volume, then $N_v\rho_{aa} = n_a$ gives the number of atoms with electrons in state "a." The left side of Equation 7.9.9 has terms of the form dn_a/dt. The commutator on the right hand side contains the matter-light interaction and therefore represents the gain term in the rate equations. The pump term resembles the current-number density term $\mathscr{J}$. The last terms start to resemble carrier recombination terms; however, the collision term requires further discussion.

Example 7.9.1

Consider a collection of 5 atoms as shown in Figure 7.9.1. Assume the electrons occupy either eigenstate $|u_1\rangle$ or $|u_2\rangle$. Suppose we allow only these two possible wavefunctions to appear in the density operator

$$\hat{\rho}(t) = \sum_{S=1}^{2} P_S|u_S\rangle\langle u_S| \tag{7.9.10}$$

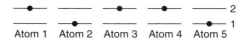

FIGURE 7.9.1
Collection of 5 atoms in various states.

where the index "S" takes on values of $S = 1,2$. For this example, we are not allowing the wavefunctions in the density operator to be coherent sums over the basis vectors. Equation 7.9.10 represents a statistical mixture. We see from the figure that the probability of wavefunction #2 is $P_2 = 3/5$ and the probability of wavefunction #1 is $P_1 = 2/5$.

Example 7.9.2

Suppose both the environmental and matter–field interactions are disabled. Assume the two-level atom has atomic energy eigenstates satisfying $\hat{H}_o |u_a\rangle = E_a |u_a\rangle$, where $\hat{H}_o$ represents the atomic Hamiltonian. Find the probability of particle occupying eigenstate $|u_a\rangle$ of a two level atom.

We recall that $\langle u_2 | \hat{\rho} | u_2 \rangle$ gives the probability of an electron occupying the second energy. So we just need to find the density operator as a function of time. Without the environmental terms, the Liouville equation becomes

$$\frac{\partial \hat{\rho}}{\partial t} = \frac{1}{i\hbar}[\hat{H}, \hat{\rho}] + \left(\frac{\partial \hat{\rho}}{\partial t}\right)_{env} \quad \rightarrow \quad \frac{\partial \hat{\rho}}{\partial t} = \frac{1}{i\hbar}[\hat{H}, \hat{\rho}]$$

Without an applied EM field, the Hamiltonian reduces to the atomic Hamiltonian $\hat{H} = \hat{H}_o + \hat{V} = \hat{H}_o$ and we can write the Liouville Equation as

$$\frac{\partial \hat{\rho}}{\partial t} = \frac{1}{i\hbar}[\hat{H}_o, \hat{\rho}] = \frac{1}{i\hbar}\left(\hat{H}_o \hat{\rho} - \hat{\rho}\hat{H}_o\right)$$

Let's operate with the projector $\langle u_a |$ and the ket $|u_a\rangle$ which are always independent of time.

$$\frac{\partial}{\partial t}\langle u_a | \hat{\rho} | u_a \rangle = \frac{1}{i\hbar}\langle u_a | \left(\hat{H}_o \hat{\rho} - \hat{\rho}\hat{H}_o\right) | u_a \rangle = \frac{1}{i\hbar}\langle u_a | \hat{H}_o \hat{\rho} | u_a \rangle - \frac{1}{i\hbar}\langle u_a | \hat{\rho}\hat{H}_o | u_a \rangle$$

To evaluate this last expression, note that $\hat{H}_o |u_n\rangle = E_n |u_n\rangle$ so that $\langle u_n |\hat{H}_o = \langle u_n | E_n$.

$$\frac{\partial}{\partial t}\langle u_a | \hat{\rho} | u_a \rangle = \frac{1}{i\hbar}\langle u_a | E_2 \hat{\rho} | u_a \rangle - \frac{1}{i\hbar}\langle u_a | \hat{\rho} E_a | u_a \rangle = 0$$

Therefore

$$\frac{\partial}{\partial t}\rho_{aa} = 0 \quad \rightarrow \quad \rho_{aa}(t) = \rho_{aa}(0)$$

We see the rate of change of the probability for an electron occupying the second level must be constant in time. Without the environmental relaxation term, the number of carriers in state 2 cannot change. For a semiconductor, we know that electrons in the conduction band will decay to the valence band by collisions with other electrons or phonons (for example). This example shows that the necessity of the relaxation term and that it accounts for those processes not normally in the Hamiltonian describing stimulated emission.

7.9.2   The Liouville Equation Using a Phenomenological Relaxation Term

Unfortunately, we don't often know the Hamiltonians representing the interaction of the collection of atoms with the environment. We can most conveniently recognize that the commutators 7.9.8 provide the rate of change of the density operator due to the effect of the pump, collisions and spontaneous emission as shown in Equation 7.9.9. Each term must inherently involves a time constant. Without knowing the Hamiltonian, we replace each term with a phenomenological term. For simplicity, let's consider the case of a single environmental term representing the effect of collisions. Equation 7.9.9 becomes

$$\frac{\partial \hat{\rho}}{\partial t} = \frac{1}{i\hbar}\left[\hat{H}, \hat{\rho}\right] + \left(\frac{\partial \hat{\rho}}{\partial t}\right)_{env} \tag{7.9.11}$$

where $\hat{\rho}$ denotes the single electron density operator and $\hat{H} = \hat{H}_o + \hat{\mathcal{V}}$ represents the atom and the EM interaction. Obviously, the Hamiltonian $\hat{H}$ does not include some physical mechanisms capable of changing the occupation probabilities. The environment causes the system to relax to some steady state value in the absence of the matter-light interaction. We will refer to this steady state as the "no-light steady state."

We can reason-out a suitable form for the phenomenological term. Consider a system with N 2-level atoms. Assume the collisions destroy the phase coherence and therefore, the density operator does not have any off diagonal terms (see 5.11.4). The next section will show how the "loss of coherence" can occur over a time period due to relaxation process. This has important implications for the spectral content of the emission or absorption.

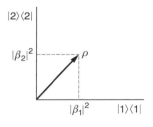

FIGURE 7.9.2
The density operator in a portion of its Hilbert space.

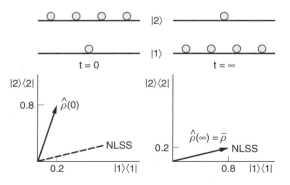

FIGURE 7.9.3
An ensemble of 2-level atoms where the density operator relaxes to a no-light steady-state value (NLSS).

We can focus our thoughts using the example in Figure 7.9.2 which shows $N = 5$ atoms with four electrons in state $|2\rangle$ and one in state $|1\rangle$ at time $t = 0$. We can write

$$\hat{\rho}(0) = 0.2|1\rangle\langle 1| + 0.8|2\rangle\langle 2|$$

Without the EM interaction, the system can relax to a no-light steady-state (NLSS) value $(t \to \infty)$ given by

$$\bar{\rho} = 0.8|1\rangle\langle 1| + 0.2|2\rangle\langle 2|$$

The no-light steady-state density operator might be given by the Fermi–Dirac distribution as appropriate for thermal equilibrium. Without light, we expect $\hat{\rho}(t) \to \bar{\rho}$ as time increases. For very large times, when $\hat{\rho}(t) \cong \bar{\rho}$, the density operator should stop changing

$$\frac{\partial \hat{\rho}}{\partial t} \cong 0 \sim \hat{\rho}(t) - \bar{\rho}$$

We can define the constant of proportionality τ and write

$$\frac{\partial \hat{\rho}}{\partial t} = -\frac{\hat{\rho}(t) - \bar{\rho}}{\tau} \tag{7.9.12}$$

The last relation includes a minus sign to indicate that the density operator must change toward the no-light steady-state value. If only the environment influences the collection of independent atoms, then we can solve Equation 7.9.12 to find

$$\hat{\rho}(t) = \bar{\rho} + [\hat{\rho}(0) - \bar{\rho}]e^{-t/\tau} \tag{7.9.13}$$

This shows that the environment causes the system to exponentially decay toward no-light steady state.

The Liouville Equation must be

$$\frac{\partial \hat{\rho}}{\partial t} = \frac{1}{i\hbar}\left[\hat{H}, \hat{\rho}\right] - \frac{\hat{\rho}(t) - \bar{\rho}}{\tau} \tag{7.9.14}$$

Technically, each matrix element can have a different time constant as we will see later.

Example 7.9.3

Suppose a beam of light excites a collection of 2-level atoms. Assume a researcher turns-off the beam at $t = 0$. The carrier population can relax by collisions using the diagonal version of the NLSS density operator. Assume atomic energy eigenstates $\{|u_a\rangle\}$ satisfying $\hat{H}_o|u_a\rangle = E_a|u_a\rangle$ for the atomic Hamiltonian. Find the density operator as a function of time and find the probability as a function of time that the electron occupies the second energy level.

Solution: We need to find the density operator as a function of time from the Liouville equation $(\partial\hat{\rho}/\partial t) = (1/i\hbar)(\hat{H}_o\hat{\rho} - \hat{\rho}\hat{H}_o) - (\hat{\rho} - \bar{\rho})/\tau$ where $\hat{H}_o$ represents the atomic Hamiltonian. Operating with the projector $\langle u_2|$ and the ket $|u_2\rangle$, which are always independent of time, produces the result

$$\frac{\partial}{\partial t}\langle u_2|\hat{\rho}|u_2\rangle = \frac{1}{i\hbar}\langle u_2|H_o\hat{\rho}|u_2\rangle - \frac{1}{i\hbar}\langle u_2|\hat{\rho}\hat{H}_o|u_2\rangle - \frac{1}{\tau}[\rho_{22} - \bar{\rho}_{22}]$$

To evaluate this last expression, note that $\hat{H}_o|u_n\rangle = E_n|u_n\rangle$ so that $\langle u_n|\hat{H}_o = \langle u_n|E_n$. Again we find

$$\frac{1}{i\hbar}\langle u_2| E_2 \,\hat{\rho}\,|u_2\rangle - \frac{1}{i\hbar}\langle u_2| \,\hat{\rho}\, E_2 \,|u_2\rangle = 0$$

Therefore

$$\frac{\partial}{\partial t}\rho_{22} = -\frac{1}{\tau}[\rho_{22} - \bar{\rho}_{22}]$$

This first order, ordinary differential equation can be solved

$$\rho_{22}(t) = \bar{\rho}_{22} + [\rho_{22}(0) - \bar{\rho}_{22}]e^{-t/\tau}$$

Keep in mind that ρ_{22} represents the probability of finding a particle in state #2. As time increases, the probability of finding an electron in the conduction band must change to the steady state value, which might be given by the Fermi Distribution.

7.10 The Liouville Equation for the Density Matrix with Relaxation

The Liouville equation describes the interaction of an atom with electromagnetic fields and the surrounding environment (defined here to be any nonlight object or influence). The environment can have complicated interactions with the atom but the effects can easily be represented by relaxation terms in the Liouville equation. Without an applied electromagnetic (EM) field, an atomic system will relax to steady state or to equilibrium (termed no-light steady state - NLSS). In this section, we find the phenomenological terms for the effects of the pump and collisions. The results apply to a collection of two level atoms but can be generalized to two semiconductor bands or to more levels.

7.10.1 Preliminaries

We look for the matrix elements of the Liouville equation

$$\frac{\partial\hat{\rho}}{\partial t} = \frac{1}{i\hbar}\left[\hat{H}, \hat{\rho}\right] + \left(\frac{\partial\hat{\rho}}{\partial t}\right)_{\text{pump}} + \left(\frac{\partial\hat{\rho}}{\partial t}\right)_{\text{coll}} + \left(\frac{\partial\hat{\rho}}{\partial t}\right)_{\text{spont}} \qquad (7.10.1)$$

where $\hat{H} = \hat{H}_o + \hat{V}$ contains the atomic Hamiltonian and the interaction energy for stimulated emission and absorption. The last three terms in the Hamiltonian $\mathcal{H} = \hat{H} + \hat{H}_{\text{pump}} + \hat{H}_{\text{coll}} + \hat{H}_{\text{spont}}$ produce the last three terms in Equation 7.10.1 as discussed in the previous section. The density operator

$$\hat{\rho}(t) = \sum_S P_S|\psi_S(t)\rangle\langle\psi_S(t)| \qquad (7.10.2a)$$

incorporates wavefunctions satisfying the Schrodinger's equation

$$\hat{\mathscr{H}}|\psi_S(t)\rangle = i\hbar \frac{\partial}{\partial t}|\psi_S(t)\rangle \tag{7.10.2b}$$

where the "s" substript represents each possible different system in the ensemble and P_s represents the probability of that particular type. We assume the extra "environmental" terms do not change the atomic energy levels but instead, induce transitions between them. As in the previous section, the set $\{|u_n\rangle = |n\rangle : n = 1, 2\}$ consists of the atomic energy eigenstates with $\hat{H}|n\rangle = E_n|n\rangle$. The commutator in Equation 7.10.1 describes the laser gain. However, we focus on the "environmental" terms

$$\left(\frac{\partial\hat{\rho}}{\partial t}\right)_{env} = \left(\frac{\partial\hat{\rho}}{\partial t}\right)_{pump} + \left(\frac{\partial\hat{\rho}}{\partial t}\right)_{coll} + \left(\frac{\partial\hat{\rho}}{\partial t}\right)_{spont} \tag{7.10.3}$$

that describe the effects of electrical or optical pumping, elastic and inelastic collisions and spontaneous emissions. This section focuses on developing an expression for the no-light steady state (NLSS) density operator $\bar{\rho}$.

We will see the similarity between Equation 7.10.3 and the nonphoton terms for the rate equations discussed Chapter 2. The development is similar to that found in Chapter 2 for the rate equations. For example, suppose we wish to find the NLSS value of the number of electrons n_2 in state #2 of a 2-level atom. We assume the EM interaction has been turned-off. The rate equation has the form

$$\frac{dn_2}{dt} = \text{gain} + J - \frac{n_2}{\tau_n} - sn_2^2 = J - \frac{n_2}{\tau_n} - sn_2^2$$

where gain term must be zero. For comparison with the results for the density operator, we set $c = \tau_n^{-1}$ and linearize the spontaneous emission term to read $sn_2^2 \to bn_2$. The rate equation becomes

$$\frac{dn_2}{dt} = J - cn_2 - bn_2$$

At no-light steady state, we require the derivative to be zero. The no-light steady state number of electrons in state #2, denoted by $\bar{n}_2$, then becomes

$$\bar{n}_2 = \frac{J}{b + c}$$

At no-light steady state, the pump-number current J increases the number of electrons in the second state while the recombination represented by "c" and "b" tends to lower it.

7.10.2 Assumptions for the Density Matrix

Let us now discuss the assumptions for the relaxation terms in the Liouville equation. We will find the quantum mechanical gain (complete with the gain saturation effect) and the emission/absorption spectrum. The spectrum has nonzero width (i.e., it is not a delta function of frequency as predicted by simple theories).

For the density matrix, we make the following assumptions.

1. There exists a total of N atoms (per volume) with each one having two energy levels. Of the total number of atoms N, N_1 atoms have electrons in state $|1\rangle$ with energy E_1 and N_2 have electrons in state $|2\rangle$ with energy E_2. The Pauli exclusion principle forbids more than one electron from occupying precisely one single state. However, because we assume the atoms to be independent, many electrons can have the same energy so long as they remain in separate independent atoms. The probability of an electron occupying state $|1\rangle$ is ρ_{11} and the probability of occupying state $|2\rangle$ is ρ_{22}.

2. The pumping current J removes electrons from state $|1\rangle$ and inserts them into state $|2\rangle$ as shown in Figure 7.10.1. Therefore, the probability for level 2 increases at the expense of level one. We use the rates $(\dot{\rho}_{22})_{\text{pump}} = aJ = -(\dot{\rho}_{11})_{\text{pump}}$, where "$a$" provides a constant of proportionality.

3. Spontaneous emission produces a photon and causes electrons to transit from energy level $|2\rangle$ to level $|1\rangle$. We have previously seen greater numbers of electrons in $|2\rangle$ increase the likelihood of a spontaneous emission event. The number of electrons in state 2 decreases in proportion to the number of electrons in that level so that $\dot{n}_2 = -bn_2$. For N atoms (i.e., N electrons), we must have

$$\dot{n}_2 = -bn_2 \quad \rightarrow \quad N\dot{\rho}_2 = -bN\rho_2 \quad \rightarrow \quad \dot{\rho}_2 = -b\rho_2$$

Therefore, the probability of the electron occupying $|2\rangle$ decreases in proportion to the number of electrons in that level (i.e., in proportion to the probability ρ_{22} of occupying level 2). We therefore assume that $(\dot{\rho}_{22})_{\text{spont}} = -b\rho_{22}$ while the probability of level one must increase $(\dot{\rho}_{11})_{\text{spont}} = b\rho_{22}$. We think of the ρ_{aa} almost as if it were the number of atoms in state "a." By the way, there isn't any such thing as spontaneous absorption.

4. Collisions between the atoms and other particles (phonons or electrons) that cause the atoms to change energy levels without radiating light produce changes in the diagonal elements of the "collision" density matrix. These processes are sometimes called "inelastic collisions" since the colliding objects carry away some of the energy and cause the atoms to change energy. Any nonradiative process decreasing the number of electrons in $|2\rangle$ must increase the number in $|1\rangle$ and therefore $(\dot{\rho}_{22})_{\text{coll}} = -c\rho_{22} = -(\dot{\rho}_{11})_{\text{coll}}$. We assume that c includes the negligible reverse effect whereby the collision remove electrons from the lower level and place them in the upper one.

5. Collisions between the atoms and other particles sometimes do not induce transitions but instead interfere with the stimulated emission and absorption process. This process actually requires some discussion since it is the cause of homogeneous broadening and effectively sets the bandwidth for optical emission and absorption. 7.10.5 is devoted to these so-called "elastic collisions." For now, we will assume that the elastic collisions affect the off-diagonal elements of the "collision" density matrix. These types of collisions affect the induced dipole moment and polarization which in-turn affects the rate of transition. We assume that

$$(\dot{\rho}_{12})_{\text{coll}} = -\frac{\rho_{12}}{T_2} \quad \text{and} \quad (\dot{\rho}_{21})_{\text{coll}} = -\frac{\rho_{21}}{T_2}$$

The time constant T_2 is called the "dephasing time" and must be related to the time between collisions. Notice that both off-diagonal collision terms decrease due to dipole dephasing. High collision rates lead to smaller T_2 and therefore larger changes in the off-diagonal terms of the density matrix.

6. Thermodynamic statistical distributions must regulate the number of electrons in each given level when both the EM fields and pump are removed. Let $f(E, T)$ represent the probability of an electron occupying energy E for temperature T at thermal equilibrium. Then in the absence of fields and pump, we must have an equilibrium density matrix of the form

$$\underline{\rho}_{\text{eq}} = \begin{bmatrix} f(E_1, T) & 0 \\ 0 & f(E_2, T) \end{bmatrix}$$

The functions must provide the relation $f(E_1, T) + f(E_2, T) = 1$. We assume random influences make the off-diagonal terms in $\underline{\rho}_{\text{eq}}$ equal to zero (see 5.11.4). Sometimes we ignore the equilibrium value since the number of carriers due to population inversion must be much larger than the number at thermal equilibrium.

7.10.3 Liouville's Equation for the Density Matrix without Thermal Equilibrium

This topic writes the Liouville equation for the density matrix taking into account the phenomenology of the relaxation terms. The procedure provides the laser rate equations stated in the language of density matrices. We first ignore the thermal equilibrium density matrix ρ_{eq}. The procedure provides the quantum mechanical *gain* for an ensemble of independent atoms (refer to the next few sections).

Now combine the results from all of the assumptions in 7.10.2 to find the Liouville equation for the density matrix. The change in the "environmental" density matrix

$$\left(\frac{\partial \rho}{\partial t}\right)_{\text{env}} = \left(\frac{\partial \rho}{\partial t}\right)_{\text{pump}} + \left(\frac{\partial \rho}{\partial t}\right)_{\text{coll}} + \left(\frac{\partial \rho}{\partial t}\right)_{\text{spont}}$$

can now be written as

$$\left(\frac{\partial \rho}{\partial t}\right)_{\text{env}} = \begin{bmatrix} -aJ & 0 \\ 0 & aJ \end{bmatrix} + \begin{bmatrix} c\rho_{22} & -\dfrac{\rho_{12}}{T_2} \\ -\dfrac{\rho_{21}}{T_2} & -c\rho_{22} \end{bmatrix} + \begin{bmatrix} b\rho_{22} & 0 \\ 0 & -b\rho_{22} \end{bmatrix}$$

The individual matrices can be combined

$$\left(\frac{\partial \rho}{\partial t}\right)_{\text{env}} = \begin{bmatrix} -aJ + c\rho_{22} + b\,\rho_{22} & -\dfrac{\rho_{12}}{T_2} \\ -\dfrac{\rho_{21}}{T_2} & aJ - c\rho_{22} - b\,\rho_{22} \end{bmatrix} \tag{7.10.4}$$

We would like to write the Liouville equation for the density matrix in the convenient form

$$\frac{\partial \rho_{ab}}{\partial t} = \frac{1}{i\hbar}\left[\hat{H}, \hat{\rho}\right]_{ab} - \frac{\rho_{ab} - \bar{\rho}_{ab}}{\tau_{ab}} \tag{7.10.5}$$

where, in principle, each of the four relaxation terms can have their own time constant denoted by τ_{ab}. As in the previous section, the quantity $\bar{\rho}_{ab}$ denotes the "no-light steady state—NLSS" value obtained when we turn-off stimulated emission and stimulated absorption. We can find these NLSS values by replacing the commutator in Equation 7.10.5 with zero and setting the derivative of the density matrix equal to zero. For the case of Equation 7.10.5, we can write $\bar{\rho}_{ab} = \rho_{ab}(\infty)$. We find

$$\left(\frac{\partial \bar{\rho}}{\partial t} \right)_{env} = 0 = \begin{bmatrix} -aJ + c\bar{\rho}_{22} + b\,\bar{\rho}_{22} & -\dfrac{\bar{\rho}_{12}}{T_2} \\[2ex] -\dfrac{\bar{\rho}_{21}}{T_2} & aJ - c\bar{\rho}_{22} - b\,\bar{\rho}_{22} \end{bmatrix} \qquad (7.10.6)$$

where as a reminder, the a, b, c, T_2 terms refer to effects of the pump, spontaneous recombination, inelastic (energy changing) collisions, and elastic collisions, respectively. Working with the quantity for level two, we obtain

$$aJ - c\bar{\rho}_{22} - b\,\bar{\rho}_{22} = 0 \quad \rightarrow \quad \bar{\rho}_{22} = \frac{a}{b+c}J \qquad (7.10.7)$$

The NLSS probability of an electron occupying the second level is directly proportional to the pump current. The value of "a" sets the balance between the pump and recombination processes. Notice that the NLSS value for the off diagonal terms must be zero. Using the relation that the sum over probabilities must be equal to 1

$$\bar{\rho}_{11} + \bar{\rho}_{22} = 1 \qquad (7.10.8)$$

we find

$$\bar{\rho}_{11} = 1 - \frac{a}{b+c}J \qquad (7.10.9)$$

We only need to make sure that both probabilities always have values within the range $[0, 1]$. We have ignored the thermal equilibrium probabilities and therefore $\bar{\rho}_{22} \rightarrow 0$ without the pump J. Furthermore in this case, we can identify $\bar{\rho}$ as a "pump" because of the J appearing in Equations 7.10.8 and 7.10.9.

The form of the Liouville equation (7.10.5) requires Equation 7.10.4 to be expressed in terms of $\bar{\rho}_{22}$, $\bar{\rho}_{11}$ and time constants as in

$$\left(\frac{\partial \rho}{\partial t} \right)_{env} = \begin{bmatrix} -aJ + c\rho_{22} + b\rho_{22} & -\dfrac{\rho_{12}}{T_2} \\[2ex] -\dfrac{\rho_{21}}{T_2} & aJ - c\rho_{22} - b\rho_{22} \end{bmatrix} = -\frac{\rho_{ab} - \bar{\rho}_{ab}}{\tau_{ab}} \qquad (7.10.10)$$

Element (2,2) can be rewritten using Equation 7.10.7, $\bar{\rho}_{22} = \frac{a}{b+c}J \rightarrow aJ = \bar{\rho}_{22}(b+c)$ and element (1, 1) can be rewritten using this last result for aJ along with $\rho_{11} + \rho_{22} = 1$ to find

$$\left(\frac{\partial \rho}{\partial t} \right)_{env} = - \begin{bmatrix} \dfrac{\rho_{11} - \bar{\rho}_{11}}{\tau} & -\dfrac{\rho_{12}}{T_2} \\[2ex] -\dfrac{\rho_{21}}{T_2} & \dfrac{\rho_{22} - \bar{\rho}_{22}}{\tau} \end{bmatrix} \qquad (7.10.11)$$

where the population relaxation time is $\tau = (b + c)^{-1}$. The full Liouville equation for the density matrix is

$$\frac{\partial \rho_{ab}}{\partial t} = \frac{1}{i\hbar}\left[\hat{H}, \hat{\rho}\right]_{ab} - \frac{\rho_{ab} - \bar{\rho}_{ab}}{\tau_{ab}} \tag{7.10.12}$$

where

$$\tau_{11} = \tau_{22} = (b + c)^{-1} \quad \text{and} \quad \tau_{12} = \tau_{21} = T_2 \tag{7.10.13}$$

For gallium arsenide semiconductor lasers, the population relaxation time has a magnitude on the order of the few nanoseconds and the dephasing time has a magnitude on the order of a tenth picosecond or less. We already know that the population relaxation time represents the amount of time required for the electron to make a *nonradiative* transition from $|2\rangle$ to level $|1\rangle$. For semiconductors, it is the average time required for an electron to recombine with a hole.

The next topic includes the distribution for thermal equilibrium. The final topic discusses the origin of the dephasing time.

7.10.4   The Liouville Equation for the Density Matrix with Thermal Equilibrium

Now we include the probability distribution for thermal equilibrium in the Liouville equation. Assumption 6 uses a diagonal form for the density matrix ρ_{eq} describing thermal equilibrium. The generalized version of the rate equation (7.10.6) for the environmental effects must have the for

$$\left(\frac{\partial \rho}{\partial t}\right)_{env} = \begin{bmatrix} -aJ + (c + b)(\rho_{22} - f_2) & -\dfrac{\rho_{12}}{T_2} \\[2ex] -\dfrac{\rho_{21}}{T_2} & aJ - (c + b)(\rho_{22} - f_2) \end{bmatrix} \tag{7.10.14}$$

where we use the shortcut notation $f_1 = f(E_1, T)$ and $f_2 = f(E_2, T)$ for the probability of an electron occupying energy E_1 and E_2, respectively, for thermal equilibrium at temperature T.

Now consider steady state described by $\dot{\rho}_{env} = 0$. The steady state values of the diagonal terms of the density operator must be

$$\bar{\rho}_{11} = \rho_{11}(t = \infty) = -\frac{aJ}{b + c} + f_1 \tag{7.10.15a}$$

$$\bar{\rho}_{22} = \rho_{22}(t = \infty) = \frac{aJ}{b + c} + f_2 \tag{7.10.15b}$$

The Liouville equation then has the form

$$\frac{\partial \rho_{ab}}{\partial t} = \frac{1}{i\hbar}\left[\hat{H}, \hat{\rho}\right]_{ab} - \frac{\rho_{ab} - \bar{\rho}_{ab}}{\tau_{ab}} \tag{7.10.16}$$

where τ_{ab} has the same definition as in Equation 7.10.13 (see the chapter review exercises). Notice that J decreases the number of electrons in the first level but increases it in the second. Without any current, the system relaxes to the thermal equilibrium values.

7.10.5   The Dephasing Time

This topic provides a conceptual picture of how elastic collisions affect the emission from a collection of atoms through the dephasing time T_2.

An incident electromagnetic field induces a dipole moment as indicated in Figure 7.10.2. Assume the atom emits light only when the oscillating dipole moment maintains a phase relation with the driving field for a time. Now imagine a time sequence for the oscillating dipole as shown in Figure 7.10.3. For times $t=1$ through $t=6$, the dipole has a definite phase relation with the driving field. At time $t=6$, an electron collides with the oscillating atom and destroys the phase relation with the driving field. The oscillating dipole loses all memory of its phase relation with the field and its oscillations must start all over. The average time between collisions is the "dephasing" time T_2.

As we shall see, the dephasing time controls the width of the emission and absorption curve for the ensemble of atoms.

7.10.6   The Carrier Relaxation Time

The carrier relaxation time $\tau = 1/(b+c)$ gives the average time required for electrons in the excited state to decay to the ground state through the processes of spontaneous emission (the "b" term) and nonradiative collisions (the "c" term). The population decays according to $e^{-t/\tau}$. If a person goes to the store, buys a GaAs LED and connects it in the circuit as shown in Figure 7.10.4 (don't forget to include some type of current limiter like a resistor). Closing the switch causes the LED to emit light. Now suppose the person suddenly opens the switch. How long does it take for the carriers to recombine? The question equivalently asks "How long does it take for the spontaneously emitted light to stop?" Recall that an LED (small ones) only spontaneously emit—there isn't any

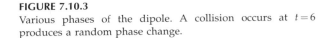

FIGURE 7.10.1
Number-current density J increases the number of electrons in state 2. Collisions decrease the number of electrons in state 2.

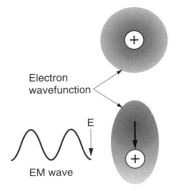

FIGURE 7.10.2
EM wave induces a dipole moment.

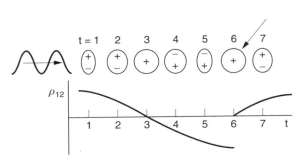

FIGURE 7.10.3
Various phases of the dipole. A collision occurs at $t=6$ produces a random phase change.

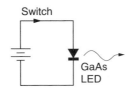

FIGURE 7.10.4
An LED will be suddenly switched off.

stimulated emission (or very little). The carrier relaxation time τ essentially equals the time required for the GaAs LED to turn-off. For GaAs, the carrier relaxation time has a magnitude on the order of $\tau = 10^{-9}$ seconds. This means that the highest possible modulation rate for an LED (without using heroic efforts) must be on the order of $\nu = 1/\tau = 10^9 = 1\ GHz$.

How is it that lasers can be modulated by a factor of 40 faster than the LED? The answer is that the stimulated emission forces the carriers to recombine in addition to carrier relaxation. A laser has stimulated emission that effectively increases the rate of recombination and thereby effectively lowers the carrier lifetime.

7.11 A Solution to the Liouville Equation for the Density Matrix

The present section shows the solution to the Liouville equation for the density matrix. The solution in this section applies to a collection of independent atoms as might be appropriate for gases (i.e., fluorescence or gas lasers); however, it can be generalized for semiconductors. The section treats the density matrix element as an unknown and finds an expression for it in terms of the perturbing electric field and the no-light steady-state (NLSS) values of the density matrix. Recall the NLSS values represent the pumping (such as bias current), effects of collisions, and thermal equilibrium. We find solutions for the diagonal elements (occupation probabilities) and off-diagonal elements (related to transitions). The next section uses the solutions to find the gain produced by the collection of independent atoms.

7.11.1 Evaluating the Commutator

Section 7.9 showed that the Liouville equation for the density matrix is

$$\frac{\partial \rho_{ab}}{\partial t} = \frac{1}{i\hbar}\left[\hat{H}, \hat{\rho}\right]_{ab} - \frac{\rho_{ab} - \bar{\rho}_{ab}}{\tau_{ab}} \tag{7.11.1}$$

for a collection of independent two-level atoms. The Liouville equation harbors the laser rate equations. The relaxation time constant $\tau_{12} = \tau_{21} = T_2$ represents the dipole dephasing time, and $\tau_{11} = \tau_{22} = \tau$ represents the population relaxation time. The quantities $\bar{\rho}_{ab}$ denote the no-light steady state (NLSS) values of the density matrix in the absence of stimulated emission and stimulated absorption (i.e., replace the commutator with 0). The NLSS values of the off diagonal terms of the density matrix are zero $\bar{\rho}_{12} = \bar{\rho}_{21} = 0$. The Hamiltonian $\hat{H}$ consists of the atomic Hamiltonian and the interaction energy

$$\hat{H} = \hat{H}_o + \hat{V} \tag{7.11.2}$$

We assume the atomic energy eigenvectors

$$\{\,|n\rangle = |u_n\rangle\,\} \tag{7.11.3}$$

satisfy

$$\hat{H}_o|n\rangle = E_n|n\rangle \tag{7.11.4}$$

We define the interaction potential in a semiclassical picture by

$$\hat{V} = -\hat{\mu}\mathscr{E}(t) \tag{7.11.5}$$

where $\hat{\mu}$ and $\vec{\mathscr{E}}$ denote the dipole operator and the applied electric field, respectively. As with Fermi's golden rule, we assume that the diagonal terms of the dipole operator are zero.

First we evaluate the quantum mechanical part consisting of the commutator.

$$\frac{1}{i\hbar}\Big[\hat{H}, \hat{\rho}\Big]_{ab} = \frac{1}{i\hbar}\Big[\hat{H}_o + \hat{V}, \hat{\rho}\Big]_{ab} = \frac{1}{i\hbar}\,\langle a|\Big[\hat{H}_o + \hat{V}, \hat{\rho}\Big]|b\rangle$$

$$= \frac{1}{i\hbar}\,\langle a|\hat{H}_o\hat{\rho} - \hat{\rho}\hat{H}_o|b\rangle + \frac{1}{i\hbar}\,\langle a|\hat{V}\hat{\rho} - \hat{\rho}\hat{V}|b\rangle \tag{7.11.6}$$

Next, insert the closure relation $1 = \sum_{c=1}^{2}|c\rangle\,\langle c|$ between the pairs of operators to find

$$[\hat{H}, \hat{\rho}]_{ab} = \sum_c \Big\{ \langle a|\hat{H}_o|c\rangle\langle c|\hat{\rho}|b\rangle - \langle a|\hat{\rho}|c\rangle\langle c|\hat{H}_o|b\rangle + \langle a|\hat{V}|c\rangle\langle c|\hat{\rho}|b\rangle - \langle a|\hat{\rho}|c\rangle\langle c|\hat{V}|b\rangle \Big\}$$

Using $\hat{H}_o|c\rangle = E_c|c\rangle \;\rightarrow\; \langle a|\hat{H}_o|c\rangle = E_c\langle a\mid c\rangle = E_c\delta_{ac}$ we find

$$\Big[\hat{H}, \hat{\rho}\Big]_{ab} = \sum_c \Big\{ E_c\delta_{ac}\langle c|\hat{\rho}|b\rangle - \langle a|\hat{\rho}|c\rangle E_b\delta_{bc} + \langle a|\hat{V}|c\rangle\langle c|\hat{\rho}|b\rangle - \langle a|\hat{\rho}|c\rangle\langle c|\hat{V}|b\rangle \Big\}$$

Next we substitute matrix notation for the inner products and use

$$\langle a|\hat{V}|c\rangle = -\mathscr{E}(t)\langle a|\hat{\mu}|c\rangle = -\mathscr{E}(t)\mu_{ac}$$

to find

$$\Big[\hat{H}, \hat{\rho}\Big]_{ab} = \sum_c \{E_c\delta_{ac}\langle c|\hat{\rho}|b\rangle - \langle a|\hat{\rho}|c\rangle E_b\delta_{bc}\} - \mathscr{E}\sum_c \{\mu_{ac}\rho_{cb} - \rho_{ac}\mu_{cb}\}$$

Next, converting to matrix notation and keeping in mind that the dipole matrix has only off-diagonal elements, we find

$$
\left[\hat{H}, \hat{\rho}\right]_{ab} = \begin{bmatrix} E_1 & 0 \\ 0 & E_2 \end{bmatrix} \begin{bmatrix} \rho_{11} & \rho_{12} \\ \rho_{21} & \rho_{22} \end{bmatrix} - \begin{bmatrix} \rho_{11} & \rho_{12} \\ \rho_{21} & \rho_{22} \end{bmatrix} \begin{bmatrix} E_1 & 0 \\ 0 & E_2 \end{bmatrix} +
$$

$$
- \mathscr{E} \left\{ \begin{bmatrix} 0 & \mu \\ \mu & 0 \end{bmatrix} \begin{bmatrix} \rho_{11} & \rho_{12} \\ \rho_{21} & \rho_{22} \end{bmatrix} - \begin{bmatrix} \rho_{11} & \rho_{12} \\ \rho_{21} & \rho_{22} \end{bmatrix} \begin{bmatrix} 0 & \mu \\ \mu & 0 \end{bmatrix} \right\} \tag{7.11.7}
$$

$$
= \begin{bmatrix} 0 & (E_1 - E_2)\rho_{12} \\ (E_2 - E_1)\rho_{21} & 0 \end{bmatrix} - \mathscr{E}\mu \left\{ \begin{bmatrix} \rho_{21} & \rho_{22} \\ \rho_{11} & \rho_{12} \end{bmatrix} - \begin{bmatrix} \rho_{12} & \rho_{11} \\ \rho_{22} & \rho_{21} \end{bmatrix} \right\}
$$

Define the angular frequency corresponding to the difference in energy of the two atomic levels

$$
E_2 - E_1 = \hbar\omega_o \tag{7.11.8}
$$

so that the last equation can be written as

$$
\left[\hat{H}, \hat{\rho}\right]_{ab} = - \begin{bmatrix} 0 & \hbar\omega_o\rho_{12} \\ -\hbar\omega_o\rho_{21} & 0 \end{bmatrix} - \mathscr{E}\mu \begin{bmatrix} \rho_{21} - \rho_{12} & \rho_{22} - \rho_{11} \\ \rho_{11} - \rho_{22} & \rho_{12} - \rho_{21} \end{bmatrix}
$$

Now we substitute into the Liouville equation

$$
\frac{\partial \rho_{ab}}{\partial t} = \frac{1}{i\hbar}\left[\hat{H}, \hat{\rho}\right]_{ab} - \frac{\rho_{ab} - \bar{\rho}_{ab}}{\tau_{ab}}
$$

and separate out the four equations. The *rate equations* for the carrier-population probability can be written as

$$
\dot{\rho}_{11} = \frac{i}{\hbar}\mathscr{E}(t)\mu(\rho_{21} - \rho_{12}) - \frac{\rho_{11} - \bar{\rho}_{11}}{\tau}
$$

$$
\tag{7.11.9}
$$

$$
\dot{\rho}_{22} = \frac{i}{\hbar}\mathscr{E}(t)\mu(\rho_{12} - \rho_{21}) - \frac{\rho_{22} - \bar{\rho}_{22}}{\tau}
$$

Notice that the commutator in the Liouville Equation (7.11.1), which describes stimulated emission and absorption, now involves the off-diagonal terms of the density matrix. Recall the off-diagonal terms can be related to the induced polarization and the susceptibility, which in turn relates to the rate of transition between levels (i.e., gain). Therefore, we shouldn't be surprised to find that the off-diagonal terms of the density matrix must be related to the gain of the laser medium. The rate equations for the off-diagonal elements are

$$
\dot{\rho}_{12} = i\omega_o\rho_{12} + \frac{i}{\hbar}\mathscr{E}(t)\mu(\rho_{22} - \rho_{11}) - \frac{\rho_{12}}{T_2}
$$

$$
\tag{7.11.10}
$$

$$
\dot{\rho}_{21} = -i\omega_o\rho_{21} + \frac{i}{\hbar}\mathscr{E}(t)\mu(\rho_{11} - \rho_{22}) - \frac{\rho_{21}}{T_2}
$$

The induced polarization apparently depends on the difference in population because the diagonal terms of the density matrix occur in these last two equations. This means that the carrier population must be responsible for the gain.

7.11.2 Two Independent Equations

As we will see in the next section, the difference in the population between the two energy levels produces the laser gain (also refer to Chapter 2). The populations of level 1 and level 2 can be written as

$$N_1 = N \rho_{11} \qquad N_2 = N \rho_{22}$$

since the diagonal elements of the density matrix represent the probability of an electron occupying the corresponding level. We consider the total number of atoms N (*per unit volume*) in the ensemble to be constant. Therefore, the population difference responsible for gain must be

$$\Delta N = N_2 - N_1 = N(\rho_{22} - \rho_{11}) \tag{7.11.11}$$

Figure 7.11.1 shows an example where $N_1 = 3$ and $N_2 = 1$ so that $N = 4$ and $\Delta N = N_2 - N_1 = -2$. We expect this material to be absorptive since $\Delta N < 0$.

The four Liouville equations 7.11.9 and 7.11.10 are not independent. Equations 7.11.9 can be reduced to a single equation by working with the population difference

$$\frac{\Delta \dot{N}}{N} = \dot{\rho}_{22} - \dot{\rho}_{11} = \frac{2i}{\hbar} \mathscr{E}(t)\mu(\rho_{12} - \rho_{21}) - \frac{(\rho_{22} - \rho_{11}) - (\bar{\rho}_{22} - \bar{\rho}_{11})}{\tau} \tag{7.11.12}$$

We can add Equations 7.11.9

$$\dot{\rho}_{11} + \dot{\rho}_{22} = \frac{1}{N}\frac{d}{dt}(N_1 + N_2) = 0$$

but we already know this result because the probabilities must add to one (i.e., because the total number of atoms $N = N_1 + N_2$ must be constant). A second independent equation can be found from Equations 7.11.10 since the density matrix must be Hermitian $\rho_{21} = \rho_{12}^*$. We therefore find a single equation related to the induced polarization and gain

$$\dot{\rho}_{21} = -i\omega_o \rho_{21} + \frac{-i}{\hbar} \mathscr{E}(t)\mu(\rho_{22} - \rho_{11}) - \frac{\rho_{21}}{T_2} \tag{7.11.13}$$

Later we will substitute the population difference ΔN (Equation 7.11.11) into the rate of change of the population difference (Equation 7.11.12) and this last equation (7.11.13) to find the laser rate equations and gain.

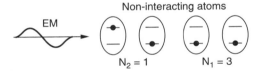

FIGURE 7.11.1
Example of population number.

To solve these differential equations as a function of time, we must substitute for the electric field. The electromagnetic (EM) field in the previous equation can be written as

$$\mathcal{E}(t) = E_o \cos(\omega t) = \frac{E_o}{2}\left[e^{i\omega t} + e^{-i\omega t}\right] \tag{7.11.14}$$

Notice this last relation has the angular frequency ω of the applied EM wave, whereas previous equations contain the angular frequency ω_o corresponding to the difference in atomic energies. The plane wave version of the electric field indicates a semiclassical theory since we are treating the light field as a classical field (we are not quantizing it). Substituting the electric field into Equations 7.11.12 and 7.11.13 provides

$$\dot{\rho}_{22} - \dot{\rho}_{11} = \frac{iE_o\mu}{\hbar}\left(e^{i\omega t} + e^{-i\omega t}\right)(\rho_{12} - \rho_{21}) - \frac{(\rho_{22} - \rho_{11}) - (\bar{\rho}_{22} - \bar{\rho}_{11})}{\tau} \tag{7.11.15}$$

and

$$\dot{\rho}_{21} = -i\omega_o\rho_{21} - \frac{iE_o\mu}{2\hbar}\left(e^{i\omega t} + e^{-i\omega t}\right)(\rho_{22} - \rho_{11}) - \frac{\rho_{21}}{T_2} \tag{7.11.15}$$

These equations are evolving into the laser rate and gain equations.

7.11.3 The Optical Bloch Equations

We now convert the rate equations (7.11.15 and 7.11.16) into the optical Bloch equations. We develop the optical Bloch equations as an intermediate step to find the solutions to the rate equations.

The driving field induces an oscillating dipole moment. We assume the dipoles oscillate at the same frequency as the driving field. We therefore write

$$\rho_{21} = \sigma_{21}e^{-i\omega t} \tag{7.11.17}$$

If we modulate the driving field with a slowly varying envelope function (Figure 7.11.2) then we expect $\sigma_{21} = \sigma_{21}(t)$. In order for the density operator to be Hermitian $\rho_{21} = \rho_{12}^*$, we conclude ρ_{12} must have the form $\rho_{12} = \sigma_{12}e^{i\omega t}$. Furthermore, we conclude the matrix representing the envelope function in Equation 7.11.17 must be Hermitian because the density matrix is Hermitian.

$$\rho_{21} = \rho_{12}^* \Rightarrow \sigma_{21}(t)e^{-i\omega t} = \left(\sigma_{12}(t)e^{i\omega t}\right)^* \Rightarrow \sigma_{21}(t)e^{-i\omega t}$$

$$= \sigma_{12}^*(t)e^{-i\omega t} \Rightarrow \sigma_{21}(t) = \sigma_{12}^*(t)$$

The slowly varying amplitude σ_{21} contains any phase difference between the driving field and ρ_{21}. As an example, the envelope function might arise from a signal modulating the laser output such as a voice or TV signal; this laser beam might be injected into an optical fiber. Once we substitute Equation 7.11.17 into the rate equations for the density matrix elements (Equations 7.11.15 and 7.11.16), we will find exponentials of the form $e^0, e^{\pm 2i\omega t}, e^{\pm\omega t}$. When we apply the Rotating Wave Approximation (RWA) to these terms, only the low frequency ones survive and not those at the optical frequencies. In this way, the rate equations for the elements of the density matrix reduce to the optical Bloch equations.

Substituting Equation 7.11.17 into Equation 7.11.15 and 7.11.16, we find

$$\dot{\rho}_{22} - \dot{\rho}_{11} = \frac{i E_o \mu}{\hbar} \left(e^{i\omega t} + e^{-i\omega t} \right) \left(\sigma_{12}(t) e^{i\omega t} - \sigma_{21}(t) e^{-i\omega t} \right) - \frac{(\rho_{22} - \rho_{11}) - (\bar{\rho}_{22} - \bar{\rho}_{11})}{\tau}$$

(7.11.18)

and

$$\dot{\rho}_{21} \equiv \frac{d}{dt} \left(\sigma_{21}(t) e^{-i\omega t} \right) = -i\omega_o \sigma_{21} e^{-i\omega t} - \frac{i E_o \mu}{2\hbar} \left(e^{i\omega t} + e^{-i\omega t} \right) (\rho_{22} - \rho_{11}) - \frac{\sigma_{21} e^{-i\omega t}}{T_2} \quad (7.11.19)$$

We first work with Equation 7.11.19 by carrying out the derivative

$$\dot{\sigma}_{21}(t) e^{-i\omega t} - i\omega \sigma_{21}(t) e^{-i\omega t} = -i\omega_o \sigma_{21} e^{-i\omega t} - \frac{i E_o \mu}{2\hbar} \left(e^{i\omega t} + e^{-i\omega t} \right) (\rho_{22} - \rho_{11}) - \frac{\sigma_{21} e^{-i\omega t}}{T_2}$$

Dividing this last equation by $e^{-i\omega t}$ gives

$$\dot{\sigma}_{21}(t) - i\omega \sigma_{21}(t) = -i\omega_o \sigma_{21} - \frac{i E_o \mu}{2\hbar} \left(e^{2i\omega t} + 1 \right) (\rho_{22} - \rho_{11}) - \frac{\sigma_{21}}{T_2}$$

where $\sigma_{21}(t)$ changes slowly compared with $e^{\pm i\omega t}, e^{\pm 2i\omega t}$. The rotating wave approximation discussed in Section 7.7 and Appendix 8 allows us to drop "fast" terms such as $e^{2i\omega t}$ (actually, this can be dropped because, to find σ_{21}, we essentially take an integral of the equation which is related to an average). Therefore, the equation for $\dot{\sigma}_{21}$ becomes

$$\dot{\sigma}_{21}(t) = i(\omega - \omega_o)\sigma_{21} - \frac{i \mu E_o}{2\hbar} (\rho_{22} - \rho_{11}) - \frac{\sigma_{21}}{T_2}$$

(7.11.20)

Next we simplify Equation 7.11.18, which is

$$\dot{\rho}_{22} - \dot{\rho}_{11} = \frac{i E_o \mu}{\hbar} \left(e^{i\omega t} + e^{-i\omega t} \right) \left(\sigma_{12}(t) e^{i\omega t} - \sigma_{21}(t) e^{-i\omega t} \right) - \frac{(\rho_{22} - \rho_{11}) - (\bar{\rho}_{22} - \bar{\rho}_{11})}{\tau},$$

by multiplying the exponentials in parenthesis and dropping terms such as $e^{\pm i 2\omega t}$ to obtain

$$\dot{\rho}_{22} - \dot{\rho}_{11} = \frac{i E_o \mu}{\hbar} \left(\sigma_{21}^* - \sigma_{21} \right) - \frac{(\rho_{22} - \rho_{11}) - (\bar{\rho}_{22} - \bar{\rho}_{11})}{\tau}$$

(7.11.21)

where we used $\sigma_{12} = \sigma_{21}^*$.

In both Equations 7.11.20 and 7.11.21, the terms at optical frequencies drop from the equations. Any change in the density operator terms (ρ_{aa}, σ_{12}) must be at the slower modulation frequency. We know from Chapter 2 and some common sense that, the difference in carrier density $N(\rho_{22} - \rho_{11})$ cannot oscillate at the optical frequency. Chapter 2 shows that the photon density depends on the lower frequency and those photon rate equations do not have the faster oscillating electric field. However, the dipoles related to σ_{12} can oscillate at these frequencies. We can see from Equation 7.11.16, we make the difference $\rho_{22} - \rho_{11}$ independent of the optical frequency by requiring the off-diagonal terms to oscillate at the optical frequencies. This observation leads to the condition $\rho_{21} = \sigma_{21}e^{-i\omega t}$ in Equation 7.11.17.

The optical Bloch equations have the form

$$\dot{\sigma}_{21}(t) = i(\omega - \omega_o)\sigma_{21} - \frac{i\mu E_o}{2\hbar}\frac{\Delta N}{N} - \frac{\sigma_{21}}{T_2} \tag{7.11.22a}$$

$$\dot{\rho}_{22} - \dot{\rho}_{11} = \frac{iE_o\mu}{\hbar}\left(\sigma_{21}^* - \sigma_{21}\right) - \frac{(\rho_{22} - \rho_{11}) - (\bar{\rho}_{22} - \bar{\rho}_{11})}{\tau} \tag{7.11.22b}$$

or

$$\frac{d(\Delta N)}{dt} = \frac{iE_o\mu N}{\hbar}\left(\sigma_{21}^* - \sigma_{21}\right) - \frac{\Delta N - \Delta\bar{N}}{\tau} \tag{7.11.23}$$

by substituting $\Delta N = N(\rho_{22} - \rho_{11})$, where N is the total number of independent 2-level atoms. To arrive at these equations, we used several assumptions to rewrite Equations 7.11.22 and 7.11.23. First the form of the interaction Hamiltonian required the dipole approximation (refer to Section 7.7 and Appendix 8). Second, we used the rotating wave approximation (i.e., energy conservation) to eliminate terms of the form $e^{\pm 2i\omega_o t}$, where ω_o denotes the atomic resonant frequency. We also assumed N independent 2-level atoms. Of course, including additional matrix elements for the density matrix can treat more than two levels. The N atoms must be independent so that we can write the number of atoms in level "a" as $N_a = N\rho_{aa}$.

7.11.4 The Solutions

We want to find a solution to the optical Bloch Equations. We could make a perturbation approximation similar to the one made for time-dependent perturbation theory. We could also perform a small signal analysis to find the modulation bandwidth for either amplitude or frequency modulation. However, in the present topic, we look for the steady state solutions similar to those in Chapter 2. The quantities σ_{21}, the dipole modulation, and $\Delta N = N(\rho_{22} - \rho_{11})$, the population difference, can depend on time but any changes must be much slower than processes represented by the time constants in Equations 7.11.22 and 7.11.23. The end result at steady state will have the form

$$\text{Re}\sigma_{21} = \frac{\Delta\bar{N}}{N}\frac{T_2^2(\omega - \omega_o)\Omega}{1 + (\omega - \omega_o)^2 T_2^2 + 4\Omega^2\tau T_2} \qquad \text{Im}\sigma_{21} = -\frac{\Delta\bar{N}}{N}\frac{T_2\Omega}{1 + (\omega - \omega_o)^2 T_2^2 + 4\Omega^2\tau T_2}$$
$$\tag{7.11.24}$$

$$\Delta N = \Delta\bar{N}\frac{1 + (\omega - \omega_o)^2 T_2^2}{1 + (\omega - \omega_o)^2 T_2^2 + 4\Omega^2\tau T_2} \tag{7.11.25}$$

where

$$\Omega = \frac{\mu E_o}{2\hbar} \tag{7.11.26}$$

and E_0 denotes the electric field amplitude. Note the use of the steady state carrier difference $\Delta \bar{N}$. Without the stimulated processes, the steady state carrier difference would be sustained by the pump or the thermal distribution. These last equations for steady state can be rewritten as

$$\text{Re}(\sigma_{21}) = -\frac{(\omega - \omega_o)T_2^2 \Omega(\bar{\rho}_{22} - \bar{\rho}_{11})}{1 + (\omega - \omega_o)^2 T_2^2 + 4\Omega^2 \tau T_2} \qquad \text{Im}(\sigma_{21}) = -\frac{\Omega T_2(\bar{\rho}_{22} - \bar{\rho}_{11})}{1 + (\omega - \omega_o)^2 T_2^2 + 4\Omega^2 \tau T_2}$$

$$(7.11.27)$$

$$\rho_{22} - \rho_{11} = (\bar{\rho}_{22} - \bar{\rho}_{11})\frac{1 + (\omega - \omega_o)^2 T_2^2}{1 + (\omega - \omega_o)^2 T_2^2 + 4\Omega^2 \tau T_2} \qquad (7.11.28)$$

Before continuing, we need to discuss the difference between "steady state," denoted by SS, and "no-light steady state," denoted by NLSS. Figure 7.11.3 provides an illustration of the process. The NLSS refers to the distribution of carriers (or state of the atoms) without an electromagnetic (EM) interaction but with the pump active or with the background thermal distribution. The "steady state" refers to the same steady state as in Chapter 2. In this case, the EM interaction maintains a carrier distribution different from the NLSS

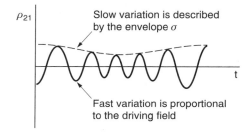

FIGURE 7.11.2
The envelope function.

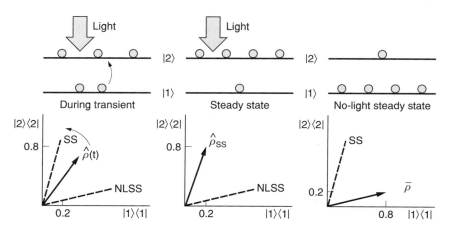

FIGURE 7.11.3
Density operator first changes to the steady state value because of the matter-light interaction. Without light, the density operator relaxes to the no-light steady state value (NLSS).

value. The steady state requires the time derivatives to be zero. If we also turn off the EM interaction then the "steady state" carrier or atom distribution must relax to the NLSS distribution.

To start the solution, we work with Equations 7.11.22 and 7.11.23 at steady state $\dot{\sigma}_{21} = 0 = \Delta\dot{N}$. Inserting the real and imaginary parts for σ_{21} into Equation 7.11.22 gives

$$(\omega - \omega_o)\,[-\text{Im}\,\sigma_{21} + i\text{Re}\,\sigma_{21}] - \frac{i\mu E_o}{2\hbar}\frac{\Delta N}{N} - \frac{1}{T_2}[\text{Re}\,\sigma_{21} + i\text{Im}\,\sigma_{21}] = 0$$

Separating the imaginary and real parts provides

$$(\omega - \omega_o)\,\text{Re}\,\sigma_{21} - \frac{\mu E_o}{2\hbar}\frac{\Delta N}{N} = \frac{\text{Im}\,\sigma_{21}}{T_2} \qquad \text{Imag} \qquad (7.11.29)$$

$$(\omega - \omega_o)\,\text{Im}\,\sigma_{21} = -\frac{\text{Re}\,\sigma_{21}}{T_2} \qquad \text{Real} \qquad (7.11.30)$$

Substituting Equation 7.11.30 into 7.11.29 for $\text{Im}\,\sigma_{21}$ provides

$$(\omega - \omega_o)\text{Re}\,\sigma_{21} - \frac{\mu E_o}{2\hbar}\frac{\Delta N}{N} = -\frac{\text{Re}\,\sigma_{21}}{(\omega - \omega_o)T_2^2}$$

Next, rearranging a little bit gives

$$\text{Re}\sigma_{21} = \frac{T_2^2(\omega - \omega_o)\frac{\Delta N}{2\hbar N}\mu E_o}{1 + (\omega - \omega_o)^2 T_2^2} \qquad (7.11.31)$$

Now substituting Equation 7.11.31 into 7.11.30 yields

$$\text{Im}\sigma_{21} = \frac{-T_2 \frac{\Delta N}{2\hbar N}\mu E_o}{1 + (\omega - \omega_o)^2 T_2^2} \qquad (7.11.32)$$

Both of the last two equations seem to be missing the saturation term in the denominator involving $\Omega \sim E_o$ the electric field amplitude. The reason is that these two equations have the *actual* (steady state) population difference ΔN and not the population difference due to the pump $\Delta\bar{N}$. We need to demonstrate Equation 7.11.25 before continuing.

We can demonstrate Equation 7.11.25 as follows. Setting the time derivative to zero in Equation 7.11.23b provides

$$\frac{\Delta N}{N} = \frac{\Delta\bar{N}}{N} + \frac{iE_o\mu\tau}{\hbar}(\sigma_{21}^* - \sigma_{21}) = \frac{\Delta\bar{N}}{N} + \frac{2E_o\mu\tau}{\hbar}\text{Im}(\sigma_{21})$$

Substituting Equations 7.11.32 and 7.11.26 into this last result provides Equation 7.11.25 as required

$$\Delta N = \Delta\bar{N}\,\frac{1 + (\omega - \omega_o)^2 T_2^2}{1 + (\omega - \omega_o)^2 T_2^2 + 4\Omega^2\tau T_2} \qquad (7.11.33)$$

Finally, we find Equation 7.11.24 by substituting Equation 7.11.33 into Equations 7.11.31 and Equation 7.11.32.

$$\mathrm{Re}\sigma_{21} = \frac{\Delta\bar{N}}{N}\frac{T_2^2(\omega - \omega_0)\Omega}{1 + (\omega - \omega_0)^2 T_2^2 + 4\Omega^2\,\tau T_2} \tag{7.11.34}$$

$$\mathrm{Im}\sigma_{21} = -\frac{\Delta\bar{N}}{N}\frac{T_2\Omega}{1 + (\omega - \omega_0)^2 T_2^2 + 4\Omega^2\,\tau T_2} \tag{7.11.35}$$

where $\Omega = (\mu E_o)/(2\hbar)$. The parameters involving N have units of "per unit volume."

Equations 7.11.34 and 7.11.35 describe the real and imaginary parts of the envelope function and shows that they depend linearly on steady-state values of the population probabilities $\bar{\rho}_{11}, \bar{\rho}_{22}$. These envelope functions will be part of the induced polarization. We think of the envelope function as representing the amplitude of the polarization. Therefore the real part must be related to the refractive index while the imaginary part must be related to the gain and absorption of the collection of N gas molecules. We also see that the induced polarization (though the envelope function) must be related to the pumping number density J. Similar comments apply to the population difference $N(\rho_{22} - \rho_{11})$. The next section shows the relation between these expressions and the classical polarization.

The next couple of sections discuss the denominator term and its relation to homogeneous broadening. The denominator depends on the optical power (inside the laser cavity) through

$$\Omega^2 \sim E_o^2 \sim \text{power}$$

As the optical power increases, the population difference and the "envelope" functions decrease. However, the envelope is related to the induced polarization and, hence, the material gain. As the optical power in the cavity increases, the material gain tends to decrease! This should remind the reader of gain saturation in optical amplifiers.

7.12 Gain, Absorption and Index for Independent Two Level Atoms

The gain, absorption and refractive index represent the primary results from the investigation of the matter–field interaction. We find these quantities by comparing classical and quantum expressions for polarization. The classical expression for polarization can be written in terms of susceptibility $\chi = \chi_r + i\chi_i$ as

$$\mathscr{P}(t) = \mathrm{Re}(\varepsilon_o\chi\mathscr{E}) = \mathrm{Re}\{\varepsilon_o\,(\chi_r + i\chi_i)\,E_o\,e^{i\omega t}\} \tag{7.12.1}$$

The polarization, as a collection of dipoles, can absorb or emit an electromagnetic (EM) field. Therefore, to find the material gain, we must find the susceptibility. However, we know from previous sections that the quantum mechanical polarization (including the ensemble average) can be related to the density matrix through

$$\mathscr{P}(t) = N\langle\mu\rangle = N\,Tr\,(\hat{\rho}\hat{\mu}) = N\mu\,(\rho_{12} + \rho_{21}) \tag{7.12.2}$$

Comparing these two expressions (7.12.1 and 7.12.2) for the polarization, produces the susceptibility in terms of the density matrix. We found expressions for the matrix elements in the previous few sections in terms of the pumping level, field amplitude E_o and the induced dipole moment μ.

7.12.1 The Quantum Polarization and the Polarization Envelope Functions

The polarization $\mathcal{P}$ can be related to the number of independent atoms (per volume) N and the average dipole moment $\langle \mu \rangle$

$$\mathcal{P}(t) = N\langle \mu \rangle = N\, Tr\,(\hat{\rho}\hat{\mu})$$

As before, we assume each atom has two levels $\{\,|n\rangle = |u_n\rangle : n = 1, 2\,\}$ where $\hat{H}_o|n\rangle = E_n|n\rangle$. We evaluate the polarization by inserting the closure relation $1 = \sum_n |n\rangle\langle n|$ between the density and dipole operators.

$$\mathcal{P}(t) = N\, Tr\,(\hat{\rho}\hat{\mu}) = N\sum_n \langle n|\hat{\rho}\hat{\mu}|n\rangle = N\sum_{mn}\langle n|\,\hat{\rho}\,|m\rangle\,\langle m|\,\hat{\mu}\,|n\rangle = N\,(\rho_{12} + \rho_{21})\,\mu \quad (7.12.3)$$

where we neglect any permanent dipole so that the dipole matrix has only off-diagonal elements

$$\underline{\mu} = \begin{bmatrix} 0 & \mu \\ \mu & 0 \end{bmatrix}$$

The off-diagonal elements of the density matrix can be related to the "slowly varying" envelope function σ by

$$\rho_{21} = \sigma_{21}(t)\,e^{-i\omega t} \quad \text{and} \quad \rho_{12} = \sigma_{12}(t)\,e^{i\omega t}$$

where both $\hat{\rho}, \hat{\sigma}$ are Hermitian. Any slow variation of the envelope function $\sigma(t)$ can be attributed to amplitude modulation on the laser optical signal. However, we assume steady-state conditions for σ_{21} so that it does not depend on time. The polarization in Equation 7.12.3 can be written as

$$\mathcal{P}(t) = N(\rho_{12} + \rho_{21})\mu = N\big(\sigma_{12}e^{i\omega t} + \sigma_{21}e^{-i\omega t}\big)\mu$$

Using Euler's theorem for the complex exponentials provides

$$\mathcal{P}(t) = N\big(\sigma_{12}e^{i\omega t} + \sigma_{21}e^{-i\omega t}\big)\mu = N\mu[(\sigma_{12} + \sigma_{21})\cos(\omega t) + i(\sigma_{12} - \sigma_{21})\sin(\omega t)] \quad (7.12.4)$$

Now use $\sigma_{12} = \sigma_{21}^*$ to find

$$\sigma_{12} + \sigma_{21} = \sigma_{21}^* + \sigma_{21} = 2\mathrm{Re}(\sigma_{21}) \quad \text{and} \quad \sigma_{12} - \sigma_{21} = \sigma_{21}^* - \sigma_{21} = -2\,i\,\mathrm{Im}(\sigma_{21})$$

Therefore, the relation between the polarization and the density matrix (Eq. 7.12.4) becomes

$$\mathcal{P}(t) = 2N\mu\,[\mathrm{Re}\,(\sigma_{21})\cos(\omega t) - \mathrm{Im}\,(\sigma_{21})\sin\,(\omega t)] \quad (7.12.5)$$

7.12.2   The Quantum Polarization and Macroscopic Quantities

The next step writes the polarization from Equation 7.12.5 in terms of accessible parameters such as the pump and relaxation times by substituting the relations derived in the previous section

$$\text{Re}\,(\sigma_{21}) = \frac{(\omega - \omega_o)\,T_2^2\,\Omega\,(\bar{\rho}_{22} - \bar{\rho}_{11})}{1 + (\omega - \omega_o)^2\,T_2^2 + 4\Omega^2\,\tau T_2}$$

$$\text{Im}(\sigma_{21}) = -\frac{\Omega T_2\,(\bar{\rho}_{22} - \bar{\rho}_{11})}{1 + (\omega - \omega_o)^2\,T_2^2 + 4\Omega^2\tau T_2} \tag{7.12.6}$$

$$\rho_{22} - \rho_{11} = (\bar{\rho}_{22} - \bar{\rho}_{11})\frac{1 + (\omega - \omega_o)^2\,T_2^2}{1 + (\omega - \omega_o)^2\,T_2^2 + 4\Omega^2\,\tau T_2}$$

where $\Omega = (\mu E_o)/(2\hbar)$ and the electric field is $\mathscr{E}(t) = E_o \cos(\omega t)$. Therefore, the polarization in Equation 7.12.5 can be rewritten as

$$\mathscr{P}(t) = 2N\mu\,\frac{(\omega - \omega_o)T_2^2\Omega(\bar{\rho}_{22} - \bar{\rho}_{11})\cos(\omega t) + \Omega T_2(\bar{\rho}_{22} - \bar{\rho}_{11})\sin(\omega t)}{1 + (\omega - \omega_o)^2 T_2^2 + 4\Omega^2\tau T_2} \tag{7.12.7}$$

We define the total number of independent atoms N (per unit volume) in the ensemble and the number in the first level N_1 and the number in the second level N_2 (per unit volume)

$$N_1 = N\,\rho_{11} \quad N_2 = N\,\rho_{22} \quad \Delta N = N_2 - N_1 \quad N_1 + N_2 = N$$

along with

$$\bar{N}_1 = N\,\bar{\rho}_{11} \quad \bar{N}_2 = N\,\bar{\rho}_{22} \quad \Delta\bar{N} = \bar{N}_2 - \bar{N}_1$$

The value ΔN denotes the population difference at operating conditions whereas the NLSS value $\Delta\bar{N}$ describes the population difference in the absence of stimulated emission and absorption (i.e., it is the pumping level taking into account the relaxation processes). Equation 7.12.7 can be rewritten in terms of the population difference $\Delta\bar{N}$ given in Equation 7.11.25 as

$$\Delta N = \Delta\bar{N}\,\frac{1 + (\omega - \omega_o)^2\,T_2^2}{1 + (\omega - \omega_o)^2\,T_2^2 + 4\Omega^2\,\tau T_2}$$

$$\mathscr{P}(t) = 2\Delta\bar{N}\mu\Omega T_2\frac{\sin(\omega t) + T_2(\omega - \omega_o)\cos(\omega t)}{1 + (\omega - \omega_o)^2 T_2^2 + 4\Omega^2\tau T_2} \tag{7.12.8}$$

The factor $2\Delta\bar{N}\mu\Omega T_2$ can be rewritten using the definition of $\Omega = \frac{\mu E_o}{2\hbar}$

$$2\,\Delta\bar{N}\,\mu\,\Omega T_2 = \frac{\mu^2}{\hbar}\,\Delta\bar{N}\,E_o T_2$$

The polarization in Equation 7.12.8 written in terms of accessible parameters provides

$$\mathscr{P}(t) = \frac{\mu^2}{\hbar} \Delta \bar{N} E_o T_2 \frac{\sin(\omega t) + T_2(\omega - \omega_o)\cos(\omega t)}{1 + (\omega - \omega_o)^2 T_2^2 + 4\Omega^2 \tau T_2} \tag{7.12.9}$$

As a note, the sine and cosine terms depend on the driving frequency because the off-diagonal elements of the density matrix have the form $\rho_{21} = \sigma_{21} e^{-i\omega t}$ with σ_{21} independent of time (because of steady state conditions). The expression for the population difference can be found from the last of Equation 7.11.25

$$\Delta N = \Delta \bar{N} \frac{1 + (\omega - \omega_o)^2 T_2^2}{1 + (\omega - \omega_o)^2 T_2^2 + 4\Omega^2 \tau T_2}$$

7.12.3 Comparing the Classical and Quantum Mechanical Polarization

Comparing the expressions for the classical and quantum mechanincal polarization provides the susceptibility, which can be related to gain. Equation 7.12.1 provides the classical polarization

$$P(t) = \mathrm{Re}(\varepsilon_o \chi \mathscr{E}) = \mathrm{Re}\{\varepsilon_o(\chi_r + i\chi_i)E_o e^{i\omega t}\} = \varepsilon_o E_o \chi_r \cos(\omega t) - \varepsilon_o E_o \chi_i \sin(\omega t)$$

Comparing this last expression with Equation 7.12.9 provides the susceptibility

$$\chi_r = \frac{\mu^2 \Delta \bar{N}}{\varepsilon_o \hbar} \frac{T_2(\omega - \omega_o)}{1 + (\omega - \omega_o)^2 T_2^2 + 4\Omega^2 \tau T_2}$$

$$\tag{7.12.10}$$

$$\chi_i = -\frac{\mu^2 \Delta \bar{N}}{\varepsilon_o \hbar} \frac{1}{1 + (\omega - \omega_o)^2 T_2^2 + 4\Omega^2 \tau T_2}$$

where $\Omega = \mu E_o/(2\hbar)$. These are the important equations.

As a note for the next topic, it is common to define the "natural line shape function" using a Lorentzian function

$$\mathscr{L}(\omega) = \frac{2T_2}{1 + 4\pi^2(\nu - \nu_o)^2 T_2^2} = \frac{2T_2}{1 + (\omega - \omega_o)^2 T_2^2} \tag{7.12.11}$$

The next topic discusses the Lorentzian line shape in more detail. For now, we restate Equations 7.12.10 in terms of the line shape function.

$$\chi_r = \frac{\mu^2 T_2(\omega - \omega_o)}{2\varepsilon_o \hbar} \Delta N \, \mathscr{L}(\omega) \quad \chi_i = \frac{\mu^2}{\varepsilon_o \hbar} \Delta N \, \mathscr{L}(\omega) \tag{7.12.12}$$

where the actual population difference ΔN (and not $\Delta \bar{N}$) appears in the last equations. However, we must eventually substitute $\Delta \bar{N}$ in place of ΔN in order to predict the susceptibility as a function of the pump level. The reader should recall that the real part of the susceptibility can be related to the refractive index and the imaginary part to the gain (or absorption).

7.12.4 The Natural Line Shape Function

The natural line shape function (Figure 7.12.1)

$$\mathscr{L}(\omega) = \frac{2T_2}{1 + (\omega - \omega_0)^2 T_2^2} \tag{7.12.13}$$

essentially provides the (normalized) emission spectrum from the ensemble of atoms (at low light levels $\Omega \cong 0$); it is also the shape of the absorption spectrum. Such a shape is characteristic of atoms that return to equilibrium through exponential relaxation process. The spectrum for a semiconductor appears different because of an integral over energy (i.e., over ω_0, the resonant frequency). The dipole dephasing time controls the width of the line distribution. Equation 7.12.13 has a full-width at half-max (FWHM) of $2/T_2$. Recall the dephasing time T_2 represents an average time-interval between collisions; these collisions interfere with the coherence between the oscillating dipole and the driving field. Apparently, large dephasing times (i.e., very few collisions) produces very sharp (i.e., narrow) spectral lines. In such a case, the semiconductor emits or absorbs at nearly a single wavelength. At the other extreme, many dephasing collisions lead to increased bandwidth (larger widths) and smaller heights. The line shape function gives the emission and absorption spectra for sufficiently small optical fields that produce negligible saturation.

The line-shape function is normalized in such a manner that its integral over frequency $\nu = \omega/2\pi$ (Hertz) equals to one.

$$\int_0^\infty d\nu \mathscr{L}(2\pi\nu) = 1 \tag{7.12.14}$$

The lower limit can be replaced by $-\infty$ to simplify the expression without changing the value of the integral since the line-shape function remains nonzero only near optical frequencies. Some books denote the line-shape function by $g(\nu) = \mathscr{L}(2\pi\nu)$. The natural line-shape function $\mathscr{L}(\omega)$ gives the shape of the absorption/emission curve for a collection of independent two-level atoms at low light levels.

7.12.5 Quantum Mechanical Gain

Let us recall from Chapter 3, the connection between the complex refractive index and the susceptibility. In this section let's use "~" to indicate complex quantities. The complex electric field is

$$\tilde{\mathscr{E}} = E_0 e^{i\tilde{k}z} = E_0 e^{ik_0\tilde{n}z} \tag{7.12.15}$$

where $\tilde{k} = k_0(n_r + in_i)$, where $k_0 = (2\pi)/(\lambda_0)$ and λ_0 is the wavelength in vacuum. Recall that the complex wave vector can also be written as

$$\tilde{k} = k_0 n_r + i\frac{\alpha}{2} \quad \text{or as} \quad \tilde{k} = k_0 n_r - i\frac{g}{2} \tag{7.12.16}$$

where "α" and "g" represent the absorption and gain (per unit length) respectively and $g = -\alpha$.

The above quantities are related to each other since $\tilde{n}$ is related to the complex permittivity $\tilde{\varepsilon}$ which is related to the complex susceptibiltiy $\tilde{\chi}$. We know from Maxwell's equations the complex permittivity $\nabla \cdot \vec{\mathscr{D}} = \rho_{\text{free}}$ and the Displacement field $\vec{\mathscr{D}}$ is

$$\vec{\mathscr{D}} = \tilde{\varepsilon}\vec{\mathscr{E}} = \varepsilon_o\vec{\mathscr{E}} + \vec{\mathscr{P}} = \varepsilon_o\vec{\mathscr{E}} + \varepsilon_o\tilde{\chi}\vec{\mathscr{E}}$$

so that the complex permittivty is identified as

$$\tilde{\varepsilon} = \varepsilon_o(1 + \tilde{\chi})$$

Therefore, the complex index of refraction is

$$\tilde{n} = \sqrt{\frac{\tilde{\varepsilon}}{\varepsilon_o}} = \sqrt{1 + \tilde{\chi}} = \sqrt{1 + \chi_r + i\chi_i} \cong \sqrt{1 + \chi_r}\left\{1 + \frac{i\chi_i}{2(1 + \chi_r)}\right\}$$

by a Taylor expansion similar to the one used in Chapter 3. The complex index can be written

$$\tilde{n} = n_r + in_i \cong \sqrt{1 + \chi_r} + \frac{i\chi_i}{2\sqrt{1 + \chi_r}}$$

Identifying the real index (responsible for optical dispersion) as

$$n_r = \sqrt{1 + \chi_r} \tag{7.12.17}$$

and the imaginary index (i.e., absorption or gain) as

$$n_i \cong \frac{\chi_i}{2n_r} \tag{7.12.18}$$

Comparing $\tilde{k} = k_o\tilde{n} = k_o(n_r + in_i)$ with $\tilde{k} = k_on_r + i\frac{\alpha}{2}$, we see

$$\alpha = 2k_on_i = 2k_o\frac{\chi_i}{2n_r} = k_o\frac{\chi_i}{n_r} \quad \text{and} \quad g = -\alpha = -k_o\frac{\chi_i}{n_r} \tag{7.12.19}$$

Using Equation 7.12.15, we see that an optical signal increases as

$$\frac{E_{out}}{E_{in}} = \exp\left(iz\tilde{k}\right) = \exp(izk_on_r)\exp(zg/2)$$

so once we know the gain per unit length g, we can calculate the exponential increase in the electric field. Of course the optical power increases as the magnitude squared of the electric field and it therefore increases as $\exp(gz)$.

Equations 7.12.10 provide the real and imaginary parts of the susceptibility

$$\chi_r = \frac{\mu^2\Delta\bar{N}}{\varepsilon_o\hbar}\frac{T_2(\omega - \omega_o)}{1 + (\omega - \omega_o)^2T_2^2 + 4\Omega^2\tau T_2}$$

$$\chi_i = -\frac{\mu^2\Delta\bar{N}}{\varepsilon_o\hbar}\frac{1}{1 + (\omega - \omega_o)^2T_2^2 + 4\Omega^2\tau T_2} \tag{7.12.20}$$

Equations 7.12.17, 7.12.18, and 7.12.19 provide equations (that have a slight problem) describing the refractive index and gain

$$n_r^2 = 1 + \chi_r = 1 + \frac{\mu^2 \Delta \bar{N}}{\varepsilon_o \hbar} \frac{T_2(\omega - \omega_o)}{1 + (\omega - \omega_o)^2 T_2^2 + 4\Omega^2 \tau T_2}$$

$$(7.12.21)$$

$$g = -\alpha = \frac{2\pi \chi_i}{\lambda_o n_r} = \frac{2\pi}{\lambda_o n_r} \frac{\mu^2 \Delta \bar{N}}{\varepsilon_o \hbar} \left(\frac{1}{1 + (\omega - \omega_o)^2 T_2^2 + 4\Omega^2 \tau T_2} \right)$$

where $\Omega = \mu E_o/(2\hbar)$, and parameters involving N have units of "per unit volume."

Although we followed all the correct steps, Equations 7.12.21 do not give correct results for a collection of N independent 2-level atoms (per unit volume). We forgot something. We can see the problem by considering the right hand side of the real part of the index. When more atoms occupy the lower energy level than the upper one (i.e., $\Delta \bar{N} < 0$), the real index must be smaller than the index for vacuum (i.e., $n_r < 1$). The problem stems from the fact that the N-atoms must be embedded in a host crystal or part of a gas mixture. Equations 7.12.21 only accounts for the change in index due to pumping and does not account for the background index (refer to Section 3.2). However, we assume that the background matter has very little absorption or gain at the optical frequency and therefore the second of Equations 7.12.21 remains correct. We can rewrite Equations 7.12.21 to include the background material as

$$n_r^2 = 1 + \chi_b + \chi_r = 1 + \chi_b + \frac{\mu^2 \Delta \bar{N}}{\varepsilon_o \hbar} \frac{T_2(\omega - \omega_o)}{1 + (\omega - \omega_o)^2 T_2^2 + 4\Omega^2 \tau T_2}$$

$$(7.12.22)$$

$$g = -\alpha = -\frac{2\pi \chi_i}{\lambda_o n_r} = +\frac{2\pi}{\lambda_o n_r} \frac{\mu^2 \Delta \bar{N}}{\varepsilon_o \hbar} \left(\frac{1}{1 + (\omega - \omega_o)^2 T_2^2 + 4\Omega^2 \tau T_2} \right)$$

where χ_b, the background susceptibility, must be real and $\Omega = \mu E_o/(2\hbar)$, and E_o represents the amplitude of the driving field. The index of the background material must be given by $n_b = \sqrt{1 + \chi_b}$. Also notice that indices of refraction do not add unlike the susceptibility. The Equations 7.12.22 provide the gain as a function of the population difference or as a function of the number of excited electrons.

7.12.6 Discussion of Results

The refractive index and gain/absorption can be plotted on the same set of axes as in Figure 7.12.2. For *N*-independent atoms, the gain curve has the Lorentzian shape with a peak at the atomic resonant frequency

$$\omega_o = \frac{E_2 - E_1}{\hbar}$$

As a note on notation, the subscript "*o*" on ω_o refers to the resonant frequency, while on E_o it refers to the amplitude of the electric field. As previously mentioned, the value $2/T_2$ represents the width of the gain curve at half the maximum height (for low light levels)

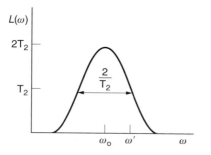

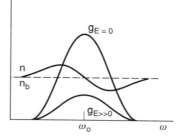

FIGURE 7.12.1
The Lorentzian line shape is zero for negative frequencies unlike the Gaussian function.

FIGURE 7.12.2
The refractive index "n" and gain "g" as a function of the EM frequency.

where T_2 denotes the dipole dephasing time. Fewer collisions lead to more narrow emission and absorption spectra. When we include the saturation effect (i.e., include Ω in the denominator of Equation 7.12.22), we find the peak gain decreases according to

$$g_{\text{peak}}(E_o \gg 0) = \frac{g_{\text{peak}}(E_o = 0)}{1 + 4\Omega^2 \tau T_2} \sim \frac{g_{\text{peak}}(E_o = 0)}{4\Omega^2 \tau T_2} \tag{7.12.23}$$

The decrease in g is sometimes called gain saturation. We can see the reason for the name based on the equations for a laser amplifier. The rate equations provide a differential equation

$$dP = P\,g(P)\,dz \tag{7.12.24}$$

for the power $P(z)$ traveling along the length of the amplifier z. Let P_o be the power at $z = 0$. The gain can be written in the simpler form

$$g = \frac{c_1}{c_2(\omega) + P/P_s} \tag{7.12.25a}$$

where P_s denotes a saturation power (refer to the next topic for alternate definitions). The constants can be found from Equation 7.12.22 to be

$$c_1(J) = \frac{2\pi}{\lambda_o n_r} \frac{\mu^2 \Delta \bar{N}}{\varepsilon_o \hbar} \qquad c_2(\omega) = 1 + (\omega - \omega_0)^2 T_2^2 \tag{7.12.25b}$$

and power P can be calculated from the Poynting vector.

$$P = IA = \frac{c n_r \varepsilon_o E_o^2}{2} A \qquad P_s = I_s A = \frac{c n_r \varepsilon_o \hbar^2}{2\mu^2 \tau T_2} A \tag{7.12.25c}$$

where A represents the beam cross section area and I represents the intensity (Watts per area). Integrating Equation 7.12.24 provides

$$Ln\frac{P}{P_o} = \frac{c_1}{c_2}z - \frac{1}{c_2}\frac{P - P_o}{P_s}$$

(7.12.26)

For large saturation threshold power P_s, Equation 7.12.26 reduces to the usual exponential form for the power in the amplifier $P = P_o e^{c_1 z/c_2}$. However, small saturation power P_s necessarily reduces the optical power P at z compared with the pure exponential case without saturation.

The top of Equations 7.12.22 shows that the refractive index depends on (1) the magnitude of the EM wave through E_o, (2) the frequency because of $(\omega - \omega_0)$ in both the numerator and denominator, and (3) the population difference through ΔN. In fact, Figure 7.12.2 shows that the pumped atoms can either increase or decrease the index depending on whether $\omega < \omega_0$ or $\omega > \omega_0$, respectively. Also, notice from Equations 7.12.20, the index is related to the real part of the susceptibility and the gain/absorption is related to the imaginary part.

7.12.7 Comments on Saturation Power and Intensity

Often another form of the saturation intensity is used so that the saturated gain has the form (for a monochromatic beam)

$$g(\omega) = \frac{g_{E=0}(\omega)}{1 + \frac{I}{I_{s\omega}(\omega)}}$$

(7.12.27a)

where, I represents the intensity found from the Poynting vector (Watts/area). The parameters are

$$g_{E=0} = \frac{2\pi}{\lambda_o n_r}\frac{\mu^2 \Delta \bar{N}}{\varepsilon_o \hbar}\frac{1}{1 + (\omega - \omega_0)^2 T_2^2}$$

(7.12.27b)

$$I = \frac{c n_r \varepsilon_o E_o^2}{2} \qquad I_{s\omega}(\omega) = \frac{c n_r \varepsilon_o \hbar^2 T_2}{\mu^2 \tau \mathscr{L}(\omega)}$$

(7.12.27c)

where, $\mathscr{L}(\omega)$ appears in Equation 7.12.13. Apparently I_s in Equation 7.12.25c agrees with 7.12.27c for $\omega = 0$ so that $I_s = I_{s\omega}(0)$.

7.13 Broadening Mechanisms

A very important distinction between lasers concerns the homogeneously and inhomogeneously broadened gain media. A homogeneously broadened laser will have atmost one lasing longitudinal mode. An inhomogeneously broadened laser can have any number of longitudinal modes. The reason for these different behaviors has to do with the way the laser produces gain within the laser cavity. Identical atoms in identical environments produce the homogeneously broadened line shapes. Atoms affected differently from one another by their environment produce the inhomogeneously broadened line shapes. For example, one atom might experience a different strain than another one. The Doppler effect can also produce inhomogeneous broadening since each atom, because of its motion, radiates at a frequency slightly different from the average-value frequency radiated by the entire ensemble.

7.13.1 Homogeneous Broadening

The emission or absorption spectrum from a collection of atoms shows homogeneous broadening when the environment relaxes for all the atoms in the same way. Each dipole exponentially relaxes with the relaxation time T_2. This type of relaxation produces the homogeneously broadened spectral line with width $\delta\omega \sim 1/T$.

The Liouville equation shows the interaction of the dipoles with the environment producing homogenous broadened spectra. Equation 7.12.22 in the previous section provides the gain for a collection of two-level atoms

$$g = -\alpha = +\frac{2\pi}{\lambda_o n_r}\frac{\mu^2 \Delta\bar{N}}{\varepsilon_o \hbar}\left(\frac{1}{1+(\omega-\omega_o)^2 T_2^2 + 4\Omega^2 \tau T_2}\right) \tag{7.13.1}$$

where $\mu, \Delta\bar{N}, \lambda_o, n_r, \varepsilon_o, \omega_o, T_2, \tau$ represent the dipole moment, the no-light steady state (NLSS) population difference (i.e., pump number), real refractive index, free-space permitivity, resonant frequency for the two levels $\omega_o = (E_2 - E_1)/\hbar$, the dipole dephasing time, and the population relaxation time, respectively. The $\Omega^2 = ((\mu E_o)/(2\hbar))^2$ has the amplitude E_o of the driving electric field oscillating at frequency ω. The driving field can be the optical field in the laser cavity or it can be the field amplified by an optical amplifier. Recall that the intensity is proportional to the square of the field amplitude $I \sim E_o^2$. As the intensity increases, the gain decreases because Ω appears in the denominator (see Figure 7.13.1). The entire gain curve decreases for homogeneous broadening. Increasing the amplitude of the electromagnetic power also increases the width of the spectral line. The full-width at half-max (FWHM) refers to the points $\omega_o \pm \delta\omega/2$, where the gain curve drops to ½ its peak value at the resonant frequency ω_o.

$$(FWHM)_{\text{Sat}} = \frac{2}{T_2}\sqrt{1 + 4\Omega^2 \tau T_2} = (FWHM)_{\text{unsat}}\sqrt{1 + \mu^2 E_o^2 \tau T_2/\hbar^2} \tag{7.13.2}$$

where "sat" refers to the nonzero driving-field case and "unsat" refers to the zero driving-field case. This last equation shows that the width of the spectrum is $\Delta\omega_{\Omega=0} = 2/T_2$ in the low driving-field limit $\Omega \sim 0$. However, larger driving fields produce a larger line width of $(\Delta\omega)_{\Omega=0}\sqrt{1 + \mu^2 E_o^2 \tau T_2/\hbar^2}$.

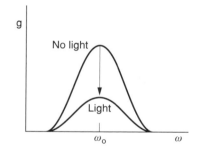

FIGURE 7.13.1
Gain saturation.

Question: Why does the emitted light have a nonzero width when we assume that the difference in energy levels $E_{21} = E_2 - E_1$ is exact? The answer comes from the fact that electrons can spontaneously relax from E_2 to E_1 which actually causes the broadening. Heisenberg's uncertainty relation gives

$$\Delta E \; \Delta t \sim \hbar \qquad\qquad (7.13.3)$$

For long relaxation time $\Delta t \sim \infty$, ΔE can be very small which implies a narrow emission curve.

It turns out that anytime a system has loss (like spontaneous decay or optical loss through mirrors) or gain there will always be some nonzero line width. This occurs because the assumed eigenstates of H_o are not exactly eigenstates for the entire system $\hat{\mathscr{H}} = \hat{\mathscr{H}}_o + \hat{V} + \hat{\mathscr{H}}_{\text{env}}$. If we knew the eigenstates of $\hat{\mathscr{H}}$ then these states would not decay. We now see that spontaneous emission common to all lasers ultimately limits the minimum width of the spectral line.

7.13.2   Inhomogeneous Broadening

Inhomogeneous broadening refers to the mechanisms responsible for line widths that are larger than the homogeneously broadened ones. The inhomogeneous broadening occurs when the environment affects each atom differently from the next. For example, defects or local stress or strain can influence the resonant frequency of the oscillating dipole. The doppler effect can also produce slightly different resonant frequency as viewed in the laboratory frame. The doppler effect arises because the atoms in the gas or solid move with respect to a stationary observer in the lab which changes the resonant frequency for that observer. Therefore each atom in random motion will a have slightly different resonant frequency.

The effect of inhomogeneous broadening appears in Figure 7.13.2. Suppose the top portion of the figure represents the *homogeneously* broadened lines of four atoms. Notice the resonant frequency differs for each atom. Adding these four spectra together produces the spectrum in the bottom portion of the figure. We can define the inhomogeneously broadened spectrum $L_I(\omega)$ by adding together all of the homogeneously

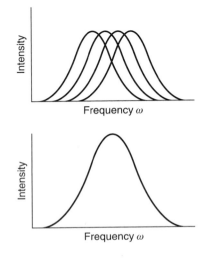

FIGURE 7.13.2
Spectra (top) add to produce the inhomogeneously broadened spectrum (bottom).

broadened lines $L_h(\omega, \omega_o)$. However, the distribution should have unit magnitude and we therefore divide by the frequency range for the resonant frequency. Alternatively, this can be viewed as the average.

$$L_I(\omega) = \frac{1}{\omega_2 - \omega_1} \int_{\omega_1}^{\omega_2} d\omega_o L_h(\omega, \omega_o) = \int_{\omega_1}^{\omega_2} d\omega_o L_h(\omega, \omega_o) f(\omega_o) \qquad (7.13.4)$$

Dividing by the frequency range assumes the resonant frequency of the dipoles is uniformly distributed. The function $f(\omega_o)$ becomes the uniform probability density.

The number of atoms with one resonant frequency does not need to be the same as the number at another resonant frequency. In fact, the limits on the frequency range do not need to be fixed at ω_1 and ω_2. We therefore take $f(\omega_o)$ to be a general probability density with the probability of finding a dipole with resonant frequency in the range (ω_1, ω_2) to be $\int_{\omega_1}^{\omega_2} d\omega_o f(\omega_o)$. The inhomogeneously broadened line has the general definition

$$L_I(\omega) = \int_{-\infty}^{+\infty} d\omega_o L_h(\omega, \omega_o) f(\omega_o) \qquad (7.13.5)$$

L_I is normalized to 1 provided L_h is normalized to 1.

The gain for the inhomogeneously broadened atoms can be treated in a similar fashion to the spectral distribution by starting with Equation 7.13.1 for the homogeneous gain

$$g_h(\omega, \omega_o) = \frac{g_{\text{peak}}(J)}{1 + (\omega - \omega_o)^2 T_2^2 + \frac{I}{I_s}} \qquad (7.13.6)$$

where the peak unsaturated gain is $g_{\text{peak}}(J) = \frac{2\pi}{\lambda_o n_r} \frac{\mu^2 \Delta \bar{N}}{\varepsilon_o \hbar}$, and the saturation intensity I_s(watts/area) is $I_s = \frac{c n_r \varepsilon_o \hbar^2}{2\mu^2 \tau T_2}$. As a reminder, the driving EM field has frequency ω. If $f(\omega_o) d\omega_o$ represents the fraction of dipoles with resonant frequency in the range $(\omega_o, \omega_o + d\omega_o)$, then the total inhomogeneous gain at the driving frequency ω must be

$$g_I(\omega) = \int_{-\infty}^{\infty} g_h(\omega, \omega_o) f(\omega_o) \, d\omega_o = \int_{-\infty}^{\infty} \frac{g_{\text{peak}}(J) f(\omega_o) \, d\omega_o}{1 + (\omega - \omega_o)^2 T_2^2 + \frac{I}{I_s}} \qquad (7.13.7)$$

In the case where the probability distribution $f(\omega_o)$ is very broad compared with the Lorentzian, the integral can be evaluated to provide

$$g_I(\omega) = g_{\text{peak}}(J) f(\omega) \int_{-\infty}^{\infty} \frac{d\omega_o}{1 + (\omega - \omega_o)^2 T_2^2 + \frac{I}{I_s}} = \frac{f(\omega) g_{\text{peak}}(J) \pi/T_2}{\sqrt{1 + \frac{I}{I_s}}} = \frac{g_{I, E=0}}{\sqrt{1 + \frac{I}{I_s}}} \qquad (7.13.8)$$

where $g_{I, E=0} = f(\omega) g_{\text{peak}}(J) \pi/T_2$. The inhomogeneous gain saturates with intensity I but not as strongly as the homogeneous gain in 7.13.1.

7.13.3 Hole Burning

"Hole burning" refers to either spatial or spectral hole burning. Spatial hole burning refers to the situation where a spatially extended gain medium has reduced or saturated gain in one location and not another. For example, a tightly focused beam in one region of the gain medium will tend to saturate the gain at that location but not in another region where the beam might have lower intensity. Spatial hole burning occurs for either homogeneously or inhomogeneously broadened spectra depending on the carrier or dipole dynamics. For example, Figure 7.13.3 shows a strong pump beam focused on one portion of a gain medium while a second weaker beam, the tickler, also passes through

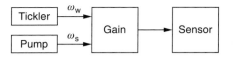

FIGURE 7.13.3

Two wavelength system to determine saturation characteristics of a gain medium.

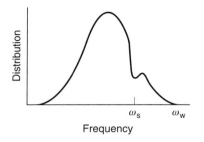

FIGURE 7.13.4

Spectral hole burning near frequency ω_S.

the material. If the weaker beam passes through the region saturated by the pump then its gain will not be as large compared to when it passes through a separate unsaturated region.

Spectral hole burning refers to the reduction in material gain near a specific wavelength rather than a location in space. Figure 7.13.3 also shows the block diagram for an experiment to determine these saturation effects. The weak "tickler" beam with optical frequency ω_w enters a gain medium and overlaps a strong pump beam with optical frequency ω_s. The frequencies need not be the same. For homogeneously broadened media, the pump beam reduces the gain for all possible tickler-beam frequencies. The whole gain curve shifts downward similar to that shown in Figure 7.13.1.

Inhomogeneously broadened media produce a tickler-beam gain curves similar to the one shown in Figure 7.13.4. The *homogeneous* gain for the dipoles with resonant frequencies near ω_s tend to saturate due to the pump beam while the gain for those with resonant frequency away from ω_s does not saturate. Therefore, the pump produces a "dip" near its frequency.

7.14 Introduction to Jaynes-Cummings' Model

The Jaynes-Cummings' Model treats the simplest system consisting of a 2-level radiator/ absorber (i.e., atom) interacting with a single-mode electromagnetic (EM) field. This fully quantized formalism incorporates the quantum theory for the atom and for the EM field in contrast to the Liouville equation where the EM field and the environmental systems remain unquantized (Sections 7.9 through 7.11). As discussed in subsequent sections, we can recover the Liouville equation from the fully quantized theory once we include the effects of quantum mechanical reservoirs. The reservoir model divides the complete system (atom plus light plus environment) into a subsystem and a number of reservoirs. The smaller system responds to the influence of the reservoirs which for lasers, account for optical loss, pumping, and collisions (etc). The reservoirs give rise to the fluctuation and relaxation effects described by the phenomenological term in the Liouville equation. In this section, we consider the atomic system interacting with a large number of modes.

The system itself consists of two parts, namely the atom and the field. The Hamiltonian $\hat{\mathscr{H}}$ includes Hamiltonians for the atom, fields and their interactions

$$\hat{\mathscr{H}} = \hat{\mathscr{H}}_a + \hat{\mathscr{H}}_f + \hat{\mathscr{H}}_{a-f}$$

The following topics explore each Hamiltonian. The atomic and EM wave functions live in separate Hilbert spaces and move independently of one another in the absence of the matter–field interaction. The motions of the two wave functions become correlated only when the atom and light interact.

7.14.1 The Pauli Operators

We will work with two level atoms and, in the subsequent topic, state the atomic Hamiltonian $\hat{H}_a$ in terms of the Pauli operators. The energy basis set must be $\{|n\rangle = |E_n\rangle$ for $n = 1, 2\}$. The Pauli operators, so termed because of their similarity with spin operators, have special symbols

$$\hat{\sigma}_{ij} = |i\rangle\langle j|$$

and

$$\hat{\sigma}_z = \hat{\sigma}_{22} - \hat{\sigma}_{11} = |2\rangle\langle 2| - |1\rangle\langle 1| \quad \hat{\sigma}_y = i(\hat{\sigma}_{12} - \hat{\sigma}_{21}) \quad \hat{\sigma}_x = \hat{\sigma}_{12} + \hat{\sigma}_{21} \tag{7.14.1}$$

The operator $\hat{\sigma}_z$ essentially gives the difference in population (i.e. probability) between the second and first energy levels for a given electron wave function. The raising and lowering operators have special significance for the atomic Hamiltonian. The raising and lowering operators are respectively defined by

$$\hat{\sigma}^+ = |2\rangle\langle 1| \quad \hat{\sigma}^- = |1\rangle\langle 2| \tag{7.14.2}$$

For example, the raising operator promotes an electron from the lower to the higher level

$$\hat{\sigma}^+|1\rangle = |2\rangle \quad \hat{\sigma}^+|2\rangle = 0 \tag{7.14.3}$$

Example 7.14.1
Suppose

$$|\psi\rangle = \frac{1}{\sqrt{2}}|1\rangle + \frac{1}{\sqrt{2}}|2\rangle$$

which says that a given atom is 50% in the first energy level and 50% in the second. The average *difference* in population must be

$$\langle\psi|\hat{\sigma}_z|\psi\rangle = 0$$

Sometimes quantum mechanical quantities (such as commutators) can be most simply computed using the matrix representation of the Pauli operators. The column vectors

$$\begin{pmatrix} 1 \\ 0 \end{pmatrix} \quad \begin{pmatrix} 0 \\ 1 \end{pmatrix} \tag{7.14.4a}$$

represent electrons in states $|1\rangle$ and $|2\rangle$, respectively. The matrix representation of the Pauli operators must be

$$\underline{\sigma}_z = \begin{pmatrix} -1 & 0 \\ 0 & 1 \end{pmatrix} \quad \underline{\sigma}_y = \begin{pmatrix} 0 & i \\ -i & 0 \end{pmatrix} \quad \underline{\sigma}_x = \begin{pmatrix} 0 & 1 \\ 1 & 0 \end{pmatrix} \tag{7.14.4b}$$

The raising and lowering operators have a matrix representation

$$\underline{\sigma}^+ = \begin{pmatrix} 0 & 0 \\ 1 & 0 \end{pmatrix} \quad \underline{\sigma}^- = \begin{pmatrix} 0 & 1 \\ 0 & 0 \end{pmatrix} \tag{7.14.4c}$$

These operators have the name "raising and lowering" even though they do not necessarily have any connection with a harmonic oscillator. The commutation relations can be easily calculated using either the operator or matrix form. For example,

$$\left[\hat{\sigma}^-, \hat{\sigma}^+\right] = -\hat{\sigma}_z \tag{7.14.5}$$

7.14.2   The Atomic Hamiltonian

The Hamiltonians for the atom, field and interactions

$$\hat{\mathcal{H}} = \hat{\mathcal{H}}_a + \hat{\mathcal{H}}_f + \hat{\mathcal{H}}_{a-f}$$

comprise the entire Hamiltonian $\hat{\mathcal{H}}$. The energy basis set is $\{\ |n\rangle = |E_n\rangle \text{for} n = 1, 2\ \}$. The atomic Hamiltonian can be written in terms of projection operators after using the closure relation for the two-dimensional space $\sum_{n=1}^{2} |n\rangle\langle n| = 1$

$$\hat{\mathcal{H}}_a = \hat{\mathcal{H}}_a \sum_{n=1}^{2} |n\rangle\langle n| = \sum_{n=1}^{2} \hat{\mathcal{H}}_a |n\rangle\langle n| = \sum_{n=1}^{2} E_n |n\rangle\langle n| = E_1 |1\rangle\langle 1| + E_2 |2\rangle\langle 2| \tag{7.14.6}$$

We rewrite the atomic Hamiltonian in terms of the Pauli z-spin operator that measures the difference in population. The Hamiltonian becomes

$$\hat{\mathcal{H}}_a = E_1 |1\rangle\langle 1| + E_2 |2\rangle\langle 2| = \frac{1}{2}(E_2 - E_1)[|2\rangle\langle 2| - |1\rangle\langle 1|] + \frac{1}{2}(E_2 + E_1)[|2\rangle\langle 2| + |1\rangle\langle 1|]$$

which simplifies to

$$\hat{\mathcal{H}}_a = \frac{E_2 - E_1}{2}\hat{\sigma}_z + \frac{E_2 + E_1}{2}\hat{1} \tag{7.14.7}$$

The energy difference between levels is usually defined in terms of an atomic resonance frequency ω_o as

$$\hbar\omega_o = E_2 - E_1 \tag{7.14.8}$$

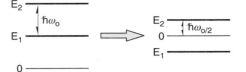

FIGURE 7.14.1
Redefining the zero of energy for the atom.

The atomic Hamiltonian can then be written as

$$\hat{\mathcal{H}}_a = \frac{1}{2}\hbar\omega_o\hat{\sigma}_z + \frac{E_2 + E_1}{2}\hat{1} \qquad (7.14.9)$$

The energy scale for the atom can be reset so that the zero-energy sits midway between E_2 and E_1 as shown in Figure 7.14.1. The last term in $\hat{H}_a$ becomes equal to zero. This last term (regardless of whether it's zero or not) hasn't any effect on the rate of change of a density operator (for example) since it commutes with all operators (and c-numbers). For example, the equation of motion for the atomic density operator $\hat{\rho}_a$ does not depend on this last term

$$\frac{\partial\hat{\rho}_a}{\partial t} = \frac{1}{i\hbar}\left[\hat{\mathcal{H}}_a, \hat{\rho}_a\right] = \frac{1}{i\hbar}\left[\frac{1}{2}\hbar\omega_o\hat{\sigma}_z + \frac{E_2 + E_1}{2}\hat{1}, \hat{\rho}_a\right] = \frac{1}{i\hbar}\left[\frac{1}{2}\hbar\omega_o\hat{\sigma}_z, \hat{\rho}_a\right]$$

Either by resetting the energy scale or for the reason of computing rates of change, the added c-number constant term can be dropped from the Hamiltonian. The final version of the atomic Hamiltonian must be

$$\hat{\mathcal{H}}_a = \frac{1}{2}\hbar\omega_o\hat{\sigma}_z \qquad (7.14.10)$$

7.14.3   The Free-Field Hamiltonian

The atoms and electromagnetic EM fields live in a product space. One constituent product space, spanned by the photon Fock states, describes the amplitude of the EM field. The quantization of the electromagnetic field requires us to replace the classical EM Hamiltonian found in Sections 3.5 and 6.5

$$\vec{\mathcal{H}}_f = \int_V dV \left(\frac{\varepsilon_o}{2}\vec{E}\cdot\vec{E} + \frac{1}{2\mu_o}\vec{B}\cdot\vec{B}\right)$$

with the quantum mechanical one

$$\hat{\mathcal{H}}_f = \sum_{\vec{k}s}\hbar\omega_k\left(\hat{b}_{\vec{k}s}^+\hat{b}_{\vec{k}s} + \frac{1}{2}\right) = \sum_{\vec{k}s}\hbar\omega_k\left(\hat{N}_{\vec{k}s} + \frac{1}{2}\right) \qquad (7.14.11)$$

The operators $\hat{b}_{\vec{k},s}^+$, $\hat{b}_{\vec{k},s}$ represent the creation and annihilation operators, respectively. Equation 7.14.11 can be viewed as the "free field" Hamiltonian in the Heisenberg representation when the operators have only the trivial time dependence. The equation can also be viewed as the total EM Hamiltonian (with or without the matter-light interaction) for the interaction representation. Recall that the "creation operator" creates a particle in the electromagnetic mode represented by $\vec{k}$ with one of two polarizations represented by "s." The creation and annihilation operators satisfy the equal-time commutation relations

$$\left[\hat{b}_{\vec{k},s}, \hat{b}^{+}_{\vec{K},S}\right] = \delta_{\vec{k},\vec{K}} \delta_{s,S} \tag{7.14.12}$$

The EM creation and annihilation operators (in either the Schrodinger or the interaction representation) commute for all times with the atomic raising and lowering operators (in either the Schrodinger or the interaction representation). As a comment, the Heisenberg representation places all of the dynamics into the operators. This implies that the creation and annihilation operators must change in a nontrivial manner when the EM field interacts with matter. The Heisenberg operators do not necessarily satisfy the commutation relations in Equation 7.14.12 for the case when the annihilation operator and creation operator are evaluated at different times.

7.14.4   The Interaction Hamiltonian

The interaction Hamiltonian links the atomic subsystem with the EM subsystem. The interaction Hamiltonian correlates the motion of the EM wave functions and the atomic wave functions in their respective Hilbert spaces. The interaction Hamiltonian drives the motion of the wave functions in the interaction representation.

   This topic discusses the fully quantized interaction Hamiltonian. We start by discussing how energy conservation leads to particular combinations of creation and annihilation operators. Next, starting from basic formulas, we show the Hamiltonian does in fact contain the expected terms. Finally, we write the fully quantized interaction Hamiltonian $\hat{\mathcal{H}}_{af}$ between the atom and fields.

   Let us consider as to how the principle of energy conservation leads to specific terms in the interaction Hamiltonian. The principle requires that any quanta of energy removed from the field must reappear in the atom and vice versa. Therefore, we expect to find terms of the form

$$\hat{b}^{-}_{k}\hat{\sigma}^{+} \quad \text{and} \quad \hat{b}^{+}_{k}\hat{\sigma}^{-} \tag{7.14.13}$$

Consider $\hat{b}^{+}_{k}\hat{\sigma}^{-}$ as an example. The operator $\hat{\sigma}^{-}$ removes a quanta of energy from the two level atom while the operator $\hat{b}^{+}_{k}$ increases the field quanta by one. Now here's an important point. While these two operators keep track of the number of quanta, they do not quite guarantee conservation of energy because we need to know that mode k has approximately the same energy as the difference in atomic levels. If it doesn't then other factors in the equation must prohibit the atom from radiating. The two terms in Equations 7.14.13 link the atomic and EM Hilbert spaces (i.e., link the two subspaces in the larger direct product space). For a *single* optical mode and a two-level atom, the basis vectors in the direct product space have the form

$$\{|a,n\rangle = |a\rangle|n\rangle \quad a = 1,2 \quad n = 0,1,2\ldots\} \tag{7.14.14}$$

where "a" stands for the atomic state and "n" stands for the number of photons in the electromagnetic mode (assuming Fock states for the EM basis set). A general wave vector in the direct product space has the form

$$|\psi(t)\rangle = \sum_{a,n} \beta_{an}(t)\,|a\rangle|n\rangle \tag{7.14.15}$$

For *multiple* optical modes, the interaction Hamiltonian $\hat{\mathscr{H}}_{af}$ must contain terms such as

$$\sum_{\vec{k},s} \hat{\sigma}^- \hat{b}^+_{ks} \quad \text{and} \quad \sum_{\vec{k},s} \hat{\sigma}^+ \hat{b}^-_{ks} \tag{7.14.16}$$

The basis vectors in the direct product space consist of the direct product of the atomic basis set with the EM Fock set (for example)

$$\{|a\rangle|m_1, m_2 \ldots\rangle = |a\rangle \, |\{m\}\rangle \quad a = 1, 2 \quad m_i = 0, 1, \ldots\} \tag{7.14.17}$$

where m_i is the number of photons in the ith mode.

Example 7.14.2

Calculate the following quantity: $\hat{b}^- \hat{\sigma}^+ |a = 1, n = 5\rangle$

Solution

$$\hat{b}^- \hat{\sigma}^+ |a = 1, n = 5\rangle = \hat{\sigma}^+ |a = 1\rangle \, \hat{b}^- |n = 5\rangle = |a = 2\rangle \sqrt{5}|n = 4\rangle$$

We can find the fully quantized form of the Hamiltonian by starting with the interaction Hamiltonian

$$\hat{H}_{af} = \hat{\mu} \cdot \hat{E} \tag{7.14.18}$$

where, $\hat{\mu}$ is the dipole moment operator and $\hat{E}$ is the electric field operator. Both operators are also vectors in the physical three-dimensional space (hence the reason for the dot product). The dipole moment operator can be written as a basis vector expansion

$$\hat{\mu} = \left(\sum_{i=1}^{2} |i\rangle\langle i| \right) \hat{\mu} \left(\sum_{j=1}^{2} |j\rangle\langle j| \right) = \sum_{ij} \vec{\mu}_{ij} \, |i\rangle\langle j| \tag{7.14.19}$$

Typically, we assume that the atom does not have a permanent dipole moment $\mu_{ii} = 0$ and that the induced dipole moment has the property that $\mu_{12} = \mu_{21}$. The dipole operator reduces to

$$\hat{\mu} = \vec{\mu}_{12}[\, |1\rangle\langle 2| + |2\rangle\langle 1| \,] = \left(\hat{\sigma}^- + \hat{\sigma}^+ \right) \vec{\mu}_{12} \tag{7.14.20}$$

We usually assume the physical size of the dipole to be small compared with the electromagnetic wavelength (dipole approximation). The matrix element appearing in Equations 7.14.19 and 7.14.20 has the form

$$\vec{\mu}_{12} = \int d^3\vec{r} \, u_1^*(\vec{r}) \, (-e\vec{r}) \, u_2(\vec{r}) \tag{7.14.21}$$

The electric field operator in the interaction Hamiltonian (Equation 7.14.18) can be written as

$$\hat{E}(\vec{r}, t) = \sum_{\vec{k}s} \sqrt{\frac{\hbar\omega_k}{2\varepsilon_0}} \left[\hat{b}^-_{\vec{k}s}(t) \, f_{\vec{k}}(\vec{r}) + \hat{b}^+_{\vec{k}s}(t) \, f_{\vec{k}}^*(\vec{r}) \right] \tilde{e}_{\vec{k}s} \tag{7.14.22}$$

where, $\tilde{e}_{\vec{k}s}$ represents a polarization vector. The mode functions $f_{\vec{k}}$ satisfy the classical Maxwell equations with specified boundary conditions (c.f., see 6.3.4) and can be normalized according to

$$\int_V dV\, f_{\vec{k}}^*(\vec{r})\, f_{\vec{K}}(\vec{r}) = \delta_{\vec{k}\vec{K}} \qquad (7.14.23)$$

For plane waves with periodic boundary conditions, the mode function is

$$f_{\vec{k}}(\vec{r}) = \frac{e^{i\vec{k}\cdot\vec{r}}}{\sqrt{V}} \qquad (7.14.24)$$

Example 7.14.3

Orthonormality of "f" for plane waves. The orthonormality can easily be shown by considering the integral for the 1-D case

$$I = \int_0^L dx\, f_k^*(x)\, f_K(x) = \frac{1}{L}\int_0^L dx\, e^{i(k-K)x}$$

where, for periodic boundary conditions, $k = 2\pi n/L$. If $k=K$ then the integral becomes

$$I = \frac{1}{L}\int_0^L dx\, 1 = 1$$

For $k \neq K$, the integral is

$$I = \frac{1}{L}\int_0^L dx\, \exp\left\{i\,\frac{2\pi(n-m)}{L}x\right\} = \frac{e^{2\pi i(n-m)} - 1}{2\pi i(n-m)} = 0$$

where the last step follows because (*n-m*) is an integer. Putting the two results together gives the orthonormality relation.

The interaction Hamiltonian can be rewritten by combining the operator expressions for the dipole moment and the electric field.

$$\mathscr{H}_{af} = \hat{\mu}\cdot\hat{E}(\vec{r},t) = \sum_{\vec{k}s}\sqrt{\frac{\hbar\omega_k}{2\varepsilon_0}}\left(\hat{\mu}_{12}\cdot\tilde{e}_{\vec{k}s}\right)\left[\hat{\sigma}^-(t) + \hat{\sigma}^+(t)\right]\left[\hat{b}_{\vec{k}s}(t)f_{\vec{k}}(\vec{r}) + \hat{b}_{\vec{k}s}^+(t)f_{\vec{k}}^*(\vec{r})\right] \qquad (7.14.25)$$

where the raising, lowering, creation and annihilation operators are all written in the interaction representation (refer to the following topics). The rotating wave approximation (RWA) allows us to drop the terms

$$\hat{\sigma}^-\hat{b} \quad \text{and} \quad \hat{\sigma}^+\hat{b}^+ \qquad (7.14.26)$$

The rotating wave approximation is equivalent to dropping terms that do not conserve energy. For example, $\hat{\sigma}^-\hat{b}$ removes a photon from the EM field and also removes a unit of energy from the atom without placing the extra energy anywhere. The RWA is usually associated with an integral over time; this turns out to be the case when we calculate the integral of the rate of change of the density operator. The final form of the interaction Hamiltonian is

$$\hat{H}_{af} = \hat{\mu} \cdot \hat{E}(\vec{r}, t) = \sum_{\vec{k}s} \sqrt{\frac{\hbar \omega_k}{2\varepsilon_o}} \left(\hat{\mu}_{12} \cdot \tilde{e}_{\vec{k}s} \right) \left[\hat{\sigma}^+(t) \, \hat{b}_{\vec{k}s}(t) \, f_{\vec{k}}(\vec{r}) + \hat{\sigma}^-(t) \, \hat{b}_{\vec{k}s}^+(t) \, f_{\vec{k}}^*(\vec{r}) \right] \qquad (7.14.27)$$

As a note, it is interesting to speculate on whether or not conservation of energy must only hold in the long-time limit. It might be possible to create a particle from the vacuum so long as it returns to the vacuum within a very short time period.

7.14.5 Atomic and Interaction Hamiltonians Using Fermion Operators

Some books (e.g., Haken) write the atomic and interaction Hamiltonian in terms of Fermion creation and annihilation operators. Rather than the two vectors used for the atomic states, we must now use the following three Fermion Fock states.

$|0, 0\rangle$ is the vacuum state

$|1, 0\rangle$ specifies an electron in the lowest energy level

$|0, 1\rangle$ specifies an electron in the highest energy level

The Fermion creation $\hat{f}_n^+$ and annihilation $\hat{f}_n^-$ operators add or subtract a particle from energy level E_n, respectively. For example, the Fermion creation operator has the following affect.

$$\hat{f}_1^+ |0, 0\rangle = |1, 0\rangle \qquad \hat{f}_1^+ |1, 0\rangle = 0$$

with similar results for $\hat{f}_2^+$ etc. The Fermion creation and annihilation operators behave similar to those for the boson creation and annihilation operators (e.g. for EM fields). The boson operators produce any number of particles in a state with a particular energy. The Fermion operators produce either 0 or 1.

The Fermion operators satisfy anticommutation relations, which allow only one electron per state (Pauli exclusion principle).

$$\left\{ \hat{f}_a^-, \hat{f}_b^+ \right\} = \delta_{ab} \qquad \left\{ \hat{f}_a^-, \hat{f}_b^- \right\} = 0 \qquad \left\{ \hat{f}_a^+, \hat{f}_b^+ \right\} = 0$$

where the anticommutator is defined by

$$\left\{ \hat{A}, \hat{B} \right\} = \hat{A}\hat{B} + \hat{B}\hat{A}$$

For example, the anticommutation relation

$$\left\{ \hat{f}_a^+, \hat{f}_b^+ \right\} = 0$$

yields the Pauli exclusion principle

$$\hat{f}_a^+ \hat{f}_a^+ |0\rangle = \frac{1}{2} \, 2\hat{f}_a^+ \hat{f}_a^+ |0\rangle = \frac{1}{2} \left\{ \hat{f}_a^+, \hat{f}_a^+ \right\} |0\rangle = \frac{1}{2} \, 0 \, |0\rangle = 0$$

So it is not possible to create two particles in a single energy state. The atomic raising and lowering operators must be replaced as follows

$$\hat{\sigma}^+ \rightarrow \hat{f}_2^+ \hat{f}_1^- \quad \text{and} \quad \hat{\sigma}^- \rightarrow \hat{f}_1^+ \hat{f}_2^-$$

This book uses the raising and lowering operators rather than the creation and annihilation operators.

7.14.6 The Full Hamiltonian

The full Hamiltonian includes terms for the atom, free fields and the matter–field interaction.

$$\hat{\mathcal{H}} = \hat{\mathcal{H}}_a + \hat{\mathcal{H}}_f + \hat{\mathcal{H}}_{a-f}$$

Combining Equations 7.14.10, 7.14.11 and 7.14.27 produces the full Hamiltonian for the full system consisting of the atoms and fields.

$$\hat{\mathcal{H}} = \frac{\hbar\omega_0}{2}\hat{\sigma}_z + \sum_{\vec{k}s}\hbar\omega_k\left(\hat{b}_{\vec{k}s}^+\hat{b}_{\vec{k}s} + \frac{1}{2}\right) +$$
$$+ \sum_{\vec{k}s}\sqrt{\frac{\hbar\omega_k}{2\varepsilon_o}}\left(\hat{\mu}_{12}\cdot\tilde{e}_{\vec{k}s}\right)\left[\hat{\sigma}^+(t)\hat{b}_{\vec{k}s}(t)f_{\vec{k}}(\vec{r}) + \hat{\sigma}^-(t)\hat{b}_{\vec{k}s}^+(t)f_{\vec{k}}^*(\vec{r})\right]$$

(7.14.28)

7.15 The Interaction Representation for the Jaynes-Cummings' Model

The Jaynes-Cummings' model, although fairly simple, provides a great deal of insight into the matter-light interaction. As with the Liouville equation discussed earlier in this chapter, we will use the density operator to predict the evolution of an electron in the two-level atom. However this time, we also want information on the evolution of the optical field. We therefore need a rate equation (a.k.a., master equation) for the direct-product density operator for the atom and the light field.

The master equation for the density operator describes the time-development of the density operator and the evolution of the system. We have already seen one example of a master equation when we discussed the Liouville equation. A master equation has the simplest interpretation when written in the interaction representation. The density operator, being similar to a wavefunction, moves through Hilbert space under the influence of an interaction. The density operator remains constant without the interaction. We will see that the ladder, creation and annihilation operators assume the form of a complex exponential in time. This section discusses the interaction representation for the wave functions, operators and Hamiltonians.

The next section includes reservoirs in the system. The reservoirs produce damping effects leading to the "phenomenological" term in the Liouville equation. We will see that they also produce fluctuations in the system.

7.15.1 Atomic Creation and Annihilation Operators

We focus on the interaction representation for both the atom and the EM field. Consider the atom first. Chapter 5 shows that the wavefunction $|\psi\rangle$ in the Schrodinger representation can be related to the wave function $|\breve{\psi}\rangle$ in the *interaction* representation by

$$|\psi\rangle = \exp\left\{\frac{\hat{H}_a t}{i\hbar}\right\} |\breve{\psi}\rangle \tag{7.15.1}$$

The operator

$$\hat{u}_a = \exp\left\{\frac{\hat{H}_a t}{i\hbar}\right\} \tag{7.15.2}$$

relating the two must be unitary.

The interaction representation of the atomic lowering operator $\breve{\sigma}^-$ can be found by requiring expectation values for the Schrodinger and Interaction representation to agree.

$$\langle\psi|\,\hat{\sigma}^-\,|\psi\rangle = \left\langle\breve{\psi}\right|\hat{u}_a^+ \hat{\sigma}^- \hat{u}_a \left|\breve{\psi}\right\rangle = \left\langle\breve{\psi}\right|\breve{\sigma}^-\left|\breve{\psi}\right\rangle$$

so that

$$\breve{\sigma}^- = \hat{u}_a^+ \hat{\sigma}^- \hat{u}_a \tag{7.15.3}$$

The operator expansion theorem provides the explicit time-dependent form of the lowering operator in the interaction representation. The operator expansion theorem is

$$e^{\hat{A}}\hat{B}e^{-\hat{A}} = \hat{B} + \left[\hat{A},\hat{B}\right] + \frac{1}{2!}\left[\hat{A},\left[\hat{A},\hat{B}\right]\right] + \cdots \tag{7.15.4}$$

The interaction representation of the lowering operator is

$$\breve{\sigma}^- = \hat{u}_a^+ \hat{\sigma}^- \hat{u}_a = \exp\left\{-\frac{\hat{H}_a t}{i\hbar}\right\}\hat{\sigma}^-\exp\left\{\frac{\hat{H}_a t}{i\hbar}\right\} = \hat{\sigma}^- - \left[\frac{\hat{H}_a t}{i\hbar},\hat{\sigma}^-\right] + \cdots \tag{7.15.5}$$

Using Equation 7.14.10 from the previous section, $\hat{H}_a = \frac{1}{2}\hbar\,\omega_o\hat{\sigma}_z$, the commutator in the last equation becomes

$$\left[\hat{H}_a,\hat{\sigma}^-\right] = \frac{\hbar\omega_o}{2}[\hat{\sigma}_z,\hat{\sigma}^-] = \hbar\omega_o\begin{pmatrix} 0 & 1 \\ 0 & 0 \end{pmatrix} = -\hbar\omega_o\hat{\sigma}^- \tag{7.15.6}$$

Performing similar computations for the other commutators in Equation 7.15.5 and adding the results together provides the interaction representation for the lowering operator

$$\breve{\sigma}^-(t) = \hat{\sigma}^-\,e^{-i\omega_o t} \tag{7.15.7a}$$

where, $\hat{\sigma}^-$ is the Schrodinger representation for the atomic lowering operator. The adjoint of Equation 7.15.7a provides

$$\breve{\sigma}^+(t) = \hat{\sigma}^+ \, e^{+i\omega_o t} \tag{7.15.7b}$$

The atomic Hamiltonian $\hat{H}_a$ has the same form in either the interaction or Schrodinger representation. Recall that a function of an operator always commutes with that operator so that we can write

$$\breve{H}_a = \hat{u}_a^+ \hat{H}_a \hat{u}_a = e^{-\frac{\hat{H}_a t}{i\hbar}} \hat{H}_a e^{\frac{\hat{H}_a t}{i\hbar}} = \hat{H}_a e^{-\frac{\hat{H}_a t}{i\hbar}} e^{\frac{\hat{H}_a t}{i\hbar}} = \hat{H}_a$$

Technically, to find the interaction representation of the atomic quantities, we should be using an evolution operator that includes both the atomic and field Hamiltonians. However, they operate on separate Hilbert space and therefore commute. As a result, we can find the interaction representations for the atom and light fields separately.

7.15.2   The Boson Creation and Annihilation Operators

We can find the interaction representation variables for the EM operators. The EM-field evolution operator can be written as

$$\hat{u}_f = \exp\left\{\frac{\hat{H}_f t}{i\hbar}\right\} \tag{7.15.8}$$

where the Schrodinger Hamiltonian has the form

$$\hat{H}_f = \sum_{\vec{k}s} \hbar\omega_{\vec{k}} \left(\hat{b}_{\vec{k}s}^+ \hat{b}_{\vec{k}s} + \frac{1}{2} \right) \tag{7.15.9}$$

and where the creation/annihilation operators must be independent of time. The operator expansion theorem

$$e^{\hat{A}} \hat{B} e^{-\hat{A}} = \hat{B} + \left[\hat{A}, \hat{B}\right] + \frac{1}{2!}\left[\hat{A}, \left[\hat{A}, \hat{B}\right]\right] + \cdots$$

provides the interaction representation of the EM-field creation and annihilation operators

$$\breve{b}_{\vec{k}s}(t) = \hat{u}_f^+ \hat{b}_{\vec{k}s} \hat{u}_f = \hat{b}_{\vec{k}s} + \frac{it}{\hbar}\left[\sum_{\vec{k'}s'} \hbar\omega_{\vec{k'}} \left(\hat{b}_{\vec{k'}s'}^+ \hat{b}_{\vec{k'}s'} + \frac{1}{2}\right), \hat{b}_{\vec{k}s}\right] + \cdots \tag{7.15.10}$$

The first commutator in Equation 7.15.10 reduces to

$$\left[\sum_{\vec{k'}s'} \hbar\omega_{\vec{k'}} \left(\hat{b}_{\vec{k'}s'}^+ \hat{b}_{\vec{k'}s'} + \frac{1}{2}\right), \hat{b}_{\vec{k}s}\right] = \sum_{\vec{k'}s'} \hbar\omega_{\vec{k'}} \left[\hat{b}_{\vec{k'}s'}^+ \hat{b}_{\vec{k'}s'} + \frac{1}{2}, \hat{b}_{\vec{k}s}\right]$$

$$= \sum_{\vec{k'}s'} \hbar\omega_{\vec{k'}} \left[\hat{b}_{\vec{k'}s'}^+, \hat{b}_{\vec{k}s}\right] \hat{b}_{\vec{k'}s'} = -\hbar\omega_{\vec{k}} \hat{b}_{\vec{k}s}$$

Working out all of the commutators in Equation 7.15.10 for $\breve{b}_{\vec{k}s}$ and collecting terms provides

$$\breve{b}_{\vec{k}s}(t) = \hat{b}_{\vec{k}s}\,e^{-i\omega_k t} \tag{7.15.11a}$$

The adjoint of this expression provides the interaction representation of the boson creation operator

$$\breve{b}_{\vec{k}s}^{+}(t) = \hat{b}_{\vec{k}s}^{+}\,e^{-i\omega_k t} \tag{7.15.11b}$$

where

$$\hat{b}_{\vec{k}s} = \hat{b}_{\vec{k}s}(0) \quad \text{and} \quad \hat{b}_{\vec{k}s}^{+} = \hat{b}_{\vec{k}s}^{+}(0)$$

The Hamiltonian for the free fields has the same form in either the Schrodinger or interaction representation

$$\breve{H}_f = \sum_{\vec{k}s} \hbar\omega_{\vec{k}} \left(\hat{b}_{\vec{k}s}^{+}(t)\,\hat{b}_{\vec{k}s}(t) + \frac{1}{2} \right) = \sum_{\vec{k}s} \hbar\omega_{\vec{k}} \left(\hat{b}_{\vec{k}s}^{+}\,e^{-i\omega_k t}\,\hat{b}_{\vec{k}s}\,e^{+i\omega_k t} + \frac{1}{2} \right)$$

$$= \sum_{\vec{k}s} \hbar\omega_{\vec{k}} \left(\hat{b}_{\vec{k}s}^{+}\,\hat{b}_{\vec{k}s} + \frac{1}{2} \right) = \hat{H}_f$$

7.15.3 Interaction Representation of the Subsystem Density Operators

We start the discussion of the interaction representation of the density operator with the atomic density operator. Let $|\psi\rangle$ be a Schrodinger wave function for an isolated atom. Let $|\breve{\psi}\rangle$ be the interaction representation of the same wave function. The two wavefunctions can be related to one another by

$$|\psi\rangle = \exp\left\{ \frac{\hat{H}_a t}{i\hbar} \right\} |\breve{\psi}\rangle \tag{7.15.12}$$

with the atomic evolution operator and the atomic Hamiltonian, respectively, given by

$$\hat{u}_a = \exp\left\{ \frac{\hat{H}_a t}{i\hbar} \right\} \quad \text{and} \quad \hat{H}_a = \frac{1}{2}\hbar\omega_o\hat{\sigma}_z$$

Keep in mind the evolution operator must be unitary with the property that $\hat{u}_a^{-1} = \hat{u}_a^{+}$. The Schrodinger atomic density operator $\hat{\rho}_a$

$$\hat{\rho}_a = \sum_{\psi} |\psi\rangle P_\psi \langle\psi| \tag{7.15.13}$$

obtains from the interaction density operator $\breve{\rho}_a$ by substituting Equation 7.15.12 into Equation 7.15.13

$$\hat{\rho}_a = \sum_{\psi} |\psi\rangle P_\psi \langle\psi| = \sum_{\psi} \left\{ \exp\left(\frac{\hat{H}_a t}{i\hbar}\right) |\breve{\psi}\rangle \right\} P_\psi \left\{ \exp\left(\frac{\hat{H}_a t}{i\hbar}\right) |\breve{\psi}\rangle \right\}^{+}$$

This equation simplifies upon use of the Hermitian property of the atomic Hamiltonian $\hat{H}_a = \hat{H}_a^+$ to find

$$\hat{\rho}_a = \exp\left(\frac{\hat{H}_a t}{i\hbar}\right) \left\{ \sum_\psi |\breve{\psi}\rangle P_\psi \langle\breve{\psi}| \right\} \exp\left(-\frac{\hat{H}_a t}{i\hbar}\right) = \hat{u}_a \, \breve{\rho}_a \hat{u}_a^+$$

The final form of the interaction density operator must be

$$\breve{\rho}_a = \hat{u}_a^+ \hat{\rho}_a \hat{u}_a \qquad\qquad (7.15.14)$$

The "free-field" density operator can be similarly found

$$\breve{\rho}_f = \hat{u}_f^+ \hat{\rho}_f \hat{u}_f \qquad\qquad (7.15.15a)$$

with

$$\hat{u}_f = \exp\left\{\frac{\hat{H}_f t}{i\hbar}\right\} \quad\text{and}\quad \hat{H}_f = \sum_{\vec{ks}} \hbar\omega_{\vec{k}} \left(\hat{b}_{\vec{ks}}^+ \hat{b}_{\vec{ks}} + \frac{1}{2}\right) \qquad (7.15.15b)$$

7.15.4   The Interaction Representation of the Direct-Product Density Operator

The next several sections use the master equation (i.e., the rate equation) for the density operator in the direct product space. We know that the density operator for the Jaynes-Cummings' model consists of two subspaces, one for the atom and another for the light. Later we will add reservoirs and enlarge the direct product space to include reservoir terms. These master equations have the simplest interpretation when written in the interaction representation. For this reason, we now make a few comments on the interaction representation of the density operator for the direct product space.

The wave functions used in the Jaynes-Cummings' model reside in a direct product space denoted by $V_a \otimes V_f$ where V_a and V_f are the Hilbert spaces for the atom and the field, respectively. For a *single* optical mode, the general wave function in the direct product space has the form

$$|\Psi\rangle = \sum_{a,n} \beta_{an}^\Psi(t) \, |a\rangle |n\rangle \qquad\qquad (7.15.16)$$

where $|a\rangle$ represents the atomic state $(a = 1, 2)$ and $|n\rangle$ represents the Fock electromagnetic state $(n = 0, 1, \ldots)$. The density operator must be given by

$$\hat{\rho} = \sum_\Psi |\Psi\rangle P_\Psi \langle\Psi| \qquad\qquad (7.15.17)$$

Only under special circumstances (e.g., without interaction between the atomic and optical subsystems) can the density operator be written as the direct product $\hat{\rho} = \hat{\rho}_a \hat{\rho}_f$. Usually, we assume that just prior to initiating the matter–field interaction ($t = 0$), that the density operator can be factored according to

$$\hat{\rho}(0) = \hat{\rho}_a(0) \, \hat{\rho}_f(0) \qquad\qquad (7.15.18)$$

However, we can always write the atomic density operator by "tracing" the full density operator over the degrees of freedom for the field (Fock states in this case).

$$\hat{\rho}_a = Tr_f(\hat{\rho}) \equiv \sum_n \langle n| \; \hat{\rho} \; |n\rangle \tag{7.15.19a}$$

We can similarly find the field density operator from the direct product one by tracing over the atomic degrees of freedom.

$$\hat{\rho}_f = Tr_a(\hat{\rho}) \equiv \sum_a \langle a| \; \hat{\rho} \; |a\rangle \tag{7.15.19b}$$

where Tr_f and Tr_a signifies the trace over the field modes and the atomic states respectively. To get a better understanding of how the trace affects the direct-product density operator, substitute Equation 7.15.17 into Equation 7.15.16 for $\hat{\rho}$ to find

$$\hat{\rho} = \sum_\Psi |\Psi\rangle P_\Psi \langle \Psi| = \sum_\Psi \sum_{a,n} \beta_{an}^\Psi(t)|a\rangle|n\rangle P_\Psi \sum_{b,m} \beta_{bm}^{\Psi*}(t)\langle b|\langle m|$$

which can be rearranged to get

$$\hat{\rho} = \sum_\Psi \sum_{\substack{a,n \\ b,m}} P_\Psi \beta_{an}^\Psi(t)\beta_{bm}^{\Psi*}(t)|n\rangle|a\rangle\langle b|\langle m|$$

where, $|a\rangle, |b\rangle$ refer to basis set for the atomic subsystem. Now tracing over the basis set for the field provides

$$Tr_f\hat{\rho} = \sum_{n'} \langle n'|\hat{\rho}|n'\rangle = \sum_{n'} \sum_\Psi \sum_{\substack{a,n \\ b,m}} \langle n'| \; \{ \; P_\Psi \beta_{an}^\Psi(t)\beta_{bm}^{\Psi*}(t)|n\rangle|a\rangle\langle b|\langle m| \; \} \; |n'\rangle$$

$$= \sum_\Psi \sum_{a,b,n} P_\Psi \beta_{an}^\Psi(t)\beta_{bn}^{\Psi*}(t)|a\rangle\langle b| \tag{7.15.20}$$

This last result can be written in the form of a density operator for the atomic system by defining the coefficients

$$\beta_{ab}'\Psi = \sum_n \beta_{an}^\Psi(t)\beta_{bn}^{\Psi*}(t)$$

so that

$$Tr_f\hat{\rho} = \sum_\Psi \sum_{a,b} P_\Psi \beta_{ab}'\Psi(t)|a\rangle\langle b| \equiv \hat{\rho}_a \tag{7.15.21}$$

This last expression has the form of a density operator for the atomic system as required for the first of Equations 7.15.19. Taking the trace in this manner should remind the reader of the probability relation for independent events A, B

$$\sum_B P(A \text{ and } B) = \sum_B P(A) \, P(B) = P(A) \sum_B P(B) = P(A)$$

The trace in Equation 7.15.21 basically removes the field subsystem from consideration and embeds its effects into a single reduced formula.

The interaction representation of the direct-product density operator can be defined analogously to that for the atom or EM density operators. The unitary evolution operator becomes

$$\hat{u} = \exp\left\{\frac{\hat{H}_o t}{i\hbar}\right\} \tag{7.15.22}$$

where the Hamiltonian $\hat{H}_o$ does not contain the matter-light interaction Hamiltonian

$$\hat{H}_o = \hat{H}_a + \hat{H}_f \tag{7.15.23}$$

Because $\hat{H}_a$ and $\hat{H}_f$ contain dynamical variables (operators) in distinct/disjoint spaces, the two Hamiltonians commute. The operator expansion theorem says

$$e^{\hat{A}+\hat{B}} = e^{\hat{A}} e^{\hat{B}} e^{[\hat{A},\hat{B}]/2}$$

so long as $[\hat{A},[\hat{A},\hat{B}]] = [\hat{B},[\hat{A},\hat{B}]]$. Therefore the evolution operator can be written as

$$\hat{u} = \exp\left\{\frac{\hat{H}_o t}{i\hbar}\right\} = \exp\left\{\frac{\hat{H}_a t}{i\hbar}\right\} \exp\left\{\frac{\hat{H}_f t}{i\hbar}\right\} = \hat{u}_a \hat{u}_f \tag{7.15.24}$$

where $\hat{u}_a$ operates only in atom-space and $\hat{u}_f$ operates only in light-space. As a note, $\hat{u}_a$ and $\hat{u}_f$ commute and can be arranged as either $\hat{u}_a \hat{u}_f$ or $\hat{u}_f \hat{u}_a$.

The interaction representation of the full density operator can be written as

$$\breve{\rho} = \hat{u}^+ \hat{\rho} \, \hat{u} = \hat{u}_a^+ \hat{u}_f^+ \hat{\rho} \, \hat{u}_a \hat{u}_f \tag{7.15.25}$$

Under special circumstances when $\hat{\rho} = \hat{\rho}_a \hat{\rho}_f$, the interaction density operator can then be written as

$$\breve{\rho} = \hat{u}^+ \hat{\rho} \, \hat{u} = \hat{u}_a^+ \hat{u}_f^+ \hat{\rho}_a \hat{\rho}_f \, \hat{u}_a \hat{u}_f = \breve{\rho}_a \breve{\rho}_f$$

7.15.5 Rate Equation for the Density Operator in the Interaction Representation

The rate equation for the density operator can be written in the interaction representation. The density operator (in the interaction representation) moves through its Hilbert space only when an interaction potential links the light fields and the matter fields. We will write the master equation without the fluctuation and damping terms characteristic of an interaction between a system and reservoir. This section demonstrates two alternative forms for the equation of motion.

In this topic, we start with the Schrodinger representation $\hat{\rho} \sim |\psi(t)\rangle\langle\psi(t)|$ and use the Schrodinger equation

$$\hat{H}|\psi\rangle = i\hbar \frac{\partial}{\partial t}|\psi\rangle \quad \text{with} \quad \hat{H} = \hat{H}_o + \hat{V} \tag{7.15.26}$$

where $\hat{V}$ is an interaction potential. Subsequent sections write $\hat{H}_o = \hat{H}_s + \hat{H}_r$ as the sum of the system and reservoir Hamiltonians. The interaction potential $\hat{V}$ is the interaction energy between the system and reservoir. Continuing with Equations 7.15.26 and taking the derivative of the density operator, we find

$$\dot{\hat{\rho}} \sim \left(\frac{\partial}{\partial t}|\psi(t)\rangle\right)\langle\psi(t)| + |\psi(t)\rangle\left(\frac{\partial}{\partial t}|\psi(t)\rangle\right)^+ = \frac{1}{i\hbar}\left[\hat{H}, \hat{\rho}\right] \tag{7.15.27}$$

This is the equation of motion of the density operator in the Schrodinger Representation.

Next, to find one form of the equation of motion in the interaction representation, differentiate the interaction representation of the density operator

$$\breve{\dot{\rho}} = \frac{\partial}{\partial t}\left(\hat{u}^+\hat{\rho}\hat{u}\right) \quad \text{with} \quad \hat{u} = \exp\left(\frac{\hat{H}_o t}{i\hbar}\right) \tag{7.15.28}$$

to get

$$\breve{\dot{\rho}} = -\frac{\hat{H}_o}{i\hbar}\hat{u}^+\hat{\rho}\,\hat{u} + \hat{u}^+\frac{\partial\hat{\rho}}{\partial t}\hat{u} + \hat{u}^+\hat{\rho}\,\hat{u}\frac{\hat{H}_o}{i\hbar} \tag{7.15.29a}$$

which reduces to

$$\breve{\dot{\rho}} = -\frac{1}{i\hbar}\left[\hat{H}_o, \breve{\rho}\right] + \hat{u}^+\frac{\partial\hat{\rho}}{\partial t}\,\hat{u} \tag{7.15.29b}$$

where $[\hat{H}_o, \hat{u}] = 0$ was used. Notice that the partial derivative of the density operator in the last term can be nonzero in the Schrodinger representation.

The second form of the equation comes from Equation 7.15.29a by rearranging terms. Factoring out the unitary operators provides

$$\breve{\dot{\rho}} = -\frac{\hat{H}_o}{i\hbar}\hat{u}^+\hat{\rho}\,\hat{u} + \hat{u}^+\frac{\partial\hat{\rho}}{\partial t}\,\hat{u} + \hat{u}^+\hat{\rho}\,\hat{u}\frac{\hat{H}_o}{i\hbar} = \hat{u}^+\left\{-\frac{1}{i\hbar}\left[\hat{H}_o, \hat{\rho}\right] + \dot{\hat{\rho}}\right\}\hat{u}$$

Substituting Equation 7.15.27, specifically $\dot{\hat{\rho}} = [\hat{H}, \hat{\rho}]/i\hbar = [\hat{H}_o, \hat{\rho}]/i\hbar + [\hat{V}, \hat{\rho}]/i\hbar$, gives the second form of the equation of motion

$$\breve{\dot{\rho}} = \frac{1}{i\hbar}\left[\breve{V}, \breve{\rho}\right] \tag{7.15.29c}$$

since $\hat{u}^+(\hat{H} - \hat{H}_o)\hat{u} = \hat{u}^+\hat{V}\hat{u} = \breve{V}$. Equation 7.15.29c clearly demonstrates how the interaction Hamiltonian completely controls the motion of the density operator in the interaction representation.

7.16 The Master Equation

The environment surrounding a system relaxes it to steady state. However, a collection of reservoirs defines the environment. The present section illustrates how a collection of reservoirs defining the environment drives the system toward steady state and how the quantum models produce the Liouville equation. One often encounters three approaches, using density operators, Heisenberg dynamical variables, or distribution functions. The density operators conveniently describe ensembles of atoms or fields. The Heisenberg approach demonstrates the fluctuation-dissipation theorem as shown in the next section. This section reproduces the Liouville equation for the density operator.

7.16.1   The System

The present section focuses on finding the relaxation terms in the Liouville equation from Section 7.11.

$$\frac{\partial \hat{\rho}}{\partial t} = \frac{1}{i\hbar}\left[\hat{H}, \hat{\rho}\right] + \frac{1}{i\hbar}\left[\hat{H}_{\text{env}}, \hat{\rho}\right] \qquad \frac{\partial \rho_{ab}}{\partial t} = \frac{1}{i\hbar}\left[\hat{H}, \hat{\rho}\right]_{ab} - \frac{\rho_{ab} - \bar{\rho}_{ab}}{\tau_{ab}} \qquad (7.16.1)$$

where $\hat{H} = \hat{H}_o + \hat{V}$ represents the atomic and matter-field Hamiltonians. The term environmental "env" term really refers to reservoirs that have many degrees of freedom as discussed in Section 2.6. The reservoirs produce damping and fluctuations, which average to zero. The damping represented by the last term in Equations 7.16.1 tends to move the system to steady state (no-light steady-state NLSS) or to equilibrium.

Consider a small system consisting of a two-level atom in order to demonstrate the relaxation terms in Equations 7.16.1. Denote the small system Hamiltonian by $\hat{H}_s$. We ignore the matter-field interaction although it can be included as necessary. The Hamiltonian $\hat{H}$ for the complete system consists of the small system Hamiltonian $\hat{H}_s$, a summation of Hamiltonians for N multiple reservoirs $\hat{H}_r$ (such as a pump and thermal reservoir), and a summation of Hamiltonians $\hat{H}_{sr}$ to describe the interaction between the small system and the many possible reservoirs.

$$\hat{H} = \hat{H}_s + \hat{H}_{sr} + \hat{H}_r \qquad (7.16.2)$$

The density operator exists in a direct product space. Normally, the interaction between one system (such as a reservoir) and another one (such as the small system) causes the wavefunctions, and hence the density operators, to become entangled. The density operator for two systems cannot be factored into separate density operators for each individual system. However, if we assume the interaction between the systems starts at $t_o = 0$, then up to and including time $t_o = 0$ we can separate and reduce the density operator. For the case of the small system interacting with reservoirs, we have

$$\hat{\rho}(0) = \hat{\rho}^{(s)}(0)\hat{\rho}^{(1)}(0)\hat{\rho}^{(2)}(0)\ldots\hat{\rho}^{(N)}(0) = \hat{\rho}^{(s)}(0) \bigotimes_{r=1}^{N} \hat{\rho}^{(r)}(0) \equiv \hat{\rho}^{(s)}(0)\, \hat{\rho}^{(1\ldots N)}(0) \qquad (7.16.3)$$

where the superscript "s" refers to the small system and the other superscripts refer to the reservoirs. However, we can write the small-system density operator in terms of the full density operator (not just $t = 0$)

$$\hat{\rho}^{(s)}(t) = Tr_1 Tr_2 \cdots Tr_N \hat{\rho}(t) \equiv Tr_{1\ldots N}\,\hat{\rho}(t) \qquad (7.16.4)$$

The rate equation for the density operator can be rewritten using the total Hamiltonian in Equation 7.16.2

$$\dot{\rho} = \frac{1}{i\hbar}\left[\hat{H}, \rho\right] = \frac{1}{i\hbar}\left[\hat{H}_s, \rho\right] + \frac{1}{i\hbar}\left[\hat{H}_{sr}, \rho\right] + \frac{1}{i\hbar}\left[\hat{H}_r, \rho\right] \tag{7.16.5}$$

where H_s does not involve any of the reservoir operators. Taking the trace of Equation 7.16.5 over the *reservoir* states then reduces $\dot{\rho}$ and $[H, \rho]$ to terms involving only the *system* density operator ρ_s. The last two terms in Equation 7.16.5 provide the "relaxation" effects

$$\dot{\rho}_s = \frac{1}{i\hbar}\left[\hat{H}_s, \rho_s\right] + \frac{\partial \rho_s}{\partial t}\bigg|_{envir} = \frac{1}{i\hbar}\left[\hat{H}_s, \rho_s\right] + \sum_i \frac{\partial \rho_s}{\partial t}\bigg|_{\substack{reserv \\ \#i}} \tag{7.16.6}$$

Equation 7.16.6 is similar to the Liouville equation discussed in the first two sections of this chapter.

7.16.2 Multiple Reservoirs

Suppose multiple reservoirs interact with a single small system. Let the operators $R^{(a)} = \{R_i^{(a)}\}$ refer to the ath reservoir. Each reservoir can be characterized differently from the other; for example, thermal reservoirs might have different temperatures. Segregate the operators for the system so that an interaction Hamiltonian links the set $S^{(a)} = \{S_i^{(a)}\}$ with the ath reservoir; i.e., the degrees of freedom represented by set $S^{(a)}$ interact with the ath reservoir (see Figure 7.16.1). The sets $S^{(a)}$ are not necessarily disjoint. A two level atom has 2 ladder operators that can be linked to any number of reservoirs. All of the reservoir operators commute with the system operators for all times in the Schrodinger and interaction representations

$$\left[\hat{S}_i^{(a)}, \hat{R}_j^{(b)}\right] = 0 \tag{7.16.7}$$

since S and R refer to separate Hilbert spaces. However, by necessity, each individual set contains noncommuting operators. For example, there exists indices "i" and "j" such that

$$\left[\hat{S}_i^{(a)}, \hat{S}_j^{(a)}\right] \neq 0 \tag{7.16.8}$$

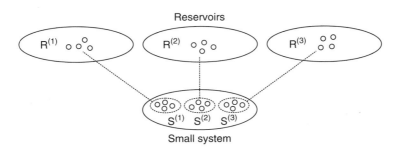

FIGURE 7.16.1
The operators for the small system link with operators for the reservoirs by the interaction Hamiltonian.

and similar considerations for the set R. Assume a total of "N" reservoirs and an interaction Hamiltonian of the form

$$\hat{H}_{sr} = \hat{V} = \sum_{a=1}^{N}\sum_{i} C_i \hat{S}_i^{(a)} \hat{R}_i^{(a)} \tag{7.16.9}$$

Notice that we don't distinguish between raising and lowering operators in this topic. The coupling constant C_i describes the strength of the particular term for the interaction. We suppress it for notational convenience.

Example 7.16.1

Suppose the system consists of an atom and an electromagnetic EM field attributed to spontaneous emission. The field interacts with the vacuum reservoir. The "free" field Hamiltonian is

$$\hat{H} = \sum_{k} \hbar\omega_k \left(\hat{b}_k^+ \hat{b}_k^- + \frac{1}{2} \right)$$

There are two operators $\hat{S}_i^{(1)}$ and $\hat{S}_j^{(1)}$ in the set "S" such that, for each k,

$$\hat{S}_i^{(1)} = \hat{b}_k^+ \quad \text{and} \quad \hat{S}_j^{(1)} = \hat{b}_k^-$$

7.16.3   Dynamics and the Perturbation Expansion

The total density operator $\hat{\rho}$ in the Schrodinger representation satisfies

$$\dot{\hat{\rho}} = \frac{1}{i\hbar}\left[\hat{H}, \hat{\rho} \right] \tag{7.16.10}$$

for the complete Hamiltonian $\hat{H}$ in Equation 7.16.2. We denote operators in the Schrodinger representation by the "caret" such as $\hat{O}$, and operators in the interaction representation by $\breve{O}$. It would be nice to use a perturbation expansion to find the density operator in Equation 7.16.10 by treating the right-hand side as small. Working with the interaction representation allows us to replace the Hamiltonian $\hat{H}$ with the interaction potential $V = \hat{H}_{sr}$ which can be made as small as necessary. In the interaction representation, this interaction Hamiltonian controls the motion of $\breve{\rho}$ in Hilbert space (see Section 7.15) according to

$$\dot{\breve{\rho}} = \frac{1}{i\hbar}\left[\breve{H}_{sr}, \breve{\rho} \right] \tag{7.16.11}$$

The interaction representation writes the density operator and Hamiltonian as

$$\breve{\rho} = \hat{u}^+ \hat{\rho}\, \hat{u} \qquad \breve{H}_{sr} = \hat{u}^+ \hat{H}_{sr}\, \hat{u}$$

where the explicit form of the unitary operator $\hat{u} = \exp(\hat{H}_o t/i\hbar)$ assumes the reservoir-system interaction starts at $t_o = 0$. The "free" Hamiltonian is $\hat{H}_o = \hat{H}_s + \hat{H}_r$, which describes the evolution of the system and reservoirs without any interaction between them. The evolution operator can be written as $\hat{u} = \hat{u}_s \hat{u}_r$ since the two Hamiltonians commute $[\hat{H}_s, \hat{H}_r] = 0$ and the exponential can therefore be divided into two pieces. Equation 7.16.11 can be integrated to give

$$\breve{\rho}(t) = \breve{\rho}(0) + \frac{1}{i\hbar} \int_o^t d\tau \left[\breve{H}_{sr}(\tau), \breve{\rho}(\tau) \right]$$

(7.16.12)

At this point, two approaches can be taken. The first approach assumes that $\breve{H}_{sr} = \breve{V}$ is small. Equation 7.16.12 can be repeatedly substituted into itself to find

$$\breve{\rho}(t) = \breve{\rho}(0) + \frac{1}{i\hbar} \int_o^t d\tau_1 \left[\breve{H}_{sr}(\tau_1), \breve{\rho}(0) \right]$$
$$+ \frac{1}{(i\hbar)^2} \int_o^t d\tau_1 \int_o^{\tau_1} d\tau_2 \left[\breve{H}_{sr}(\tau_1), \left[\breve{H}_{sr}(\tau_2), \breve{\rho}(0) \right] \right] + \cdots$$

(7.16.13)

Rearranging terms and using the course grain derivative (see 7.16.5)

$$\dot{\breve{\rho}}(t) = \lim_{t \to 0} \frac{\breve{\rho}(t) - \breve{\rho}(0)}{t}$$

Notice that the derivative would normally be defined as $\dot{\breve{\rho}}(0)$ rather than $\dot{\breve{\rho}}(t)$ but we are taking t as small. The course grained derivative uses a time t longer than the correlation time of the reservoir. Equation 7.16.14 becomes

$$\dot{\breve{\rho}}(t) = \frac{1}{i\hbar} \left[\breve{H}_{sr}(t), \breve{\rho}(0) \right] + \frac{1}{(i\hbar)^2} \int_o^t d\tau \left[\breve{H}_{sr}(t), \left[\breve{H}_{sr}(\tau), \breve{\rho}(0) \right] \right] + \cdots$$

(7.16.14)

Again, Equation 7.16.14 makes it quite clear that the interaction Hamiltonian alone causes the density operator to evolve in time. For notational convenience, we use V rather than H_{sr}

$$\dot{\breve{\rho}} \cong \frac{1}{i\hbar} \left[\breve{V}, \breve{\rho}(0) \right] + \frac{1}{(i\hbar)^2} \int_o^t dt' \left[\breve{V}(t), \left[\breve{V}(t'), \breve{\rho}(0) \right] \right] + \cdots$$

(7.16.15)

This first approach clearly shows where Equation 7.16.3 becomes important. However, later we will want the density operator to depend on time rather than having $t = 0$. Then we will need to assume the equation holds for times other than $t = 0$.

The second approach does not make an approximation for the density operator. Instead, Equation 7.16.12 is substituted back into Equation 7.16.11, specifically $\dot{\breve{\rho}} = \frac{1}{i\hbar} [\breve{H}_{sr}, \breve{\rho}]$, to find

$$\dot{\breve{\rho}} = \frac{1}{i\hbar} \left[\breve{V}, \breve{\rho}(0) \right] + \frac{1}{(i\hbar)^2} \int_o^t dt' \left[\breve{V}(t), \left[\breve{V}(t'), \breve{\rho}(t') \right] \right]$$

(7.16.16)

The derivative will again be taken to mean a course derivative. The argument of the integral has a density operator that depends on time. However in this case, we will need to assume the density operator can be factored into a form similar to Equation 7.16.3 at the time t. We will follow the first approach.

It is necessary to take the trace of Equation 7.16.16 over the reservoir states to find the equation of motion for the system density operator. There are two ways to accomplish the task. First consider that the eigenvector are always independent of time. If $\{|n\rangle\}$ denotes the set of reservoir eigenstates then

$$Tr_r\left(\dot{\bar{\rho}}\right) = \sum_n \langle n|\, \dot{\bar{\rho}}\, |n\rangle = \frac{\partial}{\partial t} \sum_n \langle n|\, \breve{\rho}\, |n\rangle = \frac{\partial}{\partial t} Tr_r\, \breve{\rho} = \dot{\breve{\rho}}^{(s)}$$

As a second method, the same expression is easily seen to hold by using the definition of derivative

$$Tr_r\, \dot{\breve{\rho}}(t) \cong Tr_r\, \frac{\breve{\rho}(t + \Delta t) - \breve{\rho}(t)}{\Delta t} = \frac{1}{\Delta t}\left\{ Tr_r\, \breve{\rho}(t + \Delta t) - Tr_r\, \breve{\rho}(t) \right\}$$

$$= \frac{1}{\Delta t}\left\{ \breve{\rho}^{(s)}(t + \Delta t) - \breve{\rho}^{(s)}(t) \right\} \cong \dot{\breve{\rho}}^{(s)}(t)$$

Either way, taking the trace over all of the reservoirs provides

$$\dot{\breve{\rho}}^{(s)} = \frac{1}{i\hbar} Tr_{1\cdots N}\left[\breve{V}, \breve{\rho}(0)\right] + \frac{1}{(i\hbar)^2} Tr_{1\cdots N} \int_o^t dt' \left[\breve{V}(t), \left[\breve{V}(t'), \breve{\rho}(0)\right]\right] + \cdots \tag{7.16.17}$$

Equation 7.16.17 calculates a "course-grain derivative" in that the times involved are longer than the correlation times of the reservoir.

Finally for this topic, the interaction representation of the density operator at $t = 0$ can be replaced by the Schrodinger representation since

$$\breve{\rho}(t) = \hat{u}^+(t)\, \hat{\rho}(t)\, \hat{u}(t) \quad \text{with} \quad \hat{u} = \exp\left(\frac{\hat{H}_o t}{i\hbar}\right) \quad \text{and} \quad \hat{H}_o = \hat{H}_s + \hat{H}_r$$

Setting $t = 0$ provides

$$\hat{u}(0) = 1 \quad \text{and therefore} \quad \breve{\rho}(0) = \hat{u}^+(0)\, \hat{\rho}(0)\, \hat{u}(0) = \hat{\rho}(0)$$

Equation 7.16.17 has an alternate form

$$\dot{\breve{\rho}}^{(s)} = \frac{1}{i\hbar} Tr_{1\cdots N}\left[\breve{V}, \hat{\rho}(0)\right] + \frac{1}{(i\hbar)^2} Tr_{1\cdots N} \int_o^t dt' \left[\breve{V}(t), \left[\breve{V}(t'), \hat{\rho}(0)\right]\right] + \cdots \tag{7.16.18}$$

7.16.4 The Langevin Displacement Term

Now we show the average $Tr_{1\cdots N}[\breve{V}, \breve{\rho}(0)]$ in the first term of Equation 7.16.18 can be neglected. The first term yields

$$Tr_{1\cdots N}\left[\breve{V}, \rho(0)\right] = Tr_{1\cdots N}\sum_{ai}\left[\breve{S}_i^{(a)}\breve{R}_i^{(a)}, \rho(0)\right] = Tr_{1\cdots N}\sum_{ai}\left[\breve{S}_i^{(a)}\breve{R}_i^{(a)}, \rho^{(s)}(0)\rho^{(1\cdots N)}(0)\right]$$

$$= Tr_{1\cdots N}\sum_{ai}\left\{\rho^{(s)}(0)\breve{S}_i^{(a)}\left[\breve{R}_i^{(a)}, \rho^{(1\cdots N)}(0)\right] + \left[\breve{S}_i^{(a)}, \rho^{(s)}(0)\right]\breve{R}_i^{(a)}\rho^{(1\cdots N)}(0)\right\}$$

The trace refers to the reservoir degrees of freedom. The terms in this last equation are all similar.

$$Tr_{1\cdots N}\breve{R}_i^{(a)}\rho^{(1\cdots N)}(0) = Tr_1\rho^{(1)}(0)\ldots Tr_{a-1}\rho^{(a-1)}(0)\left\{Tr_a\breve{R}_i^{(a)}\rho^{(a)}(0)\right\}Tr_{a+1}\rho^{(a+1)}(0)\ldots Tr_N\rho^{(N)}(0)$$

$$(7.16.19)$$

Factors without the reservoir variable $\breve{R}_i^{(a)}$ provide results similar to

$$Tr_1\,\rho^{(1)}(0) = 1$$

by definition of the density operator. The factors with the reservoir variable give

$$Tr_a\breve{R}_i^{(a)}\rho^{(a)}(0) = Tr_a\,\rho^{(a)}(0)\breve{R}_i^{(a)} = \left\langle\breve{R}_i^{(a)}\right\rangle_{res\atop a} \qquad (7.16.20a)$$

This average at most adds a constant to Equation 7.16.18. However, the operetors $\hat{R}$ represent either creation or annihilation operators so that the average becomes zero according to $\langle n|\hat{R}|n\rangle \sim \langle n \mid n\pm 1\rangle = 0$. Also notice that the cyclic property of the trace can be used to rearrange the order of "R" and "ρ" as necessary. Therefore, the Langevin displacement term becomes

$$Tr_{1\cdots N}\left[\breve{V}, \rho(0)\right] = 0 \qquad (7.16.20b)$$

where the equality follows from the cyclic property of the trace operation.

7.16.5 Reservoir Correlation Time and the Course Grain Derivative

The ensemble average of reservoir operators $\hat{R}_j$ can be found by calculating the trace $Tr_{1\cdots N}\breve{R}_i^{(a)}\rho^{(1\cdots N)}$ as shown in Equation 7.16.20a. The ensemble average must agree with the time average taken over a sufficiently long time interval (for an ergodic process). This is where the question of the interval of integration becomes important and why the time derivative in Equation 7.16.15 is usually called the "course grain derivative." If the values of the random variables fluctuate and are correlated over a small time interval, then the time average of the random variable might depend on the length of the time interval. Figure 7.16.2 for example shows that the average of some variable R over the time interval [0, 1] must be nonzero while over [0, 2] it must be zero. Presumably, with long

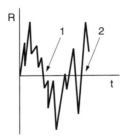

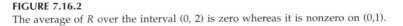

FIGURE 7.16.2
The average of R over the interval $(0, 2)$ is zero whereas it is nonzero on $(0,1)$.

enough integration time, the average will be identically zero for all times. This time average (over sufficiently long times) matches the ensemble average since every duplicate system in the ensemble is in a different possible state.

We show how the use of the course grain derivative implies Equation 7.16.20b. Consider the reservoirs. A derivative of a reservoir quantity is defined by

$$\frac{dR}{dt} = \lim_{\Delta t \to 0} \frac{R(t + \Delta t) - R(t)}{\Delta t} \qquad (7.16.21a)$$

For the case of Figure 7.16.2, the derivative is essentially nonzero for all times. However, if we take an ensemble average of the derivative, which is equivalent to an average over time "t" on the right hand side of the derivative, we find

$$\left\langle \lim_{\Delta t \to 0} \frac{R(t + \Delta t) - R(t)}{\Delta t} \right\rangle \sim \lim_{\Delta t \to 0} \frac{\langle R(t + \Delta t) \rangle - \langle R(t) \rangle}{\Delta t} = \frac{d}{dt} \bar{R}(t) \qquad (7.16.21b)$$

Averaging removes any fluctuations and the derivative would produce zero. A system is not stationary when the random variable changes on average with time $\bar{R}(t)$ (see Appendix 4). For the average, we assume a time scale long compared with the correlation time so that changes in the reservoir parameters remain constant.

Next consider a term such as $\breve{\rho}^{(s)} = \frac{1}{(i\hbar)^2} \mathrm{Tr}_{1 \cdots N} \int_o^t dt' [\breve{V}(t), [\breve{V}(t'), \hat{\rho}(0)]]$ in Equation 7.16.18. We assume large times t compared with the reservoir correlation time but small compared with any system time constants. The system density operator $\breve{\rho}^{(s)}$ slowly evolves in time. In what follows, we will remove terms from the integral that depend on "t" (but not t') to find integrals of the form

$$\sum_{ija} \breve{S}_i^{(a)}(t) \int_o^t dt' \, \breve{S}_j^{(a)}(t') \rho^{(s)}(0) \left\langle \breve{R}_i^{(a)}(t - t') \breve{R}_j^{(a)}(0) \right\rangle$$

Because the reservoir quantities are only correlated on a very small time scale, one that is ignored by course graining, the time "t" on the integral can be replaced by ∞.

Essentially, course graining a quantity means to average out any fast variations (see also Appendix 7). For example, a system quantity S averages over a time $\Delta \tau$ long compared with random fluctuations (such as a Langevin source) would have an average that can still depend on time

$$\bar{S}(t) = \frac{1}{\Delta \tau} \int_t^{t + \Delta \tau} S(\tau) \, d\tau \qquad (7.16.22)$$

7.16.6 The Relaxation Term

The second term in Equation 7.16.18 gives the relaxation effects

$$\text{Term 2} = \text{Tr}_{1\ldots N} \int_o^t dt' \left[\breve{V}(t), \left[\breve{V}(t'), \breve{\rho}(0) \right] \right]$$

Expanding the commutators provide

$$\text{Term 2} = \text{Tr}_{1\ldots N} \int_o^t dt' \left[\breve{V}(t), \breve{V}(t') \breve{\rho}(0) - \breve{\rho}(0) \breve{V}(t') \right]$$

$$= \text{Tr}_{1\ldots N} \int_o^t dt' \left[\underbrace{\breve{V}(t) \breve{V}(t') \breve{\rho}(0)}_{\leftarrow \text{Term 2.1} \rightarrow} - \underbrace{\breve{V}(t) \breve{\rho}(0) \breve{V}(t')}_{\leftarrow \text{Term 2.2} \rightarrow} - \underbrace{\breve{V}(t') \breve{\rho}(0) \breve{V}(t)}_{\leftarrow \text{Term 2.3} \rightarrow} + \underbrace{\breve{\rho}(0) \breve{V}(t') \breve{V}(t)}_{\leftarrow \text{Term 2.4} \rightarrow} \right]$$

$$(7.16.23)$$

Examine each sub-term in Equation 7.16.23. Substitute expressions for the interaction Hamiltonian in each one. For example, Term 2.1 becomes

$$\text{Term 2.1} = \text{Tr}_{1\ldots N} \int_o^t dt' \breve{V}(t) \breve{V}(t') \breve{\rho}(0) = \text{Tr}_{1\ldots N} \int_o^t dt' \sum_{aij} \breve{S}_i^{(a)}(t) \breve{R}_i^{(a)}(t) \sum_b \breve{S}_j^{(b)}(t') \breve{R}_j^{(b)}(t') \, \breve{\rho}(0)$$

$$= \text{Tr}_{1\ldots N} \int_o^t dt' \sum_{ij,ab} \breve{S}_i^{(a)}(t) \, \breve{R}_i^{(a)}(t) \, \breve{S}_j^{(b)}(t') \, \breve{R}_j^{(b)}(t') \, \breve{\rho}^{(s)}(0) \, \breve{\rho}^{(1\ldots N)}(0)$$

Recall from the discussion at the start of 7.16.2 that, for example, the operator $\breve{S}_i^{(a)}$ is linked to reservoir "(a)" through the interaction Hamiltonian. This linking cannot affect the basic commutation relations $[\hat{S}_i^{(a)}, \hat{R}_j^{(b)}] = 0$ in Equation 7.16.7, since the operators refer to separate Hilbert spaces. Likewise the position of $\breve{\rho}^{(s)}(0)$ with respect to $\breve{\rho}^{(1\ldots N)}$ can be rearranged at will. However, operators $\hat{S}_i^{(a)}$ do not necessarily commute with other operators in the set $S^{(a)}$ nor with the system density operator $\breve{\rho}^{(s)}(0)$. Similar comments apply to the reservoir operators. Moving the trace inside the summation and commuting operators as appropriate provides

$$\text{Term 2.1} = \int_o^t dt' \sum_{ij,ab} \breve{S}_i^{(a)}(t) \, \breve{S}_j^{(b)}(t') \, \breve{\rho}^{(s)}(0) \, \text{Tr}_{1\ldots N} \left[\breve{R}_i^{(a)}(t) \breve{R}_j^{(b)}(t') \breve{\rho}^{(1\ldots N)}(0) \right]$$

Assuming the reservoir variables must be stationary means that trace over the reservoir does not depend on the origin of time and last integral can be written as

$$\text{Term 2.1} = \int_o^t dt' \sum_{ij,ab} \breve{S}_i^{(a)}(t) \, \breve{S}_j^{(b)}(t') \, \breve{\rho}^{(s)}(0) \, \text{Tr}_{1\ldots N} \left[\breve{R}_i^{(a)}(t - t') \breve{R}_j^{(b)}(0) \breve{\rho}^{(1\ldots N)}(0) \right]$$

Replacing the integration variable with $\tau = t - t'$ produces

$$\text{Term 2.1} = \int_o^t d\tau \sum_{ij,ab} \breve{S}_i^{(a)}(t) \, \breve{S}_j^{(b)}(t - \tau) \, \breve{\rho}^{(s)}(0) \, \text{Tr}_{1\ldots N} \left[\breve{R}_i^{(a)}(\tau) \breve{R}_j^{(b)}(0) \breve{\rho}^{(1\ldots N)}(0) \right]$$

Following the procedure outlined in the previous subtopic (see Equation 7.16.19), the trace factor can be rewritten as

$$\text{Term2.1} = \int_0^t d\tau \sum_{ij,ab} \breve{S}_i^{(a)}(t) \, \breve{S}_j^{(b)}(t-\tau) \, \breve{\rho}^{(s)}(0) \left\langle \breve{R}_i^{(a)}(\tau) \breve{R}_j^{(b)}(0) \right\rangle \delta_{ab}$$

$$= \int_0^t d\tau \sum_{ij,a} \breve{S}_i^{(a)}(t) \breve{S}_j^{(a)}(t-\tau) \breve{\rho}^{(s)}(0) \left\langle \breve{R}_i^{(a)}(\tau) \breve{R}_j^{(a)}(0) \right\rangle$$

(7.16.24a)

since for $a \neq b$ the fluctuations average to zero

$$\text{Tr}_{1...N}\left[\breve{R}_i^{(a)}(\tau) \, \breve{R}_j^{(b)}(0) \, \breve{\rho}^{(1...N)}(0) \right] = \left\langle \breve{R}_i^{(a)}(\tau) \right\rangle \left\langle \breve{R}_j^{(b)}(0) \right\rangle = 0$$

Similar reasoning applies to the other parts of "term 2."

$$\text{Term 2.2} = \int_0^t d\tau \sum_{ij,a} \breve{S}_i^{(a)}(t) \, \breve{\rho}^{(s)}(0) \, \breve{S}_j^{(a)}(t-\tau) \left\langle \breve{R}_i^{(a)}(0) \breve{R}_j^{(a)}(\tau) \right\rangle \qquad (7.16.24b)$$

$$\text{Term 2.3} = \int_0^t d\tau \sum_{ij,a} \breve{S}_i^{(a)}(t-\tau) \, \rho^{(s)}(0) \, \breve{S}_j^{(a)}(t) \left\langle \breve{R}_i^{(a)}(\tau) \breve{R}_j^{(a)}(0) \right\rangle \qquad (7.16.24c)$$

$$\text{Term 2.4} = \int_0^t d\tau \sum_{ij,a} \rho^{(s)}(0) \, \breve{S}_i^{(a)}(t-\tau) \, \breve{S}_j^{(a)}(t) \left\langle \breve{R}_i^{(a)}(0) \, \breve{R}_j^{(a)}(\tau) \right\rangle \qquad (7.16.24d)$$

where the cyclic property of the trace is used to obtain the results for Terms 2.2 and 2.3. Also, dummy indices i, j and m, n can be interchanged at will.

All of the terms in Equations 7.16.24 can be combined into Equation 7.16.23 to produce the equation of motion for the density operator. The limit "t" on the integral is replaced by ∞ since the correlation functions are nonzero only for exceedingly small times. We also assume that the system density operator at small time t can be replaced

$$\hat{\rho}^{(s)}(0) = \breve{\rho}^{(s)}(0) \; \rightarrow \; \breve{\rho}^{(s)}(t)$$

Equation 7.16.23 becomes

$$\dot{\breve{\rho}}^{(s)}(t) = -\frac{1}{\hbar^2} \sum_{ija} \int_0^\infty d\tau \left\{ \left[\breve{S}_i^{(a)}(t), \breve{S}_j^{(a)}(t-\tau)\breve{\rho}^{(s)}(t) \right] \left\langle \breve{R}_i^{(a)}(\tau) \, \breve{R}_j^{(a)}(0) \right\rangle + \right.$$

$$\left. - \left[\breve{S}_i^{(a)}(t), \breve{\rho}^{(s)}(t)\breve{S}_j^{(a)}(t-\tau) \right] \left\langle \breve{R}_j^{(a)}(0) \, \breve{R}_i^{(a)}(\tau) \right\rangle \right\}$$

(7.16.25)

We can make the replacement $\breve{R}_j^{(a)}(0) = \hat{R}_j^{(a)}(0)$ if desired.

7.16.7 The Pauli Master Equation

Now we develop the relaxation terms in the Liouville equation for the density matrix. We modify the approach found in Weissbluth's book for a single reservoir by including multiple reservoirs. Let $\{|k\rangle\}$ be eigenvectors of $\hat{H}_s$, the system Hamiltonian.

$$\dot{\breve{\rho}}^{(s)}(t) = -\frac{1}{\hbar^2} \sum_{ija} \int_o^\infty d\tau \; \left\{ \underbrace{\left[\breve{S}_i^{(a)}(t), \breve{S}_j^{(a)}(t-\tau)\breve{\rho}^{(s)}(t) \right]\left\langle \breve{R}_i^{(a)}(\tau)\breve{R}_j^{(a)}(0)\right\rangle}_{\leftarrow \text{Term 3.1}\rightarrow} + \right.$$

$$\left. - \underbrace{\left[\breve{S}_i^{(a)}(t), \breve{\rho}^{(s)}(t)\breve{S}_j^{(a)}(t-\tau) \right]\left\langle \breve{R}_j^{(a)}(0)\breve{R}_i^{(a)}(\tau)\right\rangle}_{\leftarrow \text{Term 3.2}\rightarrow} \right\} \tag{7.16.26}$$

where $\breve{R}_j^{(a)}(0) = \hat{R}^{(a)}$. After considerable algebraic manipulation, Term 3.1 in $\dot{\breve{\rho}}^{(s)}(t)$ can be rewritten using the closure relation for the system eigenstates

$$\langle k| \left[\breve{S}_i^{(a)}(t), \breve{S}_j^{(a)}(t-\tau)\breve{\rho}^{(s)}(t) \right] |l\rangle = \sum_{nm} \langle m| \breve{\rho}^{(s)}(t) |n\rangle \left\{ \delta_{ln} \sum_r \langle k|S_i^{(a)}|r\rangle \langle r|S_j^{(a)}|m\rangle e^{-i\omega_{rm}t} + \right.$$

$$\left. - \delta_{ln} \sum_r \langle n|S_i^{(a)}|l\rangle \langle k|S_j^{(a)}|m\rangle e^{-i\omega_{km}t} \right\} e^{i(\omega_{km}+\omega_{nl})t}$$

where the complex exponential functions come from changing the interaction representation into the Schrodinger representation. Term 3.2 becomes

$$\langle k| \left[\breve{S}_i^{(a)}(t), \breve{\rho}^{(s)}(t)\breve{S}_j^{(a)}(t-\tau) \right] |l\rangle = \sum_{nm} \langle m|\breve{\rho}^{(s)}(t)|n\rangle \left\{ \langle n|\hat{S}_j^{(a)}|l\rangle \langle k|\hat{S}_i^{(a)}|m\rangle e^{-i\omega_{nl}t} + \right.$$

$$\left. - \delta_{km} \sum_r \langle n|\hat{S}_j^{(a)}|r\rangle \langle r|\hat{S}_i^{(a)}|l\rangle e^{-i\omega_{nr}t} \right\} e^{i(\omega_{km}+\omega_{nl})t} \tag{7.16.27}$$

where "l" is lower case "L." The difference in angular frequency is

$$\omega_{nl} = \frac{1}{\hbar}(E_n - E_l) \tag{7.16.28}$$

where E_n (etc.) refers to the energy of nth level of the system. Notice that ω_{nl} can be negative.

Define matrix elements

$$\Gamma_{nlkm}^{(a)+} = \frac{1}{\hbar^2} \sum_{ij} \langle n|\hat{S}_i^{(a)}|l\rangle \langle k|\hat{S}_j^{(a)}|m\rangle \int_o^\infty d\tau \; e^{-i\omega_{km}\tau}\left\langle \breve{R}_i^{(a)}(\tau)\breve{R}_j^{(a)}(0)\right\rangle \tag{7.16.29}$$

$$\Gamma_{nlkm}^{(a)-} = \frac{1}{\hbar^2} \sum_{ij} \langle n|\hat{S}_j^{(a)}|l\rangle \langle k|\hat{S}_i^{(a)}|m\rangle \int_o^\infty d\tau \; e^{-i\omega_{nl}\tau}\left\langle \breve{R}_j^{(a)}(0)\,\breve{R}_i^{(a)}(\tau)\right\rangle \tag{7.16.30}$$

and also the relaxation matrix elements

$$R_{klmn}^{(a)} = -\delta_{\ln} \sum_r \Gamma_{krrm}^{(a)+} + \Gamma_{nllm}^{(a)+} + \Gamma_{nlkm}^{(a)-} - \delta_{km} \sum_r \Gamma_{nrrl}^{(a)-} \qquad (7.16.31)$$

Substitute all Equations 7.16.27 through 7.16.31 into Equation 7.16.26 to find

$$\langle k| \dot{\breve{\rho}}_s(t) |l\rangle = \sum_{nma} \langle m| \breve{\rho}_s(t) |n\rangle R_{klmn}^{(a)} e^{i(\omega_{km}+\omega_{nl})t} \qquad (7.16.32)$$

Applying the rotating wave approximation (RWA) requires

$$\omega_{km} + \omega_{nl} = 0$$

There are three cases implied by the RWA

$$(k=m, l=n k \neq l) \quad (k=l, m=n, k \neq n) \quad (k=l=n=m)$$

From these, two separate equations can be found

$$\langle k| \dot{\breve{\rho}}_s(t) |l\rangle = \sum_a \langle k| \breve{\rho}_s(t) |l\rangle R_{klkl}^{(a)} \quad \text{and} \quad \langle k| \dot{\breve{\rho}} |k\rangle = \sum_a \sum_{\substack{n \\ n \neq k}} \langle n| \breve{\rho}_s(t) |n\rangle R_{kknn}^{(a)}$$

Define

$$R_{ijkl} = \sum_a R_{ijkl}^{(a)}$$

and combine the previous two equations to obtain

$$\langle k| \dot{\breve{\rho}}_s(t) |l\rangle = \langle k| \breve{\rho}_s(t) |l\rangle R_{klkl} + \delta_{kl} \sum_{\substack{n \\ n \neq k}} \langle n| \breve{\rho}_s(t) |n\rangle R_{kknn} \qquad (7.16.33)$$

Notice that the effect of all the reservoirs combines into a single constant. This equation clearly shows that the reservoir causes a relaxation of the density operator (the system changes state due to the reservoir interaction). If $k=l$, then the first term on the right side represents the number of electrons, for example, leaving state k, while the second term represents the number entering state k from a different state n.

Equation 7.16.33 shows how the density operator changes due to the interaction between the reservoir and system in the interaction representation. Now we write the equation of motion in the Schrodinger representation. Equation 7.15.29b in 7.15.5 is repeated here as

$$\frac{\partial \hat{\rho}_s(t)}{\partial t} = \hat{u}_s \dot{\breve{\rho}}_s(t) \hat{u}_s^+ + \frac{1}{i\hbar} [\hat{H}_s, \hat{\rho}_s(t)] \qquad (7.16.34)$$

To combine Equations 7.16.33 and 7.16.34, requires that we take the "kl" matrix element of Equation 7.16.34.

$$\frac{\partial \hat{\rho}_s(t)}{\partial t}\Big]_{kl} = \langle k| \left[\hat{u}_s \dot{\breve{\rho}}_s(t) \hat{u}_s^+ \right] |l\rangle + \frac{1}{i\hbar} [\hat{H}_s, \hat{\rho}_s(t)]_{kl}$$

$$= \langle k| \left[e^{\frac{\hat{H}_s t}{i\hbar}} \dot{\breve{\rho}}_s(t) \, e^{-\frac{\hat{H}_s t}{i\hbar}} \right] |l\rangle + \frac{1}{i\hbar} [\hat{H}_s, \hat{\rho}_s(t)]_{kl}$$

$$(7.16.35)$$

Operating with the unitary operators inside the first expectation value on the right-hand side provides

$$\langle k| \left[e^{\frac{\hat{H}_s t}{i\hbar}} \dot{\breve{\rho}}_s(t)\, e^{-\frac{\hat{H}_s t}{i\hbar}} \right] |l\rangle = e^{i\omega_{lk} t} \langle k| \dot{\breve{\rho}}_s(t) |l\rangle$$

so that Equation 7.16.34 becomes

$$\frac{\partial \hat{\rho}_s(t)}{\partial t}\bigg]_{kl} = e^{i\omega_{lk} t} \langle k| \dot{\breve{\rho}}_s(t) |l\rangle + \frac{1}{i\hbar}\left[\hat{H}_s, \hat{\rho}_s(t)\right]_{kl} \tag{7.16.36}$$

Combining Equations 7.16.33 and 7.16.36 provides

$$\frac{\partial \hat{\rho}_s(t)}{\partial t}\bigg]_{kl} = \frac{1}{i\hbar}\left[\hat{H}_s, \hat{\rho}_s(t)\right]_{kl} + e^{i\omega_{lk} t} \langle k| \breve{\rho}_s(t) |l\rangle\, R_{klkl} + e^{i\omega_{lk} t} \delta_{kl} \sum_{\substack{n \\ n \neq k}} \langle n| \breve{\rho}_s(t) |n\rangle\, R_{kknn}$$

Writing the interaction density operator in terms of the Schrodinger operator, yields

$$\frac{\partial \hat{\rho}_s(t)}{\partial t}\bigg]_{kl} = \frac{1}{i\hbar}\left[\hat{H}_s, \hat{\rho}_s(t)\right]_{kl} + e^{i\omega_{lk} t} \langle k|\hat{u}_s^+ \hat{\rho}_s(t)\, \hat{u}_s |l\rangle\, R_{klkl} + e^{i\omega_{lk} t} \delta_{kl} \sum_{\substack{n \\ n \neq k}} \langle n| \hat{u}_s^+ \hat{\rho}_s(t)\, \hat{u}_s |n\rangle\, R_{kknn}$$

The equation can be rewritten by substituting the exponential form of the unitary operators $\hat{u}_s = \exp(\hat{H}_s t/i\hbar)$ which can operate on the basis vectors. The operation yields a factor of $e^{-i\omega_{lk} t}$ to yield

$$\frac{\partial \hat{\rho}_{kl}}{\partial t} = \frac{1}{i\hbar}\left[\hat{H}_s, \hat{\rho}_s(t)\right]_{kl} + e^{i\omega_{lk} t} e^{-i\omega_{lk} t} \langle k| \hat{\rho}_s(t) |l\rangle\, R_{klkl} + e^{i\omega_{lk} t} \delta_{kl} \sum_{\substack{n \\ n \neq k}} e^{i\omega_n t - i\omega_n t} \langle n| \hat{\rho}_s(t) |n\rangle\, R_{kknn}$$

Canceling exponential terms and noting that the factor δ_{kl} results in $\omega_{lk} = \omega_l - \omega_k = 0$, we end up with

$$\frac{\partial \hat{\rho}_{kl}^{(s)}}{\partial t} = \frac{1}{i\hbar}\left[\hat{H}_s, \hat{\rho}^{(s)}(t)\right]_{kl} + \langle k| \hat{\rho}^{(s)}(t) |l\rangle\, R_{klkl} + \delta_{kl} \sum_{\substack{n \\ n \neq k}} \langle n| \hat{\rho}^{(s)}(t) |n\rangle\, R_{kknn} \tag{7.16.37}$$

Equation 7.16.37 is the Liouville equation for the density matrix as discussed in the first several sections of this chapter. The commutator provides the dynamics internal to the system itself. As a very important point, the exact composition of the "system" is unspecified; it can be composed of an atom and an electromagnetic wave interacting with the atom. Using notation previously employed, the last two terms in Equation 7.16.37, can be written as

$$\frac{\partial \hat{\rho}_{kl}^{(s)}}{\partial t}\bigg|_{\text{environ}} = \langle k|\hat{\rho}^{(s)}(t)|l\rangle R_{klkl} + \delta_{kl} \sum_{\substack{n \\ n \neq k}} \langle n|\hat{\rho}^{(s)}(t)|n\rangle R_{kknn}$$

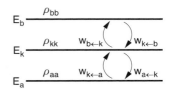

FIGURE 7.16.3
The rates of transition "W" between energy levels "E." The diagonal
density matrix elements give the population of each level.

The rate of change of the diagonal elements must be

$$\left.\frac{\partial \hat{\rho}_{kk}^{(s)}}{\partial t}\right|_{\text{environ}} = \langle k|\hat{\rho}^{(s)}(t)|k\rangle R_{kkkk} + \sum_{\substack{n \\ n \neq k}} \langle n|\hat{\rho}^{(s)}(t)|n\rangle R_{kknn}$$

The first term is the transition out of level "k" and the second term is the transition rate into level "k" from other levels "n." Figure 7.16.3 indicates the transition rates by W.

7.17 Quantum Mechanical Fluctuation-Dissipation Theorem

The laser rate equations must incorporate fluctuation and dissipation sources in accordance with the fluctuation-dissipation theorem. Usually we assume a long enough time scale that the fluctuation terms average to zero. However, the rate equations in Chapter 2 show the dissipation terms such as that for spontaneous recombination. This section demonstrate how the dissipation-fluctuations arise from quantum mechanical interactions between a system and a reservoir. In particular, the Langevin source terms appear as the Fourier transforms of the reservoir-system coupling coefficients. For the Langevin forces to be truly delta-function correlated, the coupling strengths (as a function of frequency) must be broadband (approximately independent of frequency). The jitter and relaxation effects for a single atom can be most clearly seen in the Hesienberg representation of the raising/lowering operators.

7.17.1 Some Introductory Comments

Recall that the Liouville equation comes from writing the Hamiltonian as

$$\hat{H} = \hat{H}_o + \hat{H}_{\text{environ}} = \hat{H}_a + \hat{V} + \hat{H}_{\text{environ}}$$

which leads to

$$\frac{\partial \hat{\rho}}{\partial t} = \frac{1}{i\hbar}\left[\hat{H}_o, \hat{\rho}\right] + \left(\frac{\partial \hat{\rho}}{\partial t}\right)_{\text{environ}} = \frac{1}{i\hbar}\left[\hat{H}_o, \hat{\rho}\right] + \left(\frac{\partial \hat{\rho}}{\partial t}\right)_{\text{pump}} + \left(\frac{\partial \hat{\rho}}{\partial t}\right)_{\text{coll}} + \left(\frac{\partial \hat{\rho}}{\partial t}\right)_{\text{spont}}$$

The density operator provides both a macroscopic statistical average and the microscopic quantum mechanical average. The $\hat{H}_{\text{environ}}$ term indicates that an "external" agent acts on the system. These external agents cause the relaxation effects. The external agents can be modeled as "reservoirs" (similar concept to thermal reservoirs). This section shows that a reservoir induces rapid fluctuations (Langevin noise) as well as damping. 2.6.3 discusses the reservoir and the fluctuation-dissipation theorem in more detail.

The Liouville equation for the density operator is essentially a differential equation for the energy level occupation number (i.e., $\langle n| \hat{\rho} |n\rangle$) and the induced polarization (off-diagonal terms). This section shows how a quantum mechanical reservoir gives rise to damping and fluctuation terms in the Liouville equation (a.k.a., master equation). The damping term appears as

$$\frac{\partial \hat{\rho}}{\partial t}\bigg|_{\text{other}}$$

while, as expected, the average of the fluctuations disappears. The next sections will use the trace over the reservoir states to calculate the average. The tracing operation produces a zero average for the fluctuations and removes the reservoir degrees-of-freedom from the differential equation.

The formalism can be applied to spontaneous and stimulated emission from atoms. The density operator can incorporate the various types of EM states (Fock, coherent and squeezed) and various atomic states. The density operator/matrix accounts for all possible knowledge of the system.

7.17.2 Quantum Mechanical Fluctuation Dissipation Theorem

Consider a small system composed of a single harmonic oscillator and a reservoir with a very large number of harmonic oscillators. Let $\{\hat{R}_\omega^+\}$ and $\{\hat{R}_\omega^-\}$ be the creation and annihilation operators for mode ω in the reservoir (in the Schrodinger representation). Let $\hat{S}^+$ and $\hat{S}^-$ be *raising* and *lowering* operators for a harmonic oscillator defining a small system. Alternatively, the same conclusions can be drawn for the raising $\hat{\sigma}^+$ and $\hat{\sigma}^-$, lowering operators for the Jaynes-Cummings' 2-level atoms. The Hamiltonian for the system and reservoir $\hat{H} = \hat{H}_s + \hat{H}_r + \hat{H}_{sr}$ consists of the Hamiltonians for the system $\hat{H}_s$, reservoir $\hat{H}_r$ and system-reservoir interaction $\hat{H}_{sr}$ defined by

$$\hat{H}_s = \hbar\omega_s\left(\hat{S}^+\hat{S}^- + \frac{1}{2}\right) \qquad \hat{H}_r = \sum_\omega \hbar\omega\left(\hat{R}_\omega^+\hat{R}_\omega^- + \frac{1}{2}\right)$$

$$\hat{H}_{sr} = \sum_\omega \beta_\omega \hat{S}^+\hat{R}_\omega^- + \sum_\omega \beta_\omega^* \hat{S}^-\hat{R}_\omega^+ = \hat{S}^+\sum_\omega \beta_\omega \hat{R}_\omega^- + \hat{S}^-\sum_\omega \beta_\omega^* \hat{R}_\omega^+$$

The system-reservoir interaction Hamiltonian consists of terms that explicitly conserve energy. For example, the annihilation operator $\{\hat{R}_\omega^-\}$ removes a quantum of energy from the reservoir and the raising operator $\hat{S}^+$ increase the energy of the system. The symbols β_ω represent the coupling strength of a reservoir oscillator (with angular frequency ω) to the system. The operators obey the following commutation relations.

$$\left[\hat{S}^-, \hat{S}^-\right] = 0 = \left[\hat{S}^+, \hat{S}^+\right] \qquad \left[\hat{S}^-, \hat{S}^+\right] = 1$$

$$\left[\hat{R}_i^-, \hat{R}_j^-\right] = 0 = \left[\hat{R}_i^+, \hat{R}_j^+\right] \qquad \left[\hat{R}_i^-, \hat{R}_j^+\right] = \delta_{ij}$$

$$\left[\hat{S}^\pm, \hat{R}^\pm\right] = 0$$

which also apply for equal-time commutators in the Heisenberg representation. The *Heisenberg* representation of an operator $\hat{O}$ is given by

$$\tilde{O} = \hat{u}^+ \hat{O} \, \hat{u}$$

where, for a closed system, the evolution operator must be $\hat{u} = \exp(\hat{H}t/i\hbar)$. Note the use of the symbol over the operator "$\tilde{O}$" to indicate the Heisenberg representation. The *total* Hamiltonian has the same form for either the Schrodinger or the Heisenberg representations since $\hat{u}$ and $\hat{H}$ commute

$$\tilde{H} = \hat{u}^+ \hat{H} \, \hat{u} = \hat{u}^+ \hat{u} \, \hat{H} = \hat{H}$$

However, each individual Hamiltonian such as $\hat{H}_{sr}$ does not necessarily commute with the full Hamiltonian. The Heisenberg Hamiltonian can be found just by adding the "twiddle" to each operator. The equation of motion for a Heisenberg operator $\tilde{O}$ is

$$\frac{d\tilde{O}}{dt} = \frac{i}{\hbar}\left[\tilde{H}, \tilde{O}\right] + \hat{u}^+ \frac{\partial \hat{O}}{\partial t} \hat{u}$$

Usually the last term must be zero for most operators (except possibly the density operator).

The operators of the theory therefore satisfy the following rate equations

$$\dot{\tilde{S}}^- = \frac{i}{\hbar}\left[\tilde{H}, \tilde{S}^-\right] = \frac{i}{\hbar}\left[\tilde{H}_s + \tilde{H}_r + \tilde{H}_{sr}, \tilde{S}^-\right] = \frac{i}{\hbar}\left[\tilde{H}_s + \tilde{H}_{sr}, \tilde{S}^-\right] \qquad (7.17.1)$$

The first commutator can be computed

$$\left[\tilde{H}_s, \tilde{S}^-\right] = \left[\hbar\omega_s\left(\tilde{S}^+\tilde{S}^- + \frac{1}{2}\right), \tilde{S}^-\right] = \hbar\omega_s\left[\tilde{S}^+\tilde{S}^-, \tilde{S}^-\right] = \hbar\omega_s\left[\tilde{S}^+, \tilde{S}^-\right]\tilde{S}^- = -\hbar\omega_s\tilde{S}^-$$

The second commutator in Equation 7.17.1 can also be computed

$$\left[\tilde{H}_{sr}, \tilde{S}^-\right] = \left[\tilde{S}^+ \sum_\omega \beta_\omega \tilde{R}_\omega^- + \tilde{S}^- \sum_\omega \beta_\omega^* \tilde{R}_\omega^+, \tilde{S}^-\right]$$

$$= \left[\tilde{S}^+ \sum_\omega \beta_\omega \tilde{R}_\omega^-, \tilde{S}^-\right] = \left[\tilde{S}^+, \tilde{S}^-\right]\sum_\omega \beta_\omega \tilde{R}_\omega^- = -\sum_\omega \beta_\omega \tilde{R}_\omega^-$$

Combining the commutators into Equation 7.17.1 yields

$$\dot{\tilde{S}}^- = -i\omega_s\tilde{S}^- - \frac{i}{\hbar}\sum_\omega \beta_\omega \tilde{R}_\omega^- \qquad (7.17.2)$$

The adjoint of Equation 7.17.2 provides a similar equation for the raising operator.

$$\dot{\tilde{S}}^+ = i\omega_s\tilde{S}^+ + \frac{i}{\hbar}\sum_\omega \beta_\omega^* \tilde{R}_\omega^+ \qquad (7.17.3)$$

It is also necessary to find the equations of motion for the reservoir operators because they occur in Equations 7.17.2 and 7.17.3.

$$\dot{\tilde{R}}_\omega^- = \frac{i}{\hbar}\left[\tilde{H}, \tilde{R}_\omega^-\right] = \frac{i}{\hbar}\left[\tilde{H}_s + \tilde{H}_r + \tilde{H}_{sr}, \tilde{R}_\omega^-\right] = \frac{i}{\hbar}\left[\tilde{H}_r + \tilde{H}_{sr}, \tilde{R}_\omega^-\right] \tag{7.17.4}$$

The first commutator provides

$$\left[\tilde{H}_r, \tilde{R}_\omega^-\right] = \left[\sum_{\omega'}\hbar\omega'\left(\tilde{R}_{\omega'}^+\tilde{R}_{\omega'}^- + \frac{1}{2}\right), \tilde{R}_\omega^-\right] = \sum_{\omega'}\hbar\omega'\left[\tilde{R}_{\omega'}^+, \tilde{R}_\omega^-\right]\tilde{R}_{\omega'}^- = -\sum_{\omega'}\hbar\omega'\delta_{\omega\omega'}\tilde{R}_\omega^- = -\hbar\omega\tilde{R}_\omega^-$$

The second commutator gives

$$\left[\tilde{H}_{sr}, \tilde{R}_\omega^-\right] = \left[\tilde{S}^+\sum_{\omega'}\beta_{\omega'}\tilde{R}_{\omega'}^- + \tilde{S}^-\sum_{\omega'}\beta_{\omega'}^*\tilde{R}_{\omega'}^+, \tilde{R}_\omega^-\right] = \left[\tilde{S}^-\sum_{\omega'}\beta_{\omega'}^*\tilde{R}_{\omega'}^+, \tilde{R}_\omega^-\right] = -\beta_\omega^*\tilde{S}^-$$

Combining the last two commutators into Equation 7.17.4 gives

$$\dot{\tilde{R}}_\omega^- = -i\omega\tilde{R}_\omega^- - \frac{i}{\hbar}\beta_\omega^*\tilde{S}^- \tag{7.17.5}$$

The adjoint gives

$$\dot{\tilde{R}}_\omega^+ = i\omega\tilde{R}_\omega^+ + \frac{i}{\hbar}\beta_\omega\tilde{S}^+ \tag{7.17.6}$$

An equation for the lowering operator (for the system) can be found by combining Equations 7.17.5 and 7.17.2. First, formally solve equatsion 7.17.5 by using an integrating factor. As reviewed in Appendix 1, a first order differential equation $\dot{y} - ay = f(t)$ has the solution

$$y(t) = \frac{\mu(0)y(0)}{\mu(t)} + \frac{1}{\mu(t)}\int_0^t d\tau\, \mu(\tau)f(\tau)$$

with $\mu(t) = e^{-at}$. Equation 7.17.5 has the integrating factor $\mu = e^{i\omega t}$ and the formal solution

$$\tilde{R}_\omega^-(t) = e^{-i\omega t}\tilde{R}_\omega^-(0) + \frac{i\beta_\omega^*}{\hbar}\int_0^t d\tau e^{-i\omega(t-\tau)}\tilde{S}^-(\tau) \tag{7.17.7}$$

Substituting Equation 7.17.7 into Equation 7.17.2, specifically $\dot{\tilde{S}}^- = -i\omega_s\tilde{S}^- - \frac{i}{\hbar}\sum_\omega\beta_\omega\tilde{R}_{\omega'}^-$, provides

$$\dot{\tilde{S}}^- + i\omega_s\tilde{S}^- + \frac{1}{\hbar^2}\sum_\omega|\beta_\omega|^2\int_0^t d\tau e^{-i\omega(t-\tau)}\tilde{S}^-(\tau) = -\frac{i}{\hbar}\sum_\omega\beta_\omega e^{-i\omega t}\tilde{R}_\omega^-(0)$$

which can be rewritten as

$$\dot{\tilde{S}}^- + i\omega_s\tilde{S}^- + \frac{1}{\hbar^2}\int_0^t d\tau\,\tilde{S}^-(\tau)\sum_\omega|\beta_\omega|^2 e^{-i\omega(t-\tau)} = -\frac{i}{\hbar}\sum_\omega\beta_\omega e^{-i\omega t}\tilde{R}_\omega^-(0) \tag{7.17.8}$$

The integral in Equation 7.17.8 gives the damping whereas the summation is the Langevin fluctuation. First evaluate the integral. The summation in the integrand can be evaluated. Let $g(\omega)$ be the density of states (i.e., the number of β_ω per unit frequency range).

$$\sum_\omega |\beta_\omega|^2 e^{-i\omega(t-\tau)} = \int_x^y d\omega \, g(\omega) \, |\beta_\omega|^2 e^{-i\omega(t-\tau)} \qquad (7.17.9)$$

where x, y are the smallest and largest allowed frequencies, respectively. A typical assumption is that $g(\omega)|\beta_\omega|^2$ is essentially independent of frequency, which is usually stated as the coupling strength β_ω being relatively constant. For a number of calculations, the upper limit on the integral remains finite, which avoids pesky infinities. A good example is for the Casimir effect (refer to Milonni's book on the Quantum Vacuum). We take the upper limit to be infinity for simplicity. The summation becomes

$$\sum_\omega |\beta_\omega|^2 e^{-i\omega(t-\tau)} = \int_x^\infty d\omega \, g(\omega) \, |\beta_\omega|^2 e^{-i\omega(t-\tau)} = g \, |\beta|^2 \int_x^\infty d\omega \, e^{-i\omega(t-\tau)} \qquad (7.17.10)$$

Two basic methods appear in texts and papers to calculate the integral. Some authors include negative frequencies in the integral to obtain a Dirac delta function; however, it is nonphysical to have negative frequencies since this corresponds to negative energies for the harmonic oscillator.

$$\int_x^\infty d\omega \, e^{-i\omega(t-\tau)} = \begin{cases} 2\pi \, \delta(t-\tau) & x = -\infty \\[2mm] \pi g |\beta|^2 \delta(t-\tau) - i g |\beta|^2 \dfrac{P}{t-\tau} & x = 0 \end{cases} \qquad (7.17.11)$$

The "P" in the above relation refers to the principal part. The development in this book uses the expression with the negative frequencies.

We obtain the fluctuation-damping expression for the system operators by substituting the top of Equations 7.17.11 into Equation 7.17.8.

$$\dot{\tilde{S}}^- + i\omega_s \tilde{S}^- + \frac{g|\beta|^2 2\pi}{\hbar^2} \int_0^t d\tau \, \tilde{S}^-(\tau) \delta(t-\tau) = -\frac{i}{\hbar} \sum_\omega \beta_\omega e^{-i\omega t} \tilde{R}_\omega^-(0) \qquad (7.17.12)$$

With the definition of the Dirac delta function

$$\int_0^b d\tau \delta(\tau - a) = \begin{cases} 1 & a \in (0, b) \\[1mm] 1/2 & a = 0 \text{ or } b \\[1mm] 0 & a \notin [0, b] \end{cases}$$

the equation of motion (7.17.12) becomes

$$\dot{\tilde{S}}^- + i\omega_s \tilde{S}^- + \frac{\pi g |\beta|^2}{\hbar^2} \tilde{S}^-(t) = -\frac{i}{\hbar} \sum_\omega \beta_\omega e^{-i\omega t} \tilde{R}_\omega^-(0) \qquad (7.17.13a)$$

The adjoint of this last result provides the equation of motion for the raising operator.

$$\dot{\tilde{S}}^+ - i\omega_s \tilde{S}^+ + \frac{\pi g |\beta|^2}{\hbar^2} \tilde{S}^+(t) = \frac{i}{\hbar} \sum_\omega \beta_\omega^* e^{+i\omega t} \tilde{R}_\omega^+(0) \qquad (7.17.13b)$$

Now examine the terms in Equations 7.17.13a and 7.17.13b. The second term, which is $i\omega_s \tilde{S}^-$, controls the frequency of oscillation. The third term

$$\frac{\pi g |\beta|^2}{\hbar^2} \tilde{S}^-(t)$$

controls the damping/relaxation. Define the damping coefficient as

$$\gamma = \frac{\pi g |\beta|^2}{\hbar^2}$$

The last term in Equation 7.17.13 comprises the Langevin "force" describing the random fluctuations that the reservoir induces in the system. The Langevin force term

$$\tilde{\Gamma}(t) = -\frac{i}{\hbar} \sum_\omega \beta_\omega e^{-i\omega t} \tilde{R}_\omega^-(0) \qquad (7.17.14)$$

averages to zero as shown in 7.17.3. The final topic shows how Equations 7.17.13 lead to an exponential decay.

7.17.3 The Average of the Langevin Noise Term

The average of the Langevin noise term $\tilde{\Gamma}(t) = -(i/\hbar) \sum_\omega \beta_\omega e^{-i\omega t} \tilde{R}_\omega^-(0)$ in Equation 7.17.14 can be calculated by either taking the time average or the ensemble average. For ergodic systems, the two types of averages must produce identical results. The ensemble average is

$$\langle \Gamma(t) \rangle_{\text{reserv}} = Tr(\rho_r \Gamma(t)) \qquad (7.17.15)$$

The density operator $\hat{\rho}_r$ is defined through a Boltzmann distribution as discussed in Appendix 9 with the normalization Z given by

$$\hat{\rho}_r = \frac{1}{Z} \exp\left(-\frac{\hat{H}_r}{k_{\text{B}} T}\right) \qquad Z = Tr_r \left\{ \exp\left(-\frac{\hat{H}_r}{k_{\text{B}} T}\right) \right\}$$

For this calculation, we really only require that the density operator be diagonal (a pure state). For a reservoir consisting of a very large number of Harmonic oscillators with many possible frequencies, the Hamiltonian must be

$$\hat{H}_r = \sum_\omega \hbar\omega \left(\hat{R}_\omega^+ \hat{R}_\omega^- + \frac{1}{2} \right)$$

Denote a Fock state by

$$|\{n\}\rangle = |n_1, n_2, n_3, \ldots\rangle$$

Returning to Equation 7.17.15, the average of the fluctuation term becomes

$$\langle \Gamma(t) \rangle_{\text{reserv}} = Tr(\hat{\rho}_r \Gamma(t)) = -\frac{i}{\hbar} Tr\left(\hat{\rho}_r \sum_\omega \beta_\omega e^{-i\omega t} \breve{R}_\omega(0) \right)$$

$$= -\frac{i e^{-i\omega t}}{\hbar Z} \sum_\omega \beta_\omega Tr_r \left\{ \exp\left(-\frac{\hat{H}_r}{k_B T} \right) \hat{R}_\omega^- \right\}$$

This last equation uses the notation of $\hat{R}_\omega^- = \breve{R}_\omega(0)$. We can easily calculate the trace as follows

$$Tr_r \left\{ \exp\left(-\frac{\hat{H}_r}{k_B T} \right) \hat{R}_\omega^- \right\} = \sum_{\{n\}} \langle \{n\} | \exp\left(-\frac{\hat{H}_r}{k_B T} \right) \hat{R}_\omega^- |\{n\}\rangle$$

Inserting the closure relation provides

$$Tr_r \left\{ \exp\left(-\frac{\hat{H}_r}{k_B T} \right) \hat{R}_\omega^- \right\} = \sum_{\{n\},\{m\}} \langle \{n\} | \exp\left(-\frac{\hat{H}_r}{k_B T} \right) |\{m\}\rangle \langle \{m\} | \hat{R}_\omega^- |\{n\}\rangle$$

Here's where the fact that $\hat{\rho}_r$ is diagonal becomes important.

$$Tr_r \left\{ \exp\left(-\frac{\hat{H}_r}{k_B T} \right) \hat{R}_\omega^- \right\} = \sum_{\{n\},\{m\}} \exp\left(-\frac{1}{k_B T} \sum_{\{m\}} E_{\{m\}} \right) \langle \{n\} | \{m\}\rangle \langle \{m\} | \hat{R}_\omega^- |\{n\}\rangle$$

where $E_{\{m\}}$ is an obvious notation. The orthonormality of the Fock states provides

$$\langle \{n\} | \{m\} \rangle = \delta_{\{n\},\{m\}}$$

The trace becomes

$$Tr_r \left\{ \exp\left(-\frac{\hat{H}_r}{k_B T} \right) \hat{R}_\omega^- \right\} = \sum_{\{n\},\{m\}} \exp\left(-\frac{1}{k_B T} \sum_{\{m\}} E_{\{m\}} \right) \langle \{n\} | \hat{R}_\omega^- |\{n\}\rangle$$

Now its clear that the trace operation gives zero because the annihilation operator removes a quanta of energy from the ket. For example

$$\langle n_1, n_2 \ldots | \hat{R}_1^- |n_1, n_2 \ldots\rangle = \langle n_1, n_2 \ldots | n_1 - 1, n_2 \ldots\rangle = 0$$

since Fock states with unequal occupation numbers must be orthogonal. Therefore we find the expectation value of the Langevin fluctuation to be

$$\langle \Gamma(t) \rangle_{\text{reserv}} = Tr(\rho_r \Gamma(t)) = 0$$

7.17.4 Damping of the Small System

The raising and lowering operators for the small system have the form

$$\dot{\tilde{S}}^- + i\omega_s \tilde{S}^- + \frac{\pi g |\beta|^2}{\hbar^2} \tilde{S}^-(t) = \tilde{\Gamma}(t) \tag{7.17.16}$$

$$\dot{\tilde{S}}^+ - i\omega_s \tilde{S}^+ + \frac{\pi g |\beta|^2}{\hbar^2} \tilde{S}^+(t) = \tilde{\Gamma}^+(t) \tag{7.17.17}$$

where $\Gamma(t) = -\frac{i}{\hbar} \sum_\omega \beta_\omega e^{-i\omega t} \tilde{R}_\omega^-(0)$. These equations can be solved (chapter exercises) to find

$$\tilde{N}(t) = e^{-2\gamma t} \tilde{N}(0) \tag{7.17.18}$$

where $\tilde{N}(t) = \tilde{S}^+(t)\tilde{S}(t)$ is the Heisenberg representation of the number operator and $\gamma = \pi g |\beta|^2/\hbar^2$. Notice that this solution can only hold when the system starts in a level higher than the ground state.

7.18 Review Exercises

7.1 Consider a permanent dipole with charges $\pm e$ separated by a distance vector $\vec{d}$ pointing from $-e$, located at $\vec{r}$, to $+e$, located at $\vec{r} + \vec{d}$. Show the total force (not torque) acting on an electric dipole can be written as $\vec{F} = \vec{\mu} \cdot \nabla \vec{\mathcal{E}}$, where $\vec{\mathcal{E}}$ denotes the electric field. Hint: First work the problem for $\mathcal{E}_x$ by Taylor expanding $\mathcal{E}_x(\vec{r} + \vec{d})$.

7.2 Consider a permanent dipole with charges $\pm q$ which can oscillate about its center of mass in the x-y plane. The charged particles both with mass m are separated by a distance d. Let I denote the moment of inertia of the dipole. A torque $\vec{\tau}_e = -\tilde{z} c_1 I \theta$ returns the dipole to its equilibrium orientation along the x-axis (i.e., $\vec{\mu} = qd\tilde{x}$ at equilibrium) where $\tilde{x}, \tilde{y}, \tilde{z}$ represent unit vectors. A damping torque of the form $\vec{\tau}_d = -c_1 I \dot{\theta}$ acts on the dipole. A very small electromagnetic wave $\vec{E} = \hat{y} E_o \cos(\omega t)$ incident on the dipole induces small oscillations.

1. Find the moment of inertia I.
2. Find the torque produced by the electric field.
3. Using Newton's law relating torque and angular momentum, write the equation of motion for the dipole. Explain the effect of the various terms.
4. Find the frequency response and detail any assumptions.

7.3 Show that a pure state $\hat{\rho} = |\psi\rangle\langle\psi|$ reduces $\langle\hat{\mu}\rangle = Tr(\hat{\rho}\hat{\mu})$ to $\langle\vec{\mu}\rangle = \int d^3r\,\psi^*(q\vec{r})\,\psi$.

7.4 A student is playing with a high-voltage distributor coil (30 kV) from an old car. The student is trying to make a "shock box" for a demonstration. Another student has a demonstration nearby that which has excited gas molecules enclosed in a glass jar; most of the atoms have electrons in the $n = 2$ excited state. The first student powers up the shock box and it emits a HUGE spark. With hair still smoking, the student notices that the nearby gas emits a photon. Assume the spark produces an electromagnetic field of the form

$$\mathscr{E} = \frac{\mathscr{E}_o}{\sqrt{2\pi}\,\sigma}\exp\left[-\frac{t^2}{2\sigma^2}\right]$$

at the position of the atoms. The perturbation potential is then

$$\hat{V} = \hat{\mu}\cdot\vec{\mathscr{E}} \quad \text{or} \quad \underline{V} = \mu\vec{\mathscr{E}} \quad \text{so that} \quad V_{12} = \mu_{12}\frac{\mathscr{E}_o}{\sqrt{2\pi}\,\sigma}\exp\left[-\frac{t^2}{2\sigma^2}\right]$$

Show the approximate probability of transition from state #2 to #1 has the form

$$\mathscr{E}_o\left[\frac{\mu_{12}}{i\hbar}\right]\exp\left[-\frac{\sigma^2}{2}\omega_{12}^2\right]$$

Hint: Integrate using

$$\exp\left\{-\frac{\tau^2}{2\sigma^2} + i\omega_{12}\tau\right\} = \exp\left\{-\frac{\sigma^2}{2}\omega_{12}^2\right\}\exp\left\{-\frac{1}{2\sigma^2}\left(\tau + i\sigma^2\omega_{12}\right)^2\right\}$$

$$\int_{-\infty}^{\infty}d\tau\,\frac{1}{\sqrt{2\pi}\,b}\exp\left[-\frac{(\tau\pm a)^2}{2b^2}\right] = 1$$

7.5 For the electron harmonic oscillator $\hat{H} = (\hat{p}^2/2m) + (k/2)\hat{x}^2$ with $\omega = \sqrt{k/m}$, define $\hat{Q} = (\hat{a} + \hat{a}^+)/\sqrt{2}$ and $\hat{P} = -i(\hat{a} - \hat{a}^+/\sqrt{2})$, where $\hat{a}, \hat{a}^+$ represent the ladder operators

1. Show $\hat{x} = \hat{Q}\sqrt{(\hbar/m\omega)}$ and $\hat{p} = \hat{P}\sqrt{m\omega\hbar}$
2. Show $[\hat{Q}, \hat{P}] = i$.
3. Show $\hat{H} = (\hbar\omega/2)(\hat{Q}^2 + \hat{P}^2)$

7.6 Using the Schrodinger representation of the electron coherent state

$$|\alpha(t)\rangle \equiv e^{\hat{H}t/i\hbar}|\alpha\rangle = e^{i\omega t/2}e^{-|\alpha(t)|^2/2}\sum_n\frac{\alpha(t)^n}{\sqrt{n!}}|n\rangle$$

where $\alpha(t) \equiv \alpha e^{-i\omega t} = |\alpha|e^{i\phi-i\omega t}$. Show the center position of the coherent state must oscillate according to $\bar{Q}(t) = \langle\alpha(t)|\hat{Q}|\alpha(t)\rangle = \sqrt{2}\,\text{Re}[\alpha(t)] = \sqrt{2}\,|\alpha|\cos(\phi - \omega t)$.

7.7 Starting with $|\alpha\rangle = \hat{D}(\alpha)|0\rangle = e^{\alpha \hat{a}^+ - \alpha^* \hat{a}}|0\rangle$, show all of the steps leading to

$$\psi_\alpha(Q) = \langle Q \mid \alpha \rangle = \langle Q|\hat{D}(\alpha)|0\rangle = e^{i\bar{P}Q}e^{-i\bar{P}\bar{Q}/2}\langle Q|e^{-i\bar{Q}\hat{P}}|0\rangle = e^{i\bar{P}Q}e^{-i\bar{P}\bar{Q}/2}u_o(Q - \bar{Q})$$

and

$$\psi^*\psi = \frac{1}{\pi^{1/2}}e^{-(Q-\bar{Q})^2}.$$

7.8 Suppose a density operator has the form $\hat{\rho} = \int_0^{2\pi} d\phi\, P(\phi)\, |\alpha\rangle\langle\alpha|$, where $\alpha = r\,e^{i\phi}$ and P denotes a probability density. Show, unlike the results of Problem 7.6, $\langle \hat{Q}\rangle = 0$ for a uniform distribution $P = 1/2\pi$ for ϕ.

7.9 Show the antisymmetry property of the electromagnetic field tensor $F^{\mu\nu} = -F^{\nu\mu}$.

7.10 Use $\partial_\mu F^{\mu\nu} = j^\nu$ to show the equation of continuity $\partial_t \rho + \nabla \cdot \vec{J} = 0$.

7.11 Show Ampere's law $\nabla \times \vec{\mathcal{B}} = \vec{\mathcal{J}} + (\partial \vec{\mathcal{E}}/\partial t)$ starting with $\partial_\mu F^{\mu\nu} = j^\nu$.

7.12 Show the relation $\partial_1 F^{23} + \partial_2 F^{31} + \partial_3 F^{12} = 0$ for the field tensor F.

7.13 Show $F_{\mu\nu}F^{\mu\nu} = -2(\mathcal{E}^2 - \mathcal{B}^2)$

7.14 Show all of the mathematical detail leading to the momentum density $\pi^\alpha = -\partial^0 A^\alpha + \partial^\alpha A^0$ in Section 7.5 starting with

$$\pi^\alpha = \frac{\partial}{\partial \dot{A}_\alpha}\left(-\frac{1}{4}F_{\mu\nu}F^{\mu\nu} - j^\mu A_\mu\right) = -\frac{1}{4}\frac{\partial}{\partial \dot{A}_\alpha}F_{\mu\nu}F^{\mu\nu}$$

7.15 Show the relation $\vec{p} = (im/\hbar)[\hat{H}_A, \hat{r}]$, where "$A$" denotes the atomic Hamiltonian.

7.16 For the interaction Hamiltonian $\hat{H}_{A-L} = \hat{\mu} \cdot \hat{\mathcal{E}}$, find its semiclassical form $\hat{H}_{AL} = \langle \alpha\eta|\hat{\mu} \cdot \hat{\mathcal{E}}|\alpha\eta\rangle$ and compare it with the result for the pure coherent state given at the end of Section 7.6.

7.17 Calculate the transition rate in Equation 7.8.6 for the transition from atomic level 2 to 1. Perform the calculation for the following optical parts of the transition.

1. Two different coherent states as in $\langle \beta|\hat{E}|\alpha\rangle$.

2. A squeezed state as in $\langle \alpha\eta|\hat{E}|\alpha\eta\rangle$.

7.18 Let $\hat{P}_{ab} = |a\rangle\langle b|$ and $\hat{\rho} = \sum_\psi P_\psi|\psi\rangle\langle\psi|$, respectively, be a basis vector in the direct product space and the density operator where $|\psi\rangle = \sum_n \beta_n|n\rangle$. Show $\langle \hat{P}_{aa}|\hat{\rho}\rangle = \langle|\beta_a|^2\rangle$. Explain the significance of this results.

7.19 Assume the electromagnetic interaction has been turned-off $\hat{H}_{A-L} = \hat{\mathscr{V}} = 0$. Find the density operator as a function of time using the situation depicted in Figure P7.19 and assume the Liouville equation holds.

7.20 Find the matrix elements of the solution to the Liouville equation for the density operator when the light-field and the relaxation mechanisms are switched off. Assume a collection of 2-level atoms with energy basis satisfying $\hat{H}_o|n\rangle = E_n|n\rangle$.

1. Find the diagonal and off-diagonal elements for $\hat{\rho}(t)$ in terms of the β_n in the wave function expansion.

2. Assume at $t = 0$, the density matrix $\rho(0) = \rho_o$ is diagonal. Further assume initially that any given atom has its electron definitely in atomic state $|1\rangle$ or $|2\rangle$ but not

a superposition of both. Explain with diagrams for a wave function in a 2-level Hilbert space why the following relation holds $\langle \beta_n(0)\beta_m^*(0)\rangle = P(n)\delta_{nm}$.

3. Write all elements of the density matrix as a function of time.

7.21 A small time-varying electric field is incident on a collection of two-level atoms starting at $t=0$. Assume initially that any given atom has its electron definitely in atomic state $|1\rangle$ or $|2\rangle$ but not a superposition of both, where $\hat{H}_o|n\rangle = E_n|n\rangle$. Assume the perturbation has the form $\hat{V} = E\begin{pmatrix} 0 & \mu \\ \mu & 0 \end{pmatrix}$, where the field is assumed small. Let ρ_o represent the density matrix at $t=0$. Find the lowest order terms in $\rho(t)$. Do not include the relaxation terms in the Liouville equation.

7.22 Show the algebra for the following transformation and identify the time τ. Neglect thermal equilibrium statistics.

$$\left(\frac{\partial \rho}{\partial t}\right)_{env} = \begin{bmatrix} -aJ + c\rho_{22} + b\rho_{22} & -\dfrac{\rho_{12}}{T_2} \\[2mm] -\dfrac{\rho_{21}}{T_2} & aJ - c\rho_{22} - b\rho_{22} \end{bmatrix} \rightarrow -\begin{bmatrix} \dfrac{\rho_{11} - \bar{\rho}_{11}}{\tau} & -\dfrac{\rho_{12}}{T_2} \\[2mm] -\dfrac{\rho_{21}}{T_2} & \dfrac{\rho_{22} - \bar{\rho}_{22}}{\tau} \end{bmatrix}$$

7.23 Consider the Liouville equation for the density operator

$$\frac{\partial \rho_{ab}}{\partial t} = \frac{1}{i\hbar}\left[\hat{H}, \hat{\rho}\right]_{ab} - \frac{\rho_{ab} - \bar{\rho}_{ab}}{\tau_{ab}}$$

Assume the field has been removed and neglect thermal equilibrium statistics.

1. Explain why the current density must be small in $\bar{\rho}_{22} = (aJ)/(b+c)$. If J becomes large, what must change. Explain the physical significance.

2. Find ρ_{22} as a function of time and the pump-number current J.

7.24 Consider

$$\left(\frac{\partial \rho}{\partial t}\right)_{env} = \begin{bmatrix} -aJ + (c+b)(\rho_{22} - f_2) & -\dfrac{\rho_{12}}{T_2} \\[2mm] -\dfrac{\rho_{21}}{T_2} & aJ - (c+b)(\rho_{22} - f_2) \end{bmatrix}$$

that includes thermal equilibrium statistics.

1. Show $\bar{\rho}_{11} = \rho_{11}(t=\infty) = -(aJ)/(b+c) + f_1$ and $\bar{\rho}_{22} = \rho_{22}(t=\infty) = (aJ/b+c) + f_2$

2. Show the rate of change of the density matrix due to environmental effects can be written as $(\partial \rho_{ab}/\partial t)_{env} = -(\rho_{ab} - \bar{\rho}_{ab})/\tau_{ab}$

3. Find $\rho_{ab}(t)$ for the environmental effects in terms of a, b, c, etc. You will also need to include the value at $t=0$ defined by $\rho(0) = \rho_o$.

7.25 Assume the probability of finding an electron in state E has the form given by $P \sim e^{-E/kT}$. Assume N independent 2-level atoms with energy levels $E_1 = 0$ and $E_2 > 0$. Assume $E_1 = 0$.

1. What is the probability of an electron occupying level n where $n=1$ or 2. Hint: probabilities must add to 1.

2. When $0 < E_2 \ll kT$, why is $P(E_2) = 1/2$? Show this mathematically and explain physically.

3. Show $\lim_{T \to 0} P(E_1) = 1$ and $\lim_{T \to 0} P(E_2) = 0$. Explain this result physically.

4. Show the relaxation terms for the case with thermal equilibrium reduce to those for the case without thermal equilibrium using the results of part c.

7.26 Work through the algebra in Section 7.11. Be sure to fill-in any missing steps.

7.27 Show the Lorentz line shape function $L(\omega) = (2T_2)/(1 + (\omega - \omega_o)^2 T_2^2)$ has the full-width half-max of $2/T_2$.

7.28 Suppose N two-level atoms are embedded in a host dielectric (fiber) with real susceptibility χ_b. Assume an external, monochromatic light source pumps the embedded atoms and induces susceptibility χ_p. Assuming the total polarization is linear in the EM field, show the total susceptibility must be $\chi = \chi_b + \chi_p$.

7.29 Consider an optical amplifier made from a collection of two-level atoms having having relaxation time τ and phase decoherence time T_2. A pump-number density J excites the atoms. A beam with optical power P_o enters the collection at $z = 0$. Neglect the thermal distribution and assume the real part of the index remains approximately constant.

1. Write the gain as a function of J, τ, T_2, etc. Find the constants in

$$g = \frac{c_1}{c_2(\omega) + P/P_s}$$

Be sure to show the equations for P in terms of E_o starting with the Poynting vector.

2. Show the optical power P as a function of distance z has the form

$$\mathrm{Ln}\frac{P}{P_o} = \frac{c_1}{c_2}z - \frac{1}{c_2}\frac{P - P_o}{P_s}$$

3. For small EM fields, show the exponential form for P versus z.

4. Briefly discuss the significance of c_1 and c_2 in terms of the material gain in chapter 2, the peak gain, and the frequency dependence of c_2.

7.30 Consider a collection of two-level atoms used as an optical amplifier as in the previous problem. The beam starts at $z = 0$ and propagates along z.

1. Find the asymptotic behavior of $P(z)$ for large z.

2. Suppose $c_1 = 50 \,\mathrm{cm}^{-1}$, $P_o = 0.1\,\mathrm{mW}$, $P_s = 5\,\mathrm{mW}$, with a center wavelength of $\lambda_o = 1000\,\mathrm{nm}$. Let z range from 0 to 0.1 cm and let $\omega - \omega_o = 0$. Make a plot of P vs. z. Plot the unsaturated gain (i.e., $P_s = \infty$) on the same plot. Explain any differences in magnitude and shape.

3. For the frequency $\omega - \omega_o = 1/T_2$, make a plot of the gain vs. distance for the same conditions in part b.

7.31 Consider a collection of two-level atoms used as an optical amplifier as discussed in the previous two problems. Make a rough sketch of the single-pass gain defined by

$$G = \frac{P - P_o}{P_o}$$

for the peak frequency and another at half-max.

7.32 For homogeneous broadening of two-level atoms, show

$$(FWHM)_{Sat} = \frac{2}{T_2}\sqrt{1 + 4\Omega^2 \tau T_2}$$

as discussed in Section 7.13.

7.33 Suppose a dipole p oscillates with frequency ω_o but is exponentially damped $p(t) = e^{-t/T}e^{i\omega_o t}$ for $t > 0$ and zero otherwise.

1. Find the Fourier Transform and make a sketch of the power spectrum $|p(\omega)|^2$.

2. Find the Full-Width at Half Max (FWHM).

3. Discuss how this results applies to homogeneous broadening.

7.34 Find all of the parameters in $g(\omega) = (g_{E=0}(\omega))/(1 + I/(I_{s\omega}(\omega)))$ for homogeneous broadening. Be sure to find the intensity I using the Poynting vector.

7.35 Show the inhomogeneously broadened line $\mathscr{L}_I$

$$\mathscr{L}_I(\omega) = \int_{-\infty}^{\infty} d\omega_o \, \mathscr{L}_h(\omega, \omega_o) f(\omega_o)$$

is normalized to 1 provided $\mathscr{L}_h$ is normalized to 1.

7.36 Consider the gain for inhomogeneously broadened atoms

$$g_I(\omega) = \int_{-\infty}^{\infty} \frac{g_{peak}(J)f(\omega_o)\,d\omega_o}{1 + (\omega - \omega_o)^2 T_2^2 + \frac{I}{I_s}}$$

1. What is the inhomogeneous gain for atoms with only a single resonant frequency?

2. Show for a very broad distribution $f(\omega_o)$ that the inhomogeneous gain must be

$$g_I(\omega) = \frac{f(\omega)g_{peak}(J)\pi/T_2}{\sqrt{1 + \frac{I}{I_s}}}$$

7.37 For $|\psi\rangle = (1/\sqrt{2})|1\rangle + (1/\sqrt{2})|2\rangle$, use matrices to show $\langle\psi|\hat{\sigma}_z|\psi\rangle = 0$.

7.38 For a two-level atom with operators $\hat{\sigma}_{ij} = |i\rangle\langle j|$, find all of the commutation relations $[\hat{\sigma}_{ab}, \hat{\sigma}_{cd}]$, $[\hat{\sigma}^-, \hat{\sigma}^+]$, and $[\hat{\sigma}_{ab}, \hat{\sigma}^\pm]$.

7.39 Problem on how haken or miloni handle the constant in $cf + b$- etc.

7.40 Generalize the Jaynes-Cumming's hamiltonian to a 3 level atom. Assume transition only, between adjacent levels. Assume a two mode field to match.

7.41 Starting with the definition of the atomic lowering operator in the interaction representation for a two-level atom $\breve{\sigma}^- = \hat{u}_a^+ \hat{\sigma}^- \hat{u}_a = \hat{\sigma}^- - [(\hat{H}_a t/i\hbar), \hat{\sigma}^-] + \cdots$, show $\breve{\sigma}^-(t) = \hat{\sigma}^- e^{-i\omega_o t}$, using the matrix representation.

7.42 Starting with the definition of the EM lowering operator in the interaction representation $\breve{b}_{\vec{ks}}(t) = \hat{u}_f^+ \hat{b}_{\vec{ks}} \hat{u}_f$, show $\breve{b}_{\vec{ks}}(t) = \hat{b}_{\vec{ks}} e^{-i\omega_k t}$.

7.43 Show the interaction representation of the "free-field" density operator has the form $\breve{\rho}_f = \hat{u}_f^+ \hat{\rho}_f \hat{u}_f$ where $\hat{u}_f = \exp\{\hat{H}_f t/i\hbar\}$ and $\hat{H}_f = \sum_{\vec{ks}} \hbar\omega_{\vec{k}}(\hat{b}_{\vec{ks}}^+ \hat{b}_{\vec{ks}} + 1/2)$.

7.44 Show the following relation from 7.16.4

$$\text{Tr}_{1\ldots N}\left[\breve{V}, \rho(0)\right] = \text{Tr}_{1\ldots N} \sum_{aij} \left\{ \rho^{(s)}(0)\breve{S}_i^{(a)}\left[\breve{R}_j^{(a)}, \rho^{(1\ldots N)}(0)\right] + \left[\breve{S}_i^{(a)}, \rho^{(s)}(0)\right]\breve{R}_j^{(a)} \rho^{(1\ldots N)}(0)\right\}$$

and work through to find the result $\text{Tr}_a \breve{R}_j^{(a)} \rho^{(a)}(0) = \text{Tr}_a \rho^{(a)}(0)\breve{R}_j^{(a)} = \langle \breve{R}_i^{(a)} \rangle_a^{\text{res}}$.

7.45 Work through the reasoning in Section 7.16 to show

$$\text{Term 2.2} = \int_0^t d\tau \sum_{\substack{ij, mn \\ a}} \breve{S}_i^{(a)}(t)\, \breve{\rho}^{(s)}(0)\, \breve{S}_m^{(a)}(t - \tau) \left\langle \breve{R}_j^{(a)}(0)\breve{R}_n^{(a)}(\tau)\right\rangle$$

7.46 Derive the equation below (Equation 7.16.26)

$$\langle k| \left[\breve{S}_i^{(a)}(t), \breve{S}_j^{(a)}(t - \tau)\breve{\rho}^{(s)}(t)\right] |l\rangle = \sum_{nm} \langle m| \breve{\rho}^{(s)}(t) |n\rangle \left\{ \delta_{ln} \sum_r \langle k|S_i^{(a)}|r\rangle \langle r|S_j^{(a)}|m\rangle e^{-i\omega_{rm}t} + \right.$$

$$\left. - \langle n|S_i^{(a)}|l\rangle \langle k|S_j^{(a)}|m\rangle e^{-i\omega_{km}t} \right\} e^{i(\omega_{km} + \omega_{nl})t}$$

7.47 Find the differential equations for the Jaynes-Cummings' 2-level-atom raising and lowering operators following the procedure outlined in Section 7.16. Find the equation of motion for the number operator.

7.48 Consider a Homonic oscillator interacting with a reservoir where the raising and lowering operators satisfy

$$\dot{\tilde{S}}^- + i\omega_s \tilde{S}^- + \frac{\pi g |\beta|^2}{\hbar^2} \tilde{S}^-(t) = \tilde{\Gamma}(t)\tilde{S}^+ - i\omega_s \tilde{S}^+ + \frac{\pi g |\beta|^2}{\hbar^2} \tilde{S}^+(t) = \tilde{\Gamma}^+(t)$$

as in Section 7.17.

1. Find a solution to each equation and explain why the Γ can be neglected to lowest order.
2. Show $\tilde{N}(t) = e^{-2\gamma t}\tilde{N}(0)$.

7.49 Read how it might be possible to see through solid objects and report your findings.

* Harris, Electromagnetically Induced Transparency, *Physics Today*, July 1997, p. 36.

7.50 Read and report on M. Lax's account of Laser Noise in The Theory of Laser Noise, *SPIE* Vol. 1376, pp. 1–18 (1990).

7.51 Provide a brief summary on interesting new ideas that perhaps make SciFi-a reality. Can you find any other publications along these lines? Search the internet but accept only legitimate publications. How many "crack pot" schemes do you find?

* Jaekel and Reynaud, Quantum Fluctuations and Inertia, *Proceedings of NATO-ASI Conference on Electron Theory and Quantum Electrodynamics—100 years later*, Edirne, Turkey, Sept. 5–16, 1994.
* Alcubierre, The Warp Drive: Hyper-Fast Travel within General Relativity, *Class. Quantum Grav.*, 11, L73–L77 (1994).
* Coule, No Warp Drive, *Class. Quantum Grav.*, 2523 (1998).
* Search the internet on "Faster than Light Communications".

7.52 Find and report on the Unruh effect (i.e., Unruh-Davies effect). You will find some information in *Milonni's quantum vacuum book*.

7.53 Read and report on the Casimir force from the following sources. Look up any of their references pertinent to these effects. Use the citation indices to find newer publications.

* Read Sections 2.7 on the Casimir force in *Milonni's quantum vacuum book*.
* Hawton, One Photon Operators and the Role of Vacuum Fluctuations in the Casimir Force, *Phys. Rev.*, A50, 1057 (1994).
* Maclay, Fearn, Milonni, Of some theoretical significance: implications of Casimir effects, *Eur. J. Phys.*, 22 463 (2001).

7.19 Further Reading

The following list contains well-known references for the material presented in the chapter.

General

1. Heitler W., *The Quantum Theory of Radiation*, 3rd ed., Dover Publications, Mineola, NY (1984).
2. Allen L., Eberly J.H., *Optical Resonance and Two-Level Atoms*, Dover Publications, Mineola, NY (1987).

Liouville Equation

3. Milonni P.W., Eberly J.H., *Lasers,* Wiley-Interscience, New York (1988).
4. Yariv A., *Quantum Electronics*, 3rd ed., John Wiley & Sons, New York (1989).

Matter-Fields

5. Sakurai J.J., *Advanced Quantum Mechanics*, Addison-Wesley Publishing, Reading, MA (1967).
6. Milonni P.W., *The Quantum Vacuum, An Introduction to Quantum Electrodynamics*, Academic Press, Boston (1994).
7. Jackson J.D., *Classical Electrodynamics*, 2nd ed., John Wiley & Sons, New York (1975).
8. Rohrlich F., *Classical Charged Particles*, Addison-Wesley Publishing, Reading, MA (1965).
9. Jauch J.M., Rohrlich F., *The Theory of Photons and Electrons*, Springer-Verlag, New York (1976).

Reservoir Theory

10. Haken H., *Laser Theory*, Springer-Verlag, Berlin (1970).
11. Weissbluth M., *Photon-Atom Interactions*, Academic Press, Boston (1989).
12. Yokoyama H., Ujihara K., Eds, *Spontaneous Emission and Laser Oscillation in Microcavities*, CRC Press, Boca Raton, FL (1995).
13. Berman P.R., Ed., *Cavity Quantum Electrodynamics*, Academic Press, Boston (1994).

8

Semiconductor Emitters and Detectors

Modern technology uses a variety of semiconductor devices in a myriad of applications. This chapter focuses on optoelectronic devices capable of emitting and detecting light. It completes and summarizes the material by deriving the gain and rate equations for semiconductor lasers and detectors. The study proceeds along two paths. First, we start with Fermi's golden rule. The gain curves have sharp features and not the rounding expected from the spectral broadening mechanisms. The second method uses the results for the Liouville equation, for the independent atoms. We will show why the development applies to the semiconductor laser with "vertical transitions."

The two approaches share important common features. Fermi's golden rule requires matrix elements involving the electromagnetic interaction. The matrix elements use the energy basis set of wave functions. For semiconductors, the electron basis set consists of the Bloch functions. The Liouville equation implicitly incorporates the Bloch functions by considering only vertical transitions. Both approaches, Fermi's golden rule and the Liouville equation, require the density of states. Fermi's golden rule provides the transition rate from a single initial state to a range of final states. For the semiconductor, we must develop the reduced density of states. The reduced density of states applies to all semiconductor lasers including the quantum diode, well, wire and dot lasers.

The chapter shows how the parts of the emitter and detector fit together. It shows how to compute the gain required for lasing and the range of useful wavelengths for detectors. Reverse biasing a device can emphasize effects not normally important for the forward biased regions. Changing the bias applied to a homojunction device necessarily changes the internal electric field and the shape of the band edge. Forward biased devices tend to have flat bands at least for high levels of injection. Reversed biased quantum well devices do not have flat bands; the wells become triangular shape.

8.1 Effective Mass, Density of States and the Fermi Distribution

The density of states has a primary role in determining the gain and absorption for semiconductor lasers, light emitting diodes and detectors. This distribution of extended states versus energy determines the range of optical wavelengths involved in the transitions. The bulk materials, without nanostructure, produce a density of states with an energy profile of the form $E^{(n-2)/2}$ for n-dimensions. Nanostructure embedded within a bulk crystal produces sub-bands and a density of states that differs from that of the bulk. The present section discusses the basics of the density of states for the bulk material and shows how it depends on the effective mass.

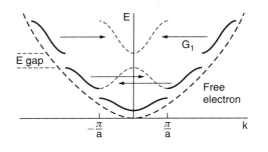

FIGURE 8.1.1
Comparing the dispersion curves for the free electron and the nearly-free electron. The symbol "*a*" represents the inter-atomic spacing so that $2\pi/a$ gives the width of the first Brillouin zone.

8.1.1   Effective Mass

The introductory material in Chapter 1 reviewed the formation of bands in bulk semiconductor material. A kinetic energy of a free electron can be plotted as a quadratic dispersion curve in terms of energy E versus momentum p or equivalently, angular frequency ω versus wave vector k (or any combination). An electron in a crystal interacts with a periodic potential. This interaction alters the topology of the dispersion curve from a single solid curve into one consisting of disjoint sections that form bands. For example, Figure 8.1.1 shows the structure of the dispersion curves in the extended zone (solid curved lines) and the reduced zone (dotted curved lines within the confines of the outer parabola representing the free electron dispersion curve).

An electron moving through a crystal experiences the periodic crystal potential due to the atomic charge distribution. The Schrodinger equation incorporates the crystal potential and the mass of the electron. The energy eigenfunctions consist of the Bloch wave functions as discussed in the next section. The crystal potential can be eliminated so long as the Schrodinger's equation uses the effective mass and the envelope part of the Bloch wave function. The effective mass describes the curvature of the band in which the electron (or hole) moves. If the band diagram has the form $E = p^2/2m_e$ where $p = \hbar k$, then the effective mass m_e must have the form

$$m_e^{-1} = \frac{\partial^2 E}{\partial p^2} \quad \text{or} \quad m_e^{-1} = \frac{1}{\hbar^2}\frac{\partial^2 E}{\partial k^2} \tag{8.1.1}$$

This equation shows that the curvature of the band determines the effective mass. As discussed in the companion volume on Solid State and Quantum Theory, the effective mass in 8.1.1 relates the force applied to an electron (not due to the periodic crystal potential) to its group velocity according to $dv/dt = F/m_e$.

The effective mass cannot be a scalar since the dispersion curves for two different directions do not necessarily have the same shape as shown in Figure 8.1.2 for bulk GaAs. The band shape depends on the direction of electron motion (i.e., E versus k). The parabolic band E versus k for the 1-D case can be generalized to

$$E = E_c + A(k_x - k_{ox})^2 + B(k_y - k_{oy})^2 + C(k_z - k_{oz})^2 \tag{8.1.2}$$

where the wave vector has the form $\vec{k} = k_x\tilde{x} + k_y\tilde{y} + k_z\tilde{z}$. The coordinates (k_{ox}, k_{oy}, k_{oz}) describe the band minimum (or maximum for the valence band) and the coefficients A, B, C describe the shape of the band along the corresponding direction. If the constants A, B, C differ then the parabola curvatures differ with direction and therefore

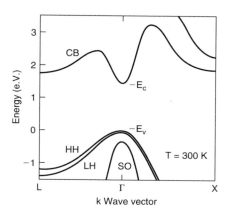

FIGURE 8.1.2
GaAs Band diagram for two different crystal directions. The bands are conduction band CB, heavy hole HH, light hole LH and split-off SO bands.

the effective mass must also differ with direction.

$$(m_x)^{-1} = \frac{1}{\hbar^2}\frac{\partial^2 E}{\partial k_x^2} = \frac{2A}{\hbar^2} \quad (m_y)^{-1} = \frac{1}{\hbar^2}\frac{\partial^2 E}{\partial k_y^2} = \frac{2B}{\hbar^2} \quad (m_z)^{-1} = \frac{1}{\hbar^2}\frac{\partial^2 E}{\partial k_z^2} = \frac{2C}{\hbar^2} \qquad (8.1.3)$$

The acceleration of an electron in the crystal due to noncrystalline forces (i.e., due to forces other than the periodic force caused by the periodic charge distribution).

$$a_x = (m_x)^{-1}F_x = \frac{2AF_x}{\hbar^2} \quad a_y = (m_y)^{-1}F_y = \frac{2BF_y}{\hbar^2} \quad a_z = (m_z)^{-1}F_z = \frac{2CF_z}{\hbar^2} \qquad (8.1.4)$$

An average effective mass often appears in formulas such as for the density of states. The average usually appears as a geometric average such as $\langle m \rangle = (m_x m_y m_z)^{1/3}$ (refer to the companion volume on the solid state).

Equation (8.1.4) indicate that the effective mass should be represented by a dyad

$$\vec{a} = \overset{\leftrightarrow}{m}{}^{-1} \cdot \vec{F} \qquad (8.1.5)$$

Although Equation (8.1.2) is useful for illustration, it does not describe the general band shapes depicted in Figure 8.1.2. At minimum, the coefficients must depend on the components of the wave vector. The companion volume shows the effective mass can be written as

$$\overset{\leftrightarrow}{m}{}^{-1} = \frac{1}{\hbar^2}\nabla_k \nabla_k E \qquad (8.1.6)$$

Example 8.1.1
Find the effective mass m_{ij} for the isotropic band

$$E = A\hbar^2 k^2 = A\hbar^2\left(k_x^2 + k_y^2 + k_y^2\right)$$

Solution: Using 7.8.12, namely $(m^{-1})_{ij} = 1/\hbar^2(\partial^2 E)/(\partial k_i \partial k_j)$ produces $(m^{-1})_{ij} = 2A\,\delta_{ij}$. Therefore the effective mass $m = 1/2A$ must be isotropic.

8.1.2 Introduction to Boundary Conditions

The previous topic treats the semiconductor bands as continuous. The energy $E_{\vec{k}}$ versus k dispersion curve represents the eigenvalues for the time-independent Schrodinger equation for the crystal. Boundary conditions over a finite region of space produce a discrete set of wave vectors $\{\vec{k}\}$ and hence a discrete set of energy eigenvalues $\{E_{\vec{k}}\}$. The energy bands consist of the collection of closely spaced points $\{(k, E_{\vec{k}})\}$ that represent the Bloch eigenfunctions. The envelope part of the Bloch wave function consists of plane waves with wavelength $\lambda = 2\pi/|\vec{k}|$ and angular frequency $\omega_k = E_k/\hbar$.

The Bloch wave functions, as eigenfunctions of the Schrodinger wave, must be normalized over some region of space. In general, either an electron can be confined to a finite region of space as for an atom or it can remain unconfined as for the case of a plane wave. In either case, the wave function must be normalized over a finite volume of space by requiring it to satisfy boundary conditions over a finite region of space. These boundary conditions place conditions on the particle wavelength and hence also on the wave vector. Finite regions of space produce discrete allowed wave vectors and therefore discrete energy values. Two types of boundary conditions, specifically fixed-endpoint and periodic, are typically applied to the electron.

The fixed-endpoint boundary conditions require the particle wave function to have a fixed value at the boundary of the finite region of space. Often the wave function must be zero at both the boundary and outside the finite region since the particle should not be found anywhere except inside the finite region. These fixed-endpoint boundary conditions produce sine and cosine standing waves for the energy eigenfunctions (i.e., the energy basis set). The wave vectors $\vec{k}$ have only positive components since negative values do not change the form of the basis function. Electrons confined to nanostructure can use this type of boundary condition.

Travelling waves usually require periodic boundary conditions whereby the wave function must repeat every macroscopic distance L. The plane wave functions comprising the basis set must be normalized over this same distance L although the wave extends across larger distances. In this case, the wave vectors $\vec{k}$ have both positive and negative components in order to account for motion along the positive and negative directions. However, the finite size of L leads to discrete allowed wavelengths, wave vectors, and therefore also energy. Very large regions L produce closely space wavelengths, wave vectors and energy and therefore approximate a continuum. It is customary to use the length L of a real crystal as the repetition length for the boundary conditions. In such a case, the size of the crystal sets a minimum spacing for allowed k.

Once we know the allowed energies for a finite system, we can count the number of allowed states. Figure 8.1.3 shows the descrete states for the conduction band. We can count the number of states ΔN in the energy range ΔE to find the density of states $g(E) \sim \Delta N/\Delta E$. The figure shows how the number of states along the energy axis must be related to the number along the k-axis. In fact, the total number of states in the range ΔE comes from the two regions marked Δk. For 2-D systems, the Δk region corresponds to an annular region between two circles.

8.1.3 The Fixed Endpoint Boundary Conditions

The fixed-endpoint boundary conditions require a wave to be zero at the edges of the bounding region. The fundamental modes fitting the region appear as sine and cosine waves as shown in Figure 8.1.4. The wavelengths cannot be larger than

$$\lambda_1 = 2L$$

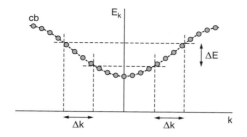

FIGURE 8.1.3
The density of energy states must be related to the density of k-states.

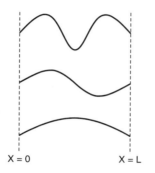

FIGURE 8.1.4
The fixed end-point boundary conditions.

In fact, the wave must exactly fit into the distance L according to the relation

$$\lambda = \frac{2L}{1}, \frac{2L}{2}, \frac{2L}{3}, \ldots, \frac{2L}{n}, \ldots$$

Therefore, the allowed wave vectors must be

$$k_n = \frac{2\pi}{(2L/n)} = \frac{n\pi}{L} \qquad n = 1, 2, 3, \ldots \tag{8.1.7}$$

The finite region of space $0 < x < L$ produces sinusoidal functions

$$B_s = \sin k_n x \tag{8.1.8}$$

Three-dimensional problems require three-dimensional wave vectors. For a cube, with sides of length L, the allowed wave vectors can be written as

$$\vec{k} = \frac{n_x \pi}{L} \tilde{x} + \frac{n_y \pi}{L} \tilde{y} + \frac{n_z \pi}{L} \tilde{z} \tag{8.1.9}$$

where $n_x, n_y, n_z = 0, +1, +2, \ldots$ for plane waves. As we will see, traveling waves most naturally use the *periodic* boundary conditions since then the waves don't need to be zero at the boundaries.

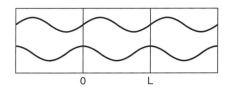

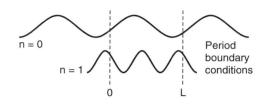

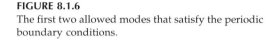

FIGURE 8.1.5
Repeating the physical crystal every distance L.

FIGURE 8.1.6
The first two allowed modes that satisfy the periodic boundary conditions.

8.1.4 The Periodic Boundary Condition

Periodic boundary conditions describe macroscopically sized real crystals. The electron wavefunction must repeat itself every distance L, which usually matches the physical size of the crystal. For free space or physically large media, the length L can be increased without bound. We have primary interest in finite physical crystals. In this case, the waves can be imagined to have infinite extent by imagining copies of the physical crystal placed next to each other as shown in Figure 8.1.5.

Two allowed modes with the longest wavelengths appear in Figure 8.1.6. The allowed wavelengths must be given by

$$\lambda_n = \frac{L}{n} \qquad n = 0, \pm 1, \ldots$$

and the allowed 1-D wave vectors must be

$$k_n = \frac{2\pi}{\lambda_n} = \frac{2\pi n}{L} \tag{8.1.10a}$$

The value of n can assume any nonzero integer value in the range $(-\infty, \infty)$ for the extended-zone band diagram.

The reduced-zone band diagrams use only the First Brillouin Zone (FBZ). If we assume an even number of atoms N spaced apart by lattice constant "a" in the 1-D crystal then we can write $L = Na$ and

$$k_n = \frac{2\pi}{\lambda_n} = \frac{2\pi}{a}\frac{n}{N} \tag{8.1.10b}$$

The longest wavelength corresponds to $L = Na$ so that the closest spacing of k-values must be $\Delta k = 2\pi/L = 2\pi/Na$. The wavelengths $\lambda_{\min} = 2a$ correspond to k-vectors $k_{\text{FBZ}} = \pi/a$ at the edge of the First Brillouin Zone (FBZ). This smallest wavelength sets the maximum integer n in Equation (8.1.10b).

$$k_n = \frac{2\pi}{\lambda_n} = \frac{2\pi}{a}\frac{n}{N} \qquad n = \pm 1, \pm 2, \ldots, \pm N/2 \tag{8.1.10c}$$

We see that each band (in the reduced band scheme) must have N states. For crystals with an atomic basis, N represents the number of unit cells.

Three-dimensional crystals use similar periodic boundary conditions

$$\vec{k} = \frac{2\pi n_x}{L_x}\tilde{x} + \frac{2\pi n_y}{L_y}\tilde{y} + \frac{2\pi n_z}{L_z}\tilde{z} \quad n_x, n_y, n_z = 0, \pm 1, \dots \quad (8.1.11a)$$

where, L_x, L_y, L_z represent the lengths of the three sides of the normalization volume. The lengths refer to the size of the normalization volume and not necessarily to the crystal. For convenience, we can assume that all of the sides have the same length $L = L_x = L_y = L_z$. Consequently, the wave function must be normalized to the volume of a cube. For N atoms along each edge of the cube (N^3 total atoms), the length L must be $L = Na$ and the allowed k-vectors become

$$\vec{k} = \frac{2\pi n_x}{aN}\tilde{x} + \frac{2\pi n_y}{aN}\tilde{y} + \frac{2\pi n_z}{aN}\tilde{z} \quad n_x, n_y, n_z = 0, \pm 1, \dots \quad (8.1.11b)$$

The size of the crystal sets the smallest spacing of the components of the wave vectors

$$\Delta k_x = \Delta k_y = \Delta k_z = \frac{2\pi}{L} = \frac{2\pi}{aN}$$

The upper limit corresponds to a wave vector k^{FBZ} at the edge of the FBZ for which the wave does not propagate. Strong reflections occur for the smallest wavelength $\lambda = 2a$ so that

$$k_x^{\text{FBZ}} = k_y^{\text{FBZ}} = k_z^{\text{FBZ}} = \frac{\pi}{a}$$

Therefore the allowed wave vectors must be

$$\vec{k} = \frac{2\pi n_x}{L}\tilde{x} + \frac{2\pi n_y}{L}\tilde{y} + \frac{2\pi n_z}{L}\tilde{z} = \frac{2\pi}{a}\frac{n_x}{N}\tilde{x} + \frac{2\pi}{a}\frac{n_y}{N}\tilde{y} + \frac{2\pi}{a}\frac{n_z}{N}\tilde{z}$$

$$(8.1.11c)$$

$$n_x, n_y, n_z = 0, \pm 1, \dots, \pm N/2$$

Again we see each k-axis has N states corresponding to the number of atoms along the axis. The number of states for the entire 3-D band must be N^3 corresponding to the total number of atoms within the solid. The total number of atoms will be very large for any physically sized crystal (on the order of Avagadro's number). Notice that the k-vectors can have positive or negative values depending on the direction of wave propagation unlike for the fix-endpoint boundary conditions.

The plane waves corresponding to these macroscopic boundary conditions have the form

$$B_{1\text{-D}} = \left\{ \frac{e^{ik_n x}}{\sqrt{L}} \right\} \quad \text{or} \quad B_{3\text{-D}} = \left\{ \frac{e^{i\vec{k}\cdot\vec{r}}}{\sqrt{V}} \right\} \quad (8.1.12)$$

where $V = L^3$ for the 3-D case. These wave functions correspond to the envelope part of the Bloch wave function. The next topics show how the macroscopic boundary conditions determine the density of states.

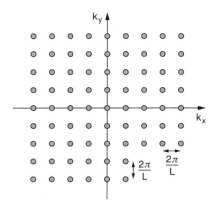

FIGURE 8.1.7
The allowed values of $\vec{k}$ as determined by periodic boundary conditions.

8.1.5   The Density of k-States

The "density of k-states" measures the number of possible modes in a given region of k-space. Consider a *two-dimensional bulk crystal*. Figure 8.1.7 shows a 2-D region of k-space with a point for each k-vector

$$\vec{k} = \frac{2\pi m}{L}\tilde{x} + \frac{2\pi n}{L}\tilde{y} \quad m,n = 0,\ \pm 1,\ \dots$$

from periodic boundary conditions. Consider the horizontal direction for a moment. The distance between adjacent points can be calculated as

$$\frac{2\pi(m+1)}{L} - \frac{2\pi m}{L} = \frac{2\pi}{L}$$

Therefore, each elemental area of k-space

$$\frac{2\pi}{L} \cdot \frac{2\pi}{L} = \left(\frac{2\pi}{L}\right)^2$$

corresponds to precisely one mode. The number of modes per unit area of $\vec{k}$-space must then be given by

$$g_{\vec{k}}^{(2D)} = \frac{1}{(2\pi/L)^2} = \frac{L^2}{4\pi^2} = \frac{A_{\text{xal}}}{4\pi^2} \tag{8.1.13}$$

where A_{xal} represents the area of the crystal. Note the use of the "vector k" as opposed to the "scalar k" as a subscript on g.

In general, n-dimensions produces a k-spaced density of states (DOS) of

$$g_{\vec{k}}^{(n\text{-}D)} = \frac{1}{(2\pi/L)^n} = \left(\frac{L}{2\pi}\right)^n \tag{8.1.14}$$

where bulk crystals can be 3-D, 2-D, or a line of atoms for 1-D.

8.1.6 The Electron Density of Energy States for a 2-D Crystal

Semiconductor-based devices require the density of states. For electronic devices, recall the combination of the Fermi-Dirac distribution $F(E)$ with the density of energy states $g(E)$ leads to the number of electrons per crystal volume in a given energy range according to

$$n = \int dE \, g(E) \, F(E)$$

This topic finds the density of energy states using a graphical technique. The interested reader should consult the companion volume on the solid state and introduction to quantum theory for more detail.

We need to clearly distinguish the bulk cases from those encountered with reduced dimensional systems such as quantum wells, wires, and dots. These latter systems still have 3-D arrangements of atoms. However, the 3-D pattern of atoms (heterostructure) produces potentials that tend to confine electrons to wells. In this topic, we discuss 2-D and 3-D arrays of atoms without regard to confining the electron to smaller wells. For simplicity, we apply the procedure to portions of the band having a parabolic shape. The density of states for the entire band requires the full dispersion curve $E = E(k)$ and not just the portion at the top or bottom of the band.

For simplicity of drawing figures, consider the 2-D case for the electronic density of energy states. We want to calculate the number N of energy states per unit energy, specifically $g(E) = dN/dE$, whereas the previous topic calculates the density of k-states with units of k-states per unit k-region. The calculation requires the energy versus wave vector k in terms of the effective mass m_e

$$E = \frac{\hbar^2 k^2}{2m_e} \quad k^2 = k_x^2 + k_y^2 \tag{8.1.15}$$

where for convenience, we start the energy scale at the bottom of the conduction band so that $E_c = 0$. For constant effective mass m_e, Equation (8.1.15) represents the conduction band only near the minimum. The number N can be related to E through Equation (8.1.15) provided we can find the number of states up to the value $k = |\vec{k}|$.

To calculate the number of states up to the value k for the 2-D case, first plot the allowed k-states and then use k as the radius of a circle in k-space (see Figure 8.1.8). The number of states up to radius k can be calculated as

$$N(k) = \frac{\# \text{ States}}{k\text{-area}} \quad k\text{-area} \tag{8.1.16a}$$

or

$$N(k) = g_k^{2D} \pi k^2 = \left(\frac{L}{2\pi}\right)^2 \pi k^2 = \frac{A_{\text{xal}}}{(2\pi)^2} \pi k^2 \tag{8.1.16b}$$

where A_{xal} represents the area of the 2-D crystal. The total number of states per unit energy then becomes

$$g^{2D}(E) = \frac{dN(k)}{dE} = \frac{dk}{dE}\frac{dN}{dk} = \frac{1}{dE/dk}\frac{dN}{dk} \tag{8.1.16c}$$

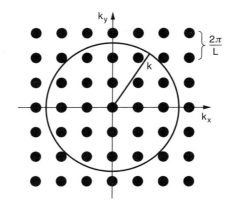

FIGURE 8.1.8
Number of modes N in circle of radius k.

Dividing out the area A_{xal} and substituting Equations (8.1.16b) and (8.1.15) provides

$$g_E^{(2D)} = \frac{1}{2\pi} \frac{m_e}{\hbar^2} \quad \text{(no spin)} \tag{8.1.17a}$$

Usually a factor of "2" should be included for the degenerate electron spin

$$g_E^{(2D)} = \frac{1}{\pi} \frac{m_e}{\hbar^2} \quad \text{(spin)} \tag{8.1.17b}$$

Notice that the density of energy states for the 2-D bulk case remains constant above the bottom of the band. In the case of a band increasing upward from a vertex at E_c, the formula must include the step function θ.

$$g_E^{(2D)} = \frac{1}{\pi} \frac{m_e}{\hbar^2} \theta(E - E_c) \quad \text{(spin)} \tag{8.1.17c}$$

The density of state must be zero for energy E smaller than E_c and constant for energy larger than E_c.

In general, an n-dimensional bulk crystal has the density of energy states (*including spin*) given by

$$g_E^{(3D)} = \frac{\sqrt{2} m_e^{3/2}}{\pi^2 \hbar^3} \sqrt{E - E_c} \theta(E - E_c) \tag{8.1.18a}$$

$$g_E^{(2D)} = \frac{1}{\pi} \frac{m_e}{\hbar^2} \theta(E - E_c) \tag{8.1.18b}$$

$$g_E^{(1D)} = \frac{\sqrt{2}}{\pi\hbar} \frac{\sqrt{m_e}}{\sqrt{E - E_c}} \theta(E - E_c) \tag{8.1.18c}$$

where the step function θ has a value of one for $E \geq E_c$ and zero otherwise, and the equations have units of # states/xal-vol/energy, # states/xal-area/energy, # states/xal-length/energy, respectively.

As an example, the 3-D density of energy states can be plotted next to the band diagram as illustrated in Figure 8.1.9. Both the conduction band and heavy-hole valence band

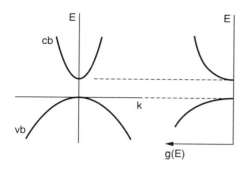

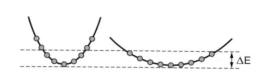

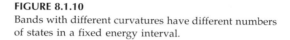

FIGURE 8.1.9

The conduction and valence band both have a density of states function.

FIGURE 8.1.10

Bands with different curvatures have different numbers of states in a fixed energy interval.

produce a density of states that increase as $\sqrt{E}$. The effective mass controls the shape of the density of states as illustrated in Figure 8.1.10. The two bands have different curvatures. The boundary conditions produce equally spaced states along the horizontal k-axis. Let ΔE represent a small fixed energy interval. The curvature of the bands produces two different numbers of states within the energy interval. The band with the larger curvature and therefore smaller effective mass has fewer states within the energy interval. The flatter band with the larger effective mass has more states within the interval.

8.1.7 Overlapping Bands

Gallium Arsenide has overlapping Heavy Hole (HH) and Light Hole (LH) valence bands as shown in Figure 8.1.11. We will find overlapping sub-bands for the reduced dimensional structures such as quantum wells. Each band must have states corresponding to the allowed discrete wave vectors k. Therefore the number of states within the energy range ΔE must include states from both the HH and LH bands.

We now discuss the method for calculating the density of states for overlapping bands. The calculation for the quantum well sub-bands will be very similar. For simplicity, let $-E \rightarrow E$ in Figure 8.1.11 and consider the two overlapping bands now with positive curvature as shown in Figure 8.1.12. We demonstrate that the density of states must

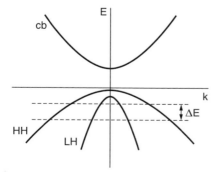

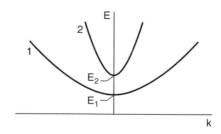

FIGURE 8.1.11

Light and heavy hole valence bands.

FIGURE 8.1.12

Two over-lapping 3-D bands (inverted for convenience).

be given by

$$g_E^{(3D)}(E) = \frac{m_1^{3/2}}{\sqrt{2}\pi^2\hbar^2}\sqrt{E-E_1}\,\Theta(E-E_1) + \frac{m_2^{3/2}}{\sqrt{2}\pi^2\hbar^2}\sqrt{E-E_2}\,\Theta(E-E_2) \qquad (8.1.19)$$

The figure shows that the density of states must be zero below E_1. As E increases, we eventually encounter band #1 starting at energy E_1 where the states start. The density of states (3-D crystal) must therefore increase as $\sqrt{E-E_1}$ according to Equation (8.1.18). At energy E_2, the number of states in band 2 must be included. The density of states in band 2 increases as $\sqrt{E-E_2}$ again according to Equation (8.1.18). To find the total number of states for energy larger than E_2, we must add the states from bands 1 and 2. Therefore, we find Equation (8.1.19).

8.1.8 Density of States from Fixed-Endpoint Boundary Conditions

The topics in the present section find the density of states using the periodic boundary conditions. The length L in Figure 8.1.6 appears to be rather arbitrary. For the fixed-endpoint boundary conditions, the length L matches the physical length of the crystal. We make the same requirement for the length L in the periodic boundary conditions as illustrated in Figure 8.1.5. However, the fixed-endpoint conditions might seem to give the more accurate density of states since electrons must surely be confined to the crystal and therefore cannot be standing wave that repeat every length L. Let's examine how the choice of the type of boundary conditions affects the density of states. We will find that both types give precisely the same density of energy state function.

The following table compares the wavelength, wave vectors and minimum wave vector spacing using periodic and fixed-endpoint boundary conditions for a 2-D crystal (for example).

Periodic BCs	Fixed-Endpoint BCs
$\lambda_x = L/m$ $\lambda_y = L/n$	$\lambda_x = 2L/m$ $\lambda_y = 2L/n$
$k_x = 2\pi m/L$ $k_y = 2\pi n/L$	$k_x = \pi m/L$ $k_y = \pi n/L$
$\Delta k_x = 2\pi/L$ $\Delta k_y = 2\pi/L$	$\Delta k_x = \pi m/L$ $\Delta k_y = \pi n/L$
Travelling waves	Standing waves
m, n can be positive and negative	m, n must be nonnegative

The periodic boundary conditions produce twice the spacing between allowed k values as does the fixed-endpoint ones. Figure 8.1.13 shows the Periodic Boundary Conditions

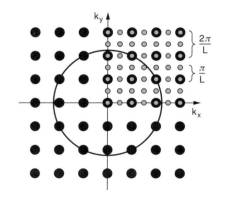

FIGURE 8.1.13
Large dots represent allowed k for periodic BC while the small dots represent the fixed-endpoint BCs.

(PBC) produce density of k-states 4 times smaller than that from the Fixed-Endpoint Boundary Conditions (FEBC)

$$g_{\vec{k}(pbc)}^{(2\text{-}D)} = \frac{g_{\vec{k}(febc)}^{(2\text{-}D)}}{4} \tag{8.1.20a}$$

Next, we see that constant energy circle covers 4 times the area for the PBCs as it does for the FEBCs.

$$A_{pbc} = 4A_{febc} \tag{8.1.20b}$$

The density of energy states can then be calculated from the product of Equations (8.1.20). We find the same result for either set of boundary condtions.

$$g(E) = g_{\vec{k}(pbc)}^{(2\text{-}D)} A_{(pbc)} = \frac{g_{\vec{k}(febc)}^{(2\text{-}D)}}{4} 4A_{(febc)} = g_{\vec{k}(febc)}^{(2\text{-}D)} A_{(febc)} \tag{8.1.20c}$$

8.1.9 Changing Summations to Integrals

The density-of-states can be used to convert summations to integrals. Suppose we start with a summation of coefficients $C_{\vec{k}}$ of the form

$$S = \sum_{\vec{k}} C_{\vec{k}}$$

The index $\vec{k}$ on the summation means to sum over allowed values of k_x, k_y, k_z; i.e., think of the two-dimensional plot in the previous topics and imagine that $C_{\vec{k}}$ has a different value at each point on the plot. For one-dimension, a plot of "C_k vs. k" might appear as in Figure 8.1.14. Suppose the allowed values of "k" are close to one another. Let Δk_i be a small interval along the k-axis but assume that it contains 4 of the "k" points. Let k_i be the center of each of these intervals. The figure shows that

$$S = (C_{1.00} + C_{1.01} + C_{1.02} + C_{1.03}) + (C_{1.04} + C_{1.05} + C_{1.06} + C_{1.07}) + \cdots$$

The sum can be recast into

$$S = 4C_{1.00} + 4C_{1.04} + \cdots = \sum [g(k)\Delta k]\, C_k \equiv \sum C(k)\, g(k)\Delta k \cong \int dk\, C(k)\, g(k)$$

where, for the figure, $\Delta k = 0.04$ and $g(k) = 4/0.04 = 100$ (states per k-length).

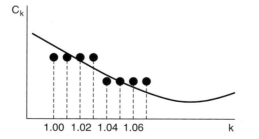

FIGURE 8.1.14
Example of closely spaced modes.

Alternatively, suppose a slowly varying function $f(x)$ is defined at the points in the set $\{x_1, x_2, \ldots\}$ where the points x_i are equally spaced and separated by the common distance Δx. The summation can be rewritten as

$$\sum_i f(x_i) = \sum_i \frac{1}{\Delta x_i} f(x_i)\, \Delta x_i$$

We recognize the quantity $1/\Delta x$ as the density of states; that is, $g = 1/\Delta x$. Recognizing the second summation as an integral for sufficiently small Δx, the summation can be written as

$$\sum_i f(x_i) \cong \int dx\, g(x) f(x) \qquad (8.1.21)$$

The last expression generalizes to a 3-D case most commonly applied to the wave vectors discussed in the preceding topics.

$$\sum_{\vec{k}} f(\vec{k}) \to \int d^3k\, g(\vec{k}) f(\vec{k}) = \frac{V}{(2\pi)^3} \int d^3k\, f(\vec{k}) \qquad (8.1.22)$$

where V represents the normalization volume coming from periodic boundary conditions. We essentially use this last integral when we find the total number of discrete states within a sphere or circle.

8.1.10   A Brief Review of the Fermi-Dirac Distribution

At thermal equilibrium the number of electrons occupying a conduction band state or a valence band state can be written as

$$F_e = \frac{1}{1 + e^{(E - E_f)/kT}} \qquad (8.1.23a)$$

where E_f represents the Fermi level. The average number of electrons per state can range between 0 and 1. Equation (8.1.23) can be interpreted as a probability. At $E = E_f$, the probability has the value of 0.5. The probability of a hole occupying a state necessarily has the form

$$F_h(E) = 1 - F_e = \frac{1}{1 + e^{-(E - E_f)/kT}} \qquad (8.1.23b)$$

If the states in either case have energy at least several kT away from the Fermi level then these two equations reduce to the Boltzman distribution for electrons in the conduction band and holes in the valence band, respectively.

$$F_e = e^{-(E - E_f)/kT} \qquad F_h = e^{+(E - E_f)/kT} \qquad (8.1.23c)$$

These do not apply to electrons in the valence band or holes in the conduction bands since otherwise the Fermi functions would be larger than 1 for sufficiently large energy.

The number of electrons (per volume) over a range of energy must be given as follows

$$n = \sum_{energy} \#e = \sum_{energy} \frac{\#e}{State} \frac{\#States}{Energy} \rightarrow \int dE\, F_e(E)\, g(E) \qquad (8.1.24)$$

As discussed in solid state and device physics books, the doping level determines the placement of E_f. The law of mass action $np = n_i^2$ holds regardless of the doping level where n_i refers to the number of electrons at thermal equilibrium without doping ($n_i = p_i$).

8.1.11   The Quasi-Fermi Levels

Thermal equilibrium requires the number of electrons and holes to satisfy the Fermi-Dirac distribution. There exists only one Fermi level E_f as shown on the left side of Figure 8.1.15. Nonthermal generation of carriers (such as photogeneration or injection) produces more carriers than should be present for temperature T as dictated by the Fermi-Dirac distribution. Therefore, the distribution does not obey the Fermi-Dirac distribution and the position of the Fermi level E_f cannot simultaneously describe the number of free electrons and holes. Instead, the Fermi level splits into two quasi-Fermi levels E_{fc} and E_{fv} for the conduction and valence electrons, respectively.

The probability of an electron occupying a state in the conduction and valence bands, respectively, becomes

$$F_c(E) = \frac{1}{1 + \exp(E - E_{fc})/kT} \qquad (8.1.25a)$$

$$F_v(E) = \frac{1}{1 + \exp(E - E_{fv})/kT} \qquad (8.1.25b)$$

The probability of a valence or conduction band state being empty must be $1 - F_v$ and $1 - F_c$, respectively.

The quasi-Fermi functions provide the number of electrons in the conduction and valence bands. The number of conduction electrons (per volume), for example, can be written as

$$n = \int_{E_g}^{\infty} dE\, g_c(E) F_c(E) \qquad (8.1.26)$$

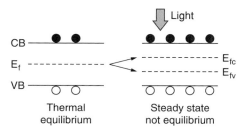

FIGURE 8.1.15
Nonthermal generation of carriers splits a single Fermi level E_f into two quasi-Fermi levels E_{fc} and E_{fv}. Both cases are maintained at the same temperature T.

The quasi-Fermi *level* fixes the number of electrons as will now be demonstrated. This is important for lasers since the separation of the quasi-Fermi levels must generally be larger than the band gap. In the parabolic band approximation for a 3-D bulk crystal, the number of electrons must be

$$n = \int_{E_g}^{\infty} dE\, C\, m_e^{3/2}\, \frac{\sqrt{E - E_g}}{1 + e^{(E - E_{fc})/kT}} \tag{8.1.27}$$

where C denotes constants associated with the 3-D bulk density of states $C = \sqrt{2}/(\pi^2 \hbar^3)$. Setting

$$\xi = \frac{E - E_g}{kT} \qquad d = \frac{E_{fc} - E_g}{kT} \tag{8.1.28}$$

where "d" signifies the difference in energy between the quasi-Fermi level E_{fc} and the band gap energy E_g. Equation (8.1.27) for the total number of electrons in the conduction band can be rearranged into either of two equivalent forms

$$n = C(kT)^{3/2} m_e^{3/2} F(d) \tag{8.1.29a}$$

$$n = C(kT)^{3/2} (m_e d)^{3/2} F_d(d) \tag{8.1.29b}$$

where the functions F and F_d are

$$F(d) = \int_0^{\infty} d\xi \frac{\sqrt{\xi}}{1 + e^{\xi - d}} \qquad F_d = \int_0^{\infty} d\xi \frac{\sqrt{\xi/d^3}}{1 + e^{\xi - d}} \tag{8.1.29c}$$

These equations differ only by the placement of the energy separation d. Equation (8.1.29b) incorporates the energy separation d with the effective mass whereas Equation (81.29a) includes it in the function F_d. Figure 8.1.16 plots both functions. In particular, notice F_d approaches 0.7 for large energy separation $d > 2$ or equivalently $E_{fc} > E_g + 2kT$. Therefore, the number of electrons can be characterized by $m_e d$ in Equation (8.1.29b) when the quasi-Fermi level is more than several kT beyond the band edge. The number of

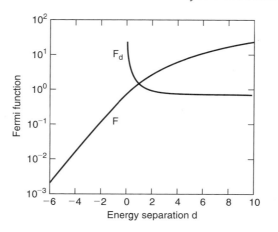

FIGURE 8.1.16
A plot of the Fermi F and F_d.

electrons becomes approximately

$$n \cong 0.7C(kT)^{3/2}(m_e d)^{3/2} \cong \frac{(kT)^{3/2}}{\pi^2\hbar^3}(m_e d)^{3/2} \qquad (8.1.30)$$

where m_e represents the effective mass of the electron and $d = (E_{fc} - E_g)/kT$.

8.2 The Bloch Wave Function

The free electron does not interact with matter or fields. The Hamiltonian produces plane wave eigenfunctions and the collection of energy eigenvalues forms a parabolic dispersion curve. The nearly-free electron weakly interacts with the periodic potential in a crystal. The Schrodinger equation includes the periodic potential and produces Bloch energy eigenfunctions rather than the pure plane waves for the free electron. The periodic potential modifies the energy eigenvalues; the dispersion curve becomes nonparabolic and develops band gaps. In either case, the macroscopic boundary conditions lead to closely-spaced discrete states. The envelope portion of the Bloch wave function has the form of a plane wave. The Hamiltonian for the nearly free electron can incorporate an externally applied potential V_E in addition to the kinetic energy and the periodic lattice potential V_L. The external potential describes situations such as the square well or the *p-n* junction.

The Schrodinger equation for the electron in a crystal can be simplified by making the effective mass approximation. In this case, the lattice potential can be removed from the Hamiltonian and the Bloch wave function can be reduced to the plane wave by using the effective mass rather than the free mass. The effective mass approximation essentially allows us to use Newton's laws without including the forces exerted by the crystal periodic potential. The curvature of the dispersion curve determines the effective mass.

8.2.1 Free Electron Model

The *free electron* model treats the motion of the electron in a crystal as if it were free of the periodic potential due to the atomic cores on the lattice. Figure 8.2.1 shows the electrostatic potential energy of the electron in the neighborhood of the charged cores for a 1-D crystal (for example). The periodic potential appears in the Schrodinger wave equation as $V_L(x)$.

$$-\frac{\hbar^2}{2m}\nabla^2\psi(x) + V_L(x)\psi(x) = i\hbar\frac{\partial}{\partial t}\psi(x) \qquad (8.2.1)$$

The potential $V_L(x)$ has the periodicity of the lattice. For sufficiently large total energy $E \gg V_L(x)$, we might consider the periodic potential energy $V_L(x)$ to be negligible.

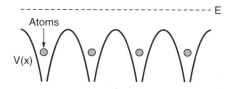

FIGURE 8.2.1
Periodic potential $V(x)$ for a 1-D monatomic crystal.

In this case, we should drop the potential energy term from the Schrodinger wave equation.

$$-\frac{\hbar^2}{2m}\nabla^2\psi(x) = i\hbar\frac{\partial}{\partial t}\psi(x)$$ (8.2.2)

$$\psi \sim e^{i\vec{k}\cdot\vec{r}-i\omega t}$$ (8.2.3)

Substituting Equation (8.2.3) in Equation (8.2.2) produces

$$\frac{\hbar^2 k^2}{2m} = \hbar\omega_k = E$$ (8.2.4)

The energy E represents the kinetic energy of the particle and provides the quadratic dispersion curve for the free particle as shown in Figure 8.2.2.

The periodic boundary conditions allow only certain wave vectors. The periodic boundary conditions require the wave to repeat itself every macroscopic distance L. For 1-D motion, the plane waves can only have the wave vectors $k_n = 2\pi n/L$ where n denotes an integer. The dispersion curve must be augmented with information on the allowed wave vectors. Figure 8.2.2 shows the allowed states corresponding to allowed plane waves. Each allowed k has a corresponding allowed energy E_k. For very large macroscopic length L, the allowed wave vectors will essentially form a continuum.

In addition to quantizing the wave vector and energy, the boundary conditions must be used to normalize the wave function in Equation (8.2.3). The finite interval of integration L produces a finite amplitude for the wave (as opposed to an infinitesimal one as $L \to \infty$). The plane waves can be normalized to describe either one electron per volume $1/V$ or to describe N electrons per volume N/V. That is, for periodic boundary conditions, the plane wave can be normalized to $\sqrt{N/V}$ or $1/\sqrt{V}$ for 3-D.

8.2.2   The Nearly Free Electron Model

The *nearly free* electron model describes electrons moving through a periodic potential. The electron although free to move, experiences the periodic potential of the crystal as shown in Figure 8.2.1 so that the periodic potential $V_L(x)$ must be included in the Schrodinger equation. The periodic potential gives rise to band structure (Figure 8.2.3) and determines the form of the wave function. Even a small infinitesimal periodic

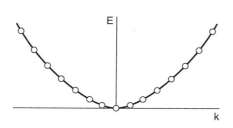

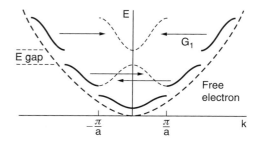

FIGURE 8.2.2
The dispersion curve for the QM free electron. The circles represent the wave vectors and corresponding energy allowed by the periodic boundary conditions.

FIGURE 8.2.3
Comparing the dispersion curves for the free electron and the nearly free electron. The symbol "a" represents the inter-atomic spacing so that $2\pi/a$ gives the width of the first Brillouin zone.

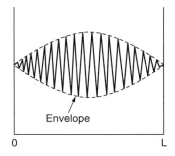

FIGURE 8.2.4
A Bloch wave-function for an electron in a semiconductor infinitely deep well. The long wavelength satisfies the macroscopic boundary conditions at $x = 0$, L. The small wavelengths provide a periodicity matching that of the crystal.

potential radically alters the topology of the dispersion curve by opening small gaps. However, for small periodic potential, the gaps become smaller and the dispersion curves for the free and nearly free electron models coincide.

As just mentioned, the allowed values of the wave vector $\vec{k}$ come from the macroscopic boundary conditions, which usually span a distance on the order of $L \sim 100$ angstroms or more. Therefore, the spacing between allowed wave vectors must be on the order of $\Delta k \sim 2\pi/L \sim 0.06$. The reciprocal lattice vectors G provide important markers for the band diagram. The set of reciprocal lattice vectors $G = 2n\pi/a$ (for 1-D) come from the inter-atomic spacing "a." For a lattice constant on the order of $a \sim 2$ angstroms, we find $\Delta G \sim 2\pi/a \sim 3$. The first value of G denotes the First Brillouin Zone (FBZ) for k-space, which has the width $2\pi/a$ (see Figure 8.2.3 for example). The wave vectors G have spacing much larger than the wave vectors k.

The band gaps in the dispersion curves occur near the Brillouin zone edges defined by the reciprocal lattice vectors. Near the zone edges, the dispersion curve has zero slope and the electron must have negligible group velocity there. We can understand this behavior as follows. Electron waves propagating through the periodic structure experience reflections at the atomic cores (i.e., the periodic potential). Those electrons having wavelengths approximately twice the inter-atomic spacing experience strong reflections that prevent forward motion of the electron wavefunction. These effects occur for wave vectors near the Brillouin zone edges.

The energy eigenfunctions form a basis for the Hilbert space so that a general solution to the Schrodinger wave equation can be represented as a time-dependent sum over these functions. The eigenfunctions have the form of a plane wave but with some modification. The Bloch wave functions resemble plane waves except they consist of a type of carrier with a modulation as suggested by Figure 8.2.4. For the electron wave function, the carrier portion of the wave has the periodicity of the lattice. The envelope function solves the free-electron Schrodinger wave equation so long as it incorporates the effective electron mass rather than the free electron mass, and neglects the periodic potential.

8.2.3   Introduction to the Bloch Wave Function

Schrodinger's equation for a single electron in the periodic potential $V_L(\vec{r})$ can be written as

$$\left[-\frac{\hbar^2}{2m} \nabla^2 + V_L + V_E \right] \Psi = i\hbar \frac{\partial}{\partial t} \Psi \tag{8.2.5}$$

where "m" represents the *actual* mass of the electron (not the effective mass). The external potential V_E denotes any potential in addition to the periodic crystal potential such as for the quantum well at heterostructure interfaces and in regions of band bending for *p-n* junctions. In order to focus on the effects of the crystal potential, set $V_E = 0$. The potential energy V_L has the periodicity of the lattice when, for $\vec{R}$ a direct lattice vector (i.e. points from one lattice point to another), the potential has the property that

$$V_L(\vec{r} + \vec{R}) = V_L(\vec{r}) \tag{8.2.6}$$

The time-independent Schrodinger equation can be found by setting $\Psi(\vec{r}, t) = \phi(\vec{r})T(t)$ in Equation (8.2.5).

$$\left[-\frac{\hbar^2}{2m} \nabla^2 + V_L \right] \phi = E \phi \tag{8.2.7}$$

As usual, we want to find the eigenfunctions and eigenvalues for the time independent Schrodinger equation in Equation (8.2.7). The eigenvalues E depend on k and produce the dispersion curves (i.e., band diagrams with gaps). The eigenfunction of Equation (8.2.7) form a basis set—the Bloch wavefunctions. The general solution to the time-dependent Schrodinger wave equation consists of the time-dependent superposition of these energy eigenfunctions.

Each energy state E_k in each band corresponds to a basis state. The allowed k values come from the periodic boundary conditions. Aside from the bandgaps caused by the periodic lattice potential, the boundary conditions impose quantization conditions on the wave vectors and hence the energy; therefore, the system supports only certain wave vectors k and energies E_k.

For simplicity, consider a two-band model as shown in the "zoomed-in" view of Figure 8.2.5. Each k value provides a state in both the conduction and valence band. We must distinguish between the two allowed energies for each wave vector $\vec{k}$. Let the integer n represent the band. The energy eigenvalue $E_{n,k}$ specifies the energy of an electron in band "n" with wave vector k. The function $E_{1,\vec{k}} = E_1(\vec{k})$ represents the energy states in the valence band (it gives the dispersion curve) and $E_{2,\vec{k}} = E_2(\vec{k})$ represents the energy states in the conduction band. Generally, semiconductors have more than just two bands, which can be labeled by the band index "n." The k-p theory and its derivatives provide the states for degenerate bands.

Having specified notation for the eigenvalues, we can now enumerate the eigenstates forming a basis for the Hilbert space. Continuing with the two-band example in Figure 8.2.5, we must specify one eigenfunction for each energy eigenstate. The single eigenstate corresponding to $E_{n,\vec{k}}$ can be written as $|E_{n,\vec{k}}\rangle = |n, \vec{k}\rangle = |n, k_x, k_y, k_z\rangle$. In the

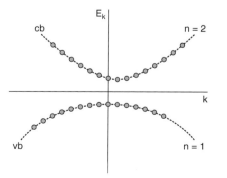

FIGURE 8.2.5
A zoomed-in view of the bands showing individual states.

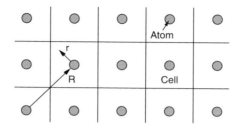

FIGURE 8.2.6
R indicates the center of the cell and r ranges over the interior of the cell.

coordinate representation, we can write $|E_{n,\vec{k}}\rangle = |n,\vec{k}\rangle \rightarrow \phi_{n\vec{k}}(\vec{r})$. These eigenfunctions are the Bloch wavefunctions and satisfy $\hat{H}|n,\vec{k}\rangle = E_{n\vec{k}}|n,\vec{k}\rangle$ where $\hat{H} = -(\hbar^2/2m)\nabla^2 + V_L$.

The Bloch wavefunctions make up the energy basis states and consist of two separate functions.

$$\phi_{n\vec{k}}(\vec{r}) = \frac{e^{i\vec{k}\cdot\vec{r}}}{\sqrt{V}}u_{n\vec{k}}(\vec{r}) \qquad (8.2.8)$$

where V represents the normalization over the region of space L characterizing the periodic boundary conditions. The product contains the plane wave $e^{i\vec{k}\cdot\vec{r}}$ and a function u having the periodicity of the lattice. That is, for $\vec{R}$ a lattice vector, the function u has the property that $u_{n\vec{k}}(\vec{r}+\vec{R}) = u_{n\vec{k}}(\vec{r})$. The subscripts indicate the possibility of different periodic functions $u_{n\vec{k}}(\vec{r})$ depending on the band. Equation (8.2.8) is the coordinate representation of the eigenvector $|E_{n,\vec{k}}\rangle = |n,\vec{k}\rangle$ where "u" represents the wave function for a single unit cell. The traveling plane wave $e^{i\vec{k}\cdot\vec{r}}$ constitutes an envelope function. The full solution $\Psi(\vec{r},t)$ to the time-dependent Schrodinger's wave equation requires a summation over the basis functions (the eigenfunctions ψ).

$$\Psi(\vec{r},t) = \sum_{n,k}\beta_{n\vec{k}}(t)\phi_{n,k}(\vec{r}) = \sum_{n,k}\beta_{n\vec{k}}(t)\frac{e^{i\vec{k}\cdot\vec{r}}}{\sqrt{V}}u_{n\vec{k}}(\vec{r}) = \sum_{n,k}\beta_{n\vec{k}}(0)\frac{e^{i\vec{k}\cdot\vec{r}-i\omega_n t}}{\sqrt{V}}u_{n\vec{k}}(\vec{r}) \qquad (8.2.9)$$

where $\omega_n = E_{nk}/\hbar$. Notice that the envelope has the form of a plane wave.

The envelope function can be used to satisfy the macroscopic boundary conditions (see Figure 8.2.4). For example, consider the infinitely deep well where the wavefunction must be zero at the boundaries, which produces sinusoidal wave functions. A solution to the time-independent Schrodinger wave equation might be expected to have the form

$$X(x) = C_1 e^{ikx}u_{2,k}(x) + C_2 e^{-ikx}u_{2,-k}(x)$$

for an electron in the conduction band ($n=2$). The portion of the wave function u periodic in the crystal potential has negligible dependence on k. Often k, in the function "u," is set to zero $k = 0$ because of the insensitivity of "u" to the macroscopic boundary conditions. Equivalently, for u symmetric in k we find

$$X(x) = (C_1 e^{ikx} + C_2 e^{-ikx})u_{2,k}(x)$$

Therefore the summation over the envelope wave functions must be zero at the boundaries so that $C_1 = -C_2 \equiv C$ and

$$X(x) = (C_1 e^{ikx} - C_1 e^{-ikx})u_{2,k}(x) \sim \sin(kx)u_{2,k}(x)$$

As discussed below, the wave function includes the periodic part of the wave function when the Schrodinger equation includes the crystal potential but not the effective mass. However, the effective mass Schrodinger equation eliminates the crystal potential V_L but includes the effective mass. In this case, the periodic part of the wave function can be dropped.

We might picture the energy eigenfunctions $\phi_{n\vec{k}}$ as shown in Figure 8.2.4; the dotted curve corresponds to the envelope. For the standing wave shown in the figure, we have assumed the electron is in the conduction band ($n = 2$) and that the envelope function consists of right-traveling and left-traveling plane wave (i.e., $k > 0$, and $-k$). Of course, the same reasoning applies to other physical situations besides the infinitely deep well. Another example might be the wave packet travelling through a semiconductor. As discussed in the companion volume on Solid State and Quantum Theory, the boundary conditions at an interface do not necessary have the simple requirements of the wave function and its spatial derivative being continuous across the interface. However, this is a fairly close approximation for two different materials with identical effective mass and lattice constants.

Equation (8.2.4) can be restated as

$$\phi_{n\vec{k}}(\vec{r} + \vec{R}) = e^{i\vec{k}\cdot\vec{R}}\,\phi_{n\vec{k}}(\vec{r}) \tag{8.2.10}$$

where $\vec{R}$ is a direct lattice vector. We can easily demonstrate this result by using Equation (8.2.8).

$$\phi_{n\vec{k}}(\vec{r} + \vec{R}) = e^{i\vec{k}\cdot(\vec{r}+\vec{R})}u_{n\vec{k}}(\vec{r} + \vec{R}) = e^{i\vec{k}\cdot\vec{R}}\,e^{i\vec{k}\cdot\vec{r}}u_{n\vec{k}}(\vec{r}) = e^{i\vec{k}\cdot\vec{R}}\phi_{n\vec{k}}(\vec{r}) \tag{8.2.11}$$

where we have used the periodicity of the function "u," namely $u_{n\vec{k}}(\vec{r} + \vec{R}) = u_{n\vec{k}}(\vec{r})$.

8.2.4   Orthonormality Relation for the Bloch Wave Functions

Now we demonstrate the normalization of the Bloch wave functions. First, these wavefunctions represent a type of plane wave throughout space—a crystal actually has infinite size based on the definition of a lattice. Therefore, the wave function must be normalized on a finite region of space with volume V that usually comes from periodic boundary conditions over the length L so that $V = L^3$.

We start with the definition of

$$\phi_{n\vec{k}}(\vec{r}) = \frac{e^{i\vec{k}\cdot\vec{r}}}{\sqrt{V}}u_{n\vec{k}}(\vec{r}) \tag{8.2.12}$$

and explicitly demonstrate the normalization for u. We want to satisfy the orthonormality relation for $|n, \vec{k}\rangle$.

$$\delta_{m\vec{\kappa},\,n\vec{k}} = \langle m, \vec{\kappa} \mid n, \vec{k}\rangle = \int_V d^3r \frac{e^{-i\vec{\kappa}\cdot\vec{r}}}{\sqrt{V}}u_{m\vec{\kappa}}^*(\vec{r})\frac{e^{+i\vec{k}\cdot\vec{r}}}{\sqrt{V}}u_{n\vec{k}}(\vec{r}) \tag{8.2.13}$$

The orthonormality in $\vec{k}$ mostly comes from the $e^{i\vec{k}\cdot\vec{r}}$ term since $\vec{k}$ corresponds to a wavelength having the size of many unit cells whereas the periodic function $u_{n\vec{k}}(\vec{r})$ has distinct values only within the unit cell (Figure 8.2.6). Therefore, we expect u to be relatively independent of $\vec{k}$.

To simplify the calculation, we make the substitution of $\vec{r} \rightarrow \vec{R}_j + \vec{r}$ where $\vec{R}_j$ gives the center of unit cell #j and we now confine $\vec{r}$ to a unit cell. Note that $u(\vec{r} + \vec{R}_j) = u(\vec{r})$

since u is periodic. This means that the integral in Equation (8.2.13) must be divided into a summation over all unit cells.

$$\delta_{m\vec{\kappa},n\vec{k}} = \sum_{j=1}^{N} \frac{e^{i(\vec{k}-\vec{\kappa})\cdot\vec{R}_j}}{V} \int_{V_j} d^3r e^{i(\vec{k}-\vec{\kappa})\cdot\vec{r}} u^*_{m\vec{\kappa}}(\vec{r}) u_{n\vec{k}}(\vec{r}) \qquad (8.2.14)$$

The wavevectors k must have very small magnitude since the electron wavelength spans many unit cells. In fact, $\vec{k}\cdot\vec{r} \sim 2\pi(|\vec{r}|/\lambda) \sim 0$ since $\vec{r}$ is now confined to a single cell. We therefore take the exponential under the integral to be unity and Equation (8.2.14) becomes

$$\delta_{m\vec{\kappa},n\vec{k}} = \sum_{j=1}^{N} \frac{e^{i(\vec{k}-\vec{\kappa})\cdot\vec{R}_j}}{V} \int_{V_j} d^3r \; u^*_{m\vec{\kappa}}(\vec{r}) u_{n\vec{k}}(\vec{r}) \qquad (8.2.15)$$

To continue, set $\vec{\kappa} = \vec{k}$ otherwise the sum over the exponentials will add approximately to zero. The functions "u" are periodic which means that their integral must be independent of the particular unit cell V_j. Therefore, as far as the summation is concerned, the integrals are constants. We have

$$\delta_{m,n} = \frac{1}{V} \sum_{j=1}^{N} \int_{V_j} d^3r \; u^*_{m\vec{k}}(\vec{r}) u_{n\vec{k}}(\vec{r}) = \frac{N}{V} \int_{V_j} d^3r \; u^*_{m\vec{k}}(\vec{r}) u_{n\vec{k}}(\vec{r}) \qquad (8.2.16)$$

Using the fact that there are N unit cells in the volume V yields $V = NV_{\text{cell}}$ and

$$\int_{V_{\text{cell}}} d^3r \; u^*_{m\vec{k}}(\vec{r}) u_{n\vec{k}}(\vec{r}) = V_{\text{cell}}\delta_{m,n} \qquad (8.2.17)$$

We can go further by noting the functions "u" are approximately independent of k. Then we have

$$\int_{V_{\text{cell}}} d^3r u^*_{m\vec{\kappa}}(\vec{r}) u_{n\vec{k}}(\vec{r}) \cong V_{\text{cell}}\delta_{m,n} \qquad (8.2.18)$$

We assume that for a given $\vec{k}$, the functions $u_{n,\vec{k}}$ form a complete set (where "n" runs over all of the bands).

We can see the normalization factor V_{cell} must be correct by using the case of the periodic potential going to zero since then $u \to 1$ and the integral then produces V_{cell}. Given that u is relatively independent of k, the exponential carries most of the orthonomality over the k variable. We can normalize the function u so that the integral does not require the extra V_{uc} factor. Making the replacement

$$u_{n\vec{k}} \to \sqrt{V_{uc}} u_{n\vec{k}} \quad \text{or} \quad u_{n\vec{k}} \to \sqrt{V} u_{n\vec{k}} \qquad (8.2.19a)$$

we then have the orthonormality relation

$$\int_{V_{uc}} d^3r' u^*_{m\vec{\kappa}} u_{n\vec{k}} = \delta_{mn} \quad \text{or} \quad \int_{V} d^3r' u^*_{m\vec{\kappa}} u_{n\vec{k}} = \delta_{mn} \qquad (8.2.19b)$$

However, this is not the usual procedure.

8.2.5 The Effective Mass Equation

The Hamiltonian for the Schrodinger equation incorporates the periodic crystal potential V_L as

$$\hat{H} = \frac{\hat{p}^2}{2m} + V_L \qquad (8.2.20a)$$

This Hamiltonian uses the electron free mass m. The energy basis states have the Bloch form

$$\phi_{n\vec{k}}(\vec{r}) = \frac{e^{i\vec{k}\cdot\vec{r}}}{\sqrt{V}} u_{n\vec{k}}(\vec{r}) \qquad (8.2.20b)$$

The periodic boundary conditions over the macroscopic length L for all three directions ($V = L^3$) determine the normalization and the number (and form) of the Bloch wave functions. In general, extra potentials V_E can be added to the Hamiltonian to account for band bending and so on.

$$\hat{H} = \frac{\hat{p}^2}{2m} + V_L + V_E \qquad (8.2.21)$$

The effective mass approximation includes the effects of the lattice potential by replacing the free mass in Equation (8.2.21) with the effective mass m_e. The potential V_L is removed from the Hamiltonian.

$$\hat{H} = \frac{\hat{p}^2}{2m_e} + V_E \qquad (8.2.22a)$$

In these cases, the periodic part of the wave function should be removed and the basis states become plane waves equivalent to those for free-particle transport.

$$\phi_{n\vec{k}}(\vec{r}) = \frac{e^{i\vec{k}\cdot\vec{r}}}{\sqrt{V}} \qquad (8.2.22b)$$

The solution found for free-space can be applied to the equivalent case with a semiconductor so long as the effective mass is used.

$$m_e = \hbar^2 \left(\frac{\partial^2 E}{\partial k^2} \right)^{-1} \qquad (8.2.23)$$

The Schrodinger equation incorporating the tensor mass can be written as

$$\hat{\mathscr{H}} = -\frac{\hbar^2}{2} \nabla \cdot \vec{\vec{m}}_e^{-1} \cdot \nabla + V_E = -\frac{\hbar^2}{2} \sum_{a,b} (m_e^{-1})_{ab} \partial_a \partial_b + V_E \qquad (8.2.24)$$

8.3 Density of States for Nanostructures

One of the most exciting areas of modern research focuses on fabricating reduced dimensional structures. These structures can incorporate dissimilar groups of atoms arranged and grouped on scales of 5 to 100 angstroms. Such small sizes induce quantum confinement effects in the system, which radically affects the band structure and most of the opto electronic properties.

In this section, we develop the density of states for these reduce dimensional structures after briefly reviewing the solution to the Schrodinger wave equation in the effective mass approximation. The companion volume on solid states contains the full mathematical treatment.

8.3.1 Envelope Function Approximation

We want to model 3-D crystals using potentials that reduce the Schrodinger wave equation to simpler 1-D or 2-D problems. To fix our thoughts, Figure 8.3.1 shows a 3-D heterostructure with varying aluminum concentration along the growth axis z. The GaAs forms a 2-D reduced dimensional structure, namely a quantum well. The crystal atoms produce a periodic potential V_L (L for lattice) and the interfaces produce the "macroscopic" confining potential V. The 3-D character of the structure leads to a 2-D equation for the x-y directions and a 1-D equation for the z direction. All three directions generally use the Bloch wavefunctions. Figure 8.3.2 shows the Bloch wavefunction for the z-direction.

The Schrodinger wave equation for the heterostructure can be written as

$$-\frac{\hbar^2}{2m}\nabla^2\Psi + (V + V_L)\Psi = i\hbar\frac{\partial}{\partial t}\Psi \tag{8.3.1}$$

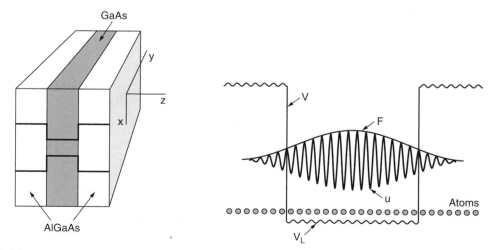

FIGURE 8.3.1

The band offset produces quantum wells in a heterostructure.

FIGURE 8.3.2

The wave function for the finitely deep well.

where m denotes the free mass of the electron. The wave function has the form

$$|\Psi(t)\rangle = \sum_{\vec{k}} \beta_{n\vec{k}}(t)|n,\vec{k}\rangle = \sum_{\vec{k}} \beta_{n\vec{k}}(0)|n,\vec{k}\rangle e^{-iE_{n\vec{k}}t/\hbar} \qquad (8.3.2)$$

where the eigenfunctions have the form

$$|n,\vec{k}\rangle \sim \psi(\vec{r}) = \frac{1}{\sqrt{V}} e^{i\vec{k}\cdot\vec{r}} u_{n,\vec{k}}(\vec{r}) \qquad (8.3.3)$$

and we confine our attention to the conduction band. A similar expression can be used for the valence bands so long as the light and heavy hole bands have sufficient separation in energy (nondegenerate bands). The basis functions for the Hilbert space of envelope functions

$$\phi_{\vec{k}}(\vec{r}) = \frac{1}{\sqrt{V}} e^{i\vec{k}\cdot\vec{r}} \qquad (8.3.4a)$$

satisfy the orthonormality relation

$$\langle \phi_{\vec{K}} | \phi_{\vec{k}} \rangle = \delta_{\vec{k}\vec{K}} \qquad (8.3.4b)$$

The Bloch functions $u_{n,\vec{k}}$ are periodic on the crystal so that the values of $u_{n,\vec{k}}$ repeat from one unit cell to the next. As discussed in the previous section, the Bloch wave function $u_{n,\vec{k}}$ are normalized so that they satisfy an inner product over the unit cell of the form

$$\langle u_{n\vec{k}} | u_{m\vec{k}} \rangle_{uc} = \int_{uc} dV \, u^*_{n\vec{k}} u_{m\vec{k}} = V_{\text{cell}} \delta_{mn} \qquad (8.3.5)$$

We consider only the conduction band ($n = 2$) and redefine the notation as $u_{2,\vec{k}} = u_{\vec{k}}$. The general vector in the space spanned by the basis set

$$|n,\vec{k}\rangle \sim \psi_{n,\vec{k}}(\vec{r}) = \frac{1}{\sqrt{V}} e^{i\vec{k}\cdot\vec{r}} u_{n,\vec{k}}(\vec{r}) \qquad (8.3.6a)$$

has the form

$$\Psi(\vec{r},0) = \sum_{\vec{k}} \beta_{\vec{k}} \psi_{n,\vec{k}}(\vec{r}) = \sum_{\vec{k}} \beta_{\vec{k}} \phi_{\vec{k}} u_{n,\vec{k}}(\vec{r}) \qquad (8.3.6b)$$

The envelope approximation allows Equation (8.3.6b) to be simplified. For this approximation, the functions $u_{n,\vec{k}}(\vec{r})$ are assumed to be relatively independent of the wave vector $\vec{k}$ since it corresponds to a wavelength having the size of many unit cells whereas $u_{n,\vec{k}}(\vec{r})$ has distinct values only within the unit cell. Therefore, writing $u_{n,\vec{k}}(\vec{r}) \sim u_{n,0}(\vec{r}) \equiv u_n(\vec{r})$, we can write

$$\Psi(\vec{r},0) = \sum_{\vec{k}} \beta_{\vec{k}} \phi_{\vec{k}} u_{n,\vec{k}}(\vec{r}) \cong \left[\sum_{\vec{k}} \beta_{\vec{k}} \, \phi_{\vec{k}}(\vec{r}) \right] u_n(\vec{r}) = F(\vec{r}) u_n(\vec{r}) \qquad (8.3.6c)$$

The envelope function $F(\vec{r})$ resides in the Hilbert space spanned by the envelope basis set $\{\phi_{\vec{k}}(\vec{r})\}$. We therefore see that the solution (8.3.6c) to the Schrodinger wave equation must have the form of a modulated carrier. The envelope function must satisfy the boundary conditions for the microstructure. We can find the density of states using either the full basis set $\{|2, \vec{k}\rangle\}$ or those for the envelope functions $\{\phi_{\vec{k}}(\vec{r})\}$ since for each $\vec{k}$, there exists a basis vector in either set.

The effective mass Schrodinger equation eliminates the periodic potential V_L in Equation (8.3.1) but replaces the free mass with the more complicated effective mass m_e.

$$-\frac{\hbar^2}{2m_e}\nabla^2\Psi + V\Psi = i\hbar\frac{\partial}{\partial t}\Psi \tag{8.3.7a}$$

The solution has the form

$$|\Psi(t)\rangle = \sum_{\vec{k}} \beta_{2\vec{k}}(t)\left|\phi_{\vec{k}}\right\rangle \tag{8.3.7b}$$

As usual, the functions $\phi_{\vec{k}}$ satisfy the eigenvector equation

$$\hat{H}\phi_{\vec{k}} = E_{\vec{k}}\phi_{\vec{k}} \tag{8.3.8}$$

where $E_{\vec{k}} = E_{2,\vec{k}}$ and $\hat{H} = -(\hbar^2/2m_e)\nabla^2\Psi + V\Psi$

8.3.2 Summary of Solution to the Schrodinger Wave Equation for the Quantum Well

Consider an electron confined to a well. To find the energy basis states for the quantum well, separate variables in the time-independent Schrodinger wave equation

$$-\frac{\hbar^2}{2m_e}\nabla^2\psi(x, y, z) + V(z)\psi(x, y, z) = E\psi(x, y, z) \tag{8.3.9}$$

where E represents the *total* energy of the electron. The total kinetic energy consists of the motion perpendicular and parallel to the interfaces in the heterostructure. The total energy inside the quantum well where $V(z) = 0$ for $0 \leq z \leq L$ must be the same as the total kinetic energy. A standing wave describes the electron motion along the confinement direction z similar to the one shown in Figure 8.3.2. As usual, we separate the equation by substituting

$$\psi = X(x)Y(y)Z(z) \tag{8.3.10}$$

and then divide by ψ to find

$$\underbrace{-\frac{\hbar^2}{2m_e}\frac{1}{X}\frac{\partial^2}{\partial x^2}X(x)}_{E_x} \underbrace{-\frac{\hbar^2}{2m_e}\frac{1}{Y}\frac{\partial^2}{\partial y^2}Y(y)}_{E_y} \underbrace{-\frac{\hbar^2}{2m_e}\frac{1}{Z}\frac{\partial^2}{\partial z^2}Z(z) + V(z)}_{E_z} = E \tag{8.3.11}$$

The total energy consists of the sum of the energies for motion in the x, y, z directions

$$E = E_x + E_y + E_z \tag{8.3.12}$$

We already know the eigenfunctions and eigenvalues for motion in the x and y directions.

$$X_{k_x} = e^{ik_x x} \quad Y_{k_y} = e^{ik_y y} \quad E_x = \frac{\hbar^2 k_x^2}{2m_e} \quad E_y = \frac{\hbar^2 k_y^2}{2m_e} \tag{8.3.13}$$

These last two equations represent dispersion curves for the x and y directions; the electron acts as a free electron so long as the effective mass replaces the free mass. Equation (8.3.13) assumes spherical bands but the effective masses can be replaced with m_x and m_y as necessary. The allowed values of k_x and k_y come from macroscopic periodic boundary conditions as usual. The equation for the z-direction takes the form of

$$-\frac{\hbar^2}{2m_e}\frac{\partial^2}{\partial z^2}Z + V(z)Z = E_z Z \tag{8.3.14}$$

We need to find the eigenfunctions and eigenvalues for this last equation.

For the infinitely deep well, we assume that the envelope wave function must be zero outside the well as given by the fixed-endpoint boundary conditions $z = 0$ and $z = L_z$. The basis set has the form

$$Z(z) = \sqrt{\frac{2}{L_z}}\sin(k_z z) \quad k_z = n\pi/L_z \quad n = +1, +2, \ldots \tag{8.3.15a}$$

and the energy eigenvalues can be written as

$$E_{(z)n} = \frac{\hbar^2 k_z^2}{2m_e} = \frac{n^2\pi^2\hbar^2}{2m_e L_z^2} \tag{8.3.15b}$$

The solution applies to the infinitely deep well, the density of states only needs the dependence of energy on k_z and not the exact form of the wave function.

Now that we have the allowed energies and the eigenfunctions for the z-direction, the general solution to the original Schrodinger *time-dependent* equation can be determined. The total energy consists of the quantum well energy E_z plus the energy due to motion parallel to the interfaces.

$$E = E_z + \frac{\hbar^2}{2m_e}(k_x^2 + k_y^2) \tag{8.3.16}$$

and the general wavefunction must have the form

$$\Psi = \sum_{k_x k_y k_z} C_{k_x k_y k_z} X_{k_x} Y_{k_y} Z_{k_z} e^{-itE/\hbar}$$

$$= \sum_{k_x k_y k_z} C_{k_x k_y k_z} X_{k_x} Y_{k_y} Z_{k_z} e^{-itE_z/\hbar} e^{-itE_{xy}/\hbar} \tag{8.3.17}$$

where

$$E = E_z + \frac{\hbar^2}{2m_e}(k_x^2 + k_y^2) \quad E_{xy} = \frac{\hbar^2}{2m_e}(k_x^2 + k_y^2) \tag{8.3.18}$$

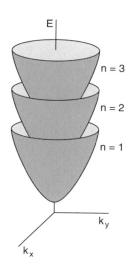

FIGURE 8.3.3
The energy sub-bands from Equation (8.3.16).

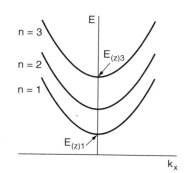

FIGURE 8.3.4
Sub-bands for the quantum well in the 3-D crystal.

The dispersion relation $E_{xy}(k_x, k_y)$ applies to directions parallel to the plane of the quantum well. It describes the motion of a free particle with mass m_e.

The sequence of parabaloids in Figure 8.3.16 represents the total energy in Equation (8.3.16). The vertex of each one increases in energy according to the energy E_z in Equation (8.3.15b). Electron motion in the x-y plane of Figure 8.3.1 is similar to the motion of free electrons because of the parabolic dispersion relations. Each parabaloid in Figure 8.3.3 corresponds to the portion of the electron motion parallel to the layers. However the parabaloids must be displaced from the origin along the energy axis because of the additional discrete energy levels due to the quantum well. For simplicity, the dispersion curves in Equation (8.3.16) are usually plotted as shown in Figure 8.3.4.

8.3.3 Density of Energy States for the Quantum Well

We now calculate the density of states inside the quantum well using the results for the bulk case. A more rigorous treatment can be found in the companion volume for Solid State and Quantum Theory.

As indicated in Figure 8.3.1, the quantum well essentially consists of a 2-D plane embedded within a crystal. Electrons can freely move parallel to the plane but the well produces widely separated wave vectors k_z and the corresponding energy E_z values. Figure 8.3.5 shows the sub-bands produced by a quantum well. Each sub-band corresponds to motion along the 2-D plane. We therefore expect the density of states for each sub-band to correspond to the 2-D bulk case.

$$g^{2D}(E) = \frac{m_e A_{\text{xal}}}{2\pi\hbar^2} \quad \text{(no spin)} \tag{8.3.19}$$

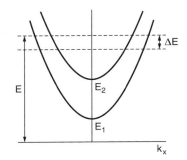

FIGURE 8.3.5
The sub-bands produced by a quantum well.

where $A_{\text{xal}} = L_x L_y$ represents the crystal area and L_x, L_y represent the macroscopic lengths for the periodic boundary conditions along the x and y directions, respectively. The area will be removed later.

The density of states for the quantum well can now be found. For the region of energy $0 \leq E < E_1$, the density-of-states for the well g_{well} must be zero. For the energy range $E_1 < E < E_2$, only one sub-band must be considered. The density of states must correspond to a single 2-D parabaloid given by equation 8.3.19. Finally the energy range $E_2 < E < E_3$ requires us to count states in two sub-bands in the energy range ΔE. We therefore expect to include Equation (8.3.19) twice. In general then, the density of states for the quantum well must be

$$g_{\text{well}}(E) = \frac{m_e}{2\pi\hbar^2} \sum_{E_n} \Theta(E - E_n) \quad \text{per xal area} \quad \text{no spin}$$

(8.3.20)

$$g_{\text{well}}(E) = \frac{m_e}{\pi\hbar^2 L_z} \sum_{E_n} \Theta(E - E_n) \quad \text{per xal vol} \quad \text{with spin}$$

where we have divided out the well volume $A_{\text{xal}}L_z$ in the second equation. The units of "per crystal volume" allow us to compare bulk and quantum well density of states.

Typically, the literature shows the density of energy states for the quantum well next to a plot of the sub-bands as in Figure 8.3.6. Each 2-D plane of $\vec{k}$ vectors leads to a constant density of energy states (independent of energy). Larger energy E requires more sub-bands be included in the energy state counting. A step occurs in the well density of

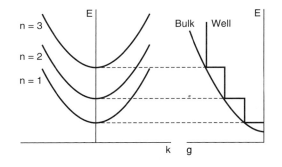

FIGURE 8.3.6
The density of energy states for the quantum well and its relation to the sub-band diagram. Both DOS are normalized to crystal volume.

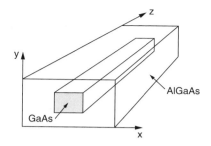

FIGURE 8.3.7
The quantum wire confines electrons along the x and y direction.

states at the start of each sub-band as shown. By the way, the step-like form for the well density of states means that thermal electrons occupy narrower range of energy than for the bulk material. This also means that the population inversion required for lasing can also occupy more narrow range.

8.3.4   The Density of Energy States for the Quantum Wire

The quantum wire confines the electron in two directions, say the x and y directions. For example, Figure 8.3.7 shows a GaAs "wire" embedded within AlGaAs. We assume the wire length along z has macroscopic size compared with the microscopic size along either x or y.

$$L_x, L_y \ll L_z$$

The lengths along x or y are approximately 50–100 angstroms.

We can solve Schrodinger's equation in the effective mass approximation for the *infinitely* deep well. The solutions along x and y must have a sinusoidal form while those for the "z" direction can be taken as traveling waves. The allowed wave vectors can be written as

$$k_m^{(x)} = \frac{m\pi}{L_x} \quad k_n^{(y)} = \frac{n\pi}{L_y} \quad k_q^{(z)} = \frac{2q\pi}{L_y} \quad m, n = +1, +2, \ldots \quad q = \pm 1, \pm 2, \ldots \tag{8.3.21}$$

with the spacing between states

$$\Delta k_x = \frac{\pi}{L_x} \quad \Delta k_y = \frac{\pi}{L_y} \quad \Delta k_z = \frac{2\pi}{L_z} = \text{small} \tag{8.3.22}$$

and the energy

$$E = E_x + E_y + E_z = \frac{\hbar^2 k_x^2}{2m_e} + \frac{\hbar^2 k_y^2}{2m_e} + \frac{\hbar^2 k_z^2}{2m_e} = \frac{\hbar^2 k^2}{2m_e} \tag{8.3.23}$$

where $k^2 = k_x^2 + k_y^2 + k_z^2$ and k_z is essentially continuous.

The density of states can be calculated using a diagram very similar to those in Figures 8.3.3 and 8.3.4. In this case, the parabola represents free motion along the z-direction. The parabolas have vertices given by the various energy levels in the wire for the x and y directions.

$$E_{mn} = E_x + E_y = \frac{\hbar^2 k_x^2}{2m_e} + \frac{\hbar^2 k_y^2}{2m_e} = \frac{m^2 \pi^2 \hbar^2}{2m_e L_x^2} + \frac{n^2 \pi^2 \hbar^2}{2m_e L_y^2} \tag{8.3.24}$$

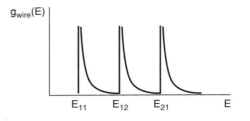

FIGURE 8.3.8
The density of energy states for the quantum wire.

The density of states of a single sub-band has the same form as for the bulk 1-D crystal

$$g^{1D}(E) = \frac{L_z}{2\pi}\sqrt{\frac{2m_e}{\hbar^2}}\frac{1}{\sqrt{E}} \tag{8.3.25}$$

The crystal scale factor L_z will be separately handled. If the energy E in question is larger than the vertex energy for several sub-bands, then the density of states must include several terms similar to equation 8.3.25 to give

$$g_{\text{wire}}(E) = \frac{L_z}{2\pi}\sqrt{\frac{2m_e}{\hbar^2}}\sum_{E_{mn}}\frac{1}{\sqrt{E-E_{mn}}}\Theta(E-E_{mn}) \quad \text{per length, no spin}$$

$$\tag{8.3.26}$$

$$g_{\text{wire}}(E) = \frac{1}{\pi L_x L_y}\sqrt{\frac{2m_e}{\hbar^2}}\sum_{E_{mn}}\frac{1}{\sqrt{E-E_{mn}}}\Theta(E-E_{mn}) \quad \text{per volume, with spin}$$

The units of "per volume" allow the quantum wire and 3-D bulk density-of-states to be compared.

The density of energy states for the quantum wire appears in Figure 8.3.8. The density of states rapidly falls off after each confinement energy E_{mn}. Combining this density of states with the Fermi-Dirac distribution produces very sharp electron (vs. energy) distributions.

As an important note, we have assumed an infinitely deep well. The energy levels for the finitely deep well have different values than those for the finitely deep ones. Therefore we expect a density of states function that appears similar to Figure 8.3.8 except that the steps must occur at different values of E. Further, the finitely deep well only binds the electron for a fixed number of states; the remaining states correspond to plane waves. Therefore only a finite number of steps appear in the density of states plot.

8.3.5   The Quantum Box

The quantum box confines the electron in all dimensions and therefore produces a series of spikes for the density of states. For a box with sides of length L_x, L_y, L_z, the density of states must have the form

$$g_{\text{box}}(E) = \sum_{m,n,p}\delta(E - E_{mnp}) \quad \text{no spin}$$

8.4 The Reduced Density of States and Quasi-Fermi Levels

The laser gain and rate equations for semiconductor lasers can be deduced using a variety of methods. The phenomenological approach in Chapter 2 provides the rate equations and the gain using basic energy conservation but does not include any wavelength dependence nor broadening mechanisms. Fermi's golden rule incorporates the quantum mechanics of the transition process but not the collision broadening and the steady state conditions. The Liouville equation includes all of these and describes the rounding of the spectral features due to the homogeneous broadening.

All of the approaches require the reduced density of states and the Bloch wave functions. The reduced density of states provides the number of pairs of states (between the conduction and valence bands) that contribute to the emission process at a particular wavelength. In this way, the transition process includes the density of states for each band.

This section starts with Fermi's golden rule and shows how to include the reduced density of states and the effects of the pump through the quasi-Fermi levels. Section 7.3 shows the transition rate from an initial state with energy E_i to a group of final states centered on energy E_f.

$$R_{i \to f} = \frac{d}{dt} P_V = \frac{\pi}{2\hbar}(\mu E_o)^2 \rho_f(E_f = E_i \pm \hbar\omega) \tag{8.4.1}$$

This rate provides the number of transitions per second per crystal volume. The symbol ρ denotes the density of states to distinguish it from the gain g. The symbols μ, E_o represent the induced dipole moment and the electric field amplitude, respectively. The section replaces the density of states ρ_f with the reduced density of states and explains the reason for including the Fermi distributions as an extra factor.

8.4.1 The Reduced Density of States

The number of final and initial states partly determines the transition rate. If there were only one final state but many electrons in initial states, then we could have only one transition to a final state. Increasing the number of final states (receptors) means many more of the electrons in the initial states can make transitions. Therefore the transition rate must increase with greater numbers of final states. Likewise, a single initial state (source) can have only a single electron. Therefore there can be only one upward transition. Increasing the number of initial states (assuming they have electrons) must therefore increase the transition rate. It should also be apparent that the Fermi distribution must play a role in the transition rate. Initial states without electrons do not contribute to upward transitions. Likewise, filled final states cannot receive another electron and therefore cannot contribute to the transition rate.

A complication arises since Equation (8.4.1) has a single density of states but the conduction and valence bands each have their own density of states. How do we combine the two density of states into a single one for Fermi's golden rule? The paradox can be resolved by realizing that the density of states in Equation (8.4.1) actually refers to the transition energy $E_T = \hbar\omega$. We actually want the number of states per unit transition energy. Figure 8.4.1 shows a conduction and valence band. The vertical lines between the two bands indicate the possible transitions. Each electron wave vector $\vec{k}$ has a different transition energy associated with it. The transitions appear vertical because the photon has very little momentum compared with the energy difference it provides to the

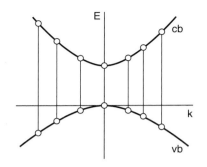

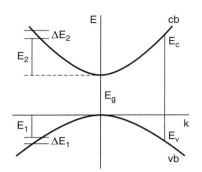

FIGURE 8.4.1
Vertical transitions.

FIGURE 8.4.2
The transition energy $E_T = |E_2| + E_g + |E_1|$.

electrons. An electron in the conduction band can emit a photon and drop into the valence band. The energy difference between the initial and final states for the electron exceeds the gap energy E_g. If phonons were involved with the transition, then the transitions would no longer appear approximately vertical. Consequently, each transition can be uniquely associated with an electron wave vector $\vec{k}$.

We can find the reduced density of states. Figure 8.4.2 shows the transition energy can be written as a sum of two energies so long as the sum adds to $\hbar\omega$ where ω denotes the optical angular frequency. We want the quantity

$$\rho_r(E_T) = \frac{dN}{dE_T} \tag{8.4.2}$$

where N denotes the number of pairs of electron-hole states separated by the transition energy E_T.

$$E_T = |E_2| + E_g + |E_1|$$

Then we can write

$$dE_T = dE_2 + dE_1$$

We can now write

$$\frac{1}{\rho_r} = \frac{dE_T}{dN} = \frac{dE_2}{dN} + \frac{dE_1}{dN} = \frac{1}{\rho_c} + \frac{1}{\rho_v} \quad \text{or} \quad \rho_r = \frac{\rho_c \rho_v}{\rho_c + \rho_v} \tag{8.4.3}$$

where ρ_c, ρ_v represent the density of states in the conduction and valence bands respectively. These last equations use the fact that N represents the number of pairs of electron-hole states; it therefore must also represent the number of states in the conduction band in ΔE_2 and the number of states in the valence band in ΔE_1. The second of Equations (8.4.3) shows the reason for the name "reduced;" it has a form resembling the reduced mass from classical mechanics. The reduced density of states must be smaller than either the conduction or valence band density of states.

The method of calculating the reduced density of states can perhaps be best understood from the simple case of equal numbers of states for the conduction and valence bands as plotted in Figure 8.4.3. If $E_T < E_g$, then the transition energy E_T does not connect states in the conduction and valence band so that the reduced density of states must be zero. In other words, there does not exist any pairs of states separated by energy E_T. If $E_T = E_g$, then the states at the band edges can participate in transitions; these transitions can involve the absorption or emission of photons with energy E_T. In this case, the

FIGURE 8.4.3
The reduced density of states (right) and its relation to the conduction and valence band density of states (left). For equal valence and conduction band density of states, the "transition bar," denoted by E_T is symmetrically placed about the gap.

reduced density of states begins to increase from zero. For $E_T > E_g$ as shown in the figure, a large number of states in the valence band can be connected with a large number of states in the conduction band by the transition energy. The figure shows equal density of states for both bands. Therefore, the reduced density of states must be $\rho_r = \rho_c/2 = \rho_v/2$. In this case, the reduced density of states is easy to calculate just by drawing a horizontal bar representing the length E_T; this bar is symmetrical about the band gap.

We need a method for calculating the reduced density of states when the conduction and valence bands do not have equal numbers of states. We need to find the location (i.e., the energy) of the states in the conduction band and the location of those in the valence band that are connected by the transition energy E_T. Assuming a bulk 3-D crystal with parabolic bands and direct band gap, the functional form of the reduced density of states can be found from Equations (8.4.3) and the relation found in Problems 8.15, namely $m_c E_2 = -m_v E_1$ (for $E_v < 0$). The reduced density of states in Equations (8.4.3) requires E_2 and E_1 to be written in terms of the constant e.g., and E_T.

$$E_2 = \frac{m_v}{m_c + m_v}(E_T - E_g) \qquad E_1 = -\frac{m_c}{m_c + m_v}(E_T - E_g) \qquad (8.4.4a)$$

Equivalently, using $E_c = E_g + E_2$ and $E_v = E_1$ we find

$$E_c = \frac{m_v E_T + m_c E_g}{m_c + m_v} \qquad E_v = -\frac{m_c}{m_c + m_v}(E_T - E_g) \qquad (8.4.4b)$$

Using Equation (8.4.4b), the reduced density of states in Equation (8.4.3) becomes

$$\frac{1}{\rho_r(E_T)} = \frac{1}{\rho_c(m_v E_T + m_c E_g)/(m_c + m_v)} + \frac{1}{\rho_v - (m_c/(m_c + m_v))(E_T - E_g)} \qquad (8.4.5)$$

The last equation for the reduced density of states indicates the "transition energy bar" does not have a symmetrical position relative to the band gap nor is it parallel to the energy axis in a plot of the density of states versus energy (unlike Figure 8.4.3). The relation $m_c E_2 = -m_v E_1$ with unequal effective masses, rather than $E_2 = -E_1$, sets the position of the bar relative to the band gap. One of the chapter review exercises has an example for applying this equation.

8.4.2 Quantum Well Reduced Density of States

An infinitely deep quantum well of width L_z has the density of states

$$\rho(E) = \frac{m_e}{\pi \hbar^2 L_z} \sum_{E_n} \Theta(E - E_n) \quad \text{per xal vol, with spin} \qquad (8.4.6)$$

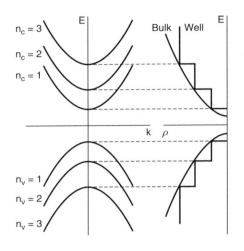

FIGURE 8.4.4
The sub-bands and density of states for an infinitely deep quantum well.

where m_e denotes the effective mass. This equation includes spin degeneracy. The sub-bands produce the step functions Θ in Equation (8.4.6). A step occurs in the well density of states at the start of each sub-band as shown in Figure 8.4.4. The density of states for the well is smaller than that for the bulk. This lowers the total number of thermal electrons and holes for the well.

The reduced density of states must include both the conduction and valence sub-bands as shown in Figure 8.4.4. Notice that the valence sub-bands do not start at $E = 0$ unlike the bulk density of states. This occurs because the quantum well pushes the vertex energy away from the typical minimum or maximum.

The reduced density of states can be written using

$$\frac{1}{\rho_r} = \frac{1}{\rho_c} + \frac{1}{\rho_v}$$

where

$$\rho_c(E) = \frac{m_c}{\pi\hbar^2 L_z} \sum_{E_{cn}} \Theta(E - E_{cn}) \quad \text{per xal vol} \tag{8.4.7a}$$

$$\rho_v(E) = \frac{m_v}{\pi\hbar^2 L_z} \sum_{E_{vn}} \Theta(E - E_{vn}) \quad \text{per xal vol} \tag{8.4.7b}$$

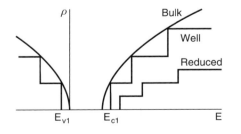

FIGURE 8.4.5
The density of states for the 3-D bulk, quantum well and the reduced density of states. The reduced DOS must be a function of E_T.

For simplicity, assume identical effective masses for valence and conduction bands. The reduced density of states appears as in Figure 8.4.5. Notice how the reduced DOS has half the height of the other curves and it's displaced toward larger energy. The energies E_{c1} and E_{v1} refer to the first well energy levels for the conduction and valence bands, respectively. The displacement occurs since the energy E_T must be larger than $E_{c1} - E_{v1}$ for transitions.

8.4.3 The Quasi-Fermi Levels

Replacing the density of states with the reduced density of states in Equation (8.4.1) does not complete the prescription for the rate of transition. We must include the number of filled or empty initial and final states. As discussed in Section 8.1, photogeneration or charge injection produces more carriers in some states than should be present for temperature T as dictated by the Fermi-Dirac distribution. Therefore, the distribution does not obey the Fermi-Dirac distribution and the Fermi level E_f splits into two quasi-Fermi levels E_{fc} and E_{fv} for conduction and valence electrons, respectively.

The laser necessarily involves nonequilibrium statistics by virtue of the pumping mechanism. The probability of an *electron* occupying a state in the conduction and valence bands, respectively, becomes

$$F_c(E) = \frac{1}{1 + \exp((E - E_{fc})/kT)} \quad F_v(E) = \frac{1}{1 + \exp((E - E_{fv})/kT)} \tag{8.4.8}$$

The probability of a valence or conduction band state being empty must be $1 - F_v$ and $1 - F_c$, respectively. In order for a laser to oscillate, a population inversion must exist. In the language of Sections 7.9 and 7.11, we require the population difference $\Delta N = N_2 - N_1$ to satisfy $\Delta N > 0$. This means a large number of electrons must occupy the upper state E_2 and only a few occupy the lower state E_1. For the semiconductor laser, generally the conduction quasi-Fermi level E_{fc} must be above the conduction band edge and the valence quasi-Fermi level E_{fv} must be near or below the valence band edge as shown in Figure 8.4.6. The separation of the quasi-Fermi levels must generally be larger than the band gap.

The number of conduction electrons and valence holes (per unit volume) can be written as

$$n = \int_{E_g}^{\infty} dE \rho_c(E) F_c(E) \qquad p = \int_{-\infty}^{0} dE \rho_v(E) F_v(E) \tag{8.4.9}$$

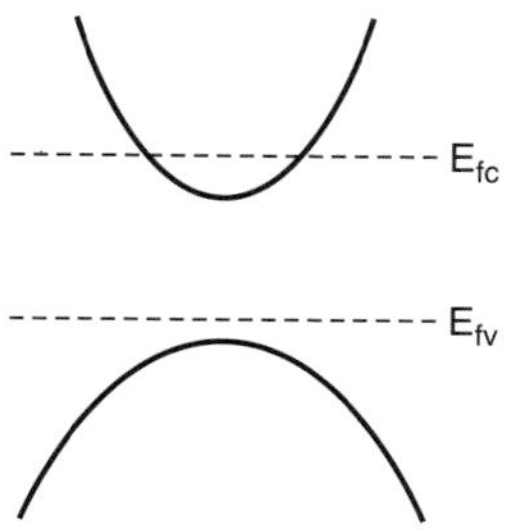

FIGURE 8.4.6
A possible position of the quasi-Fermi levels for lasing.

The quasi-Fermi level fixes the number of electrons as demonstrated in Section 8.1. Set $d_c = (E_{fc} - E_g)/kT$ and $d_v = E_v/kT$ similar to that done for Topic 8.1.11. For a quasi-Fermi level more than several kT beyond the band edge, the number of conduction *electrons* can be *characterized* by $m_c d_c$ and the number of valence *holes* by $m_v d_v$ where m_c and m_v denote the conduction and valence band effective masses, respectively. Specifically, the number of conduction and valence holes can be approximated by Equation (8.1.29b) when the quasi-Fermi level is more than several kT beyond the band edge. The number of electrons becomes approximately Equation (8.1.30) provides the approximate number of electrons and holes

$$n \cong \frac{(kT)^{3/2}}{\pi^2 \hbar^3} (m_c d_c)^{3/2} \quad p \cong \frac{(kT)^{3/2}}{\pi^2 \hbar^3} (m_v d_v)^{3/2} \tag{8.4.10}$$

The characterization of the number of electrons in the conduction band or the holes in the valence band is conceptually important for lasers. For sufficient gain to initiate lasing, the quasi-Fermi level separation $E_{fc} - E_{fv}$ must exceed the band gap e.g., We can approximately position the quasi-Fermi levels in the bands by requiring the number of holes to be the same as the number of electrons $p = n$. Therefore, Equations (8.4.10) give $m_c d_c = -m_v d_v$ where the minus sign occurs because the valence band has negative energy. This can be equivalently written as $m_c(E_{fc} - E_g) = -m_v E_{fv}$. However, for the most part, this simple relation must be modified for energy near the quasi-Fermi level where the approximation breaks down.

These quasi-Fermi levels should appear in Fermi's golden rule. For downward transitions, we need a large number of electrons in the conduction band and very few in the valence band. Absorption requires the reverse. Therefore for emission, we need to know the number of filled conduction states and the number of empty valence states. Fermi's golden rule should be written as follows.

$$R = \frac{\pi}{2\hbar} (\mu E_o)^2 \rho_r(E_T) \times \begin{cases} F_c(1 - F_v) & \text{Emission} \\ F_v(1 - F_c) & \text{Absorption} \end{cases} \tag{8.4.11}$$

Keep in mind that the transition energy refers to the difference in energy for the semiconductor but will be later linked to the photon energy as $E_T = E_\gamma = \hbar\omega$. The factors F_c and F_v must be evaluated at E_c and E_v, respectively, for which $E_T = E_c - E_v$ where $E_v < 0$.

8.5 Fermi's Golden Rule for Semiconductor Devices

Fermi's Golden Rule provides the rate of transition from a single filled state to a single final state (or group of states. Section 7.3 states Fermi's Golden Rule for an electromagnetic interaction as

$$R = \frac{\pi}{2\hbar} (\mu E_o)^2 \delta(E_f = E_i \pm \hbar\omega) \quad \text{(single final state)} \tag{8.5.1a}$$

$$R = \frac{\pi}{2\hbar} (\mu E_o)^2 \rho_f(E_f = E_i \pm \hbar\omega) \quad \text{(group of final states)} \tag{8.5.1b}$$

where the "+" refers to absorption and "−" refers to emission of a photon. The electromagnetic interaction has the form

$$\hat{\mathcal{V}} = \hat{\mu}E_o\cos(\omega t) = \hat{H}'e^{-i\omega t} + \hat{H}'e^{+i\omega t} \tag{8.5.2}$$

where $\hat{\mu}$ denotes the dipole operator and E_o denotes the amplitude of the incident electric field. Equation (8.5.2) relates the time-independent interaction energy $\hat{H}'$ to the dipole operator and the electric field according to $\hat{H}' = \hat{\mu}E_o/2$. In such a case, Fermi's Golden rule can be restated as

$$R = \frac{2\pi}{\hbar}\left|H'_{fi}\right|^2 \rho_f(E_f = E_i \pm \hbar\omega) \tag{8.5.3a}$$

where i, f represent the initial and final states, respectively. Taking into account the reduced density of states and the Fermi factor as in Equation (8.4.11), we can write

$$R = \frac{2\pi}{\hbar}\left|H'_{fi}\right|^2 \rho_r(E_T)\, F(E_T) \tag{8.5.3b}$$

where we define

$$F(E_T) = \begin{cases} F_c(1 - F_v) & \text{Emission} \\ F_v(1 - F_c) & \text{Absorption} \end{cases} \tag{8.5.3c}$$

and where ρ_r, F_c, F_v denote the reduced density of states, the probability of finding an electron in a conduction state, and the probability of finding an electron in a valence state, respectively.

Previous sections have examined the reduced density of states and the Fermi factor. The present section shows how to calculate the remaining factor for the transition matrix element. For semiconductors, the states have the form of Bloch states. We will find the gain for a semiconductor material (unsaturated) using the vector potential form $\mathcal{A} \cdot \hat{p}$ for the Hamiltonian rather than $\hat{H}' = \hat{\mu}E_o/2$. This alternate form proves most useful when calculating the effects of multiple bands. The matrix element describes emission or absorption between the valence and conduction band and therefore involves the Bloch wave functions for the valence and conduction bands. The calculation proceeds by using $\hat{p}$ in differential form and dividing the resultant integral into a portion specific to the unit cell and another applicable to the entire crystal without reference to the unit cell. The integral over the unit cell produces a type of dipole moment and harbors the transition selection rules. The integral over the entire crystal provides the vertical transitions expected for the interaction between matter and light. The results apply to emitters and detectors.

The next section puts together all of the pieces in Equation (8.5.3b) to calculate the gain. It shows a graphical method to calculate the gain for a semiconductor laser.

8.5.1 Vector Potential Form of the Interaction

Before calculating the transition matrix element, we first state the vector potential form of the interaction. Rather than use the $\hat{\mu} \cdot \vec{\mathcal{E}}$ form of the matter–light interaction, we use the alternative form given in Section 7.6

$$\hat{\mathcal{H}} = \frac{(\hat{p} - q\vec{\mathcal{A}})^2}{2m} + \hat{V} \tag{8.5.4}$$

where $\hat{V}$ represents the internal potentials of the atom (such as the Coulomb potential due to the nucleus), and $\mathscr{A}$ represents the vector potential (Section 3.12). The charge q has the value $q = -e$ for an electron where $e = 1.6 \times 10^{-19}$ coulombs. The vector potential satisfies the coulomb gauge condition $\nabla \cdot \mathscr{A} = 0$. Neglecting the square of the vector potential, we can identify the EM interaction Hamiltonian as

$$\hat{V} = \frac{e}{2m}(\hat{p} \cdot \vec{\mathscr{A}} + \vec{\mathscr{A}} \cdot \hat{p}) \tag{8.5.5}$$

where $\hat{p} = \hbar/i\nabla$.

The interaction Hamiltonian (Equation (8.5.5)) can be simplified by allowing it to operate on a wave function

$$\left(\hat{p} \cdot \vec{\mathscr{A}} + \vec{\mathscr{A}} \cdot \hat{p}\right)\psi = \hat{p} \cdot \left(\vec{\mathscr{A}}\psi\right) + \vec{\mathscr{A}} \cdot \hat{p}\psi = \left(\hat{p} \cdot \vec{\mathscr{A}}\right)\psi + \vec{\mathscr{A}} \cdot \left(\hat{p}\psi\right) + \vec{\mathscr{A}} \cdot \hat{p}\psi \tag{8.5.6a}$$

where the derivative of a product has been used (since $\hat{p}$ involves a derivative). The first term on the right hand side must be zero due to the Coulomb gauge condition

$$\hat{p} \cdot \vec{\mathscr{A}} = \frac{\hbar}{i}\nabla \cdot \vec{\mathscr{A}} = 0$$

Therefore, Equation (8.5.6a) becomes

$$\left(\hat{p} \cdot \vec{\mathscr{A}} + \vec{\mathscr{A}} \cdot \hat{p}\right)\psi = 2\vec{\mathscr{A}} \cdot \hat{p}\psi \tag{8.5.6b}$$

and Equation (8.5.5) becomes

$$\hat{V} = \frac{e}{m}\vec{\mathscr{A}}(\vec{r}, t) \cdot \hat{p} \tag{8.5.7}$$

The vector potential $\vec{\mathscr{A}}(\vec{r}, t)$ must be real. It has the form

$$\vec{\mathscr{A}}(\vec{r}, t) = \text{Re}\left\{\vec{A}(\vec{r})e^{-i\omega t}\right\} = \frac{1}{2}\left[\vec{A}e^{-i\omega t} + \vec{A}^*e^{+i\omega t}\right] \tag{8.5.8}$$

where we have used $z + z^* = 2\,\text{Re}(z)$ and $A(\vec{r}) = A_o e^{i\vec{k}_\gamma \cdot \vec{r}}$. Therefore, Equation (8.5.7) becomes

$$\hat{V} = \frac{e}{2m}\vec{A} \cdot \hat{p}e^{-i\omega t} + \vec{A}^* \cdot \hat{p}e^{+i\omega t} = \frac{e}{2m}\left[A(\vec{r})\tilde{e} \cdot \hat{p}e^{-i\omega t} + A^*(\vec{r})\tilde{e} \cdot \hat{p}e^{+i\omega t}\right] \tag{8.5.9}$$

where $\tilde{e}$ gives the direction of the vector potential and hence also the direction of the electric field. Recall from Section 6.2 that the vector potential must be perpendicular to the direction of the wave vector $\vec{k}$ because of the Coulomb gauge condition. Comparing this last result with Equation (8.5.2) shows the time independent Hamiltonian must be

$$\hat{H}' = \frac{e}{2m}A(\vec{r})\,\tilde{e} \cdot \hat{p} \tag{8.5.10}$$

where $\tilde{e}$ gives the polarization direction of the vector potential.

8.5.2 The Matrix Elements for the Homojunction Devices

Homojunction devices include lasers, detectors and LEDs made from a single semi-conductor but with different types of doping. The matrix element provides the quantum mechanics of the transition.

For simplicity, assume we want to calculate the transition matrix element for an emission event. Assume an electron can make a transition from a state in the conduction band to one in the valence band as shown in Figure 8.5.1. The electron initially in the conduction band state $|2k_2\rangle$ transitions to the valence band state $|1k_1\rangle$. We will be able to find the transition matrix element and also demonstrate $k_1 = k_2$ which produces the "vertical transition."

We want to calculate the transition matrix element

$$H_{fi} = \left\langle 1\vec{k}_1 \middle| \hat{H}' \middle| 2\vec{k}_2 \right\rangle = \frac{e}{2m} \left\langle 1\vec{k}_1 \middle| A(\vec{r})\tilde{e} \cdot \hat{p} \middle| 2\vec{k}_2 \right\rangle \tag{8.5.11}$$

in Equation (8.5.3b). Each band has a state labeled by k. The Bloch wave function $\Phi_{n,\vec{k}} \sim |n,k\rangle$ distinguishes between these states by the band index n. Recall that the Bloch wavefunction has the form

$$\Phi_{n,\vec{k}}(\vec{r}) = \frac{e^{i\vec{k}\cdot\vec{r}}}{\sqrt{V}} u_{n,\vec{k}}(\vec{r}) \tag{8.5.12}$$

as suggested by Figure 8.5.2. The function $u_{n\vec{k}}$ has the periodicity of the lattice while the envelope function corresponds to a wavelength very large compared with the unit cell. The large normalization volume V can be written as $V = L_x L_y L_z$ and the wave vector $\vec{k}$ becomes

$$\vec{k} = \frac{2m\pi}{L_x}\tilde{x} + \frac{2n\pi}{L_y}\tilde{y} + \frac{2q\pi}{L_z}\tilde{z}$$

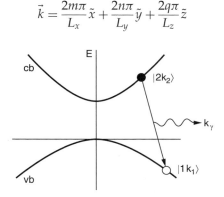

FIGURE 8.5.1
A hypothetical optical transition from the conduction to the valence band.

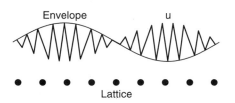

FIGURE 8.5.2
The envelope and periodic function.

as a result of periodic boundary conditions. The envelope function applies to either the valence or the conduction band states and depends only on the electron wave vector $\vec{k}$. The envelope portion satisfies the following inner product

$$\left\langle \frac{e^{i\vec{k}_1 \cdot \vec{r}}}{\sqrt{V}} \,\middle|\, \frac{e^{i\vec{k}_2 \cdot \vec{r}}}{\sqrt{V}} \right\rangle = \delta_{\vec{k}_1, \vec{k}_2} \tag{8.5.13a}$$

where the integral covers only the volume V. The periodic part satisfies

$$\left\langle u_{m,\vec{k}} \,\middle|\, u_{n,\vec{k}} \right\rangle = V_{uc}\delta_{mn} \tag{8.5.13b}$$

where the integral covers only the volume of the unit cell V_{uc}. The plane wave envelope functions apply to a homojunction whereas a quantum well requires a linear combination of the plane waves.

Now we can evaluate the transition matrix element in 8.5.11.

$$H_{fi} = \left\langle 1\vec{k}_1 \middle| \hat{H}' \middle| 2\vec{k}_2 \right\rangle = \frac{e}{2m} \left\langle 1\vec{k}_1 \middle| A(\vec{r})\, \tilde{e} \cdot \hat{p} \middle| 2\vec{k}_2 \right\rangle$$

where $A(\vec{r}) = A_o e^{i\vec{k}_y \cdot \vec{r}}$. The inner product becomes

$$\left\langle 1\vec{k}_1 \middle| \hat{H}' \middle| 2\vec{k}_2 \right\rangle = \frac{e}{2m} \int_V d^3 r\, \frac{e^{-i\vec{k}_1 \cdot \vec{r}}}{\sqrt{V}} u^*_{1,\vec{k}_1}(\vec{r}) \{A(\vec{r})\, \tilde{e} \cdot \hat{p}\} \frac{e^{i\vec{k}_2 \cdot \vec{r}}}{\sqrt{V}} u_{2,\vec{k}_2}(\vec{r}) \tag{8.5.14}$$

The operator $\hat{p} = (\hbar/i)\nabla$ requires us to differentiate the two functions on the right hand end to find

$$\left\langle 1\vec{k}_1 \middle| \hat{H}' \middle| 2\vec{k}_2 \right\rangle = \frac{e\,\tilde{e} \cdot \hbar\vec{k}_2}{2m} \int_V d^3 r\, \frac{e^{-i\vec{k}_1 \cdot \vec{r}}}{\sqrt{V}} A(\vec{r}) \frac{e^{i\vec{k}_2 \cdot \vec{r}}}{\sqrt{V}} u^*_{1,\vec{k}_1} u_{2,\vec{k}_2}$$

$$+ \frac{e}{2m} \int_V d^3 r\, \frac{e^{-i\vec{k}_1 \cdot \vec{r}}}{\sqrt{V}} A(\vec{r}) \frac{e^{i\vec{k}_2 \cdot \vec{r}}}{\sqrt{V}} u^*_{1,\vec{k}_1} \tilde{e} \cdot \hat{p}\, u_{2,\vec{k}_2} \tag{8.5.15}$$

We can identify $s = e^{-i\vec{k}_1 \cdot \vec{r}} A(\vec{r}) e^{i\vec{k}_2 \cdot \vec{r}}/V$ as a slowly varying function since the EM wave usually has wavelength much larger than 1000 angstroms and the electron wave vector (near the *cb* minimum or *vb* maximum) has a large size compared with the lattice constant. The electron wavelength is larger than 10 lattice constants for the typical quantum well. The functions $f = u^*_{1,\vec{k}_1} u_{2,\vec{k}_2}$ and $f = u^*_{1,\vec{k}_1} \tilde{e} \cdot \hat{p} u_{2,\vec{k}_2}$ must be rapidly varying functions since they must repeat over each unit cell. Therefore, we can write the integrals in Equation (8.5.15) using the results of Appendix 7 as

$$I = \int_V dV\, s(\vec{r})\, \langle f(\vec{r}) \rangle$$

where the average must be taken over the unit cell located at position $\vec{r}$. Because the function "u" in f must be periodic in the crystal lattice, the average must be independent of location of the unit cell. The integral can now be written as

$$I = \langle f \rangle \int_V dV\, s(\vec{r}) \tag{8.5.16}$$

The same result can be found using a cellular procedure similar to that in Topic 8.2.4 (refer to the chapter review exercises).

The first integral in Equation (8.5.15) can now be written as

$$
\int_V d^3r \, \frac{e^{-i\vec{k}_1 \cdot \vec{r}}}{\sqrt{V}} A(\vec{r}) \frac{e^{i\vec{k}_2 \cdot \vec{r}}}{\sqrt{V}} u^*_{1,\vec{k}_1} u_{2,\vec{k}_2} = \left\langle u^*_{1,\vec{k}_1} u_{2,\vec{k}_2} \right\rangle_{uc} \int_V d^3r \, \frac{e^{-i\vec{k}_1 \cdot \vec{r}}}{\sqrt{V}} A(\vec{r}) \frac{e^{i\vec{k}_2 \cdot \vec{r}}}{\sqrt{V}} \tag{8.5.17}
$$

where $A(\vec{r}) = A_o e^{i\vec{k}_\gamma \cdot \vec{r}}$. Using the typical definition of the average from elementary calculus $\langle f \rangle_{uc} = 1/V_{uc} \int_{V_{uc}} dV \, f(\vec{r})$, the average in Equation (8.5.17) becomes

$$
\left\langle u^*_{1,\vec{k}_1} u_{2,\vec{k}_2} \right\rangle_{uc} = \frac{1}{V_{uc}} \int_{V_{uc}} dV \, u^*_{1,\vec{k}_1} u_{2,\vec{k}_2} = \delta_{12} = 0
$$

by virtue of the orthogonality relation in Equation (8.5.13b). Therefore, Equation (8.5.15) reduces to

$$
\left\langle 1\vec{k}_1 \left| \hat{H}' \right| 2\vec{k}_2 \right\rangle = \frac{e}{2m} \int_V d^3r \, \frac{e^{-i\vec{k}_1 \cdot \vec{r}}}{\sqrt{V}} A(\vec{r}) \frac{e^{i\vec{k}_2 \cdot \vec{r}}}{\sqrt{V}} u^*_{1,\vec{k}_1} \tilde{e} \cdot \hat{p} \, u_{2,\vec{k}_2} \tag{8.5.18}
$$

We can find the matrix element in Equation (8.5.18) by again applying the results of Appendix 7 or using the technique illustrated in Topic 8.2.4. The slow function can be identified as $s = e^{-i\vec{k}_1 \cdot \vec{r}} A(\vec{r}) e^{i\vec{k}_2 \cdot \vec{r}} / V$. This time, the fast function must be

$$
f = u^*_{1,\vec{k}_1} \tilde{e} \cdot \hat{p} \, u_{2,\vec{k}_2}
$$

Using the results 8.5.16, the integral in Equation (8.5.18) can be written as

$$
\left\langle 1\vec{k}_1 \left| \hat{H}' \right| 2\vec{k}_2 \right\rangle = \frac{e}{2m} \left\langle u^*_{1,\vec{k}_1} \tilde{e} \cdot \hat{p} \, u_{2,\vec{k}_2} \right\rangle_{uc} \int_V d^3r \, \frac{e^{-i\vec{k}_1 \cdot \vec{r}}}{\sqrt{V}} A(\vec{r}) \frac{e^{i\vec{k}_2 \cdot \vec{r}}}{\sqrt{V}} \tag{8.5.19a}
$$

Notice the matrix element of the Bloch function can be written in either of two ways

$$
\left\langle u^*_{1,\vec{k}_1} \tilde{e} \cdot \hat{p} \, u_{2,\vec{k}_2} \right\rangle_{uc} = \frac{1}{V_{uc}} \int_{V_{uc}} dv \, u^*_{1,\vec{k}_1} \tilde{e} \cdot \hat{p} \, u_{2,\vec{k}_2} = \frac{\left\langle u_{1,\vec{k}_1} \left| \tilde{e} \cdot \hat{p} \right| u_{2,\vec{k}_2} \right\rangle}{V_{uc}}
$$

where the middle term comes from the elementary calculus formula for averages given by $\langle f \rangle_{uc} = V_{uc}^{-1} \int_{V_{uc}} dV \, f(\vec{r})$.

Next, examine the envelope matrix element in Equation (8.5.19a). The homojunction uses the plane wave envelope functions while the quantum well requires a linear combination of plane waves.

$$
\left\langle 1\vec{k}_1 \left| \hat{H}' \right| 2\vec{k}_2 \right\rangle = \frac{e}{2m} \frac{\left\langle u_{1,\vec{k}_1} \left| \tilde{e} \cdot \hat{p} \right| u_{2,\vec{k}_2} \right\rangle_{uc}}{V_{uc}} \int_V d^3r \, \frac{e^{-i\vec{k}_1 \cdot \vec{r}}}{\sqrt{V}} A(\vec{r}) \frac{e^{i\vec{k}_2 \cdot \vec{r}}}{\sqrt{V}} \tag{8.5.19b}
$$

The integral in this last equation requires the transitions to be vertical, as we will see. Equation (8.5.8) uses $A(\vec{r}) = A_o e^{i\vec{k}_\gamma \cdot \vec{r}}$ where $\vec{k}_\gamma$ denotes the photon wave vector. Substituting we find

$$\int_V d^3r \, \frac{e^{-i\vec{k}_1 \cdot \vec{r}}}{\sqrt{V}} A(\vec{r}) \frac{e^{i\vec{k}_2 \cdot \vec{r}}}{\sqrt{V}} = A_o \int_V d^3r \, \frac{e^{-i\vec{k}_1 \cdot \vec{r}}}{\sqrt{V}} \frac{e^{i(\vec{k}_2 + \vec{k}_\gamma) \cdot \vec{r}}}{\sqrt{V}} = A_o \delta_{\vec{k}_2 + \vec{k}_\gamma, \vec{k}_1} \qquad (8.5.20)$$

The Kronecker delta function comes from the fact that the integral has the form of an inner product between two basis vectors. Therefore combining Equations (8.5.19a) and 8.5.20, we find the transition matrix element has the form

$$\langle 1\vec{k}_1 | \hat{H}' | 2\vec{k}_2 \rangle = \frac{eA_o}{2m} \frac{\langle u_{1,\vec{k}_1} | \tilde{e} \cdot \hat{p} | u_{2,\vec{k}_2} \rangle}{V_{uc}} \delta_{\vec{k}_2 + \vec{k}_\gamma, \vec{k}_1} \quad \text{(homo junction)} \qquad (8.5.21a)$$

The delta function requires the final momentum of the electron $\hbar\vec{k}_1$ to equal the initial momentum $\hbar\vec{k}_2$ plus the photon momentum $\hbar\vec{k}_\gamma$ so that $\vec{k}_1 = \vec{k}_2 + \vec{k}_\gamma$. However, we set $\vec{k}_1 = \vec{k}_2$ because the photon momentum is very small. Therefore the transition matrix element must be

$$\langle 1\vec{k} | \hat{H}' | 2\vec{k} \rangle = \frac{eA_o}{2m} M_T \quad \text{(homo junction)} \qquad (8.5.21b)$$

where we have eliminated the unnecessary subscript on the k vector and defined the matrix element $M_T = \langle u_{1,\vec{k}} | \tilde{e} \cdot \hat{p} | u_{2,\vec{k}} \rangle / V_{uc}$

Using the transition matrix element in Equation (8.5.21b), Fermi's golden rule for the homojunction laser can be written as

$$R = \frac{2\pi}{\hbar} \left| H_{fi} \right|^2 \rho_r(E_T) \, F(E_T) \qquad (8.5.22a)$$

where, for emission, the marix element has the form

$$H_{fi}' = \langle 1\vec{k} | \hat{H}' | 2\vec{k} \rangle = \frac{eA_o}{2m} M_T = \frac{eA_o}{2mV_{uc}} \langle u_{1,\vec{k}} | \tilde{e} \cdot \hat{p} | u_{2,\vec{k}} \rangle \qquad (8.5.22b)$$

The Fermi factor depends on whether the transition produce emission or absorption

$$F(E_T) = \begin{cases} F_c(1 - F_v) & \text{Emission} \\ F_v(1 - F_c) & \text{Absorption} \end{cases} \qquad (8.5.22c)$$

8.5.3   The Quantum Well System

The quantum well falls within the class of reduced dimensional structures or nanostructure. Combining dissimilar materials such as GaAs and AlGaAs produce wells owing to the difference in band gap. Fermi's Golden rule can be applied to the quantum well devices (emitters or detectors) with very little modification. Only the reduced density of states and the matrix elements for the envelope wave function need to be re-examined. Section 8.4 discusses the reduced density of states for the quantum well laser. We now discuss the matrix elements for the quantum well laser.

For band #n, the general time-independent electron wave function has the form

$$|\Psi\rangle = \sum_{\vec{k}} \beta_{n\vec{k}} \left| n\vec{k} \right\rangle \tag{8.5.23a}$$

Writing the expansion in terms of the coordinate representation of the Bloch functions provides

$$\Psi = \sum_{\vec{k}} \beta_{n\vec{k}} \frac{e^{i\vec{k}\cdot\vec{r}}}{\sqrt{V}} u_{n\vec{k}}(\vec{r}) \tag{8.5.23b}$$

The periodic part of the Bloch wave function is essentially independent of the wave vector $\vec{k}$. We can write

$$\Psi = u_{n,0}(\vec{r}) \sum_{\vec{k}} \beta_{n\vec{k}} \frac{e^{i\vec{k}\cdot\vec{r}}}{\sqrt{V}} = u_{n,0}(\vec{r})\,\psi(\vec{r}) \tag{8.5.23c}$$

where

$$\psi(\vec{r}) = \sum_{\vec{k}} \beta_{n\vec{k}} \frac{e^{i\vec{k}\cdot\vec{r}}}{\sqrt{V}} \tag{8.5.23d}$$

The basis set $|n\vec{k}\rangle$ satisfies the Hamiltonian with the crystal potential V_L according to

$$\left(\frac{\hat{p}^2}{2m} + V_L \right) \left| n, \vec{k} \right\rangle = E_{n,\vec{k}} \left| n, \vec{k} \right\rangle$$

The envelope wavefunction satisfies the effective mass equation

$$\left(\frac{\hat{p}^2}{2m_n} + V_E \right) \psi(\vec{r}) = E_{\vec{k}} \psi(\vec{r}) \tag{8.5.24}$$

where m_n denotes the effective mass for band #n and V_E represents the potential beyond the periodic crystal potential. Assuming the effective mass is continuous across a heterostructure interface, the boundary conditions lead to solutions similar to those for a particle in a nonsemiconductor well. Infinitely deep quantum wells have sinusoidal waves for the envelope basis states. The conduction electrons and valence holes must have their own wave functions as indicated in Figure 8.5.3. This means that the boundary conditions select the appropriate coefficients β in Equation (8.5.23d) to produce the sinusoidal functions.

Let $\psi^{(c)}$ and $\psi^{(v)}$ denote the envelope wave functions for electrons in the conduction and valence bands, respectively. Following the steps leading to Equation (8.5.19b), the transition matrix elements must be

$$\left\langle 1, \vec{k}_1 \left| \hat{H}' \right| 2, \vec{k}_2 \right\rangle = \frac{e}{2m} M_T \int_V d^3 r \; \psi^{(v)}_{\vec{k}_1} A(\vec{r}) \psi^{(c)}_{\vec{k}_2} \tag{8.5.25a}$$

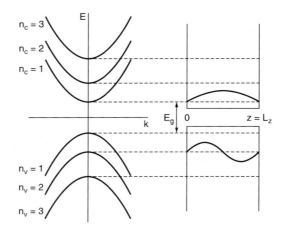

FIGURE 8.5.3
The subbands and the infinitely deep well.

where

$$M_T = \frac{\left\langle u_{1,\vec{k}_1} \left| \tilde{e} \cdot \hat{p} \right| u_{2,\vec{k}_2} \right\rangle_{uc}}{V_{uc}} \tag{8.5.25b}$$

As we will see, the integral leads to momentum conservation again.

We must evaluate the inner product for the envelope wave functions in Equation (8.5.25b); this inner product provides selection rules. We can see this by using the results of Topic 8.3.2. The well confines the electron in the z-direction and produces quantized energy for k_z direction but leaves the energy parabolic in k_x and k_y. This fact leads to the notion of sub-bands as depicted on the left side of Figure 8.5.3. The energy levels have the form

$$E = E_z + \frac{\hbar^2}{2m_e}\left(k_x^2 + k_y^2\right)$$

The general wave function has the form

$$\Psi = \sum_{k_x k_y k_z} C_{k_x k_y k_z} X_{k_x} Y_{k_y} Z_{k_z} e^{-itE/\hbar} = \sum_{k_x k_y k_z} C_{k_x k_y k_z} X_{k_x} Y_{k_y} Z_{k_z} e^{-itE_z/\hbar} \exp\left[-\frac{it}{\hbar}\frac{\hbar^2}{2m_e}\left(k_x^2 + k_y^2\right)\right] \tag{8.5.26}$$

where X and Y must satisfy periodic boundary conditions and Z satisfies the fixed endpoint boundary conditions appropriate for the infinitely deep well.

$$X_{k_x} = \frac{e^{ik_x x}}{\sqrt{L_x}} \qquad Y_{k_y} = \frac{e^{ik_y y}}{\sqrt{L_y}} \qquad Z_{k_z} = \sqrt{\frac{2}{L_z}}\sin(k_z z) \tag{8.5.27}$$

The basis function live in a direct product space and we can write

$$\psi_{\vec{k}} = X_{k_x} Y_{k_y} Z_{k_z} = \frac{e^{ik_x x}}{\sqrt{L_x}}\frac{e^{ik_y y}}{\sqrt{L_y}}\sqrt{\frac{2}{L_z}}\sin(k_z z) = \sqrt{\frac{2}{V}}e^{ik_x x + ik_y y}\sin(k_z z) \tag{8.5.28}$$

Now we evaluate the inner product of the envelope wave functions in Equation (8.5.25). Setting

$$A(\vec{r}) = A_o e^{i\vec{k}_\gamma \cdot \vec{r}} \tag{8.5.29}$$

we find

$$\left\langle \psi_{\vec{k}_1}^{(v)} \mid \psi_{\vec{k}_2}^{(c)} \right\rangle = \frac{A_o}{2} \left\langle \frac{e^{ik_x^{(v)}x}}{\sqrt{L_x}} \mid \frac{e^{i(k_x^{(c)}+k_x^{(\gamma)})x}}{\sqrt{L_x}} \right\rangle \left\langle \frac{e^{ik_y^{(v)}y}}{\sqrt{L_y}} \mid \frac{e^{i(k_y^{(c)}+k_y^{(\gamma)})y}}{\sqrt{L_y}} \right\rangle$$
$$\times \left\langle \sqrt{\frac{2}{L_z}}\sin(k_z^{(v)}z) \mid e^{ik_z^{(\gamma)}z} \sqrt{\frac{2}{L_z}}\sin(k_z^{(c)}z) \right\rangle \tag{8.5.30}$$

where the integrals in the first two inner products on the right hand side integrate over the macroscopic lengths L_x and L_y while the last inner product integrate over the width of the well L_z. The first two inner products provide

$$k_x^{(v)} = k_x^{(c)} + k_x^{(\gamma)} \quad \text{and} \quad k_y^{(v)} = k_y^{(c)} + k_y^{(\gamma)} \tag{8.5.31a}$$

Small photon momentum requires

$$k_x^{(v)} = k_x^{(c)} \quad \text{and} \quad k_y^{(v)} = k_y^{(c)} \tag{8.5.31b}$$

which can be conveniently summarized by

$$\vec{k}_\perp^{(c)} = \vec{k}_\perp^{(v)} \tag{8.5.31c}$$

where $\vec{k}_\perp = \tilde{x} k_x + \tilde{y} k_y$ and $\tilde{x}, \tilde{y}$ denote unit vectors. The third inner product in Equation (8.5.30) simplifies by noting the wavelength of the EM wave is much larger than the size of the well. Assume the well is located near $z = 0$. The argument $k_z^{(\gamma)}z$ must be approximately zero over the width of the well. Therefore, the exponential in the third inner product becomes equal to 1. Returning to Equation (8.5.30), assuming $e^{ik_z^{(\gamma)}z}$ is constant over the width of the well, we find

$$\left\langle \psi_{\vec{k}_1}^{(v)} \mid \psi_{\vec{k}_2}^{(c)} \right\rangle = \delta_{\vec{k}_\perp^{(c)}, \vec{k}_\perp^{(v)}} \frac{A_o}{2} \left\langle \sqrt{\frac{2}{L_z}}\sin(k_z^{(v)}z) \mid \sqrt{\frac{2}{L_z}}\sin(k_z^{(c)}z) \right\rangle \tag{8.5.32}$$

The matrix element in 8.5.25 becomes

$$\left\langle \Psi_{1,\vec{k}_1} \mid \hat{H}' \mid \Psi_{2,\vec{k}_2} \right\rangle = \frac{eA_o}{2m} M_T \, \delta_{\vec{k}_\perp^{(c)}, \vec{k}_\perp^{(v)}} \left\langle \sqrt{\frac{2}{L_z}}\sin(k_z^{(v)}z) \mid \sqrt{\frac{2}{L_z}}\sin(k_z^{(c)}z) \right\rangle \tag{8.5.32}$$

This last equation provides one of the most basic selection rules. We see that $k_z^{(v)} = k_z^{(c)}$ and given that $k = n\pi/L$ for n an integer, we find $n^{(c)} = n^{(v)}$. Therefore the electron can make transitions only between energy levels with the same quantum numbers (i.e., the envelope wavefunctions must have the same shape). The transition from $n^{(c)} = 1$ to $n^{(v)} = 2$ in Figure 8.5.3 cannot occur.

We can write a convenient form for Fermi's golden rule

$$R = \frac{2\pi}{\hbar} \left| \langle f | \hat{H}' | i \rangle \right|^2 \rho_r(E_T)\, F(E_T) \qquad F(E_T) = \begin{cases} F_c(1 - F_v) & \text{Emission} \\ F_v(1 - F_c) & \text{Absorption} \end{cases} \tag{8.5.33a}$$

where the matrix element H_{fi} for emission is

$$H'_{fi} = \left\langle 1, \vec{k}_1 \middle| \hat{H}' \middle| 2, \vec{k}_2 \right\rangle = \frac{eA_o}{2m} M_T \left\langle \psi^{(v)}_{\vec{k}_1} \middle| \psi^{(c)}_{\vec{k}_2} \right\rangle \tag{8.5.33b}$$

and recalling $E_o = i\omega A_o$ and $\vec{\mathscr{E}} = -\dot{\vec{\mathscr{A}}}$ from Section 6.2, we have

$$R = \frac{\pi}{2\hbar} \left(\frac{e}{m} \right)^2 \left(\frac{E_o}{\omega} \right)^2 |M_T|^2 \left| \left\langle \psi^{(v)}_{\vec{k}_1} \middle| \psi^{(c)}_{\vec{k}_2} \right\rangle \right|^2 \rho_r(E_T)\, F(E_T) \tag{8.5.33c}$$

where

$$M_T = \frac{\langle u_{1,\vec{k}} | \tilde{e} \cdot \hat{p} | u_{2,\vec{k}} \rangle_{uc}}{V_{uc}} \tag{8.5.33d}$$

8.6 Fermi's Golden Rule and Semiconductor Gain

The rate equations in Chapter 2 require an accurate expression for the gain. An expression incorporating the transition energy or wavelength makes it possible for the rate equations to describe wavelength-dependent effects. For example, the effect of a DBR mirror or the effects of two beams of light with two different wavelengths interacting with a gain medium. Fermi's golden rule provides the gain without the intraband relaxation effects that otherwise broaden the spectrum and "smear out" the gain versus wavelength profile. The gain describes various light emitters or photo-detectors depending on whether the gain is positive or negative in a wavelength range of interest. Semiconductor lasers can operate with a gain curve simultaneously having both positive and negative values. The laser emits light in the positive regions while it can absorb light as a pumping mechanism in the negative regions.

8.6.1 Homojunction Emitters and Detectors

Fermi's Golden rule provides the gain for the semiconductor laser rate equations. The rate R has units of transitions per second per unit volume and therefore accounts for the number of photons emitted or absorbed per second per volume $d\gamma/dt$ in the rate equations. The gain takes into account the difference between stimulated emission $R_{c \to v}$ and absorption events $R_{v \to c}$ where "c" and "v" refer to the conduction and valence bands, respectively. The difference $\mathscr{R} = R_{c \to v} - R_{v \to c}$ can be calculated from Fermi's golden rule using the nonequilibrium Fermi-Dirac distributions.

To start, consider the rate of transition from Section 8.5

$$R = \frac{\pi}{2\hbar} \left(\frac{e}{m} \right)^2 \left(\frac{E_o}{\omega} \right)^2 |M_T|^2 \rho_r(E_T)\, F(E_T) \tag{8.6.1}$$

where M_T refers to a momentum matrix element. The rate of emission from the conduction to valence band must be

$$R_{c \to v} = \frac{\pi}{2\hbar} \left(\frac{e}{m}\right)^2 \left(\frac{E_o}{\omega}\right)^2 |M_T|^2 \rho_r(E_T) \, F_c(1 - F_v) \tag{8.6.2a}$$

and the rate of absorption must be

$$R_{v \to c} = \frac{\pi}{2\hbar} \left(\frac{e}{m}\right)^2 \left(\frac{E_o}{\omega}\right)^2 |M_T|^2 \rho_r(E_T) \, F_v(1 - F_c) \tag{8.6.2b}$$

found by interchanging the "c" and "v" subscripts. The net rate of emission must be

$$\mathscr{R} = R_{c \to v} - R_{v \to c} = \frac{\pi}{2\hbar} \left(\frac{e}{m}\right)^2 \left(\frac{E_o}{\omega}\right)^2 |M_T|^2 \rho_r(E_T) \, (F_c - F_v) \tag{8.6.3}$$

Therefore the net rate of photon emission per volume becomes

$$\frac{d\gamma}{dt} = \mathscr{R} \tag{8.6.4}$$

where γ denotes the number of photons per unit volume.

Let's rewrite the emission rate $\mathscr{R}$ in a more familiar form by changing the electric field on the right-hand side of Equation (8.6.3) into the photon density. The electric field term $|E_o|^2$ can be related to the power in the cavity. The Poynting vector gives power per unit area

$$S = \frac{1}{2} |E \times H^*| = \frac{n_r |E_o|^2}{2\mu_o} \tag{8.6.5a}$$

where μ_o denotes the permeability of free space. Assuming a uniform distribution of EM energy in a material with refractive index n_r, the Poynting vector and the *energy density* ρ_E (energy per volume) must be related by

$$S = \rho_E \frac{c}{n_r} \tag{8.6.5b}$$

The energy density can be written as

$$\rho_E = \frac{n_r^2 |E_o|^2}{2\mu_o c} \tag{8.6.5c}$$

Therefore, the photon density must be given by

$$\gamma = \frac{\#\text{Phots}}{\text{Vol}} = \frac{\text{Energy}}{\text{Volume}} \left(\frac{\text{Energy}}{\text{Photon}}\right)^{-1} = \rho_E \frac{1}{\hbar\omega} = \frac{n_r^2 |E_o|^2}{2\mu_o c \hbar\omega} \tag{8.6.5d}$$

Substituting this last result for $|E_o|^2$ in Equation (8.6.3) produces

$$\mathscr{R} = R_{c \to v} - R_{v \to c} = \frac{\pi}{2\hbar} \left(\frac{e}{m}\right)^2 \left\{\frac{2\mu_o c \hbar}{\omega n_r^2} \gamma\right\} |M_T|^2 \rho_r(E_T) \, (F_c - F_v) \tag{8.6.6a}$$

where the electron occupation probability for the conduction and valence band is F_c and F_v, respectively. Canceling terms and substituting $\mu_o = 1/\varepsilon_o c^2$ provides

$$\frac{d\gamma}{dt} = \mathcal{R} = \left\{ \frac{\pi e^2}{\omega n_r^2 m^2 \varepsilon_o c} |M_T|^2 \rho_r(E_T) \, (F_c - F_v) \right\} \gamma \tag{8.6.6b}$$

where m denotes the free mass of the electron (not the effective mass).

We can identify the semiconductor gain in Equation (8.6.6b) as

$$g = \left\{ \frac{\pi e^2 \hbar}{(\hbar \omega) n_r^2 m^2 \varepsilon_o c} |M_T|^2 \rho_r(E_T) \, (F_c - F_v) \right\} \tag{8.6.7}$$

Recall from Section 8.4 that the occupation probability F_c must be evaluated at energy E_c and F_v at E_v in such a way that $E_T = E_c - E_v$. Notice we must have $F_c(E_c) - F_v(E_v) > 0$ to have positive gain. This relation along with properties for the reduced density of states requires $E_g < E_T < E_{fc} - E_{fv}$ for the system to have positive gain. A system in thermal equilibrium has one Fermi energy $E_f = E_{fc} = E_{fv}$ so that $F_c(E_c) - F_v(E_v) < 0$ and the gain satisfies $g \leq 0$ regardless of the temperature $T \geq 0$. High temperatures do not produce population inversions.

Three terms in Equation (8.6.7) account for the energy dependence (i.e., the wavelength dependence) of the gain. The dependence is primarily determined by the product $\rho_r(E_T) \, (F_c - F_v)$, which is often highly peaked over a relatively narrow energy band. The photon energy term $\hbar \omega$ can often be taken as relatively constant for sufficiently narrow peaks.

As an example, we can calculate the primary factor $\rho_r(E_T) \, (F_c - F_v)$ that determines the gain using a graphical approach. Suppose we apply a bias voltage to a homojunction laser that has equal conduction and valence band masses. The top portion of Figure 8.6.1 shows the density of states for the conduction and valence bands ρ_c, ρ_v, respectively, and the nonequilibrium Fermi distributions F_c, F_v. The bias voltage separates the two quasi-Fermi levels as shown. To calculate the gain, imagine a horizontal line of length E_T

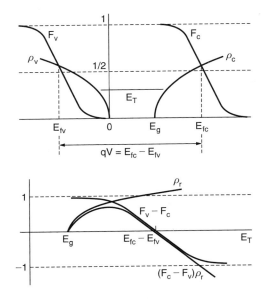

FIGURE 8.6.1
The applied voltage V sets the quasi-Fermi levels and therefore the gain.

connecting states in the valence band with those in the conduction band. For example, at $0 \le E_T < E_g$ the density of states in both the conduction and valence band is zero and therefore the reduce density of states must likewise be zero in this region as shown in the figure. Next, suppose $E_T = E_g$ so that the line E_T starts at $E_v = 0$ and ends at $E_c = E_g$. At the value $E_c = E_g$, the quasi-Fermi function F_c has a value approximately equal to one while at $E_v = 0$ the valence quasi-Fermi function F_v has a value approximately equal to zero. Consequently, the Fermi factor provides

$$F(E_T = E_g) = F_c - F_v = 1$$

The other values can be similarly calculated. For example, consider E_T such that $F_c = F_v = 1/2$. In this case, the transition energy is $E_T = E_c - E_v = E_{fc} - E_{fv}$. The Fermi factor produces zero as illustrated since

$$F = F_c(E_c) - F_v(E_v) = 0.5 - 0.5 = 0$$

Figure 8.6.1 shows the Fermi factor F as a function of the transition energy. It also shows the reduced density of states must be approximately half of either the valence or conduction band density-of-states.

Figure 8.6.1 shows the gain

$$g \sim \rho_r(E_T)\,(F_c - F_v)$$

must be positive where the Fermi factor $F_c - F_v$ is positive. The electron density reaches its peak value within this region of the plot. Therefore, the figure verifies that emitted photons will have an energy $\hbar\omega$ in the range $E_g < \hbar\omega < E_{fc} - E_{fv}$. The gain is small near the band edge e.g., since the density of states is small there. The gain is also small for large energy since the Fermi factor decreases at higher energy.

8.6.2 The Gain for Quantum Well Materials

We now develop the gain for semiconductor quantum well lasers. The previous section provides the transition rate as

$$R = \frac{\pi}{2\hbar} \left(\frac{e}{m}\right)^2 \left(\frac{E_o}{\omega}\right)^2 |M_T|^2 \left|\left\langle \psi_{\vec{k}_1}^{(v)} \,\middle|\, \psi_{\vec{k}_2}^{(c)} \right\rangle\right|^2 \rho_r(E_T)\,F(E_T) \tag{8.6.7a}$$

where

$$M_T = \frac{\left\langle u_{1,\vec{k}} \,\middle|\, \tilde{e} \cdot \hat{p} \,\middle|\, u_{2,\vec{k}} \right\rangle_{uc}}{V_{uc}} \tag{8.6.7b}$$

and $\langle \psi_{\vec{k}_1}^{(v)} \,|\, \psi_{\vec{k}_2}^{(c)} \rangle$ represents the overlap of the envelop wavefunctions for the basis states. It consists of sinusoidal functions for the quantum well and complex exponentials for the nonconfined directions. The amplitude E_o gives the electric field amplitude in $E = E_o \cos(\omega t)$ where ω represents the angular frequency of the EM wave.

For the quantum well laser with "z" as the confinement direction, the rate reduces to

$$R = \frac{\pi}{2\hbar} \left(\frac{e}{m}\right)^2 \left(\frac{E_o}{\omega}\right)^2 |M_T|^2 \delta_{n^{(v)}, n^{(c)}} \rho_r(E_T)\,F(E_T) \tag{8.6.8a}$$

where $n^{(v)}, n^{(c)}$ denote the valence and conduction sub-bands, respectively. Proceeding similarly to Equation (8.6.7), we can write the gain as

$$g = \left\{ \frac{\pi e^2 \hbar}{(\hbar\omega) n_r^2 m^2 \varepsilon_0 c} \delta_{n^{(v)}, n^{(c)}} |M_T|^2 \rho_r(E_T) (F_c - F_v) \right\} \tag{8.6.8b}$$

where m denotes the free mass of the electron. The primary difference between the homojunction and quantum well lasers appears to be the reduced density of states. However, the Kronecker delta function also forces the quantum well material to make transitions between envelope states with the same principal quantum number.

Figure 8.6.2 shows a plot of the sub-bands (left), the density of states (middle) and the normalized gain curve $\rho_r(E_T) (F_c - F_v)$ (right). The spacing of the sub-bands depends on the quantized energy levels due to electron and hole confinement along the z direction. We assume an infinitely deep well for convenience. The discrete levels have the energy

$$E_{zn} = \frac{\hbar^2 k_{zn}^2}{2m_e} = \frac{\pi^2 \hbar^2 n^2}{2m_e L_z^2} \tag{8.6.9}$$

where m_e denotes the effective mass for the band and n denotes an integer. The sub-bands have a parabolic form since the electrons can freely move along the unconfined x and y directions. This can be seen from Schrodinger's equation that gives the dispersion relation as $E = \hbar^2 k^2 / 2m_e$ where $k^2 = k_x^2 + k_y^2 + c$ and the constant c represents the quantized k_{zn} component in Equation (8.6.9). Each value of $E_n = \hbar^2 k_{zn}^2 / (2m_e)$ therefore determines the position of the corresponding parabola along the energy axis. Different effective masses between the conduction and valence bands lead to different positions for the sub-bands along the energy axis. The collection of states in all of the sub-bands represents the allowed energy states for the system.

The density of states and the quasi-Fermi distributions appear in the middle graph of Figure 8.6.2. The figure shows both the bulk and quantum well density of states (refer to Section 8.4). The position of the conduction sub-band minimums (or maximums

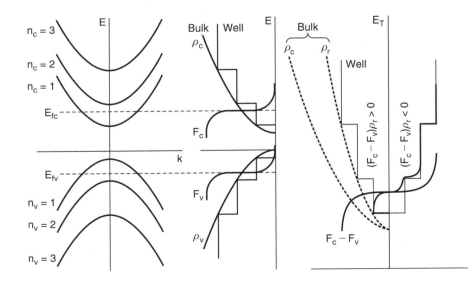

FIGURE 8.6.2
Hypothetical quantum well structure with equal electron and hole masses. Left: Sub-bands. Middle: Density of States. Right: the Fermi factor multiplied by the reduced density of states.

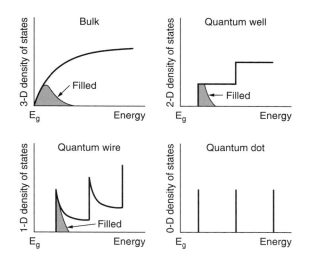

FIGURE 8.6.3
Density of states plots.

for the valence sub-band) determines the position of the steps for the well density of states. The minimum of the quantum well density-of-states occurs at a larger energy than the minimum for the bulk because the quantum-well energy levels sit above the bottom of the well, which corresponds to the bottom of the conduction band.

Briefly, recall why the well produces a step-like structure for the density of states. Consider the conduction states. For energy E in the interval $0 < E < E_{z1}$, none of the states have energy smaller than the bottom of the $n = 1$ sub-band. For energy E between the bottoms of the $n = 1$ and $n = 2$ sub-bands in $E_{z1} < E < E_{z2}$, the density of states comes from only the $n = 1$ sub-band. Section 8.4 shows the density of states is a constant for a single sub-band. Next consider energy between the bottoms of the $n = 2$ and $n = 3$ sub-bands. In this case, the total density of states at energy E must be the sum of the number of states in the $n = 1$ sub-band plus the number of states in the $n = 2$ sub-band. This total density-of-states must be given by the sum of two constants and therefore the density of states plot must reach a new, higher plateau. The other plateaus occur for similar reasons.

The right-hand graph of Figure 8.6.2 shows the reduced density of states and the term $(F_c - F_v)\rho_r$ related to the gain in Equation (8.6.8b). The quasi-Fermi levels E_{fc} and E_{fv} occupy positions within the sub-bands above the first quantized energy levels. Therefore, the gain $g \sim (F_c - F_v)\rho_r$ must be positive for $E < E_{fc} - E_{fv}$ and negative for $E > E_{fc} - E_{fv}$. Basically, the material emits light for those energies where the gain is positive. It absorbs light for those energies where the gain is negative. Keep in mind that the energy width for regions of positive or negative gain depend on the pumping level of the material (i.e., the separation of the quasi-Fermi levels). Those regions where the gain is negative can be used for photodetectors or for pumping a laser. The gain curve in Figure 8.6.3 therefore gives the spectral response of the photodetector.

8.6.3 Gain for Quantum Dot Materials

The quantum dot materials confine the electron in three dimensions. In this case, Fermi's golden rule once again provides a relation of the form

$$g = \frac{\pi e^2 \hbar}{(\hbar\omega) n_r^2 m^2 \varepsilon_o c} |M_T|^2 \rho_r(E_T) (F_c - F_v)$$

for allowed transitions (i.e., between envelope wavefunctions with the same set of quantum numbers). The reduced density of states appears as a set of spikes as in Figure 8.6.3. Without collision broadening mechanisms, we would expect the emission and absorption spectra to have the form of a sequence of very narrow spikes. The reduced density of states would equip an emitter or detector sharp filtering characteristics.

8.6.4   Example of a Homojunction 2-D Laser with Unequal Band Masses

Consider a 2-D bulk homojunction laser with a bias voltage set by a battery that separates the quasi-Fermi levels according to $qV_b = E_{fc} - E_{fv}$ where $q < 0$ for an electron. Note that a bias current does not completely determine the separation of the Fermi levels since the recombination terms reduce the number of carriers while the pump current increases it; consequently, a balance between pump and recombination determines the separation. Assume for the example that the effective valence band mass is twice that of the conduction band $m_v = 2m_c$, the energy gap is $E_g = 1.7\,\text{eV}$ and $qV_b = E_{fc} - E_{fv} = 3.3$.

The 2-D bulk cyrstal has conduction and valence bands described by

$$\rho_v = \frac{1}{\pi}\frac{m_h}{\hbar^2}\theta(-E) \quad \rho_c = \frac{1}{\pi}\frac{m_e}{\hbar^2}\theta(E - E_g) \tag{8.6.10}$$

where θ is the step function. Figure 8.6.4 shows the conduction, valence and reduced density of states. To calculate the gain $g \sim \rho_r(E_T)(F_c - F_v)$, we need to know how to place $E_T = \hbar\omega$ and $(F_c - F_v)$ in a diagram similar to Figure 8.6.4. The position of E_T depends on the effective mass of the bands. The Fermi factor can be positioned using $n = p$ (number of electrons per unit volume equals number of holes per unit volume) and by knowing the shape of the bands (effective mass).

First consider E_T. The relation for the transition energy

$$E_T = E_c - E_v \tag{8.6.11}$$

requires placement of the energy E_c and E_v (Figure 8.6.5). The energy E_v has a minus sign since the top of the valence band is taken to be zero. Section 8.4 indicates the energies E_v and E_c must be related by

$$m_v E_v = -m_c (E_c - E_g) \tag{8.6.12}$$

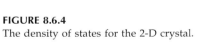

FIGURE 8.6.4
The density of states for the 2-D crystal.

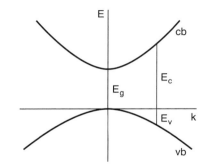

FIGURE 8.6.5
Relation between E_c and E_v.

where m_v and m_c represent the valence and conduction band effective masses. As demonstrated in Equation (8.4.4b),

$$E_c = \frac{m_v E_T + m_c E_g}{m_c + m_v} \quad E_v = -\frac{m_c}{m_c + m_v}\left(E_T - E_g\right) \tag{8.6.13}$$

For this example, the relation $m_v = 2m_e$ produces $E_c = (2/3)E_T + (1/3)E_g$ and $E_v = -1/3$ $(E_T - E_g)$. For example, when $E_T = E_g$, then $E_c = E_g$ and $E_v = 0$ so that $\rho_r = 0$. When $E_T = 2$ then $E_c = 1.9$ and $E_v = -0.1$. Equations (8.6.10) indicate density of states of $\rho_c = m_c$ and $\rho_v = m_v$ in units of $1/(\pi\hbar^2)$ for the values of $E_c = 1.9$ and $E_v = -0.1$. The reduced density of states for $m_v = 2m_e$ must be $\rho_r = m_c m_e/(m_c + m_v) = 2/3$ as shown in Figure 8.6.4.

Next, find the position of E_{fv} and E_{fc} for the Fermi factor $F_c - F_v$. The number of electrons equals the number of holes $n = p$ when they remain approximately confined to an intrinsic region. Relating n to the density of conduction states and to the quasi-Fermi distribution produces

$$n = \int_{E_g}^{\infty} dE\, \rho_c(E)\, F_c(E) = \int_{E_g}^{\infty} dE\, \frac{1}{\pi}\frac{m_c}{\hbar^2}\frac{1}{1 + e^{(E-E_{fc})/kT}} = \frac{kT}{\pi}\frac{m_c}{\hbar^2} Ln\left[1 + e^{(E_{fc}-E_g)/kT}\right] \tag{8.6.14a}$$

Therefore, for $E_{fc} - E_g > 2kT = 0.05\,\text{eV}$ at room temperature, the density of electrons (per area in this case) becomes

$$n = \frac{m_c}{\pi\hbar^2}\left(E_{fc} - E_g\right) \tag{8.6.14b}$$

This approximation holds since the number of states not completely filled near E_{fc} becomes small in comparison to those filled below E_{fc}. Figure 8.6.6 show the relation. A similar expression for the number of holes in the valence band must hold

$$p = -\frac{m_v}{\pi\hbar^2} E_{fv} \tag{8.6.14c}$$

The conditions $E_{fc} - E_g > 2kT$ and $E_{fv} < -2kT$ should be verified at the end. The condition $n = p$ produces

$$m_e\left(E_{fc} - E_g\right) = -m_v E_{fv} \tag{8.6.15}$$

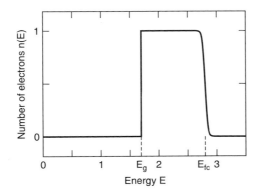

FIGURE 8.6.6
The number of electrons at energy E. The number is normalized to the density of states.

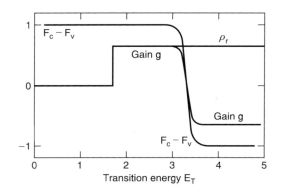

FIGURE 8.6.7
Gain for 2-D crystal.

Combining 8.6.15 with $qV_b = E_{fc} - E_{fv}$ produces

$$E_{fc} = \frac{m_c E_g + m_v qV_b}{m_c + m_v} \qquad E_{fv} = \frac{m_c E_g - m_c qV_b}{m_c + m_v} \qquad (8.6.16)$$

For this example with $m_v = 2m_e$, the quasi-Fermi levels become $E_{fc} = 2.8$ $E_{fv} = -0.53$. Figure 8.6.4 shows the relation between the reduced density of states and the quasi-Fermi levels.

Finally, the gain can be determined as

$$g \sim \rho_r(E_T)\,[F_c(E_c) - F_v(E_v)]$$

where $E_T = E_c - E_v$. Figure 8.6.7 shows the gain versus the transition energy. To obtain the plot, first select a transition energy E_T and calculate the corresponding E_c and E_v using Equations (8.6.13). Calculate the density of states from Equations (8.6.10) using E_c and E_v. The Fermi factor $F_c(E_c) - F_v(E_v)$ comes from Equations (8.4.8), (8.6.16), and (8.6.13). The results appear in Figure 8.6.7.

8.7 The Liouville Equation and Semiconductor Gain

The second approach to semiconductor gain and absorption starts with the results from the Liouville equation for the density matrix. In this approach, the Liouville equation includes the relaxation effects similar to the rate equations in Chapter 2 so that the solution incorporates saturation and gain-broadening effects. We start with the susceptibility, substitute an expression for the population difference based on the density of states, and then develop the gain and absorption. Unlike the results from Fermi's golden rule, the gain incorporates the line broadening.

8.7.1 Homojunction Devices

The Liouville equation for the density matrix

$$\dot{\rho}_{ab} = \frac{1}{i\hbar}\left[\underline{H}, \rho\right]_{ab} - \frac{\rho_{ab} - \bar{\rho}_{ab}}{\tau_{ab}} \qquad (8.7.1)$$

The Hamiltonian $\underline{H}$ includes the atomic Hamiltonian $\underline{H}_A$ and the EM interaction $\hat{V}$ according to $\underline{H} = \underline{H}_A + \underline{V}$. The relaxation term includes the population relaxation time constant $\tau = \tau_{11} = \tau_{22}$ and the dipole dephasing time $T_2 = \tau_{12} = \tau_{21}$. Recall the applied voltage essentially sets the "no light steady state (NLSS)" density operator $\bar{\rho}_{ab}$. The quasi-Fermi levels determine the NLSS density operator. The Liouville equation produces an expression for the susceptibility (Topic 7.12.3)

$$\chi_r = \frac{\mu^2 \Delta \bar{n}}{\varepsilon_o \hbar} \frac{T_2(\omega - \omega_o)}{1 + (\omega - \omega_{21})^2 T_2^2 + 4\Omega^2 \tau T_2}$$

$$\chi_i = -\frac{\mu^2 \Delta \bar{n}}{\varepsilon_o \hbar} \frac{1}{1 + (\omega - \omega_o)^2 T_2^2 + 4\Omega^2 \tau T_2}$$

(8.7.2)

where $\Omega = \mu_o E_o/(2\hbar)$, E_o denotes the electric field amplitude, ω denotes the angular frequency of the driving EM field, and $\omega_o = [E(\vec{k}_c) - E(\vec{k}_v)]/\hbar = (E_c - E_v)/\hbar = E_T/\hbar$ and E_T represents the transition energy (since $E_v < 0$ as shown in Figure 8.7.1). These equations depend on the NLSS value of the population difference $\Delta \bar{n} = \bar{n}_c - \bar{n}_v$ (units of "per unit volume"). We want to know the susceptibility under actual operating conditions. We need to use the actual population difference $\Delta n = n_c - n_v$ for the susceptibility and not the NLSS population difference $\Delta \bar{n} = \bar{n}_c - \bar{n}_v$. The rate equations use a gain that depends on the number of electrons at time t and not at steady state, and they include the NLSS population difference through the pump-number current density. In fact, the rate equation for dn/dt can be solved for steady state to find a result reminiscent of Equations (8.7.2) with the saturation term.

Let the quantity n_c represents the number of electrons (per unit volume) in the conduction band at energy $E_c = E(k_c)$ and n_v represents the number of electrons in the valence band at energy $E_v = E(k_v)$. The differential form of the susceptibility comes from Topic 7.12.3

$$d\chi(E) = \frac{\mu_o^2 T_2(dn)[(E - E_T)T_2 - i]}{\varepsilon_o \hbar^2 [1 + (E - E_T)^2 T_2^2/\hbar^2]}$$

(8.7.3)

where $dn = dn_c - dn_v$ has units of "per unit volume." The real and imaginary parts of the susceptibility contribute to the refractive index and gain respectively. Section 3.2 (Equation (3.2.17)) shows the relations

$$g = -\frac{k_o}{n_r}\left(\mathrm{Im}(\chi) + \frac{\sigma}{\varepsilon_o \omega}\right) = -\alpha \quad n_r = \sqrt{1 + \mathrm{Re}(\chi) + \chi_b}$$

(8.7.4)

where in this case, χ_b is approximately real and represents the susceptibility of the host crystal at the frequency of the EM field. Recall the gain refers to the power in the electric field according to

$$|E|^2 = \left|E_o \exp[i(k_o n - ig/2)z)]\right|^2 = |E_o|^2 \exp(gz)$$

(8.7.5a)

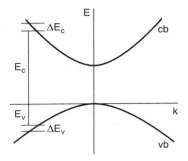

FIGURE 8.7.1
The band diagram.

so that the rate equation has the form

$$\frac{dP}{dt} = gP \quad \text{or} \quad \frac{d\gamma}{dt} = g\gamma \tag{8.7.5b}$$

where the optical power $P \sim |E|^2$ and γ denotes the photon density. We assume negligible conductivity σ and write

$$g = -\frac{k_o}{n_r} \text{Im}(\chi) = -\alpha \tag{8.7.6}$$

where k_o denotes the wave vector for the EM wave in vacuum. To find the gain in Equation (8.7.6), we must find the susceptibility in 8.7.3 and the population difference.

The population difference must depend on the position of the quasi-Fermi levels (under operating conditions) and on the position of the states within the band. The population difference can be written in terms of the reduced density of states and the quasi-Fermi distributions as in Topic 8.1.11. First starting with the density of states for the conduction and valence bands, we have

$$dn_v = (\rho_v dE_v)F_v \qquad dn_c = (\rho_c dE_c)F_c \tag{8.7.7}$$

where $\rho_v dE_v$ and $\rho_c dE_c$ give the number of states within the energy range dE_v and dE_c respectively. The quasi-Fermi distributions F_v and F_c describe the probability of an electron occupying a valence band or conduction band state at energy E_v or E_c, respectively.

$$F_{v,c} = \frac{1}{1 + \exp\{(E - E_{fv,fc})/(k_b T)\}} \tag{8.7.8}$$

where E_{fv} and E_{fc} denote the valence and conduction quasi-Fermi levels respectively. Furthermore Section 8.4 shows that the "reduced density of states ρ_r" describes the number of pairs of states at energy $E_T = E_c - E_v$, which must be the same as the number of states at E_c in range dE_c and also the same as the number of states at E_v in range dE_c

$$\rho_r dE_T = \rho_c dE_c = \rho_v dE_v \tag{8.7.9}$$

Now we can combine Equations (8.7.7) and (8.7.9) to find the population difference

$$dn = dn_c - dn_v = (\rho_c dE_c)F_c - (\rho_v dE_v)F_v = \rho_r dE_T(F_c - F_v) \tag{8.7.10}$$

The susceptibility can now be written in terms of the reduced density of states as

$$d\chi(E) = \frac{\mu_o^2 T_2(dn)[(E-E_T)T_2 - i]}{\varepsilon_o \hbar^2 [1 + (E-E_T)^2 T_2^2/\hbar^2]} = \frac{\mu_o^2 T_2[(E-E_T)T_2 - i]}{\varepsilon_o \hbar^2 [1 + (E-E_T)^2 T_2^2/\hbar^2]} \rho_r dE_T (F_c - F_v) \qquad (8.7.11)$$

The gain in Equation (8.7.6) becomes

$$g = \int_0^\infty \frac{k_o}{n_r} \frac{\mu_o^2 T_2}{\varepsilon_o \hbar^2 [1 + (E-E_T)^2 T_2^2/\hbar^2]} \rho_r dE_T (F_c - F_v) \qquad (8.7.12)$$

The dephasing collisions described by T_2 broaden the gain curve. The expression can be compared with the results of Fermi's golden rule in Section 11.4, which does not include any relaxation effects.

Example 8.7.1

Find the gain for large dephasing time T_2 (very few collisions).

Solution: The integrand is nonzero only for small energy differences. We can make the substitution (see Appendix 5)

$$\lim_{T_2 \to \infty} \frac{T_2}{\pi[1 + (E-E_T)^2 T_2^2/\hbar^2]} = \delta(E - E_T) \qquad (8.7.13)$$

The gain becomes

$$g = \frac{\pi k_o \mu_o^2}{\varepsilon_o \hbar^2 n_r} \rho_r (F_c - F_v) \qquad (8.7.14)$$

We can see that the results of Equation (8.7.14) must be equivalent to the results in Equation (8.6.7), specifically

$$g = \left\{ \frac{\pi e^2 \hbar}{(\hbar\omega) n_r^2 m^2 \varepsilon_o c} |M_T|^2 \rho_r(E_T) (F_c - F_v) \right\}$$

by using the relation $\mu_{cv}\mathscr{E} = e/m\mathscr{A} M_T$, where $E, \mathscr{A}$ represent the electric field and vector potential, respectively.

8.7.2   Quantum Well Material

The results from the Liouville equation remain independent of the type of material. Equation (8.7.12) provides

$$g = \int_0^\infty \frac{k_o}{n_r} \frac{\mu_o^2 T_2}{\varepsilon_o \hbar^2 [1 + (E-E_T)^2 T_2^2/\hbar^2]} \rho_r dE_T (F_c - F_v) \qquad (8.7.15)$$

We just need to replace the reduced density of states with the one appropriate for the quantum well.

The lineshape function

$$\mathscr{L}(E - E_T) = \frac{T_2}{\pi\left[1 + (E - E_T)^2 T_2^2/\hbar^2\right]} \tag{8.7.16}$$

provides a convolution type of average. We can see this as follows. For very large T_2, the lineshape function becomes a Dirac delta function as indicated in Example 8.7.1. Taking the delta function limit, we find a result similar to that for Fermi's golden rule in Section 11.4.

$$g_{T_2\to\infty} = \frac{\pi\mu_o^2 k_o}{n_r\varepsilon_o\hbar^2}\rho_r(E_T)(F_c - F_v) \tag{8.7.17}$$

Next consider the case $0 \ll T_2 < \infty$.

For simplicity consider a region where $(F_c - F_v) = +1$ and where $0 \ll T_2 < \infty$. The lineshape function has its largest value at $E_T = E$ where E is the given energy of the photons and E_T is the integral in Equation (8.7.15). The gain for photons at energy E has the form

$$g(E) \sim \int_a^b \mathscr{L}(E - E_T)\rho_r(E_T)dE_T \tag{8.7.18a}$$

It is clear from this last equation that the gain (or absorption) is the convolution between the reduce density of states and the lineshape function. The lineshape function rounds-off the gain curves. This can be seen as follows. Convert the integral in 8.7.18a to a summation to find

$$g(E) \sim \sum_i \mathscr{L}(E - E_{Ti})\rho_r(E_{Ti})dE_i \tag{8.7.18b}$$

The prescription in 8.7.18b then says to multiply the density of states by the sequence of lineshape functions shown in Figure 8.7.2 and sum. Those lineshape functions for $E_T < E$ multiply a constant density of states and reproduce essentially the value of ρ_r near E. However, the lineshape functions for $E_T > E$ multiply a larger density of states and therefore add values somewhat larger than the value of ρ_r near E. For this reason, the gain $g(E)$ increases near E even though the density of states does not increase at E. Figure 8.7.3

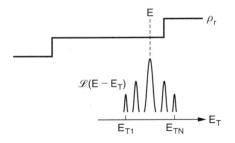

FIGURE 8.7.2
The density of states and the lineshape function.

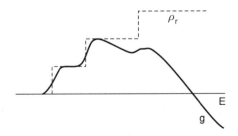

FIGURE 8.7.3
The lineshape function rounds off the corners of the gain.

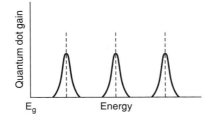

FIGURE 8.7.4
Gain broadening for the quantum dot.

shows how the convolution integral therefore rounds off the corners of the density of states to produce $g(E)$.

The collisional broadening leads to broad linewidths rather than the delta function type that might be expected. The broadening also leads to the rounding off of corners. If the phonon is responsible for broadening, then we would expect to find more narrow linewidths for lower temperatures.

8.7.3 Quantum Dot Material

Similar to the development of the gain and absorption for the quantum well, we expect the broadening mechanisms to produce similar effects in the quantum dot material. Equation (8.7.18a) provides

$$g(E) \sim \int_a^b \mathcal{L}(E - E_T)\rho_r(E_T)dE_T \tag{8.7.19}$$

For delta function density of states

$$\rho_r = \sum_i C_i \delta(E - E_{Ti}) \tag{8.7.20}$$

the gain has the form of a sequence of lineshape functions

$$g(E) = \sum_i C_i \mathcal{L}(E - E_{Ti}) \tag{8.7.21}$$

Figure 8.7.4 shows the broadening.

Quantum dot gain curves can also become broadened as a result of nonuniform sizes. Slightly different sized dots produce slightly different energy levels. This would result in gain curves similar to Figure 8.7.4.

8.8 Review Exercises

8.1 Find the effective mass m_{ij} for the band $E = A\hbar^2 k^2 = \hbar^2(Ak_x^2 + Bk_y^2 + Ck_z^2)$ and find an expression for the force given the components a_x, a_y, a_z.

8.2 Find the effective mass matrix for $E - E_c = 3(k_x - 1)^2 + 3(k_y - 2)^2$. Be sure to discuss the effective mass m_{zz}.

8.3 Find the effective mass for $E = E_c + \sum_{a,b=1}^{3} C_{ab}k_ak_b$ where E_c represents the bottom of the conduction band and C_{ab} is a constant. Using a plane wave $\psi \sim e^{i\vec{k}\cdot\vec{r}}$, show the Hamiltonian $\hat{\mathscr{H}} = -\hbar^2/2\nabla \cdot m_e^{-1} \cdot \nabla + V_E$ can be written as $\hat{\mathscr{H}} = -\hbar^2/2\sum_{a,b}(m_e^{-1})_{ab}k_ak_b + V_E$

8.4 For a bulk 3-D crystal, show the density of energy states must be

$$g_E^{(3D)} = \frac{\sqrt{2}\,m_e^{3/2}}{\pi^2\hbar^3}\sqrt{E - E_c}\,\theta(E - E_c)$$

8.5 For a bulk 1-D crystal, show the density of energy states must be

$$g_E^{(1D)} = \frac{\sqrt{2}}{\pi\hbar}\frac{\sqrt{m_e}}{\sqrt{E - E_c}}\theta(E - E_c)$$

8.6 Show the law of mass action $np = n_i^2$ holds regardless of the doping lever where n_i refers to the number of electrons at thermal equilibrium without doping ($n_i = p_i$).

8.7 Explain why the electron wave does not propagate for wave vectors at the edge of the FBZ. Show the wave functions $B_{3D} = \{e^{i\vec{k}\cdot\vec{r}}/\sqrt{V}\}$ are correctly normalized when $\vec{k} = (2\pi n_x/L_x)\hat{x} + (2\pi n_y/L_y)\hat{y} + (2\pi n_z/L_z)\hat{z}$ and $V = L_xL_yL_z$.

8.8 Show for a conduction band for a 3-D crystal with nonspherical constant energy surfaces that the density of states must be $g(E) = V_{xal}m_e^{3/2}\sqrt{E}/(\sqrt{2}\,\pi^2\hbar^3)$ where the *effective mass is* $m_e = (m_xm_ym_z)^{1/3}$ as shown in Figure P8.8. Assume the dispersion relation has the form $E = -\hbar^2/2\sum_{a,b}(m_e^{-1})_{ab}k_ak_b$ where $(m_e^{-1})_{ab} = \begin{pmatrix} 1/m_x & 0 & 0 \\ 0 & 1/m_y & 0 \\ 0 & 0 & 1/m_z \end{pmatrix}$. Hint: you will need the volume enclosed by an ellipsoid $V = 4\pi/3abc$ where a, b, c are the intercepts with the x, y, z axis.

8.9 Consider a line of atoms space by lattice constant "a." Suppose $f(x)$ is a function periodic on the lattice $f(x) = f(x + R)$ for *all* direct lattice vectors $R = na$ where "n" denotes an integer. Using the Fourier expansion of $f(x)$ of the form $f(x) = \sum_m B_me^{iG_mx}$ to find the allowed reciprocal lattice vectors G.

8.10 Suppose someone suggests the following wave function for an electron traveling along a line of atoms separated by a distance "a."

$$\psi = \frac{1}{(2\pi)^{1/4}\sqrt{\sigma}}\,e^{-\frac{(x-\bar{x})^2}{4\sigma^2}}A\sin(Cx)$$

where $\bar{x}$ represents the center of the packet and depends on time, and σ describes the width of the packet.

1. If n deBroglie wavelengths fit within the length "a," then find C.

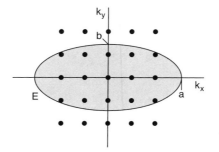

FIGURE P8.8
An ellipse in k-space as a constant energy surface.

2. In terms of the previous problem, explain why the n deBroglie wavelengths must fit within the distance a.

3. Find the value of A to normalize this wavefunction over all space x. Assume $\sigma \gg a$.

8.11 Find the eigenfunctions for an infinitely deep quantum well where the effective mass has the form

$$\underline{m}_e = \begin{pmatrix} m_x & 0 & 0 \\ 0 & m_y & 0 \\ 0 & 0 & m_z \end{pmatrix}$$

and the confinement direction is along z similar to that in Section 8.3.

8.12 Explain how the density of states differs between the finitely and infinitely deep quantum wells.

8.13 Consider the conduction band for a GaAs heterostructure. The width of the finitely deep well is L. Assuming the effective electron mass is constant across the interfaces, apply boundary conditions for which the wave function and its derivative is continuous across the interfaces. Find the conditions to find the wave vector k and the corresponding energy of an electron in the well. You should find that the wave vector k in the well can be written implicitly as

$$\tan(kL) = \frac{2k\sqrt{\alpha^2 - k^2}}{2k^2 - \alpha^2}$$

where $\alpha^2 = 2m_e V_o/\hbar^2$, m_e denotes the electron effective mass, and V_o represents the height of the well.

8.14 Consider an electron in a finitely deep GaAs quantum well embedded in an $Al_xGa_{1-x}As$ as in the previous problem. Find the allowed values of k and $E = \hbar^2 k^2/2m_e$ and find an expression for the density of states. Assume the band offset is 0.3 eV, the electron effective mass is $m_e = 0.0665$ and the width of the well is $L = 70$ angstroms. One method of finding k from the results of the previous problem is to plot the left side and right side as two separate curves on the same set of axes and find the intersection point.

8.15 The density of states (Figure P8.15) for either the conduction or valence band has a form similar to

$$\rho = \frac{m^{3/2}}{\sqrt{2}\,\pi^2\hbar^2}\sqrt{E}$$

where $m = m_v$ represents the valence band effective mass and $m = m_c$ represents the conduction band effective mass. Assume a parabolic band near $k = 0$. Note that $E_1 < 0$.

Show $m_c E_2 = -m_v E_1$ for vertical transitions defined by $E_T = E_g + |E_1| + |E_2|$. Hint, draw a band diagram, identify a vertical transition and use $E = \hbar^2 k^2/(2m)$ where m refers to the effective mass of either band.

8.16 Consider the density of states plot and thermal equilibrium Fermi Function F shown in Figure P8.16. Assume the hole and electron effective masses are identical. Use the definitions of E_1 and E_2 in the previous problem. Assume an emission event.

1. Using a ruler and the results of problem 3, plot the reduced density of states as a function of energy E_T. Hint: use the results of the last problem to identify E_1 in terms of E_T and then E_2 in terms of E_1.

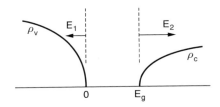

FIGURE P8.15
Density of states for the conduction and the valence bands.

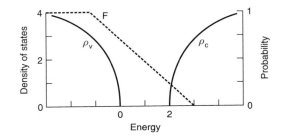

FIGURE P8.16
The density of states and thermal equilibrium Fermi Function.

2. Plot the Fermi Factor $F_c(1 - F_v)$ assuming the semiconductor remains in thermal equilibrium.
   Hint: use F and $(1 - F)$ in the figure.

3. Explain why a laser composed of a semiconductor in thermal equilibrium can never have gain regardless of the temperature.

4. Draw on rough sketch of the gain curve.

8.17 For parabolic bands, show the transition energy E_T versus wave vector k has the form $E_T = \hbar^2 k^2 / 2m_r$ where $1/m_r = 1/m_c + 1/m_h$ gives the reduced mass and m_c, m_v are the effective mass of the conduction and valence electron respectively.

8.18 Consider the reduced density of states discussed in Section 8.11 for parabolic bands.

1. For $m_v \to \infty$, make a rough drawing of the conduction and valence density of states versus energy. Calculate and draw the reduced density of states.

2. Show all steps leading to the following relations found in Topic 8.4.1

$$E_c = \frac{m_v E_T + m_c E_g}{m_c + m_v} \qquad E_v = -\frac{m_c}{m_c + m_v}(E_T - E_g)$$

3. For parabolic bands with $m_c = m_v$, find the reduced density of states in terms of the conduction band density of states.

8.19 A 3-D bulk semiconductor with parabolic bands has a valence band density of states twice as large as for the conduction band. For simplicity, normalize constants such as $\sqrt{2}/(\pi^2 \hbar^3)$ to one. Assume $m_c = 1$ and $E_g = 1$.

1. Find the effective mass for the valence band in terms of the effective mass for the conduction band.

2. On a piece of graph paper or using computer software, plot the density of states for the conduction and valence bands.

3. For transition energy $E_T = 2E_g$, find the energy E_1 and E_2 from the band edge to the states participating in the transition for the valence and conduction bands, respectively. Refer to Section 8.4 for notation if necessary.

4. Draw or generate a plot for the reduced density of states.

8.20 Repeat all steps in the previous problem when the valence band density of states is 3 times larger than for the conduction band.

8.21 Explain any changes for finding the reduced density of states when the bands are nonparabolic (i.e., the effective mass depends on k) but produce a direct band gap. What happens for indirect band gaps?

8.22 Show the 3-D bulk and quantum well density-of-states have the same value at the vertex of each sub-band as indicated in Figure 8.4.5.

8.23 Suppose a quantum well is placed in a bulk 3-D crystal with a conduction band and single valence band. Suppose the valence band density-of-states for the bulk 3-D crystal is a factor of C times larger than that for the conduction band. Assume the confined electrons and holes in the well have the same effective mass as they do for the 3-D case.

1. Show the energy levels for the confined holes and electrons must be related by $E_{vn} = E_{cn}/C^{2/3}$ where "n" indicates the sub-band.

2. Show $\rho_v^{3\text{-}D}(E_{vn}) = C^{2/3}\rho_c^{3\text{-}D}(E_{cn})$

8.24 Use the information from Problem 8.23 and find an expression for the reduced density of states (Figure P8.24). In this case, the transition energy E_T must be $E_T = -E_1 + E_2 + E_g - E_{vz}^{(n)} + E_{cz}^{(n)}$ where $E_1 < 0$ is the energy from the top of valence sub-band #n to some point in the sub-band, $E_2 > 0$ is the energy from the bottom of conduction sub-band #n to some point in the band, E_g represents the usual 3-D bulk band gap, $E_{vz}^{(n)} < 0$ is the energy to the top of the valence sub-band #n, and $E_{cz}^{(n)} > 0$ is the energy to the bottom of conduction sub-band #n.

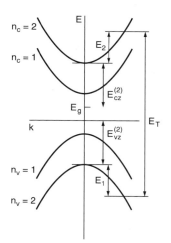

FIGURE P8.24
The sub-bands for the quantum well.

8.25 Start with Equation (8.5.11) in Section 8.5,

$$\left\langle 1\vec{k}_1 \middle| \hat{H}' \middle| 2\vec{k}_2 \right\rangle = \frac{e\,\tilde{e} \cdot \hbar \vec{k}_2}{2m} \int_V d^3r \, \frac{e^{-i\vec{k}_1 \cdot \vec{r}}}{\sqrt{V}} A(\vec{r}) \frac{e^{i\vec{k}_2 \cdot \vec{r}}}{\sqrt{V}} u^*_{1,\vec{k}_1} u_{2,\vec{k}_2}$$

$$+ \frac{e}{2m} \int_V d^3r \, \frac{e^{-i\vec{k}_1 \cdot \vec{r}}}{\sqrt{V}} A(\vec{r}) \frac{e^{i\vec{k}_2 \cdot \vec{r}}}{\sqrt{V}} u^*_{1,\vec{k}_1} \tilde{e} \cdot \hat{p}\, u_{2,\vec{k}_2}$$

and use the cellular method in Topic 8.2.4 to show

$$\left\langle 1\vec{k}_1 \middle| \hat{H}' \middle| 2\vec{k}_2 \right\rangle = \frac{eA_o}{2m} \frac{\left\langle u_{1,\vec{k}_1} \middle| \tilde{e} \cdot \hat{p} \middle| u_{2,\vec{k}_2} \right\rangle}{V_{uc}} \delta_{\vec{k}_2 + \vec{k}_\gamma, \vec{k}_1}$$

for the homojunction.

8.26 Consider the quantum well discussed in Topic 8.5.3. Let $\psi^{(c)}$ and $\psi^{(v)}$ denote the envelope wave functions for electrons in the conduction and valence bands, respectively. Following the steps leading to Equation (8.5.19b), show the transition matrix elements must be

$$\left\langle 1,\vec{k}_1 \middle| \hat{H}' \middle| 2,\vec{k}_2 \right\rangle = \frac{e}{2m} M_T \int_V d^3r \, \psi^{(v)}_{\vec{k}_1} A(\vec{r}) \psi^{(c)}_{\vec{k}_2} \quad \text{where} \quad M_T = \frac{\left\langle u_{1,\vec{k}_1} \middle| \tilde{e} \cdot \hat{p} \middle| u_{2,\vec{k}_2} \right\rangle_{uc}}{V_{uc}}.$$

8.27 Based on the results of Section 8.5.3, can an electron make a transition from one conduction sub-band to another conduction sub-band in an infinitely deep well? Assume an electromagnetic interaction. What about a finitely deep well? Explain and back up your argument with calculations.

8.28 Prove the relation $E_g < E_T < E_{fc} - E_{fv}$ must hold for the gain to be positive at the transition energy E_T, where E_g, E_{fc}, E_{fv} represent the gap energy, conduction quasi-Fermi energy and the valence quasi-Fermi energy.

8.29 Show that a system at thermal equilibrium can only potentially exhibit gain for negative temperatures in °K.

8.30 A researcher fabricates a homojunction laser using a direct bandgap semiconductor with parabolic bands.

1. Show the largest "reduced density of states" occurs when the effective masses for the conduction and valence band are equal $m_c = m_v$.

2. Show that the gain must also be maximized by this choice for the effective masses.

8.31 A student fabricates a new type of homojunction LED in the MERL fabrication center. The LED can simultaneously emit two colors of light: Blue 410 nm (3 eV) and Red 620 nm (2 eV). Figure P8.31.1 shows a side view of the device. The device has 3 semitransparent electrodes. The voltage V_b controls the blue LED and voltage V_r controls the red LED. Assume the device is maintained at a temperature near 0°K. Assume the density of states for the bands are step functions.

1. Figure P8.31.2 shows the density of states and the thermal equilibrium Fermi-Dirac distribution for the blue emitter. Draw a plot of the gain as a function of

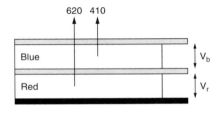

FIGURE P8.31.1
The two-color light emitting diode (LED).

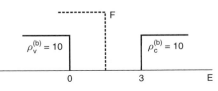

FIGURE P8.31.2
The blue emitter at thermal equilibrium.

transition energy E_T. Include the reduced density of states. Explain why this device will work as a photodetector and what range of energies can be detected.

2. Consider Figure P8.31.3. Draw pictures for the gain for both LEDs and include a diagram of the reduced density of states (draw to scale). Explain what happens to the wavelength range of the emitted light as the bias voltage V to either LED increases. Assume the quasi-Fermi levels in each material satisfy a relation of the form $qV = E_{fc} - E_{fv}$ and the relation $n = p$ for electrons and holes causes the quasi-Fermi levels to have equal energy to their respective band edge.

3. What voltage should be applied to the red LED so that its emission spectrum does not overlap the emission spectrum from the blue LED?

8.32 Consider a pump-number current density $\mathscr{I}$ (carrier pairs/volume/second) that removes electrons from the valence band and places them in the conduction band for a 3-D crystal. Assume $\mathscr{I}$ starts at $t = 0$. Suppose the carrier recombination rate is characterized by a constant τ. Assume the same number of holes and electrons (per unit volume) $n = p$.

1. For n_o initial electrons, find the number of electrons versus time $n(t)$.

2. Find the difference in quasi-Fermi levels $E_{fc} - E_{fv}$.

8.33 The chapter found a condition on the quasi-Fermi levels for equal numbers of holes and electrons $n = p$ in a 3-D crystal when $|E_c - E_{fc}| > 2kT$ and $|E_v - E_{fv}| > 2kT$. Consider the case of $-2kT < E_c - E_{fc} < 2kT$ and $-2kT < E_v - E_{fv} < 2kT$. Assume the band edges are within 2kT of the respective quasi-Fermi level. Making a linear approximation of the quasi-Fermi functions, find a relation between E_{fc} and E_{fv} for the condition $n = p$. The valence and conduction effective masses do not need to be identical.

8.34 Use a ruler and Figure P8.34 to calculate the reduced density of states and the normalized gain given by $g' = \rho_r(F_c - F_v)$.

8.35 For Figure P8.35, draw the quasi-Fermi function F_v based on the quasi-Fermi function F_c. Use a ruler and Figure P8.35 to find the reduced density of states and normalized gain given by $g' = \rho_r(F_c - F_v)$.

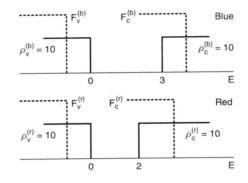

FIGURE P8.31.3
The diagrams for both the red and blue LEDs.

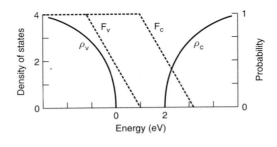

FIGURE P8.34
The density-of-states and quasi-Fermi functions.

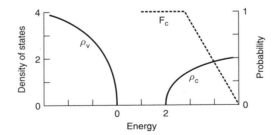

FIGURE P8.35
Quasi-Fermi functions and bands with unequal effective mass.

8.36 Based on Figure P8.34, find the peak gain and the corresponding wavelength. Repeat the procedure for $qV = (E_{fc} - E_{fv}) = 3, 4, 5$. Make a plot of your findings.

8.37 Suppose a grating mirror has peak reflectivity at 413 nm. Based on Figure P8.34 and P8.35, what peak gain can be expected in both cases?

8.38 Explain why the saturated gain from the Liouville equation differs from the unsaturated gain for the rate equations. Prove Equation (8.7.3)

$$d\chi(E) = \frac{\mu_o^2 T_2 (dn)[(E - E_T)T_2 - i]}{\varepsilon_o \hbar^2 [1 + (E - E_T)^2 T_2^2 / \hbar^2]}$$

Why don't the rate equations incorporate the saturated gain? Find the saturated gain from the rate equations in Chapter 2 (Hint: refer to the Chapter 2 review exercises).

8.39 First prove the relation

$$\mu_{cv}\mathscr{E} = \frac{e}{m}\mathscr{A}\,M_T$$

in Section 8.7 and then show the relations

$$g = \frac{\pi k_0 \mu_o^2}{\varepsilon_o \hbar^2 n_r}\rho_r(F_c - F_v)g = \left\{\frac{\pi e^2 \hbar}{(\hbar\omega)n_r^2 m^2 \varepsilon_o c}|M_T|^2 \rho_r(E_T)\,(F_c - F_v)\right\}$$

are equivalent. Discuss any assumptions.

8.9 Further Reading

The following list contains well-known references for the material presented in the chapter.

1. Zory P.S., Ed., *Quantum Well Lasers*, Academic Press, Boston (1993).
2. Chuang S.L., *Physics of Optoelectronic Devices*, John Wiley & Sons, New York (1995).
3. Yariv A., *Quantum Electronics*, 3rd ed., John Wiley & Sons, New York (1989).
4. Coldren L.A., *Diode Lasers and Photonic Integrated Circuits*, John Wiley & Sons, New York (1995).
5. Klingshirn C.F., *Semiconductor Optics*, Springer-Verlag, Berlin (1997).
6. Vedeyan J.T., *Laser Electronics*, 2nd ed., Prentice Hall, Englewood Cliffs, NJ (1989).

Appendix 1

Review of Integrating Factors

This appendix provides a quick review of integrating factors as a method of solving first order differential equations. Suppose we want to solve the equation

$$\dot{y} - ay = f(t) \tag{A1.1}$$

where $y = y(t)$ and the dot indicates the first derivative with respect to time. Suppose we multiply through out by a function $\mu(t)$, the integrating factor,

$$\mu\dot{y} - a\mu y = \mu f(t) \tag{A1.2}$$

with the particular property that the left-hand side is an exact derivative

$$\frac{d}{dt}(\mu y) = \mu\dot{y} - a\mu y \tag{A1.3}$$

Then we could write Equation A1.2 as

$$\frac{d}{dt}(\mu y) = \mu f(t)$$

If the forcing function $f(t)$ starts at $t = 0$, we can integrate both sides of the equation with respect to time to obtain

$$\mu(t)y(t) = \mu(0)y(0) + \int_0^t d\tau\,\mu(\tau)f(\tau)$$

or

$$y(t) = \frac{\mu(0)y(0)}{\mu(t)} + \frac{1}{\mu(t)}\int_0^t d\tau\,\mu(\tau)f(\tau) \tag{A1.4}$$

Once we know the integrating factor $\mu(t)$ then we also know the form of the solution even when the exact form of the forcing function has not been specified. This is the property that makes the integrating factor useful for our purposes.

How do we find the integrating factor? Use Equation A1.3 and expand the derivative

$$\frac{d}{dt}(\mu y) = \mu\dot{y} + \dot{\mu}y \tag{A1.5}$$

Combining Equations A1.3 and A1.5 we find

$$\mu \dot{y} + \dot{\mu} y = \mu \dot{y} - a\mu y$$

to arrive at

$$\dot{\mu} = -a\mu$$

By separating variables, this simple first order differential equation has the solution

$$\mu(t) = e^{-at}$$

Notice that constants of integration are unimportant for integrating factors—they cancel out of the final equation.

Appendix 2

Rate and Continuity Equations

Two rate equations can be obtained from two "equations of continuity" which have the form

$$\nabla \cdot \vec{J} + \frac{\partial \rho}{\partial t} = S - R \qquad \text{(A2.1)}$$

where S and R represent the source and sink. First consider an equation of continuity for the charged carriers. The source and sink, S and R respectively, denote the carrier generation and recombination rates (units of #/vol/sec). J is the current density (amps per unit area), $\rho = n$ is the density of carriers (number per volume). Equation A2.1 shows that the rate of change of the number of carriers in the volume (time-derivative term) decreases as carriers leave the volume (divergence term) or recombine (R), but increases as more carriers are generated (S) within the volume by absorbing light. For a semiconductor, ρ should be interpreted as the number of carrier pairs (one pair consists of one hole and one electron) since they always recombine or generate as pairs.

$$\rho = n$$

If none of the charges leave the volume, then the divergence term is negligible.

$$\nabla \cdot \vec{J} = 0$$

The source term describes the amount of carrier pairs created in the semiconductor. In this case, carriers appear as a result of pump and the stimulated absorption process. The recombination term corresponds to both stimulated recombination (i.e., stimulated emission) and to the spontaneous recombination. Notice that the gain terms $g_t \gamma$ already includes the stimulated recombination and absorption. S can therefore be written as

$$S = \mathscr{J} - g_t \gamma$$

Substituting all the terms together gives the carrier rate equation found in the previous topic.

$$\frac{dn}{dt} = -g_t \gamma + \mathscr{J} - R$$

It is important to realize that the divergence term is not always negligible. For example, carriers in a given region might diffuse away. In such a case, the divergence of the current density must be related to the carrier diffusion. The current number density J is given by nv where v is the speed of the carrier.

Next, consider an equation of continuity for the photons.

$$\nabla \cdot \vec{J} + \frac{\partial \gamma}{\partial t} = S - R \qquad \text{(A2.2)}$$

J is the "photon current density" which is defined similarly to the usual current density

$$J = \rho_\gamma v_g = \gamma v_g$$

"Photon currents" can be useful for finding the photon density in a beam after it strikes a mirror. Also notice how the speed of the photons in the medium multiplies the photon density similar to that found in the photon rate equation. The rate of change of the photon density $\partial \gamma / \partial t$ measures the rate of increase of photons in a small region of space even though photons are never at rest (they are either moving at the speed of light or they are absorbed). The source S of photons consists of stimulated and spontaneous emission. The sink R of photons consists of stimulated absorption and the optical losses embodied in the cavity lifetime τ_γ. The photon equation of continuity can be rewritten as

$$v_g \cdot \nabla \gamma + \frac{\partial \gamma}{\partial t} = \Gamma g_t \gamma - \frac{\gamma}{\tau_\gamma} + \beta R_{\text{spont}} \tag{A2.3}$$

where the confinement factor Γ and the geometry factor β are included.

Example A2.1
Demonstrate the photon rate equation from Equation A2.3.
Assume a "lumped" model with $\nabla \gamma = 0$. Then Equation A2.3 reduces to the usual photon rate equation

$$\frac{\partial \gamma}{\partial t} = \Gamma g_t \gamma - \frac{\gamma}{\tau_\gamma} + \beta R_{\text{spont}}$$

Example A2.2
Show the rate equation that involves the z coordinate and the gain "g" can be found from Equation A2.3.
Assume steady-state in Equation A2.3 so that $\partial \gamma / \partial t = 0$. For changes along the z direction, we have

$$v_g \cdot \nabla \gamma = v_g \frac{\partial \gamma}{\partial z}$$

For simplicity, define $g_t = v_g g$ and $1/\tau_\gamma = v_g \alpha$. Substituting into Equation A2.3 provides

$$\frac{\partial \gamma}{\partial z} = \Gamma g \gamma - \alpha \gamma + \beta R_{\text{spont}} / v_g$$

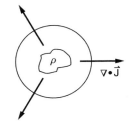

FIGURE A2.1
Schematic representation of the equation of continuity.

Appendix 3

The Group Velocity

The *group* velocity describes a type of average speed of a wave packet. A wave packet consists of many sinusoidal waves with each having a specific wavelength and frequency. That is, a wave packet consists of the superposition of multiple plane waves. *Phase* velocity describes the speed of a single sinusoidal wave with a single frequency. The phase velocity of the plane wave

$$\psi(x, t) = A_k e^{ikx - i\omega t} \tag{A3.1}$$

can be found by watching the motion of a single point of the wave. Focus on the point initially at $x = 0$ at $t = 0$. Setting the phase to zero

$$kx - \omega t = 0 \tag{A3.2}$$

provides the *phase* velocity

$$v_p = \frac{x}{t} = \frac{\omega}{k}$$

The *group* velocity describes the average speed of "wave-packets" travelling in a dispersive medium. Plane waves with different frequencies travel with different phase velocities in a dispersive medium. For optics, this means that the index of refraction depends on wavelength. Wave packets can represent photons, electrons, holes, and phonons (etc.). These wave packets can perhaps be most conveniently pictured as travelling Gaussian waves $f(z, t)$ as indicated in Figure A3.1 although they can have any arbitrary form. As we will shortly discuss, these Guassian waves are "envelope" wavefunctions.

The Fourier transform of the wave packet appears in Figure A3.2 that shows the amplitude $\phi(k)$ of the various spectral components plotted against the wave vector. For an optics example, the wave packet and its Fourier transform might describe a pulse of light. Suppose the center wavelength corresponds to green and the smaller amplitudes on either side of the center correspond to red and blue (see Figure A3.3).

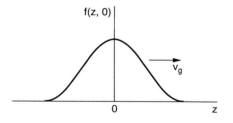

FIGURE A3.1
A wave packet moving to the right with group velocity v_g.

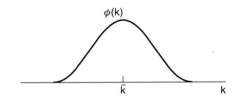

FIGURE A3.2
The Fourier transform of the wave packet *f*.

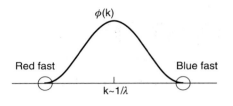

FIGURE A3.3
Various colors of light travel faster or slower than the average. Note that "*k*" refers to the carrier wavevector.

Obviously, the average wave vector $k = 2\pi/\lambda$ denoted by $\bar{k}$ cannot be anywhere near zero! Also, some pulses have narrow Fourier transforms $\phi(k)$ (unlike the one shown in Figure A3.3).

For a nondispersive medium, the wave packet shown in Figure A3.1 does not spread because all of the constituent components travel at the same speed. On the other hand, a dispersive medium (such as glass) requires the components to travel at different speeds. This means that the wave packet will spread out with time. A dispersive medium does not require the various components making up the pulse to interact with each other. Two spectral components can interact with each other in a nonlinear medium. For example, a blue component might get larger at the expense of two nearby infrared components.

One issue concerns the motion of a packet as compared with the motion of a plane wave. This is especially important for dispersive media where $\omega = \omega(k)$ or equivalently, $E = E(k)$. For optics, the relations are especially easy to picture. Consider the speed of the wave. If we write the phase velocity of a given plane wave as $v = \omega(k)/k$, we see that different colors travel at different speeds (this is dispersion). For example, blue light interacts more with a piece of glass than red light; therefore blue light runs slower (some materials are the reverse of this behavior). It is also blue light that is most deflected from its straight-line path by a glass prism (the index of refraction is larger for blue). As an example, consider Figure A3.3 showing that certain colors of light travel faster than the average while others travel slower. We might expect the width of the Gaussian to change as some of the wave run slower than an average while others run faster. The issue becomes one of describing the motion of the wave packet (the envelope) in spite of the fact that the various components travel at different speeds. The phase velocity is not the correct measure. Usually, people describe the wave packet as consisting of a slowly varying envelope function superimposed on the fast moving carrier waves. The function in Figure A3.1 provides one example of the envelope, and Figure A3.4 provides another for two superimposed sine waves with nearly identical frequencies and wavevectors (discussed in para A3.1). The envelope function is very long compared with the small wavelength carrier. The figure shows the group velocity v_g describing the speed of the envelope.

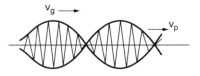

FIGURE A3.4
Envelope and phase velocity.

A3.1 Simple Illustration of Group Velocity

We can easily understand how the envelope can travel slower (much slower) than the plane waves by considering a simple example of adding two traveling sine waves together. We will work the same example in two ways that both lead to the same conclusion. First, assume that κ, κ' and ω, ω' are wave vectors and angular frequencies and that they are very close together in value. Assume two sine waves travel parallel to each other.

$$y = A \sin(\kappa x - \omega t) + A \sin(\kappa' x - \omega' t)$$

$$= 2A \cos\left(\frac{\kappa - \kappa'}{2}x - \frac{\omega - \omega'}{2}t\right)\sin\left(\frac{\kappa + \kappa'}{2}x - \frac{\omega + \omega'}{2}t\right) \quad \text{(A3.3)}$$

These last equations show that the summation of the two sine waves can be viewed as another sine wave with modulated amplitude. We can identify the carrier as

$$\sin\left(\frac{\kappa + \kappa'}{2}x - \frac{\omega + \omega'}{2}t\right) \quad \text{(A3.4)}$$

having approximate wave vector and frequency of

$$\frac{\kappa + \kappa'}{2} \cong \kappa \quad \text{and} \quad \frac{\omega + \omega'}{2} \cong \omega$$

(since $\kappa \cong \kappa'$ and $\omega \cong \omega'$). The envelope (modulation) function must be

$$\cos\left(\frac{\kappa - \kappa'}{2}x - \frac{\omega - \omega'}{2}t\right) \quad \text{(A3.5)}$$

The envelope function has a very long wavelength encompassing many cycles of the sine term since

$$\kappa - \kappa' \ll \kappa \quad \rightarrow \quad \lambda_{\text{env}} = \frac{2\pi}{(\kappa - \kappa')/2} \gg \lambda = \frac{2\pi}{\kappa}$$

As far as Fourier series and transforms are concerned, the results seems a little unfamiliar because we are adding two high-frequency waves whereas we normally add two low frequency waves (with equal speed) to get a square wave etc. Anyway, to continue, the speed of the carrier wave is approximately $v_p = \omega/k$ and the speed of the envelope is

$$v_{\text{env}} \cong \frac{\omega - \omega'}{\kappa - \kappa'} = \frac{\Delta\omega}{\Delta k} \cong \frac{d\omega}{dk} \quad \text{(A3.6)}$$

Notice we only required two waves (at high frequency) with slightly different phase velocities $v_p = \omega/k$. So the wave packet motion is really the motion of the beat wave.

There is another way to see this result that perhaps better illustrates the role of the different speeds of the two individual waves. Figure A3.4 shows the sum of two waves

$$y = y_1 - y_2 = A \sin(\kappa x - \omega t) - A \sin(\kappa' x - \omega' t) \quad \text{(A3.7)}$$

near $x=0$ and $t=0$. The minus sign for the second term is chosen so that the envelope function crosses zero near $x=0$ for convenience. The point where the envelope crosses through zero depends on the relative positions of the two waves y_1 and y_2. If one wave moves faster than the other one then the zero point of $y_1 - y_2$ must move. Near the origin $x=0$ and $t=0$ both $y_1 - y_2$ and the envelope crosses zero. To find the group velocity, consider x and t to be very small but not necessarily zero. Focus on the zero point crossing by setting the sum of the two waves $y_1 - y_2$ to zero to find

$$0 = y = y_1 - y_2 = A\sin(\kappa x - \omega t) - A\sin(\kappa' x - \omega' t) \cong A[(\kappa x - \omega t) - (\kappa' x + \omega' t)]$$

from the lowest order Taylor approximation. Solving for $v_g = x/t$ provides

$$v_g = \frac{x}{t} = \frac{\omega - \omega'}{k - k'} = \frac{\Delta\omega}{\Delta k} \cong \frac{d\omega}{dk} \qquad (A3.8)$$

similar to the previous result. Figure A3.5 illustrates the results. The top portion of the figure shows the superposition of two sine waves at $t=0$. The wave vectors and angular frequencies are $k_1 = 1$, $k_2 = 1.03$, $\omega_1 = 10$, and $\omega_2 = 10.1$ which gives two slightly different phase velocities nearly equal to 10. The slight difference in the wave vectors yields a group velocity three times smaller than the phase velocity. Focus on the point in the top portion of Figure A3.5 where the envelope passes through zero. The bottom portion shows a close-up view for three different times: $t=0$, $t=0.03$ and $t=0.06$. Notice how the zero-point crossing moves to the right; this motion corresponds to the envelope

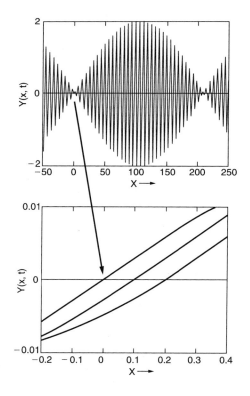

FIGURE A3.5

Focus on the point where the envelope crosses zero. The velocity that it moves to the right is the same as the group velocity.

(wave packet) moving toward the right (top portion). You can measure directly from the lower potion of the figure or calculate $v_g = (\omega_2 - \omega_1)/(k_2 - k_1)$ to find a group velocity of $v_g = 3.3$.

A3.2 Group Velocity of the Electron in Free-Space

The above considerations apply equally well to the wave motion of electrons. This is especially true for free-space since the free-space dispersion relation is

$$E = \frac{p^2}{2m} = \frac{\hbar^2 k^2}{2m} \tag{A3.9}$$

Using $\omega = E/\hbar$, we see that the phase velocity $v_p = \omega/k$ depends on k according to

$$v_p = \frac{\hbar k}{2m} \tag{A3.10}$$

(note the extra factor of 2). The reason for the k-dependence of the phase velocity in Equation A3.10 is that $\hbar k$ is related to the particle momentum (however, infinitely long plane waves don't *intuitively* represent particles very well). The point of Equation A3.10 is that the phase velocity of the electron depends on the wavevector (i.e., wavelength) even for a free particle. The free photon propagating through free space behaves completely different. The speed of light in free space is independent of the wavevector since the speed of light c $= \omega/k$ is constant for all EM waves. The previous section shows that the group velocity for a dispersion relation such as A3.9 must be

$$v_g = \frac{\partial \omega}{\partial k} = \frac{\partial}{\partial k} \frac{\hbar k^2}{2m} = \frac{\hbar k}{m} \tag{A3.11}$$

A3.3 Group Velocity and the Fourier Integral

Now is a good time to talk about the mathematics for group velocity. Suppose $f(x, t)$ is a wave packet made up of a discrete set of spectral components—this is a good illustration of converting summations to integrals.

$$f(z, t) = \sum_j c_j e^{i(k_j z - \omega_j t)} \tag{A3.12}$$

For each j, there is a k, so relabel the sum as

$$f(z, t) = \sum_k c_k e^{i(kz - \omega_k t)} \tag{A3.13}$$

We are considering a one-dimensional problem in k-space. Assume that the sums over an extremely large number of k-values. In fact, left $\rho(k)\,dk$ be the number of k-values in the length dk. The summation $\sum_k \ldots$ can be changed to the integral

$$f(z,t) = \int_{-\infty}^{\infty} dk\; c(k)\rho(k)e^{i\{kz-\omega_k t\}} = \int_{-\infty}^{\infty} dk\; c(k)\rho(k)e^{i\{kz-\omega(k)t\}}$$

where ρ is the density of states (#k—values per unit length of k). Next defining the Fourier amplitude for $f(x,0)$ as $\phi(k) = \sqrt{2\pi}\,c(k)\,\rho(k)$ we find the expansion

$$f(z,t) = \frac{1}{\sqrt{2\pi}} \int_{-\infty}^{\infty} dk\; \phi(k)\, e^{i\{kz-\omega(k)t\}} \tag{A3.14}$$

The wave packet f and its Fourier transform $\phi(k)$ appear in Figures A3.2 and A3.3. We could have started with Equation A3.14 directly, but sometimes it's nice to see how the individual modes make up the wave packet.

An average wave vector $\bar{k}$ and angular frequency $\bar{\omega}$ characterize the wave packet (as in Figure A3.2). For a wave packet with a very narrow spread in frequency and wave vector, we can write a Taylor expansion for the angular frequency (keeping only two terms)

$$\omega(k) \cong \omega(\bar{k}) + \left.\frac{\partial\omega}{\partial k}\right|_{\bar{k}} (k-\bar{k}) \equiv \bar{\omega} + \bar{\omega}' \cdot (k-\bar{k}) \tag{A3.15}$$

Substituting this last result into $f(z,t)$ in Equation A3.14, we find

$$f(z,t) = \frac{1}{\sqrt{2\pi}} \int_{-\infty}^{\infty} dk\; \phi(k)e^{i\bar{k}z}e^{i(k-\bar{k})z}e^{it\{\bar{\omega}+\bar{\omega}'\cdot(k-\bar{k})\}} = e^{i\bar{k}z-i\bar{\omega}t}\frac{1}{\sqrt{2\pi}} \int_{-\infty}^{\infty} dk\; \phi(k)e^{i(k-\bar{k})z}e^{-it\bar{\omega}'(k-\bar{k})}$$

Defining a new variable that shows the deviation between the wave vector and its average as $k' = k - \bar{k}$ we find

$$f(z,t) = \underbrace{e^{i\bar{k}z-i\bar{\omega}t}}_{\text{phase-factor}} \underbrace{\frac{1}{\sqrt{2\pi}} \int_{-\infty}^{\infty} dk'\; \phi(k'+\bar{k})\, e^{ik'z}e^{-it\bar{\omega}'k'}}_{\text{Envelope}} = e^{i\bar{k}z-i\bar{\omega}t}f(z-t\bar{\omega}',0) \tag{A3.16}$$

The leading phase factor is unimportant for our purposes. Equation A3.16 defines $\phi(k'+\bar{k}) = \phi(k')$ to replace the original function $f(z,t)$ by the envelope function

$$f(z-\bar{\omega}'t,0) = f\left(z - t\frac{\partial\bar{\omega}}{\partial k}, 0\right) \tag{A3.17}$$

where

$$f\left(z - t\frac{\partial\bar{\omega}}{\partial k}, 0\right) = \frac{1}{\sqrt{2\pi}} \int_{-\infty}^{\infty} dk'\; \phi(k')\, e^{ik'\{z-t\bar{\omega}'\}}$$

To interpret this Equation A3.17, if the wave packet had the value f_o at point $z_o = 0$ at time $t = 0$, then (to lowest approximation) it has the same value at the point $z = z_o + \overline{\omega}'t$ at time t. On average, the wave packet moves with speed (group velocity)

$$v_{\text{group}} = \frac{\partial \omega}{\partial k} \qquad (A3.18)$$

As a note, if "f" above is the electric field (for EM) or the probability amplitude (for QM), then the power or the probability becomes

$$f^*f = \left[e^{i\overline{k}z - i\overline{\omega}t} f(z - t\overline{\omega}', 0) \right]^* \left[e^{i\overline{k}z - i\overline{\omega}t} f(z - t\overline{\omega}', 0) \right] = \left| f(z - t\overline{\omega}', 0) \right|^2$$

For electromagnetics and quantum theory, it is the modulus-squared that has physical significance and the phase factor $e^{i(\overline{k}z - t\overline{\omega})}$ drops out. Equation A3.17 shows that the wave packet does not change shape as it moves to the right with the group speed

$$v_g = \frac{\partial \omega}{\partial k}$$

All of the manipulations used for the Fourier Transform also hold for the periodic discrete case.

A3.4 The Group Velocity for a Plane Wave

Consider a single frequency component

$$\varphi(k) = \delta(k - k_o)$$

where

$$\delta(k - k_o) = \begin{cases} \infty & k = k_o \\ 0 & k \neq k_o \end{cases} \quad \text{such that} \quad \int_{-\infty}^{\infty} dk \, \delta(k - k_o) = 1$$

(refer to the Dirac Delta Function). Equation A3.14

$$f(z, t) = \frac{1}{\sqrt{2\pi}} \int_{-\infty}^{\infty} dk \, \varphi(k) \, \exp(ikz - i\omega_k t)$$

reduces to

$$f(z, t) = \frac{\exp(ik_o z - i\omega_{k_o} t)}{\sqrt{2\pi}}$$

so that the phase velocity must be identical with the group velocity.

Appendix 4

Review of Probability Theory and Statistics

The present appendix reviews selected topics from probability and statistics. Most of the examples focus on optics and noise process. We first introduce the probability density, cumulative probability and the average.

![line separator]

A4.1 Probability Density

The probability density function ρ measures the probability per unit "something" such as per unit length or per unit volume. If $\rho(x)\, dx$ is the probability of finding a particle in the infinitesimal interval dx centered at the position x, then the probability of finding the particle in the interval $[a, b]$ is given by

$$P(a \leq x \leq b) = \int_a^b dx\, \rho(x) \tag{A4.1}$$

The integral presumes the random variable x is continuous. Discrete random variables reduce the integral in Equation A4.1 to a summation. As is typical for classical probability theory, the integral of density function ρ must be one

$$\int_{-\infty}^{\infty} dx\, \rho(x) = 1 \tag{A4.2}$$

The fact that the integral over all space equals unity is a reflection of the fact that the particle must be found somewhere in space (i.e., the total probability equals one for finding the particle somewhere). For volume rather than length, the probability density is $\rho(x, y, z)$. The probability of finding a particle in a volume V of space is then

$$P(a \leq x \leq b, c \leq y \leq d, e \leq z \leq f) = \int_a^b \int_c^d \int_e^f dx\, dy\, dz\ \rho(x, y, z) = \int_V dV\, \rho \tag{A4.3}$$

The average of a real-valued random variable "x" can be symbolized several ways

$$\bar{x} = \langle x \rangle = E[x] \tag{A4.4}$$

where $E(\ )$ is the expectation operator from probability theory. These averages are calculated as usual

$$\langle f(x) \rangle = \int_{-\infty}^{\infty} dx\, f(x)\, \rho(x) \tag{A4.5}$$

As will be elsewhere in the book for quantum mechanics, the state must be specified before the average can be taken. This is essentially the same as saying for Equation A4.5, the probability distribution must be specified before the average can be computed (which is the usual way of doing things).

The variance of a real-valued random variable "x" can be written

$$\sigma^2 = \langle (x - \bar{x})^2 \rangle \tag{A4.6}$$

where σ is the standard deviation. The term $(x - \bar{x})$ measures the deviation between "x" and its average. The average of all of the terms $(x - \bar{x})$ gives zero since, by definition of average, "x" is larger than $\bar{x}$ as often as it is smaller. Therefore taking the square $(x - \bar{x})^2$ makes the term always positive and it still tends to measure the deviation between x and $\bar{x}$. We are not interested in a point-by-point difference $(x - \bar{x})^2$ but instead, we want the expected behavior over all the possible values. Therefore the variance is defined with the average in Equation A4.6.

For a complex-valued random variable z, the average is given similar to Equation A4.5. The average can have real and imaginary parts. The variance must be real (as a measure of total deviation) and is given by

$$\sigma_z^2 = \langle (z - \bar{z})^* (z - \bar{z}) \rangle = \langle |z - \bar{z}|^2 \rangle \tag{A4.7}$$

The probability density leads to a probability for discrete random variables rather than those having a continuous range as appropriate for the probability density. We convert the integral for the average of an arbitrary function $\langle f \rangle = \int_0^L dx\, f(x)\, \rho(x)$ into a discrete summation. First divide the region of integration $(0, L)$ into very small intervals δx so that $L = N\, \delta x$. Let x_i be a point in interval #i. Assume the interval δx centered on the interval #i small enough that a measurement of x produces value x_i with probability Π_i. The probability Π_i must be $\Pi_i = \rho(x)\, \delta x$. Notice the units of Π_i and ρ differ. We can now write

$$\langle f \rangle = \int_0^L dx\, f(x)\, \rho(x) \cong \sum_{i=1}^{N} f(x_i)\, \Pi_i \tag{A4.8}$$

This suggests writing the discrete form of the probability density as

$$\rho(x) = \sum_{i=1}^{N} \Pi_i\, \delta(x - x_i) \tag{A4.9}$$

A4.2 Processes

Making a series of measurements of a quantity Y produces a set of discrete points $\{y\}$. Each measurement takes place at a separate time t_i. For example, we might measure

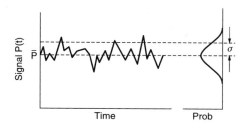

FIGURE A4.1

A signal as a function of time. A large amount of noise is superimposed on the average signal.

slight fluctuations in optical power $P(t)$ from a laser as a function of time (Figure A4.1). Each sequence of points (i.e., each possible graph like Figure A4.1) produces a realization of a random process. For a given value of the parameter t, the quantity $Y(t)$ represents a random variable. The collection of random variables $\{Y\}$, with one such Y for each parameter t, constitutes the random process. The set $\{y_i : i = 1, 2, \ldots\}$ provides a representation of the random process. These sets might be so dense as to approximate a continuous set.

Consider an example for the power $P(t)$ from a laser or light emitting diode. Measurements at time $\{t_i\}$ produce results $\{P_i\}$ that can be plotted as points on a graph. The time t serves as an "index" for the points (t_1, P_1), (t_2, P_2) and so on. Each sequence of points $P(t)$ (i.e., each possible graph like Figure A4.1) represents a realization of the random process. Let t_1 be a specific time. The power $P(t_1)$, a random variable, can assume any number of values. That is, for a given fixed time t_1, the value of P can assume a range of values. For example, P might be in the range $[-1, 1]$ or it might assume a set of discrete values in that range. Therefore, for every time t_1, t_2, and so on, there exists a random variable $P_1 = P(t_1)$, and another random variable $P_2 = P(t_2)$, and so on. The collection of all random variables forms the random process $P(t)$. Sometimes people refer to quantities such as $P(t)$ as a time dependent random variable rather than a process.

A probability density ρ describes the distribution of possible values at a given value of the parameter. The probability density $\rho(y_1, t_1)$ refers to a single random variable $Y_1 = Y(t_1)$ indexed by a particular value of the parameter t_1. The quantity $\rho(y_1, t_1)$ represents the probability (density) for finding the random variable Y_1 and has the value y_1 at the specific time t_1. The joint probability density $\rho(y_1, t_1, y_2, t_2)$ gives the probability of finding the value of $Y_1 = y_1$ at t_1 *and* $Y_2 = y_2$ at t_2. Sometimes people refer to $\rho(y_1, t_1, y_2, t_2)$ as the "two-time probability density." Notice that the two-time probability refers to two-separate values of the parameter for the same process. The multi-time probability density provides more information than does the single-time probability density. As will be seen momentarily, the multi-time probabilities contain information on correlation.

A4.3   Ensembles

An ensemble consists of the collection of *all possible* realizations of the random process. To understand this statement, consider an experiment to measure the optical power $P(t)$ at times t_1, t_2, ... from a semiconductor laser. Suppose the experiment starts at 2 pm on December 3, ends at 3 pm, and produces the results given by plot #1 in Figure A4.2. Now suppose the experimenter goes back in time to 2 pm on December 3rd and

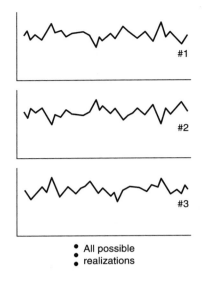

FIGURE A4.2
The ensemble consists of all possible realizations.

repeats the experiment. Realization #2 in Figure A4.2 shows this data set. In fact, suppose that the experimenter goes back an infinite number of times and collects all possible realizations. That collection represents the ensemble. Of course, we only imagine going back in time and obtaining the ensemble; we can't really collect the information. Sometimes we focus on a single time t such as t_1. An ensemble might consist of all possible values $P(t_1)$. The average power $\bar{P}(t) = \langle P(t) \rangle$ can be found by averaging all of the possible realizations of the process at time t_1. Using the density function, the average becomes

$$\bar{P}_1 = \bar{P}(t_1) = \int dP_1 \, P_1 \, \rho(P_1, t_1)$$

That is, the average is found by a point-by-point average over the infinite number of possible points at time t_1.

A4.4 Stationary and Ergodic Processes

A process receives the designation of "stationary" when its characteristics do not change with time. For example, the average and the standard deviation do not depend on time. The time-dependence of the probability distribution determines the stationary character of a process. Consider again the power from a laser. The single-time probability distribution $\rho(P_1, t_1)$ cannot depend on time for a stationary process. However, a multi-time probability distribution $\rho(P_1, t_1; P_2, t_2; P_3, t_3; \ldots)$ describing for example, the power P_i in an optical beam at time "t_i," depends only on a difference in time.

$$\rho(P_1, t_1; P_2, t_2; P_3, t_3 \ldots) = \rho(P_1, 0; P_2, t_2 - t_1; P_3, t_3 - t_1 \ldots) \tag{A4.10}$$

Some stationary processes have the designation of "ergodic" when the average such as $\bar{P} = \langle P(t) \rangle$ can be found by either (1) the ensemble average or by (2) a time average. The two averages must produce identical results for the process to be ergodic. The time-average has the usual definition

$$\langle P \rangle_T = \frac{1}{N} \sum_{i=1}^{N} P(t_i) \qquad \langle P(t) \rangle = \frac{1}{T} \int_0^T dt\, P(t) \qquad \text{(A4.11a)}$$

while the ensemble average uses only a single time t_i and calculates

$$\bar{P}_i = \bar{P}(t_i) = \sum P_i\, \rho(P_i, t_i) \qquad \langle P_i \rangle = \int dP_i\, P_i\, \rho(P_i, t_i) \qquad \text{(A4.11b)}$$

The distinction will become clear in the following examples. Strictly speaking, a process can only be ergodic if every realization contains exactly the same statistical information as the ensemble. In this case, the realizations don't all need to start at the same time.

Example A4.4.1
Nonstationary process

Figure A4.3 shows two examples of nonstationary processes. The first one shows that the standard deviation of the noise $a(t)$ decreases with time. The second one shows that the average value of $b(t)$ decreases with time.

Example A4.4.2
Nonergodic Process

Figure A4.4 shows a nonergodic process because (for some reason) the standard deviation differs for two different realizations (perhaps taken at widely different times).

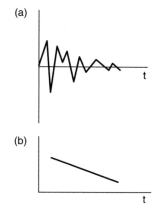

FIGURE A4.3
A nonstationary processes.

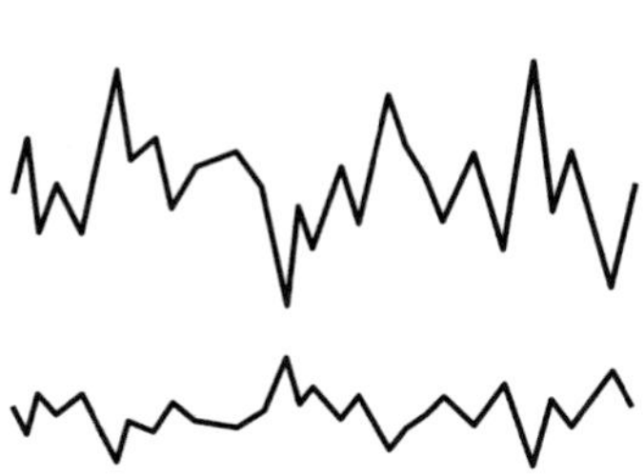

FIGURE A4.4
Two realizations of a nonergodic process.

A4.5   Correlation

This section discusses the meaning of correlation for a single random variable and cross correlation for two random variables. Two random variables X and Y are correlated if the values of one are "linked" (to some extent) with the values of the other. Probability and statistics courses define the covariance. We freely interchange the name correlation and covariance. The correlation (or perhaps more properly the covariance) of two random variables X and Y is defined by

$$\Gamma_{XY} = \text{cov}(X, Y) = \left\langle (X - \overline{X})(Y - \overline{Y})^* \right\rangle \tag{A4.12}$$

The complex conjugate only applies to complex-valued random variables; the correlation function generally has real and imaginary parts. Sometimes Γ_{XY} is interpreted as an element of a matrix (the covariance matrix); the elements are Γ_{XX}, Γ_{XY}, Γ_{YX}, Γ_{YY}. The *correlation coefficient* is defined as

$$\frac{\left\langle (X - \overline{X})(Y - \overline{Y})^* \right\rangle}{\sigma_X \sigma_Y} \tag{A4.13}$$

where σ_X and σ_Y are the standard deviations for X and Y respectively. The complex conjugate only applies to complex-valued random variables. Both the correlation function and the correlation coefficient measure the linkage between two random variables. However, the correlation coefficient removes arbitrary scaling factors. *As an important note*, if $X = Y$ then the correlation function $\Gamma_X = \Gamma_{X,Y}|_{X=Y}$ reduces to the usual variance according to

$$\Gamma_{XY} = \left\langle (X - \bar{X})(Y - \bar{Y})^* \right\rangle = \left\langle |X - \bar{X}|^2 \right\rangle = \sigma_X^2 \tag{A4.14}$$

Figure A4.5, as an example, shows two sets of measured values exhibiting positive and negative correlation, and a third exhibiting negligible correlation. The values x_i and y_i have positive cross correlation for the set marked "pos" because, as the values of one increase, so do the values of the other. For example, the set of points might represent the x–y position of an ant as it follows a scent across a tabletop. The subscript "i" represents the time (in seconds) on a clock. For the eight points, the cross correlation

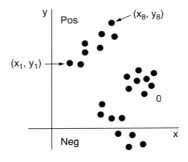

FIGURE A4.5
Three types of cross correlation.

between x and y is

$$\langle (x(t) - \bar{x})(y(t) - \bar{y}) \rangle = \frac{1}{8} \sum_{i=1}^{8} (x_i - \bar{x})(y_i - \bar{y})$$

The cross correlation is positive for the "pos" case in Figure A4.5.

Next consider the autocorrelation function defined by

$$\Gamma_X = \langle X(t) X^*(t + \tau) \rangle = \Gamma_X(t, t + \tau) \tag{A4.15}$$

Equation A4.14 looks similar to the cross correlation function in Equation A4.12. In some sense, the term $X^*(t + \tau)$ acts like a new random variable $Y(t)$. This brings us back to interpreting t and $t + \tau$ as indices in a sequence of measured values; the symbol τ is then similar to an offset. The autocorrelation function measures the similarity between two subsets of a single string of numbers.

The following set of example lead the reader to the meaning of the correlation and autocorrelation of the Langevin noise sources.

Example A4.4.3

Correlation (for illustration purposes)

Consider a discrete process with realization given by

$$\underbrace{x_0, x_1, \ldots, x_i, \ldots}_{\leftarrow \ n \ \rightarrow}$$

that is, $x(t_0) = x_0$ and so on. The correlation between the set $x_0, x_1, \ldots$ and the set $x_i, x_{i+1}, \ldots$ must be given by

$$\Gamma = \frac{1}{N} \sum_{i=1}^{N} (x_i - \bar{x})(x_{i+n} - \bar{x}')$$

where a string of N numbers is taken for each subset. This is the same as the auto-correlation. Notice that if the offset $n = 0$ then the autocorrelation becomes the variance

$$\Gamma = \frac{1}{N} \sum_{i=1}^{N} (x_i - \bar{x})(x_{i+n} - \bar{x}') = \frac{1}{N} \sum_{i=1}^{N} (x_i - \bar{x})(x_i - \bar{x}) = \sigma_x^2$$

We have not been careful to properly define estimators, which would require N to be replaced by $N-1$.

Example A4.4.4

Autocorrelation

Suppose a coin has sides labeled with $+1$ and -1. Suppose 22 tosses of the coin yields the following string of numbers.

$$x_0 = -1, +1, -1, -1, +1, +1, -1, +1, -1, +1, -1, +1, -1,$$

$$+1, +1, +1, -1, -1, +1, +1, +1, -1 = x_{22}$$

Consider two small subsets with $N=7$ elements. Suppose the first subset starts at x_0 and the second one starts at x_{12}.

$$x_0 = -1, +1, -1, -1, +1, +1, -1 = x_6 \quad x_{12} = -1, +1, +1, +1, -1, -1, +1 = x_{18}$$

The correlation between these two sets (assuming $\bar{x} \cong 0$ for convenience) is therefore $-3/7$. This is the autocorrelation because the two subsets are from the same initial string. For the coin toss, the seven-number sets could produce a correlation value anywhere between -1 and $+1$ (with 0 as the expected outcome so long as the sets are different). For this case, the offset is $n=12$.

Example A4.4.5
The Kronecker-Delta Correlation

For the previous example, what is the autocorrelation for $n=0$? The answer is 1. It's not too hard to imagine a situation where the correlation is 0 for $n \neq 0$. The correlation as a function of "n" is then

$$\Gamma_x(n) = \sigma^2 \delta_{n,0} = \begin{cases} \sigma^2 & n = 0 \\ 0 & n \neq 0 \end{cases}$$

where $\delta_{a,b}$ is the Kronecker delta function which is 1 when $a=b$ and 0 otherwise.

Appendix 5

The Dirac Delta Function

The Dirac Delta function (also called the impulse function) arises in many fields of engineering and physics. In many respects, the Dirac Delta function can be thought of as a function. The Dirac delta function departs from classical mathematical theory and must be defined as the limit of a sequence of functions. Distribution function theory provides a firm basis for the Dirac delta function. This section provides a number of representations of the Dirac delta function. We will find that every basis set of functions provides another representation. This section also discusses the idea of principal part.

A5.1 Introduction to the Dirac Delta Function

We often think of the Dirac delta function $\delta(x - x_o)$ as a function with exactly one infinite value at the point x_o and zero everywhere else (Figure A5.1).

$$\delta(x - x_o) = \begin{cases} \infty & x = x_o \\ 0 & x \neq x_o \end{cases} \tag{A5.1}$$

The function must be infinitely large at x_o but infinitely narrow so that the area under the function equals to one. Apparently, integrals over the delta function have wonderful properties.

We might also consider an alternate definition of the Dirac delta function by the effect it has on integrals. Define the delta function by

$$\int_a^b dx\, f(x)\delta(x - x_o) = \begin{cases} f(x_o) & x_o \in (a, b) \\ \frac{1}{2}f(x_o) & x_o = a \text{ or } b \\ 0 & \text{else} \end{cases} \tag{A5.2}$$

Notice that if $f(x) = 1$ then Equation A5.2 provides

$$\int_a^b dx\, \delta(x - x_o) = \begin{cases} 1 & x_o \in (a, b) \\ 1/2 & x_o = a \text{ or } b \\ 0 & \text{else} \end{cases} \tag{A5.3}$$

The integral of the delta function has the value of one when the point of discontinuity x_o appears entirely inside the integration interval. When you encounter a delta function in an equation, you should consider it an "invitation to integrate." The next topic shows how the Dirac delta function really comes from the limit of a sequence of functions, which substantiates Equations A5.2 and A5.3.

Example A5.1

What is $\displaystyle\int_{10}^{50} \frac{\sin x}{\sqrt{1+3x^2}} \delta(x - 45°)\, dx$?

$$\int_{10}^{50} \frac{\sin x}{\sqrt{1+3x^2}} \delta(x - 45°)\, dx = \left.\frac{\sin x}{\sqrt{1+3x^2}}\right|_{x=45°} = \frac{\sin 45}{\sqrt{1+3(45)^2}}$$

A5.2   The Dirac Delta Function as Limit of a Sequence of Functions

The Dirac delta function should really be defined as the limit of a sequence of functions S_n according to the definition

$$\int_{-\infty}^{\infty} dz\, \delta(z - z_0) = \underset{n\to\infty}{\text{Lim}} \int_{-\infty}^{\infty} dz\, S_n(z - z_0) \qquad (A5.4)$$

The order of the limit, integral and function S_n should be carefully noted. Many different sequences of functions S_n will work even those that can not be differentiated everywhere.

Example A5.2.1

Figure A5.2 shows a sequence of functions $S_n(z - z_0)$ given by

$$S_1 = 1/6 \qquad x \in (0, 6)$$
$$S_2 = 1/2 \qquad x \in (2, 4)$$
$$S_3 = 1 \qquad x \in (2.5, 3.5)$$
$$\vdots \qquad\qquad \vdots$$

Notice that the area under each function S_n equals to one. We can then trivially write

$$\underset{n\to\infty}{\text{Lim}} \int_{0}^{9} dz\, S_n(z - z_0) = \underset{n\to\infty}{\text{Lim}} 1 = 1 = \int_{0}^{9} dz\, \delta(z - z_0)$$

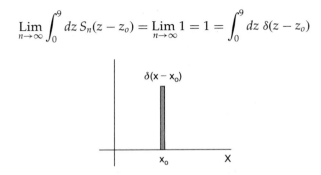

FIGURE A5.1
Representation of the delta function as a narrow spike.

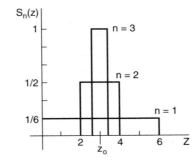

FIGURE A5.2
A sequence of functions with a "limit" that represents the Dirac delta function.

This last example brings up an important point regarding the definition for the integral of the delta function in terms of the limit of a sequence of functions.

$$\operatorname*{Lim}_{n\to\infty} \int_a^b dx\, f(x) S_n(x - x_o) \equiv \int_a^b dx\, f(x)\delta(x - x_o)$$

The integral of *each* function S_n does not need to equal unity; however, at the very least, the integral of S_n should approach "1" as "n" becomes large. In many cases, we require each function in the sequence S_n to be everywhere differentiable. For example, S_n might be Gaussian-shaped functions.

Many books use a shorthand notation for the Dirac delta function. For example, looking at the defining relation

$$\int_{-\infty}^{\infty} dz\, \delta(z - z_o) = \operatorname*{Lim}_{n\to\infty} \int_{-\infty}^{\infty} dz\, S_n(z - z_o) \tag{A5.5}$$

we might be tempted to make the identification

$$\delta(z - z_o) = \operatorname*{Lim}_{n\to\infty} S_n(z - z_o) \tag{A5.6}$$

However, this can only be correct when interpreted as in Equation A5.5. We can easily see the problem with directly integrating Equation A5.6. Setting the Dirac delta function δ directly equal to the limit of a sequence of functions produces a limit function equal to zero everywhere except at one point. This limit function matches the intuitive view of the Dirac delta function. Taking the integral of this limit function must produce zero because a Riemann integral is insensitive to a single point. The integral of the limit function does not produce a value equal to unity contrary to the definition of the Dirac delta function.

Now let's discuss why the first integral property in Equation A5.2 holds, namely

$$\int_a^b dx\, f(x)\delta(x - x_o) = \begin{cases} f(x_o) & x_o \in (a, b) \\ \frac{1}{2} f(x_o) & x_o = a \text{ or } b \\ 0 & \text{else} \end{cases} \tag{A5.7}$$

Figure A5.3 shows a sequence of functions S_n all enclosing unit area. The first graph shows that $f(z)$ varies along the nonzero portion of S_1. The middle picture shows a case with $f(z)$ almost constant over the width of S_2. Finally, the last graph shows a function $S_3(z - z_o) \cong \delta(z - z_o)$ sufficiently narrow to provide a very good approximation

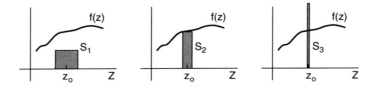

FIGURE A5.3

Making n sufficiently large makes S_n sufficiently narrow so that $f(z)$ does not vary along the nonzero portion of S_n. In this case, we can take $\delta(z - z_o) \cong S_3(z - z_o)$.

$f(z) \cong f(z_o)$ over the nonzero width of $S_3(z - z_o)$. As a result of this intuitive approach, we can write

$$\int_{-\infty}^{\infty} dz\, f(z)\, \delta(z - z_o) \cong \int_{-\infty}^{\infty} dz\, f(z) S_3(z - z_o) \cong \int_{-\infty}^{\infty} dz\, f(z_o)\, S_3(z - z_o)$$

$$= f(z_o) \int_{-\infty}^{\infty} dz\, S_3(z - z_o) = f(z_o)$$

which demonstrates the first of the integrals. This last approximation also works for functions that aren't delta functions so long as they are very sharply peaked; however, the result must be multiplied by a constant equal to the integral over the function.

Now what about the property in Equation A5.2 for $z_o = a$, namely

$$\int_{a}^{b} dz\, f(z)\delta(z - z_o) = \frac{1}{2} f(z_o)$$

This property holds because the integral covers only half of the delta function. Using Figure A5.4 and a fairly narrow S_n (as shown), we can again write

$$f(z)\, S_n(z) \cong f(z_o)\, S_n(z)$$

and the integral becomes

$$\int_{a}^{b} dz\, f(z)\, S_n(z - z_o) \cong \int_{a}^{b} dz\, f(z_o)\, S_n(z - z_o) \quad \text{or}$$

$$\int_{a}^{b} dz\, f(z)\, S_n(z - z_o) = f(z_o) \int_{a}^{b} dz\, S_n(z - z_o)$$

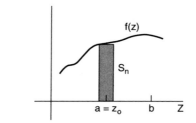

FIGURE A5.4

The integral covers only 'half' of the delta function.

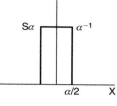

FIGURE A5.5
Sequence of rectangles.

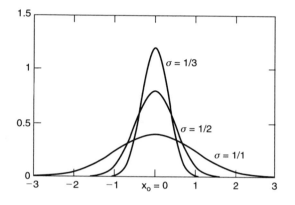

FIGURE A5.6
The limit of the Gaussian probability distribution approaches the Dirac delta function.

Now, because $a = z_0$, the integral covers only half of the width of S_n, and the integral becomes

$$\int_{a=z_0}^{b} dz\, S_n(z - z_0) = 1/2$$

Finally, including $f(z)$

$$\int_{z_0}^{b \gg z_0} dz\, f(z)\delta(z - z_0) = \frac{1}{2} f(z_0)$$

A5.3 The Dirac Delta Function from the Fourier Transform

The Dirac Delta function is most often first encountered with Fourier transforms. The following derivation shows how this comes about. Start with the Fourier integral

$$f(x) = \int_{-\infty}^{\infty} dk\, \frac{e^{ikx}}{\sqrt{2\pi}} f(k)$$

and then substitute the Fourier transform for $f(k)$

$$f(k) = \int_{-\infty}^{\infty} dX\, \frac{e^{-ikX}}{\sqrt{2\pi}} f(X)$$

to find

$$f(x) = \int_{-\infty}^{\infty} dk \, \frac{e^{ikx}}{\sqrt{2\pi}} \, f(k) = \int_{-\infty}^{\infty} dk \, \frac{e^{ikx}}{\sqrt{2\pi}} \int_{-\infty}^{\infty} dX \, \frac{e^{-ikX}}{\sqrt{2\pi}} \, f(X)$$

$$= \int_{-\infty}^{\infty} dX \int_{-\infty}^{\infty} dk \, \frac{e^{-ik(X-x)}}{2\pi} \, f(X) = \int_{-\infty}^{\infty} dX \, f(X) \int_{-\infty}^{\infty} dk \, \frac{e^{-ik(X-x)}}{2\pi}$$

Comparing both sides of the equation we see that the second integral must be related to a Dirac Delta function in order that $f(X)$ becomes $f(x)$. Therefore

$$\delta(x - X) = \int_{-\infty}^{\infty} dk \, \frac{e^{-ik(x-X)}}{2\pi}$$

and similarly

$$\delta(k - K) = \int_{-\infty}^{\infty} dx \, \frac{e^{-i(k-K)x}}{2\pi}$$

which can be proved in the same manner as for x-Delta function but starting with $f(k)$ instead of $f(x)$.

A5.4   Other Representations of the Dirac Delta Function

This topic lists some common sequences for the Dirac delta function.

1. The previous topic discusses the sequence of rectangles defined by

$$S_\alpha = \begin{cases} 1/\alpha & |x| \leq \alpha/2 \\ 0 & |x| \geq \alpha/2 \end{cases}$$

Note that $S_\alpha(x - x_o)$ is obtained by replacing x with $x - x_o$ in the formula.

2. The Gaussian probability density function

$$g_\sigma(x - x_o) = \frac{1}{\sqrt{2\pi}\sigma} \exp\left[-\frac{(x - x_o)^2}{2\sigma^2} \right]$$

represents a delta function when the standard deviation σ approaches zero. These distribution functions can be written in terms of the integer "n" by setting $\sigma = 1/n$ for example. The delta function can be written as

$$\lim_{\sigma \to 0} g_\sigma(x - x_o) = \delta(x - x_o)$$

with the understanding that this means

$$\lim_{\sigma \to 0} \int_a^b dx \, f(x) \, g_\sigma(x - x_o) \equiv \int_a^b dx \, f(x) \, \delta(x - x_o)$$

Without the integral, the limit of the sequence of distribution functions g_σ would be zero at all points except at x_o where the limit of the distribution is infinite. The point x_o is at the center of the distribution and σ is the standard deviation.

3. $\delta(x) = \text{Lim}_{\varepsilon \to 0} \, S_\varepsilon(x) = \text{Lim}_{\varepsilon \to 0} \dfrac{1}{\pi} \dfrac{\varepsilon}{x^2 - \varepsilon^2}$

4. The theory of Fourier transforms provides an integral representation (see Topic A5.3 above)

$$\delta(x) = \text{Lim}_{\kappa \to \infty} \int_{-\kappa}^{\kappa} dk \, \frac{e^{ikx}}{2\pi} = \int_{-\infty}^{\infty} dk \, \frac{e^{ikx}}{2\pi} \tag{A5.8}$$

which can be written in two other forms

$$\delta(x) = \text{Lim}_{\kappa \to \infty} \int_{0}^{\kappa} dk \, \frac{\cos(kx)}{\pi} \tag{A5.9}$$

and

$$\delta(x) = \text{Lim}_{\kappa \to \infty} \frac{\sin(\kappa x)}{\pi x} \tag{A5.10}$$

Equation A5.9 is related to the "sinc" function. Figure A5.7 shows how increasing the value of "κ" causes the function "$\sin(\kappa x)/\pi x$" to become sharper and more narrow; the height of the function is "κ/π" and the distance from $x=0$ to the first zero is "π/κ." Equation A5.9 follows from Equation A5.8

$$\delta(x) = \int_{-\infty}^{\infty} dk \, \frac{e^{ikx}}{2\pi} = \int_{-\infty}^{0} dk \, \frac{e^{ikx}}{2\pi} + \int_{0}^{\infty} dk \, \frac{e^{ikx}}{2\pi}$$

$$= \int_{0}^{\infty} dk \, \frac{e^{-ikx}}{2\pi} + \int_{0}^{\infty} dk \, \frac{e^{ikx}}{2\pi} = \int_{0}^{\infty} dk \, \frac{\cos(kx)}{\pi}$$

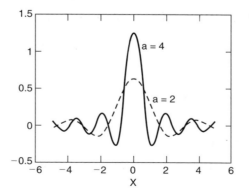

FIGURE A5.7
A plot of Equation A5.10 for two values of κ.

where the integral is divided into two (one over negative k and the other over positive k), replacing k with $-k$ in one of them (the one for negative k) and then recombining the two integrals using one of Eulers' equations $\cos(kx) = [e^{ikx} + e^{-ikx}]/2$. Equation A5.10 follows from Equation A5.8 as follows

$$\delta(x) = \lim_{\kappa \to \infty} \int_{-\kappa}^{\kappa} dk \frac{e^{ikx}}{2\pi} = \lim_{\kappa \to \infty} \left[\frac{e^{i\kappa x} - e^{-i\kappa x}}{2\pi i x} \right] = \lim_{\kappa \to \infty} \frac{\sin(\kappa x)}{\pi x}$$

Note that the $\sin(\kappa x)/x$ appears as a sequence in "κ" just like the previous examples while Equations A5.8 and A5.9 have the parameter as the bounds on an integral.

A5.5 Theorems on the Dirac Delta Functions

There are some useful theorems on the Dirac delta function that allow a person to simplify expressions. G. Barton's book "Elements of Green's Functions and Propagation" published by Oxford Science Publications in 1989 provides a good reference.

1. $\delta(x - \xi) = \delta(\xi - x)$

2. $\delta(ax) = \dfrac{1}{|a|} \delta(x)$

3. If $g(x)$ has real roots x_n (that is, $g(x_n) = 0$) then

$$\delta[g(x)] = \sum_n \frac{\delta(x - x_n)}{|g'(x_n)|} \quad \text{where} \quad g'(x) = dg/dx$$

4. For $\xi \in (a, b)$,

$$\int_a^b dx\, f(x)\, \delta'(x - \xi) = -f'(\xi)$$

This property is important because it allows for a weak identity that is exceedingly useful

$$f(x)\, \delta'(x - \xi) = -f'(\xi)\, \delta(x - \xi)$$

A5.6 The Principal Part

If half the range of "k" is left off the integral in Equation A5.8 then a function $\zeta(x)$ can be defined by

$$\zeta(x) = -i \lim_{\kappa \to \infty} \int_0^{\kappa} dk \frac{e^{ikx}}{2\pi}$$

where an extra $i = \sqrt{-1}$ is added for later convenience. Integrating provides

$$\zeta(x) = \frac{1}{2\pi} \underset{\kappa \to \infty}{\mathrm{Lim}} \left(\frac{1 - e^{i\kappa x}}{x} \right) = \frac{1}{2\pi} \underset{\kappa \to \infty}{\mathrm{Lim}} \left(\frac{1 - \cos(\kappa x)}{x} - i\, \frac{\sin(\kappa x)}{x} \right)$$

where the last step obtains using $e^{i\kappa x} = \cos(\kappa x) + i \sin(\kappa x)$. Half the range of the integral in Equation A5.8 is removed to obtain the expression for $\zeta(x)$. The reader should realize that for Equation A5.9, half the range of the integral was *not* removed from Equation A5.8; the range was folded up (so to speak) into the cosine term. Now for $\zeta(x)$, define the principal part $\wp(1/x) = \wp/x$

$$\frac{\wp}{x} = \underset{\kappa \to \infty}{\mathrm{Lim}} \frac{1 - \cos(\kappa x)}{x}$$

as the principal part of $1/x$. The imaginary part of $\zeta(x)$ is related to the Dirac delta function as shown in #4 above. Now it is possible to write an alternate expression for $\zeta(x)$ as

$$\zeta(x) = \underset{\kappa \to \infty}{\mathrm{Lim}} \frac{1 - \cos(\kappa x)}{2\pi x} - i \underset{\kappa \to \infty}{\mathrm{Lim}} \frac{\sin(\kappa x)}{2\pi x} = \frac{\wp}{2\pi x} - i\, \frac{\delta(x)}{2}$$

Restricting the range of "k" for the integral is therefore seen to give something that differs from the delta function by the value of the principal part.

What is $P(1/x) = \mathrm{Lim}_{\kappa \to \infty}(1 - \cos(\kappa x))/(2\pi x)$? As a function of x, taking the limit literally, only $x = 0$ is defined since $\cos(\kappa x)$ does not have a limit (with κ as the limit variable) where $x \neq 0$. At $x = 0$, the limit becomes (by Taylor expanding the cosine function)

$$P(1/x) = \underset{\kappa \to \infty}{\mathrm{Lim}} \frac{1 - \cos(\kappa x)}{2\pi x} \cong \underset{\kappa \to \infty}{\mathrm{Lim}} \underset{x \to 0}{\mathrm{Lim}} \frac{1 - \left[1 - \frac{(\kappa x)^2}{2!} + \cdots \right]}{2\pi x} = 0$$

by L'Hospital rule. Now, because the principal part occurs in the same equation as the Dirac delta function, the reader should anticipate that the principal part has special integral properties. The integral of the terms in $\zeta(x)$ is found before taking the limit (the limit is understood to be outside the integral).

The integral of $P(1/x)$ requires some explanation. Consider two cases for the integration interval of $[a, b]$. First assume that $a > 0$ and $b > 0$ and second, assume that $a < 0$ and $b > 0$. Consider case 1 for $a > 0$ and $b > 0$. Figure A5.8 shows a plot of $[1 - \cos(\kappa x)]/x$ (solid curve) for a *fixed* κ and also a plot of $1/x$ (dotted curve). *Notice how $1/x$ appears as a "local" average for the curve.* To evaluate the integral, divide the interval $[a, b]$ into smaller intervals $[a_i, b_i]$ such that

1. $[a, b] = \overset{n}{\underset{i=1}{\cup}} [a_i, b_i]$ where $a_i = b_{i-1}$ and $\overset{n}{\underset{i=1}{\cup}} [a_i, b_i]$ means the union of the sub-intervals.
2. the function $1/x$ does not vary appreciably over $[a_i, b_i]$,
3. $[1 - \cos(\kappa x)]$ passes through many cycles over each $[a_i, b_i]$; this is certainly the case for large κ when $b_i - a_i \gg \lambda$. (see λ in Figure A5.8).

Using the first property, the integral can be rewritten as

$$\int_a^b = \sum_{i=1}^n \int_{a_i}^{b_i}$$

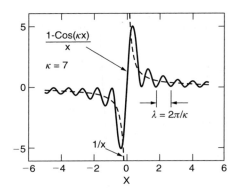

FIGURE A5.8
The function "$1/x$" is an average of $[1 - \cos(\kappa x)]/x$.

We also need the mean value theorem from calculus, which can be written as

$$\int_{a_i}^{b_i} dx\, f(x) = \langle f(x)\rangle (b_i - a_i)$$

Now, applying the mean value theorem to $[1 - \cos(\kappa x)]/x$ keeping in mind that $1/x$ is a local average, we find

$$\int_{a_i}^{b_i} dx\, \frac{\wp}{x} = \operatorname*{Lim}_{\kappa \to \infty} \int_{a_i}^{b_i} dx\, \frac{1 - \cos(\kappa x)}{x} = \operatorname*{Lim}_{\kappa \to \infty} \left\langle \frac{1 - \cos(\kappa x)}{x} \right\rangle (b_i - a_i)$$

$$= \operatorname*{Lim}_{\kappa \to \infty} \frac{1}{x}(b_i - a_i) = \int_{a_i}^{b_i} dx\, \frac{1}{x}$$

The third and last terms were found by applying the mean value theorem. The limit in the fourth term doesn't matter and can be dropped. How is $\langle (1 - \cos(\kappa x))/x \rangle$ found? This can be seen in two ways. For the first way, $1/x$ was already noted to be the average of $[1 - \cos(\kappa x)]/x$ for small enough intervals. For the second way, we can write

$$\int_{a_i}^{b_i} dx\, \frac{1 - \cos(\kappa x)}{x} \simeq \frac{1}{x} \int_{a_i}^{b_i} dx\, [1 - \cos(\kappa x)] = \frac{b_i - a_i}{x} - \left. \frac{\sin(\kappa x)}{\kappa} \right|_{a_i}^{b_i} \frac{1}{x} \simeq \frac{b_i - a_i}{x}$$

Thus for case 1, we can make the replacement

$$\int_{a_i}^{b_i} dx\, f(x) \frac{\wp}{x} \Rightarrow \int_{a_i}^{b_i} dx\, \frac{f(x)}{x}$$

so long as $f(x)$ is slowly varying. The original integral can be written as

$$\int_{a}^{b} dx\, f(x) \frac{\wp}{x} = \sum_{i=1}^{n} \int_{a_i}^{b_i} dx\, \wp \frac{f(x)}{x} = \sum_{i=1}^{n} \int_{a_i}^{b_i} dx\, \frac{f(x)}{x} = \int_{a}^{b} dx\, \frac{f(x)}{x}$$

for $a, b > 0$. For this case, the principal part has no effect. Also notice that the sine term (i.e. the delta function) in

$$\zeta(x) = \underset{\kappa \to \infty}{\text{Lim}} \frac{1 - \cos(\kappa x)}{2\pi x} - i \underset{\kappa \to \infty}{\text{Lim}} \frac{\sin(\kappa x)}{2\pi x} = \frac{\wp}{2\pi x} - i \frac{\delta(x)}{2}$$

is approximately zero since the point of discontinuity is outside the interval (i.e., $a > 0, b > 0$).

Consider the second case of $a < 0$ and $b > 0$. Again divide up the interval into small subintervals satisfying the properties on the previous page. Those subintervals that don't contain zero are handled just like case 1. Therefore consider the subinterval $[-\varepsilon, \varepsilon]$ where ε is a small number. As discussed above $P(1/x) \cong 0$ for "x" near zero. The integral over the ε subinterval becomes

$$\int_{-\varepsilon}^{\varepsilon} dx \, f(x) \, \wp \left(\frac{1}{x} \right) = f(x) \, \wp \left(\frac{1}{x} \right) \Bigg|_{-\varepsilon}^{\varepsilon} \cong 0$$

The smaller the value of ε, the better the approximation. The original integral becomes

$$\int_{a}^{b} dx \, f(x) \, \wp \left(\frac{1}{x} \right) = \int_{a}^{-\varepsilon} dx \, f(x) \, \wp \left(\frac{1}{x} \right) + \int_{-\varepsilon}^{\varepsilon} dx \, f(x) \, \wp \left(\frac{1}{x} \right) + \int_{\varepsilon}^{b} dx \, f(x) \, \wp \left(\frac{1}{x} \right)$$

$$= \int_{a}^{-\varepsilon} dx \, \frac{f(x)}{x} + 0 + \int_{\varepsilon}^{b} dx \, \frac{f(x)}{x}$$

Some people define the principal part of the integral as

$$\wp \int_{a}^{b} = \int_{a}^{-\varepsilon} + \int_{\varepsilon}^{b}$$

A5.7   Convergence Factors and the Dirac Delta Function

In many cases, the form of the Dirac delta function (for a given Hilbert space) is surmised from the closure relation. This topic discusses one method of showing that the area under a Dirac delta function is equal to one. Consider the Fourier representation of the Dirac delta function $\delta(k - 0)$ given by

$$I(k) = \int_{-\infty}^{\infty} dx \, \frac{e^{ikx}}{2\pi} \tag{A5.11}$$

The integral can be evaluated by including a "convergence" factor $e^{\pm \alpha x}$ with $\alpha > 0$. The "positive" sign in $e^{\pm \alpha x}$ is used when "x" is negative and the "negative" sign in $e^{\pm \alpha x}$ is used when "x" is positive. Including the appropriate integrating factor forces the integrand in Equation A5.11 to approach zero near $\pm\infty$. After the calculation is complete, the parameter α is set to zero.

$$I(k) = \int_{-\infty}^{\infty} dx \, \frac{e^{ikx}}{2\pi} = \int_{0}^{\infty} dx \, \frac{e^{ikx}}{2\pi} + \int_{-\infty}^{0} dx \, \frac{e^{ikx}}{2\pi} = \int_{0}^{\infty} dx \, \frac{e^{-\alpha x + ikx}}{2\pi} + \int_{-\infty}^{0} dx \, \frac{e^{\alpha x + ikx}}{2\pi}$$

Notice that integrating factors are included in the integrals. Carrying out the integrals provides

$$I(k) = \int_{-\infty}^{\infty} dx \, \frac{e^{ikx}}{2\pi} = -\frac{1}{2\pi(-\alpha + ik)} + \frac{1}{2\pi(\alpha + ik)} = \frac{1}{2\pi} \frac{2\alpha}{(k - i\alpha)(k + i\alpha)} \qquad \text{(A5.12)}$$

Notice that if $k = 0$ then as $\alpha \to 0$ the integral becomes infinite $I(k) \to \infty$. On the other hand, if $k \neq 0$ then as $\alpha \to 0$ the integral becomes zero $I(k) \to 0$. This behavior matches that for a Dirac delta function $\delta(k - 0)$.

Now to evaluate the integral of $I(k)$

$$\int_{-\infty}^{\infty} dk \, I(k)$$

a contour integration can be performed on A5.12. The contour can be closed in either the lower half plane or the upper half plane. A closed contour in the upper half plane encloses a pole at $k = i\alpha$. The basic formula for residues can be used

$$\oint dz \, \frac{f(z)}{z - z_0} = 2\pi i \sum \text{residues} = 2\pi i f(z_0)$$

to find

$$\oint dk \, I(k) = \oint dk \, \frac{1}{2\pi} \frac{2\alpha}{(k - i\alpha)(k + i\alpha)} = 2\pi i \left\{ \frac{1}{2\pi} \frac{2\alpha}{(k + i\alpha)} \right\}_{k = i\alpha} = 1$$

Appendix 6

Coordinate Representations of the Schrodinger Wave Equation

This appendix illustrates how the Schrodinger wave equation such as for the Harmonic Oscillator

$$\left(\frac{-\hbar^2}{2m}\frac{\partial^2}{\partial x^2} + \frac{1}{2}kx^2\right)\Psi(x,t) = i\hbar\frac{\partial}{\partial t}\Psi(x,t) \tag{A6.1}$$

can be found from operator-vector form of the equation

$$\left(\frac{\hat{p}^2}{2m} + \frac{1}{2}k\hat{x}^2\right)|\Psi(t)\rangle = i\hbar\frac{\partial}{\partial t}|\Psi(t)\rangle \tag{A6.2}$$

We use the harmonic oscillator as an example with the understanding that other Hamiltonians can be similarly treated.

We begin with Equation A6.2 by operating on both sides using the x-coordinate projection operator $\langle x|$ to get

$$\langle x|\left(\frac{\hat{p}^2}{2m} + \frac{1}{2}k\hat{x}^2\right)|\Psi(t)\rangle = i\hbar\frac{\partial}{\partial t}\langle x|\Psi(t)\rangle$$

where the x-coordinate operator moves past the time derivative. On the left-hand side, insert the unit operator

$$1 = \int |x'\rangle dx'\langle x'|$$

between the Hamiltonian operator and the ket $|\Psi(t)\rangle$. We obtain

$$\langle x|\left(\frac{\hat{p}^2}{2m} + \frac{1}{2}k\hat{x}^2\right)\left(\int |x'\rangle dx'\langle x'|\right)|\Psi(t)\rangle = i\hbar\frac{\partial}{\partial t}\langle x|\Psi(t)\rangle$$

The x-terms can be moved under the integral since they do not depend on x'.

$$\int dx'\langle x|\left(\frac{\hat{p}^2}{2m} + \frac{1}{2}k\hat{x}^2\right)|x'\rangle\langle x'||\Psi(t)\rangle = i\hbar\frac{\partial}{\partial t}\langle x|\Psi(t)\rangle \tag{A6.3}$$

The momentum and position operators are diagonal in "x" so that

$$\langle x|\hat{p}^2|x'\rangle = \langle x|x'\rangle \, [\hat{p}(x')]^2 = \delta(x'-x)\left[\frac{\hbar}{i}\frac{\partial}{\partial x'}\right]^2 \quad \text{and} \quad \langle x|\hat{x}^2|x'\rangle = \delta(x'-x)\,[x']^2$$

since $\hat{x}|x'\rangle = x'|x'\rangle$. Therefore Equation A16.3 becomes

$$\int dx' \, \delta(x'-x)\left(\frac{-\hbar^2}{2m}\frac{\partial^2}{\partial x'^2}+\frac{1}{2}kx'^2\right)\langle x'|\Psi(t)\rangle = i\hbar\frac{\partial}{\partial t}\langle x|\Psi(t)\rangle$$

Integrating over the delta function yields

$$\left(\frac{-\hbar^2}{2m}\frac{\partial^2}{\partial x^2}+\frac{1}{2}kx^2\right)\langle x|\Psi(t)\rangle = i\hbar\frac{\partial}{\partial t}\langle x|\Psi(t)\rangle$$

and, using $\langle x|\Psi(t)\rangle = \Psi(x,t)$, gives the desired results

$$\left(\frac{-\hbar^2}{2m}\frac{\partial^2}{\partial x^2}+\frac{1}{2}kx^2\right)\Psi(x,t) = i\hbar\frac{\partial}{\partial t}\Psi(x,t)$$

Appendix 7

Integrals with Two Time Scales

Consider an integral of the form

$$I = \int_a^b f(t)\, s(t)\, dt \tag{A7.1}$$

where $f(t)$ and $s(t)$ are "fast" and "slow" functions, respectively, in terms of their variations along the t-axis. The figure shows an example of the functions $f(t)$ and $s(t)$. The period of $f(t)$ is small compared with all time scales of interest. The time axis is divided into intervals Δ_s that are small compared to the length of any variation of the slow function $s(t)$. We show the following properties

1. $I = \int_a^b d\tau\, s(\tau) \langle f(\tau) \rangle$

   where the average $\langle f(t) \rangle$ is over the interval Δ_s. The "average" is the type that every reader has learned in calculus. The average $\langle f(t) \rangle$ might still depend on time since the average of $f(t)$ over each interval Δ_s might depend on the location of that interval.

2. Nom the average $\langle f(t) \rangle$ is independent of time, the integral in Equation A7.1 can be written as

   $$I = \langle f \rangle \langle s \rangle (b - a)$$

3. If the average $\langle f(t) \rangle = 0$ then $I = 0$

4. $\int d\tau \left[f(\tau) + s(\tau) \right] = \int d\tau\, s(\tau)$ if $\langle f(t) \rangle = 0$

5. The *Rotating Wave Approximation* shows that an integral can be approximated as

   $$\int_0^t d\tau \left[e^{i(\omega_{ni} - \omega)\tau} + e^{i(\omega_{ni} + \omega)\tau} \right] \cong \int_0^t d\tau\, e^{i(\omega_{ni} - \omega)\tau}$$

   for an angular frequency $\omega \cong \omega_{ni}$.

To prove the properties starting with the first, divide the small intervals Δ_s into smaller intervals δ_{sf} which are small compared to the length of the variations for the fast

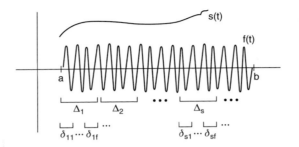

FIGURE A7.1
Slow $s(t)$ and fast $f(t)$ functions.

function $f(t)$. Using the basic definition from calculus, the integral in Equation A7.1 can be written as

$$I \cong \sum_{sf} f(t_{sf}) s(t_{sf}) \delta_{sf}$$

where as usual t_{sf} is a point in the small interval δ_{sf}. Over each interval Δ_s, the function $s(t)$ is constant. Let t_s be a point in the interval Δ_s so that $s(t_s) \cong s(t_{sf})$. The function $s(t_{sf})$ can be moved outside of one of the summations as follows

$$I \cong \sum_{s} s(t_s) \sum_{f} f(t_{sf}) \delta_{sf}$$

Multiplying and dividing by the larger interval Δ_s produces

$$I \cong \sum_{s} s(t_s) \sum_{f} f(t_{sf}) \delta_{sf} = \sum_{s} s(t_s) \Delta_s \left[\frac{1}{\Delta_s} \sum_{f} f(t_{sf}) \delta_{sf} \right] \tag{A7.2}$$

The term in brackets provides the average of the function $f(t)$ over the interval Δ_s. The summation in the brackets provides an integral

$$\frac{1}{\Delta_s} \sum_{f} f(t_{sf}) \, \delta_{sf} = \frac{1}{\Delta_s} \int_{\Delta_s} d\tau \, f(\tau) = \langle f(t) \rangle = g(t)$$

Notice that the average of the fast function in this last equation $g(t) = \langle f(t) \rangle$ can depend on time since the average over each of the subintervals Δ_s might not be the same. Equation A7.2 can now be written as

$$I = \int_a^b d\tau \, s(\tau) \, g(\tau) = \int_a^b d\tau \, s(\tau) \langle f(\tau) \rangle$$

which proves the first property.

If the average over the interval Δ_s, namely $g(t) = \langle f(t) \rangle$, does not depend on the location of the interval Δ_s, then the average must be independent of time $\langle f(t) \rangle = \langle f \rangle$ and can be removed from the integral

$$I = \langle f \rangle \int_a^b d\tau\, s(\tau)$$

Using the definition of an average from calculus

$$\langle s \rangle = \frac{1}{b-a} \int_a^b d\tau\, s(\tau)$$

The integral becomes

$$I = \langle f \rangle \int_a^b d\tau\, s(\tau) = \langle f \rangle \langle s \rangle (b-a)$$

which proves the second property.

Obviously, if as shown in the figure, the average of the function $f(t)$ is zero $\langle f(t) \rangle = 0$ then the integral is zero. The fourth and fifth properties follow.

Appendix 8

The Dipole Approximation

The dipole approximation treats the wavelength of a traveling electric field (plane wave) as large compared with the size of the atom. Consider the integral

$$\langle 1|f(\vec{r})\,\mathrm{e}^{\pm i\vec{k}\cdot\vec{r}}\,|2\rangle = \underset{\text{all space}}{\iiint} dV\; u_1^*(\vec{r})f(\vec{r})\,\mathrm{e}^{\pm i\vec{k}\cdot\vec{r}}u_2(\vec{r})$$

where the volume V is centered at $\vec{r}=0$ and the wave functions u_1 and u_2 are essentially confined to the volume V (they have tails that extend slightly beyond the volume V). Therefore, the integral must be zero for regions of space outside the volume V since the wave functions in the integrand are zero outside the volume V. However, the spatial part $\mathrm{e}^{\pm i\vec{k}\cdot\vec{r}}$ of the plane wave is constant over the volume V; it has the value of $\mathrm{e}^{\pm i\vec{k}\cdot\vec{r}}\big|_{\vec{r}=0}=1$. Therefore the integral can be written as

$$\langle 1|f(\vec{r})\,\mathrm{e}^{\pm i\vec{k}\cdot\vec{r}}\,|2\rangle = \mathrm{e}^{\pm i0}\underset{\text{all space}}{\iiint} dV\; u_1^*(\vec{r})f(\vec{r})\,u_2(\vec{r}) = \underset{\text{all space}}{\iiint} dV\; u_1^*(\vec{r})f(\vec{r})\,u_2(\vec{r})$$

The approximation is called the "dipole approximation" because (classically) the volume V is considered to be composed of dipoles (polarized atoms) that absorb and emit electromagnetic radiation. These dipoles are small compared with the wavelength of the electromagnetic wave. It should be obvious that the position of the atom can be at any location (say $\vec{r}\,'$) besides the origin. We would then have

$$\langle 1|f(\vec{r})\mathrm{e}^{\pm i\vec{k}\cdot\vec{r}}|2\rangle = \underset{\text{all space}}{\iiint} dV\; u_1^*(\vec{r})\,f(\vec{r})\mathrm{e}^{\pm i\vec{k}\cdot\vec{r}}\,u_2(\vec{r}) = \mathrm{e}^{\pm i\vec{k}\cdot\vec{r}\,'}\underset{\text{all space}}{\iiint} dV\; u_1^*(\vec{r})\,f(\vec{r})\,u_2(\vec{r})$$

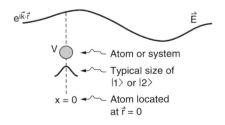

FIGURE A8.1
The electron wavefunction is nonzero over a region that is small compared with the wavelength of light.

Appendix 9

The Density Operator and the Boltzmann Distribution

We can define the density operator $\hat{\rho}_r$ through a Boltzmann distribution

$$\hat{\rho}_r = \frac{1}{Z} \exp\left(-\frac{\hat{H}_r}{k_B T}\right)$$

where Z denotes the normalization (partition function)

$$Z = Tr_r\left\{\exp\left(-\frac{\hat{H}_r}{k_B T}\right)\right\}$$

Consider the average of an operator $\hat{O}$

$$\langle\hat{O}\rangle = Tr\left(\hat{\rho}_r\hat{O}\right) = \sum_n \langle n| \hat{\rho}_r\hat{O} |n\rangle = \sum_{n,m} \langle n| \hat{\rho}_r |m\rangle\langle m| \hat{O} |n\rangle$$

where the closure relation for the *energy* basis set $\{ |n\rangle \}$ has been inserted between the two operators. The energy eigenstates are chosen for the basis since the density operator is diagonal in that basis set. First, evaluate the matrix elements of the density operator.

$$\langle n| \hat{\rho}_r |m\rangle = \frac{1}{Z} \langle n| \exp\left(-\frac{\hat{H}_r}{k_B T}\right) |m\rangle = \frac{1}{Z} \langle n|m\rangle \exp\left(-\frac{E_m}{k_B T}\right)$$

where the factor $1/Z$ can be removed from the inner product by virtue of it being a c-number and where the last term obtains by operating with the Hamilton on the ket $|m\rangle$. Using the orthogonality of the basis provides

$$\langle n| \hat{\rho}_r |m\rangle = \frac{\delta_{nm}}{Z} \exp\left(-\frac{E_n}{k_B T}\right)$$

and the average of an operator becomes

$$\langle\hat{O}\rangle = Tr\left(\hat{\rho}_r\hat{O}\right) = \sum_n \frac{1}{Z} \exp\left(-\frac{E_n}{k_B T}\right) O_{nn}$$

Notice that this last expression only requires the diagonal matrix elements O_{nn} of the operator $\hat{O}$. The partition function can be similarly evaluated. The expectation value of the operator $\hat{O}$ shows that the density operator for the reservoir gives rise to the Boltzmann probability distribution. The energy levels E_n are expected to be populated according to the thermal distribution.

Index

Note: Italicized page numbers refer to illustrations